TEMPORARY STRUCTURES IN CONSTRUCTION

TEMPORARY STRUCTURES IN CONSTRUCTION

Robert T. Ratay
Editor in Chief

Third Edition

New York Chicago San Francisco Lisbon London Madrid
Mexico City Milan New Delhi San Juan Seoul
Singapore Sydney Toronto

The McGraw·Hill Companies

Cataloging-in-Publication Data is on file with the Library of Congress.

Temporary Structures in Construction, Third Edition

1 2 3 4 5 6 7 8 9 0   DOC/DOC   1 9 8 7 6 5 4 3 2

ISBN 978-0-07-175307-4
MHID 0-07-175307-9

The Second Edition was published by McGraw-Hill in 1996 with the title *Handbook of Temporary Structures in Construction*.

This book is printed on acid-free paper.

Sponsoring Editor
Larry S. Hager

Acquisitions Coordinator
Bridget Thoreson

Editorial Supervisor
David E. Fogarty

Project Manager
Sapna Rastogi,
Cenveo Publisher Services

Copy Editor
Manish Tiwari,
Cenveo Publisher Services

Proofreader
Akhilesh Gupta,
Cenveo Publisher Services

Indexer
Robert Swanson

Production Supervisor
Richard C. Ruzycka

Composition
Cenveo Publisher Services

Art Director, Cover
Jeff Weeks

*To **Elyse** who couldn't care less about engineering,*

and

*to **Scott** for whom there is still hope*

ABOUT THE EDITOR

Robert T. Ratay, Ph.D., P.E., is a structural engineer in private practice in New York with nearly 50 years of design, construction, and teaching experience, and an Adjunct Professor at Columbia University. His current practice and teaching are focused on the investigation and litigation support of structural and construction failures. Dr. Ratay has been an expert consultant/witness for engineers, contractors, owners, attorneys, and insurance companies in the investigation and litigation support of nearly 100 cases of temporary structure and other construction failures. Earlier in his career, he had worked with such prominent engineering firms as Le Messurier Associates, Severud Associates, and HNTB, and had been Professor and Chairman of the Civil Engineering Department, then Dean of the School of Engineering at Pratt Institute, and Professor at Polytechnic University (formerly Brooklyn Polytech). Dr. Ratay has lectured extensively at conferences and seminars in the US and abroad, and published in US and international journals on temporary structures in construction and on structural and construction failures. In addition to the three editions of the *Temporary Structures in Construction*, he is also the Editor in Chief of two other books: *Structural Condition Assessment* and *Forensic Structural Engineering Handbook*, Second Edition, published in 2005 and 2010, respectively. Dr. Ratay has served on and chaired several technical committees, including an advisory committee of the Federal Highway Administration to tighten regulations for temporary structures. He is the originator of the effort and chairman of the committee that developed and maintains the *SEI/ASCE 37, Design Loads on Structures during Construction* Standard. In 2003, he was named by *Engineering News-Record* as one of the *Top Newsmakers 2003* who "made a mark in the construction industry"; in 2010, he received the Palotás Award of the *Federation International du Beton* from the Hungarian Group of *fib* "for his very valuable achievements in the field of structural concrete"; in 2011 he was named by *Structural Engineer* magazine as one of the *Power List* of 10 "individuals who bring progress and passion to the structural engineering field." Dr. Ratay is a Life Member and Fellow of the American Society of Civil Engineers (ASCE), former member of the Board of Governors of its Structural Engineering Institute (SEI), and a Fellow of the International Association of Bridge and Structural Engineering (IABSE) where he formed and chairs the Working Group on Forensic Structural Engineering, and he serves on the Editorial Advisory Panel of the *Forensic Engineering Journal* of the British Institution of Civil Engineers (CE).

CONTENTS

Part 3   Failures

CONTRIBUTORS

Francis J. Arland, P.E. *Partner, Mueser Rutledge Consulting Engineers, New York, New York* (CHAP. 11)

Mohammad Ayub, P.E., S.E. *Director, Office of Engineering Services, Directorate of Construction, Occupational Safety and Health Administration (OSHA), U.S. Department of Labor, Washington, D.C.* (CHAP. 4)

Wilson C. Barnes, Ph.D., AIA, FCIOB *Professor and Dean, Southern Polytechnic State University, Marietta, Georgia* (CHAP. 19)

Sarah B. Biser, Esq. *Partner, McCarter & English, New York, New York* (CHAP. 5)

John A. Brain, P.E. *Director of Engineering, Harsco Infrastructure—a Division of Harsco Corporation, Fair Lawn, New Jersey* (CHAP. 16)

Arthur B. Corwin, P.E. *President and Chief Executive Officer, Moretrench American Corporation, Rockaway, New Jersey* (CHAP. 9)

Michael S. D'Alessio P.E. *President, Michael D'Alessio Engineering, Bayside, New York* (CHAPS. 14, 15)

John S. Deerkoski, P.E., S.E., P.P. *President and Owner, John S. Deerkoski, P.E. & Associates, Warwick, New York* (CHAP. 1)

John F. Duntemann, P.E., S.E. *Principal, Wiss, Janney, Elstner Associates, Inc., Northbrook, Illinois* (CHAP. 2)

Patrick E. Durnal, P.E., MEng. *Consulting Engineer, San Francisco, California; formerly, Lead Foundation Engineer, Ben C. Gerwick Inc. Consulting Engineers, San Francisco, California; and Project Manager/Engineer, Wagner Construction Shoring and Foundation Co., Brisbane, California* (CHAP. 6)

Donald O. Dusenberry, P.E. *Senior Principal, Simpson Gumpertz & Heger Inc., Waltham, Massachusetts* (CHAP. 3)

Chris Evans, P.E. *Chief Operating Officer, Universal Builders Supply (UBS), Inc., New Rochelle, New York (CHAP. 22)*

Ben C. Gerwick, Jr., P.E. *(deceased) formerly, Chairman, Ben C. Gerwick Inc. Consulting Engineers, San Francisco, California; and Professor Emeritus of Civil Engineering, University of California, Berkeley, California* (CHAP. 6)

Robert G. Lenz, P.E. *Groundwater Control Consultant, Port Saint Lucie, Florida; formerly Chairman and C.E.O., Moretrench American Corporation, Rockaway, New Jersey* (CHAP. 9)

Bernard Monahan, Sr., P.E. *Consultant, Rockaway Point, New York; formerly, Vice President, Crow Construction Company—a subsidiary of J.A. Jones Construction Company, New York, New York* (CHAPS.12, 13)

Richard C. Mugler, Jr. *(deceased) formerly, President, Richard C. Mugler Co., Inc., Bronx, New York* (CHAP. 21)

Richard C. Mugler III *President, Richard C. Mugler Co., Inc., Bronx, New York* (CHAP. 21)

Kevin M. O'Callaghan, *President, Universal Builders Supply (UBS), Inc., New Rochelle, New York* (CHAP. 19)

John J. Peirce, P.E., D.GE *Owner, Peirce Engineering, Inc., Phoenixville, Pennsylvania* (CHAP. 7)

Robert T. Ratay, Ph.D., P.E. *Consulting Structural Engineer, and Adjunct Professor at Columbia University, New York, New York* (CHAP. 23)

Robert A. Rubin, Esq., P.E. *Partner, McCarter & English, New York, New York* (CHAP. 5)

GianCarlo Santarelli, Civil Engineer *President, Bencor Corporation of America, Inc., Dallas, Texas* (CHAP. 8)

James Scheld, P.E. *Associate, Howard I. Shapiro and Associates Consulting Engineers, Lynbrook, New York* (CHAP. 17)

Paul C. Schmall, P.E., D.GE *Vice President and Chief Engineer, Moretrench American Corporation, Rockaway, New Jersey* (CHAP. 9)

Lawrence K. Shapiro, P.E. *Principal, Howard I. Shapiro & Associates Consulting Engineers, Lynbrook, New York* (CHAPS. 17, 20)

Gary Strong, Esq. *Lewis Brisbois Bisgaard & Smith, LLP, New York, New York* (CHAP. 5)

Vincent Tirolo, Jr., P.E. *Geotechnical Consultant, Brooklyn, New York; formerly, Chief Engineer, Skanska USA Civil Northeast, Whitestone, New York* (CHAP. 10)

Thomas J. Tuozzolo, P.E. *Vice President, Moretrench American Corporation, Rockaway, New Jersey* (CHAP. 7)

Jake van Baarsel, P.E. *Senior Structures Estimator, Skanska USA Civil, Inc., Riverside, California* (CHAP. 18)

Rudi J. van Leeuwen, P.E. *Consulting Engineer, Fredericksburg, Virginia; formerly, President, Spencer, White and Prentis, Inc. New York, New York* (CHAP. 11)

PREFACE

Twenty-six years after the publication of the original *Handbook of Temporary Structures in Construction*, and 15 years after the publication of the second edition, this third edition is still the only comprehensive book on the subject. The book presents much-needed practical information on professional practice, codes and standards, design, erection, maintenance, and failures of temporary support and access structures used in the execution of construction work. It fills the gap between books on structural design and books on construction, in that it deals with those *systems that are used to construct the designed facilities*. These systems have primary influence on the quality, safety, speed, and financial success of all construction projects. It has been said that the temporary structures provide the "competitive edge" for contractors.

Responses to the first and second editions have been most complimentary: reviews by reviewers and communications from users have been congratulatory; several printings sold out over the years; the book has been used as a valuable and authoritative reference by engineers, contractors, inspectors, building officials, and construction lawyers; it has been adopted as the text or reference book in construction engineering courses at several universities.

Time has come to once again update the book in order to reflect the new developments and advancing technologies in construction, as well as some lessons learned from construction failures. In particular, three new chapters "Environmental and Construction Loads," "Cranes," and "Failures of Temporary Structures in Construction" have been added; the chapters "Earth-Retaining Structures," "Underpinning," "Bracing and Guying for Stability," and "Bridge Falsework," among others, have been heavily revised and updated; and a special note "Suggestions to the Instructor" has been added for those using the book as the text for a course.

The available information on many types of temporary structures and procedures is not as extensive, well developed, standardized, and well documented as on most other subject of civil engineering design and construction. Consequently, many parts of this book are original with little or no reliance on works published elsewhere. On topics where there are no standard methods, the more widely used practices and/or those based upon the authors' own experiences and judgments are presented. Design methods ranging from rule of thumb, through approximate calculations, to "exact" engineering analyses are given according to what the generally accepted and warranted degree of reliability is for the particular problem. The reader is advised to use his/her professional judgment in applying information set forth in the book.

The book is divided into three major parts: Part 1—"Business, Safety, and Legal Aspects"; Part 2—"Design, Construction, Maintenance," and Part 3—"Failures."

The opening chapter, "Professional and Business Practices," explains how the business of procuring, furnishing, installing, and maintaining temporary structures and procedures should be conducted in order to ensure technical and financial success. Chapter 2, "Standards, Codes, and Regulations," introduces national standards, design codes and specifications, model building codes, and state and local codes applicable to temporary works and structures during construction. Chapter 3, "Construction and Environmental Loads," is new to this third edition; it explains some of the reasons forces on temporary structures differ from those on permanent structures and gives some guidance on where to find information on temporary loads and how to calculate them. Chapter 4, "Construction Site Safety," presents data on the number and nature of construction failures, injuries, and fatalities; describes the OSHA Construction Standards; and points to guidance on the causes and prevention of construction failures. The fifth and last chapter of Part 1, "Legal Aspects," deals with the rights, responsibilities, and liabilities of the involved parties as defined by common law, statute, and the signed contract.

The first six chapters of Part 2 deal with below-ground temporary works. Chapter 6, "Cofferdams," is confined to structural cofferdams and cribs as temporary installations, explaining in step-by-step detail the proper and safe methods and materials to be used. Chapter 7, "Earth-Retaining Structures," introduces the types of retaining structures and bracing systems, and explains and illustrates methods of design and construction in much detail. Chapter 8, "Diaphragm/Slurry Walls," is an amply illustrated presentation of the types, materials, uses and methods of construction of diaphragm walls, and their bracing systems, built by the slurry trench method, as well as the various procedures for their design and analysis. Chapter 9, "Construction Dewatering and Groundwater Control," presents a detailed discussion of the methods, design considerations, system installation, operation, and removal, environmental applications, and cost elements of construction dewatering, together with an introduction to various methods of groundwater cutoff and exclusion. Chapter 10, "Underground/Tunneling Supports," is a rather extensive treatment of the types of supports in tunneling and shaft construction, their design concepts, materials, methods of installation, and risk management. When to use underpinning below existing foundations, the traditional and newest means and methods of its installation, its consequences, and safety precautions are discussed in Chap. 11, "Underpinning." Two chapters address the maintenance of traffic and site access. Chapters 12 and 13, "Roadway Decking" and "Construction Ramps, Runways, and Platforms," respectively, present several ways of providing access into, out of, over, and within the work area; the texts are amply illustrated by photographs. Temporary works for primarily building construction are covered in the next eight chapters. Scaffolding and shoring are the most frequent temporary structures in building construction and are treated in considerable detail in Chap. 14, "Scaffolding," and Chap. 15, "Falsework/Shoring." This may well be the most complete treatment of these two subjects in one book and includes design, hardware, installation, purchase and rental costs, safety requirements, and numerous illustrations. An equally common temporary structure is "Concrete Formwork," discussed in Chap. 16, with an eye on the contractor's concerns, that is, types of forms, hardware, details, and dimensions, as well as ACI code requirements in both design and construction. The critical importance and the basic engineering principles of providing lateral stability of structures during construction are discussed under "Bracing and Guying for Stability" in Chap. 17. Temporary supports in bridge construction are addressed in Chap. 18, "Bridge Falsework," which includes design considerations, such as loads and stability, as well as installation, inspection, maintenance, and removal practices. Repair, restoration, and retrofitting of buildings and other structures have been receiving increased attention in recent years; some of the temporary structures used in that work are discussed in Chap. 19, "Temporary Structures in Repair and Restoration," with consideration of the existing conditions of the permanent structure under modification and care for the safety of the public. New to this third edition of the book is Chap. 20, "Cranes"—prompted by the rather large numbers of construction crane failures in recent years—discussing the various lifting machines and their accessories, installations, operations, and safety practices for risk management. Chapter 21 of the book deals with the critical subject of "Protection of Site, Adjacent Areas, and Utilities," including signs, barricades, sidewalk sheds, and other protective installations, as well as pointers on avoiding damage to adjacent facilities, and references to OSHA and other safety regulations. Chapter 22, "Leading Edge Vertical Containment Systems" is the publication for the first time of an innovative temporary system for the protection of workers and the public at high-rise construction sites in congested urban areas.

Part 3 of the book consists of one chapter only: Chap. 23, "Failures of Temporary Structures in Construction." It is new in the third edition for the purpose of illustrating the errors in temporary structure work, their consequences, the risks involved, and lessons to be learned from the failures.

The 23 chapters of the book have been written by 30 invited authors who were selected for their eminence in their fields of expertise. They prepared manuscripts of high quality and value to the profession. I am grateful to them for their effort and cooperation.

The contributing authors and I thank the numerous engineers, contractors, educators, attorneys, and other professionals who have used the first two editions of the book and who communicated their comments and suggestions over the years to us that helped us make this third edition an even better book. I also owe thanks to several of the McGraw-Hill people: particularly to Larry Hager for his valuable assistance and continuing cooperation in producing both the second and this third edition, and to David Fogarty for his effective direction of the copyeditors, illustrators, compositors, and others whom I never met but whose work I much appreciated.

But foremost, I thank you, the reader, for buying and using the book!

Robert T. Ratay, Ph.D., P.E.
Consulting Structural Engineer, New York, and
Adjunct Professor, Columbia University, New York

SUGGESTIONS TO
THE INSTRUCTOR

When in the late 1970s I started getting involved with temporary structures in construction, I realized that neither the designer and contractor, nor the student had one good comprehensive book to consult. There were, and are, isolated journal articles and conference presentations discussing the design, construction, performance, and failure of temporary works and the causes and lessons learned from individual failures. That is to say, in a diligent search one can find useful literature on many aspects of temporary structures used in construction. One could use one or a group of these to be references for teaching isolated topics but several of them would have to be acquired and assembled for a comprehensive course. After all, in order to learn to deal intelligently with temporary structures, the student not only has to have an understanding of loads, strength, and stability, but also has to learn about the philosophy and processes of their design, construction and use, safety, natural as well as code-specified limitations, standard industry practices, and even the prevailing business practices and legal issues.

In an attempt to fill the apparent need for a comprehensive and authoritative reference, with a group of contributing authors I created the first edition of this book, published by McGraw-Hill in 1984. The book was well received by engineers, contractors, regulators, and attorneys, so a second edition was prepared and published in 1996. A number of university professors and continuing education instructors also recognized its value for their construction courses, and assign it as the standard text book. While not a dedicated textbook—mainly because of the absence of numerical examples and homework problems—the book has proven to be a valuable, comprehensive, and convenient text.

This third edition is an improved version of the first two, including new methods, codes, regulations, and practices that have come about in the past 15 years.

A special feature of this book is its coverage in appropriate order and detail of the range of the topics that are necessary for the study and successful practice of temporary works in construction. This is a "what I need to know," "what I need to do," and "how I need to do it" book.

Upon completing the course with the use of this book, students claim to have a broader view of the construction engineering profession; better understanding of the importance, behavior, misbehavior, and vulnerability of temporary structures; more appreciation of the consequences of their design errors and construction defects; at least some awareness of the related legal issues; and a fuller realization of their own personal responsibilities.

In my view, they turn out to be better informed, more reliable, and more successful engineers.

Robert T. Ratay, Ph.D., P.E.
Consulting Structural Engineer, New York, and
Adjunct Professor, Columbia University, New York

TEMPORARY STRUCTURES IN CONSTRUCTION

P · A · R · T · 1

BUSINESS, SAFETY, AND LEGAL ASPECTS

CHAPTER 1
PROFESSIONAL AND BUSINESS PRACTICES*

John S. Deerkoski, P.E., S.E., P.P.

---

*This chapter is updated and revised by John S. Deerkoski, P.E., S.E., P.P., from the one titled *Technical and Business Practices*, contributed by Robert F. Borg, Esq., P.E. in the second edition.

INTRODUCTION

Temporary structures in construction are those structures that are erected and used to aid in the construction of permanent projects. They are used to facilitate the construction of buildings, bridges, tunnels, and other above- and below-ground facilities by providing access, support, and protection for the facility under construction, as well as to ensure the safety of the workers and the public. They are either dismantled and removed when the permanent works become self-supporting or completed, or they are incorporated into the finished work.

Many aspects of temporary structures in construction are different from those of permanent structures, thus requiring somewhat different design, erection, and maintenance philosophies. They also have a primary influence on the quality, safety, speed, and profitability of all construction projects.

Construction safety and, hence, temporary structures are the concern of the owner, the designer, the contractor, the insurer, the Authority Having Jurisdiction, as well as of the workers at the site and the general public.

PROFESSIONAL PRACTICES

Sequence of Construction

Enough cannot be said about the importance of the sequencing of the construction operation on a project. Among other things, it requires meetings, coordination, and planning between all the parties on the site. There are times when a minor change in the final structure can have a major impact on the construction sequence and schedule. All parties have to be open minded and not be defensive of their particular position.

The loads that a temporary structure may have to sustain are directly related to the construction sequence.

The owner's goal is to have a completed structure meeting his requirements at the lowest cost while minimizing the duration of the construction. All parties are answerable to the owner in this regard.

The goal of the coordination and construction sequence is to allow many operations to occur in parallel rather than sequentially. For example, some of the items that affect the schedule are access ramp interference, crane locations, and storage on site to service multiple subcontractors.

Design

The details of who does the design, who prepares the drawings, and how general or detailed the specifications should be all vary depending on the temporary structure under consideration. Being means and methods, it is the responsibility of the contractor to develop the details of the design and perform them in construction. There are specialty contractors with licensed engineers who make it their business of doing specific types of temporary structures.

If the design of the structure is such that it requires shoring of elements such as composite girders where the deflection under the wet concrete would be excessive or if the member cannot sustain the weight of the poured concrete and shoring is required, as a minimum the designer of the permanent structure must provide the loads to be sustained. On occasion, because of unique restrictions, the designer of the permanent structure may provide a full design of the shoring. Generally, however, the Structural Engineer of Record, architect, and owner of the new structure rightly want to distance themselves from liability resulting from the means and methods of temporary structure and construction activities.

Rules, Regulations, and Industry Practices

Technical as well as administrative rules, regulations, and practices are generated by the Authority Having Jurisdiction. The following is a reasonable hierarchy of authority:

• Occupational Safety and Health Administration (OSHA)
• Local Authority Having Jurisdiction
• Contract documents
• Industry consensus standards

When the requirements by different governmental authorities differ, the more restrictive standard must be followed. When industry consensus standards are more restrictive than those of the Authority Having Jurisdiction, the Authority having jurisdiction requirements apply unless the owner of the property chooses to use the more restrictive standard. The owner's input in this decision is necessary because there is a fiduciary responsibility to the owner under state professional licensing laws, and a more restrictive specification could increase the cost of the temporary structure as well as construction time which may result in back-charges to the owner.

Specifications

The specifications for the temporary structure are usually drawn up by the temporary structure contractor. In most cases, architects or engineers designing a complete structure will not provide a design or specifications for the temporary structure work. For example, a temporary structure contractor working on concrete formwork will have to specify the type of plywood, framing, form ties, snap ties, braces, and other form accessories that are required for the work. These materials are often ordered from a catalog or from a supplier that has supplied materials to the contractor in the past. The responsibility for the adequacy of the specified material and for its quality will be of the temporary structure contractor.

In the case of a design that is prepared by a temporary structure contractor for the approval of the architect or engineer, the specifications that are part of that design must also be the responsibility of the temporary structure contractor. These specifications may have to be reviewed by the engineer or architect for their adequacy and durability and strength. Should there then be a question about any of these qualities, the temporary structure contractor will have to justify the specified materials as being proper.

Permits

The permit requirements are a matter of the local Authority Having Jurisdiction and/or the contract. If the general contractor is filling out the permits, he or she would enter his information and engineering submittals, if required, and submit them to the owner or the owner's expediter. Virtually all permits require the owner or his or her designated agent to be one of the signatories to the permits. If a subcontractor is preparing the permit with technical information for temporary structures, he or she would submit it to the general contractor and then the procedure noted previously would be followed.

Shop Drawings

In a temporary structure, the shop drawing and its preparation are somewhat different from a shop drawing prepared by a subcontractor or supplier for a permanent part of the finished work. The shop drawing for a temporary structure may, in fact, be a completely original design that does not appear in any way on the general contractor's contract drawings. An example of this might be a shop drawing prepared for the dewatering system using well points. In this case, it is the responsibility of the subcontractor for the dewatering work not only to prepare the details of where the dewatering system will be located and the type of piping and pumping equipment that would be used, but also to incorporate in the shop drawings or its specifications the calculations made by the subcontractor to ensure the adequacy of the system being installed.

This then brings up the question of the architect's or engineer's checking and approval of the shop drawings. Since the temporary structure subcontractor has provided the calculations for the work to be done, it is therefore a question of who will be responsible for those calculations. Normally an engineer or architect will check the shop drawings only for the compliance with the original contract drawings. In this case, since there are no original contract drawings of the material or equipment being shown on the shop drawings, it is necessary for the engineer or architect to decide whether the shop drawings have to be inspected and approved at all. In most cases, the engineer or architect will merely review the shop drawings for interferences or for conditions that might affect the finished design.

In the case of a dewatering system, for example, the removal of fine material from a sandy substrate during pumping could cause later problems in the foundations. However, in the end it would have to be the temporary structure contractor's responsibility not only to make the calculations for the design of the temporary structure but also to be responsible for the adequacy of the calculations.

The difficulty sometimes encountered is when a shop drawing is submitted for approval by a temporary structure subcontractor and is, in fact, checked and approved by the architect or engineer. Is the temporary structure subcontractor then relieved of any further responsibility for the adequacy of this temporary structure?

The answer to this type of problem will be found in Chap. 2. In any event, it is clear that shop drawings for temporary structures have an important difference in the legal sense from shop drawings for portions of the work that are going to be left in the final structure. It is also evident that shop drawings for temporary structures sometimes require a great deal of more design work than is normally required for shop drawings for work that will be left in the finished structure.

Supervision

Supervision of a temporary structure installation includes not only supervision of its initial installation but supervision of the adequacy of the structure during the entire period that it is performing its function. The contractor must maintain adequately trained superintendents and supervisors who are on the site at all times when work is under way.

A qualified supervisor is one who has training and experience in the type of work that is being done. Obviously, a supervisor who has only a general background in construction but has never been directly involved in the supervision of a temporary structure, such as the one which is being built, is not as qualified to do this work as one who has many years of experience, directly or indirectly related to the work on hand.

Concrete formwork for multistoried high-rise reinforced-concrete buildings, for example, requires many intricate details of bracing, posting, struts, and ties so that they will not only be adequate to sustain the load of the wet concrete and the embedded reinforcing bars but also will be able to undergo the impact of motorized buggies for distributing the concrete and of temporary hoppers, sustain the weight of workers, and have the necessary clearances required for cranes and hoists.

Even after the concrete is poured, it is necessary to observe proper supervision for such activities as winter heating, time constraints before stripping, and reporting of the fresh concrete so that it can attain its required strength for the support of loads of succeeding doors.

Improperly trained or inexperienced supervisors can endanger workers' lives, and the neglect of good practices can cause structural failures.

Inspection

The responsibility for inspection of the work of the contractor for temporary structures is shared by three entities:

1. The temporary structure subcontractor whose own supervisors or engineers must inspect the work as it is being installed, for example, when a temporary structure contractor is installing shoring. As the work is being placed, the supervisor and supervisory personnel of the shoring contractor must be alert to all the requirements of safe and adequate construction practice and be certain that shop drawings or permits which specified the methods of performing the work are being adhered to.

2. The general contractor who is responsible for the entire finished structure, both temporary and permanent. The second required inspection by the general contractor is to see that the installation of the temporary structure contractor is being performed safely and adequately, that it is not interfering with other portions of the work that may be a part of the finished structure, and that it is not interfering with other contractors or subcontractors who may be working simultaneously or subsequent to the work of the temporary structure contractor. The inspection will also cover such things as quality of the materials being installed and compliance with plans and specifications and shop drawings to the extent that they pertain to the temporary structure contractor.

3. The party whose responsibility is to perform the ultimate inspection of the work. This third required inspection may be made by either an architect or an engineer who has the inspection responsibility. It may also be made by a government official such as a member of a building department or government financing agency. This third inspector is the one whose inspection must be the most comprehensive, painstaking, and flawless. This inspection ultimately determines whether, if there is a failure of the temporary structure, the work has been truly reviewed and found to be "approved" by the inspecting party. It is the responsibility of this inspector to be absolutely sure that all necessary compliance is being taken with building codes, safe structural practices, OSHA requirements, and any other procedure for good construction which is made a part of the contract, as well as with the plans, specifications, and shop drawings for the temporary structure.

Estimation

Estimates for the cost of temporary structures entail a three-step process:

1. A layout or design of the structure must be prepared.

2. An estimate of the cost of construction, using standard estimating practices, must be made.

3. An estimate of the cost for maintenance and eventual removal of the temporary structures, such as the labor involved in keeping the temporary structure operative, must be made.

The first part of the estimate, the design phase, is best handled by consultation between the temporary structure contractor, the general contractor or the general contractor's field forces (if general contractors propose to erect the same for themselves with their own forces), and the engineer.

In a typical example involving dewatering by well points, the design would first have to be made showing the location of all the well points, piping, pumps, settling basins, and disposal facilities. Once the layout is completed, step 2, the estimate, using standard methods, comes next. This, of course, will involve quantity takeoffs of materials, equipment, and labor, and then pricing, using unit prices that have been produced from past experience or from an analysis of the probable size of the work gangs, and the probable output per day or per shift in the production of the gangs.

Step 3, which would cover the operation of the system, would estimate the labor and fuel required to maintain the system after it is installed and for the length of time that it is expected to be in operation, and the cost of later removal.

BUSINESS PRACTICES

Contract Negotiations

The negotiation of the contract between the temporary structure contractor and the general contractor, or the owner, is closely related to the concepts set forth in estimating previously. One of the important considerations is the type of design that has been prepared by the temporary structure contractor for the proposed work. In the case where the temporary structure contractor is competing with others for the same or similar services, differences in the design could, of course, have a marked effect on the estimate and greatly affect the price. If the owner is not aware of the possible advantages and disadvantages of the proposed methods of design being presented with various prices, the temporary structure contractors should endeavor to make clear the salient features of their designs in the negotiations. In the example given for the estimating section previously, the spacing of the well points might vary between two proposals. While the temporary structure contractors in each proposal may guarantee to dewater the site, a difference in the spacing, which could depend on factors of judgment, could also affect the adequacy of the dewatering system. These factors must be made apparent to the buyer.

Once the buyer and the temporary structure contractor agree in concept to the design, the hard bargaining begins. Here all the factors that go into the purchase of any commodity come into play. Such matters as wage rates required to be paid, sales tax requirements, insurance requirements, the matter of scheduling and timing, method of payment, and all the other intangible and tangible factors that affect the price will have to be discussed.

The final, agreed-upon price will reflect all these factors.

Who Pays for the Unexpected in Underground Construction?

The problem of who shall bear the risk of unexpected subsurface conditions in construction has plagued the engineering profession and construction industry for many years. If a contractor is engaged in construction for an owner in accordance with a previously prepared design, and the contractor encounters subsurface conditions materially different from what might be expected, resulting in increased costs, should the contractor absorb the increased cost or should the owner pay it?

Many owners or engineers attempt to pass on these unexpected costs to the contractor under the theory that a "grandfather" clause, making the contractor responsible for anything and everything that happens in the way of unexpected contingencies, will amply protect them. Needless to say, when disputes arise, contractors will point to many unusual or extraordinary circumstances that will tend to contradict such "grandfather" clauses. On the other hand, by including such a clause in the agreement, the owner has requested the contractor to assume certain risks for which the contractor must provide contingency allowances in the estimate. These contingencies may or may not arise, and it is therefore an added cost burden because the owner is paying for something that may not come about.

The best view, and the one that is understood by most enlightened engineers, contractors, and owners, is that the contractor should provide in the estimate only for certain agreed-upon assumptions which all of the parties stipulate are the basis for the estimated cost and for the contract. The concept of "agreed-upon assumptions" was originated by the writer in an article published in *Civil Engineering* magazine in June 1961, entitled, "Who Pays for the Unexpected in Subsurface Construction?" There is a recommended contract clause in the article that introduces the phrase. Later, in September 1963, this concept was expanded at a seminar held by the Committee on Contract Administration of the American Society of Civil Engineers in New York City on April 25, 1963. The seminar proceedings were published in the *Journal of the Construction Division, ASCE*, in September 1963. In a discussion published in the same journal, the writer expanded on the concept of the "agreed assumptions."

Any situations or eventualities that are outside of these agreed-upon assumptions, which would have resulted in an additional cost to the owner in the original bid or estimate, will be considered as an unexpected cost which should be reimbursed.

A recommended contract clause for such a provision is as follows:

> The contract documents indicating the design of the portions of the work below the surface are based upon available data and the judgment of the engineer. The quantities, dimensions, and classes of work shown on the contract documents are agreed upon by the parties as embodying the assumptions from which the contract price was determined.
>
> As the various portions of the subsurface are penetrated during the work, the contractor shall promptly, and before such conditions are disturbed, notify the engineer and owner in writing, if the actual conditions differ substantially from those which were assumed. The engineer shall promptly submit to the owner and contractor a plan or description of the modifications he proposes should be made in the contract documents. The resulting increase or decrease in the contract price, or the time allowed for the completion of the contract, shall be estimated by the contractor and submitted to the engineer in the form of a proposal. If approved by the engineer, he shall certify the proposal and forward it to the owner with recommendation for approval. If no agreement can be reached between the contractor and the engineer, the question shall be submitted to arbitration as provided elsewhere herein. Upon the owner's approval of the engineer's recommendation, or receipt of the ruling of the arbitration board, the contract price and time of completion shall he adjusted by the issuance of a change order in accordance with the provisions of the section entitled, "Changes in the Work" and "Extensions of Time."

The foregoing is a business consideration. For a legal consideration, see Chap. 5.

Extras

A contractor's claim for extras under a temporary works contract is more difficult to justify and is more likely to be resisted than in a contract for the permanent structure. The reason is that the temporary structure is generally not fully detailed or designed on the contract drawings, and it is often on the basis of a performance type of specification that the contract is awarded for the temporary structure. Nevertheless, there are many instances in which an extra or change order in favor of the temporary structure contractor is justified. Some instances are changed conditions, delays, interferences by the owner, interferences by separate prime contractors, changes or redesign to the permanent structure, or subsurface conditions materially different from those that are contemplated.

The legal aspects of the claims for extras are dealt with in Chap. 5.

After it has been established that a change order is justified, it will probably be in one of the following forms.

Unit Prices

Unit prices may have been established, prior to entering into the agreement, for possible changes or revisions to the work. These unit prices are generally different for added work than for deducted work. The deducted work unit prices are usually somewhat lower than the unit prices for added work. Should the quantity of work applied under any one unit price be radically different from that which could conceivably have been contemplated, a renegotiation of the unit price is sometimes justified.

Negotiation by Lump Sum

This is the most common type of method of arriving at the amount of an extra or of a change order. Generally speaking, the negotiation will take place on the basis of an estimate which has been prepared by the contractor showing what is believed to be the probable cost of the work. Often, the change order will be negotiated after the extra work is done. Here again, the negotiation may start with the contractor's estimate of the cost of the work as it is performed or with the contractor's actual time sheets or material sheets of the work as it progressed. A final price might be agreed upon after these estimates or actual time sheets or material bills are examined for correctness and proper allocation of costs.

Cost of Labor and Material Plus Markup

This type of change order can result when there is no agreement between the parties as to how best to resolve the value of a contemplated change. When work is done in this manner, costs are kept contemporaneously and daily as the work progresses and the records are signed by representatives of both the contractor and the engineer or owner.

Contractor's Insurance

Insurance requirements fall under three categories:

1. Insurance required by contract, such as liability insurance or builder's risk insurance.
2. Insurance is mandated by law, such as workers' compensation.
3. Insurance is neither required by contract nor mandated by law, but that the contractor may feel is prudent to have as additional protection for good business practice.

Insurance costs have risen spectacularly in the past few years because of the increasing awareness of juries of the high human cost of injuries to persons and the high material costs of damage to property. In addition, many jurors are also aware of the fact that in most instances, recoveries for such damages are covered by insurance and there seems to be a tendency on the part of jurors to award higher recoveries when they are of the opinion that these will be paid for by an insurance company.

Because of these high present-day insurance costs, it is most important that contractors for temporary structures be particularly alert to constant reexamination of their insurance programs. The broker or agent who handles the contractor's insurance is particularly charged in this instance with using intelligence in reappraising the policy each time it is being renewed. The contractor may attempt to be "self-insured" for some of the less likely eventualities that are capable of being covered by insurance. On the other hand, instances where there are frequent losses and where coverage has been inadequate should be boosted up and greater amounts of insurance should always be considered.

Examples of excess insurance are when the contractor is paying for an insurance limit in an amount that is not covered under the policy anyway. This situation can exist, for example, in insurance on buildings under construction under a builder's risk form. This type of policy excludes coverage for demolition work or work below the lowest basement slab and landscaping. Since these are not covered, it is unnecessary to insure such a building for the full contract price.

Liability Insurance

General liability insurance is often purchased to cover the owner and contractor for the risks assumed in temporary construction. The designer should also maintain professional liability [also known as "errors and omissions" (E & O)] insurance to cover these risks. However, it should be noted that E & O liability arising from plans and designs is a common exclusion to most owners' and contractors' liability insurance. Hence, if the contractor is to prepare a design for temporary construction, it is important that the contractor procure Professional Liability E & O insurance coverage for the plans and designs. It should also be noted that general liability insurance only covers the claims of third parties for personal injury or property damage and ordinarily does not cover damage to the owner's own property or the contractor's work.

Thus, if a scaffolding collapses, injuring a pedestrian and damaging the building being worked on, the contractor's general liability insurance in most instances would only cover a claim by the pedestrian and not the cost of repairing the building and replacing the scaffolding.

State workers' compensation laws require employers to provide specific insurance coverage for injury to their employees. However, workers' compensation benefits have not kept pace with inflation and in many instances are grossly inadequate. Injured construction workers are barred by law from suing their employers beyond workers' compensation benefits, but they are not barred from suing others, such as the owner or the designer, for negligence or breach of statutory or contractually imposed duty. For example, in New York State, building owners have an absolute statutory duty to provide workers with a safe place to work, even if they are the employees of an independent contractor hired by the building owner. For that reason, and also because of the owner's vicarious liability to third parties such as adjacent property owners, the owner will often insert a contract clause requiring the contractor to indemnify and hold the owner harmless from all claims and liabilities arising from the contractor's activities, even if the claim is based on the owner's breach of a statutory duty. Sometimes the indemnity is expanded to include the owner's own negligence. Likewise, contractors can seek contractual indemnity from their subcontractors, and designers from contractors.

State laws vary on the enforceability of indemnity for one's own negligence and for defects in a design or plan. Nevertheless, an indemnity is no better than the indemnitor's ability to pay. For that reason, indemnitors are frequently required to provide specific "contractual liability" insurance to cover the added risks assumed by a hold-harmless agreement. This coverage is generally provided in a rider or supplement to the contractor's liability policy. Contractual indemnity and hold-harmless provisions, unless backed by adequate contractual liability insurance coverage, may be of little or no intrinsic value.

Insurance coverage (and exclusions) for temporary, construction are highly complex. The advice of a competent insurance professional should be sought.

Bonding

Performance and payment bonds, when issued by a fidelity company, are a contract in which the company agrees to complete the project if the contractor fails to do so, and agrees to pay all bills for material and labor if the contractor should not do so.

In order to obtain performance and payment bonds, temporary structure contractors must demonstrate to a bonding company that they have adequate finances and experience to perform and complete the jobs being bonded, Generally speaking, a bonding company will favor issuing bonds to those contractors with whom it has previously established a track record, and those who have been able to show continuing financial stability and profits over a period of years. Bonding company relationships are very much like bank relationships. They depend on confidence on the part of the surety in the ability of the contractor, and they depend on experience that has been trouble-free and satisfactory in the past for future business to be conducted.

Cost Analysis

A contractor's cost-keeping system should provide the contractor with information for

Estimating on future contracts

Checking the efficiency of operations to determine whether a completed contract was more or less costly than had been estimated

Controlling the costs of jobs as they are in progress to pinpoint areas where costs are running over, so that corrective action can be taken

An adequate cost system is more than just allocating the total costs of the labor and material for each contract to a specific job number. This is, of course, the simplest type of cost system and is elementary in the contracting business. However, an adequate system will attempt to provide information in all of the above three categories.

Such an adequate cost system would be based on the following two types of reports:

1. Daily work reports, which will show all types of work performed, the quantity of work performed, and the number of work hours and hourly pay rates that are being committed

2. Unit cost reports, which will be prepared periodically to provide data on production, a check on costs at an early date, and records for estimating for the future

Purchasing

In issuing a purchase order for materials to be delivered on a temporary construction contract, the purchase order must take into account the scope of the work and the quantities and unit prices or lump-sum price, and must list inclusions properly, note exceptions or exclusions, and, where practicable, state unit prices for added or deleted work. There must also be listed on the purchase order the time of performance and the dates of delivery.

In purchasing materials for temporary construction contracts, one must also take into account the availability of equipment to perform the work, the adequacy of the plant or supplier to deliver on time, and the reliability of the supplier's previous promises for materials that were purchased.

Purchase orders should also contain a provision for field measurements by the vendor, if this is required, and should indicate whether delivery and transportation charges and sales tax are included in the prices.

Subcontracting

In many instances, the temporary structure contractor will be a subcontractor to a general contractor on the job. This will be the case, for example, in a temporary dewatering subcontract. The negotiation and award of a subcontract will proceed along the following lines:

Price

Time of performance

Availability of labor, material, and equipment exceptions, inclusions, and exclusions

Form and language of subcontract

Terms of payment

Other subcontract clauses that will be important and will be dealt with during the course of the subcontract negotiation are

Subcontractor to be bound by plans and specifications for the general contract Submission of payment application and breakdown with each request for payment

Payment by subcontractor of all monies owed for materials and labor previously paid for by the general contractor on past requisitions

Making of claims promptly

Protection of other trades

Safety requirements

Assignment of subcontract not made without consent of contractor

Guarantees, whether personal or corporate

Pursuit of work diligently

Use of the general contractor's equipment only at the general contractor's discretion reports on material in fabrication

Changes

Coordination

Shop drawings

Laws and permits

Work subject to approval

Hold-harmless clauses

Arbitration clause

A common form of standard subcontract is the AIA Form A401, "Contractor-Subcontractor Agreement."

Financial Controls and Banking

Proper financial controls and banking practices on the part of the contractor include safeguards for issuance of checks and payment of bills. These include, first, that a purchase order or subcontract be issued for each liability that is being incurred by the contractor. When invoices are received, it is important that they be accompanied by a delivery ticket or that a responsible individual who is familiar with the work being invoiced issues an approval to accompany the invoice before it is processed for payment. When the invoice is approved for payment, checks that are issued should be, if possible, on an account that requires two signatures. This will provide for a further safeguard and duplicate scrutiny of the invoice.

Often, it is advantageous to open a separate bank account for each major project. This will aid in cost keeping and will also provide for a stricter control of the expenditure of funds versus the income received for a particular job.

It is also good practice to have banking relationships with more than one bank. These can then be drawn upon, for example, when borrowing or when requesting the issuance of such bank accommodations as letters of credit, with greater strength and facility than if only one bank is relied upon.

SAFETY CODES AND REGULATIONS

The federal government and many states and municipalities have established safety codes concerning construction activities. Failure to comply with the federal code can subject the contractor to substantial civil penalties by the Occupational Health and Safety Administration (OSHA). Moreover, willful disobedience can subject a contractor to criminal penalties. The federal regulations created a duty extending from contractors to their own employees. This duty may not be delegated by the contractor to another. Violation of an OSHA safety provision does not, in and of itself, give an injured employee the right to bring a case in federal court against the employer or anyone else. Rather, this right will depend on the circumstances of each case.

OSHA requires, for example, that there be an engineer-designed system of support where the excavation will not have sloped earth banks. Further, OSHA requires an inspection of the shoring system on a daily basis and after every endangering storm. OSHA will permit any shoring system within accepted engineering standards.

The extent of safety requirements regarding construction activities varies greatly. Generally, state requirements are considerably less extensive than OSHA requirements. In some states, however, the duties owed are broader than under OSHA. In New York State, for example, the duty extends from all contractors and owners connected with the construction or demolition of a building to employees of subcontractors. A violation has been held to entitle a contractor, sued by an employee of a subcontractor, to contribution from the subcontractor to the extent of the subcontractor's fault. Other typical provisions contained in the New York code are the requirement for daily inspection of shoring systems, and prohibitions against placing excavated material within 2 ft (60 cm) of the edge of the open excavation.

Municipal codes are generally somewhat between the state and the federal codes in length but deal with local problems not usually considered in the state or OSHA codes.

For example, under the New York City Administrative Code, abandoned excavations must be filled: controlled inspection must be provided by a licensed architect or engineer for all construction relating to the support of adjacent properties or buildings; excavated material and other superimposed loads must be kept at a certain distance from the edge of the cut in proportion to the depth of the cut, unless the excavation is in rock or the sides of the cut have been sloped or sheet-piled and shored to withstand such loads; and shoring or other protection must be provided for earth cuts below a certain depth unless the sides are of a specified gradual slope.

Because three sets of regulations have been issued at the federal, state, and local levels, contractors are best advised to follow the strictest requirement when the codes merely supplement each other. When there is a direct conflict between state, federal, or city regulation, the federal code should take precedence, followed by the state, and lastly the municipal regulations.

The code will, of course, outline whose responsibility it is for the design, what permits are necessary, which shop drawings are needed, and how supervision and inspection are to be performed.

What happens, however, if there is no code in the particular location where the structure is to be erected, or if the code is silent on the particular type of structure? In the absence of a code, the contractor for the temporary structure has even more responsibility for the observance of good practices. If inexperienced in building the type of structure being provided or in doubt as to the adequacy of the design, the contractor should get supplemental help from those who have expertise in these areas. In addition, the general contractor should be more alert to the possibility of a design or construction failure.

An extended treatment of standards, codes, and regulations for the design and construction of temporary structures is included in Chap. 2.

PROFESSIONAL AND BUSINESS ADVICE

In the following chapters, experts in many of the fields of temporary structures in construction have described their specialties. If readers find themselves involved with temporary structures that are not discussed in this book, perhaps because they are too novel, a few words of advice may be in order.

As a practicing general contractor with five decades of involvement in the civil engineering and construction industry, the writer still feels the inadequacy of his background and experience. Daily, in the conduct of his profession and business, he is assailed with problems that are new and have never been dealt with by him before. In each case, he attempts to obtain a thorough understanding of the issues and problems involved before attempting a solution. However, most importantly, he is constantly seeking the advice and help of others in finding these solutions. There are, in a broad practice of this profession, experts who are dealing with every conceivable problem. When they are sought out, they are usually more than anxious and willing to offer their advice. The writer is not ashamed to say that the advice of others is always given great weight in the conduct of his affairs, but the principles of careful investigation, adequacy of preparation and design, and careful checking of all work will help any practitioner in overcoming the most difficult problems.

CHAPTER 2
STANDARDS, CODES, AND REGULATIONS

John F. Duntemann, P.E., S.E.

GENERAL OVERVIEW OF STANDARDS, CODES, AND REGULATIONS

Numerous standards, codes, and regulations provide guidance for construction in the United States. These publications include national standards, design codes and specifications, model building codes, and state and local codes. Many of the existing standards were developed for permanent construction, and few are specifically written to address temporary structures in construction.

With the exception of the Occupational Safety and Health Administration (OSHA) regulations, as well as related state and local safety regulations, most standards for temporary structures in construction are voluntary. These voluntary standards, or guidelines, are generally produced by a consensus procedure through organizations such as the American National Standards Institute (ANSI), the American Concrete Institute (ACI), and the American Society of Civil Engineers (ASCE). In some cases, provisions of voluntary standards are adopted either in part or as a whole by regulatory agencies, thereby becoming mandatory standards. Many of the ANSI standards, for instance, have been incorporated into the OSHA regulations.

The objective of this chapter is to familiarize the reader with the available national codes, standards, and guidelines, and discuss their applicability to temporary structures in construction.

NATIONAL STANDARDS, SPECIFICATIONS, AND CODES

National Standards

For the purpose of this discussion, *national standards* are defined as standards developed by national organizations or organizations representing a wide variety of regional interests. Included in this group are the American National Standards Institute and the American Society of Civil Engineers. There are also several design codes, such as the American Concrete Institute's *Building Code Requirements for Structural Concrete and Commentary* (ACI 318-08), the American Institute of Steel Construction's *Code of Standard Practice for Steel Buildings and Bridges*, and the American Forest & Paper Association's *National Design Specification for Wood Construction*, that are subject-specific and nationally recognized.[1-3] Many of these documents are developed and written in a form that allows them to be adopted by reference in a general building code. For example, the introduction to ACI 318-08 notes the following:

> The code has no legal status unless it is adopted by government bodies having the police power to regulate building design and construction. Where the code has not been adopted, it may serve as a reference to good practice even though it has no legal status.

Similarly, the AISC *Code of Standard Practice for Steel Buildings and Bridges* states that:

> In the absence of specific instructions to the contrary in the Contract Documents, the trade practices that are defined in this Code shall govern the fabrication and erection of Structural Steel.

American National Standard for Construction and Demolition Operations, ANSI A10.   In 1931, the American Standards Association (ASA), which is now known as the American National Standards Institute (ANSI) Committee on Standards for Safety in the Construction Industry, issued the *American Safety Code for Building Construction*. Since then, it has been updated as a series of standards known as *American National Standard for Construction and Demolition Operations*, ANSI A10. The topics addressed by ANSI A10 include the following.[4-6]

ANSI Standard A10.8—*Scaffolding* covers a broad range of scaffold types, including general requirements and provisions for platforms, tube and coupler scaffolds, and fabricated tubular frame scaffolds commonly used for construction access.

ANSI Standard A10.9—*Concrete and Masonry Work* contains sections on concrete placement, vertical shoring, formwork, prestressed concrete, precast concrete, lift slab operations, and masonry construction. General guidelines in regard to construction means and methods are discussed. Minimum design loads and safety factors are also specified.

ANSI Standard A10.13—*Steel Erection* contains general guidelines on temporary flooring, bolting and "fit-up," building and bridge erection, dismantling, mill work, and working over water. Like the other ANSI standards, ANSI A10.13 tends to be relatively prescriptive and cross-references a variety of related standards.

At present, there are 46 existing or proposed standards under development in the A10 series for safety requirements in construction and demolition operations. A list of the A10 series can be found in the foreword of most of these standards.

Minimum Design Loads for Buildings and Other Structures, ASCE 7-10. A report of the Department of Commerce Building Code Committee, entitled "Minimum Live Loads Allowable for Use in Design of Buildings," was published by the National Bureau of Standards in 1924. The recommendations contained in that document were widely used in revision of local building codes. These recommendations, based upon the engineering data available at that time, represented the collective experience and judgment of the committee responsible for drafting this document.

The ASA Committee on Building Code Requirements for Minimum Design Loads in Buildings subsequently issued a report in 1945 that represented a continuation of work in this field. This committee took into consideration the work of the previous committee and expanded on it to reflect current knowledge and experience. The end result was *American Standard Building Code Requirements for Minimum Design Loads in Buildings and Other Structures*, A58.1-1945.

The A58.1 standard has been revised nine times since 1945, the latest revision corresponding to ASCE 7-10, *Minimum Design Loads for Buildings and Other Structures*.[7] Subsequent to the 1982 edition of ANSI A 58.1, the American National Standards Institute (ANSI) and the ASCE Board of Direction approved ASCE rules for the standards committee to govern the writing and maintenance of the ANSI A58.1 standard. The current document prescribes load combinations, dead loads, live loads, soil and hydrostatic pressures, wind loads, snow loads, rain loads, and earthquake loads. Like earlier editions of the ANSI standard, ASCE 7 has significantly influenced the development and revision of other building codes.

Design Loads on Structures During Construction, SEI/ASCE 37-02. In 2002, ASCE published the SEI/ASCE 37-02, *Design Loads on Structures During Construction*.[8] (At the time of this writing, ASCE 37 is in balloting for the next edition intended for publication in 2012. Since balloting is not complete, there might be some differences between content described herein and content in the final version of ASCE 37-12.)

The objective of the standard is to establish performance criteria, design loads, load combinations, and safety factors to be used in the analysis and design of structures during their transient stages of construction as well as temporary structures used in construction operations. The standard is composed of six sections corresponding to a general introduction identifying the purpose and scope of the document, loads and load combinations, dead and live loads, construction loads, lateral earth pressures, and environmental loads.

The construction loads, load combinations, and load factors were developed to account for the relatively short duration of load, variability of loading, variation in material strength, and the recognition that many elements of the completed structure that are relied upon implicitly to provide strength, stiffness, stability, or continuity are not present during construction. The load factors are based on a combination of probabilistic analysis and expert opinion. The concept of using maximum and arbitrary point-in-time (APT) loads and corresponding load factors is adopted to be consistent with ASCE 7.

The basic reference for the computation of environmental loads is also ASCE 7. However, modification factors have been adopted to account for reduced exposure periods. Furthermore, certain loads may be disregarded due to the relatively short reference period associated with typical construction projects, and certain loads in combinations may effectively be ignored because of the practice of shutting down job sites during these events, for example, excessive snow and wind.

Design Codes and Specifications

While many of the national design codes can also be applied to temporary structures in construction, specific commentary on design criteria and construction methods is limited. Some notable exceptions, however, are the AISC *Specification for Design, Fabrication and Erection of Structural Steel for Buildings*, the AISC *Code of Standard Practice for Steel Buildings and Bridges*, the American Association of State Highway and Transportation Officials (AASHTO) *Guide Design Specifications for Bridge Temporary Works*, and the AASHTO *LRFD Bridge Construction Specifications*.[9-11] The latter two publications are undergoing revisions at the time of this writing and will be republished in 2013.

AISC Specifications and Code of Standard Practice.   As evident by the title, the AISC *Specification for Design, Fabrication and Erection of Structural Steel for Buildings* applies to both temporary and permanent construction. As in earlier editions, many of the general provisions, such as stability and slenderness, are also applicable to temporary structures. The specifications also contain a section "Fabrication, Erection and Quality Control," designated Chapter M.

The practices defined in the AISC *Code of Standard Practice for Steel Buildings and Bridges* have been adopted by the AISC as the commonly accepted standards of the structural steel fabricating industry. In the absence of other instructions in the contract documents, the trade practices defined in this code govern the fabrication and erection of structural steel. Specific sections related to temporary structures in construction include Section 6, "Shop Fabrication and Delivery," and Section 7, "Erection." Basic fabrication tolerances are stipulated in Section 6.4 of the code.

The erection tolerances defined in Section 7.13 of the code have been developed through long-standing usage as practical criteria for the erection of structural steel. The current requirements were first published in the October 1, 1972, edition of the code. The basic premise that the final accuracy of location of any specific point in a structural steel frame results from the combined mill, fabrication, and erection tolerances, rather than from erection tolerances alone, remains unchanged.

Section 7.10 of the code, entitled *Temporary Support of Structural Steel Frames*, presents some basic guidelines on temporary support. Temporary supports, such as guys, braces, falsework, cribbing, or other elements required for erection, are specified to be determined, furnished, and installed by the erector.

AASHTO Guide Design Specifications for Bridge Temporary Works.   The AASHTO *Guide Design Specifications for Bridge Temporary Works* was developed for use by state agencies to update their existing standard specifications for falsework, formwork, and related temporary structures and was first published in 1995. The *Guide Design Specifications* provides unified design criteria that reflect the state of practice at the time the specifications were developed. The specifications were prepared in a format similar to the AASHTO *Standard Specifications for Highway Bridges*.

In this document, "falsework" is defined as temporary construction used to support the permanent structure until it becomes self-supporting. "Shoring" is generally considered a component of falsework, such as horizontal or vertical support members, but often used interchangeably with falsework. "Formwork" is a temporary structure or mold used to retain plastic or fluid concrete in its designated shape until it hardens. "Temporary retaining structures" are both earth-retaining structures and cofferdams.

The *Guide Specifications* contains four sections: Introduction, Falsework, Formwork, and Temporary Retaining Structures. A brief description of the falsework, formwork, and temporary retaining structures sections is as follows:

Falsework. The falsework provisions include four general topics: materials, loads, design considerations, and construction. Allowable stress provisions for steel and timber, as well as modification factors for salvaged (used) materials, are identified. Safety factors and limitations of manufactured (proprietary) components are also as specified. Four general load categories including environmental loads are defined. The basic reference for computation of wind load is the *Uniform Building Code*.[12] General design topics such as load combinations, stability against overturning, traffic openings, and foundations are addressed. Presumptive soil-bearing values are also provided. Construction topics include foundation protection; erection tolerances; and clearances of traffic openings, adjustment methods, and removal.

The specification is supplemented with commentary and appendices, which include design values for ungraded structural lumber, provisions for steel beam webs and flanges under concentrated forces, design wind pressure from selected model codes, and foundation investigation and design.

Formwork. ACI 347-88 along with ACI SP-4, *Formwork for Concrete*, served as the principal reference documents for this section.[13,14] Formwork includes materials, loads, and construction. Requirements for sheathing, form accessories, prefabricated formwork, and stay-in-place formwork are specified, as are minimum vertical and horizontal loads. The ACI equations for lateral pressure of fluid concrete are adopted, and the limitations of these equations are discussed in the commentary. Construction topics such as form removal, placement of construction joints, and tolerances are also discussed.

Temporary Retaining Structures. Although developed primarily to address earth-retaining systems more common to bridge construction, this section also applies to temporary cofferdams. General requirements and types of excavation support are identified. Federal standards of OSHA and other regulations are referenced.[15] Empirical methods for determining design lateral pressures in various soils and their limitations are identified. The simplified earth pressure distributions presented in the AASHTO 1991 *Interim Specifications for Highway Bridges* are adopted.[16]

AASHTO LRFD Bridge Construction Specifications. In 1991, the AASHTO *Interim Specifications* contained a newly created section entitled "Temporary Works," which included subsections on falsework and forms, cofferdams and shoring, temporary water controls systems, and temporary bridges. This section was developed, in part, to update Division II and consolidate information found in other parts of the AASHTO *Standard Specifications for Highway Bridges*. The section on falsework and forms includes design criteria as well as guidelines for removal of these temporary structures. The other sections tend to be more general in content.

With the adoption of the AASTHO *LRFD Bridge Construction Specifications* in 1998, AASHTO moved the Division II—Construction provisions that appeared in the *Standard Specifications for Highway Bridge Structures* to the AASHTO *LRFD Bridge Construction Specifications*. Further discussion regarding the evolution of these specifications can be found in NCHRP Research Results Digest No. 198 available from the Transportation Research Board.[17]

The AASHTO *LRFD Bridge Construction Specifications* consists of 32 sections. Section 3—"Temporary Works" references the documents that were developed by the Federal Highway Administration (FHWA) Bridge Temporary Works Research Program including the *Guide Design Specification for Bridge Temporary Works* and the *Construction Handbook for Bridge Temporary Works*.[18,19] This section includes general requirements regarding working drawings, design and removal and specific requirements regarding falsework and forms, cofferdams and shoring, temporary water control systems, temporary bridges and measurement and payment. The section on falsework and forms contains similar provisions to the construction provisions found in the *Guide Design Specifications for Bridge Temporary Works*.

The AASHTO *LRFD Bridge Construction Specifications* requires that the design of temporary works be in accordance with the AASHTO *LRFD Bridge Design Specifications* or the *Guide Design Specifications for Bridge Temporary Works* unless another recognized specification is accepted by the engineer (presumably the state bridge engineer or governing authority). The AASHTO *LRFD Bridge Construction Specifications* notes that the design of access scaffolding is subject to the OSHA regulations. The AASHTO *LRFD Bridge Construction Specifications* also requires the working drawings for falsework to be prepared and sealed by a registered professional engineer when the height of the falsework exceeds 14 ft. The specifications require the design of formwork to conform with ACI 347-04, "Guide to Formwork for Concrete."[20]

State and Local Codes

Until recently, most local building codes in the United States were patterned after the so-called model building codes, which included the *National Building Code* by the Building Officials and Code Administrators (BOCA), the *Uniform Building Code* by the International Conference of Building Officials (ICBO), and the *Southern Standard Building Code* by the Southern Building Code Congress (SBCC).[21,22]

In 2000, the International Code Council (ICC) published the first edition of the *International Building Code*.[23] This code was the culmination of an effort initiated in 1997 by the ICC that included five drafting subcommittees appointed by the ICC and consisted of representatives of BOCA, ICBO, and SBCC. The intent was to draft a comprehensive set of regulations for building systems consistent with the existing model codes. The technical content of the latest model codes promulgated by BOCA, ICBO, and SBCC was utilized as the basis for the development. The 2006 edition presents the code as originally issued, with changes approved through 2005. A new edition of the code is produced every 3 years.

With the development and publication of the *International Building Code* in 2000, the continued development and maintenance of the model codes individually promulgated by BOCA, ICBO, and SBCC was discontinued. The 2000 *International Building Code*, and subsequent editions of the code, was intended to be the successor building code to those codes previously developed by BOCA, ICBO, and SBCC.

Discussion related to construction safety and temporary construction in these model codes is generally minimal or nonexistent. The model codes, however, adopt many of the national design standards developed by organizations such as the American Concrete Institute (ACI), the American Institute of Steel Construction (AISC), and the National Forest Products Association (NFPA), all of which have application to temporary structures. The model codes also adopt by reference many of the American Society for Testing and Materials (ASTM) Standards as the recognized test procedures to ensure construction quality.

Related Foreign Standards

Several countries, such as Canada and Britain, have comprehensive national standards that address the more common temporary structures used in construction, such as falsework and scaffolding. These standards serve as good reference documents and are briefly discussed herein.

Canada. In 1975, the Canadian Standards Association (CSA) published a national standard entitled *Falsework for Construction Purposes,* designated CSA Standard S269.1-1975.[24] This standard provides rules and requirements for design, fabrication, erection, inspection, testing, and maintenance of falsework. The falsework standard was prepared by the Technical Committee on Scaffolding for Construction Purposes and is one of the first national standards developed on this subject. CSA Standard S269.1-1975 was reaffirmed in 2003.[25]

CSA Standard S269.1-1975 R2003 adopts the *National Building Code of Canada* and existing CSA Standards by reference. Materials that cannot be identified as complying with specified standards are not allowed for falsework construction.

In addition to material and design standards, CSA S269.1-1975 R2003 specifies design loads and forces, analysis and design methods, erection procedures, and test procedures for steel shoring systems and components. Vertical loads are generally prescribed in terms of a uniformly distributed load. Loads due to special conditions such as impact, asymmetrical placement of concrete, and overpressures due to pneumatic pumping are discussed but not quantified. Horizontal loads are specified as either the lateral wind force found in the *National Building Code of Canada,* or 2 percent of the total vertical load, whichever is greater. Design capacity is determined by existing CSA design codes or, where proprietary components are used, based upon test results with prescribed factor(s) of safety. Additional requirements for tubular scaffold frames and wood falsework are specified.

Tolerances for vertical load-carrying members are also specified, and general inspection guidelines are discussed and illustrated. Test procedures and safety factors for welded tubular frame scaffolding, tube and coupler scaffolding and components, vertical shores, and horizontal shores are prescribed.

Great Britain. The British *Code of Practice for Falsework* (BS 5975) was originally published in 1982 and is similar in format to the Canadian falsework standard in that the content is organized under the general headings of procedures, materials and components, loads, foundations and ground conditions, design, and construction.[26] However, it also contains a considerable amount of in-depth commentary and several detailed appendices, which include properties of components in tube and coupler falsework, design of steel beams at points of reaction or concentrated load, effective lengths of steel members in compression, and so forth. The most recent edition of BS 5975 was published in 2008.[27]

One of the unique features of the British code is the distinction made between maximum wind force during the life of the falsework, which represents an extreme condition, and a maximum allowable wind force during construction operations. Forces from both of these conditions are used to check the stability of the falsework at appropriate stages of construction.

The British *Code of Practice for Falsework* is relatively complete with respect to foundations and ground conditions for temporary works. Pile foundations are addressed in a separate British standard on foundations. BS 5975 includes allowable bearing pressures for a wide range of rock and soil types. The British *Code of Practice* includes modification factors which—depending upon the reliability of site information, magnitude of anticipated settlement, and fluctuations in groundwater level—are applied to the prescribed bearing pressure. The code also contains some specific guidelines for the protection of foundation areas.

In addition to the *Code of Practice for Falsework*, the British Standards Institute publishes BS 5973 *Code of Practice for Access and Working Scaffolds and Special Scaffold Structures in Steel* and BS1139 *Metal Scaffolding*.[28,29] Although the latter standards are primarily descriptive in terms of materials and tolerances, they also provide guidance on bracing arrangements, effective lengths, joint eccentricity, and allowable loads for couplers and fittings.

OSHA AND STATE REGULATIONS

The regulations most profoundly affecting temporary structures in construction are those of the Occupational Safety and Health Administration (OSHA) of the U.S. Department of Labor. The Occupational Safety and Health Act of 1970 was enacted to provide a safe workplace and requires employers to provide a job environment that is free from hazards that can cause serious physical harm or death. The OSHA act applies to virtually all employers, but specifically to those engaged in construction and commerce. Failure to meet standards or comply with the provisions of the act can subject the contractor to substantial civil penalties. Moreover, willful disobedience can subject a contractor to criminal penalties.

The inspection and enforcement of OSHA regulations is performed by regional offices. OSHA inspectors are generally allowed to enter any construction site at reasonable times and without undue delay. An OSHA inspector who is refused admission can obtain a search warrant. Inspections can also be initiated by accidents, collapses, or employee complaints. In addition to federal OSHA, compliance checks of construction sites are often made by state or local agencies, most insurance carriers, and some private consultants.

Although most states administer their own occupational safety and health programs, they generally adopt the federal OSHA regulation or similar requirements. Because three sets of regulations may apply at the federal, state, and local levels, contractors are advised to follow the strictest requirement when the codes merely supplement each other. When there is a direct conflict between state, federal, or city regulation, the federal code should take precedence, followed by the state and finally the municipal regulations.

OSHA Regulation 29CFR, Part 1926, defines mandatory requirements to protect employees from the hazards of construction operations. Part 1926 has 30 subparts, or subdivisions, which include Subpart L—*Scaffolding*, Subpart P—*Excavations*, Subpart Q—*Concrete and Masonry Construction*, Subpart R—*Steel Erection*, and Subpart S—*Underground Construction, Caissons, Cofferdams and Compressed Air*. Some of the more relevant OSHA regulations to temporary structures in construction are as follows:

Subpart L—Scaffolding. The OSHA regulations are patterned after ANSI A10.8—*Scaffolding*, and organized into general and type-specific provisions. The general provisions apply to scaffolding systems of all types, while the type-specific provisions can vary depending upon the particular type of scaffolding.

OSHA adopts the light-, medium-, and heavy-duty definitions found in ANSI A10.8, and described as follows:

Light duty: Scaffold design for a 25 lb/ft² maximum working load to support workers and tools only. Equipment or material storage on the platform is not allowed.

Medium duty: Scaffold designed for a 50 lb/ft² maximum working load for workers and material, often intended for bricklayers' and plasterers' work.

Heavy duty: Scaffold designed for a 75 lb/ft² maximum working load for workers and material storage, often intended for stone masonry work.

Similar load ratings are presented for platform units, such as wood scaffold planks, scaffold decks, and fabricated planks and platforms. Specific provisions are presented for the wide variety of scaffold types, including the tube and coupler; fabricated tubular frame; and suspended, bracket, and form scaffolds common on most construction sites.

Subpart P—Excavations. This subpart applies to all open excavations including trenches, which are defined as narrow excavations where the depth is greater than the width and the width is not greater than 15 ft. Section 1926.652 prescribes the requirements for protection systems. Specific topics include protection of employees in excavations; design of sloping and benching systems; design of support systems, shield systems, and other protective systems; materials and equipment; installation of removal and support; sloping and benching systems; and shield systems. For excavations greater than 20 ft in depth, OSHA requires the protective system be designed by a registered professional engineer.

Subpart Q—Concrete and Masonry Construction. Subpart Q contains requirements for cast-in-place concrete, precast concrete, lift-slab construction operations, and masonry construction. The provisions of ANSI Standard 10.9-1983 are non-mandatory guidelines referenced in an appendix to Section 1926.703, *Requirements for Cast-in-Place Concrete.* Similarly, App. A to Subpart Q lists a variety of related ACI, ANSI, and ASTM standards, including *Building Code Requirements for Structural Concrete and Commentary* (ACI 318-83), "Guide to Formwork for Concrete" (ACI 347-78), and *Formwork for Concrete* (ACI SP-4). The documents in App. A are also identified as non-mandatory references and do not necessarily reflect their most recent editions.

RECOMMENDED PRACTICE AND GUIDELINES

Perhaps some of the best sources of information related to the design, erection, and construction of temporary structures can be found in the publications produced by various industry groups. Several of these groups have already been identified in this chapter. Others include the Precast/Prestressed Concrete Institute (PCI), the American Institute of Timber Construction (AITC), the Truss Plate Institute (TPI), and the Scaffolding, and Shoring and Forming Institute (SSFI). Private industry groups, such as U.S. Steel Corporation, also produce related publications. In addition, there are several federal and state organizations, such as the FHWA, AASHTO, the Department of Commerce, and the Army and the Navy that publish some noteworthy manuals. A selective list of these publications is as follows: ACI 347-04 "Guide to Formwork for Concrete," ACI Publication SP-4, *Formwork for Concrete*, the *PCI Design Handbook*, PCI *Recommended Practice for Erection of Precast Concrete*, AITC *Timber Construction Manual*, TPI *Commentary and Recommendations for Handling and Erecting Wood Trusses*, the *Steel Sheet Piling Design Manual* by U.S. Steel Corporation, the *California Falsework Manual*, and the AASHTO *Construction Handbook for Bridge Temporary Works.*[30-37]

ACI 347-04 "Guide to Formwork for Concrete" is the basic source document for many other codes and standards and has been adopted in its entirety as an ANSI standard. The standard describes various design and construction considerations, and includes special guidelines for shoring and reshoring multistory structures, slipforming, and bridge construction. ACI Publication SP-4, *Formwork for Concrete*, serves as a commentary to ACI 347-04, and includes design aids and illustrative examples and figures. Although ACI 318-08 *Building Code Requirements for Structural Concrete and Commentary* includes some general provisions for design of formwork and removal of forms and shores, it references ACI 347-04 in the *Commentary*.

The *PCI Design Handbook* contains several chapters, addressing general topics such as product information, analysis and design of precast prestressed concrete structures, design of components, and design of connections. Also included are chapters on product handling, erection bracing, and tolerances. The section on erection bracing discusses recommended loads used for erection design, suggested factors of safety, bracing equipment and related materials, and erection analysis and sequencing. This particular chapter is a unique and comprehensive treatise on erection procedures. PCI also publishes another related document, entitled *Recommended Practice for Erection of Precast Concrete*. This document is more comprehensive than the passages found in the *PCI Design Handbook* and is well illustrated.

The AITC *Timber Construction Manual* is principally a design manual, applicable to a wide variety of temporary structures. Related documents include AITC 104—*Typical Construction Details*, AITC 108—*Standard for Heavy Timber Construction*, and AITC 112—*Standard for Tongue-and-Groove Heavy Timber Roof Decking*.

The Truss Plate Institute publishes two very good guidelines related to wood truss construction, entitled *Guide to Good Practice for Handling, Installing, Restraining & Bracing of Metal Plate Connected Wood Trusses* and *Bracing Wood Trusses: Commentary and Recommendations*.

The *Steel Sheet Piling Design Manual* was prepared by U.S. Steel Corporation to be used by engineering design professionals to design steel sheet pile-retaining structures. Emphasis is placed on step-by-step procedures for estimating the external forces on the structure, evaluating the overall stability, and sizing the sheet piling and other structural elements. Three basic types of sheet pile structures are considered: (1) cantilevered and anchored retaining walls, (2) braced cofferdams, and (3) cellular cofferdams. Consideration is also given to the design of anchorage systems for walls and bracing systems for cofferdams. Graphs and tables are included as design aids.

The California Department of Transportation (Caltrans) *Falsework Manual* is one of the more authoritative documents written on the subject of falsework used in bridge construction. The first edition of the *Falsework Manual* was published in 1977 to fill a long-recognized need for a comprehensive design and construction manual devoted to bridge falsework. Topics include design considerations and stability. The *Falsework Manual* also includes some specific guidelines and design procedures for pile foundations. The majority of the discussion on pile foundations relates to timber pile bents and an empirical method of analysis based on a modified combined stress equation. The derivation of the equation is described in detail and is based upon field testing and analytical studies. General guidelines with respect to required penetration, point-of-fixity, and soil relaxation are also discussed.

The AASHTO *Construction Handbook for Bridge Temporary Works* was developed for use by contractors and construction engineers involved in bridge construction on federal highway projects. This document supplements information found in the *Guide Design Specifications for Bridge Temporary Works*. The content is construction-oriented, focusing primarily on standards of material quality and means and methods. The handbook contains chapters on falsework, formwork, and temporary retaining structures.

Chapter 2, "Falsework" identifies material standards, the assessment and protection of foundations, construction-related topics, loading considerations, and inspection guidelines. Methods for in situ testing of foundations are identified. General guidelines regarding timber construction, proprietary shoring systems, cable bracing, bridge deck falsework, and traffic openings are also discussed.

Chapter 3, "Formwork" identifies and describes the various components and formwork types commonly used in bridge construction. Information on load considerations and design nomographs is provided. General guidelines relating to formwork construction and form maintenance are also discussed.

Chapter 4, "Temporary Retaining Structures" focuses primarily on cofferdams and their application to bridge construction. As indicated by the chapter title, however, general topics relating to a wide range of temporary retaining structures are also addressed. Specific topics include classification of construction types, relative costs, sealing and buoyance control, seepage control, and protection. The construction of timber sheet pile cofferdams, soldier pile and wood lagging cofferdams, and steel sheet piles cofferdams is reviewed. Methods of internal bracing and soil and rock anchorage are also discussed.

While the publications identified in this chapter are some of the more widely distributed documents, this information by no means represents the full extent of available literature on the subject of temporary structures in construction. Other related publications and guidelines are identified in a bibliography at the end of this chapter. Some of these publications are referenced and further discussed in subsequent chapters.

REFERENCES

National Standards and Codes

1. ACI Committee 318: *Building Code Requirements for Structural Concrete and Commentary* (ACI 318-08), American Concrete Institute, Detroit, Mich., 2008.

2. American Institute of Steel Construction: *Code of Standard Practice for Steel Buildings and Bridges* (AISC 303-10), Chicago, Ill., 2010.

3. American Forest & Paper Association: *ANSI/AF&PA NDS-2005 National Design Specification (NDS) for Wood Construction*, American Wood Council, Leesburg, Va., 2005.

4. American National Standards Institute: *Safety Requirements for Scaffolding* (ANSI/ASSE A10.8-2001) New York, 2001. (Withdrawn, for historical purposes only.)

5. American National Standards Institute: *American National Standard for Construction and Demolition Operations: Concrete and Masonry Work—Safety Requirements* (ANSI A10.9-1997 R2004), New York, 2004.

6. American National Standards Institute: *Safety Requirements for Steel Erection (ANSI/ASSE A10.13-2011), American National Standards Institute*, New York, 2011.

7. American Society of Civil Engineers: *Minimum Design Loads for Buildings and Other Structures* (ASCE 7-10), New York, 2010.

8. ASCE Standards Committee: *Design Loads on Structures During Construction* (ASCE 37-02), American Society of Civil Engineers, Reston, Va., 2002.

9. American Institute of Steel Construction: *Specification for Structural Steel Buildings* (ANSI/ AISC 365-05), Chicago, Ill., 2005.

10. American Association of State Highway and Transportation Officials: *Guide Design Specifications for Bridge Temporary Works*, Washington, D.C., 1995, p. 89.

11. American Association of State Highway and Transportation Officials: AASHTO *LRFD Bridge Construction Specifications*, 2d ed., Washington, D.C., 2004.

12. International Conference of Building Officials: *Uniform Building Code*, 1991 ed., Whittier, Calif., 1991.

13. American Concrete Institute: "Guide to Formwork for Concrete (ACI 347R-88)," *ACI Manual of Concrete Practice, Part 2*, Detroit, Mich., 1990.

14. Hurd, M. K and ACI Committee 347: *Formwork for Concrete* (SP-4), 5th ed., American Concrete Institute, Detroit, Mich., 1989.

15. Occupational Safety and Health Administration Safety and Health Standards, CFR Part 1926—Safety and Health Regulations for Construction, U.S. Department of Labor, Washington, D.C., 2011.

16. American Association of State Highway and Transportation Officials: *Interim Specifications for Highway Bridges*, Washington, D.C., 1991.

17. Transportation Research Board, National Cooperative Highway Research Program, "Development of Comprehensive Bridge Specifications and Commentary." *NCHRP Research Results Digest*, no. 198, Washington D.C., May 1998.

18. Duntemann, J. F., L. E. Dunn, S. Gill, R. G. Lukas, and M. D. Kaler: *Guide Design Specification for Bridge Temporary Works*, FHWA Report No. FHWA-RD-93-032, Federal Highway Administration, Washington, D.C., November 1993, p. 87.

19. Duntemann, J. D, F. Calabrese, and S. Gill: *Construction Handbook for Bridge Temporary Works*, Report FHWA-RD-93-034, FHWA, U.S. Department of Transportation, November 1993.

20. ACI Committee 347: "Guide to Formwork for Concrete (ACI 347-04)," *ACI Manual of Concrete Practice, Part 3*, American Concrete Institute, Detroit, Mich., 2011.

21. Building Officials and Code Administrators International, Inc.: *The BOCA National Building Code/1999*, Country Club Hills, Ill., 1999.

22. Southern Building Code Congress International, Inc.: *Standard Building Code*, Birmingham, Ala., 1999.

23. *International Building Code*, International Code Council, Country Club Hills, Ill., 2000.

24. Canadian Standards Association: *Falsework for Construction Purposes* (CSA Standard S269.1-1975), Rexdale, Ontario, Canada, 1975.

25. Canadian Standards Association: *Falsework for Construction Purposes* (CSA Standard S269.1-1975 R2003), Rexdale, Ontario, Canada, 1975 (Reaffirmed 2003).

26. British Standards Institution: *Code of Practice for Falsework* (BS 5975:1982), London, England, 1982.

27. British Standards Institution: *Code of Practice for Falsework* (BS 5975:2008), London, England, 2008.

28. *Code of Practice for Access and Working Scaffolds and Special Scaffold Structures in Steel*, BS 5973. London, England: British Standards Institution, 1993.

29. *Metal Scaffolding*, BS1139-2.2:2009. London, England: British Standards Institute, 2009.

Recommended Practice and Guidelines

30. Precast/Prestressed Concrete Institute: *PCI Design Handbook*, 7th ed., Chicago, Ill., 2010.

31. Precast/Prestressed Concrete Institute: *Recommended Practice for Erection of Precast Concrete*, Chicago, Ill., 1985.

32. American Institute of Timber Construction: *Timber Construction Manual*, 5th ed., Wiley, New York, 2005.

33. WTCA and Truss Plate Institute: *Guide to Good Practice for Handling, Installing, Restraining & Bracing of Metal Plate Connected Wood Trusses*, WTCA, Madison, Wisc., 2008.

34. WTCA and Truss Plate Institute: *Bracing Wood Trusses: Commentary and Recommendations*, WTCA, Madison, Wisc., 2008.

35. U.S. Steel Corporation: *Steel Sheet Piling Design Manual*, Pittsburgh, Pa., 1984.

36. California Department of Transportation: *California Falsework Manual, Rev. 35*, Division of Structures, California Department of Transportation, Sacramento, Calif., 2010.

37. American Association of State Highway and Transportation Officials: *Construction Handbook for Bridge Temporary Works*, 1st ed., with 2008 Interim Revisions, American Association of State Highway and Transportation Officials, Washington, D.C., 1995.

BIBLIOGRAPHY

American Association of State Highway and Transportation Officials, Washington, D.C., and National Steel Bridge Alliance, Chicago, *Steel Bridge Fabrication Guide Specification*, S2.1-2002.

American Association of State Highway and Transportation Officials: *Guidelines for Design for Constructability*, Washington, D.C., 2003.

American Plywood Association: *Concrete Forming*, Form C345P, American Plywood Association, Tacoma, Wash., 1988.

American Plywood Association: *Concrete Forming*, Form No. V345U, Tacoma, Wash., 2004.

California Department of Transportation: *Trenching and Shoring Manual*, Division of Structures, California Department of Transportation, Sacramento, Calif., 1977.

Dayton-Superior Corporation: *Bridge Deck Forming Handbook*, Miamisburg, Ohio, 1985 (Rev. 6-88A).

Department of the Navy, Naval Facilities Engineering Command: *Foundations and Earth Structures,* NAVFAC DM-7, Alexandria, Va., May 1982.

Goldberg, D. T., W. E. Jaworski, and M. D. Gordon: *Lateral Support Systems and Underpinning* Vols. I, II, III, Federal Highway Administration Report FHWA-RD-75-128, 129, 130, Washington, D.C., 1976.

Harris, F.: *Ground Engineering Equipment and Methods*, McGraw-Hill, New York, 1983.

National Cooperative Highway Research Program, *Steel Bridge Erection Practices NCHRP Synthesis 345*, Washington, D.C., 2005.

Peck, R. B., W. E. Hanson, and T. H. Thornburn: *Foundation Engineering*, 2d ed., Wiley, New York, 1974.

Peurifoy, R. L.: *Formwork for Concrete Structures*, 2d ed., McGraw-Hill, New York, 1976.

Scaffolding, Shoring, and Forming Institute, Inc.: *Guidelines for Safety Requirements for Shoring Concrete Formwork*, Publication SH306, Cleveland, Ohio, 1990.

Scaffolding, Shoring, and Forming Institute, Inc.: *Recommended Procedure for Compression Testing of Welded Frame Scaffolds and Shoring Equipment*, Publication S102, Scaffolding, Shoring, and Forming Institute, Inc., Cleveland, Ohio, 1989.

Tomlinson, M. J.: *Foundation Design and Construction*, 3d ed., Wiley, New York, 1975.

Yokel, F. Y.: NBS Building Science Series 127—Recommended Technical Provisions for Construction Practice in Shoring and Sloping of Trenches and Excavations, U. S. Department of Commerce, Washington, D. C., June 1980.

CHAPTER 3

CONSTRUCTION AND ENVIRONMENTAL LOADS

Donald O. Dusenberry, P.E.

PURPOSE AND SCOPE

Temporary structures, in the broadest sense, are built to perform a specific purpose for a limited time. Whether they are special-purpose structures that are constructed to be disassembled at the end of their useful life or they are temporary configurations of portions of structures that are being built for a long-term function, they must sustain loads from the environment and from the construction process. In any case, the loads that temporary structures in construction are likely to experience have little relationship to the forces that the related finished structure is likely to encounter over its life.

This chapter is intended to explain some of the reasons forces on temporary structures differ from those on permanent structures and gives some guidance on where to find information on temporary loads and how to calculate them.

GENERAL CONSIDERATIONS FOR LOADS

Loads on temporary structures are transient and temporal. They move from place to place, having dynamic components. Some induce impact. Usually construction loads, which might be applied for only a brief time, are substantially different in magnitude and configuration than loads that are likely in a finished structure.

When taken in the context of a building under construction, intended for an occupancy that is entirely different from the construction environment, these characteristics of construction loads imply that careful planning is needed to accommodate construction loads on temporary structures and structures in interim configurations. The planning needs to verify that temporary configurations have adequate strength to support construction loads, whether additional strength must be provided in the design of the structure under construction to accommodate construction loads, or procedural measures need to be implemented to limit loads during construction.

One of the principal challenges when designing temporary structures for construction is to plan the construction process sufficiently to be able to anticipate and establish structural configurations that provide the best platform for support of construction loads. For complicated construction projects, this usually requires early development of a construction sequence plan. That plan should identify materials handling needs, loads associated with those processes, site planning for the provision of loads handing equipment, loads associated with erection processes, and protocol for safely implementing the plan. The plan also requires anticipation of environmental loads that are likely to occur during the construction period. With that information in hand, engineers designing for construction loads can anticipate loads associated with interim structural configurations and provide responsive designs.

Many load-generating activities do not impact a structure under construction. Supports for construction loads can be on structures that are entirely separate from the structure under construction. As an obvious example, a mobile crane parked on grade next to a building without a basement usually does not induce loads on the building.

More commonly though, temporary structures interact with the structure under construction, relying on the completed portion of the structure for partial support. For example, scaffolding erected next to a structure usually supports its own weight and the weight of materials placed on it. However, for lateral stability and for support of horizontal loads from wind, tall scaffolding (Fig. 3.1) normally is attached to the structure under construction and relies on it for resistance.

In these cases, the resulting loads on the structure under construction may be similar during the construction phase to loads that are anticipated on the finished structure, or they may be entirely different. Differences occur, for example, when the temporary structure has a substantially different silhouette to the wind as compared to the finished structure. Clearly, wind load on a temporary structure will be less than that on the finished structure if the temporary structure is only in place during seasons when winds are relatively calm.

Then there are temporary structures that rely entirely on the structure under construction. In construction, it is routine to support concrete formwork on finished floors as a new floor is added to a building. Depending on the relative magnitudes of the loads used for design of the finished structure and the weight of the construction materials to be supported, it could be possible to rely entirely on the design for in-service loads to support construction loads.

FIGURE 3.1 Towers for movable scaffold are attached to the building for lateral support.

For construction loads that do not compare favorably to the loads anticipated for the finished structure, one option is to modify the design to provide the required strength, over and above that needed for service loads, in the structure under construction. Another option is to shore the finished portion of the structure (Fig. 3.2). Creating a temporary system to engage several floors in support of loads on one floor reduces the demand on the finished structure by spreading the load across a larger portion of the structure.

When relying on the structure under construction to support construction loads, designers need to consider the timing and duration of the construction loads in the context of the materials and age of the structure under construction. Structures that might be able to support construction loads when their materials have achieved full strength might not be able to do so at the time during construction when they are put under load. For instance, concrete structures gain strength and stiffness with time. Loads applied too early in a concrete structure can cause unrecoverable creep deflections.

In most cases, buildings are more vulnerable to instability while they are under construction. This is primarily because the components of the system that provides structural stability are installed in a sequence, meaning that the full system is not complete until the last component is in place.

Failures occur when the erection sequence does not adequately provide for stability, or the loads induced by the environment and construction activities are not matched with the capacity at the time. In addition, there is activity during the erection of a structure. Sometimes that activity has potential to create loads that are not well controlled. Loads

FIGURE 3.2 Shoring for temporary support of a floor.

may be placed with impact, they might shift, structural components might be bumped, and other human and environmental effects might come to play in ways they do not in finished structures.

Should there be an initial failure in a building under construction, there is heightened potential for the failure to spread in the structure, causing disproportionate collapse. This is because connections might not be complete, critical components often have not yet been installed, systems have low ductility, and other construction-related loads on the structure might be impacting stability.

Sometimes erection sequences for structures require construction of one or more floor levels before all the lateral bracing elements are in place. This sometimes happens in steel-framed structures that rely on concrete or masonry shear walls for lateral support. In these cases, the steel might be erected before the walls are built. For the interim condition, the building might require temporary lateral bracing for stability. That bracing sometimes is cables installed between floor levels, which might create a vulnerable system because it does not have the ductility of completed structures.

Designers should anticipate that vertical loads applied with eccentricities with respect to their supporting elements require corresponding base moment capacity or horizontal bracing. Sufficient column anchor bolts need to be installed to keep cantilevered columns from toppling, or the columns need to be braced until such time as a competent frame is established.

A beam recently placed might be able to support only a fraction of its full capacity because the floor slab that provides bracing against lateral buckling is not in place yet.

Tall scaffolding usually derives overall lateral stability from connection to the structure under construction. In such structures, stability issues still can arise from the adequacy of the foundation. Scaffolding erected on soft or shifting ground can become unstable because forces in vertical elements change or components tip. These structures often do not have ductility that will allow them to resist local overstresses or to sustain unexpected behaviors the same way as do structures designed for long life following building codes.

From these observations, it is clear that stability and overall structural integrity of temporary structures and structures under construction often need careful attention. The reaction forces required to maintain stability usually need to be determined by analysis. However, one common rule of thumb suggests that lateral bracing needs to support on the order of 2 percent of the vertical load, when the vertical load is applied without significant eccentricity on a plumb element.

The designer of a construction process that involves creating loads that are unique to the construction process on structures that do not have the robustness, interconnectivity, and ductility of a complete building system needs to consider

- The extent of the predictability of construction loads.
- The differences in the character of temporary structures and structures under construction as compared to permanent structures.
- The real possibilities that loads induced during construction can exceed the capacity of the interim condition of the permanent structure unless they are specifically controlled when feasible.
- Analyses of the circumstances of the construction stage and the magnitude and arrangement of construction loads to avoid damage and failure.

LOAD FACTORS AND LOAD COMBINATIONS

The construction process involves ever-changing gravity, environmental, and process-induced loads. The effects of these loads usually are additive, so they must be combined for evaluation. The specific load combinations need to represent realistic conditions that are likely to occur during the life of a temporary structure or the duration of a particular interim structural configuration. They also need to reflect the reliability with which we can anticipate the magnitudes of specific load types and the need to have appropriate factors of safety against failure states.

Certain loads, such as in-place building dead load and construction dead load, are fixed loads that are constant at their full value. Load combinations with a single variable load should reflect this, and the maximum effects of the variable loads should be combined directly with the fixed loads.

Recognizing that the maximum credible magnitudes of multiple uncorrelated variable load types are unlikely to occur simultaneously (e.g., maximum wind loads are unlikely to occur when maximum live load is in a structure), designers need to consider combinations that account for this low probability during the construction process. This is accomplished by reducing the maximum combined effects of variable loads when it is assumed that there will be some simultaneous effects of two or more variable loads.

Designers also need to understand that the reliability with which they know certain loads often is a function of the loading type. For instance, the weight of a structure can be estimated with relatively high accuracy. Live load estimates normally are not as precise, at least in part because they change from time to time, but also because they often are not controlled as closely as are dead loads. Hence, to provide reliability in the design of structures that support both dead and live loads, we should acknowledge the difference by adding a larger margin on the effects of live load as compared to dead load.

Substantial research into the variability of loads in permanent structures has led to the development of load combinations that are expressed in *Minimum Design Loads for*

Buildings and Other Structures, ASCE 7 (ASCE, 2010). The loading combinations in ASCE 7 have been further extended in *Design Loads on Structures During Construction*, ASCE 37 (ASCE, 2012)[*] to reflect conditions during a construction project, in which there are loading scenarios that do not exist in permanent structures.

Loading Types

Drawing upon ASCE 7 and ASCE 37, loads readily identifiable for the construction phase and symbols to represent them are as follows:

Final loads
   D—dead load
   L—live load
Construction loads
   Weight of temporary structures
      C_D—construction dead load
   Material loads
      C_{FML}—fixed material load
      C_{VML}—variable material load
   Construction procedure loads
      C_P—personnel and equipment loads
      C_H—horizontal construction load
      C_F—erection and fitting forces
      C_R—equipment reactions
      C_C—lateral pressure of concrete
Lateral earth pressures
   C_{EH}—lateral earth pressures
Environmental loads
   W—wind
   T—thermal
   S—snow
   E—earthquake
   R—rain
   I—ice

There might be other phenomena (Fig. 3.3), in addition to those listed above, that cause loads on structures. Examples include internal forces developed by foundation settlement or concrete shrinkage. To the extent that the designer anticipates that such loads could contribute significantly to overall stresses in building elements, the effects of these loads should be evaluated in combination with other foreseeable load types.

Strength Design Load Combinations

To address the probabilities that several of these loads are likely to occur simultaneously, designers must consider combinations that represent the anticipated magnitudes of the expected loads. For strength designs, this is accomplished by assigning load factors to the

---

[*]At the time of this writing, ASCE 37 is in balloting for the edition intended for publication in 2012. Since balloting is not complete, there might be some differences between content described herein and content in the final version of ASCE 37-12

FIGURE 3.3   Structure designed for loads due to dismantling an enclosure tent.

load types, to amplify them to establish an appropriate factor of safety against overload given the variability of each load type.

The concept involves assigning load factors based on the variability of the load magnitude and the likelihood that more than one time-variable load will occur at maximum values at the same time. This requires primary load factors for the dominant loads that one might consider to be the focus of a particular load combination, and arbitrary-point-in-time factors for uncorrelated loads that could occur simultaneously.

Algebraically, these loads are assumed to combine as follows:

Combined design load = dead and/or material loads + variable loads at their maximum values + uncorrelated variable loads at their arbitrary-point-in-time reduced values in accordance with the equation:

$$U = \sum_k c_{D,k} D_{n,k} + \sum_k c_{D,k} C_D + \sum_i c_{\max,i} Q_{n,j} + \sum_j c_{\mathrm{APT},j} Q_{n,j}$$

where c_D = dead load factor, $c_{\max}$ = load factor for the maximum value of variable load, c_{APT} = load factor for the arbitrary-point-in-time value of variable load, D_n = nominal dead load, C_D = nominal construction material load, Q_n = nominal variable load, k = all dead and construction material loads, i = all loads occurring at maximum value, and j = all relevant simultaneously occurring uncorrelated variable loads at arbitrary-point-in-time values.

Correlated variable loads, which therefore have interdependence, should not be combined using reduced arbitrary-point-in-time values since they are not subject to randomness in there probability of simultaneous occurrence. Such variable loads, when they are the focus of a load combination, should be combined using their primary load factors.

While the variabilities of most loadings in permanent structures have been studied to the level necessary to provide scientifically justified load factors, similar research generally has not been performed for many loads that are unique to the construction process. For this reason, load factors on certain construction-related loads need to be set at conservative levels to both account for their variability and also address the general precision with which they can be set at this time. Again, drawing upon ASCE 7 and ASCE 37, load factors that are appropriate under these circumstances for strength design load combinations are given in Table 3.1.

Examples of load combinations using these load factors are as follows.

$$1.4D + 1.4C_D + 1.2C_{FML} + 1.4C_{VML}$$

$$1.2D + 1.2C_D + 1.2C_{FML} + 1.4C_{VML} + 1.6C_P + 1.6C_H + 0.5L$$

$$1.2D + 1.2C_D + 1.2C_{FML} + 1.0W + 1.4C_{VML} + 0.5C_P + 0.5L$$

$$1.2D + 1.2C_D + 1.2C_{FML} + 1.0E + 1.4C_{VML} + 0.5C_P + 0.5L$$

$$0.9D + 0.9C_D + (1.0W \text{ or } 1.0E)$$

The last of the loading cases above is intended to apply when the designer will be relying on the counteracting effect of dead loads to reduce the consequences of wind or earthquake. In these cases, the designer must acknowledge that the actual dead load in place could be less than the nominal dead load, such that the benefit gained will

TABLE 3.1 Load Factors for Certain Construction-Related Loads

Load	Load factor (c_{max})	Arbitrary-point-in-time load factor (c_{APT})
D	0.9 (when counteracting wind or seismic loads)	—
	1.4 (when combined with only construction and material loads)	—
	1.2 (for all other combinations)	
L	1.6	0.5
C_D	0.9 (when counteracting wind or seismic loads)	—
	1.4 (when combined with only construction and material loads)	—
	1.2 (for all other combinations)	—
C_{FML}	1.2	—
C_{VML}	1.4	*by analysis*
C_P	1.6	0.5
C_C	1.3 (full head)	—
	1.5 (otherwise)	—
C_{EH}	1.6	—
C_H	1.6	0.5
C_F	2.0	*by analysis*
C_R	2.0 (unrated)	0
	1.6 (rated)	0
W	1.0	0.5
T	1.2	—
S	1.6	0.5
E	1.0	—
R	1.6	—
I	1.6	—

not actually be realized. Hence, load factors on the relevant dead loads need to be less than 1.0.

An example might be erection of a building that has cantilever girder (also called "Gerber") framing. In this system, stubs of girders are attached rigidly to columns, with the span-end of those stubs positioned near to the anticipated inflection point of the girder in the finished structure. The connections of the girders that span the ends of adjacent stubs normally are assumed to be pinned. In framing such as this, when a concrete slab is placed on one side of the column before the other, essentially all of the negative moment delivered to the column-girder connection by the stub being loaded is resisted by the column. In this case, any benefit from the dead load of the framing on the opposite side of the column should be reduced.

Given the unlikely coincidence of an earthquake and high winds at the same time during a particular construction operation, it is generally not required to consider these events simultaneously. In effect, their arbitrary-point-in-time load factors for a simultaneous occurrence of these two events can be taken as zero.

These might not be the only loading cases that need to be considered. Engineers designing for temporary conditions need to take reasonable care to anticipate the types of loads that are likely to occur and to use prudent judgment to address the consequences. For instance, combinations with live load at its full value (i.e., $c_{max} = 1.6$) may be required. A load factor of 2.0 is generally prudent for load types when specific values cannot be developed reasonably from available data.

The dynamic component of loads from equipment, such as some types of pumps, that induces significant cyclic or impact loads should be evaluated separately from weight and generally applied with a load factor of 1.3.

There might be instances when the uncertainty surrounding the magnitude of the load justifies using load factors that are higher than those presented previously. An example of the application of this concept is the load factor suggested above for equipment reactions. When the equipment is rated (the estimate of reactions has relatively high reliability), a load factor of 1.6 is appropriate. However, when the equipment is not rated (the estimate of reactions might be less reliable), a load factor of 2.0 is more prudent.

Consideration should also be given to the values of capacity reduction factors (Φ factors) when components of temporary structures are subject to damage or deterioration due to repetitive use over time.

Allowable Stress Load Combinations

Some material standards and some industry approaches evaluate adequacy based on allowable stress design. When applying allowable stress approaches, the following basic combinations, as a minimum, should be considered:

$$D + C_D + C_{FML} + C_{VML}$$
$$D + C_D + C_{FML} + C_{VML} + C_P + C_H + L$$
$$D + C_D + C_{FML} + C_{VML} + 0.6W + C_P + L$$
$$D + C_D + C_{FML} + C_{VML} + 0.7E + C_P + L$$
$$0.6D + C_D + (0.6W \text{ or } 0.7E)$$

As with load combinations for strength design, dead, live, wind, earthquake, and other loads that also affect the finished structure should be evaluated for the anticipated magnitude during construction, instead of for the magnitude associated with service conditions. Also, other load combinations might be necessary to assess the effects of additional sources of load.

Of course, there can be loadings that are not covered specifically by these allowable stress load combinations. The designer of a structure for a temporary condition needs to use appropriate judgment to assess loads that are likely and to consider the reliability of magnitude of the load value when developing allowable stress load combinations. When unusual uncertainty surrounds a load magnitude, the designer might consider increasing the estimate of the magnitude to increase the reliability of the resulting design.

When designers are evaluating the influence of two or more simultaneous uncorrelated variable loads using allowable stress approaches, the effects of these loads can be reduced to adjust for the low probability that the loads will occur at their maximum values at the same time. Standard practice is to reduce the combined effects of uncorrelated variable loads to 75 percent of their effects when combined at their maximum values, and to combine that result with the effects of dead load at full value. Correlated variable loads should not be reduced when their full magnitudes are likely to coincide.

When this approach is used to reduce combined effects to 75 percent of its full magnitude, there should be no further reduction in load or increase in effective allowable stress to address the probability of simultaneous occurrence. Of course, strength increases for short duration loads, such as are commonly applied for wood construction, can still be included in the analyses.

Additional Considerations

In both strength design and allowable stress design, the designer should evaluate all foreseeable load combinations and design for those that create the highest demand consistent with the construction process and the ability to reasonably predict the impacts. In any approach to assessing load effects, designs for individual components should be based on the conditions that create the highest demand on that component. Overall stability should be evaluated for the combinations of effects that are most severe.

In addition, there can be requirements within the jurisdiction governing the construction that demand compliance with more restrictive standards. Further, materials standards, guidance from manufacturers of load-handling equipment, federal regulations, or other authoritative and reliable resources could influence a designer's approach to combining loads.

The Occupational Safety and Health Administration (OSHA, 2011) has specific requirements for the design of scaffolds. These structures must be able to support, without failure, their own weight and at least four times the maximum intended load applied or transmitted to them. Ropes suspending scaffolds need to support at least six times the maximum intended load supported or transmitted to the rope. These requirements, and other such requirements mandated by code, regulation, or ordinance, are not superseded by the load combinations listed in this chapter.

Although designers often have the option to use either strength design load combinations or allowable stress design combinations, they should not combine approaches in a search for justification to reduce the design capacity of individual components. Each approach is internally consistent, whereas selectively using each in an effort to minimize overall costs is not.

DEAD LOADS

Dead loads in the context of design of temporary structures and permanent structures in interim configurations are the in-place weight of the permanent construction. The only conceptual difference between dead loads during construction and those in the finished structure is the amount and distribution of the dead load that is in place at any particular time. As a structure is assembled, dead load is constantly being added, continuously changing the demands on components installed previously. Generally, demands will increase over time, but there are circumstances in which incremental addition of dead load will create effects that are inconsistent with the corresponding effects in the finished structure.

When using the load combinations contained herein, dead load should not include the weight of temporary structures supported by the structure or the weights of permanent construction that has not yet been installed in its final configurations.

Weights of temporary structures, which can include weight of scaffolding, formwork, ramps, and other works associated with the construction process, are treated separately because their potential variability does not match that of the permanent structure. Further, these weights change with the construction process and often change in ways not anticipated by the designer, so there are different levels of confidence in their magnitudes at any particular time.

Weights associated with permanent construction that is not yet in its final configuration also needs to be separated from dead load. When concrete is poured for a supported slab, the process usually involves placing and spreading concrete, creating relatively concentrated loads during the construction process when distributed loads will be the end result. The variability of the load magnitude under these circumstances is different, and likely higher, than that for the loads in the finished structure.

The American Concrete Institute (ACI, 2005) provides guidance for addressing loads from shoring and reshoring concrete structures.

Dead loads most commonly are estimated from accepted values of the density of materials and the estimated volume that will be installed (see Tables 3.2 and 3.3 for densities and component weights of some common building materials). Additional guidance can be found in references by the American Institute of Steel Construction (AISC, 2011) and the American Society of Civil Engineers (ASCE, 2010; ASCE, 2012). For interim configurations of permanent structures, it is a relatively simple matter to compile these estimates from the construction documents.

LIVE LOADS

Live loads are derived from the occupancy of the structure while it is under construction. In most construction processes, the live load during construction will differ substantially from that assumed for the in-service design. New buildings under construction generally are not occupied as they will be when the building is finished. Sometimes existing buildings are renovated in a significant way while portions are in service. While some portions of these buildings have in-service live loads, the immediate area of the construction usually is vacated for the work. In these cases, live loads for the construction area and the rest of the building need to be evaluated separately.

To the extent that the structural components are influenced by ongoing occupancy elsewhere, the designer might address live loads outside of the construction area by considering code-imposed design values to apply. Alternatively, surveys of live load can be performed to estimate actual values in place to determine whether code-imposed design values are realistic for the interim conditions under consideration.

If a designer of temporary works intends to rely on surveys of live loads as the basis for design, those loads should be monitored and controlled throughout the duration of the construction process. Further, primary load factors, rather than arbitrary-point-in-time load factors, should be applied because the arbitrary-point-in-time factors are determined probabilistically and would reduce the load unrealistically, whereas survey loads are deterministic and are intended to account for a specific configuration.

TABLE 3.2 Weights and Specific Gravities

Substance	Weight (lb/ft^3)	kg/m^3
Ashlar, Masonry		
Granite	165	2650
Limestone	160	2560
Sandstone	140	2240
Mortar Rubble Masonry		
Granite	155	2480
Limestone	150	2400
Sandstone	130	2080
Dry Rubble Masonry		
Granite	130	2080
Limestone	125	2000
Sandstone	110	1760
Brick Masonry		
Pressed brick	140	2240
Common brick	120	1920
Soft brick	100	1600
Concrete Masonry		
Cement, stone, sand	144	2310
Cement, slag	130	2080
Cement, cinder	100	1600
Various Building Materials		
Cement, Portland, loose	90	1440
Cement, Portland, set	183	2930
Reinforced normal-weight concrete	150	2400
Reinforced lightweight concrete	90–115	1440–1840
Lime, gypsum, loose	53–64	850–1030
Mortar, set	103	1650
Earth, etc., Excavated		
Clay, dry	63	1010
Clay, damp, plastic	110	1760
Clay and gravel, dry	100	1600
Earth, dry, loose	76	1220
Earth, dry, packed	95	1520
Earth, moist, loose	78	1250
Earth, moist, packed	96	1540
Earth, mud, flowing	108	1730
Earth, mud, packed	115	1840
Sand, gravel, dry, loose	90–105	1440–1680
Sand, gravel, dry, packed	100–120	1600–1920
Sand, gravel, wet	120	1920
Minerals		
Gneiss	159	2550
Granite	175	2800
Gypsum	159	2550
Limestone, marble00	165	2640
Sandstone, bluestone	147	2400
Shale, slate	175	2800

TABLE 3.2 Weights and Specific Gravities (*Continued*)

Substance	Weight (lb/ft³)	kg/m³
Stone, Quarried, Piled		
Granite, gneiss	96	1500
Limestone, marble, quartz	95	1500
Sandstone	82	1300
Shale	92	1500
Metals, Alloys, Ores		
Aluminum, cast, hammered	165	2600
Brass, cast, rolled	534	8900
Bronze, 7.9–14% Sn	509	8200
Bronze, aluminum	481	7700
Copper, cast, rolled	556	8900
Copper ore, pyrites	262	4200
Iron, cast	450	7200
Iron, wrought	485	7800
Iron, ferrosilicon	437	7000
Iron ore	325	5200
Iron ore, loose	130–160	2100–2600
Iron slag	172	2800
Lead	710	11000
Timber, U.S. Seasoned		
Moisture content by weight:		
Seasoned timber 15–20%		
Ash, white, red	40	640
Cedar, white, red	22	350
Chestnut	41	660
Cypress	30	480
Fir	25–32	400–510
Maple, hard	43	690
Maple, white	33	530
Oak	41–54	660–860
Pine	25–32	400–510
Pine, yellow	38–44	610–700
Redwood, California	26	420
Spruce, white, black	27	430
Walnut, black	38	610
Walnut, white	26	420
Various Liquids		
Alcohol, 100%	49	780
Acids	75–110	1200–1760
Oils, vegetable	58	930
Oils, mineral, lubricants	57	910
Snow, fresh fallen	8	130
Snow, accumulated	13–19	210–300
Water, 4°C max density	62.4	96
Water, ice	56	900
Water, sea water	64	1000

(*Continued*)

TABLE 3.3   Weights of Building Materials

Materials	Weight (lb/ft^2)	N/m^2
Ceilings		
Channel suspended system	1	50
Lathing and plastering	See Partitions	
Acoustical fiber tile	1	50
Floors		
Steel deck	See Manufacturer	
Concrete, reinforced, 1 in		
Stone	12.5	600
Slag	11.5	550
Lightweight	6–10	290–480
Concrete, plain, 1 in		
Stone	12	580
Slag	11	530
Lightweight	3–9	140–430
Fills, 1 in		
Gypsum	6	290
Sand	8	380
Cinders	4	190
Finishes		
Terrazzo, 1 in	13	620
Ceramic or quarry tile, 3/4 in	10	480
Linoleum, 1/4 in	1	50
Mastic, 3/4 in	9	430
Hardwood, 7/8 in	4	190
Softwood, 3/4 in	2.5	120
Roofs		
Copper or tin	1	50
Corrugated steel	See Manufacturer	
3-ply ready roofing	1	50
3-ply felt and gravel	5.5	260
5-ply felt and gravel	6	290
Shingles		
Wood	2	100
Asphalt	3	140
Clay tile	9–14	430–670
Slate, 1/4 in	10	480
Sheathing		
Wood, 3/4 in	3	140
Gypsum, 1 in	4	190
Insulation, 1 in		
Loose	0.5	20
Poured	2	100
Rigid	1.5	70
Partitions		
Clay tile		
3 in	17	810
4 in	18	860
6 in	28	1300
8 in	34	1600
10 in	40	1900

TABLE 3.3 Weights of Building Materials (*Continued*)

Materials	Weight (lb/ft²)	N/m²
Partitions (*Continued*)		
Gypsum block		
2 in	9.5	460
3 in	10.5	500
4 in	12.5	600
5 in	14	670
6 in	18.5	890
Wood studs 2 × 4		
12–16 in o.c.	2	100
Steel partitions	4	190
Plaster, 1 in		
Cement	10	480
Gypsum	5	240
Lathing		
Metal	0.5	20
Gypsum board, 1/2 in	2	100
Walls		
Brick		
4 in	40	1900
8 in	80	3800
12 in	120	5800
Hollow concrete block (heavy aggregate)		
4 in	30	1400
6 in	43	2100
8 in	55	2600
12½ in	80	3800
Hollow concrete block (light aggregate)		
4 in	21	1000
6 in	30	1400
8 in	38	1800
12 in	55	2600
Clay tile (load bearing)		
4 in	25	1200
6 in	30	1400
8 in	33	1600
12 in	45	2200
Stone, 4 in	55	2600
Glass block, 4 in	18	860
Window, glass, frame, and sash	8	380
Curtain walls	See Manufacturer	
Structural glass, 1 in	15	720
Corrugated cement asbestos, 1/4 in	3	140

The occupancy in buildings that are not in service and, in any event, in the immediate area of the construction needs to be evaluated specifically. In general, assessments of live loads in areas impacted by construction need to consider that many of construction loads are included in other load types that are separately defined. To the extent that construction materials are estimated and combined separately with live loads and other load effects, they obviously need not be considered redundantly. Hence, live loads are normally applied as an allowance for the typical weights on workers and incidental materials and equipment in a construction area, when such loads are not considered otherwise. Specific heavy equipment and construction materials placed or stored in an area under consideration should be evaluated separately.

CONSTRUCTION LOADS

Construction loads are those loads that are unique to the construction process and can include loads imparted by equipment, materials, and construction processes; construction dead loads; and personnel loads when they represent a significant load on a particular structure.

Equipment Loads

Equipment placed on a structure (Fig. 3.4) obviously will impart loads on the structure. Depending on the type of equipment, these loads can be static, dynamic, or impact. Loads can have vertical and horizontal components.

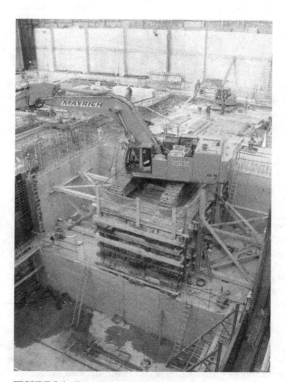

FIGURE 3.4   Excavator on stack of cribbing.

Generally, construction equipment is for demolition, processing within the structure, or moving materials. Some equipment is stationary and imparts insignificant loads other than its own weight. In these cases, designers need to determine the weight of the equipment, how the equipment with be moved within the structure, and where it will be positioned while in operation.

In general, the weight of equipment can be determined relatively easily from the manufacturer. Some data on equipment reactions are contained in other references (e.g., Jahren, 1996; Caterpillar, 2008). To the extent that the equipment is large relative to the spans of the systems on which it will rest, it might be necessary to further evaluate the distribution of the weight over the footprint of the equipment so that the effects on the structure can be determined with accuracy. If the footprint of the equipment is relatively small compared to the spans of the structural elements on which it rests, normally the weight can be represented as a concentrated load.

Other equipment specifically imparts loads other than its own weight. Motorized wheelbarrows, hoists, and cranes are of this character. In these cases, designers need to be aware of the rated capacity of the equipment, and design the supporting structure accordingly. This will involve determining the reactions that the equipment will impart on the structure when operating at the extremes of its capability. Establishing the appropriate reactions might involve investigating several configurations, such as lifted load limits as a function of reach for a crane.

Suggestions on minimum loads for wheeled equipment are as follows:

- 500 lb (2.2 kN) for the wheel of a manually powered vehicle
- 2000 lb (8.9 kN) for the wheel of powered equipment.

Of course, actual loads should be used when they are higher than these values. The areas over which these loads should be assumed to apply are the load divided by the tire pressure for pneumatic tires and 1 in (25.4 mm) times the width of the tire for hard rubber tires.

Operations that cause impact need to be assessed, and related dynamic loads should be considered separately. In addition, certain equipment induces horizontal loads. When these loads are significant, they must be coupled with vertical loads.

Except when lifted loads will be strictly controlled, reactions from cranes and similar equipment should be based on the rated capacity of the equipment. This addresses the possibility that an equipment operator will not understand the load limits assumed by the designer and inadvertently overload supporting elements. If a designer chooses to design a supporting structure for equipment reactions that are less than those associated with the rated capacity of the equipment, sufficient safeguards need to be implemented to control loads to the limits assumed in the design.

Loads beyond those associated with normal operation of certain equipment can occur when, for instance, swing staging or a forklift catches on a fixed object, causing stall loads in the lifting devices. Unless specific safeguards are imposed, designers should assume that equipment such as a forklift can be loaded to cause it to tip onto its front axle because, should that happen, the full weight of the forklift plus the tipping load is supported on one axle.

Motorized wheelbarrows can impart significant horizontal forces through braking or turning, depending on the mass of the equipment and payload, speed of operation, and the nature of the structure supporting the equipment. As a lower bound, wheeled vehicles transporting materials should be assumed to impart lateral loads equal to 20 percent of the combined weight of the vehicle and its payload. When two or more vehicles are operating together, the design should also be checked for lateral load equal to 10 percent of the combined weight of the equipment and their payloads. Tipping loads and horizontal loads can also occur if motorized load-handling equipment is driven into an obstacle that suddenly stops the forward motion.

FIGURE 3.5   Inadequate support for manlift reactions.

For equipment that is driven onto a structure, it will be necessary to assess the impacts of wheel loads on all structural components that the machine crosses (Fig. 3.5). Working with the contractor, the designer needs to understand the path of a crane to its operating position and any ramps or other structures that the contractor plans to use to facilitate access. With that information and an understanding of the structural system in its interim condition while the equipment is being placed, the designer of the interim condition can assess the effects on all the structural elements that the equipment will cross. In many cases, the structure might need to be protected by shoring or placement of cribbing for the wheel loads.

Due consideration should be given to the placement of outriggers, favoring placement over strong elements such as beams rather than on a slab between beams. Of course, such decisions need to be coordinated with decisions about the required position and orientation of a crane when it is making the lift. In that context, the placement of outriggers to protect the structure should not override procedures, such as proper extension of the outriggers for stability, that are established for safe operation of the crane. To accomplish both goals— stability of the crane and safe loads on structural components—it might be necessary to use secondary structures, such as cribbing, to properly distribute outrigger loads to structural elements that are strong enough for the purpose.

Sometimes designers encounter equipment for which they cannot identify rated capacity or equipment weight. Under these circumstances, the designer needs to determine equipment reactions through independent analysis of reactions or by measurement. It is possible to drive forklift vehicles onto portable truck scales to establish wheel loads. With that information and the basic geometry of the vehicle, designers can estimate axle loads corresponding to specific payload sizes and weights. To the extent that uncertainty remains after analyses and measurement, the designer may proceed with appropriate caution by treating the data as unconservative and increasing design loads to reduce the risk of overload.

Erection and Fitting Forces

It is necessary sometimes in construction to impart forces on structures to bring parts together for interconnection or to bring construction to plumb. This often happens even for construction that is within normal tolerances when the extremes of the ranges oppose each other. When this is the case, the forces that must be induced, and which remain in the structure once it is brought into alignment, generally are small enough that they will not have detrimental impact on the performance of the completed structure.

However, there could be temptation to bring into alignment components that are out of tolerance or that must mate with structure that was previously constructed with errors. Normally, such actions should not be tolerated in competent construction. However, when there is reason to proceed with consideration of forced correction as an option, careful evaluation is needed to assess the implications on the performance of the completed structure both in terms of serviceability (e.g., residual out-of-plumb) and the forces that are induced in the structure and that may diminish load-carrying capacity in service. The decision process usually engages the contractor, the engineer consulting for the contractor, the engineer-of-record for the completed structures, the owner, and potentially other stakeholders.

In any event, the forces that are induced to fit up a structure must be resisted somewhere, usually within the structure itself. Those reaction locations need to be strong enough to support the resulting forces without inducing localized or overall damage, and there has to be a continuous, adequate load path between the two load application points to transmit the reactions. In addition to evaluating the strength of the structure between the load application points, due consideration has to be given to the possibility that loads, such as horizontal loads to bring a column to plumb, might induce instability in any of the components engaged in the process.

Material Loads

Material loads derive from the weights of portions of the structure that are stored (Fig. 3.6) or being put into place, and the construction-related materials (other than equipment loads and other loads that are separately defined) necessary to facilitate construction.

Material loads can be either fixed or variable. As the term indicates, the magnitudes of fixed material loads do not vary over time. An example is erected scaffolding installed on the roof of a low level of a completed structure to allow construction of the façade at higher levels or to enclose a building (Fig. 3.7). Once in place, the scaffold weight usually remains constant.

An example of a variable load, with magnitude that changes over time, might be concrete blocks placed on scaffolding (Fig. 3.8) before they are installed into the structure. This is a variable material load for two reasons: (1) The blocks are not yet in its final position, so they are not yet a dead load, and (2) the amount of block on the scaffolding changes with time and location, increasing the uncertainty in what might be its maximum effect.

Sometimes variable material loads become fixed material loads over the course of construction. Before becoming a fixed material load, the scaffolding in the preceding

FIGURE 3.6   Materials stored on pallets in a building.

FIGURE 3.7   Scaffold being installed to create an enclosure.

FIGURE 3.8 Concrete blocks on scaffolding before installation.

example was a variable load if it was lifted and temporarily placed on the low roof where it was ultimately erected. While stacked or being moved on the structure, the components are a variable load. Also, the variable material loads represented by the blocks on the scaffold in the preceding example ultimately become dead loads when they are installed in the structure.

The determination of whether a material load is fixed or variable is important because the variabilities of these load, and therefore the load factors that should be used in strength design, differ. Variable material loads require a larger load factor than do fixed material loads because usually the uncertainties are greater in their magnitude and placement (Fig. 3.9).

It should be clear enough to all involved in the process that it does not take very much material load (e.g., pallets of brick) to severely exceed the in-service live load limits, which often are in the range of 40 psf (6.3 kN/m^2) for completed structures.

As with other loads applied during construction, the needs of the workers handling materials on a structure need to be communicated to the designer of the temporary works, and the assumptions of the designer for the material loads need to be communicated back to the workers handling those loads.

The effects of material loads can be controlled by restricting amounts and locations where materials can be placed. Depending on the nature of the structural system, transporting materials close to girder lines and storing them close to columns, as compared to uncontrolled handling and placement, can reduce certain effects on the structure. When loads that potentially threaten to damage a structure must be introduced during the construction process, plans need to be developed and communicated. Those plans might include marking travel lands and storage locations and posting load limits, in addition to clear verbal communication between the involved workers and designers.

FIGURE 3.9   Placement of concrete using a motorized wheelbarrow.

Worker and Personal Equipment Loads

Usually it is unnecessary to specifically consider weights of workers and their personal equipment on completed portions of structures under construction. Over large enough areas of completed structures, such loads are considered to be within live load allowances. However, for certain types of structures and in other atypical situations, it is necessary to consider these loads specifically.

Work platforms erected for the sole purpose of providing workers, with their personal equipment and some materials of construction, access to work locations often have as their primary loading the weight of the workers themselves (Fig. 3.10). Scaffolding erected for light construction on the outside of a building is a prime example.

When workers and their personal equipment constitute significant loads on a structure, it is important to anticipate the number of workers on the structure at a time and their potential positions. Once the number and arrangement are determined and when those factors are important to the design of the temporary structure, these parameters can be reflected in the design. As with limitations on the handling of material loads, when a structure is designed for limited worker loads, the bases for the design need to be communicated back to the workers.

When the weight of workers dominates the design of a work surface, the designer normally should assume that the workers can be positioned in the locations that create the highest demand. For instance, when designing specifically for three workers on a scaffold, the maximum moment should be calculated assuming that the workers are side by side in very close proximity (normally assumed to be 2 ft apart) at midspan. This is prudent because it is not clear that the positions of the workers in practice can be reliably controlled even though the activities planned for the working surface may imply certain positions for the workers. There might be circumstances when a design can be based on workers in positions that do not create peak demand, particularly if their movements are specifically constrained.

FIGURE 3.10   Scaffold for worker access.

It is common practice (ANSI, 2001) to take the weight of a worker to be 250 lb (0.89 kN) plus personal equipment weighing 50 lb (0.22 kN). Of course, the designer should consider higher load if such loads are reasonably anticipated.

OSHA defines classes of working surfaces and associated uniform loads to be used for design (OSHA, 2011). The intent is to provide default strength for platforms such as scaffolding designed for specific purposes. The categories and their associated equivalent uniform applied loads are given in Table 3.4.

Examples of light-duty working surfaces are platforms for residential framing activities, finishing operations with hand tools, and transportation of light construction materials.

TABLE 3.4   Classes of Working Surfaces and Associated Uniform Loads

	Uniform load*	
Operational class	psf	kN/m²
Light Duty: sparsely populated with personnel; hand-operated equipment; staging of materials for lightweight construction	25	1.20
Medium Duty: concentrations of personnel; staging of materials for average construction	50	2.40
Heavy Duty: material placement by motorized buggies; staging of materials for heavy construction	75	3.59

*Loads do not include dead load, D; construction dead load, C_D; or fixed material loads, C_{FML}

Medium-duty activities include transportation of moderate-weight construction materials (concrete transported by buckets or handcarts); masonry construction with hollow, light-weight masonry units; and placement of reinforcing steel. Working surfaces for heavy duty are intended to handle transportation of concrete by motorized wheelbarrow and masonry construction using heavy concrete masonry units. Of course, these are guidelines; designers should consider the activities that are planned for the working surface and verify that these uniform loads will suffice.

There might be circumstances when the loads associated with the OSHA operational classes exceed the design base for the structure under construction. For instance, few occupancies in finished structures require floor live loads of 75 psf (3.59 kN/m^2). Sometimes it is necessary to enhance the strength of the finished structure to support heavy construction loads or to shore floors so they are not overloaded. At the other end of the scale, there might be working surfaces, such as joists that ultimately will be uninhabited attics in completed structures, that cannot support even the equivalent light-duty loads. In these cases, specific plans can be developed to limit the amount of load (e.g., the number and proximity of workers) or to provide shoring so that the capacity of a very light framing system is not overloaded.

When scaffolds and similar working surfaces are designed for loads that are less than full occupancy, their use should be controlled procedurally. Barriers should be in place to keep wheeled vehicles from entering onto surfaces that cannot support their weight. When the number of workers who can enter onto a scaffold exceeds the number for which it was designed, the entrance should be posted or monitored and/or the workers should be clearly informed of the limitations. An example is a work platform that has relatively small area [say, 40 ft^2 (3.72 m^2) or less] and, hence, might be designed for relatively light total load. Such structures should be rated for the number of workers allowed at a time. Similarly, work platforms designed for relatively low uniform load [say, less than 25 psf (1.2 kN/m^2)] should be rated for the number and position of workers allowed at a time.

Horizontal loads from workers should be considered on platforms and on guards at edges of platforms and around openings. Unless other information is available, designers should consider that workers can apply horizontal loads of 50 lb (220 N) in any direction at platform level. This value should not be used if the platform is known to the designer to be intended for operations for which workers will be pushing or pulling with forces likely to exceed this value. Loads on guards are normally taken as 50 lb/ft (730 N/m) or 200 lb (890 N) as a concentrated load, whichever creates the higher demand.

Reduction in Construction Loads

As in finished structures in service, it is unlikely that the full design load will exist over large areas simultaneously. To address the possibility that overly conservative designs will follow from the basic assumptions about load magnitudes, it is permissible to reduce certain construction loading assumptions when the area of the supporting structure is relatively large.

For personnel, general equipment, and live loads, such reductions generally are permissible when the influence area impacted is greater than 400 ft^2 (37.2 m^2).

The reduced loads are as follows:

$$C_p = L_o(0.25 + 15/\sqrt{A_I})\,\text{psf}$$

or

$$C_p = L_o(0.25 + 4.57/\sqrt{A_I}) \text{kN/m}^2$$

where C_p = reduced design uniformly distributed personnel, equipment, and live load per square foot (m²) of area supported by the member

L_o = unreduced uniformly distributed personnel, equipment, and live design load per square foot (m²) of area supported by the member

A_I = influence area, in square feet (m²). The influence area A_I is normally four times the tributary area for a column, two times the tributary area for a beam, and equal to the panel area for a two-way slab.

In any event, when the unreduced uniformly distributed personnel, equipment, and live loads are more than 25 psf (1.2 kN/m²), the reduced uniformly distributed design load should neither be less than 50 percent of the unreduced load on elements supporting one level, regardless of influence area, nor be less than 40 percent of the unreduced design load for elements supporting more than one level. When the unreduced uniformly distributed personnel, equipment, and live load is 25 psf (1.2 kN/m²) or less, the reduced load should not be less than 60 percent of the unreduced design load, unless justified by an analysis of the construction operations.

Reductions of this type do not apply to fixed and variable materials loads except to the extent that they are incidental transportation or staging loads that are included in personnel loads. They also do not apply when a platform is designed for a combination of specific loads. To the extent that designs are to be for reduced material loads, they must represent specific arrangements of loads that will be reasonably controlled to be consistent with the design assumptions.

When roof slopes exceed 4 in 12 (33.3 percent), roof loading could also be reduced because it is unlikely that such roofs will be used to store or handle large loads. A suggested reduction factor is

$$R = 1.2 - 0.006S$$

where S is the roof slope expressed as a percent.

ENVIRONMENTAL LOADS

Temporary structures and structures in interim configurations need to withstand environmental loads along with the loads that are unique to the construction process. These environmental loads include wind, thermal, snow, earthquake, rain, and ice.

Perhaps the best reference on environmental loads on buildings is the American Society of Civil Engineers *Minimum Design Loads on Buildings and Other Structures*, ASCE 7-10 (ASCE, 2010). This document has been developed following a consensus process in which provisions are vetted by a committee with interests that are balanced in the construction industry. Input comes from research and experience of qualified experts in relevant fields and technical observations of building performance during environmental events.

ASCE 7 addresses loads on buildings that are designed for long life. In that context, loads specified in that document are intended to establish low probabilities of failure for structures with design lives on the order of 50 years or more. The probability that the loads in ASCE 7 will be exceeded during the life of a typical building is on the order of a few percent. As such, loads specified in ASCE 7 generally would be conservative for a temporary structure or a structure in an interim configuration that will last for days or weeks, or even longer, because the events that form the bases for loads in ASCE 7 would be highly unlikely during such short intervals.

Further, completed structures are expected to withstand environmental events without intervention. Construction sites are different. For environmental events, such as wind storms, that have warning, contractors can take proactive steps to secure the site and reduce the impact of loads. They can also vacate the site, reducing or entirely eliminating the risk to human life in many cases.

For environmental loads that accumulate over time, such as snow, contractors can also intervene. Contractors can remove snow before loads accumulate to critical levels to reduce the demand on structures.

For these reasons and others, the American Society of Civil Engineers prepared the standard *Design Loads on Structures During Construction*, ASCE 37-12 (ASCE, 2012). This standard, also developed through the consensus process, considers how the duration of interim structural configurations or the life of temporary structures impacts the probable magnitudes of environmental loads and adjusts the loads in ASCE 7 accordingly.

The intent in the development of ASCE 37 is to recognize the anticipated lifespan of temporary structures and interim conditions of structures under construction and to establish reasonable adjustment factors that provide what the committee developing ASCE 37 deems to be acceptable risk for construction operations. This is accomplished by defining a construction period, which is the length of time from first erection to structural completion, including installation of cladding, of each independent structural system.

The intent is not to increase the risk of injury to workers or the general public. In deliberations leading to the duration-related coefficients in ASCE 37, the committee recognized the visible nature of construction elements, the ability of contractors to provide bracing in anticipation of storms, protocols for reducing risk, the knowledge and training of contractor personnel, and the ability to vacate the site in response to approaching hazards that have advance warning.

Clearly, there are circumstances when construction needs to be designed for loads that are higher than those in ASCE 37. These might include construction of a high-rise building in urban environments when intervention by the contractor cannot adequately protect the general public. It might include partially occupied buildings undergoing renovation, if there are unusual risks to occupants that cannot be otherwise controlled. Very long-duration interim conditions should have higher loads, as ASCE 37 recognizes, because the structure's exposure is higher than for construction with typical durations. It would also include construction in jurisdictions that have requirements that exceed the reduced environmental loads published in ASCE 37 or in other documents.

Designers for structures in interim configurations and for temporary structures have to be aware that construction workers spend their working lives in these structures. As such, construction workers' exposure to the effects of environmental loads on such structures is continuous during the working hours. In that context, workers on construction sites that are designed for reduced loads are exposed to higher lifetime risk than is held by the general public for environmental loads for events (e.g., earthquakes) that cannot be anticipated reasonably.

Nevertheless, it is very common, if not the norm, to use reduced environmental loads for the design of temporary works for construction even when those loads are likely to occur without warning. In fact, it is probable that most temporary works and interim conditions are developed without consideration of seismic loads, given their very low probability of occurrence during any particular construction phase.

It is in this context that the provisions of ASCE 37 are summarized herein.

Risk Category

ASCE 7 (ASCE, 2010) refers to risk categories that effectively modify the expected return period for various environmental events on the basis of the occupancy of the structure. The approach effectively adjusts the level of risk to human life and the consequences to regional health and safety from a failure by modifying the basic design loads accordingly. These adjustments come through importance factors, which assume values ranging from 0.8 to 1.5 depending on load type and risk category.

The importance factors for loads on structures with the most common occupancy, corresponding to structures such as office buildings with a moderate number of occupants, are set at 1.0 for all loading types. The loads on these buildings, which are assigned risk category II, are the base loads calculated by the formulas in ASCE 7.

For structures with very few human occupants and which have no essential function for the health and safety of the regional population, the risks of significant consequences from a failure are deemed to be low. The structures are assigned to risk category I and are to have an acceptable risk of failure that is higher than for more densely occupied structures. This is accomplished by specifying importance factors with values less than 1.0 for some loads, to adjust downward the corresponding design loads.

At the other end of the spectrum, high-occupancy buildings and facilities that have essential functions for the health and safety of a population (i.e., buildings in risk categories III and IV) are designed for higher-value importance factors that effectively increase the return period and therefore the design loads.

In general, structures under construction are occupied by a significant number of workers. Also in general, structures under construction are neither low hazard nor high hazard and/ or essential facilities. In that context, structures under construction are deemed in ASCE 37 (ASCE, 2012) to have risk category that is the same as a typical building, such as an office building, during construction, regardless of the risk category it will assume when finished. Hence, importance factors according to ASCE 7 are taken as 1.0, meaning that the basic loads are not modified either upward or downward for risk.

There might be occasions when this approach is not appropriate. Jurisdictions might require more conservative interpretation of risk. There might also be projects that create unacceptable risk to workers or the general public if they are not designed for higher loads. A higher risk category might also be appropriate if a temporary structure or a structure under construction would imperil an essential facility if it were to fail. In these instances, loads otherwise determined in accordance with ASCE 37 can be modified either by increasing ASCE 37 reduction factors or by applying importance factors consistent with a higher risk category.

Wind Loads

Wind loads are a function of region, location, surrounding features, and time of year. ASCE 7 (ASCE, 2010) has detailed procedures for adjusting historical data on wind to account for these influences to result in design wind pressures on the facades and components of completed buildings. ASCE 37 (ASCE, 2012) allows for reductions in these wind pressures based on the duration of a construction period, which is taken as the time interval between first erection to completion of each independent structural system, including cladding.

The reduction is accomplished by applying one of the factors in Table 3.5 to the design wind speed specified in ASCE 7.

For construction periods less than 6 weeks, factors less than 0.75 can be used if they are justified by a statistical analysis of wind data for the locale and season of construction. Literature (Boggs and Peterka, 1992; Rosowsky, 1995) provides a basis for some of the

TABLE 3.5   Wind Speed Reduction Factors

Construction period	Factor
Less than 6 weeks	0.75
From 6 weeks to 1 year	0.8
From 1 year to 2 years	0.85
From 2 years to 5 years	0.9

construction period factors above. In general, factors for periods less than 1 year are based on judgment.

In many parts of the country, the season of the year has an important influence on the likely wind speed. A clear example is the U.S. hurricane coast along the Gulf of Mexico and the Eastern Seaboard. For risk category II structures, ASCE 7 specifies high wind speeds [up to 180 mi/h (80 m/s)] for the hurricane coast, as compared to wind speeds [115 mi/h (51 m/s)] in most of the rest of the United States. However, hurricanes only occur during the summer and fall of the year (the hurricane season is taken in ASCE 37 as July 1 through October 31, when the most severe storms are likely). As such, construction that takes place during the hurricane season needs to be designed considering the hurricane potential, whereas construction that is confined to occur outside of the hurricane season does not.

ASCE 37 allows risk category II temporary structures and structures under construction in the hurricane region to be designed for the same wind speed [115 mi/h (51 m/s) 3-s gust] as is specified for the rest of the country when the construction period occurs outside of the hurricane season. If a structure is designated with a higher risk category, the base wind speed should be 120 mi/h (54 m/s) 3-s gust. It is also permissible to design for this wind speed when construction occurs during the hurricane season, if additional bracing is provided and can be installed in time to support the full, unmodified hurricane wind speed.

Even when it is permissible to design for a basic wind speed of 115 mi/h (51 m/s) 3-s gust in hurricane regions, there remains a difference between the design speeds during and outside of hurricane seasons. For structural configurations during the hurricane season, the design wind speed cannot be adjusted by construction period factors, whereas the design wind speed outside of the hurricane season can. Hence, design loads for temporary conditions still will be higher in the hurricane season than they will be outside of the hurricane season.

Certain operations at a construction site can be completed in very short time (i.e., hours) under conditions that can be monitored continuously. An example is the lifting of a precast panel from a truck to its permanent position on a structure. The weather report for the day is available, ambient wind speed can be assessed immediately prior to the lift, and emergency steps can be taken if something unanticipated occurs. Under these circumstances, the design wind speed can be based on the wind speed predicted by reputable forecasters for the continuous work period.

When specific data for the day of the activity are used to calculate wind pressures, the subject operation needs to be monitored continuously, and must be completed in a timely fashion, such as by the end of the work day. Interim conditions designed based on monitored conditions should not be left unattended, nor should they be maintained for lengthy periods by repeatedly reassessing the evolving wind speeds.

Care needs to be taken when interpreting the predicted wind speed. Procedures in ASCE 7 and ASCE 37 are based on 3-s gust speeds, which might not be the basis for broadcast weather reports. Further, ASCE 7 applies a load factor of 1.0 on wind pressures because, for risk category II structures, ASCE 7 uses a return period of approximately 700 years. In order to achieve a corresponding factor of safety using predicted wind speeds, broadcast wind speeds need to be adjusted to 3-s gust speeds, if necessary, and then multiplied by 1.26 before calculating wind pressures. The value of 1.26 is the square root of 1.6, which is the load factor applied to wind pressures in earlier version of ASCE 7, when pressures were determined based on a nominal return period of 50 years. The adjustment factors for various wind speed averages are published in a chart in the commentary to ASCE 7.

With buildings being erected in pieces, first as open structures without facades, then becoming partially enclosed, their silhouettes to the wind are changing continuously through various configurations that are different than when the structures are finished. One particularly different configuration occurs when the frame of a building is erected, but the cladding is not yet in place. In this configuration (Fig. 3.11), the building receives wind loading on exposed vertical and horizontal elements across the full width and depth of the building, and on any materials that are resting on the structure.

FIGURE 3.11   Exposed framing during construction activities.

There is little guidance in the literature about calculating pressures on columns and beams from wind moving through a floor of a partially completed building. It is possible that elements in a repeating pattern will create some shielding that will reduce the wind pressures on downwind elements. It might also be possible that a building that is long in the direction of the wind, with many internal elements, might actually have a higher total silhouette to the wind as compared to the enclosed structure when it is finished. Hence, there is a dilemma in the evaluation of wind loads on an open structure such as a building.

ASCE 37 allows for shielding in some open frames of buildings that ultimately will be enclosed. The premise is that the first three rows of repeating elements in the direction of the wind should be assumed to receive full, unshielded wind pressures. Subsequent frames are assumed to receive some benefit from the breakup of the wind stream from upwind elements, and may be treated as having loads calculated for the design wind pressures reduced by 15 percent in addition to any other reductions for construction duration.

FIGURE 3.12 Temporary enclosure surrounding a church steeple.

Loads must also be assigned to materials stored in the building, internal walls, temporary enclosures (Fig. 3.12), signs, and similar objects likely to be in place.

The ASCE 37 procedure further requires simultaneous consideration of wind loads in both principal directions of the framing. The cross-wind direction wind loads are 50 percent of the wind load in the principal wind direction. This further accounts for consideration of the open nature of the structure and the complicated wind environment inside the building.

It seems imprudent to consider any shielding effects on structures such as scaffolding and trussed towers built to facilitate building erection. These structures tend to have few rows of repeating patterns of elements so shielding, if present, would be minimal. For these structures, wind loads should be calculated based on the sum of the loads independently calculated for each element or by following ASCE 7, which provides guidance for the calculation of wind loads on some structures of this type. The loads based on ASCE 7 can be reduced by construction period factors.

Further information on shielding is contained in the literature (Vickery et al., 1981; Nix et al., 1995; MBMA, 2006; Shapiro and Shapiro, 2011).

Calculation of wind loads on scaffolding erected next to a building is further complicated by its close proximity to the building. The façade of a building is an obstacle to the movement of air. When the wind streams reach the face of a building, they turn causing trajectories that are altered locally from the free field. Wind speeds are also altered and sometimes air separation at building corners creates high suctions.

This phenomenon is addressed in ASCE 37 by referring to the ASCE 7 coefficients for building cladding. The ASCE 7 coefficients are developed to represent the effects of local wind speeds, and include the influences of wind speed-up or separation. Hence, wind speeds to be used for the calculation of loads on scaffolding near to a building can be estimated by multiplying the basic speed by the square root of the cladding coefficient.

The square root is used to estimate wind speeds from pressure coefficients because pressure is proportional to velocity squared.

The estimate is imprecise. Pressure coefficients in ASCE 7 include the influence of internal pressure in buildings. Since internal pressures inside a building have no relevance to the pressures on components of scaffolding outside of buildings, wind speeds calculated following this procedure are approximate.

With drag factors for individual components of scaffolding and the modified wind speed, designers can calculate loads on the components of scaffolding. Values of drag factors for various shapes of structural elements are available (Simiu and Scanlan, 1986), but usually can be taken as 2.0 without compromising the accuracy of this approximate approach.

The designer needs to be aware that wind directions can be perpendicular (sideways or upward) to the face of the façade, and design accordingly.

Additional information on wind loads on structures during construction can be found in the literature (Ratay, 1987; Boggs and Peterka, 1992; Rosowsky, 1995).

Thermal Loads

Under some unusual circumstances, thermal effects on structures under construction can induce forces for which structures were not designed. Two primary factors contribute: (1) Building frames are sometimes erected at temperatures that differ substantially from their in-service temperatures, and (2) components that otherwise would be shielded from direct sunlight are exposed while under construction.

In many cases, the largest excursion of temperature that a structure will experience occurs between the time it is erected and the time it was put into service. For instance, in a cold climate, winter erection can be at steel temperatures that are on the order of 40°F (22°C) less than operational temperatures when the building is heated. Depending on how the resulting expansion is accommodated in the framing system and in nonstructural elements, this temperature change can cause damage. Similarly, floor slabs that normally would be shielded from the sun in the completed structure might be exposed to thermal radiation in a hot climate. This might cause more expansion than can be accommodated by supporting structures.

Problems are most common in structural components when there is high restraint against movement and in structural frames that are in-filled with nonstructural elements. Damage to the frame itself is most likely when it is rigidly connected to a stiff foundation. In these configurations, the foundations tend to restrain the frame, and forces develop in the members. Damage is also most likely at the ground level of multistory structures because the upper levels can expand or contract equally from floor to floor, whereas at the lowest level the slab above behaves differently than the slab on grade. Frames in buildings with long distances between expansion joints, particularly if there are multiple stiff elements for lateral support in a single column line, can develop high stresses.

It is usually not practical to control temperatures sufficiently to limit stresses. Theoretically, stresses due to even relatively moderate temperature changes can be very high. The more practical approach is to build the structure with recognition of the potential for damage due to thermal loads and accommodate the expected distortions by sequencing the erection so the movement can occur while inducing minimum harm.

ASCE 37 (ASCE, 2012) provides guidance on when thermal distortions might create high stresses or nonstructural damage, on the premise that most structures can accommodate distortions on the order of ½ in (13 mm) between expansion joints. The guidance identifies erection conditions that theoretically induce distortions that exceed that limit, recommending that provisions be made for thermal distortions of the structure and architectural components when structures are erected when the product of the following quantities exceeds 7000 ft·°F (1185 m·°C):

1. The largest horizontal dimension between expansion joints of the erected structures and
2. The largest of the differences between the following temperatures for the months when the portion of the structure is erected and exposed temporarily to ambient temperatures:
   a. The highest mean daily maximum temperature and the lowest mean daily minimum temperature
   b. The expected average temperature of the structure when it is in its end use and the highest mean daily maximum temperature
   c. The expected average temperature of the structure when it is in its end use and the lowest mean daily minimum temperature

Information about mean daily temperatures is available from the National Climatic Center (NCC, n.d.a; NCC, n.d.b) and the National Academy of Sciences (NAS, 1974).

In addition, solar radiation can cause exposed surfaces to heat and expand, causing damage (Martin, 1971; Ho and Liu, 1989; ACI, 1992; PCI, 1992; Chrest et al., 2001). ASCE 37 recommends accommodation of thermal distortions when portions of the structure that will be shielded when the structure is completed are subjected to direct solar radiation during hot weather, and whenever temperature changes create distortions that could damage structural or architectural components.

Snow Loads

The need to consider snow loads depends on the regional climate and the season of the year when construction is underway. When snow can reasonably be expected during the construction period, designs of temporary structures need to accommodate the snow that is likely to accumulate, contractors need to make plans for the removal of snow, or some combination of options needs to be implemented. Of course, if construction occurs entirely during the time of year when significant snow is unlikely, snow loads need not be considered in the design of temporary works.

ASCE 7 (ASCE, 2010) and ASCE 37 (ASCE, 2012) provide good references for the calculation of snow load. ASCE 7 contains maps that indicate the ground snow load that is appropriate for design throughout the United States, and provides methods to convert the ground snow load to roof snow load. Factors adjust the ground snow depending on the slope of the roof, whether the building is heated, and the exposure of the roof to wind.

The information in ASCE 7 is for long-term construction. During any particular winter construction project, the probability that snow will accumulate to the levels indicated in ASCE 7 is small. For that reason, ASCE 37 specifies that the ground snow load of ASCE 7 can be reduced by applying a factor of 0.8 when the duration of the construction period is 5 years or less.

Unheated buildings have potential to accumulate more snow on the roof than do heated buildings, because snow tends to melt and drain from heated buildings. Recognizing that buildings under construction in the winter are likely to be unheated at times, thermal factors that might apply to completed construction might not apply during construction. Therefore, for winter construction periods when the building under construction is not heated, thermal factors should be for unheated occupancies.

Snow melt on buildings under construction might not drain away because partially completed buildings often do not have functional drainage systems on the surfaces that are exposed to snow accumulation. Floors that will be enclosed do not have dedicated drainage systems. Roof drains might not be functional until the roofing system is complete. Furthermore, standpipes for roof drains might freeze because they are in unheated spaces. It is important to consider these possibilities when selecting snow loads for design of temporary conditions.

Often, the practical solution to deal with snow loads on temporary structures and structures in interim conditions is to have in place an understanding of the amount of snow that the structural systems can support safely based on calculating demands when combined with other loads, establishing a means to measure and weigh the snow on the structure and providing a process for snow removal should it accumulate and approach the safe limit. This is often a practical approach because contractors usually want to clear their work areas so they can proceed with construction. Contractors generally do not let snow accumulate on active projects, simply because it gets in the way.

Earthquake Loads

It is common in construction to ignore earthquake loads for the design of temporary structures and interim conditions. This is because the probability of a significant earthquake—one that will generate effective forces higher than those for wind—during a particular construction activity is very small.

Nevertheless, it is prudent to include earthquake loads in the analyses of temporary structures and interim conditions in structures under construction when the site is earthquake-prone, and certainly when the consequences of failure are significant.

ASCE 7 (ASCE, 2010) describes how to calculate earthquake loads on structures. The procedures rely on mapped values of the risk-targeted maximum considered earthquake (MCE_R) and the maximum considered earthquake geometric mean (MCE_G). Depending on the site and the nature of response under consideration (structural response or soil-related issues), the values selected from these tables are used together with response modification factors, R, and overstrength factors, Ω_0, in formulas to determine the effective seismic forces to be combined with other loads to evaluate performance of structures.

ASCE 37 (ASCE, 2012) provides a mechanism to adjust the procedure in ASCE 7 to account for regions of the country where the earthquake risk is low, and to account for the low probability that a severe earthquake will occur during the life of a temporary structure or the duration of a particular interim configuration. It says that earthquake loads need not be considered in regions where S_S, (i.e., the mapped MCE_R, 5 percent damped spectral response acceleration parameter at short periods) is 0.4 or less.

Furthermore, earthquake loads need not be considered even when S_S is greater than 0.4 unless the design short period response acceleration parameter, S_{DS}, is greater than 0.6. With this requirement, when S_S is equal to 0.4, only structures on sites with soft clay soil and soil vulnerable to failure or collapse (e.g., vulnerable to liquefaction, quick and highly sensitive clays, and collapsible weakly cemented soils) need to be analyzed for earthquake loads. However, as S_S increases, structures on a broader spectrum of soil types need analyses. When S_S is 0.75 or greater, buildings on any soil type should be evaluated for earthquake loads.

Once a designer determines that earthquake loads should be considered and the MCE_R values are selected, the corresponding loads can be reduced to account for the relatively short life of temporary structures and interim conditions in structures under construction. ASCE 37 suggests that S_S may be multiplied by a coefficient as small as 0.2. This value is derived from the observation that the ratios of the ground motion response accelerations for an earthquake with a 2 percent chance of exceedance in 1 year (representing construction duration) to one with a 2 percent chance of exceedance in 50 years (representing the life of the completed structure) ranges from approximately 0.05 to 0.2. Using the higher value, 0.2,

obviates the need for detailed analyses and adds a measure of conservatism to this imprecise effort to match risks between a structure under construction and one in service.

Temporary structures and structures in interim conditions often do not have levels of
ductility and redundancy normally associated with completed structures. Bracing sometimes is accomplished with cabling (Fig. 3.13). Failure modes in scaffolding sometimes are
not ductile. For these reasons, temporary structures and structural response that relies on
temporary bracing systems should be analyzed with the response modification factor not
greater than 2.5. Since this coefficient is used to reduce equivalent earthquake forces, limiting the value to 2.5 will result in higher design loads than might otherwise be consistent
with a structure's structural system in its final configuration.

FIGURE 3.13   Cables forming temporary bracing.

Rain Loads

The amount of rainfall is usually a function of season. In many regions of the country
with dry seasons, there is little need to design for rain for construction durations that are in
months of low rainfall. However, when substantial rain is common during the construction
period, rain loads should be considered.

ASCE 37 (ASCE, 2012) recommends following the procedures in ASCE 7 (ASCE,
2010) for the calculation of rain loads in months when historical average rainfall exceeds 1
in (25 mm). ASCE 7 recommends calculating rain load as the amount of water that can accumulate to a depth determined from the height of the secondary drain inlet plus a height that
corresponds to the flow depth through the secondary drainage system at the design flow rate.

In structures under construction, horizontal surfaces that later will be enclosed usually do not have drainage systems. However, they also usually do not have barriers to relatively free drainage: water can run off the edge of slabs or into stairwells. Hence, rain loads on surfaces that later will be enclosed usually do not control designs.

Roofs of structures under construction might have drainage systems that are not fully functional at the time of a storm. It is possible that membranes could be in place but roof drains not yet connected. It is also possible that roof drains could become blocked by building materials. For these reasons, roofs of structures under construction need to be scrutinized to determine whether rain can accumulate. If it can, calculations in the spirit of those in ASCE 7 should be performed.

Ice Loads

Ice accumulation can be a significant load on open structures such as scaffolding. These structures, with their many vertical, horizontal, and diagonal members where ice can accumulate, can receive ice loading that is more severe than the loads that are placed on working surfaces during use. For this reason, such structures should be designed for ice loads when they are in regions susceptible to icing—as identified in ASCE 7 (ASCE, 2010)—during seasons when icing is possible.

Structures that will be enclosed when finished normally are designed for substantial live load. As such, they generally have inherent capacity that will accommodate icing should it occur during construction. In particular, ASCE 37 (ASCE, 2012) allows structures designed to be enclosed, having a design live load of at least 20 psf (0.96 kN/m^2), to be excluded from consideration of ice loads. Most enclosed structures will be in this category. Nevertheless, snow melt and rain on snow can cause accumulation of ice when drainage is poor (Fig. 3.14).

FIGURE 3.14   Ice accumulation due to poor drainage.

ACKNOWLEDGMENT

Figures in the chapter are printed with permission from Simpson Gumpertz & Heger Inc.

REFERENCES

ACI: *Analysis and Design of Reinforced Concrete Guideway Structures*, ACI 358.1R-92, Farmington Hills, Mich.: American Concrete Institute, 1992.

ACI: (2005). *Guide for Shoring/Reshoring of Concrete Multistory Buildings*, ACE 347.2R-05, Farmington Hills, Mich.: American Concrete Institute, 2005.

AISC: *Steel Construction Manual*, AISC 325-11, Chicago, Ill.: American Institute of Steel Construction, 2011.

ANSI: *Scaffolding Safety Requirements*, ANSI/ASSE A10.8-2001, New York: American National Standards Institute, 2001.

ASCE: *Minimum Design Loads for Buildings and Other Structures*, Reston, Va.: Structural Engineering Institute of the American Society of Civil Engineers, 2010.

ASCE: *Design Loads on Structures During Construction*, Reston, Va.: Structural Engineering Institute of the American Society of Civil Engineers, anticipated 2012.

Boggs, D. W. and J. A. Peterka: *Wind Speeds for Design of Temporary Structures*, Proceedings of the ASCE Tenth Structures Congress, San Antonio, Tex., Reston, Va.: American Society of Civil Engineers, 1992.

Caterpillar: *Caterpillar Performance Handbook*, Peoria, Ill.: Caterpillar, Inc, 2008.

Chrest, A. P., M. S. Smith, S. Bhuyan, D. R. Monahan, and M. Iqbal: *Parking Structures: Planning, Design, Construction, Maintenance and Repair*, Boston, Mass.: Kluwer Academic Publishers, 2001.

Ho, D. and C-H Liu: "Extreme Thermal Loadings in Highway Bridges." *Journal of Structural Engineering*, vol. 115, no. 7, 1989.

Jahren, C. T.: "Loads Created by Construction Equipment," Chapter 6. In R. T. Ratay (ed.), *Handbook of Temporary Structures in Construction*, 2d ed., New York: McGraw-Hill, 1996.

Martin, I.: "Effects of Environmental Conditions on Thermal Variations and Shrinkage of Concrete Structures in the United States." In *Designing for Effects of Creep, Shrinkage, and Temperature in Concrete Structures*, SP-27, Farmington Hills, Mich.: American Concrete Institute, 1971.

MBMA: *Metal Building Systems Manual*, Cleveland, Ohio: Metal Building Manufacturers Association, 2006.

NAS: Expansion Joints in Buildings, Building Research Advisory Board, Federal Construction Council Technical Report No, 651974. Washington, D.C.: National Academy of Sciences, 1974.

NCC: *Climatological Summary of the US*. Asheville, N.C.: National Climatic Center, n.d.a.

NCC: *Monthly Normals of Temperature, Precipitation, Heating and Cooling Degree Days*, Asheville, N.C.: National Climatic Center, n.d.b.

Nix, H. D., C. P. Bridges, and M.G. Powers: *Wind Loading on Falsework, Part I*, Sacramento, Calif.: Caltrans publication, 1995.

OSHA: *OSHA Standards for the Construction Industry*, 29 CFR 1926, Chicago, Ill.: Occupational Health and Safety Administration, 2011.

PCI: *Precast Prestressed Concrete Parking Structures: Recommended Practice for Design and Construction*, Chicago, Ill.: Prestressed Concrete Institute Committee on Parking Structures, 1992.

Ratay, R. T.: *To Mitigate Wind Damage during Construction: Codification?* Proceedings of the NSF/ Wind Engineering Research Council Symposium on High Winds and Building Codes, Kansas City, Mo.: National Science Foundation/Wind Engineering Research Council, 1987.

Rosowsky, D.V.: Estimation of Design Loads for Reduced Reference Periods. Structural Safety, 1995.

Shapiro, L.K, and J. P. Shapiro: *Cranes and Derricks*, New York: McGraw-Hill, 2011.

Simiu, E. and R. H. Scanlan: *Wind Effects on Structures*, New York: John Wiley & Sons, 1986.

Vickery, B. J., P. N. Georgiou, and R. Church: *Wind Loading on Open Framed Structures.* Third Canadian Workshop on Wind Engineering, Toronto, Canada, 1981.

CHAPTER 4
CONSTRUCTION SITE SAFETY

Mohammad Ayub, P.E., S.E.

INTRODUCTION

Construction remains one of the most hazardous occupations though the fatalities and injuries in construction are on the decline. Recent figures from the Bureau of Labor Statistics (BLS) indicate that construction accounted for more fatal work injuries than any other industry in 2010. BLS reports that the number of fatal work injuries in the private construction sector declined by 10 percent in 2010. Fatal work injuries in construction have declined every year since 2006 and down nearly 40 percent over that time. Economic slowdown and financial meltdown could explain much of this decline with total hours worked having declined another 6 percent in construction in 2010, after declines in both 2008 and 2009. The rate of fatalities in construction workers per 100,000 full-time employees, however, remained at 0.5 percent, a miniscule decline. Even with the lower fatal injury total, construction accounted for more fatal work injuries than any other industry in 2010. The rate of fatalities of all workers in all industries was 3.5 percent. The construction rate of fatalities is approximately 275 percent that is higher than all industries combined. The direct and indirect cost of construction injuries is enormous and run in billions of dollars.

CONSTRUCTION ACCIDENTS, INJURIES, AND FATALITIES

Figure 4.1 presents the actual number of fatalities in construction from 2000 through 2010. Figure 4.2 shows the rate of fatalities per 100,000 workers for construction and all industries combined from 2001 through 2010. Clearly, the rate of fatalities in construction is among the highest.

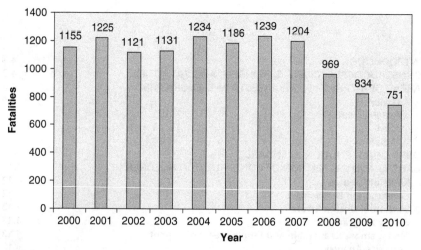

FIGURE 4.1   Fatalities in construction. (*Source: BLS CFOI Data.*)

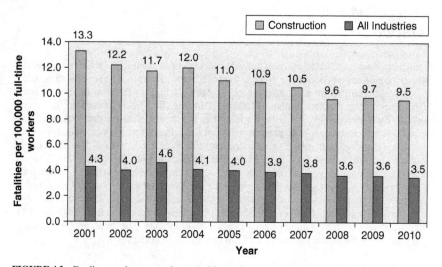

FIGURE 4.2   Fatality rates in construction and all industries combined. (*Source: BLS CFOI Data.*)

The causes of fatalities in construction remain the same, for example fall, electrocution, struck by, and caught in/between. Figure 4.3 represents the causes of construction fatalities for the year 2009. In previous years, the order of distribution among the four leading causes varied, but the fall always remained at the top.

The number of injuries in construction also remains high, though the injury numbers are not as reliable as those of fatalities. Figure 4.4 shows the distribution of injuries for the year 2009.

Fall from varying heights by far remains the single cause of largest fatalities, 34 percent of all fatalities in construction. Thirty percent of fatalities occur due to fall from a height of 11 to 20 ft. Nineteen percent of fatalities occur from a height of 20 ft or under. Figure 4.5 is a chart of varying height of fall in 2009.

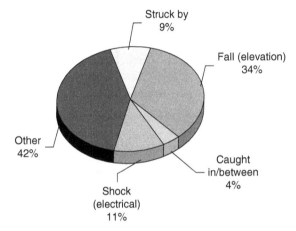

FIGURE 4.3 Causes of construction fatalities. (*Source: BLS 2009 CFOI Data.*)

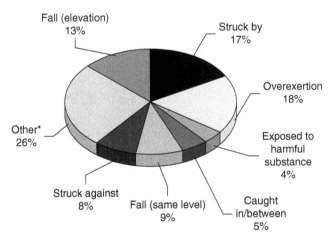

FIGURE 4.4 Causes of construction injuries. (*Source: BLS 2009 Occupational Injuries and Illness Data.*)

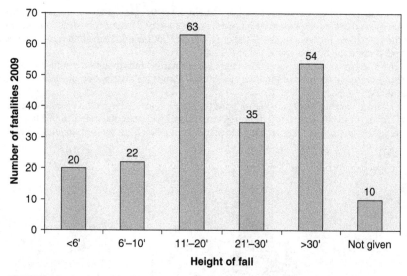

FIGURE 4.5   Analysis of fatalities caused by falls. (*Source: OSHA 2009 IMIS Data.*)

Falls occur due to various activities at the construction sites. Figure 4.6 is the chart for activities that resulted in fatal falls during 2001 through 2009.

The Occupational Safety and Health Administration (OSHA) under the Department of Labor is taking a leading role in improving construction safety. OSHA has embarked in outreach, alliances, and partnership programs with various organizations to enhance safety at the site. Labor and contractors trade groups are also actively involved in improving construction site safety. Major construction companies have expanded their safety and health plans, and have reemphasized training for their employees and have demanded the

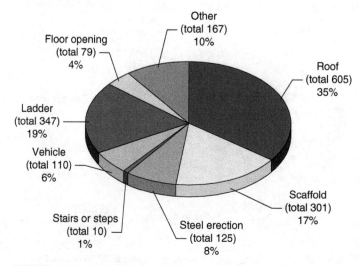

FIGURE 4.6   Activities that resulted in fatal falls during 2001 through 2009. (*Source: BLS CFOI Data.*)

same from their subcontractors. Several studies have shown that such initiatives on the part of labor and employers have resulted in savings to their contracts. It improves morale and productivity. Challenges, however, remain due to language barriers and immigration issues.

The concept of Prevention through Design (PtD) is gaining ground in the United States that calls for measures to be taken during early planning and design stage to enhance site safety and facilitate safety measures to be taken during construction. European countries have made significant advances in this regard and had taken an early lead. The challenge is to incorporate the PtD concept in codes and consensus standards.

OSHA encourages design professionals to incorporate safety in design to facilitate safe construction, as far as possible. PtD is a concept that is gaining ground in the United States. Multinational construction companies and other US governmental agencies, for example, U.S. Army Corps of Engineers, U.S. Navy, and U.S. Department of Energy are adopting PtD in their designs. It has been in practice to varying degrees in the United Kingdom and Germany. PtD has a potential of reducing injuries and fatalities at construction sites by incorporating features during the design phase that will enhance constructability. PtD will result in fewer delays in construction and savings in workers' compensation premiums. Though, architects and engineers take a "hands-off" view of methods and means of construction, they must review and examine the constructability aspect of the project they design.

AREAS OF INTEREST TO ARCHITECTS AND ENGINEERS

Particular areas of interest for the architects and engineers are

General
- Emphasize that the general contractor and its subsidiaries shall strictly follow OSHA regulations, including the training requirements for its employees as covered in 29 CFR 1926.
- Conduct constructability reviews early in the design phase. Include the constructor and maintenance personnel in the reviews.
- Ensure that structure remains stable during all phases of construction.

Fall Protection
- Provide inserts for anchorage of fall protection for horizontal and vertical lifelines during construction, which could resist minimum an ultimate load of 5000 lb per insert.
- Increase parapet wall height up to 42 in (±3 in) to act as railings.
- Provide inserts for railings instead of field drilling.
- Provide holes in columns for tie lines at 21 and 42 in above floor slab.

Steel
- Follow OSHA steel erection standard to provide minimum four anchor bolts for columns to enhance flexural capacity at the column base.
- Locate column splices at standing height from the floor.
- Maintain slenderness of columns during construction.
- Provide shop assembly of the structural elements instead of field assembly as far as practicable.
- Avoid overhead steel connections or welding except for hanger design.
- Specify shop-welded connections instead of bolts or field welds for easy erection where feasible.

- Maintain the structural integrity at all times, that is, during construction and demolition, by providing temporary bracings, guy wires, shores, underpinning, etc.
- Have approved steel erection procedure prior to erection.
- Install stairways early to avoid the need for temporary access.
- Design structural steel beam, joist, or wood trusses such that during the erection phase it can resist the weight of employees and stack of materials.
- Provide proper lateral bracing (including lateral load transfer system) from erection until final completion to steel, concrete and wooden beams, joists, girders, roof trusses, etc.
- Provide proper lateral load-resisting system for sudden crane stoppage during load testing of crane.
- Verify the structure for construction loads per SET/ASCE 37 in all stages of construction
- Maximize prefabrication.
- Ensure the spacing of purlins to allow building component to be lowered down through clear space.

Concrete

- Locate electrical conduits away from field drilling for hangers.
- Specify minimum levels of shoring and reshoring for casting of the floor slabs or beams.
- Have approved formwork drawings for any formwork erection.
- Do not allow to mix and match other manufacturer's components during the formwork and scaffold erection.
- Specify concrete test results to be verified before removal of the forms and shoring.
- Design shoring and scaffolds for inertia forces due to sudden stopping of operating vehicles over it.

Well-designed and constructed temporary structures are vital for the success of any construction project. Temporary structures could be classified in two categories. The first group provides support to the permanent structures, for example, temporary bracings to trusses, walls, framing beams, shoring to support concrete slabs, and beams and columns. The second group contains structures that support construction workers while performing work on the permanent structures, for example, scaffolds, ramps, and ladders. Failures of both groups of temporary structures would be catastrophic and would result in multiple injuries and deaths. However, falls from elevated surfaces present the greatest hazards and continue to be the topmost cause of construction fatal injuries.

OSHA CONSTRUCTION STANDARDS

OSHA construction standards are contained in 29 Code of Federal Regulations (CFR) 1926, containing regulations for personal protective equipment to concrete, masonry and steel construction, subpart A through Z, and part CC through DD. OSHA Compliance Safety and Health Officers (CSHO) inspect construction sites and, in case of violations, issue citations to the employers exposing their employees to construction hazards. Table 4.1 that has been taken from an upcoming OSHA study on the frequently cited standards presents 100 most frequently cited OSHA construction standards in the federally managed states.

TABLE 4.1 100 Most Frequently Cited OSHA Construction Standards and Their Relative Ranking to 2010

Standards (1926 Unless noted)	Description	Relative ranking to 2010				
		Years				
		2010	2009	2008	2007	2006
501(b)(13)	Fall Protection					
	Residential construction	1	1	1	1	1
1053(b)(1)	Stairways and Ladders					
	Ladders use	2	5	7	7	8
501(b)(1)	Fall Protection					
	Unprotected sides and edges	3	2	2	2	2
100(a)	Personal Protective Equipment					
	Head protection	4	4	3	3	3
503(a)(1)	Fall Protection					
	Training program—Fall hazards	5	3	5	6	5
102(a)(1)	Personal Protective Equipment					
	Eye and face protection	6	10	13	13	14
453(b)(2)(v)	Scaffolds					
	Aerial lifts—Specific requirements	7	6	4	5	7
451(g)(1)	Scaffolds					
	Fall protection	8	7	6	4	4
451(e)(1)	Scaffolds					
	Access	9	8	8	8	6
451(b)(1)	Scaffolds					
	Scaffold platform construction	10	9	9	9	9
20(b)(2)	General Safety and Health					
	Accident prevention responsibilities—Inspections	11	12	15	15	15
501(b)(10)	Fall Protection					
	Roofing work on low-slope roofs	12	15	17	18	16
454(a)	Scaffolds					
	Training requirements	13	11	11	12	13
652(a)(1)	Excavations					
	Protection of employees in excavations	14	14	10	10	10
20(b)(1)	General Safety and Health					
	Accident prevention—Employer responsibilities	15	17	14	11	11
21(b)(2)	General Safety and Health					
	Safety training and education—Employer responsibility	16	13	12	14	12
1910.1200(e)(1)	General Industry					
	Written hazard communication program	17	16	16	17	17
1053(b)(4)	Stairways and Ladders					
	Ladders use	18	27	28	33	36
404(f)(6)	Electrical					
	Grounding path	19	18	18	22	19
1060(a)	Stairways and Ladders					
	Training program	20	23	25	32	31
451(c)(2)	Scaffolds					
	Criteria for supported scaffolds	21	21	20	16	18

(*Continued*)

TABLE 4.1 100 Most Frequently Cited OSHA Construction Standards and Their Relative Ranking to 2010 (*Continued*)

Standards (1926 Unless noted)	Description	Relative ranking to 2010				
		Years				
		2010	2009	2008	2007	2006
451(g)(1)(vii)	Scaffolds					
	Fall protection—Personal fall arrest systems	22	19	19	19	20
451(g)(4)(i)	Scaffolds					
	Fall protection—Guardrail systems	23	22	22	20	23
501(b)(11)	Fall Protection					
	Duty to have fall protection—Steep roofs	24	26	33	30	25
95(a)	Personal Protective Equipment					
	PPE application	25	24	21	27	29
451(f)(7)	Scaffolds					
	Use—Erection, moving, dismantling, etc.	26	20	23	23	22
651(c)(2)	Excavations					
	Access and egress—Egress from trench excavations	27	25	24	21	21
403(b)(2)	Electrical					
	General Requirements—Installation and use of equip.	28	38	38	41	35
5(a)(1)	General Duty Clause					
	Serious hazard not covered by specific standards	29	30	30	31	28
651(k)(1)	Excavations					
	Inspections— Daily inspections of excavations	30	31	27	25	26
1910.1200(h)(1)	General Industry					
	Hazard communication—Information and training	31	28	32	37	42
404(b)(1)(i)	Electrical					
	Branch circuits—Ground fault protection	32	32	26	24	24
405(g)(2)(iv)	Electrical					
	Flexible cords and cable—Strain relief	33	34	36	43	39
651(j)(2)	Excavations					
	Protection of employees from loose rock or soil	34	35	29	28	27
451(f)(3)	Scaffolds					
	Use -Scaffolds and scaffold components inspection	35	29	34	29	33
1053(b)(13)	Stairways and Ladders					
	Ladders use	36	41	43	40	45
416(e)(1)	Electrical					
	Cords and cables—Worn or frayed electrical cords	37	39	45	47	46
1904.29(b)(1)	OSHA Injury and Illness Recordkeeping					
	Forms Implementation—OSHA 300 log	38	83	*	*	*
25(a)	General Safety and Health					
	General Housekeeping	39	43	35	39	32
701(b)	Concrete and Masonry Construction					
	Reinforcing steel—Guarding Protruding Steel Rebars	40	37	37	34	34

TABLE 4.1 100 Most Frequently Cited OSHA Construction Standards and Their Relative Ranking to 2010 (*Continued*)

Standards (1926 Unless noted)	Description	Relative ranking to 2010 — Years				
		2010	2009	2008	2007	2006
1052(c)(1)	Stairways and Ladders Stair rails and handrails	41	36	31	26	30
1910.1200(g)(1)	General Industry Hazard communication—Material safety data sheets	42	33	42	45	44
1910.1200(g)(8)	General Industry Hazard communication—Material safety data sheets	43	45	47	50	54
1053(b)(16)	Stairways and Ladders Use—Portable ladders with structural defects	44	53	52	56	59
1051(a)	Stairways and Ladders General Requirements— Stairway/ladder requirement	45	44	44	42	37
501(b)(4)(i)	Fall Protection Holes—Protection from falling through holes	46	47	40	46	43
501(b)(4)(ii)	Fall Protection Holes—Protection from tripping or stepping	47	52	50	44	47
452(c)(2)	Scaffolds Fabricated frame scaffolds—Bracing	48	40	41	35	41
760(a)(1)	Steel Erection Fall Protection—Protection from fall hazards	49	46	39	38	38
501(b)(14)	Fall Protection Wall Openings—Protection from falling	50	42	46	36	40
451(b)(2)	Scaffolds Scaffold platform construction	51	62	53	58	52
503(b)(1)	Fall Protection Certification of training	52	55	58	66	75
451(g)(1)(i)	Scaffolds Fall protection—Personal fall arrest system	53	48	51	53	51
453(b)(2)(iv)	Scaffolds Aerial Lifts—Specific Requirements	54	49	48	49	55
416(a)(1)	Electrical Protection of employees—Electric shock	55	60	74	62	72
405(b)(2)	Electrical Cabinets, boxes, and fittings— Covers and canopies	56	51	56	59	57
405(a)(2)(iii)	Electrical Temp. wiring—Guarding. temp. wiring over 600 V	57	57	64	57	62
1910.134(c)(1)	General Industry Respiratory protection program	58	58	68	71	77

(*Continued*)

Standards (1926 Unless noted)	Description	Relative ranking to 2010				
		Years				
		2010	2009	2008	2007	2006
404(b)(1)(ii)	Electrical Branch circuits—Ground fault circuit interrupter	59	68	55	52	49
1910.134(e)(1)	General Industry Respiratory protection— Medical evaluation	60	61	73	90	88
1910.1200(h)	General Industry Hazard communication— Employee info. and training	61	56	71	70	67
300(b)(1)	Tools—Hand and Power Guarding—Power operated tools	62	75	70	88	99
1910.178(l)(1)(i)	General Industry Powered industrial trucks— Operator training	63	63	72	93	78
403(b)(1)	Electrical Examination—Elect. equip. free of recognized hazards	64	71	63	86	82
451(c)(2)(v)	Scaffolds Criteria for supported scaffolds	65	54	49	63	53
454(b)	Scaffolds Training requirements	66	50	54	54	48
1052(c)(1)(i)	Stairways and Ladders Stair rails and handrails	67	64	89	64	70
451(f)(6)	Scaffolds Use—Clearance from power lines	68	72	77	72	71
403(i)(2)(i)	Electrical 600 V, nominal or less—Guarding of live parts	69	78	65	74	68
503(c)(3)	Fall Protection Retraining	70	70	91	68	60
1904.32(b)(3)	OSHA Injury and Illness Recordkeeping Implementation—OSHA 300 log certification	71	*	*	*	*
28(a)	General Safety and Health PPE—Employer responsibility	72	66	60	69	63
451(b)(1)(i)	Scaffolds Scaffold Platform Construction	73	59	59	51	56
451(h)(2)(ii)	Scaffolds Falling object protection	74	67	83	73	76
1053(b)(6)	Stairways and Ladders Ladders use	75	88	82	96	92
405(g)(2)(iii)	Electrical Flexible cords and cables—Splices	76	80	87	89	98
1053(b)(22)	Stairways and Ladders Ladders Use	77	87	94	99	*
405(b)(1)	Electrical Cabinets, boxes, and fittings	78	77	88	80	74
451(c)(2)(iv)	Scaffolds Criteria for supported scaffolds	79	73	57	76	86

TABLE 4.1 100 Most Frequently Cited OSHA Construction Standards and Their
Relative Ranking to 2010 (*Continued*)

Standards (1926 Unless noted)	Description	Relative ranking to 2010				
		Years				
		2010	2009	2008	2007	2006
501(b)(15)	Fall Protection Walking/working surfaces	80	69	69	61	61
503(a)(2)	Fall Protection Training program	81	*	95	100	73
150(c)(1)(i)	Fire Protection and Prevention Portable firefighting equipment— Fire extinguisher	82	90	79	78	83
405(a)(2)(ii)(E)	Electrical General requirements for temporary wiring—Lamps	83	82	75	*	80
502(d)(16)(iii)	Fall Protection Personal fall arrest systems	84	*	*	*	*
1053(b)(5)(i)	Stairways and Ladders Ladders Use—Non-self-supporting ladders	85	*	*	98	*
451(c)(2)(ii)	Scaffolds Criteria for supported scaffolds	86	79	61	48	50
451(a)(6)	Scaffolds Capacity—Design	87	74	67	67	66
350(a)(9)	Welding and Cutting Transporting and storing compressed gas cylinders	88	*	78	87	69
451(h)(1)	Scaffolds Falling object protection	89	84	66	60	96
200(g)(2)	Signs, Signals and Barricades Traffic Signs—Traffic control signs	90	*	*	*	*
1060(a)(1)(iii)	Stairways and Ladders Training program	91	*	*	*	*
452(j)(2)	Scaffolds Pump jack scaffolds	92	93	93	91	*
451(b)(5)(ii)	Scaffolds Scaffold platform construction	93	100	81	75	97
1910.134(d) (1)(iii)	General Industry Respiratory protection— Selection of respirators	94	95	*	*	*
304(d)	Tools—Hand and Power Woodworking tools—Guarding	95	*	*	*	*
1052(c)(1)(ii)	Stairways and Ladders Stair rails and handrails	96	81	90	83	90
651(k)(2)	Excavations Inspections	97	85	85	81	64
1053(b)(8)	Stairways and Ladders Use—Securing ladders	98	94	*	*	*
502(d)(15)	Fall Protection Personal fall arrest systems—Anchorages	99	98	*	*	*
1903.19(d)(1)	Inspections, Citations, and Proposed Penalties Abatement documentation	100	*	*	*	*

*Particular standard was not one of the 100 most frequently cited in the reference year.

CAUSES AND PREVENTION OF CONSTRUCTION FAILURES

Kinds of Failures

In addition to fall, electrocution, struck by, and caught in/between; construction workers are also killed by collapses during construction of buildings, bridges, towers, etc. Investigation of such incidents must be a part of a comprehensive safety and health program. Determination of the cause and nature of collapses and contributing factors should provide a basis of corrective actions and elimination of similar occurrences in the future. OSHA investigated some 96 incidents from 1996 to 2008. Of these, 80 percent of the collapses occurred due to construction errors and deficiencies in means and method of construction but rarely due to defective materials. The remaining 20 percent of the incidents are a result of flawed structural design. Either the structural engineer of record or the structural engineer retained by contractors during construction was responsible for the flaws. Such collapses occurred in all types of framing, steel, concrete, masonry, or timber.

The largest group of structural collapses involved 60 steel structures:

- 14 structural steel frames
- 14 scaffolds
- 18 special steel structures and cranes
- 5 television antenna towers
- 3 cofferdams
- 6 steel roof trusses and joists

The second largest group involved 29 concrete and masonry structures:

- 3 concrete frames
- 12 shorings supporting freshly placed concrete
- 4 demolitions involving concrete structures
- 5 precast concrete structures
- 5 masonry walls

The third group consisted of wood structures:

- 7 wood frames and roof trusses.

Construction Deficiencies

In 47 cases, contractor did not generally follow the installation procedures prescribed and recommended by the manufactures and designers, such as providing temporary bracings, lateral bracings, diagonal bracing, bridging, anchoring, guy cables, lateral supports, and proper welding connections. In 15 cases, contractor overloaded certain structural members beyond their ultimate capacities. In nine cases, contractor did not provide temporary bracings during construction of steel frames, concrete, or masonry walls. As a result, wind pressures caused their collapse. In seven cases, contractor began to demolish existing structures without regard to structural stability and capacity of existing structural members.

Structural Design Errors

Out of 96 incidents, 19 construction incidents were related to structural design errors. These occurred in 13 steel structures, 5 concrete structures, and 1 masonry structures. The structural design deficiencies stemmed from lack of peer review; lack of consideration of all anticipated gravity and lateral loads during construction, stability of walls during construction, lateral torsional buckling, and special design requirement of round members, vortex shedding, etc.; and poor design of beam-column joints.

Safety Program Elements

There are four parts of a well-planned and structured safety and health plan. The basics of the plan are to recognize and understand all the hazards and potential hazards of the jobsite, prevent or control those hazards and potential threats of the jobsite, and train employees at all levels so they understand the hazards and know how to help protect themselves and others.

- Management leadership and employee involvement
- Worksite analysis
- Hazard prevention and control
- Safety and health training.

Management Leadership and Employee Involvement

The company must lay down clear path of responsibility for safety and establish a culture of safety. It must state company policy, goals and objectives, assignment of responsibility, authority, accountability, and program evaluation. Safety must not be relegated below productivity. The company must aim for zero tolerance for injuries. A number of companies have achieved it. Line management and supervision for safety and health must be clear. Monitoring employees' performance, on-the-job training, and keeping employees informed of all potential hazards are functions of line management. There must be continuous improvement program where corrective measures are required.

Worksite Analysis

Individuals trained in potential hazard identification must perform worksite analysis. The individual must also be familiar with applicable safety standards. Safety planning and hazard analysis will improve safety. The frequency of site inspection will lead to continuous improvement. Employees must be encouraged to freely report hazards immediately for correction. In the event of an incident, a thorough investigation must be conducted to determine causal factors and must be shared with all employees to prevent future recurrence.

Hazard Prevention and Control

Hazards should be controlled through elimination or preventing employee exposure. Pre-job planning is crucial. Where hazard control is not practical, alternative safe work practices must be studied and implemented. Planning and preparing for emergencies is a part of an effective safety and health program. For large projects, such functions can be handled through contract or local hospitals or emergency care facilities. Employees should be trained in first aid and cardiopulmonary resuscitation (CPR).

Safety and Health Training

Training is the keystone of effective safety and health program. All employees must be familiar with hazard recognition and safe work practices to avoid or control their and their fellow workers' exposure. A significant percent of injuries occur among workers who had less than 1 month at the work site. All new employees, regardless of their construction experience, should receive orientation of the job site and work conditions, and be familiarized with safety and health rules of the worksite. Safety demands constant attention. Periodic safety and health training, safety awareness sessions, toolbox talks, and emergency drills should be provided. Special training should be provided to supervisors and managers in view of their leadership role and broader perspective of the overall work situation.

CHAPTER 5
LEGAL ASPECTS

Robert A. Rubin, Esq., P.E.
Sarah B. Biser, Esq.
Gary Strong, Esq.

LEGAL RELATIONSHIPS

Temporary structures present as many legal pitfalls as they do construction challenges. This section deals with the myriad rights, responsibilities, and liabilities of the parties with temporary structures in construction.

There are three sources of these rights, responsibilities, and liabilities: common law, statutes, and the contract, if any, between the parties. Common law is best understood as judge-made law, which is developed over the years on a case-by-case basis. Statutory law is the legislative enactments of the U.S. Congress, state legislatures, and municipal bodies, many of which are supplemented by the rules and regulations of administrative agencies charged with the enforcement of the legislative enactments. Examples are noise restrictions, building codes, and safety requirements. Contract law emerges from the agreements between the parties, setting forth their respective rights and obligations. So long as a contract provision does not contravene a statute or public policy, it will be enforced. Where a contract is vague or silent, common law often will provide the needed interpretation.

Construction people share concern about their liability to others, both on and off the jobsite. Liability arises when individuals fail to exercise the "standard of care" imposed on them. Temporary construction activities generally impose one or the other of two distinct standards of care upon the individual, depending on the circumstances of the work and the type of activity being performed. The negligence standard, usually imposed for work that is not deemed ultrahazardous, requires that an individual use reasonable care under the circumstances. The absolute standard, as the name suggests, imposes strict liability, without fault, should damage result from engaging in a particularly hazardous activity. The rationale for this severe standard is that one who subjects the public to highly dangerous conditions must be legally responsible for the full consequences, even though the individual exercised reasonable care. Blasting is the prime example of an ultrahazardous activity for which strict liability has been imposed by the courts. Strict liability for the safety of workers has been imposed by statute in some states.

Absolute or strict liability can also be imposed by statute on one party for the acts or omissions of another party. This is often referred to as vicarious liability which will be explained in detail later.

Apart from negligence or strict liability standards, the rights and duties of parties involved in temporary construction work are most often detailed in the contract documents, that is, the general, supplementary, and special conditions. These generally start with one of several standard forms, such as those provided by the American Institute of Architects (AIA), the Engineers Joint Contract Documents Committee (EJCDC), the Associated General Contractors of America (AGC), or Consensus Doc, and then include modifications tailored to meet the specific needs of a particular project. In the event of a dispute, the court or arbitrator will look to the contract documents to determine what was required or to ascertain what the actual intent of the parties was. Consequently, all parties on a construction project must thoroughly familiarize themselves with the requirements of contract documents.

RESPONSIBILITIES

All Parties

All parties on a construction project have responsibility both to the other contracting parties and to a wide range of third persons as well, such as workers, pedestrians, and adjoining property owners. As regards the contracting parties, their responsibilities to each other are governed by the contract. These include the duty not to interfere with the work of the trades and the duty to cooperate with all other parties to progress the work.

All contracting parties potentially can be held liable to third persons who are injured or damaged as a result of the contracting parties' failure to exercise the standard of care required by either common or statutory law. Sometimes, even third parties can be held to be the beneficiaries of a contract provision intended for their benefit, for instance, pedestrians protected by a sidewalk bridge. Similarly, an excavation contractor who fails to adequately underpin an adjacent structure will be liable to the owner of that structure for any damage caused by the contractor's neglect. In some states the violation of a statutory duty (e.g., to adequately underpin) is, in and of itself, conclusive proof of negligence. The practical effect of this is that the injured party has but to establish the amount of damages in order to recover; there can be no issue as to fault or liability. In other states, violation of a similar statute is considered merely some evidence of negligence to be considered with all other evidence, so that the injured party's prospect for recovery is far less certain.

The Owner

The owner is concerned with temporary structures only insofar as they are necessary to the construction of the finished product. Therefore, the construction contract documents will often simply provide what must be done rather than detailed instruction on how to do it; for example, "Earth banks will be adequately shored"; "Adjoining structures will be adequately protected." Under this performance-type specification, the involved contractors are responsible for meeting the required result, and they, not the owner, are responsible for the means and methods used to achieve it. This is in contrast to the technical or detailed specification wherein the owner or the designer specifies the method to be used and is deemed impliedly to warrant its adequacy. The difference in the owner's responsibilities can be seen as follows: If a proper performance specification is being used, the owner bears no responsibility if the contractor's work fails unless the owner was negligent in hiring the contractor or, notwithstanding the written documents, the owner undertook direction or control of the contractor's work. If, however, a detailed specification is utilized, the owner bears full responsibility for any deficiencies in the contractor's work, as long as the contractor performs that work as specified. An owner's responsibility to third persons is usually secondary, in that most lawsuits arising out of temporary construction work, while naming the owner as a party, are attributable to negligent performance by the contractors or design deficiencies by the responsible architect or engineer. The doctrine of vicarious liability traces its roots to the theory of "respondeat superior" or "Let the Master Answer." 2 *Dan B. Dobbs*, The Law of Torts 905 (2000). Typically, "liability in negligence is . . . premised on a defendant's own fault, not the wrongdoing of another person." *Feliberty v. Damon*, 72 N.Y.2d 112, 117, 531 N.Y.S.2d 778, 780 (1988). An exception to the general rule of negligence is the doctrine of vicarious liability that imputes negligence to another because of the relationship between the two individuals. *Prosser & Keeton on the Law of Torts* 69 at 499 (5th ed. 1984) (The negligence of A is to be charged to B although B played no part in A's conduct). This doctrine:

> [r]ests in part on the theory that—because of an opportunity for control of the wrongdoer, or simply as a matter of public policy loss distribution—certain relationships may give rise to a duty of care, the breach of which can indeed be viewed as the defendant's fault. *Feliberty*, 72 N.Y.2d at 118, 531 N.Y.S.2d at 782.

The intent behind vicarious liability theory is to make the damaged party whole, even if it means imputing liability to someone who may not in fact be the wrongdoer. *Paul Brothers v. New York State Electric and Gas Corporation*, 11 N.Y.3d 251, 253, 869 N.Y.S.2d 356 (2008); *Prosser & Keeton on the Law of Torts* 69 at 498 (5th ed. 1984).

Thus, even in the absence of a contractual indemnity provision, the owner can shift the liability to the party ultimately responsible, assuming that party has the financial resources to shoulder the responsibility. In other words, although the owner is primarily liable to the injured party, the owner has a right of indemnity or contribution from the party who was negligent.

The Contractor

The ambit of the contractor's responsibility is far-reaching. The contractor is responsible for performing all required work in accordance with the contract documents, generally accepted construction practices, and governing legislative enactments. The AIA 201 2007 General Conditions state that unless otherwise provided in contract documents, "the Contractor shall secure and pay for the building permit as well as other permits, fees, licenses, and inspections by government agencies" (3.7.1). When dealing with temporary structures, normally some kind of permit or license is required before the structure can be erected and/or moved on the project site.

It is important to stress that it is not necessarily enough that the contractor has performed to the best of its own ability. When assessing liability, the courts do not simply accept at face value what the contractor did, but measure that against what an ordinary contractor of similar experience would have done under similar circumstances. Consequently, when determining whether a contractor exercised ordinary or reasonable care in doing the work, the court will look at whether that work favorably compares to that degree of skill then prevalent in the industry.

The Designer

The design professional is responsible for the accuracy and adequacy of his or her plans and specifications. By distributing these documents for use, the design professional impliedly warrants to both the owner and contractor that a cohesive structure will result, provided the plans and specifications are followed. But the designer does not guarantee perfection, only that he or she exercised ordinary care. When a contractor who properly follows the designer's documents experiences a problem, liability will ultimately rest with the architect or engineer. The designer would be responsible for all additional costs of construction attributable to faulty design. For example, if an access ramp to a construction site proved unable to sustain the weight of heavy construction equipment, the engineer who designed the ramp would be required to redesign the ramp, or bear the cost of redesign, and would be liable for all resultant damages to the heavy equipment. In a more drastic case, the design professional would be liable for all damages resulting from a retaining wall collapse attributable to deficient design. This includes liability to third parties for injuries, death, and property damage.

A design professional has no duty to supervise, inspect, or observe the performance of the work designed, absent a duty which the design professional has assumed by contract, by statute, or by having filed plans with a local building department. Once some duty has been assumed, however, state courts are in wide disagreement as to the extent of responsibility and liability. At the very least, the architect or engineer will be responsible for making occasional visits to the site to verify technical compliance with the contract documents. In marked contrast, some courts have found that design professionals have a nondelegable duty to regularly supervise the means and methods of construction. Designers should be wary of so-called routine sign-offs of the contractor's work. The sign-off may include a certification by the designer that the contractor has complied with the plans and specifications and with applicable codes. Such a certification could impose on the designer liability equal to the contractor's in the event of a failure.

An engineer of record should be particularly careful in requiring submission of and in "approving" the contractor's detailed plan for temporary construction. Exculpatory language is often placed on the engineer's "approval" stamps to the effect that the engineer has assumed no responsibility by having reviewed the submission. In the event the design was not adequate and there was injury to person or property, it is likely that the engineer's stamp would not be sufficient to exculpate him or her from liability to injured third parties. If the engineer of record does not intend to carefully review each and every submission, the submission should not be required at all.

The Construction Manager

The construction manager, although not involved in the performance of any actual construction work, may assist the owner in choosing the contractors who will perform work, coordinating the work of these trades, and in approving requisitions for payment. The manager ordinarily does not assume any design responsibility, which remains with the owner's architect or engineer. However, construction managers should also be wary of contractual assumption of responsibility for the contractor's performance or sign-offs or certifications related to the work.

Responsibility for Design Adequacy

It is well established that the contractor who builds in accordance with the design that is provided by another is not responsible for the adequacy of the design or for bringing about the result for which the design was intended, provided that the design is followed. On the other hand, the contractor which provides the design is responsible for both the adequacy of the design and its fitness for the intended use.

In the provision of a temporary structure for sheeting and shoring of an embankment, for example, the design engineer or architect-engineer for the completed structure might indicate that sheeting or shoring is required for a particular embankment. This general directive may be incorporated on the plans or required by specification or code. This may also require that the contractor submit the design for the proposed sheeting and shoring to the engineer. On the other hand, the engineer may provide the design on the contract drawings.

In the first instance, the contractor will be asked to submit a detailed plan for the proposed sheeting and shoring and in all likelihood will be required to have the detailed plan checked and signed by a licensed professional engineer. The adequacy of the sheeting and shoring will be the responsibility of contractors together with their licensed professional engineer.

If, on the other hand, the designer of the completed structure shows the details of the sheeting and shoring and the contractor follows this design in the execution of the work, the responsibility for adequacy will be of the original designer.

Should the contractor deviate from this design and should it be shown that the deviation from the design has resulted in a failure of the sheeting and shoring to properly perform its function, the contractor has the responsibility of accepting the consequence of the damages that have resulted from this deviation.

Additional criteria that the designer must use in designing temporary structures are the requirements of the Occupational Safety and Health Act (OSHA) of the U.S. government (Title 29 Code of Federal Regulations, Part 1926, Government Printing Office).

OSHA can impose fines for on-the-spot violations, and regulations are not only enforced by federal or state inspectors but can be enforced as a result of complaints to the federal or state inspectors by employees or third persons not connected with the construction job.

The Occupational Safety and Health Administration (OSHA) recently announced new fall protection requirements for Residential Construction. The new directive requires all residential builders to comply with 29 Code of Federal Regulations 1926.501(b)(13). Construction and roofing companies will have until June 16, 2011, to comply with the new directive. Under

29 CFR 1926.501(b)(13) workers engaged in residential construction over six (6) feet above the ground level are to be protected by conventional fall protection. For roofers, the 25 foot, ground-to-eave height threshold no longer applies, nor do slide guards as an acceptable form of fall protection, regardless of the roof pitch or height of the roof eave. These new requirements replace the Interim Fall Protection Compliance Guidelines for Residential Construction, Standard 03-00-001 that have been in effect since 1995 and allowed residential builders to bypass fall protection requirements. Construction and roofing companies will have up to six months (June 16, 2011) to comply with the new directive.

One instance of the failure of design to conform to engineering and construction standards was cited in a report of the South Carolina Department of Labor. The department blamed faulty design as the cause of the collapse of a steel sheet-pile cofferdam in about 40 ft (12 m) of water in Lake Keowee, where a water intake structure and a pumping station were built for the city of Greenville, S.C. According to the report of Engineering News-Record on April 19, 1979, a state labor department citation specifically condemned "incorrect basis of design of the cofferdam, in that the method of calculating the requirement for elastic stability and calculating actual stress levels . . . was not adapted to the function of a cofferdam required to return free water." An independent review of the cofferdam design and of the accident was performed by Law Engineering Testing Co., Marietta, Ga. They found that "the probable cause of the cofferdam failure was an incorrect basis of design. It appears that the cofferdam was designed as though it were an excavation in earth or rock rather than a cofferdam required to retain free water."

Because of the importance of safety in overall conduct of the business of a temporary construction contractor, the contractor should appoint an individual in its office as the safety officer who is well versed in all requirements of safety regulations and of good construction practices relating to safety. This person should be given authority to order changes in methods or in construction procedures to bring about safe work practices.

Licensed Site Professionals and Controlled Inspection

In recent years, some state and local governments have initiated requirements for licensed site professionals or controlled inspection. These regulations have imposed the requirement that a licensed professional certify compliance with statutory and regulatory enactments for all or specified portions of the construction. These requirements gained favor in the aftermath of construction failures such as the one that occurred at L'Ambiance Plaza in Connecticut on April 23, 1987.

INSURANCE PROBLEMS

The requirement for the contractor to provide the design of temporary construction facilities such as sheeting and shoring has also raised coverage problems vis-a-vis the contractor's insurance.

Two companion cases arising out of the same construction project illustrate these problems: *Harbor Insurance Co. v. Omni Construction, Inc.*, 912 F. 2d 1520 (D.C. Cir. 1990) and *Harbor Insurance Co. v. Schnabel Foundation Co.*, 946 F. 2d 930 (D.C. Cir. 1991). In these cases, Omni was the general contractor for the construction of a new building owned by Sears, Roebuck & Co. Omni subcontracted to Schnabel the design and construction of a sheeting and shoring system to provide temporary support of the earth walls during excavation. During excavation of the project, settlement and damage occurred to the adjacent Sears, Roebuck building, which Omni repaired at a cost of $978,000. Omni claimed reimbursement from Harbor Insurance, its CGL insurer. Harbor, in turn, commenced an action against Omni for a declaratory judgment that the claim was barred by an endorsement in the insurance policy excluding coverage for losses caused by the "rendering of or the failure to render any professional services," "including the preparation or approval of . . . designs" and "engineering services" "by or for" Omni as the named insured. The court agreed with the insurer's interpretation of the professional liability exclusion, but ordered a new trial on the issue of whether there was coverage for negligent construction under the contractual liability endorsement.

Harbor Insurance then apparently settled with Omni and sued Schnabel and several other subcontractors, as Omni's subrogee, seeking recovery under both negligence and contract theories. Schnabel defended, in part, on the ground that Omni was contributory negligent in relying on the Schnabel engineer's shoring design. Rejecting Schnabel's defense, the court ruled that when a contractee hires an expert and relies on its advice or services, "the expert cannot blithely turn around and claim that the contracting party was negligent in retaining or listening to the expert." In coming to this decision, the court relied, in part, on the fact that Omni had no engineering personnel to review the subcontractor's design.

Thus it would appear that, on the basis of Schnabel, a contractor who properly delegates design responsibility assumes no liability if the design proves deficient.

Two issues linger, however. First, without insurance coverage, a contractor would have to bear the cost of defending a lawsuit which, in and of itself, can be very costly. Second, owing to the paucity of decisional authority, a contractor cannot be assured that other courts will follow the result in Schnabel.

A solution to this apparent dilemma is "wrap-up insurance programs."

> With a wrap-up program, the owner furnishes a single insurance program for all parties involved in the project for the duration of the project term. This insurance relates to the exposures of the project and protects the project owner, contractor, and all tiers of subcontractors. Most wrap-ups include workers compensation, general and excess liability, and builders risk coverages (auto liability and contractors equipment are not included). Wrap-ups can include project architects/engineers errors and omissions coverage and other optional coverages.
>
> Wrap-ups on large construction projects can be either Owner Controlled (OCIP) or Contractor Controlled (CCIP). OCIPs comprise about 90 percent of the wrap-up programs currently being performed in the U.S. Either wrap-up entitles the owner to reduce risks and provide a comprehensive insurance program for all participants in the project.
>
> The strongest factor in favor of a wrap-up is the potential for significant savings in the overall costs of the project. The use of wrap-ups can provide substantial savings through dividends or return premiums payable to the owner. These are granted based on favorable project loss experience and consolidated premium volume generated under the wrap-up. In addition to cost savings the owner also experiences other benefits from a wrap-up program. The owner has the benefit of a single, coordinated loss control and claims handling program, leading to a safer job site, which can result in a project completed on time and under budge. OCIPs do have several disadvantages for the owner. These include the additional administrative, safety and accounting burdens placed on the owner by the program, as well as the potential financial risk for the owner for premium cost increases and/or coverage reductions if the insurance market hardens.

CONTRACTS

Forms of Contracts

The temporary structure contractor may enter into a contract directly with the owner or ultimate user of the facility, or may enter into a subcontract with a general contractor or construction manager who subcontracts on behalf of the owner. If the contract is to be entered into with the owner, one of several forms of contract is most likely to be used.

1. A **lump-sum** contract between the owner and the temporary structure contractor. For example, AIA Document A101, "Standard Form of Agreement between Owner and Contractor for a Stipulated Sum." (All of the referenced AIA forms can be obtained from the American Institute of Architects, 1735 New York Ave., NW, Washington, D.C. 20006.) Also EJCDC Document 1910-8-A-1, "Standard Form of Agreement between Owner and Contractor on the Basis of a Stipulated Price." (All of the referenced EJCDC forms can be obtained from the American Society of Civil Engineers, 345 East 47th St., New York, N.Y. 10017.)

2. A **cost-plus-a-fee** type of arrangement, such as the AIA Form A111, "Standard Form of Agreement between Owner and Contractor, Cost of the Work plus a Fee." Also EJCDC Document 1910-8-A-2, "Standard Form of Agreement between Owner and Contractor on the Basis of Cost-Pius." The fee could be either as a percentage of the cost of the work or as a lump-sum added to the cost of the work.

3. A **unit price** form of contract between the temporary structure contractor and the owner. This type of contract would be based on an assumed estimated quantity for each of the classes or types of work to be performed, such as a unit price for steel, a unit price for concrete, and a unit price for lumber. When the work is completed, an accurate measurement can be made of each class or type of work that was actually performed and the unit prices previously agreed to can be applied to the quantities for a calculation of the amount of money payable to the contractor. The AIA does not have a standard form for this type of agreement. EJCDC Document 1910-8-A-1 can be used for this purpose. An example of a bid form for one can be found in Richard H. Clough, *Construction Contracting*, 4th ed., Wiley Interscience, New York, pp. 435–437.

On the other hand, the temporary structure contractor, rather than being in a contractual relationship directly with the owner, may enter into a subcontract with a general contractor under an arrangement similar to the AIA Form A401, "Standard Form of Subcontract between General Contractor and Subcontractor," or AGC Form 640, AGC/ASA/ASC "Standard Form Construction Subcontract." (This form can be obtained from the Associated General Contractors of America, 1957 E St., NW, Washington, D.C. 20006.) In this arrangement, which is most commonly the case, the temporary structure contractor will have no direct relationship with the owner but will look to the general contractor for whatever obligations are required to be performed and for payment of any money due. Subcontracts can be entered into between the temporary structure contractor and the general contractor on a fixed-fee, cost-plus, or unit-price basis.

In September 2007, a broad-based consortium of construction "industry" trade and other organizations jointly released a new family of model contract documents—as competitors to the AIA documents—called ConsensusDOCS.

The ConsensusDOCS are, to some degree, a reaction to the standard construction contract documents produced for over a century by the American Institute of Architects (AIA), and, indeed, the AIA is notably missing from those entities that participated in and endorsed the ConsensusDOCS. Some have contended that contractors demanded documents that ended a perceived bias in favor of design professionals allegedly inherent in the AIA's A201, the *General Conditions of the Contract for Construction*, most commonly used by owners for their contractors.

Perhaps the most important distinction between the AIA B102 and ConsensusDOCS 240 is in the area of resolving disputes. Although both contracts provide for a multistep approach to dispute resolution—placing mediation before arbitration or litigation—and allow the parties to choose whether they will arbitrate or litigate as their method of binding dispute resolution, that is where the similarities end.

As established by its predecessors, the 2007 AIA B102 makes it difficult to join all three parties (the owner, the architect, and the contractor) into the same binding dispute resolution forum, because no party may be joined without its consent. Specifically, the AIA B102 provides that:

[e]ither party, at its sole discretion, may include by joinder persons or entities substantially involved in a common question of law or fact whose presence is required if complete relief is to be accorded in arbitration, provided that the party sought to be joined consents in writing to such joinder.*

This is actually a change from the 1997 version of the AIA owner-architect agreement, which required the consent of all parties, even those already contractually compelled participants in the arbitration forum ("the Owner, Architect, and any other person or entity sought to be joined"). Now, for example, according to the 2007 version the owner may join the contractor in arbitration against the architect's wishes so long as the contractor gives its consent. Nonetheless, the AIA B102 retains the requirement that no party may be joined without its consent.

In contrast, the ConsensusDOCS 240 explicitly requires joinder of all necessary parties:

The Owner and the Architect/Engineer agree that all parties necessary to resolve a claim shall be parties to the same dispute resolution procedure. Appropriate provisions shall be included in all other contracts relating to the Project to provide for the joinder or consolidation of such dispute resolution procedures.†

The emphasis in the ConsensusDOCS 240 placed on resolving disputes without litigation can be seen in the range and number of steps established in the dispute resolution section. Unlike the AIA B102, which provides only for mediation, arbitration, and litigation, the ConsensusDOCS 240 adopts a more elaborate stepped approach. The process begins with "direct discussions" between the parties:

If the Parties cannot reach resolution on a matter relating to or arising out of the Agreement, the Parties shall endeavor to reach resolution through good faith direct discussions between the Parties' representatives, who shall possess the necessary authority to resolve such matter and who shall record the date of first discussions. If the parties' representatives are not able to resolve such matter within five (5) business Days of the date of first discussion, the Parties' representatives shall immediately inform senior executives of the Parties in writing that resolution was not affected. Upon receipt of such notice, the senior executives of the Parties shall meet within five (5) business days to endeavor to reach resolution. If the dispute remains unresolved after fifteen (15) Days from the date of first discussion, the Parties shall submit such matter to the dispute mitigation and dispute resolution procedures selected herein.‡

Thus, the "direct discussions" provision requires that representatives—and then senior executives—of the parties meet in person (presumably, a telephone conference or written correspondence would not satisfy the requirement that the parties "meet"), and that negotiations proceed quickly toward resolution. If the dispute remains unresolved after only 2 weeks, it moves to the next phase of resolution as chosen by the parties and reflected in the contract.

After direct discussions, the next phase in the dispute resolution process offered in the ConsensusDOCS 240 is a "mitigation" provision, which allows for the appointment of a project neutral or dispute review board to issue nonbinding findings on disputed project

---

*2007 AIA B102 §4.3.4.2.
†ConsensusDOCS 240 §9.6.
‡ConsensusDOCS 240 §9.2.

issues. (The use of the neutral is not mandatory, and the owner and/or architect may elect to proceed immediately to the more formal ADR process the parties have chosen by the contract.) The scope of a neutral's decision-making power is subject to the agreement of the parties, and any decisions made by the neutral are nonbinding. Although it appears that the use of a neutral or review board could substantially reduce the cost of dispute resolution by encouraging parties to reach resolution informally and early on in the process, there may also be significant costs associated with this provision. In particular, the parties must be prepared to share the costs and expenses of the neutral, who must "make regular visits to the Project so as to maintain an up-to-date understanding of the Project progress and issues and to enable the Project Neutral/Review Board to address matters in dispute between the Parties promptly and knowledgeably."[*]

Somewhat similar to the project neutral in the ConsensusDOCS 240, the 2007 edition of the AIA contracts includes a new provision for an "Initial Decision Maker" (IDM) to review and decide disputes that arise between the owner and contractor. Claims between owners and contractors will be submitted to the IDM, who then will have 30 days during which to render a written decision. Upon the IDM's rendering a written decision or the passage of 30 days without a decision, the owner and contractor may proceed to mediation. Although this initial decision-making role, traditionally performed by the architect, may now be performed by any person who is not a party to the agreement between the owner and contractor. It is expected that the architect will commonly be chosen to be the IDM or will be so designated by default. (The 2007 AIA A201 provides that "[t]he Architect will serve as the Initial Decision Maker, unless otherwise indicated in the Agreement".)[†] Even if the architect does not serve as the IDM, it is likely that the architect will be asked to provide assistance to the IDM in evaluating claims. The extent to which the architect's services will be requested by the IDM cannot be predicted, and for that reason the 2007 AIA B201 provides that "assistance to IDM, if other than the Architect" is an additional service entitling the architect to additional compensation.[‡] Unlike the ConsensusDOCS 240 provision for use of a "neutral," submission of claims to the IDM in the AIA scheme is a mandatory prerequisite for the preservation of claim rights.

The AIA B102 and ConsensusDOCS 240 both allow for binding arbitration to resolve disputes that cannot be resolved through negotiation or mediation. Like previous AIA editions, the 2007 AIA B102 requires mediation before arbitration. Specifically, it requires that "[a] demand for arbitration shall be made no earlier than concurrently with the filing of a request for mediation ..."[§] The ConsensusDOCS 240 also appears to require that the parties either mediate or "mitigate" (through a project neutral or review board, as described previously) a dispute before submitting it to binding resolution: "If the matter remains unresolved after submission of the matter to a mitigation procedure or to mediation, the Parties shall submit the matter to the binding dispute resolution procedure selected herein: ... arbitration [or] litigation."[¶]

As noted, the ConsensusDOCS 240 permits parties to choose between arbitration and litigation at the time they negotiate the contract. Now, the AIA B102 also allows that choice. In perhaps the most significant change in the 2007 edition of the AIA contracts, the AIA B102 eliminates arbitration as the default contractually stipulated binding dispute resolution procedure. For nearly 100 years, AIA construction documents provided that disputes and claims between parties would be resolved through binding arbitration. However, in the 2007 forms, including the B102, if parties do not expressly designate in

---

*ConsensusDOCS 240 §9.3.1.
†2007 AIA A201 Section 15.2.1.
‡2007 AIA B201 Section 3.3.1.11.
§2007 AIA B102 Section 4.3.1.1.
¶ConsensusDOCS 240 Section 9.5.

the contract that disputes are to be resolved through arbitration, litigation will be the default dispute-resolution procedure. Specifically, it provides:

> If the Owner and Architect do not select a method of binding dispute resolution below, or do not subsequently agree in writing to a binding dispute resolution method other than litigation, the dispute will be resolved in a court of competent jurisdiction.[*]

Although both contract systems allow for a choice between arbitration and litigation, the ConsensusDOCS 240 contains one provision that is certain to make parties consider carefully the costs of litigating a dispute. Referring to both arbitration and litigation, the ConsensusDOCS 240 states that "[t]he costs of any binding dispute resolution processes shall be borne by the non-prevailing Party, as determined by the adjudicator of the dispute."[†] "Costs" are not expressly defined, but it appears that this provision requires that the loser pay the winner's attorney's fees.

Legal Problems in Contracts

Standard forms of contract, such as the AIA, EJCDC, AGC, and Consensus Docs have been in use for many years and most contractors are familiar with their clauses. If a standard form is used, normally it is not necessary for an operating business person to send contracts, about to be entered into, to an attorney for review. The exception occurs when a standard form has been extensively modified, as is often the case, or when an unusual contract form is being used which seems to include particularly onerous clauses.

Some of the most significant cost-additive items that may affect the financial or legal outcome of the agreement for the temporary structure contractor are

Indemnification or hold-harmless clauses

Extraordinarily high insurance limits

Exculpatory clauses in which the contractor agrees not to make claims for changes, extras, or unexpected work

Waiver of lien

Fixed completion dates with liquidated damages for failure to complete

Requirements for furnishing performance and payment bonds

Warranties and guarantees

Many other clauses which impose special expenses or requirements appear in contracts, and these have to be dealt with on a case-by-case basis.

SUBCONTRACTING

In many instances the temporary structures contractor will be a subcontractor to a general contractor on the job. This will be the case, for example, in a temporary dewatering subcontract. The negotiation and award of a subcontract will proceed along the following lines:

Price

Time of performance

---

[*]2007 AIA B102 Section 4.2.4.
[†]ConsensusDOCS 240 Section 9.5.1.

Availability of labor, material, and equipment

Exceptions, inclusions, and exclusions

Forms and language of subcontract

Terms of payment

Other subcontract clauses which will be important and which will be dealt with during the course of the subcontract negotiation are

Subcontractor to be bound by plans and specifications for the general contract, as well as other business terms (general conditions) such as dispute resolution and the decision of the architect-engineer.

Submission of payment application and breakdown with each request for payment

Payment by subcontractor of all monies owed for materials and labor previously paid for by the general contractor on past requisitions

Making of claims promptly

Protection of and coordination with other trades

Removal of debris

Safety requirements

Invalidity of assignments of subcontract made without consent of contractor

Guarantees, whether personal or corporate

Pursuit of work diligently

Use of the general contractor's equipment only at the general contractor's discretion and at subcontractor's expense

Adherence to the schedules

"Pay when paid" (payment by the owner to the general contractor is a condition precedent to the general contractor's obligation to pay the subcontractor)

"No damage for delay" (subcontractor waives any compensation for delay, by whomsoever caused, and is entitled only to an extension of time to complete its work)

Reports on material in fabrication

Changes

Coordination

Shop drawings

Laws and permits

Work subject to approval

Indemnity and hold-harmless clauses

Arbitration clause

Common standard subcontract forms are the AIA Form A401, "Contractor Subcontractor Agreement," and the AGC/ASC/ASA "Standard Form Construction Subcontract."

CLAIMS: THE UNEXPECTED IN UNDERGROUND CONSTRUCTION

The problem of who shall bear the risk of unexpected conditions under the ground surface or in existing structures has plagued the engineering profession and construction industry for many years. If a contractor is engaged in construction for an owner in accordance with a previously prepared design and the contractor encounters subsurface conditions or conditions in an existing structure which are materially different from what might normally be expected, resulting in increased costs, should the contractor absorb the increased costs or should the owner pay it?

Many owners or engineers attempt to pass on these unexpected costs to the contractor, under the theory that a "grandfather" clause making the contractor responsible for anything and everything that happens in the way of unexpected contingencies will amply protect them. Needless to say when disputes arise, contractors will point to many unusual or extraordinary circumstances that will tend to contradict such "grandfather" clauses. On the other hand, by including such a clause in the agreement, the owner has requested the contractor to assume certain risks for which the contractor must provide contingency allowances in its bid estimate. These contingencies may or may not arise, and it is therefore an added cost burden, for which the owner is paying, for something which may not come about.

The best view, and the one which is endorsed by most enlightened engineers, contractors, owners, and attorneys specializing in construction, is that the contractor should provide in its bid estimate only for certain agreed-upon assumptions which all of the parties stipulate are the basis for the estimated costs on which the contract is based. Any situations or eventualities which are outside of these agreed-upon assumptions, which would have resulted in an additional cost to the owner in the original bid estimate, will be considered as an unexpected condition which should be reimbursed.

A widely used contract clause for such a provision is found at Para 3.7.4 of the AIA Document A201, "General Conditions of the Contract for Construction" (2007 edition), which reads as follows:

> If the Contractor encounters conditions at the site that are (1) subsurface or otherwise concealed physical conditions which differ materially from those indicated in the Contract Documents or (2) unknown physical conditions of an unusual nature, which differ materially from those ordinarily found to exist and generally recognized as inherent in construction activities of the character provided for in the Contract Documents, the Contractor shall promptly provide notice to the Owner and the Architect before the conditions are disturbed and in no event later than 21 days after first observance of the conditions. The Architect will promptly investigate such conditions and, if they differ materially and cause an increase or decrease in the Contractor's cost of, or time required for, performance of any part of the Work, will recommend an equitable adjustment in the Contract Sum or Contract time, or both. If the Architect determines that the conditions at the site are not materially different from those indicated in the Contract Documents and that no change in the terms of the Contract is justified, the Architect shall so notify the Owner and Contractor in writing, stating the reasons.

Also, the standard clause for U.S. government construction contracts is found in Federal Acquisition Regulations (FAR) Section 52.236-2, which reads as follows.

DIFFERING SITE CONDITIONS

(a) The Contractor shall promptly, and before the conditions are disturbed, give a written notice to the Contracting Officer of (1) subsurface or latent physical conditions at the site which differ materially from those indicated in this contract, or (2) unknown physical conditions at the site, or an unusual nature, which differ materially from those ordinarily encountered and generally recognized as inhering in work of the character provided for in the contract.

(b) The Contracting Officer shall investigate the site conditions promptly after receiving the notice. If the conditions do materially so differ and cause an increase or decrease in the Contractor's cost of, or the time required for, performing any part of the work under this contract, whether or not changed as a result of the conditions, an equitable adjustment shall be made under this clause and the contract modified in writing accordingly.

(c) No request by the Contractor for an equitable adjustment to the contract under this clause shall be allowed, unless the Contractor has given the written notice required; provided, that the time prescribed in (a) above for giving written notice may be extended by the Contracting officer.

(d) No request by the Contractor for an equitable adjustment to the contract for differing site conditions shall be allowed if made after final payment under this contract.

Contractors are well advised to insist that such a differing site condition clause be included in their contract documents.

INFORMAL DISPUTE RESOLUTIONS

Informal dispute resolution takes many forms. The parties can do it themselves or rely on third parties to help. Typically, when faced with a dispute, most parties try to work it out themselves before involving third parties. When parties are unable to resolve their disputes, there are several structured, yet informal means, through which solutions may be offered and resolutions reached; these are discussed as follows. None of the informal dispute resolution procedures discussed here is final or binding on the parties.

Arbitration

Arbitration is different from court suits in that parties must agree in advance to this procedure by including an arbitration clause in the contract, or they can agree to arbitrate after a dispute has arisen. In the absence of such an agreement, either beforehand or afterward, one party cannot compel another to arbitrate a dispute.

The procedure involved is quasi-legal. An impartial panel of arbitrators hears both sides of the dispute. Parties may be represented by counsel. Witnesses may be called to testify and other formal courtroom procedures may be incorporated. Yet it must be remembered that the forum is not a court of law, and many significant protections, such as the rules of evidence, pretrial disclosure, and mandatory consolidation and joinder, are lost in arbitration. Also, the arbitration panel's decision is final and binding and cannot be appealed. This precludes going to a higher level unless fraud, refusal to accept relevant evidence, an undisclosed conflict of interest on the part of an arbitrator, or some other serious procedural infirmity can be proved.

The American Arbitration Association (AAA) oversees the arbitration of disputes in many fields, including engineering and construction. The AAA is a private entity that has developed rules governing arbitration and maintains facilities in numerous states to administer the proceedings.

Under the AAA rules, arbitration is initiated by serving a demand for arbitration upon the other party and mailing a copy to the AAA. An initial filing fee is paid by the party initiating the arbitration. The AAA then will provide the parties with a list of about 10 to 20 potential arbitrators who, based on background information, appear to have some expertise in the subject area of the dispute. The parties strike the names of unacceptable arbitrators and list the remaining names in order of preference. One to three arbitrators may be assigned to a case, depending on the size of the claim. If the arbitration clause does not specify the number of panelists and the amount of the claim is less than $100,000, the AAA usually will designate a single arbitrator.

The experience of the arbitrators is one advantage of this method of dispute resolution compared to others. In many courts, a judge will hear an automobile accident case one day, a divorce case the next, and a construction case the next. The judge cannot be expected to know the intricacies of the industry or be knowledgeable about common practices. The members of the AAA's construction panel of arbitrators have backgrounds that are representative of all segments of the industry.

Arbitrators have their expenses paid and receive approximately $500 per day for their services. However, parties have been known to stipulate to a higher fee so that they can get the best arbitrators. These people are often very busy and deserve higher remuneration for a case that may be complex and prolonged.

Mandatory consolidation and joinder is a protection available in court that is not necessarily available in arbitration. If, for example, an owner demands arbitration claiming defective work, the architect-engineer would probably wish to question subcontractors, if their work is at issue. In arbitration, this cannot be done as a matter of right, creating the problem of multiple proceedings and the risk of inconsistent results.

Pretrial disclosure is another protection available in court but not necessarily in arbitration. In pretrial disclosure, one party can obtain, in advance, the adverse party's evidence supporting its claims and that which would support the party's defenses or counterclaims. This can save a lot of time at trial, prevent surprise, and lead to settlement in advance of trial when all the facts are on the table. Pretrial disclosure can only occur in arbitration if both parties agree. Even the arbitrators do not have the power to compel pretrial disclosure by an unwilling party.

The arbitration hearings will most likely be scheduled at great intervals of time, owing mainly to the volunteer nature of the arbitrators' duties. Also, arbitrators, as well as the parties, are busy people and schedules must be coordinated. This is one disadvantage of arbitration. Whereas once a trial in court begins, it proceeds without interruption; arbitration hearings can be sporadic and take many months or even years to complete.

Preparation for arbitration is often more difficult than for a traditional trial. Because arbitration has more relaxed procedures, a witness is made more vulnerable and must be prepared for any eventuality and line of questioning that might otherwise be precluded in court.

The decision arrived upon by the arbitrators is often quite different from the usual opinion rendered by a court. A court opinion is usually extensive and particularizes facts and law. An arbitration award can be simple, as "X owes Y $50,000" with no detailed explanation. As noted earlier, an arbitration panel's decision is very rarely overturned, because it is difficult to show fraud or some other substantial procedural infirmity.

Parties should take a hard look at the advantages and disadvantages of arbitration before including this clause in their contract. If the project could be a source of complex disputes involving a large amount of money and multiple parties, it is possible that formal litigation may be the better forum for dispute resolution.

Mediation

Unlike arbitration, which provides a means of adjudication through adversary proceedings, mediation is the attempt of one or more individuals to assist the parties in reaching a settlement of their dispute by direct negotiations. The mediator is an outside party, usually someone with expertise in the industry. Both parties agree to listen to the mediator, whose main function is to provide an atmosphere of reason and impartiality. The mediator participates in the negotiations and may or may not recommend a settlement. The mediator's recommendation is not binding upon the parties.

Mediation programs are available through the AAA, the Center for Public Resources, JAMS, Endispute, Inc., and a growing number of other entities. Unlike arbitration and other means of final adjudication, the parties retain complete control. They and their attorneys develop the procedures for mediation. Proceedings are usually informal, and the parties do not submit evidence or produce witnesses, although they may exchange information and have witnesses available. Communication is made easier if the parties agree in advance that all disclosures to the mediator are confidential and privileged.

Mediation is often useful where there is a high level of distrust or hostility among the parties that would make direct negotiations very difficult. Also, if parties have reached an impasse, a mediator often can provide a fresh look and explore options that the parties might have overlooked. Mediation can be viewed as a way of promoting settlement and avoiding full-scale litigation if it is entered into by both parties with a good faith effort to settle.

Minitrial

A minitrial is a structured negotiation where the parties present summaries of their position to principals of each party who have the authority to negotiate and settle the dispute. The theory behind the minitrial is that by presenting the facts of the case to the principals and educating them on the strengths and weaknesses of the case, they will ultimately resolve the matter.

The proceedings usually are presided over by a neutral party who has been selected by the parties involved. This person's role is to moderate the hearings, ask questions, and narrow the issues. The lawyers for each party make a short, informal presentation of their cases.

The hearings can last anywhere from 1 to 7 days or longer, depending on the number of parties involved and the complexity of the case. At the conclusion of the hearings, the optimum result would be that the principals settle their dispute by direct negotiation. If this is not possible, the parties may request that the neutral adviser render an opinion. This opinion is nonbinding and would be used by the parties in further negotiations.

The most frequently touted advantage of the minitrial is the ability to save time and money. If the parties decide to use a minitrial, an agreement should be negotiated and drafted that encompasses those terms dealing with format, discovery, participants, presentations, and scheduling.

Two provisions of a minitrial agreement often turn out to be particularly valuable. The first reassures active participation by the CEOs of each party; the second provides potential sanctions for unreasonably rejecting a final settlement offer.

Dispute Review Boards

A growing number of construction contracts provide for the establishment of a *dispute review board* that reviews referred disputes and issues nonbinding recommendations. Typically, the board comprises one member selected by the owner, one selected by the contractor, and a third selected jointly by the other two members. The owner's member must be approved by the contractor, and vice versa, and the third by both parties.

The contractor and owner enter into a contract with the board members by which the members agree to consider impartially all disputes regarding the contract work placed before them and to issue recommendations to resolve these disputes. Board findings may be either admissible or inadmissible as evidence in subsequent court or arbitration proceedings, depending upon the parties' agreement. Board members meet regularly to review the project status and potential problems. They visit the site regularly and, if possible, when specific problems arise.

Dispute review boards have had the effect of encouraging cooperation and reasonableness by owners and contractors, even when no disputes are referred to the board. The mere existence of a board has provided the incentive needed to get the parties to resolve their disputes themselves. Further, when a dispute has been resolved, parties have tended to abide by it.

Dispute review boards have two great advantages to more formal methods of dispute resolution. First, they determine the facts contemporaneously with the problem and are able to see the problem firsthand and to discuss it with the parties while the facts are well known and several potential solutions may be available. Second, they tend to resolve disputes quickly, avoiding the intense adversarial relationship and exacerbations of other contract problems that can develop when a dispute is dragged out and allowed to interfere with the work.

Board members are paid for their time and expenses by the owner and the contractor. The cost of paying three well-qualified individuals to review disputes and keep abreast of a project's status has tended to limit the use of such boards to major public works projects and large private projects. In practice, however, the cost of boards on large projects has been only $30,000 to $70,000, a range clearly affordable and practical for not-so-large projects. The reluctance to take on the expense of a board may save pennies now, but cost dollars later. If litigation can be avoided by using a dispute review board, its expense is minimal in the long run. Indeed, the effectiveness of dispute review boards cannot be questioned. Of over 100 projects studied that had dispute review boards, none had disputes that ended up in court an outstanding success rate.

On smaller private jobs, owners and contractors may wish to investigate establishing a board where their designees volunteer, perhaps in exchange for the parties' promise to do the same of the designees, and only the third member must be paid. If all parties work in good faith and nominate members who will honestly and objectively perform their duties, a smaller job may have the benefits of a review board without its costs.

Recommendations of a review board are typically not final or binding on the parties.

Observations

New methods of dispute resolution are appearing as alternatives to formal litigation. The attraction of such methods is that, at first glance, they appear to be less costly and time-consuming than litigation. Yet, before an alternative method is chosen, one should review the particular case carefully and decide what will be best. How complex is the case? Will there be numerous other claims and possibly conflicting decisions resulting from some parties not being bound by an arbitration agreement? Are all parties attempting to settle in good faith, or will mediation be futile and eventually lead to litigation anyway?

Even more important, it must be remembered that the resolution of a dispute involves more than a legal or mechanical decision about who is right or wrong. It involves psychological and emotional issues as well.

One commentator, in writing about alternative dispute resolution, termed it "plastic justice." His concern was that present-day arbitration and adjusters often resolve disputes by "splitting the baby" without really weighing or evaluating the claim. He also said that the participants did not feel that they had their day in court. This does not have to happen. Alternative dispute resolution forums can provide the opportunity for the parties to tell their story. That is one purpose behind arbitration and minitrials.

The analysis of alternative methods should be ongoing so that methods can be improved by experience. One should not view litigation and alternative dispute resolution as mutually exclusive. Each method of dispute resolution, including litigation, can play a role in the administration of justice.

Deciding how one wants to resolve a dispute is like deciding how one wants to be put to death. It is the type of decision that none of us should ever have to make.

It is important to remember that any formal dispute resolution mechanism takes the decision-making process out of the hands of the parties and puts it into the hands of the third party, that is, the outsider. This represents a failure of the system. The parties are much better off if they can prevent a problem from escalating into a formal dispute or resolve the dispute on their own.

PROTECTION OF UTILITIES

Excavation near or in a public street invariably involves the protection of utility vaults, pipes, and ducts located under the street. Shoring and other means of support must be provided to prevent damage to these utilities. However, whether the owner of the utility company is required to pay for the protection often depends on whether the construction project is privately or publicly owned. Generally, where the purpose of the owner's construction is purely private, the burden of protection is on the owner. On the other hand, where the construction work is public or quasi-public, the utility company is required to bear the costs of shoring, supporting, or relocating its structures. Thus, for example, when a contractor laying city sewers was forced to provide support to an elevated railroad, the railroad company was required to reimburse the contractor (*Necaro Co., Inc. v. Eighth Avenue R. Co.*, 220 App. Div 144, 221 N.Y. Supp. 276). In addition, a public utility company which has been given a franchise to install its facilities in public streets has the correlative duty to relocate or protect its facilities when changes are required by public necessity; the company takes the risk of the location of its facilities if public inconvenience or security demands their removal or protection.

Statutes generally have not changed these rules but rather have specified procedures for giving notice of impending excavation that will affect the utilities.

Many jurisdictions require that a contractor or person causing excavation to be done near a public street give specified written notice to the affected public utility companies. In the case of private construction, usually an engineer from the utility company will aid the excavator in locating the utility structures to be protected. If notice is given and the utility company fails to advise the excavator of the location of its subsurface structures, the utility company may not thereafter sue the contractor for negligent damage to the utility's lines while providing required support and protection. In the case of public construction, the utility company has the obligation of protecting, removing, or replacing the affected structures. If it fails to do so, the contractor will protect or remove the utility structures, as determined by the public entity, at the utility company's expense.

A good example of such a statute is the New York General Business Law Section 761. Each town and city (and each borough of the city of New York) must now maintain a registry of the underground facilities located in the municipality. The actual excavator (not the owner or general contractor) must notify the utility listed in the registry in advance of proposed excavation. The utility companies must then advise the excavator of any underground facilities that might be affected by the proposed excavation or demolition. Thereafter, the excavator must support or protect the utility company's lines in the manner set forth in the Rules and Regulations of the State Board of Standards and Appeals. An excavator, in violating the statute, will be liable for any damages caused thereby. Violation of the statute is a crime, and will subject the excavator to civil penalties as well.

Thus every excavator must contact the utilities listed in the registry. If any damages occur to the utility lines, the utility must be notified immediately. No repairs may be made in such a case, except at the direction of the utility company. Similarly, no backfilling may be attempted without notification of the utility company. This statute fairly allocates the financial burden for location and protection of utility lines and structures during construction.

PROTECTION OF ADJACENT PREMISES

Since construction can affect adjoining land and structures, the law has historically imposed certain duties of protection upon the constructing owner, differentiating, however, between protection of the adjoining land itself and protection of structures on that land. The constructing owner has a strict and absolute duty to provide lateral support to adjoining land in its natural condition, and the constructing owner's duty is limited to the use of reasonable care with respect to the buildings located on the land. Thus the constructing owner will be held liable for any and all damage to adjacent land regardless of whether the owner, or the owner's contractor, was negligent. However, the owner would be liable for any damage to buildings on that land only if it is proved that the owner was negligent. If the contractor were negligent, the owner would be liable for damage to the adjoining building only if the owner retained supervision over the contractor or negligently prescribed the methods and procedures followed by the contractor.

An exception is made to the latter rule regarding buildings where "ultrahazardous" activities are conducted such as blasting and pile driving. In such circumstances, strict and absolute liability can be imposed regardless of whether negligence can be proved.

In many jurisdictions, such as New York City, statutes have supplanted the general rules of law referred to previously. The New York City Administrative Code (Section 28-3309.4) contains what is commonly known as the "10-foot" (3-meter) rule relating to the protection of adjacent premises during excavation. Interestingly, the historical distinction between the protection of the land itself and the structures thereon is retained in that statute. Thus, with respect to adjoining land, regardless of the depth of the excavation, there is an absolute requirement that the "person causing an excavation to be made" shall provide adequate sheet piling, bracing, and other support as is necessary to prevent the sides of the excavation from caving in before the new structure is constructed. However, that person must be afforded the right to enter the adjoining property to perform such work as may be required. If the adjoining owner refuses to give such right of entry to the person causing an excavation to be made, the adjoining owner must then protect his or her own property, and in turn, is given a right to enter upon the property where the excavation is to be made to do so.

With respect to the protection of adjoining buildings, the New York City ordinance differentiates between excavations that are shallower than 10 ft (3 m) below the legally established curb level and excavations that exceed 10 ft (3 m) below such level. For excavations less than 10 ft (3 m) in depth, the owners of adjoining buildings are required to protect their own buildings, provided they are given the right to enter the property where the excavation is to be made to perform such protective work. If excavating owners refuse such right of entry, they must assume the duty to protect the adjoining buildings. They, in turn, are granted a right to enter upon adjoining property for such purposes.

Where the excavation is to be carried to a depth of more than 10 ft (3 m) below the legally established curb level, the duty to protect an adjoining building falls upon the person causing such excavation to be made, provided he or she is given the right to enter on the adjoining property to do so. Otherwise, the duty falls upon the owner of the adjoining building, who is given the right to enter on the property where the excavation is to be made in order to do so.

Additionally, the New York City Building Code requires developers to "at his or her own expense, underpin the adjacent building" provided the developer is afforded "a license ... to enter and inspect the adjoining buildings and property, and to perform such work thereon as may be necessary for such purpose" (Building Code Section 3309.5). This brings us to the question: what can a developer do if the adjacent property owner refuses or imposes unreasonable conditions to a license to underpin?

In general, when an adjacent property owner refuses to grant a developer necessary access, Section 881 of the Real Property Actions Law (RPAPL) allows the developer to commence a special proceeding to obtain a license to enter the adjacent property. The developer must demonstrate (1) it "seeks to make improvements or repairs to real property," (2) the improvements are "so situated that such improvements or repairs cannot be made by the owner ... without entering the premises of an adjoining owner," (3) "permission so to enter has been refused," and (4) the dates on which entry is needed. RPAPL Section 881.

Significantly, Section 881 evenhandedly protects the neighboring property owner by making the developer "liable to the adjoining owner or his lessee for actual damages occurring as a result of the entry." Id. It also allows the court to impose "such terms as justice requires." Id. Thus, once a developer demonstrates that entry onto an adjacent property is necessary for improvement to real property, it would appear that the only issue for the court to decide is what conditions, if any, "justice requires" to be imposed on the developer. The statute thus calls for the court to balance the competing interests of both the developer and the adjacent owner—allowing construction to proceed while protecting the adjacent owner's property rights with safeguards such as insurance, indemnity, and temporal and spatial restrictions on the incursion. (*Rosma Development, LLC v. South*, 5 Misc.3d 1014(A) (Supreme Court Kings Co., 2004).

Further requirements may be imposed such as the necessity for a permit to perform excavation work and fencing, filling, and other protection requirements, including notice to adjoining landowners prior to commencing operations.

TRESPASS

Some jurisdictions have enacted legislation providing that adjacent owners are permitted to enter upon adjoining property for the purpose of protecting their property and structures. This, however, is most unusual and contrary to the general common law rules concerning trespass. Normally, an unauthorized entry onto adjacent property subjects the invader to criminal liability as well as civil damages. Moreover, owners would be liable for such damage if their plans require the contractors to trespass on adjacent land.

Subsurface encroachments are generally treated similarly to surface encroachments regardless of whether the subsurface encroachment will impair any present use of the land or any foreseeable future use. Nevertheless, it has been held that an unauthorized subterranean trespass occasioned by construction of a municipal sewer 150 ft (45 m) below grade is not actionable unless some present or prospective use of the land is interfered with (*Boehringer v. Montalio*, 142 Misc. 560, 254 N.Y.S. Supp. 276). If the subsurface encroachment would interfere with the prospective use of the land, the adjoining owner would be entitled to recover the difference in the value of the land resulting from the encroachment.

A builder who seeks to place a foundation wall, fence, scaffold, or equipment shed directly on the property line must be aware that necessary entry upon adjoining premises will constitute a trespass. An easement from the adjoining property should be obtained.

However, some jurisdictions have legislated certain relief with respect to one's right to enter onto the adjoining land in order to repair or improve one's property. For example, a New York statute permits an owner or lessee to apply to the court for temporary authority to enter on the adjoining land for the purposes of repair or improvement (N.Y. Real Prop. Actions Law Section 881).

In one case under that statute, a waterproofing contractor was permitted by court approval to erect scaffolding temporarily on adjacent land in order to waterproof the wall located on the building line (*Chase Manhattan v. Broadway Whitney Co.*, Misc. 2d 1091, 294 N.Y.S. 2d 416). The party going onto the adjacent land for such purpose is subject to liability for any resulting damages. Furthermore, the courts will not automatically grant authority to enter the adjoining property where objection is made by the adjoining property owner. The courts will weigh the relative inconvenience and expense to each landowner in determining whether authority should be granted. Thus a request for permission to excavate on adjacent property to waterproof a basement may be denied in the absence of showing that the waterproofing was impractical or unduly costly without entry.

DESIGN, CONSTRUCTION, AND MAINTENANCE

CHAPTER 6
COFFERDAMS

Ben C. Gerwick, Jr., P.E. (dec.)
Patrick E. Durnal, P.E., MEng.

INTRODUCTION

Cofferdams are temporary enclosures to keep out water and soil so as to permit dewatering and the construction of the permanent facility (structure) in the dry. This chapter confines itself to the structural cofferdams used to permit underwater and underground constructions of bridge piers, intake structures, pump houses, and similar structures.

A cofferdam involves the interaction of the structure, soil, and water. The loads imposed include the hydrostatic forces of the water, as well as dynamic forces due to current and waves. Soils impart loads against the walls of the cofferdam and provide passive (internal) support during construction as well as after completion of the cofferdam. The structure itself interacts with the water and the soil, keeping out the soil and water while exerting global forces on the soil.

The failure of a cofferdam, should it occur, would not only be catastrophic from the point of view of the work and workers inside the cofferdam but could also precipitate disruption of the surrounding area, with damage to adjoining structures. Such a failure could even make it impracticable to reconstruct a replacement cofferdam in the same location. Therefore, more than usual precaution has to be taken to prevent failure or collapse.

For a cofferdam in water, and also in weak soil, the load follows the deformation; hence yielding does not relieve the forces. Therefore, redundancy needs to be provided within the structural system, so as to ensure that partial failure does not lead to progressive collapse. Large and deep cofferdams should be subjected to an analysis of failure modes and effects to verify that progressive collapse will not occur.

Since cofferdams are constructed at the site of the work, operations must be carried out stage by stage, often under adverse conditions. The loads acting on the cofferdam are often more severe during intermediate stages of construction than after the completion of the cofferdam. Each stage must be analyzed and evaluated to ensure its stability and safety.

As the cofferdam is constructed, stresses and deformations are built in, which are not fully released during subsequent stages. Thus residual stresses also need to be considered.

Because cofferdams are typically constructed under adverse conditions in a marine environment and because significant deformations of elements may occur at various stages of construction, it is difficult to maintain close tolerances. Ample provision must be made for deviations in dimensions so that the finished structure may be constructed according to plan.

The loads imposed on the cofferdam structure by construction equipment and operations must be considered, both during installation of the cofferdam and during construction of the structure itself. Equipment is generally located high up, hence induces critical loads at adverse locations and often involves a dynamic component.

Removal of the cofferdam must be planned and executed with the same degree of care as its installation, on a stage-by-stage basis. The effect of the removal on the permanent structure must also be considered. For this reason, sheet piles extending below the permanent structure are often cut off and left in place, since their removal may damage the foundation soils adjacent to the structure.

Safety is a paramount concern, since workers will be exposed to the hazard of flooding and collapse. Safety requires good design, proper construction, verification that the structure is being constructed as planned, monitoring of the behavior of the cofferdam and surrounding area, provision of adequate access, light, and ventilation, and attention to safe practices on the part of all workers and supervisors.

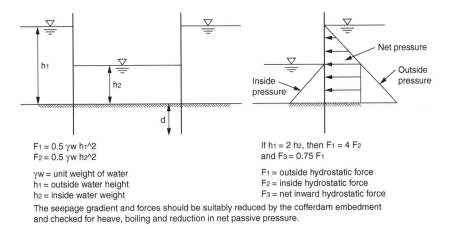

$F_1 = 0.5\ \gamma_w\ h_1{}^2$
$F_2 = 0.5\ \gamma_w\ h_2{}^2$

γ_w = unit weight of water
h_1 = outside water height
h_2 = inside water weight

If $h_1 = 2\ h_2$, then $F_1 = 4\ F_2$
and $F_3 = 0.75\ F_1$

F_1 = outside hydrostatic force
F_2 = inside hydrostatic force
F_3 = net inward hydrostatic force

The seepage gradient and forces should be suitably reduced by the cofferdam embedment and checked for heave, boiling and reduction in net passive pressure.

FIGURE 6.1 Hydrostatic forces on partially dewatered cofferdam.

DESIGN CONSIDERATIONS

Hydrostatic Pressure

The maximum probable water height outside the cofferdam during construction and the water heights inside the cofferdam during various stages of construction need to be considered. These result in the net design pressures shown in Fig. 6.1.

Determination of the probable maximum external water level needs to consider such phenomena as rise in water level due to flood, high tide, storm surge, and waves and wave crest to trough differentials on either side of the cofferdam. While the season of the year in which the structure is planned to be constructed can be taken into account, it must be recognized that cofferdam work is often delayed owing to causes both within and beyond the contractor's control, so the cofferdam may be subjected to higher levels arising later in the year.

The inside water level is that to which the cofferdam will be pumped down so that work may be executed in the dry. This is usually 0.5 to 1.0 m (1.6 to 3.2 ft) below the elevation of the work in progress, for example, the installation of bracing. If there is no seal course, the water will be pumped down below the excavation level using wells or sumps. The phreatic surface profile can then be estimated, and steps taken to ensure that it always stays a meter or so below the soil surface.

If there is a concrete seal course, the inside level may be assumed as the level of the top of the underwater concrete. Since concrete placed underwater typically slopes downward toward the edges, this elevation will always be somewhat lower than that at the location of discharge.

Forces Due to Soil Loads

The soils impose forces, both locally on the walls of the cofferdam and globally upon the structure as a whole. These forces are additive to the hydrostatic forces.

Addressing first the global loading, in the most common case where the bottom is relatively level, the forces balance. When, however, a cofferdam is constructed on a slope, such as the bank of a river, there will be a net force trying to slide and overturn the cofferdam.

A cofferdam is a relatively poor structure to resist such unbalanced loads because it has little weight and often is not founded on competent soils. Further, the necessary structural rigidity can be obtained only with additional bracing and connections. At one stage, just prior to placing the tremie seal, there is no bottom to the structure and no support from the bearing piles.

However, satisfactory solutions can often be obtained by providing one or more of the following:

- Increased penetration of sheet piles so as to develop shear resistance, bearing, and uplift (skin friction).
- Increased stiffness of sheet piles.
- Diagonal bracing (in vertical plane) to ensure rigid truss action of bracing frame.
- Adequate connections of bracing frame to sheet piles so as to make sheet-pile walls act like flanges of a truss.
- Installation of batter piles and vertical piles to support the bracing frame. (These are sometimes driven through pipe sleeves in the bracing frame.)
- Installation of batter piles, connected to sheet piles at the top.
- Installation of ground anchors or anchors to deadmen on shore.
- Temporary placement of sand and gravel surcharge on low side of cofferdam.
- Minimization of vibration, heave, and lateral movement by use of suitably spaced non-displacement foundation and hold-down pile.

Local soil forces are a major component of the lateral force on sheet-pile walls, causing bending in the sheets and wales and axial compression in the struts (see Fig. 6.2). These forces are proportional to the effective unit weight of the soil, so that submerged soils contribute only about half the force that the same soils would contribute if located above water. Critical cases may occur when surcharges are placed or soils are exposed at low tide.

A detailed analysis of soil forces on sheet-pile walls requires the use of advanced principles of soil mechanics applied to carefully determined values of the various properties. However, preliminary values may be computed on a semiempirical basis.

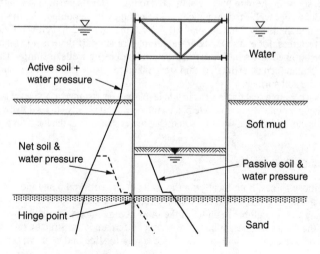

FIGURE 6.2   Soil forces in typical weak muds and sands.

For sands, a unit weight of 16 to 18 kN/m³ (100 to 115 lb/ft³) can be assumed. The angle of internal friction ϕ can be taken at 30 to 35°, giving an active pressure coefficient K_a of 0.33.

For cohesive soils, such as clays and silts, typical of marine sea floors, ϕ can be taken at 15°, the shear coefficient c varies from 14 to 50 kN/m² (300 to 1000 lb/ft²), and the unit weight is 18 to 20 kN/m² (115 to 125 lb/ft²).

Because the softer of these clay soils such as "bay mud" will tend to creep as the sheet piles deflect, at-rest pressures may develop, giving K_o as high as 0.8. At the same time, K_p may be as low as 1.2. With firmer clays, the values of K_o and K_p approach 0.5 and 1.4, respectively. For submerged soils, the unit weight should, of course, be reduced by the buoyant effect of the water, 10 kN/m³ (62.4 1b/ft³).

Satisfactory approximate computations of soil forces acting on the structure may usually be carried out by the method of equivalent fluid pressure. This, of course, gives a triangular distribution of pressure, which is approximately valid for determining total loads on a cofferdam and for the distribution of loads on the sheet piles. However, it is inadequate for proportioning the loads on the bracing, particularly at the top level. Therefore, for the bracing it is usually best to compute the total load by the method of equivalent fluid pressure, then redistribute it on a trapezoidal or rectangular pattern, from which the loads acting on the upper levels of bracing may be found (Fig. 6.2).

For fully submerged soils, the following values of equivalent fluid pressure may be assumed for preliminary calculations of the loads acting on the wall. These include both hydrostatic and soil pressures.

Medium dense sands	13 kN/m³ (80 lb/ft³)
Firm clays	12 kN/m³ (70 lb/ft³)
Soft clays and silts	14 kN/m³ (90 lb/ft³)

Firm clays may temporarily exert much lower forces owing to their sealing off the water head. However, progressive deflection of the sheet piles may subsequently allow the full water head to act.

Soft clays and silts are very sensitive to surcharge loads as, for example, sloping bottom or an adjacent riverbank. For these cases a more in-depth geotechnical analysis is essential. Loose sands may liquefy under shock or vibration (e.g., pile driving) and act as a heavy fluid, exerting a lateral pressure equal to their total density.

Current Force on Structure

The drag force D due to current is given by

$$D = AC_d\rho\frac{V^2}{2g}$$

where g = acceleration of gravity
ρ = density of water
A = projected area normal to the current
V = current velocity
C_d = drag coefficient

In metric (SI) units, ρ is approximately 10 kN/m³ and g is 10 m/s²; hence $D = AC_d \cdot V^2/2$, where D is in kN. (Similarly, in English units, ρ is approximately numerically equal to $2g$, so we can write $D = A \cdot C_d V^2$ where A is in ft², V is in ft/s, and D is in lb force.) The value of C_d depends on the overall shape of the structure and the roughness of the surface.

With a typical cofferdam, the current force consists not only of the force acting on the normal projection of the cofferdam but also on the drag force acting along the sides.

With flat sheet piles, the latter may be relatively small, whereas with Z piles it may be substantial since the current will be forming eddies behind each indentation of profile, as shown in Fig. 6.3. As a practical approach, therefore, the use of a drag coefficient $C_d = 2.0$ will conservatively include the effect of the sheet piles; hence the drag force $D = AV^2$ in SI units, and $D = 2AV^2$ in English units.

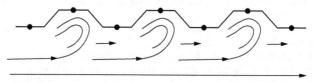

FIGURE 6.3   Current flow along sheet piles.

During construction the current force also acts on bracing and support piles and on the sheets which are partially installed. The latter may pose a critical condition, requiring special temporary support or the use of current deflectors.

If the cofferdam will be subject to accumulating debris or ice floes during a flood, consideration must be given to the current forces acting on the debris. In fact, the accumulation of debris is often the most serious factor. Debris and ice have been responsible for the destruction of a number of cofferdams in rivers.

Wave Forces

Waves acting on a cofferdam are usually the result of local winds acting over a restricted fetch and hence are of short wavelength and limited to height. Waves can also be produced by passing boats and ships, especially in a restricted waterway.

In most cases the waves acting against the face of a cofferdam are nonbreaking; that is, they reflect from the face of the sheet piles, forming a clapotis or standing wave. The effect is essentially hydrostatic in character. Attention is directed to the fact that since the water level inside the cofferdam is generally constant, the trough of the clapotis may lead to a temporary net outward hydrostatic force on the sheet piles; hence they must be well tied at the top.

Cyclic wave loads, particularly if allowed to cause play in the connections, may lead to fatigue. Methods of calculating these wave forces are given in the *Shore Protection Manual* of the U.S. Army Engineers Waterway Experiment Station. The manual also gives a method of calculating the impact of breaking waves, which may develop a dynamic component owing to the compression on entrapped air. Figure 6.4, which is from that manual, shows the typical profile and pressure distribution of waves acting on a cofferdam. Especially in the surf zone, with breaking waves, the dynamic forces can be much larger because of the high velocity of the breaking or plunging wave.

In certain exposed and open sea areas, long-period swells may develop significant inertial forces, acting on the structure as a whole. In this case, the added mass (a mass of water equivalent to that displaced) must be added to the displaced mass in calculating the inertial component. Such forces can be very large. Evaluation is best made by diffraction analysis, in a manner similar to that used on offshore caissons.

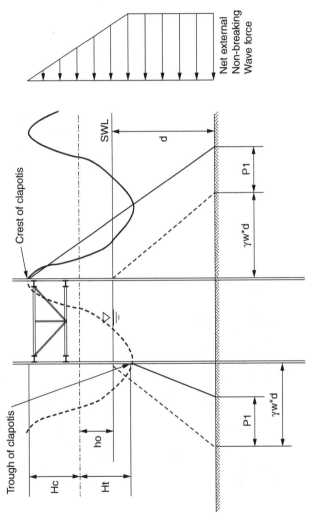

P1 = 0.5 (1 + x)γw*Hi/cosh(2πd/L)   ho = to orbit center of clapotis = f(Hi/d, Hi/(gT²))
Hi = incident wave heights   Hc = Ht = 0.5 (1 + x)Hi
d = depth of still water level SWL   T = wave periods
L = wave lengths   g = acceleration of gravity
x = 1.0 for smooth vertical surfaces

FIGURE 6.4 Pressure distribution: nonbreaking waves.

Ice Forces

These are of two types: the force exerted by the expansion of a closed-in solidly frozen-over area of water surface, and the forces exerted by the moving ice on breakup.

Static Ice Force.   This depends on the thickness of the ice sheet, the coefficient of thermal expansion of ice and the elastic modulus of ice, as well as the degree of confinement and temperature differentials. As an example, a value of 200 kN/m^2 (4000 lb/ft^2) has been used on cofferdams and structures on the Great Lakes.

Dynamic Ice Forces.   The values depend on the thickness of the ice sheet, the mode of failure of the ice against the structure, the strength of the ice, planes of fracture, etc.

Average values acting on a cofferdam-type structure are often taken at 570 to 670 kN/m^2 (12,000 to 14,000 lb/ft^2) of contact area, although extreme values may reach three times these values.

While this accounts for the normal force acting on the cofferdam, there is also an additional drag force shear along the long sides. On the favorable side, a shear or cutwater nose may serve to break the ice, reducing shear to about one-third of the value due to crushing.

Seismic Loads

These have not been normally considered in design of temporary structures in the past. For very large, important, and deep cofferdams in highly active areas, seismic evaluation should be performed. The goal is to ensure that collapse will not occur during the construction period, although local damage may be acceptable.

For the portion embedded in the soil, an imposed total value equal to the passive pressure sets an upper limit. This is approximately valid for soft sediments such as those normally encountered in harbors but is too conservative for stiff clays and sands.

For dense sands and for the firmer clays, an allowance of a 50 percent increase in active soil pressure is often assumed for an earthquake. Since earthquake forces are of a temporary nature, allowable stress levels can be increased 33 percent. This is safe for wales and sheet piles, but struts may have to be treated more conservatively because of their buckling mode of failure.

For large cofferdam structures in deep water in a highly seismic area, the stability of the structure as a whole can be computed by applying an added mass coefficient to reflect the acceleration of adjacent water by structure. For the typical cofferdam configuration, an added mass of 0.5 times the mass of displaced water will be found approximately valid.

Earthquake responses and effects are minimized for highly flexible structures such as most sheet-pile cofferdams. For the more rigid cofferdams now being considered for some deep-water bridge piers, there will be greater responses.

In loose sands, there is the possibility of liquefaction under earthquake. This can be minimized in a practical way by a blanket of crushed rock placed around the cofferdam site.

Accidental Loads

These are the loads usually caused by construction equipment working alongside the cofferdam and impacting on it under the action of waves. One approach is to assume a local velocity, say 0.6 m/s (2 ft/s), a distortion of the impacting vessel of 0.3 m (12 in), and an added mass coefficient of 2.0 to be applied to the mass of the vessel. The impact can be significantly reduced by fendering.

Logs and ice blocks can impact at the ambient current velocity. They can hit end-on and hence require consideration of local punching shear.

Scour

Scour of the river bottom or seafloor around the cofferdam may take place owing to river currents, tidal currents, or wave-induced currents. Some of the most serious and disastrous cases have occurred when these currents have acted concurrently.

Although the bottom of a river or channel may be stable before the cofferdam is constructed, the very act of constructing a cofferdam, along with moored barges, etc., may cause blockage of a significant portion of the cross-sectional area of the stream, resulting in a dramatic increase in bottom current. The sediments on the bottom were previously in a condition of dynamic stability, so any increase in the bottom current leads to erosion and scour.

Scour can cause a rapid lowering of the seafloor over the entire area adjacent to the cofferdam or local pockets at the corners. The latter are due to eddy currents and can be minimized by streamlining, that is, rounding of the upstream corners and providing a fin or tail to prevent vortices at the downstream corners. Scour is, of course, a greater potential when the bottom is fine sand or silt.

A very practical method of preventing scour, suitable in many cases, is to deposit a blanket of crushed rock or heavy gravel around the cofferdam, either before or immediately after the cofferdam sheet piles are set. Since scour can take place very rapidly, a delay in the latter case may not be acceptable when the bottom is very loose sand or silt.

A more sophisticated method is to lay a mattress of filter fabric, covering it with rock to hold it in place. A fabric made of two layers, one fine, random-oriented sheet, underlain by a layer of strong, coarse-mesh fibers, all having a specific gravity greater than 1.0 (e.g., 1.05), is best. This is laid on the bottom before starting construction. The support piles and sheet piles are then driven through it. When it will be necessary to later drive piles through scour protection rock, the size of stable rock may be minimized by using high-density rock such as iron ore.

Protection in Flood and Storm

When a flood or storm exceeds the limits assumed in design, radical steps may have to be taken promptly in order to prevent loss of the cofferdam. These limits should be established by the designer of the cofferdam and be shown in the plans in writing so there can be no confusion or delay.

Practical steps have consisted of the following: flooding the cofferdam; preinstallation of a log boom to prevent debris from lodging against the cofferdam; provisions for clearing debris during the flood or storm; installation of extra ties diagonally across the bracing, so as to ensure that it will act as a space frame (wire rope plus turnbuckles can be quickly installed); tying sheet piles to top wale, if not previously done as part of the design; and removing all barges from alongside the cofferdam.

Ice floes can exert a very heavy and potentially damaging impact on cofferdams. In the past it has been common to attempt to deflect them by a boom or barge moored independently of the cofferdam. However, the possibility that the barge or boom may break loose under the ice floe or storm, etc., and itself impact the cofferdam must be considered. Moorings for such a barge or boom must therefore be carefully designed to ensure against such a failure.

Similarly, when attempting to construct a cofferdam in a swift river current or where wind-driven waves are coming from one direction, the mooring of a large barge can effectively still the surface water. The moorings must, of course, be adequate for the design storm or flood, or else the floating breakwater must be removed at the onset of extreme conditions.

Flow deflectors can also be used; one type consists of steel frame jackets made of tubular members, through which pin piles are then installed. Sheet piles or stop logs can then be installed across the front of the jacket.

COFFERDAM CONCEPTS

The typical cofferdam for a compact structure, such as a bridge pier, intake structure, or pump house, consists of sheet piles set around a bracing frame and driven into the soil sufficiently far to develop vertical and lateral support and to cut off the flow of soil and, in some cases, the flow of water (Fig. 6.5). The structure inside may be founded directly on rock or firm soil or may require pile foundations. In the latter case, these generally extend well below the cofferdam.

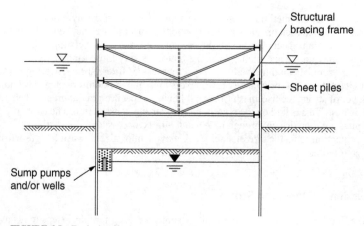

FIGURE 6.5   Typical cofferdam, without seal or piles.

In order to dewater the cofferdam, the bottom must be stable and able to resist hydrostatic uplift. Placement of an underwater concrete seal course is the safest and most common method. However, in hard and relatively impermeable soils, or where dewatering can be carried out well below the excavated bottom, the concrete seal may be omitted. In the latter case, the sheet piles must penetrate a sufficient distance to cut off water flow and to ensure stability against a "plug-type" shear failure.

An underwater concrete seal course may then be placed prior to dewatering in order to seal off the water, resist its pressure, and also to act as a slab to brace against the inward movement of the sheet piles. The concrete seal may be locked by adhesion and shear to the piles in order to mobilize their resistance to uplift under the hydrostatic pressure (Fig. 6.6).

For larger areas, such as locks and dams, cellular cofferdams or double-walled cofferdams are often used. These develop their resisting force by acting as a gravity retaining wall. In this concept, parallel walls of sheet piles are tied together, with either sheet-pile diaphragms or tie-rods, and the space between is filled with granular material. In such cases, the dewatering is usually carried out by pumping.

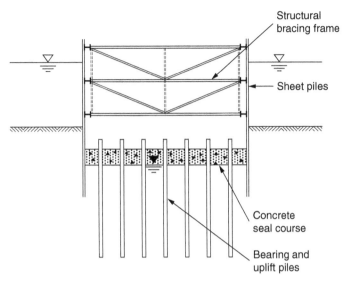

FIGURE 6.6 Typical cofferdam (with seal).

Cellular cofferdams may be built on bare rock, since penetration is not necessarily required for stability although it is required in sands and clays. The underlying soils must be capable of providing the required bearing support and lateral shear resistance.

COFFERDAM CONSTRUCTION STAGES

Typical Cofferdam Construction Sequence

For a typical cofferdam for a bridge pier or intake structure, the construction sequence involves the following stages of work:

1. Predredge to remove soft sediments and to level the area of the cofferdam (Fig. 6.7a).
2. Install temporary support piles (Fig. 6.7b).
3. Set a prefabricated bracing frame and hang on the support piles (Fig. 6.7b).
4. Set steel sheet piles, starting at all four corners and meeting at the center of each side (Fig. 6.7c).
5. Drive sheet piles to grade (Fig. 6.7c).
6. Block between bracing frame and sheets; as necessary, provide ties for sheet piles at the top (Fig. 6.7c).
7. Excavate inside the grade or slightly below grade, while leaving the cofferdam full of water (Fig. 6.8a).
8. Drive piling (Fig. 6.8b).
9. Place rock fill as a leveling and support course (Fig. 6.8b).

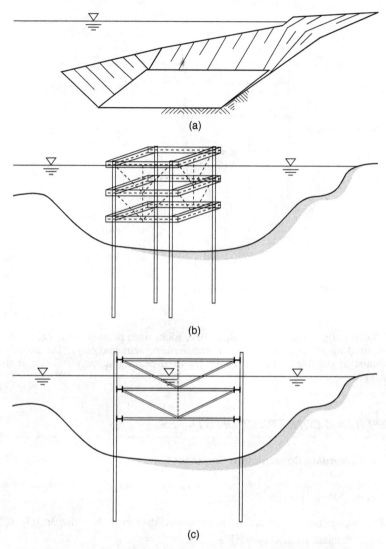

FIGURE 6.7 Cofferdam construction sequence (I). (*a*) Predredge. (*b*) Drive support piles; set prefabricated bracing frame and hang from support piles. (*c*) Set sheet piles; drive sheet piles; block and tie sheet piles to top wale.

10. Pour tremie concrete seal (Fig. 6.8*c*).
11. Check blocking between bracing and sheets (Fig. 6.9*a*).
12. Dewater (Fig. 6.9*a*).
13. Construct new structure (bridge pier or intake structure) (Fig. 6.9*a* and *b*).
14. Flood cofferdam (Fig. 6.9*c*).
15. Remove sheet piles (Fig. 6.9*c*).

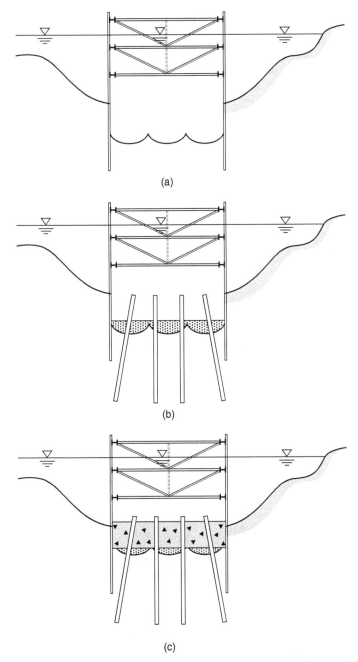

FIGURE 6.8 Cofferdam construction sequence (II). (*a*) Excavate inside to final grade. (*b*) Drive-bearing piles in place. (*c*) Pour tremie concrete.

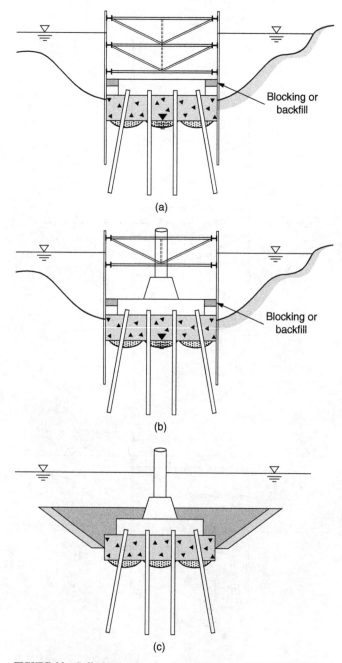

FIGURE 6.9   Cofferdam construction sequence (III). (*a*) Check blocking; dewater; construct footing block; block between footing and sheet piles. (*b*) Remove lower bracing; construct pier pedestal; construct pier shaft. (*c*) Flood cofferdam; pull sheets; remove bracing; backfill.

16. Remove bracing (Fig. 6.9c).

17. Backfill (Fig. 6.9c).

Predredging (Stage 1).   Prior removal of soil to form a deep basin means that the cofferdam will not have to resist temporary soil pressures in addition to water pressure. When the seafloor is sloping, prior dredging removes unbalanced soil loads acting on the, structure as a whole. This predredging, to be effective, must extend well back from the cofferdam wall, so as to leave a berm at the toe of the slope.

Predredging also permits the setting of a completely prefabricated bracing frame. Such a frame comprises two or more sets of horizontal bracing, tied together with vertical and diagonal trussing and hence acting as a space frame. Finally, it is generally significantly less costly to predredge than to dig inside, even though the quantity is obviously greater.

When predredging all the way to grade, the subsequent outward pressure of the tremie concrete seal course must be considered. Otherwise the sheet piles may progressively deflect outward as the seal concrete is poured. Often, it will be necessary to backfill outside the cofferdam with granular material to at least the height of the seal prior to placing the tremie concrete.

Predredging also facilitates pile driving inside the cofferdam. If it is possible to predredge to grade, it may be possible to drive the piles before the cofferdam bracing frame and sheet piles are installed, moving stage 8 up to 2.

On the negative side, predredging obviously increases both dredging and backfill quantities. It may extend the cut too close to existing structures. Dredging in the open water may stir up bottom sediments into the water, although this normally will be of little real ecological significance unless the sediments are contaminated.

Predredging can also facilitate the removal of surface or subsurface layers of hard material, such as boulders or limestone layers. In fact, if heavy excavation is required, such as blasting or use of very heavy buckets, or removal of obstructions or old piles is required, these can only be done efficiently by preexcavating before the cofferdam is started. Performance of such heavy excavation inside a cofferdam structure is usually excessively difficult owing to restrictions of space between bracing and the need to exercise care to prevent damage to the bracing.

When the site conditions and requirements of the project do not permit predredging to grade, excavation will, of course, have to be carried out inside the cofferdam, and the sheet piles and bracing will have to be designed to resist the forces from the soil during all stages of construction.

In competent soils, such as sands and firm clays, this can be accomplished in a practical manner by selecting sheet piles of adequate length and strength to resist the soil pressures during excavation, transferring the load to the bracing above and to the passive pressure zone of the soil below. Then, when the excavation is completed, the piles can be driven and the underwater concrete seal can be constructed.

If the depth of inside excavation is too great for practical spanning of the sheet piles between the bracing and the soil, another level of bracing may be lowered into place and blocked to the sheet piles after partial excavation (Fig. 6.10). Such a lower level of bracing may be particularly useful if the driving of piles requires jetting, or the vibration may otherwise temporarily destroy the passive resistance of the "passive-zone" soils below the excavation limits.

When constructing cofferdams in very weak silts and muds, there may be inadequate passive resistance. In such a case the sheet piles must be very long and very strong, often exceeding the limits of practicability and economy. This becomes particularly critical when relatively shallow excavations are to be carried out in very deep muds. Many failures have occurred when there was inadequate recognition of this problem. The failure may be characterized by either excessive rotation or excessive deflection, as shown in Fig. 6.11a and b.

FIGURE 6.10   Large and deep cofferdam for bascule pier of the 4th Street Bridge over Mission Creek Channel, San Francisco, Calif.

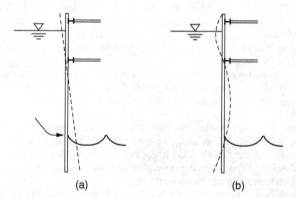

(a)   (b)

FIGURE 6.11   Deflections of sheet piles in soft soils (*a*) rotational, (*b*) bending.

Economical and practical solutions have been devised. In one, the site is predredged several feet below grade, and an artificial "stratum" of sand and gravel is dumped to act as support for the sheet piles and to develop lateral support for the sheet piles (Fig. 6.12).

Then the sheet piles need only be long enough to penetrate through the sand-gravel stratum. Such a system was successfully employed for 24 shallow cofferdams for the west side of the San Mateo-Hayward Bridge near San Francisco, reducing the required length of sheet piles from 24 m (80 ft) to 11 m (35 ft).

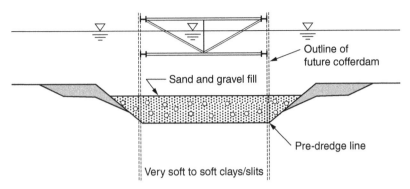

FIGURE 6.12   Use of sand and gravel mat to provide improved passive resistance.

However, the above solution assumes that predredging can be carried out. If predredging is not permissible (the initial premise), a sand-gravel "stratum" may be created by injection, that is, by using sand-drain equipment and an excess of air pressure to force the sand bulbs out at the desired elevation (Fig. 6.13).

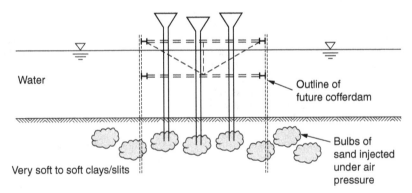

FIGURE 6.13   Creation of firm stratum to provide temporary support for cofferdam.

A more conventional solution is to cantilever the sheet piles, using a two-level bracing frame. The sheet piles are tied back at the top level and cantilevered below the lower level. The efficacy of this system depends on using high-strength sheet piles (high section modulus) and yield strength and on maximizing the distance between top and lower bracing levels. In some cases, it may be desirable to raise the top level above that which otherwise would be required (Fig. 6.14). The lower level of bracing must resist the force from the soil plus that of the tie. The sheet piles need to penetrate below the inside excavation only far enough to prevent run-in of sand or a local shear failure of the soil underneath.

Since deflections in cantilever will be relatively large, the possible remolding effect on soft clays, and thus an increase in active pressure, must be considered.

Temporary Support Piles and Bracing Frames (Stages 2 and 3).   These must provide sufficient lateral support to resist current and wave forces during the installation of the

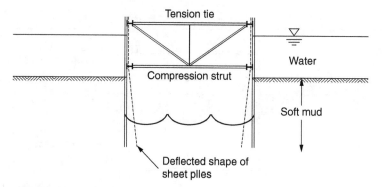

FIGURE 6.14 Use of cantilevered cofferdam concept for shallow cofferdam in weak material.

sheet piles. In calm water this can usually be done by installing vertical piles, using the bracing frame to establish a fixed-end condition and short moment arm. In strong currents or waves, batter piles or similar lateral support may be required. Recent practice has been to employ large-diameter (1 to 2 m) vertical piles to act as spuds. The bracing frame generally consists of wales, struts, posts, and diagonals. Today, cofferdam bracing frames are almost always made of steel structural shapes and tubes.

Bracing frames are generally made up of a combination of welding and bolting. Bolts facilitate subsequent disassembly if required for removal. At intersections of struts and wales, sufficient stiffeners must be provided to prevent local distortion. Since field welds are very unsatisfactory owing to wet conditions and vibrations of the elements, bolting is preferred for field connections. A typical framing arrangement is shown in Fig. 6.15.

Wales are subject to combined stresses, that is, normal loads applied horizontally caus-ing bending, plus bending due to dead weight in the vertical direction, plus axial com-pression from the ends and from diagonals. Standard structural design procedures provide methods for analyzing the stresses due to these combined forces.

Struts are subjected primarily to axial load as horizontal columns. They also have bend-ing in the vertical plane due to dead weight, frame action, accidental dredging or pile driv-ing impacts, and any equipment or frames/platforms supported on them. As columns, they must be supported in both the horizontal and vertical planes to keep the length to radius of gyration ratio (Kl/r) within allowable limits so as to prevent buckling.

Struts are also subjected to temperature loads. In cofferdams in waters, these are usually negligible since the top level (where temperature differentials are greatest) is generally free to expand and contract. However, they must be considered in land cofferdams where the walls are rigidly restrained.

In some cases, the lowest bracing frame must be installed below the level of predredging;

That is, it is not practicable to predredge to a depth that will enable the entire frame to be set as a single prefabricated unit. In such a case, the third (lower) frame is made up separately and hung immediately beneath the upper prefabricated space frame. After fur-ther excavation inside, the frames can be lowered to grade. Frame lowering is facilitated by outward inclination of the sheets using guide blocking added to the lower brace levels. Over flooding the interior a few feet will offset unbalanced dredged soil loads that can press the sheets inward and may cause friction on the wales to impede the frame lowering.

For such a frame, having bracing in the horizontal plane only, the column effects (Kl/r) will usually control the size of the struts. The lower frame must be well supported, either by piles or from the upper frame. It should be truly horizontal and well blocked to the sheet piles.

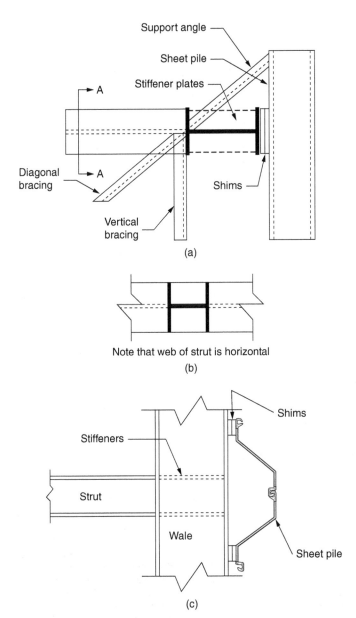

FIGURE 6.15 Typical framing arrangement: (*a*) Elevation, (*b*) intersection of struts, and (*c*) plan view.

These precautions are important to prevent any tendency for the frame to rotate under load. Therefore, maximum prefabrication of as complete a space frame as possible is to be preferred. To transport and place such a huge and heavy frame requires careful planning.

The initial space frame is usually assembled on a barge. In extreme cases it may be assembled on skidways on shore, like an offshore platform jacket, and later skidded out onto a barge. Alternatively, the space frame may be assembled on location, supported above water on large spud piles, and progressively lowered down as each level, and its next set of vertical bracing are completed. In such cases, the spud piles, acting in cantilever, must be adequate to provide lateral support.

If transported by barge, the frame, often weighing several hundred tons, can be lifted by a floating shear-leg crane barge or derrick barge and set over the support piles. Some of the bracing members may be made buoyant by closing the ends. Then the frame may be reassembled, launched, and floated into exact location where spud or support piles may be driven through sleeves. The load is then transferred to these support piles. Once again, consideration must be given to lateral forces, including an accidental impact from work barges.

With very deep cofferdams, two or more prefabricated space frames may be set on top of one another, joined by pin piles (long pipe piles running through vertical pipe sleeves). Funnel-shaped guides are used to help position the sectional frames over the spud piles as they are set.

Bracing by means of a series of circular compression rings is often an attractive solution because it eliminates cross bracing which would interfere with excavation and subsequent pier construction. However, such bracing must be designed with particular care to prevent the possibility of buckling of the ring wales. Causes of buckling are the following:

1. Out-of-round tolerances in compression-ring prefabrication and installation
2. Uneven bearing of sheet piles on compression rings. Poor blocking
3. Unbalanced soil loads, often due to equipment or unequal excavation
4. Failure to consider each stage of excavation in the design; the ring which is currently the lowest at one stage may carry much more load at that stage than subsequently when a lower ring is set
5. Accidental damage to a ring, as by a bucket
6. Installation at an angle with the horizontal
7. Inadequate web stiffeners leading to web buckling

Because the mode of failure of a ring wale is sudden and transfers its load plus impact to an adjoining set, progressive collapse is possible and has occurred. Thus each ring should be evaluated on the basis of its ability to arrest an adjacent local collapse.

Steel Sheet Piles (Stages 4 and 5).   Many different types are used. For cofferdams, where bending predominates, the Z sections, WF, or pipe Combi wall sections predominate.

Whatever section is adopted, the sheet pile should be sufficiently rigid and of sufficiently thick metal, 12 mm (1/2 in) or greater, so as to withstand driving stresses and local twisting. Figure 6.16 illustrates some of the various shapes of commercially available piles.

Use of 350 MPa (50,000 lb/in^2) yield steel is desirable to prevent local damage, increase interlock strength, and provide greater bending resistance and hence greater height between horizontal bracing levels. It is subject to much less damage and distortion than the steels of lower yield strength.

It is essential in water cofferdams (and most land cofferdams as well) that the sheet piles are continuously interlocked. To ensure this, they must be set accurately and driven progressively. By setting the corners first and working in both directions toward the middle, minor adjustments can readily be made. As a control and guide to accurate setting,

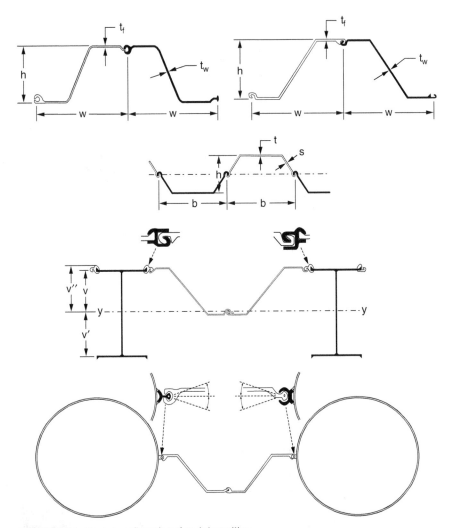

FIGURE 6.16   Typical configuration of steel sheet piling.

sheet-pile locations should be marked on the wale. An allowance should be made for interlock play [usually about 3 mm (1/8 in) per pile].

Sheet piles must be set against falsework or guides, giving enough support at a high elevation to hold them accurately in alignment and giving support against wind during the initial phases. A partially set sheet-pile wall is very susceptible to wind damage, being a large, flat wall of little strength. For this reason, the wall may be tied to the upper guide (by temporary bolting) to give it restraint in both directions. Also, it is not advisable to leave a large exposed panel of sheet piles undriven overnight or over weekends. Generally a panel should be set and at least partially driven before the end of the day's work. Some not so amusing stories are told about returning to work the morning after a day of setting, only to find the sheet piles bent flat on the ground.

Threading of sheet piles has historically been done by hand, but this resulted in many mashed fingers and was slow. New devices have been developed, both hydraulic and mechanical, by which the second sheet pile can be engaged to the first pile at ground then run up and automatically entered into the interlock of the first pile. In the case of multiple cofferdams and reuse of sheet piles, it is often practicable to handle, set, and remove the sheet piles in pairs, using a short weld along the interlock at the top to mate each pair.

In driving, no sheet pile should have its top extending more than a short distance, for example, 1.5 m (5 ft), below its neighbor at any time. Otherwise, it may tend to deviate from true alignment, and the subsequent driving of adjoining piles will result either in high friction or worse in driving out of the interlock.

Sheet piles are best driven by a vibratory hammer, as long as a satisfactory rate of penetration is being achieved (Fig. 6.17). Vibratory hammers work well in silts and sands, satisfactorily in soft clays, and poorly in gravels, peat, and sandy clay hardpan. For the latter soils either a diesel or steam hammer is used. The impact hammer should not be too large or it will lead to tip and/or butt deformations. A cast-steel driving head [cap block] should be employed, properly grooved for the sheets selected. In some very hard driving, for example, when the sheet piles must toe into cobbles, etc., cast-steel sheet-piletip protectors may be effectively used to prevent tip damage.

Tie and Block Sheet Piles (Stage 6).   Ties are usually installed at the top because the sheet piles may not bear on the top wale after the cofferdam is fully dewatered. Rather, owing to continuity effects, they may tend to move away from the top wale. To prevent vibration

FIGURE 6.17 Driving steel sheet piles in succession for Skyway pier foundation SF-Oak, Bay Bridge, Calif.

and minimize bending moments in the sheets, provision of J-bolt ties is useful. Bolts are superior to welds, since the latter often fracture under vibratory loads.

Excavation (Stage 7). Inside excavation is usually done with a clamshell bucket. It is essential to prevent hooking of the bucket under a brace. In addition to using care, temporary setting of one or more vertical sheets against the internal bracing (so as to act as guides) may be useful.

In stiff clays it is hard to excavate under bracing, since a large ridge will be left. This can be knocked down by a diver and jet or by swinging the bucket (taking care not to hit or snag the bracing).

After gross excavation, usually to 0.5 to 1.0 m (1.65 to 3.3 ft) below final grade, the clay that is still adhering to the sheet piles, especially in the indented profiles, must be removed. This is usually done by using an H-beam as a long chisel, augmented by a jet. This is run up and down along the sheet piling. The completeness of cleaning underwater should be verified by a diver. This cleaning is done to ensure that the tremie concrete seal will properly seal the entire inner perimeter. Obviously, in those relatively few cases where no seal is used, the sheet piles do not have to be cleaned. Later, after piles have been driven, but prior to placing rock backfill or tremie concrete, silt and sediments are cleaned from the bottom by means of an airlift pump. This is effective in removing soft siltation material from the pockets left by the clamshell bucket.

Pile Installation (Stage 8). Unless the pilings were driven from floating equipment right after the predredging, they are driven now (Fig. 6.18). Since their tops or cutoffs are generally in or just above the tremie concrete seal, it follows that the heads may have to be driven a substantial distance underwater. Also, in the case of bridge piers, batter piles are often

FIGURE 6.18 Driving 100-ft open end 42-in pipe pile with drive shoes in flooded cofferdam to bedrock, Wakota Bridge I-494 over Mississippi River, Minneapolis, Minn.

required, directed radially out on all four sides, as well as vertical piles, and these too may require that the heads terminate underwater.

The bearing piles, both vertical and battered, should be carefully arranged so as to avoid running into each other or into the tips of the sheet piles, during both setting and driving, and below ground. This may necessitate minor readjustment of the specified plan layout, which, of course, must be approved by the design engineer.

Piles can be set through the bracing and supported in templates affixed to the top level bracing. There are three methods used to drive such piling: (1) with an underwater hammer; (2) with a follower; (3) with an above-water hammer, acting on a longer pile than required, followed by a cutoff and removal for reuse of the extra length.

With the underwater hammer, method 1, the difficulty lies with the physical dimensions of the hammer and leads and consequent interference from bracing and previously driven piles. If a hammer accidentally lands on a previously driven "high" pile, the hammer will be badly damaged. Hence piles may have to be cut off before driving adjoining piles. If it hits on bracing, it may cause serious distortion or the collapse of the brace.

Underwater steam hammers use air pressure in the hammer casing so as to exclude water. This air pressure reduces the efficiency of the hammer, particularly with depths greater than 18 m (60 ft). The new hydraulic hammers do not use air and hence do not lose as much energy underwater.

The use of a follower, method 2, was discouraged or prohibited until recently because of bad experience with poorly made followers. Today's followers have a machined cast-steel driving head (socket) at their base and a properly reinforced head. They are fitted with guides so as to be supported in the leads and maintained in axial alignment. Followers can be designed for impedance compatibility with the pile and heat-treated after fabrication so as to anneal the welds and prevent brittle fracture under repeated impacts.

Both methods 1 and 2 encounter serious difficulties if any piles do not bring up to required bearing. Underwater splices are really not practicable for most piles, although they can be implemented with pipe piles and, in some cases, with prestressed concrete piles if proper detailing has been made beforehand. In any event, they are very costly and time-consuming.

Method 3 involves the handling and setting of a pile long enough to keep its head above water when its tip is at the designated elevation. Hence, if it becomes necessary to splice, the splice can be made above water.

Full hammer efficiency is available. The piles can be threaded past each other with minimum clearance. Unless the total pile is prohibitively long, the setting and driving rate is significantly faster.

However each pile must be subsequently cut off. In the case of batter piles, underwater cutting may be required in order to set and drive adjoining piles. As many as possible, however, will not be cut until after dewatering when they can be cut "in the dry."

These long "cutoffs" can now be taken back to the fabrication yard and spliced onto the next piles. Thus there is no cutoff loss, except on the last pier driven.

Batter piles generally extend under the edge of the cofferdam; hence they may limit the penetration of the sheet piles. In some cases, it may be necessary to increase the plan size of the cofferdam so as to obtain adequate penetration of the sheet pile prior to intersection with the batter pile.

Where solid or closed-end bearing piles are used inside the cofferdam, their displacement may cause excessive heave or lateral movement of the sheet piles and previously installed foundation pile. Open-ended pipe pile or H-piles may be preferable, especially in dense soils.

Jetting is seldom done in cofferdam piers but may occasionally be necessary with displacement piles. In at least one case, for bridge piers in sand, the use of jetting led to a liquefaction and a loss in passive resistance for the sheets, resulting in a flow of sand underneath and collapse of the cofferdam even before dewatering.

Both jetting and driving vibrations can cause local liquefaction of the sand and loss of support. Various steps to overcome this include placing of a lower set of bracing at the bottom of the excavation, so as to provide support in lieu of the soil's passive resistance, or placement of a thick layer of crushed rock at the bottom of the excavation to prevent local liquefaction of the sand and permit ready escape of the pore water.

Leveling Course (Stage 9).   A sand-and-gravel leveling course is used to bring the excavation to grade, ready to receive the tremie concrete seal. This rock subbase also gives temporary support to the seal during placement and prevents intermixing of mud, sand, and concrete.

Underwater Concrete Seal (Stage 10).   The purpose of the seal is to (1) prevent upward flow of water, (2) act as a lower strut for the sheet piles, (3) tie to the driven piles that resist uplift, (4) provide weight to offset the uplift due to differential head, and (5) provide support for subsequent construction of the pier or intake structure. In many modern piers, the tremie concrete seal course is also designed to function as part of the structural footing block. In the latter case, it may be reinforced.

Underwater concrete may be placed, either by the tremie (pipe) method or by the grout-intruded aggregate method. For tremie concrete, the mix selected should be cohesive and highly workable. A typical mix is as follows:

Coarse aggregate: gravel, rounded, 20 mm (¾ in), maximum 55 percent of total aggregate

Fine aggregate: sand, coarse gradation, 45 to 50 percent of total aggregate

Cement: 320 kg/m^3 (500 lb/yd^3)

PFA (pozzolan): 80 kg/m^3 (130 lb/yd^3)

Water-reducing admixture (WRA) not high-range water-reducing admixture (HRWRA)

Plasticizing or air-entraining admixture

Retarding admixture (as required)

Anti-washout admixture (as required)

Water and WRA, to give slump of 160 to 185 mm (8 to 10 in)

Water-cement ratio, 0.42 to 0.45

PFA (pozzolan), as a partial replacement for the cement, improves flow and reduces heat. The pozzolan should be tested for compatibility with cement and admixtures. Normally, a 15 to 20 percent replacement is used, although a higher percentage may be used (up to 50 percent), provided the delay in gain of strength is recognized.

Admixtures are commercially available with combined water-reducing, plasticizing, and retarding effects. HRWRAs, often called superplasticizers, are not suitable for mass concrete placements such as those typical of cofferdams. The high mass generates heat, and the heat causes the HRWRA to "go off" prematurely and erratically so that the slump suddenly decreases. This can result in a disastrously unsatisfactory placement of tremie concrete. Further, not all conventional admixtures are compatible and suitable for underwater concrete. A trial batch of several yards, placed in a box or pit, may indicate the degree of segregation and the workability and flow performance (Fig. 6.19).

Placement of tremie concrete is best initiated in a sealed tube or tremie pipe of 250 to 300 mm (10 to 12 in) in diameter. A plate, with gasket, is tied to the end of the tremie pipe, which is then lowered to the bottom (Figs. 6.20 and 6.21). The hydrostatic pressure holds the plate tight, so the pipe stays empty to the bottom. Obviously the empty pipe must have sufficient weight so as to not be buoyant. Concrete is then introduced to fill the pipe. It is

FIGURE 6.19 A trial batch is essential to verify the cohesiveness, slump, slump flow and washout of the concrete mix. (*a*) Trial batch, (*b*) initial 8- to 10-in slump, (*c*) initial 21- to 26-in slump flow, and (*d*) washout tests.

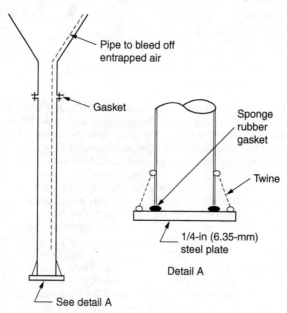

FIGURE 6.20 Arrangement of tremie pipe for start of operations.

FLGURE 6.21   Fitting a closure plate on tip of tremie pipe.

desirable to first place 1 m³ (1 yd³) of concrete from which the coarse aggregate has been omitted, then follow with the regular mix.

The upper end of the tremie pipe is capped with a hopper. Concrete should be delivered to the hopper, at atmospheric pressure, in a continuous stream, such as that obtained by pumping, conveyor belt, or a bucket with an air-controlled closure device.

The pipe is first placed on the bottom, empty and sealed as described. It is filled just above the balancing point at midheight of the pipe. Note that with a 250 to 300-mm (10 to 12-in) tremie pipe, there is very little friction head loss. Then the pipe is raised 150 mm (6 in) or so off the bottom, the seal is automatically broken by the excess concrete head, and the concrete flows out. In normal pours, the concrete exiting from the pipe flows up around the pipe to the surface, then slowly cascades down the surface. With admixtures which reduce internal shear, the flow may push out under the previously placed concrete. The pipe must always be kept embedded in the fresh concrete, a minimum distance of 1 m (3 ft) (Fig. 6.22).

The joints in the tremie pipe should be bolted and gasketed to be watertight. Otherwise the downward flow of the concrete will suck in water and mix it by the venturi effect.

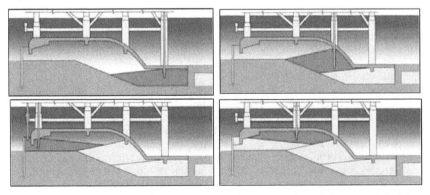

FIGURE 6.22   Advanced tremie concrete placement sequence, Olmsted Dam Tainter Gate shells filling, Ohio River, Ill.

Leaking joints have caused disastrous washing out of the mix in some extreme cases. A slightly inclined tremie pipe, where practicable, will allow the air to escape along the upper side.

If the tip of the pipe is accidentally raised above the surface of the concrete, the seal will be lost, the flow rate will increase, and water will be mixed with the concrete, causing severe segregation and laitance formation.

Any mixing of the tremie concrete and water leads to the formation of laitance (leached-out cement), a soft material of low specific gravity that rises to the top of the concrete. Segregation can also occur in cases of severe washing. It is marked by a substantial increase in volume so that the cofferdam fills with less concrete placed than calculated. Any yield greater than a 3 or 4 percent gain should be viewed as an indication of probable severe segregation.

The slope of the tremie concrete surface will be between about 6 to 1 and 10 to 1. Tremie concrete has flowed up to 20 m (70 ft) from a single tremie pipe without excessive segregation; however, 10m (35 ft) is a more reasonable spacing.

In moving a tremie pipe, it should be raised clear of the water, resealed, and set back into the fresh concrete, in other words, a complete restart. Dragging the tremie pipe through the fresh concrete inevitably produces segregation and excessive laitance.

Use of a rubber ball (volleyball) as a go-devil is widespread to start and restart a tremie pour, but it is not good practice. A volleyball is inflated with less than 1 atm (15 lb/in^2) air pressure. Once the ball is pushed deeper than 10 m (30 ft), it collapses and loses all sealing ability. A noncollapsible "pig" such as the sack of straw used in the 1900s was actually better.

This problem is exacerbated even in shallow placements, when restarting a pour after relocating the pipe or a loss of seal. The use of a go-devil or pig of any type pushes a column of water down through the recently placed concrete, washing out the concrete locally. Hence the resealing method is much to be preferred.

On very deep placements, over 15 to 21 m (50 to 70 ft), the plate on the tip may no longer be appropriate and a special pig of rubber, with squeegees, such as used in pipelines, may be needed. This can give excellent results on the initial start but is not suitable for restart into a previously placed but unhardened mass. In such a case, it is best to let the previous concrete harden, jet off any laitance, and start anew with the pig. The delay, while undesirable, may be far less than the time required to remove excessive laitance.

Ingenious valves have been developed in the United States (1920–1940), Europe (1970), and Japan (1980) but have not proved fully satisfactory because the material being placed is not a homogeneous mix but contains coarse aggregate, which can jam or wedge open a valve. Their use is not recommended for general practice.

The tremie-pipe layout and sequence should be such as to maintain acceptable flow distances and prevent cold joints (Figs. 6.23 and 6.24). In many cases, the latter requires that a relatively high rate of pour be maintained. Typical rates are 38 m^3 (50 yd^3) per hour for small- and medium-sized cofferdams, 103 m^3 (135 yd^3) per hour for larger cofferdams. Retarding admixtures will usually be needed.

As noted, tremie concrete will flow on a slight slope, about 1 on 10. Thus the spacing of pipes and points of placement depends on depth of seal. The longer the flow distance, the greater the exposure of fresh concrete to washing. A typical flow distance limit is 11 m (35 ft) as in (Fig. 6.25), although satisfactory underwater concrete has been placed in thick seals with flow distances up to 15 to 21 m (50 to 70 ft).

Placement of tremie concrete directly by pump has been often tried, with generally unsatisfactory results except in confined tubes such as drilled shafts. One problem is the formation of a vacuum when the concrete flows downhill faster than the pump can supply it; this then results in the coarse aggregate segregating and jamming the pipe. An air-relief valve at the top will help. So will an inclined pipe.

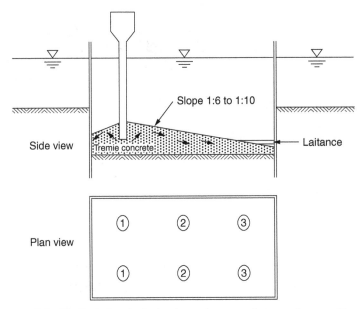

FIGURE 6.23   Typical tremie pipe locations and sequence of concrete placement. Start placement through tremie pipes 1. Place to full height. Insert pipes at 2 and continue placement until full height is reached. Insert pipe at 3 and complete placement. (Above is typical for moderate seal thicknesses.)

FIGURE 6.24   Layout of tremie pipes on one of main piers of Richmond-San Rafael Bridge, Calif.

FIGURE 6.25   Tremie concrete seal course at bottom of a medium depth cofferdam. GIWW gate structure IHNC Floodwall, New Orleans, La.

A greater problem is that the pump delivers the concrete in pulses that are not conducive to the smooth flow essential in tremie concrete. More laitance is thus formed. It is better practice to use pumps (Fig. 6.26) or steel slick lines (Fig. 6.27) to deliver the tremie concrete to the hopper and then allow the concrete to flow downward under gravity head only, that is, through an open-top tremie pipe. An airlift pump may be operated in a far corner of the cofferdam to remove silt and laitance and to prevent its being trapped by subsequent concrete.

The tremie concrete surface will be irregular, with a mound at the location of the pipe and valley in between pipe locations (Figs. 6.28 and 6.29). In attempting to fill the valleys near the end of the pour, great care must be taken to properly embed the tremie pipe; otherwise the concrete will just flow over the surface laitance. In many cases, therefore, it is best to just leave the valley low. After dewatering, the valleys can be filled with concrete placed in the dry, usually as a part of the footing block pour.

After the tremie seal has reached grade and initial set has occurred, a diver can remove the laitance by use of a jet. Later after the cofferdam is dewatered, the laitance may have hardened so as to require a jackhammer to remove it.

The typical seal course develops a substantial increase in temperature owing to the hydration of the cement and mass of concrete, which prevents rapid dissipation of heat. Therefore, for cofferdams where the seal must also serve important structural functions, precooling by batching ice or by use of liquid nitrogen will have beneficial effects and reduce thermal cracking. Covering the surface with a thermal blanket will also reduce thermal cracking (Fig. 6.30).

Horizontal lifts are undesirable in a tremie concrete placement, as they will have seams of laitance on the joint surface. If a large cofferdam is to be subdivided, this should preferably be accomplished with a vertical bulkhead. After the first placement is made, the bulkhead can be removed and the second pour made. If bulkhead is of precast concrete,

FIGURE 6.26 Transporting concrete by barged ready-mix trucks, to concrete boom pumps and into gravity tremie pipe hoppers, controlled in frames on bracing supported skids. I-205 Bridge, Ore.

(a)   (b)

FIGURE 6.27 (*a*) Transporting concrete by steel slick line to pumps and barge-mounted hoppers. (*b*) Placement of concrete from hopper barge, to boom pumps, then into gravity feed tremie pipes. Olmsted Dam, Ohio River, Tainter Gate Shell Filling.

FIGURE 6.28 Surface of tremie concrete seal immediately after dewatering bypass gate structure IHNC Floodwall, New Orleans, La.

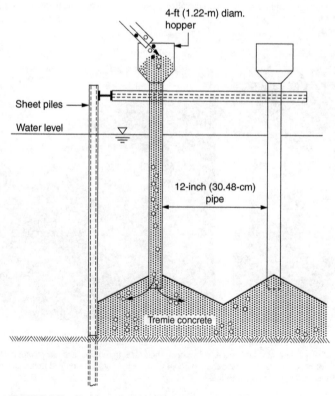

FIGURE 6.29 Tremie method for underwater concrete.

FIGURE 6.30   Covering the surface of the tremie concrete with an insulating blanket immediately after dewatering, to reduce thermal gradients and strains. Hope Creek Nuclear Power
Plant, N.J.

it can be left in place and the tremie will bond to it. Steel sheet piles may also be used as an
intermediate bulkhead left in place.

The seal resists hydrostatic uplift by its weight and its bond to the piling, which then
resist the uplift in skin friction against the soil. Typical values of ultimate bond between
tremie concrete and steel are 0.55 to 1.0 MPa (80 to 145 lb/in^2), allowing the use of a service
level design bond stresses of 0.14 to 0.22 MPa (20 to 32 lb/in^2). In U.S. practice, a few
state departments of transportation limit the allowable bond stresses to 0.05 to 0.07 MPa
(7 to 10 lb/in^2). The DOTs may also limit the bond between the soil and the sheet pile or
uplift piles. Their specifications are requirement, likely based on past adverse experiences
related to poor-quality tremie concrete and steel surface cleaning. However, tests indicate
that higher working bonds on only 1.5B pile penetrations are achievable when tremie
concrete pours are properly mixed and executed. Typically, pile penetrations are greater
than 1.5B so judgment is needed in applying the test results to a seal and hold-down
connection design. In cases where large-diameter pile holds down relatively thin seals, it
is prudent to add shear rings in the bond zone to develop bonds equivalent to the punching
shear strength of the tremie concrete. Properly placed tremie concrete will be homogeneous
and of strengths comparable to those attainable in the dry (Fig. 6.31).

Between piles, the seal concrete must be thick enough to act as an unreinforced thick
plate. Due to potential nonuniformity of section and laitance formation, the design concrete
section is reduced by at least 0.15 m (0.5 ft) for ballast and design purposes. The allowable
bending, punching shear and bond to the uplift pile and sheet pile may all be considered
in design of the tremie seal. If the cofferdam sheet pile and bracing are considered in the
ballast determination, the slab bond and strength of the seal need to be sufficient to transfer
the cofferdam weight. In some designs angles are preinstalled on the sheet pile in the seal
zone to increase the bond and ensure that the cofferdam weight is transferred as needed.

History has recorded many poor performances and failures of tremie concrete when
proper mixes and procedures have been ignored (Fig. 6.32). Conversely, high quality can
be achieved by the methods outlined here, even under extreme difficulties such as deep

FIGURE 6.31   High-quality cores of tremie concrete from Lower Monumental Dam stilling basin repair, Wash.

FIGURE 6.32   Unsatisfactory results with tremie concrete placement due to improper mix and procedure. Note segregation.

water and high currents (Fig. 6.33). The weight of tremie concrete in place will typically run 21 to 22 kN/m³ (133 to 140 lb/ft³). Underwater, the net buoyant weight will thus be 11 to 12 kN/m³ (70 to 76 lb/ft³). Hence excavating deeper in order to place a thicker seal is relatively inefficient. Such a procedure may also require deeper penetration of the sheet piles. The use of additional piles may be a more effective way of resisting uplift.

Concrete seals should normally be stopped about 150 mm (6 in) low, provided this can be done safely. The difference can be made up in the dry pour of the footing block. By stopping low, there will be less removal of humps and high points, in order to place the lower layer of reinforcing bars for the footing block.

In general, it must be assumed that full hydrostatic pressure will act on the underside of the concrete seal. The exception is when a positive cutoff is achieved by deep penetration of sheet piles and a filtered seepage relief system (dewatering system) has been installed below the seal. This is very difficult to ensure in practice.

The second method of placement of a concrete seal is by the use of grout-intruded aggregate (Fig. 6.34). In this case, coarse aggregate [12 mm (½ in)] from which all fines have been removed is placed in the cofferdam. Vertical pipes for grout injection are installed on 1.5 to 2.0 m centers. Other vertical slotted pipes are installed to serve as indicators.

FIGURE 6.33 Satisfactory results with tremie concrete placement using proper mix and procedures. (This is a second pier on the same project as that shown in Fig. 6.32, after corrective measures were instituted.)

FIGURE 6.34 Typical floating batch plant barge for tremie concrete or mortar placement.

Grout is now pumped through the pipes, so as to fill the voids and interstices in the rock with mortar. The grout must be very fluid and have high resistance to segregation.

Usually the mix will consist of fine sand and cement in equal parts, plus admixtures designed to retard set, reduce water-cement ratio, reduce bleed, promote fluidity, and prevent segregation.

If the seal thickness is too great for the first set of pipes (a typical lift is 1.5 to 2 m), a second set is installed, with its injection points set so as to enable the new grout to be injected into that which has been placed earlier, before the earlier grout has set.

The keys to successful placement of grout-intruded aggregate concrete are the cleanliness of the coarse aggregate (no fines, no silt) and proper control of the grout consistency. The rate of injection should be as rapid as possible.

The surface of a grout-intruded aggregate-concrete seal will require cleaning to remove the excess gravel, etc., much as does a tremie-concrete seal. For grout-intruded aggregate, the sheet-pile interlocks must be fully engaged, and there must be no holes; otherwise the fluid grout will flow out into the surrounding waters. A major problem which has occurred with grout-intruded aggregate is blockage of grout flow. For example, fines have inadvertently collected in lenses or strata, preventing grout flow. Marine organisms may grow, even in a short time, and block the flow of grout. Another frequent problem has been excessive bleed, resulting in small voids under the particles of coarse aggregate.

Check Blocking (Stage 11).   It is always prudent to have a diver check that the sheet piles are properly blocked against the wales and that the blocks have not floated up or been displaced during construction operations.

Dewatering (Stage 12).   During all previous steps, the water levels inside and out will have been equalized by a floodgate in the sheet piles, so as to prevent water flow through the interlocks and through the fresh concrete.

After the seal has gained sufficient strength (usually 3 to 7 days), the cofferdam is ready for dewatering. The floodgate is closed. It is often difficult to get the water level to lower initially, as water flows in through the newly driven interlocks as fast as it is pumped out. Approximately, initial flows may be on the order of 0.025 gallons per minute per square foot wall per foot of head. Later, after a head differential has been established, the sheet-pile interlocks will tighten.

Therefore, it is desirable to start dewatering with as much pumping capacity as possible.

High-volume low-head (irrigation) pumps can be used to augment the high-head submersible pumps. Start at low tide when there is less total surface exposed (Fig. 6.35b).

If an initial start proves infeasible, there are a number of ways of sealing interlocks. Sand cinders, sand and sawdust, and sand and manure have all been used. However, subsea vibration in current and waves may cause the material to fall out. Oakum caulking seams can be run by divers.

A final expedient is to drape weighted canvas or plastic sheets over the outside face of the sheet piles. These will be sucked tight against the piling by the inflow.

Unfortunately, in not a few cases it has been found that the leakage is through a handling hole in a sheet pile that has been spliced.

Construct New Structure (Stage 13).   The initial element of the structure is usually the footing block, a thick and massive block of reinforced concrete. Typically, this will develop high heat of hydration that will not dissipate for as long as 15 to 30 days. When the concrete finally cools, it will shrink. By this time, it is bonded to the tremie concrete. As a result, it will crack. Many experienced engineers and contractors have learned to place additional reinforcement in the top and even at mid-depth, so as to constrain the cracking.

An insulating blanket placed over the top of the footing block will help to reduce thermal strains and cracking. Another useful step is to precool the concrete mix by mixing

(a)

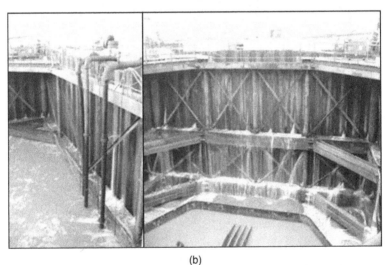

(b)

FIGURE 6.35 (*a*) Filling navigable lagoon through levee breach cofferdam and permanent double-walled sheet pile, Bethal Island, Calif. (*b*) Initial dewatering of Skyway pier foundation cofferdam SF-Oak. Bay Bridge, Calif.

with ice or by use of liquid nitrogen. Using blast furnace slag-cement or replacing Portland Cement in part by PFA will minimize this problem.

Flood Cofferdam (Stage 14). This can be done by opening a tide gate in the sheet piles at the waterline. Some cofferdams are used to dewater vast areas that eventually require re-watering. In these cases a carefully planned siphon and scour protection apron are needed to optimize the duration and efficiency of the cofferdam and levee breaching operations. (Fig. 6.35*a*)

Removal of Sheet Piling (Stage 15).   Usually a vibratory hammer is most efficient in removal. Piles are often removed in pairs to facilitate their reuse. It is often very difficult to pull the first pile. Once the first one is pulled, the others come more easily. For the first pile, there is double interlock friction and general binding of the entire cofferdam. Usually a pile in the middle of a side is easiest. Try driving (vibrating) down first to break the interlock friction. Sometimes a steam or diesel hammer must be used to drive the first pile down.

Impact extractors are made with powerful capacity to free a pile. Their clamping to the pile requires careful detailing (e.g., use of multiple high-strength bolts); otherwise they may rip out a section of steel from the sheet pile. Another "last resort" is to rig jacks and break the binding by jacking one pile up against its neighbor.

Removal of Bracing (Stage 16).   If possible, this should be designed to be lifted over the top of the partially completed structure. This is expeditious and facilitates reuse. This may not be feasible owing to weight or interference, so removal in sections may be required.

The removal, in sections, is practical if well conceived and detailed during initial design. Generally speaking, provision must be made to relieve the built-in stress on a member to be disconnected before it is cut free. Bolted joints facilitate removal and cause less damage than burning.

Upon release of a member, the remaining forces must go elsewhere, for example, into the backfill or into other bracing. Care must be taken to ensure that a second brace is not overloaded so as to fail in buckling. Blocking may be used to transfer the force to the structure.

When struts must pass through a massive portion of the new member, they are often best treated by casting them into the structure. Blockouts can be provided at the edges to permit cutting back of the surface, if required, and filling the pocket with concrete. Corrosion of steel underwater can be considered minimal because of the nonavailability of oxygen. For struts that penetrate thinner walls, sleeves may be provided so that the struts can be later removed by pulling through.

Backfill (Stage 17).   Since the backfill will be placed underwater, sand or sand and gravel should be used. In some cases where the engineer so specifies, in order to provide passive lateral support to the pier, he/she may require the backfill to be densified, for example by vibration.

Special steps may also be required to protect against both local and general scour, such as placement of filter fabric and riprap.

COFFERDAMS FOUNDED ON ROCK OR THROUGH BOULDERS

In many cases, bridge piers and pump houses are founded directly on hardpan or rock. In such cases it may be extremely difficult to achieve sufficient penetration with the sheet piles to develop adequate shear resistance at the tip. Such penetration may be aided in specific soils by one or more of the following steps:

1. Use of hardened-steel tip protectors for sheet piles.
2. Continued driving at refusal with a relatively small hammer, so as not to damage the tip, but rather chip into the rock.

3. Predrilling (line drilling) along the line of sheet piles. One way to align the drill and piles properly is to first set and drive the sheet piles to a point just above grade, taking care not to bend the tips by driving into rock. Then a 150 to 300 mm (6 to 12 in) diameter bit is used to drill within the outfacing arch of each sheet or pair of sheets. In the case of pipe sheet piles, the drill may be run down the center.

When adequate penetration below the tip cannot be reliably and positively ensured, it is important to provide shear resistance at the lowest level possible by means of a bracing frame. This ensures the structural adequacy of the cofferdam, although it does not prevent run-in of sand under the tip. Underwater concrete may be placed to seal tips.

Where a concrete seal must be placed directly on rock and its thickening is limited, anchors may be required to resist uplift. It must generally be assumed that hydrostatic pressure can eventually penetrate through seams in the rock. Anchors may consist of steel piles or ground anchors drilled and grouted into the rock, either before or after the seal is placed, but before dewatering. Alternatively, a 0.5-m layer of crushed rock may be placed to serve as an underdrain.

Cofferdams that must penetrate overlying strata of hard materials or through boulders may present major difficulties. The sheet piles are easily deformed, split, or driven out interlock. In the case of an upper stratum of hard material, predrilling (line drilling) with a drill or high-pressure jet may permit penetration, especially if the piles fitted with hardened-steel tip protectors.

If blasting is indicated (e.g., when a coral or limestone layer must be penetrated), this is usually best done after the bracing frame has been set, but before setting the sheet piles. Blasting adjacent to already installed sheet piles often causes distortion, which later prevents driving them to grade.

Boulders embedded in running sands present perhaps the worst difficulties. The hard boulders cannot easily be penetrated; they are extremely difficult to predrill because of sand run-in, and they do not always fracture by blasting. The following methods have had some success:

1. Installing a third (or lower) set of bracing as deep as possible, just above the boulder elevation.

2. Keeping the water head inside the cofferdam above that outside, or filling the inside with bentonite slurry (drilling mud).

3. Injecting chemical grouts outside (silicates or polymers) to give cohesion to the sand. (This step is not always necessary.)

4. Progressively driving and excavating inside below the tips of the sheet piles. The driving should be performed with a relatively light hammer or vibratory hammer.

A steel spud pile may be driven down in the inward-facing arch formed by the sheet piles to displace a boulder into the excavated area. A high-pressure jet may be similarly used, although this is likely to aggravate the tendency for the sand to run in.

CELLULAR COFFERDAMS

Sheet-pile cellular cofferdams are utilized to dike off relatively large areas where the use of internal bracing and/or tiebacks would be unsuitable owing to length or interference with the new structure.

Cellular cofferdams are practicable for dewatered depths up to 23 m (75 ft). They form a continuous gravity wall; hence they impose high bearing and overturning pressures as well as lateral shear on the soil beneath. For this reason, they are best used when founded on rock, hardpan, or dense sands.

The concept for the cellular cofferdam is that a circumferential ring wall is constructed progressively by setting and driving interlocking flat sheet piles to form a closed cell (Figs. 6.36 and 6.37). The cell is then filled with granular material, such as coarse sand, sand and gravel, or crushed rock. These fill materials, of course, develop tensile forces in the sheet-pile ring. The successful performance of this system depends on the internal shear strength of the fill, its friction against the sheet piles, and the tensile interlocking of the sheet piles.

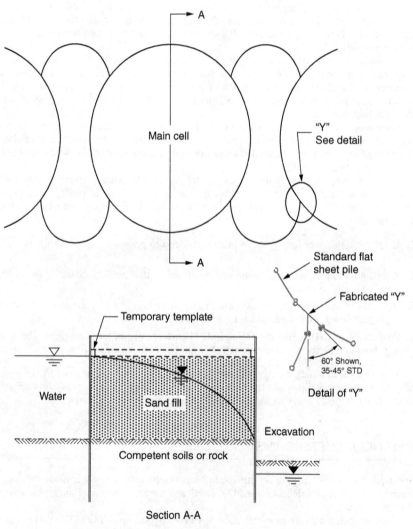

FIGURE 6.36 Typical layout for steel sheet-pile cellular cofferdam.

(a)   (b)

FIGURE 6.37 (*a*) Cellular cofferdam enclosure for construction of a new lock for Olmsted Dam on the Ohio River. (*b*) Due to deep soft clays, single-wall, combi-wall, and 54-in-diameter batter pile for new sector gate and pump station was needed in lieu of cellular cofferdam, for the IHNC levee system, New Orleans, La.

When sheet-pile cells are constructed on sedimentary soils, the sheet piles penetrate below the internal excavation level, acting to seal off water flow beneath to some degree and to develop passive lateral resistance. If there is soft material in the sheet-pile cell (e.g., mud), it must be removed before the fill is placed. Otherwise the soft material may develop high lateral fluid pressures against the sheet piles, equal to the surcharge pressure acting on it.

In order for the sheet-pile cell to function properly, the general geometry of the cell must be maintained at all horizontal cross sections, within limited tolerances. Further, the sheet piles must be stretched in tension by the fill in order to develop the necessary friction in the interlocks. Finally, the entire ring of steel-sheet piles must be interlocked continuously.

To accomplish this, it is necessary that the sheet piles be accurately set. Therefore, false-work guides must be provided, consisting of at least two and preferably more ring wales. Their supporting falsework piles are usually steel pipe spud piles driven through sleeves in the lateral bracing of each ring wale. The procedure is, therefore, to set the group of ring wales, then drive the vertical spud piles through the sleeves. Then the ring wales are raised to proper grade and pinned off.

Steel sheet piles are then set around the ring. The Ys (or Ts) should be set first, carefully positioned in exact horizontal location and in vertical alignment. Standard flat sheets are then set progressively from the Ys toward a meeting point at the center. The Ys and an occasional pile may be driven into the bottom a few feet only, during the setting, in order to enhance lateral stability. The junctions are always made at a central point between Ys so that if there is any stretching or "shaking-out" to fit, it is done as far away from the stiff Ys as possible. To aid in accurate setting and joining, the position of each pile should be premarked on the top ring wale. It should be marked as the theoretical distance, center-to-center, of each interlock, plus a small increment, say 3 mm (1/8 in), to provide for stretch. The final fit-up should remove all slack, but not be so tight as to prevent a free-running entry of the last sheet.

Driving then proceeds in incremental fashion, starting at the Ys and working toward the centers. Each pile should be driven not more than 1 m beyond the tip of its neighbor. Excessively hard driving of sheets usually indicates inaccurate setting, which in turn develops excessive interlock friction at a later stage. (The field crew usually reports it as "encountering hardpan or obstructions"; and, of course, once in a while they are right!)

Vibratory hammers are ideal for driving the sheet piles in cells, since if excessive friction does develop, a group of piles can be raised back up and redriven in the same incremental fashion. Continued hard driving can drive sheet piles out of interlock, destroying the tensile capacity of the ring.

Most cellular cofferdams are circular cells, with short arcs connecting adjacent cells.

In many cases, these short arcs can be set along with the adjacent circular cells, and thus set and driven vertically.

If, owing to the need to stabilize each cell as it is set by filling with rock (as, for example, when working in a swift river), the arc has to be set later between two completed and filled cells, then the sheet piles will have to be set on a slight outward-leaning angle so as to keep the same arc distances at top and bottom.

Consolidation of fill within the cells helps to improve the internal friction and prevent liquefaction under earthquake. Recently it has become the practice to consolidate the fill by vibration. Such vibration is very effective in a cell because of its confinement, provided the water can escape vertically, rather than just build up high pore pressure. In one case, intensive internal vibration caused the bursting of a cell where a layer of impermeable fill prevented escape of the displaced water. Therefore, the construction of several vertical drains or wells, combined with internal vibration, will expedite consolidation.

The Ys are the weak points structurally, that is, each Y and the three sheet piles connecting to it. Each leg on the Y is subject to high tension, combined with bending and moment in the flat portion. The details of the Y, its angle, and its method of fabrication are of primary importance. Many experienced designer-builders of cellular cofferdams use heavier-section steel, higher-yield steel, or both for these critical piles. Whereas in the past many connectors were 90° Ts, the present improved practice is to use 35 or 45° Ys so as to minimize the moments.

Where cellular cofferdams are founded on rock, it is very difficult to seat the sheet piles so as to prevent large underflows of water. Among the methods used are the following:

1. Cleaning off to rock and pre-placing a graded layer of relatively low-permeability sand and clay into which the sheet piles are toed. This can control but not eliminate underflow.

2. Using cast-steel "protectors" on the tips of the sheet piles to permit toeing into rock.

3. Preexcavating the rock surface so as to remove boulders, cobbles, ridges, and protuberances before the sheet piles are set.

4. Pouring a slab of tremie concrete inside the cell before filling it.

5. Pouring tremie concrete or placing sacked concrete on the outside of the cell.

6. Placing an extensive blanket of clay on the outside of the cells.

The bursting effect of fill within the cells of cofferdams and the tensile stresses in the sheet piles may be reduced by draining the cell fill through weep holes in the sheets as the cofferdam proper is dewatered. Such slowly drained fill will also consolidate and have a higher angle of internal friction.

CONCRETE SLURRY WALL COFFERDAMS

Continuous-ring (circular) walls of concrete constructed by the slurry method have been extensively used to construct shafts for tunnel access. In this method trenches have been excavated in increments of 4 to 5 m (12 to 16 ft), with a width of 1 m (3 ft), using bentonite slurry to prevent caving. Tremie concrete is then placed in the "slot." Adjoining increments are constructed so as to form a continuous ring that can then resist the external earth pressures. Such a cofferdam was constructed for the access shaft on the French side of the Channel Tunnel.

This process has been extended to the construction of very large and deep cofferdams for bridge piers, underground power plants, and ventilation structures for tunnels. Diameters of

100 m (328 ft) and more, wall depths up to 120 m (394 ft), and excavated and dewatered depths of 80 m (262 ft) have been achieved in Japan.

Notable among these are the Kobe anchorage pier for the Akashi Strait suspension bridge, the world's longest, and the Kawasaki ventilation shaft for the Trans-Tokyo Bay crossing (Fig. 6.38*a* and *b*).

(a)

(b)

FIGURE 6.38 (*a*) Slurry wall for cofferdam for Kawasaki Island Ventilation Structure, Trans-Tokyo Bay Bridge, Japan. (*b*) Circular wall is 100 m (328 ft) in diameter and 124 m (406 ft) deep, for a dewatered head of 80 m (262 ft).

One of the advantages of this method is that the wall can be constructed through any material from soft clays to sands to boulders and rock. Extreme accuracy is required in order to ensure a geometrically accurate circular ring at the bottom as well as the top.

Wall thickness must consider not only the ring compression but the moments induced by deviations in tolerances as well as any unbalanced loads due to variations in the soils or to equipment.

Vertical reinforcement can be provided by cages of reinforcing steel or by steel beams set in the slots before placement of the tremie concrete. However, it is very difficult to provide continuity in the circumferential direction. The conventional way is to install internal ring beams as excavation proceeds inside, but this is expensive. Therefore, some very clever schemes have been developed. For example, tension capacity on the inner face can be attained by steel bands anchored into the wall at close spacing and grouted or wedged so as to take up all play. On the outside, the passive soil pressure resists bulging. Other schemes are based on sheet-pile interlocks or overlapping headed bars.

BIBLIOGRAPHY

Anderson, H. V.: *Underwater Construction Using Cofferdams*, Best Publishing Company, Flagstaff, Arizona, 2001.

Anderson, P.: "Single-Wall Cofferdams," Chapter 8. In *Substructure Analysis and Design*, 2d ed., The Ronald Press Company, New York, 1956.

Dept. of Army: "Structural Design: Physical Factors," Chapter 7. In *Shore Protection Manual* 4th ed., Vol. II, U.S. Gov. Printing Office, Washington, D.C., 1984.

Gerwick, B. C., Jr.: *Construction of Marine and Offshore Structures*, 3d ed., CRC Press LLC, Boca Raton, Florida, 2007.

Leonards, G. A.: "Caissons and Cofferdams," Chapter 10. In *Foundation Engineering*, McGraw-Hill, New York,1962.

O'Brien, J., J. Havers, and S. Stubbs, Jr.: "Cofferdams and Caissons," Part D Chapter 4. In *Standard Handbook of Heavy Construction*, 3d ed., McGraw-Hill, New York, 1996.

Winkerton, H. and H. Y. Fang: *Foundation Engineering Handbook*, Van Nostrand-Reinhold, New York, 1975.

CHAPTER 7
EARTH-RETAINING STRUCTURES

John J. Peirce, P.E., D.GE
Thomas J. Tuozzolo, P.E.

INTRODUCTION

The purpose of an earth-retaining structure is to create a safe excavation into which a permanent structure can be built or to be the permanent earth-retaining structure itself. This chapter focuses on temporary earth-retaining structures. The type of earth-retaining structure and method of installation are dependent upon many factors: the principal factors being the type of soil encountered, the elevation of groundwater, the depth of excavation supported, the proximity of existing structures that might be affected, and, if more than one system can be used, their relative cost and time required for installation.

Special procedures and equipment have been developed in localized geographical areas to take advantage of local conditions, thus permitting economical installation of retaining structures under those special conditions. Contractors and engineers should be particularly careful in analyzing these special methods to be sure that the conditions at their site will permit them to be used.

Temporary retaining methods can be discussed as two separate items: the structure in contact with the soil and the bracing system that holds the structure in place. Shallow excavations may allow cantilevered construction; deeper excavations will require some form of lateral support or bracing. There are also methods such as freezing and grouting, which change the soil into structural elements and can often be designated as self-supporting systems.

Before any system can be designed, there is certain information to be obtained and reviewed and preliminary steps to be taken:

1. Analysis of soil boring data. This would include determination of groundwater level; soil stratification; soil properties such as soil weight, internal friction angle, cohesion, and corrosion potential; presence of underground obstructions (natural or man-made); and proximity of rock, its shear strength, and its bedding. Careful review of the project Geotechnical Report and soil borings is a must when choosing and designing an earth-retaining structure.

2. Survey of existing structures, including underground utilities.

3. Review of data available from libraries, utility companies, building departments, U.S. Geological Survey, etc.

4. Examination of site and existing conditions, including drainage of area, type of vegetation, etc.

5. Preparation of sketches, to scale, indicating the relationship of existing utilities and structures to the proposed excavation, soil stratification, groundwater level, and proposed new structure.

A determination must be made of the amount of deformation that would be tolerable in the retaining structure so that no damage is caused to adjacent structures or utilities. This will indicate if the adjacent structures require underpinning or if the retaining structure must be designed to support the surcharge loads of these structures. However, as a word of caution, most earth-retaining structures are considered as flexible structures and, as such, are not suitable for supporting adjacent structures. Flexibility means wall deflection; deflection means ground movement; ground movement often means lateral or vertical structure movement that may damage the structure.

Total loads acting on the temporary retaining structures can then be determined and should include soil and water pressures and surcharge loads from traffic, construction equipment, material storage, and adjacent structures. A check should also be made of the base stability of the proposed excavation. Checking base stability includes checking bottom heave of excavations in soft cohesive soils and piping or quick conditions in granular soils due to unbalanced, hydrostatic, groundwater elevations.

TYPES OF RETAINING STRUCTURES

Steel Sheet Piling

Steel sheet piling is hot-rolled or cold-formed in a variety of shapes and weights with special joints to act as interlocks between sections of the sheeting. For use as temporary retaining structures, Z sections or deep-web sections with large section moduli are most commonly used (see Figs. 7.1 and 7.2).

Greatest advantages are obtained in using steel sheet piling when there is a high groundwater level and the soil to be retained cannot easily be predrained, or if it is desirable to maintain the groundwater outside the proposed excavation at its existing level. Additionally, the use of steel sheet piling is often desirable when the soils to be retained are soft or loose.

Steel sheet piling is generally driven to its final position with conventional pile driving equipment before the excavation commences. To most effectively control groundwater and bottom heave, the sheet piling should be driven into a competent, impermeable soil layer, or with sufficient penetration below the excavation subgrade to effectively minimize excessive

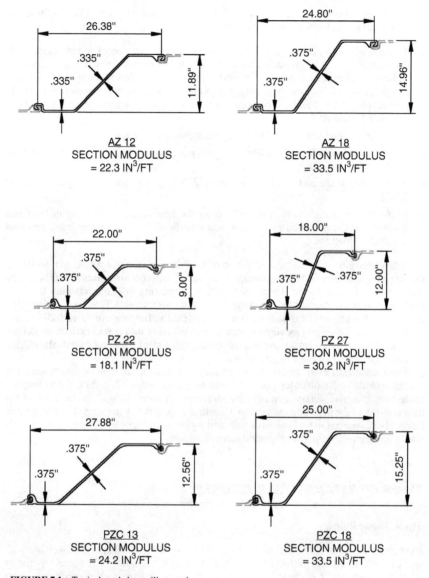

FIGURE 7.1   Typical steel sheet piling sections.

flow and pressure of groundwater under the sheet piling. When sheet piling is used to retain groundwater, some leakage through the interlocks must be anticipated. Where controlling or eliminating interlock leakage is desired, special interlock joint sealants can be installed prior to installation of the sheet piling. However, use of these sealants is relatively rare, can be expensive, and is time consuming.

FIGURE 7.2   Braced steel sheet piling for soil remediation excavation. (*Photograph courtesy of Peirce Engineering, Inc.*)

Steel sheet piling may lose its effectiveness when the interlocks are torn or if all of the sheets are not driven to the proper penetration. These conditions can be due to obstructions in the ground, hard layers of soil to be penetrated, or improper driving of the sheets. Loose sand or fill can undergo considerable consolidation during installation of the sheet piling. Therefore, the use of this method should be evaluated when driving close to existing utilities or structures that are resting on these types of soils.

Because steel sheet piling material costs are relatively high, it is often desirable to pull or extract the sheet piling after the temporary earth-retaining structure is no longer needed.

Proper installation of the sheet piling is important in ensuring the effectiveness of the system. A proper driving template should be set to provide correct alignment and horizontal spacing of the sheet piling before driving. The top wale of the bracing system can often be used as a template. When necessary to slightly adjust the overall length of a sheet-pile wall, such as to "close" the corner of a multisided sheet-pile cofferdam, the theoretical width of one or more sheet-pile doubles can be stretched wider or compressed narrower to ensure that the sheet-pile wall is installed to the proper length.

The sheet piling should be driven in "waves" or "steps" where the toe of any single or double sheet pile is driven no more than 5 to 6 ft (1.5 to 1.87 m) below the adjacent sheets. In order to prevent clogging of the socket end of the sheet piling, the ball end of the sheet should be driven as the leading edge.

Slide Rail Systems

A slide rail system is a modular excavation support system comprising heavy-duty shoring panels supported by vertical, multitrack rails into which one or more panels are placed and are advanced or slid downward as the excavation progresses toward subgrade. The vertical rails are placed at the corners of the shored excavation and at intermediate locations along the sides of the excavation. The location of the intermediate rails depends on the horizontal length of the shoring panels. Panels are available in heights of 4 to 8 ft (1.2 to 2.4 m) with lengths of 8 to 20 ft (2.4 to 6.1 m). The vertical rails, and therefore the panels, are supported by two or more levels of horizontal, steel, cross braces or by larger, fabricated cross struts (see Fig. 7.3).

FIGURE 7.3   Slide rail system for trench support. (*Photograph courtesy of Efficiency Production, Inc.*)

Slide rail systems are used for relatively narrow excavations for structures such as pipe lines, pump stations, underground vaults and tanks, utility shafts, boring and receiving pits, and similar structures. Larger excavation areas can be shored using multiple, rectangular, systems connected by double and triple corner rails or posts. Slide rail systems may not be practical where obstructions such as underground utilities cross the excavation and would interfere with sliding of the panels.

Whereas steel sheet piling is driven into the ground before the excavation is made, slide rail systems are installed as the excavation is made. This is referred to as the "dig-and-push" method where the panels are set into the vertical rails and are then slid downward, along with the rails, as the excavation progresses deeper. When a panel has been installed to its full height, a second panel is inserted into a second vertical track in the rail. This second panel is then pushed down, below the first panel, as the excavation continues. For deeper excavations, the process may be repeated as required.

Slide rail systems are proprietary systems with unpublished design parameters that make it difficult for users to prepare their own design. The design for a slide rail earth-retaining system is usually provided by the system's manufacturer, supplier, or their consulting engineer. Slide rail systems are most often rented by contractors rather than purchased.

Soldier Pile Walls with Horizontal Lagging

This earth-retaining system consists of vertical members called soldier piles (or soldier beams), generally 6 to 10 ft (1.8 to 3.0 m) on center, installed to or below subgrade in advance of the excavation, with horizontal members spanning between the soldier piles to retain the soil as the excavation progresses. For excavations up to approximately 12 ft (3.7 m), soldier pile and lagging systems may not need any extra lateral support as long as the soldier pile is strong enough and is installed deep enough in competent soil or rock below excavation subgrade. For deeper excavations, the soldier piles will need to be laterally braced at one or more levels. Retaining wall bracing systems are discussed later in this chapter.

Soldier piles are usually rolled, steel beams (H-pile or wide flange sections) but can be fabricated sections, pipe piles, or, more rarely, precast concrete elements. They are usually installed with conventional pile driving equipment but can be installed in drilled or augered holes, set in small excavated pits, or be installed by other methods as required by site conditions. While it is usually more economical to install soldier piles by pile driving methods, it is quite common and may be more economical to install them in drilled or augered holes on smaller projects and in areas of the country where soil conditions permit the holes to be drilled and remain open. When this method of installation is specified or attempted in soil that is not suited for augering, the costs could be prohibitive and progress extremely slow. Where soil conditions require the use of temporary casing or drilling mud, some of the economy is lost. The presence of boulders or large obstructions may make the use of augering equipment uneconomical, in which case, a different type of soldier pile or retention system may be more appropriate.

Using augered holes permits greater flexibility in design than when driving soldier piles. The holes can be filled with concrete (using reinforcing steel as required) to form concrete soldier piles, or steel sections and be installed, either fabricated or unfabricated. Where tieback anchors are used to brace the soldier piles, seats and/or sleeves for the tieback anchors can be installed on the soldier piles before or after placing the piles into the holes. During fabrication, before placing structural shapes into the holes, spacers are installed on the piles to help align them in the holes, and the holes are filled *entirely* with lean concrete—say one 94 lb (43 kg) sack of cement per cubic yard (0.76 m^3) of concrete. This helps keep the soldier piles in place as excavation proceeds and earth loads are imposed on the piles. Backfilling drilled or augered soldier pile holes with sand or crushed stone is not recommended. Structural concrete fill, if required, should be limited to placement below excavation subgrade so as not to interfere with installation of the horizontal, lagging.

Although often unnecessary, it is common to see engineers specify use of augered holes for projects at urban sites to eliminate noise and reduce vibrations that are present when driving soldier piles. One must be extremely careful when analyzing site conditions on such projects. Old foundations, boulders, etc. could seriously interfere with hole drilling and make this procedure extremely difficult, if not impossible.

The horizontal members spanning between soldier piles are usually wood lagging that is installed between and behind, or is attached to, the soldier pile flanges closest to the excavation. Lagging can also be attached to the front flanges of the soldier piles by various methods such as mechanical clips; welded, threaded studs with nuts and bearing plates; or welded angles or tees (see Fig. 7.4). Soldier pile and lagging walls that are being constructed in a cut situation are constructed from the original grade down toward subgrade, that is, top-down construction. After the soldier piles are installed, a vertical cut of approximately 5 ft (1.5 m) is made at the front of the soldier piles. The first lagging board is then placed at the bottom of the cut and is usually placed behind or is attached to the front flange of the pile. Next, additional lagging boards are then successively placed above the previously placed board until the initial cut is fully lagged. Then, the next cut of the same depth is made and

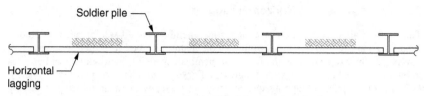

Horizontal lagging behind front flange (plan view)

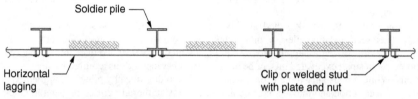

Horizontal lagging attached to in front of front flange (plan view)

FIGURE 7.4   Soldier piles and horizontal lagging—typical detail plans.

the process repeats until the entire wall is lagged to subgrade or until the excavation is lagged to the brace or tieback level. After the braces or tieback anchors are installed, the process repeats as needed. Unless being installed in a fill situation, lagging should not be installed as preassembled, multi-board panels where the intent is to install and slide the lagging panels down from the top of the soldier piles as excavation progresses. Installation of lagging in this manner usually requires overexcavation behind the lagging and results in insufficient contact or voids between the back of the lagging and the ground to be supported.

When installing horizontal, wood lagging, a vertical space of 1.5 to 2 in (3.8 to 5.1 cm) is normally left between lagging boards. These spaces are called *louvres* and are maintained by nailing short, wooden blocks to the top, narrow edges of the lagging (see Fig. 7.5).

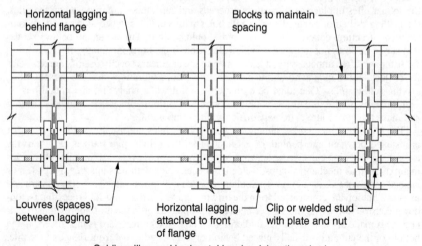

Soldier piling and horizontal lagging (elevation view)

FIGURE 7.5   Soldier piles and horizontal lagging—typical detail elevation.

These spaces allow visual inspection of the tightness or contact of the soil against the back of the lagging and permit packing of soil between the lagging boards to replace any soil that might be lost during excavation for the boards or during the service life of the retaining structure. It is a common misconception that installing louvres is done to save on lagging or to allow drainage. The amount of lagging "saved" by installing louvres is insignificant and is probably offset by the cost of louvre installation. Also, any groundwater behind the lagging wall will drain between the rough-cut lagging boards with or without louvres. However, if water flows from between the lagging boards, the louvres allow the spaces to be filled with salt hay or a geotextile that will maintain drainage while acting as a filter to prevent loss of soil from behind the lagging.

Steel plates, shotcrete, or precast concrete lagging, although usually more expensive, are sometimes used in lieu of wood lagging.

There is some disagreement regarding the design of horizontal lagging. Most designers agree that the full earth pressure is carried by the soldier piles and that the soil arches between the piles. With arching, the load on the lagging is imposed by the soil under the arch and, for braced or tiedback retaining walls, does not greatly increase as the excavation depth increases. This theory has continually proved itself on projects where 3 in (75 mm) nominal thickness, wood lagging has been used for excavation depths of 50 to 60 ft (15 to 18 m) or more. For this reason, wood lagging is often not designed but is based on past local practice or on Federal Highway Administration (FHWA) recommendations (see Fig. 7.6). However, for projects where lagging must be designed, consider the following. The New York City Transit Authority *Field Design Standards* assumes that the entire lateral load is carried by the lagging and that the lagging transfers the load to the soldier piles; however, this design manual allows a 50 percent increase in the allowable stresses for the wood lagging. Increasing the allowable lagging stress by 50 percent has the same effect as designing the lagging for 50 percent of the theoretical, lateral pressures. In cases where the lagging is installed behind the back flange of the soldier piles, it appears that the arching action of the soil is destroyed and large movements of the ground behind the wall may occur. Without the benefit of arching, the lagging takes essentially the full lateral pressure and should be designed accordingly. In addition, installing lagging behind or immediately in front of the rear flanges of the soldier piles will reduce or eliminate the lateral support of the soldier piles in their weak direction (in the plane of the wall). This, in turn, will reduce the allowable bending stress of the soldier piles that are normally considered to be fully supported along their weak axis. In cohesive soils, the excavated soil face may arch between the soldier piles without any support; however, it is imperative to install some type of support (lagging, wire mesh, chain link fence, shotcrete, etc.) to protect workers from falling pieces of soil.

Micropile Walls

Micropiles are small-diameter [4 to 12 in (100 to 300 mm)], reinforced load transfer elements capable of supporting design loads in excess of 200 tons (1800 kN). The piles are designed to function individually although they may be installed in groups. Micropiles can withstand axial and/or lateral loads and are routinely used as foundation elements for underpinning of existing structures, foundation upgrade to withstand increased static or seismic loading, or new construction.

In recent years, micropile walls, composed of a "network" of reticulated piles, have been increasingly used for slope stabilization and for earth support where access and right-of-way are limited. For these applications, the structural loads are applied to the entire reinforced pile mass rather than to the individual piles (*Micropile Design and Construction Guidelines*, 2000).

	Soil Description	Unified Classification	Depth	Recommended Thicknesses of Lagging (roughcut) for Clear Spans of:					
				5'	6'	7'	8'	9'	10'
Competent Soils	Silts or fine sand and silt above water table	ML SM-ML	0' to 25'	2"	3"	3"	3"	4"	4"
	Sands and gravels (medium dense to dense).	GW, GP, GM, GC, SW, SP, SM							
	Clays (stiff to very stiff); non-fissured.	CL, CH	25' to 60'	3"	3"	3"	4"	4"	5"
	Clays, medium consistency and $\frac{\gamma H}{S_u} < 5$.	CL, CH							
Difficult Soils	Sands and silty sands, (loose).	SW, SP, SM	0' to 25'	3"	3"	3"	4"	4"	5"
	Clayey sands (medium dense to dense) below water table.	SC							
	Clays, heavily overconsolidated fissured.	CL, CH	25' to 60'	3"	3"	4"	4"	5"	5"
	Cohesionless silt or fine sand and silt below water table.	ML; SM-ML							
*Potentially Dangerous Soils	Soft clays $\frac{\gamma H}{S_u} > 5$.	CL, CH	0' to 15'	3"	3"	4"	5"	--	--
	Slightly plastic silts below water table.	ML	15' to 25'	3"	4"	5"	6"	--	--
	Clayey sands (loose), below water table.	SC	25' to 35'	4"	5"	6"	--	--	--

Note:
* In the category of "potentially dangerous soils," use of lagging is questionable.

FIGURE 7.6 FHWA-recommended thicknesses of wood lagging.

Individual micropiles can also be used in lieu of H-piles where bouldery soils preclude conventional, pile and lagging excavation support.

Advantages and Limitations. For earth support/stabilization applications, micropile walls offer a number of advantages:

- Micropiles can be installed through almost any type of subsurface condition, from sands to clays to bouldery soils to rock.
- Orientation of the micropile element can be vertical or at any angle from horizontal.
- Low-vibration drilling techniques make micropiling well suited to sensitive sites.
- Relatively small, lightweight equipment allows micropile walls to be constructed on steeply sloping sites.

Disadvantages include

- Higher cost relative to more conventional earth support methods
- Specialized design procedures required for micropile walls

Types of Micropile Walls

Reticulated micropile walls

A-frame walls

Anchored micropile and lagging walls

Cantilevered micropile and lagging walls

Equipment and Materials. Small- to medium-sized rotary and rotary-percussion drill rigs are typically used for micropile installation. These are similar to the rigs used for tieback anchor installation. When obstructions or boulders are present, down-the-hole hammers and overburden drilling systems are typically employed.

When micropiles are used as foundation elements, the pile is typically designed to resist compressive loads and possibly small lateral loads. In addition, these piles are sometimes installed in limited access or restricted headroom conditions, and as such, threaded casing is installed in 3-, 5-, 10-, and 15-ft (1-, 1.5-, 3-, and 4.6-m) sections. When micropiles are installed as soldier piles for earth retention, it is important to try to install the casing in full lengths and to avoid the use of sectional threaded casing. Because the casing joints are often threaded, they lose some of their bending capacity and are typically the weak link in the design.

The micropile earth retention casing is either flush-joint or welded type and shall be of appropriate size to withstand the bending moments and deflections imposed on it by the ground. The casing will conform to ASTM A252. The casing is either new or used American Petroleum Institute (API) or "Mill Secondary" free from defects such as dents, cracks, or tears.

If the casing pipe is flush-joint, the pipe joint shall be completely shouldered and without stripped threads. If welded casing is used, all welds shall be full penetration welds and be able to develop the structural capacity as calculated by the design. The yield strength of the casing is most times either 45 or 80 ksi (310 or 550 MPa).

Installation Procedures. The construction of the system begins with installing the micropiles to or below subgrade prior to the excavation commencing. The spacing of the micropiles will vary depending on the overall excavation height and on whether the system will be braced, anchored, and/or cantilevered. The typical center to center spacing of the micropiles for these systems is 4 to 7 ft (1.2 to 2.1 m). Once the micropiles are installed, the excavation can proceed in lifts as horizontal lagging is secured to the face of the piles with welded studs and plates.

For excavations typically over 10 ft (3 m), the micropiles will need to be restrained laterally by either bracing or tiebacks at one or more levels. Steel wale beams are typically used to transfer the load from the tieback/brace to the micropile (see Fig. 7.7). The excavation proceeds while lagging is installed and final subgrade is reached.

Secant Pile Walls

Introduced in the 1950s, secant pile walls are an economical method of providing temporary ground support or permanent structural walls for applications such as top-down tunnel construction, cofferdams, subway structures, and retaining walls (see Fig. 7.8). The installed wall may also function as an effective groundwater cutoff, if required (Powers et al., 2007).

FIGURE 7.7 Lagged, tiedback micropile wall. (*Photograph courtesy of Moretrench American Corporation*)

Secant pile walls can be installed in a range of ground conditions, including dense soils and variable ground containing cobbles, boulders, or other obstructions. Depending on the drilling method and verticality tolerances required to ensure pile overlap at depth, secant pile walls can be installed to depths of up to 100 to 130 ft (30 to 40 m).

Advantages and Limitations.   Secant pile walls offer a number of advantages over other earth retention systems such as steel sheet piling, including

- Increased construction alignment flexibility.
- Increased wall stiffness.
- Capability of being installed in difficult ground conditions.

FIGURE 7.8 Tiedback secant pile wall. (*Photograph courtesy of Moretrench American Corporation*)

- Capability of being installed in tight working conditions that preclude the use of large equipment necessary for slurry trenches or structural slurry (diaphragm) walls.
- Low-vibration drilled installation makes secant pile walls an attractive option in dense urban areas.

However, it should be noted that there are limitations to the technique, including

- Verticality tolerances may be hard to achieve at depth.
- Total waterproofing between joints can be very difficult to obtain.
- Secant pile walls are costlier to install than steel sheet-pile walls.
- Complete replacement of the excavated soil is required; thus spoil volume is greater than that generated by sheet piling or deep soil mixing that either reuse or mix the soil in place.

Construction Sequence. Secant pile walls are constructed by drilling and concreting the piles in a primary-secondary sequence. Pile diameters can range from 16 to 60 in (410 to 1500 mm) but are more typically in the range of 16 to 36 in (410 to 900 mm). Primary piles are spaced at slightly less than the nominal pile diameter so that secondary piles will cut into and interlock with the adjacent primary piles to form a continuous cutoff wall. Secondary piles are drilled between the primary piles before the concrete achieves full strength. For ground support applications, the piles are reinforced with either structural steel sections (WF and H shapes) or reinforcing cages to provide the necessary bending strength and wall stiffness. Typically, only the secondary piles are reinforced to avoid the risk of reinforcing members being cut during secondary pile construction if the reinforcement is displaced from position.

Types of Secant Pile Walls. Several types of wall construction are in use. These include hard/soft, hard/firm, and hard/hard walls. Selection of wall type is dictated primarily by application and cost.

In hard/soft walls, the primary piles are filled with a relatively weak cement/bentonite or a sand/cement/bentonite mixture. Hard/soft walls offer the advantage of increased productivity in secondary pile construction, and relative economy since the primary piles are of lower strength and excavation into structural concrete is avoided. However, the low strength of the primary pile produces a wall with lower strength and stiffness. The long-term durability of hard/soft walls is also a concern, particularly in permanent applications where the low strength of the primary piles makes them susceptible to degradation upon repeated exposure to wetting/drying and freeze/thaw cycles. Hard/soft walls are therefore typically limited to use in temporary shallow excavations where high bending stresses do not develop. Hard/firm walls have primary piles that are constructed with low-strength concrete. Hard/hard walls have primary piles that are constructed with either unreinforced or reinforced structural concrete and are used for deep excavations and permanent walls where increased bending strength, stiffness, and durability are required.

Equipment. Standard rotary drilling equipment is typically used in hard/soft wall construction. For hard/firm or hard/hard wall construction and in dense soils and soils with obstructions, high torque rotary drilling equipment and crane-mounted or stand-alone casing rotators and oscillators are used. Drilling tools may consist of drilling buckets, soil or rock augers, coring buckets, or down-the-hole hammers.

Both standard and high torque continuous flight augers are also used in secant pile wall construction. Standard auger equipment is limited to use in hard/soft wall construction and soils without obstructions. High torque equipment uses heavy-duty augers that have a stiffened stem to improve vertical alignment in pile construction.

Cased, continuous-flight augers advance a temporary steel casing simultaneous with the penetration of the augers. The casing helps prevent overexcavation of soil from the sides of the borehole due to excessive rotation ("flighting") of the augers that can occur when upper soil strata are relatively loose or soft and the auger advance rate is slowed in the underlying, stiffer soil strata. The casing also increases system stiffness and provides improved verticality in pile construction compared to standard auger equipment. Both high torque and cased auger equipment are used in hard/soft and hard/firm wall construction.

Concrete Mix Design. Concrete mixes must provide a controlled rate of strength gain such that the concrete is soft enough to permit subsequent drilling yet is strong enough to avoid damage to primary piles and reduce vertical deviations during construction of the secondary pile. Admixtures that retard concrete strength are often used in the concrete of the primary piles. A concrete strength of between 300 and 1000 psi (2000 to 7000 kPa) at 2 to 7 days is desired in primary piles in hard/firm wall construction. In secondary piles, the 28-day strength is typically specified as 3000 to 4000 psi (21,000 to 28,000 kPa). Cement/bentonite with a 56-day strength between 150 and 700 psi (1000 to 5000 kPa) has been used in primary piles in hard/soft walls.

Wall Construction Methods. The basic wall installation sequence is as follows:

- Install a guide wall to properly position the secant pile wall (see Fig. 7.9).
- Install temporary casing for primary column construction.
- Drill and concrete primary columns.
- Drill overlapping secondary columns.
- Install reinforcement in secondary holes.
- Concrete the secondary columns.

Guide walls are essential to ensure proper horizontal alignment and minimize initial deviations between piles that could cause problems in overlap as deviations increase with

FIGURE 7.9   Guide walls for secant pile wall construction. (*Photograph courtesy of Moretrench American Corporation*)

wall depth. When high torque equipment is used, the guide walls provide restraint during initial penetration. They must therefore be reinforced and sufficiently embedded in dense soil to provide resistance and maintain alignment.

Drilled Wall Construction.   Track-mounted drill rigs with fixed leads are used to rotate a temporary steel casing into the ground using conventional rotary drilling methods with an internal drill string and bit to remove the soil as the casing is advanced. When drilling in soils below the groundwater table, a positive head of water or slurry is maintained in the casing to enhance stability. Upon reaching the necessary depth of the wall, concrete is pumped using tremie methods as the casing is simultaneously withdrawn. For secondary piles, reinforcement is inserted inside the casing prior to concreting.

Augered Wall Construction.   In this method, the soil is loosened by the auger tip and conveyed to the surface by the auger flights, with the borehole wall supported by the auger filled with drill spoil. Upon reaching the final depth of the wall, concrete is pumped through the hollow stem of the auger as the auger is withdrawn. Reinforcing cages are installed in the fluid concrete following casing withdrawal. This can be problematic, especially in granular soils above the groundwater table where water loss can cause the concrete to stiffen. The depth of the wall constructed with augers is therefore limited by the length of reinforcement cage that can be installed through the fluid concrete.

Quality-Control Measures.   Quality-control measures that are critical to the integrity of the completed wall include

Concrete Mix.   Quality control in proportions and concrete mixing is essential. Deviations in the maximum specified strength are just as important as deviations in minimum strength.

Vertical Alignment of Piles.   Vertical alignment should be checked by plumbing the drilling mast or leads and monitoring the auger or casing above ground using optical survey methods to ensure that individual secant piles do not deviate from plumb or fail to achieve continuous overlap.

Deep Soil Mix Walls

Deep soil mixing (DSM) is a process whereby in situ soils are mechanically mixed with cement and/or other cementitious materials to construct panels of overlapping soil-cement columns exhibiting higher strength, lower compressibility, and lower permeability. For excavation support, reinforcement (typically soldier beams) is inserted in completed panels, before the cement grout sets, to resist bending (Fig. 7.10). DSM walls are particularly suited to excavations in urban environments since the low-vibration, in-place mixing process creates only minimal disturbance to surrounding structures.

DSM is viable in a wide range of soils ranging from sands and gravels to clays. Predrilling of the wall alignment has been used to improve the range of applicable ground conditions to glacial tills and even weak rock; however, this significantly increases costs.

Technological advances in the proficiency of the mixing equipment in varied ground have allowed the method to become more economically competitive and more widely accepted in the United States in the last 10 years.

Advantages and Limitations.   Advantages of DSM walls include

- The soil is mixed in place, resulting in reduced material and spoil volume compared to structural slurry (diaphragm) wall and secant pile wall methods.
- Wall construction is faster than other excavation support methods such as structural diaphragm walls, steel sheet piling, soldier beams and lagging, secant pile walls, and micropile walls.
- Equipment and plant produce limited vibration and relatively low noise.

FIGURE 7.10   Deep soil mix wall. (*Photograph courtesy of Schnabel Foundation Co.*)

Limitations of the DSM method include

- There may be difficulty in penetrating and providing uniform mixing of very dense granular deposits, stiff clays or soils containing boulders, or other obstructions.
- Uniform mixing and treatment of soft silts and clays frequently requires re-stroking of soil-cement columns and increased mixing time to ensure uniform treatment.
- A large work area with no overhead restrictions is required to accommodate plant.

Equipment. DSM equipment consists of a drill rig, mixing tools, and mixing plant. DSM is performed using single and multiple shaft mixing tools and soil cutter (hydromill) systems to penetrate and mix grout into the soil to construct overlapping soil-cement columns or panels. The soil cutter techniques are a more recent DSM innovation where cutting wheels rotate around a horizontal axis to mix soils and produce rectangular panels of treated soil rather than the circular columns produced by the vertical rotating single- or multiple shaft mixing tools (see Fig. 7.11).

FIGURE 7.11   Soil mix drill rig. (*Photograph courtesy of Schnabel Foundation Co.*)

The mixing tools, consisting of a cutting head followed by discontinuous auger flights and mixing paddles, are typically mounted on a crawler crane with fixed leads for work on land. The single-shaft equipment provides less mixing action and has limited ability to disaggregate cohesive soil into small sizes and uniformly blend the grout with soil.

Multiple shaft mixing equipment is more efficient in cohesive soils and is generally more widely accepted.

Current, land-based, multiple shaft mixing equipment can penetrate to depths up to about 140 ft (43 m) and can create soil-cement columns ranging from 24 to 60 in (600 to 1500 mm) in diameter, but a range of 24 to 36 in (600 to 900 mm) in diameter is more the norm (Powers et. al., 2007).

Soil-Cement Mix Design.   Grout mix, cement dosage, grout injection ratio, mixing speed, and mixing time (penetration/withdrawal rate) are typically decided by the DSM contractor to cater to the particular tools and equipment being used.

Ordinary Portland cement is the principal grout material. Additives such as bentonite, gypsum, fly ash, slag, and other proprietary admixtures may be combined with the cement to delay setting time of the grout and improve the strength and permeability of the treated soil. Cement is injected into the soil as a slurry, which provides better mixing and more uniform soil treatment. Water/cement (w/c) ratios can range from less than 1 to greater than 2 but are more usually in the range of 1 to 1.5.

The grout injection ratio is the ratio of the total volume of grout injected into the ground to the volume of soil to be treated. It typically ranges from about 20 to 40 percent. Silts and clays require more cement than granular soils and increased mixing energy to disaggregate clay blocks and produce a uniform soil-cement product.

A laboratory testing program to demonstrate the capability of the proposed soil-cement mix to meet specified performance requirements is essential before wall construction begins.

Wall Construction Method.   Laboratory mix preparation and testing is only an index of the parameters that will actually develop in the ground. Trial panels should therefore be prepared, using field samples obtained at various depths, and visually inspected to ensure that mixing is uniform within individual soil strata.

DSM columns can be constructed sequentially or in a primary/secondary pattern. Columns are overlapped for wall continuity. During wall construction, the mixing shafts are rotated into the ground as cement grout is simultaneously delivered to the cutting head via the hollow shaft of the mixing tool. Grout injection must be carefully coordinated with the rate of penetration/withdrawal to provide proper and even distribution of grout. When final depth is reached, it is good practice to raise the mixing shafts 10 ft (3 m) and then reinsert the augers to final depth once more to ensure adequate mixing time at bottom of the wall. As the mixing tools are withdrawn, grout is usually pumped at a reduced rate. Steel H-piles are typically inserted in every other column prior to the initial set of the soil-cement mix. Interlocking steel sheet piling has also been used in lieu of H-piles.

Since a volume of grout typically equal to between 20 and 40 percent of the volume of treated soil is injected into the ground, an equal volume of waste spoil is created at the surface. This spoil must be contained and removed. After initial set, the soil-cement spoil can be handled as a solid waste or used elsewhere on site.

Quality Control.   In addition to preconstruction laboratory testing, test panel installation, and inspection, quality-control measures during the work include

Unconfined compressive strength (UCS) and hydraulic conductivity testing: UCS and hydraulic conductivity testing should be conducted using samples obtained from freshly mixed soil-cement columns to verify mixing uniformity and compliance with performance requirements.

Coring: Coring should be conducted after the soil-cement column has aged for 28 days to verify strength.

Verticality verification: During penetration, verticality of the mixing tools should be verified by inclinometers to ensure continuity of the cutoff wall, with optical survey used to confirm inclinometer accuracy. Verticality is typically specified as 1 percent of wall depth, depending on the wall depth, overlap, and geometry.

Soil mixing parameters: Monitoring and coordination of mixing rotation speeds, mixing tool penetration/withdrawal rates, and grout injection volumes are essential for uniform mixing and grout injection.

Soil Nail Walls

Since its development in Europe in the early 1970s, soil nailing has become a widely accepted method of providing temporary and permanent earth support and slope stabilization on many civil projects in the United States. In the early years, soil nailing was typically performed only on projects where specialty geotechnical contractors offered it as an alternative to other conventional systems. More recently, soil nailing has been specified as the system of choice due to its overall acceptance and effectiveness.

Soil nailing is an economical, top-down construction technique that increases the overall shear strength of unsupported soils in situ through the installation of closely spaced reinforcing bars (nails) into the soil/rock. Typically, a structural concrete facing (shotcrete) is sprayed against the excavated earth face to connect the nails and reduce deterioration and sloughing (see Fig. 7.12). A common misconception is that this structural facing is the major element of the soil nail system. In fact, the nails do the work.

FIGURE 7.12   Applying shotcrete over reinforcing mesh and drainage panels. (*Photograph courtesy of Moretrench American Corporation*)

Soil nails are passive elements and, unlike tieback anchors, are typically not mechanically pretensioned after installation. Rather, they become forced into tension when the soil they are supporting deforms laterally as the depth of the excavation increases. Without soil movement, the nails remain in a passive state. This is a fundamental difference between soil nails and tieback anchors, and one that is commonly not examined closely enough or considered during determination of proper use and monitoring.

The movements required to mobilize the nail forces are very small and generally correspond to the movements that occurs in a braced system (Chassie, 1993). In some cases, specifications require that the soil nails be tensioned and locked off, as a tieback would. This is not correct. Some specifications even require production soil nails to have a free, or unbonded, length. This is also not appropriate. In some cases, a small pretensioned load may be applied to limit the amount of deflection of the soil that is required to mobilize the nail. However, a major posttensioning effort defeats the purpose and effectiveness of the nail.

Construction Methods
Typical Construction Sequencing and Components

Excavate in Lifts.   The execution of soil nailing consists of making a 4- to 6-ft (1.2- to 1.8-m) vertical cut extending for a horizontal length to be stabilized and shotcreted the same day.

After the lift has been excavated to the proper elevation, a level work area, or bench, is constructed in front of the wall to accommodate the drilling equipment. Depending on the type of equipment used, this bench can be as small as 10 ft (3 m) or as large as 40 ft (12 m), although 25 ft (7.6 m) is typical.

Install Nail.   Typically, the next step is to install the soil nails (see Fig. 7.13). It is important to note that the shotcrete can be applied prior to installing the nails if there is a concern with the standup time of the soil and the possibility of sloughing of the soil.

FIGURE 7.13   Drilling soil nails through previously installed shotcrete. (*Photograph courtesy of Moretrench American Corporation*)

In the United States, common practice is to use rotary and rotary/percussion drill rigs for the installation of the nail, although it is quite common to utilize augers in cohesive soils. The holes are either cased or uncased, depending on the type of soil, and are, on average, 4 to 8 in (100 to 200 mm) in diameter.

Once the hole is drilled, the nail is installed and the hole is tremie-grouted with Type I, II, or III Portland cement and water. The water/cement ratio is normally on the order of 0.45 to 0.50. In the United States, it is common to use no. 8 to 11 (no. 25 to 36), grade 60 or 75 (grade 400 or 500) threaded bar as the nail. The bar is centered in the drill hole by means of plastic centralizers spaced every 10 ft (3 m) along the length of the bar. The length of the bar is based on the design, but a common rule of thumb is that the length of the first level of nails is typically 0.6 to 1 times the overall cut height. Soil-to-grout bond values are used in the design to determine the final length of the nail and the nail spacing. These values are verified by performing some verification (preproduction) and proof (production) testing as the wall is constructed.

Place Reinforcing and Drainage. Once the excavation is made and the soil nails are installed, reinforcing material, typically a welded wire mesh, may need to be placed along the face of the excavation to reinforce the concrete facing. It is not uncommon, however, to use reinforcing bars for the length of the wall in lieu of the wire mesh. As an alternative to placing the reinforcing prior to shotcreting, the reinforcing material can actually be added to the ready-mix concrete at the plant. This is known in the industry as fiber-reinforced concrete. Some common types of reinforcing material are steel or synthetic fibers.

When the bearing plate and nut are installed and the nail becomes loaded as the soil deforms, this load is transmitted to the shotcrete facing. Because of this, it is important to ensure that the facing can handle the punching shear induced from the nail load. It may be necessary to strengthen this area directly behind the bearing plate with reinforcing. A typical punching shear detail is shown in Fig. 7.14.

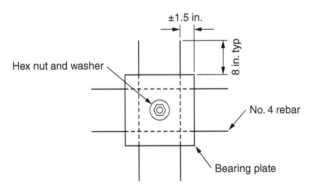

FIGURE 7.14 Typical soil nail punching shear detail.

If drainage behind the soil nail wall is a concern, a prefabricated geotextile drain mat, typically 24 in (0.6 m) wide, can be applied in strips to the excavated earth face of the wall between the nails prior to applying the shotcrete facing (see Fig. 7.12). In most cases, drainage material is not installed for temporary soil nail walls unless drainage is a major concern.

Shotcreting and Installing Bearing Plates. Once the reinforcing and any required drainage medium are placed, the next step is to apply the shotcrete facing. For most temporary shotcrete walls, this is accomplished by applying a 3- to 4-in (75- to 100-mm)-thick layer of 3000 psi (21 MPa) concrete. In a temporary application, shotcreting is merely a method of tying the system together, providing a cover for the soil, and retaining the soil between the nails. The shotcrete can, however, become more critical if the nail spacing increases. Immediately after the shotcrete is applied, the soil nail bearing plates are installed on the fresh shotcrete facing and their nuts are hand-tightened. It is very important to make sure that the bearing plates and nuts are installed as soon as possible, but definitely before the next cut is made since this is what provides the lock-off of the nail as the soil deforms.

Repeat Steps to Final Subgrade.   After the shotcrete has been installed and the soil nails have cured for 3 days, the process is repeated in lifts until the predetermined subgrade elevation is reached (see Fig. 7.13). Typical temporary soil nail details are presented in Fig. 7.15.

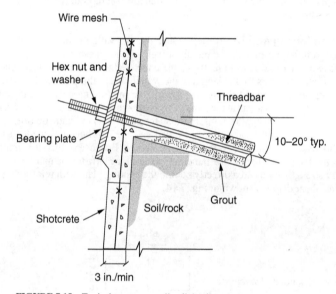

FIGURE 7.15   Typical temporary soil nail details.

Underground Conditions That Are Conducive.   In general, soil nailing can be used in any soil where a vertical cut can remain stable for at least 24 hours. These cuts should be on the order of 4 to 6 ft (1.2 to 1.8 m), which is standard practice for soil nail lifts. Cuts less than this reduce the economical advantage of soil nailing. Normally, soils that have some binder in them are considered the most favorable, such as

1. Natural cohesive materials, such as silts and low-plasticity clays not prone to creep
2. Glacial till
3. Cemented sand with little gravel
4. Fine to medium sand with silt to act as a binder
5. Weathered rock
6. Residual soils

Underground Conditions That Are Not Conducive. The soil nailing system is *not* recommended if any of the following conditions is present:

1. Urban fills and loose, natural fill material
2. Soft, cohesive soils that will not provide a high pullout resistance
3. Highly plastic clays, since they are susceptible to excessive creep
4. Loose, granular soils with N-values lower than 10 to 15
5. Soils without any apparent cohesion and with high gravel content
6. Expansive clays
7. The presence of groundwater, which presents a problem with maintaining face stability during the cut between lifts, and which must be properly controlled in the construction of a soil nail wall

Vertical Wood Sheeting Walls

Vertical wood sheeting is one of the oldest methods of ground support and, under certain conditions, is still the most economical. Most common uses are in excavations for utility or sewer trenches and manhole construction where these excavation support walls are opposite from each other but are close enough to each other to be cross-braced. However, for wide excavations with multilevel cross bracing, vertical wood sheeting may not be possible or practical. Although it is possible to install deep, multilevel, braced, vertical wood sheeting, the excavation process will be slow and the contractor may not be able to efficiently excavate with its preferred equipment. Narrowness of the shored excavation or trench and the multitude and closeness of the cross bracing may dictate that the contractor perform the excavation by hand or by crane with a clam bucket instead of by a high production rate, hydraulic backhoe. This reduced excavation rate is mitigated by the fact that vertical wood sheeting projects are usually not high-volume excavation projects. Installation of vertical wood sheeting is also a relatively low production rate process that cannot support high production excavation equipment.

Vertical wood sheeting is generally 2 to 3 in (5.1 to 7.6 cm) nominal thickness, timber planks (see Fig. 7.16). The bottom or lead edges of the planks are cut at approximately a 45° angle to aid in penetrating the soil. A timber wale and cross-bracing system is most commonly used. However, where larger horizontal spacing is needed between cross braces in a bracing level, steel wales may be used.

The vertical sheeting is usually driven with a light, air-driven hammer or is driven by the excavation equipment. Great care must be used during driving so that the sheets are not damaged. The vertical sheeting should be driven in waves with tips penetrating only 2 to 3 ft (60 to 90 cm) below adjacent excavation levels. As excavation is carried to approximate bottoms of sheets, they are redriven an additional 2 to 3 ft (60 to 90 cm) below the excavation level and the cycle is repeated. At the required depths, the wales and bracing are installed. When groundwater is present, tongue-and-groove sheeting planks can be used to try to minimize infiltration of water and prevent loss of soil.

Where ground conditions permit, excavation can be carried down to the level of the top wale, say 3 to 5 ft (0.9 to 1.5 m), before placing the top, braced wale and then setting up the vertical wood sheeting. The excavation, sheeting, and bracing then proceed as described previously. Because the full earth pressure is carried by the sheeting, the rows of wales and braces are rather close together. This increases the cost of sheeting and excavation for deeper cuts.

FIGURE 7.16 Vertical wood sheeting for pipe trench. (*Photograph courtesy of Peirce Engineering, Inc.*)

Most designs of sheeting and bracing of shallow cuts are empirical, and many codes furnish tables indicating these designs [e.g., OSHA (Occupational Safety and Health Administration) rules and regulations and New York City EPA specifications both give designs indicating sheeting thickness, wale and brace sizes, and spacing for various soil conditions and excavation widths and depths.]

Vertical wood sheeting is often used when excavating for pits or deep footings below subgrade of a larger, general excavation. When excavating small holes, say 6 ft^2 (0.6 m^2), vertical wood permits rapid excavation at minimal cost. Bracing can be relatively light because the soil arches around the holes and full earth pressure does not need to be supported. For larger holes, the arching action is lost and the bracing must be considerably stronger. It may then be advantageous to use lightweight steel sheeting or soldier piles with horizontal lagging. When excavating through soil with a high water table, dewatering is required within the excavation. This dewatering may also lower the water table behind the sheeting. However, if an unbalanced hydrostatic condition exists, there will be the possibility of a "blow" occurring under the sheeting. This possibility should be carefully evaluated before using vertical wood sheeting.

Trench Boxes and Shielding

In moderate-depth trenches, typically in the range of 5 to 20 ft (1.5 to 6.1 m), prefabricated shield and shoring systems are often found to be more effective and efficient than "custom" systems. Prefabricated systems include fixed and adjustable trench shields, trench boxes, and various jacked shoring arrangements. The latter may be used as skeleton sheeting (spot bracing) where there are suitable soil conditions, or may be combined with timber or other sheeting. Typical basic units are illustrated in Fig. 7.17. Trench boxes are

FIGURE 7.17 Stacked trench boxes for deep utility trench support. (*Photograph courtesy of Peirce Engineering, Inc.*)

most frequently used in trenches excavated in soil that stands open when first cut, but that cannot be relied upon to remain stable throughout the time needed to complete the planned work, or that does not meet code requirements for an unbraced excavation. Use of these units allows workers to safely enter a trench, within the trench box, to do the required work with a minimum of preparatory effort. The trench box can often be moved along the trench as the work progresses. Trench boxes can be used as the sole protection device or in combination with sloped excavations. Because trench boxes are usually set into an open, unsupported excavation and are usually moved along the trench in an extended, open cut, they are usually not considered sufficient protection for adjacent structures, roadways, and underground utilities. A trench box protects personnel who are working within the trench box; it does not adequately protect anything outside of the box. If protecting adjacent structures, roadways, and underground utilities is required, a shoring system other than a trench box or shielding system should be used.

There is a variety of proprietary shoring systems on the market that use screw, hydraulic, or pneumatic bracing jacks. Typically, the jacks are used in pairs, attached to two metal rails or channel-type members (see Fig. 7.18). The members may be positioned vertically as spot braces or with timber or plywood to positively retain greater areas of the trench walls. If vertical timber or metal sheeting is used, the members may be used as wales to restrain the sheeting. While these systems are not "portable" as a single unit, they are readily disassembled and reassembled, and hence can be economical.

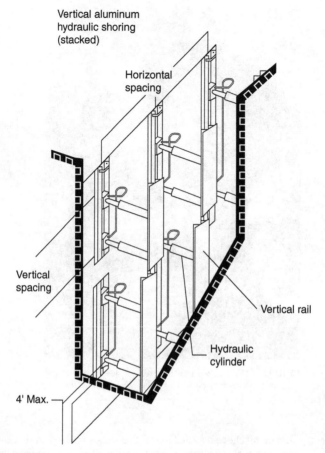

FIGURE 7.18 Hydraulic or pneumatic trench shoring system (OSHA).

The selection of the proper unit for any particular situation has been facilitated by the manufacturers who supply charts or tables that provide guidance regarding the load-carrying capacities of the units or systems and the range of soil conditions for which a particular type or model of unit may be used.

As with trench boxes, these shoring systems are placed into a trench after the excavation has been made. Therefore, until these shoring units are in place, the trench is essentially an unsupported, vertical excavation that is unsafe and possibly illegal for workers to enter. For this reason, these shoring units are placed into the trench by workers standing along the top side of the open, unsupported trench. This operation, in itself, may present an illegal, safety hazard.

Diaphragm Walls

The term *slurry wall* can mean any trench installed by the slurry method. However, for the purpose of this chapter, we will use the term *diaphragm wall* to describe a reinforced concrete wall constructed by excavation utilizing the bentonite slurry method and backfilling with tremie concrete (see Fig. 7.19*a*). A diaphragm wall is an earth-retaining structure. Although we are discussing temporary structures, the true economy of this type of construction is its use as a part of the permanent structure.

The general procedure for constructing a slurry wall is as follows:

1. Construct a "pre-trench." This consists of building two guide walls, generally of reinforced concrete. The spacing between the guide walls is about 1 in (2.5 cm) greater than the width of the diaphragm wall. The pre-trench is constructed for the entire alignment of the wall and helps guide the excavating equipment.

2. Fill the pre-trench with bentonite slurry. It is important that the level of the slurry be kept at approximately the original ground level and at least 2 ft (60 cm) above the groundwater level.

3. Commence excavation (usually with a clamshell bucket), continually filling the trench with bentonite slurry to replace the excavated material (see Fig. 7.19*b*). The length of wall dug at one time varies up to a maximum of 25 to 30 ft (7.6 to 9.1 m). Each section is called a panel. Several panels may be excavated at the same time as long as the space between panels does not permit bentonite to flow from one panel into another.

4. After the panel is excavated to the proper depth, the trench is cleaned of excess material and an "end pipe" or steel beam is placed at each end of the panel to form a joint.

5. A prefabricated, reinforcing steel cage is placed into the excavated, slurry-filled trench.

6. Concrete is placed through tremie pipes, displacing the bentonite slurry. If end pipes are used, the pipes are withdrawn as the concrete receives initial set. If steel beams are used at the panel ends, they usually remain in place.

7. Continue with steps 2 through 6 until the diaphragm wall is complete. If end pipes are used, the adjacent panels require only one end pipe since the joint has already been formed at one end. When hard soil or rock is encountered; chopping bits, churn drills, or hydromills may be required to remove this material.

The design and construction of slurry-trenched diaphragm walls are based on some theory and a lot of empirical relationships and experience. For our purpose, it should suffice to know that slurry-trenched diaphragm walls do work and that they are an excellent, but very expensive, tool in foundation construction.

Some things to consider when constructing diaphragm walls are the following:

1. When excavating a trench through fill or material with large voids, the bentonite slurry can run out through the voids and lead to a possible collapse of the trench. When this type of material is encountered, it may be necessary to alter the soil by compaction, filling, grouting, etc. to permit completion of the diaphragm wall.

2. If obstructions, boulders, etc. are removed during the trench excavation and these items extend outside the theoretical face of the wall, the void created during excavation will be filled with tremie concrete. If the excess concrete is on the inside face, it may be necessary to trim the wall as it is exposed during structure excavation.

3. Setting up a slurry plant is expensive, making the cost of the diaphragm wall rather expensive. For this reason, diaphragm walls are predominately used on larger projects with difficult ground conditions, high groundwater table, or adjacent structures.

The thickness of diaphragm walls is generally in the range of 2 to 3 ft (60 to 90 cm). This makes a rather stiff wall when compared to steel sheeting or soldier piles. This permits larger cantilevers before installing top tiers of bracing and greater vertical spans between

(a)

(b)

FIGURE 7.19   (a) Completed, tiedback diaphragm wall; (b) clamshell bucket. (*Photographs courtesy of Mueser Rutledge Consulting Engineers*)

bracing levels. Diaphragm walls also are beneficial when, in addition to earth retention, there is a need to provide cutoff of groundwater in difficult soil conditions where installation of steel sheet piling is not possible or practical. New techniques and construction equipment are constantly being developed and should be considered for use on future projects.

RETAINING WALL BRACING SYSTEMS

The purpose of a bracing system is to provide support for and prevent movement of the retaining elements that are in direct contact with the retained soil. In trenches, the bracing usually consists of cross-trench struts or jacks. Wider excavations typically utilize struts or tieback anchors that are commonly installed in horizontal rows or parallel to the ground surface.

Deciding whether to use internal braces or tieback anchors depends on several factors. Some soils are not suitable for using earth tieback anchors. If rock is at exceedingly great depth, rock anchors are not economical. High groundwater and/or underground utilities behind an earth-retaining structure may preclude the use of either earth or rock tieback anchors. The geometry of the proposed excavation (depth of excavation in relation to its width) determines if cross-lot bracing or inclined, raker braces bearing on footings or "heel blocks" should be used. If temporary penetrations through a proposed structure are not allowed, internal braces may not be an option. In cases where either tieback anchors or internal braces may be used, the determining factors should be the cost and time of installation and removal. In general, the cost of a tieback installation is greater than that of an internal bracing system. However, this initial cost is usually offset by savings in the cost of excavation, construction and backfilling of the perimeter walls, cutoff and removal of the internal bracing, and final patching and waterproofing of the penetrated walls. Note that tieback anchors extend outside the excavation. If the tieback anchors extend beyond the property line, special permits or underground easements may be required. Building codes for a few cities in the United States also require removal of temporary tieback anchors as the new structure is backfilled.

Another important aspect of selecting and implementing a suitable bracing system is consideration of the possible consequences to structures or utilities outside the excavation if deflections of the retaining structure become excessive. Most earth-retaining structures are defined as flexible structures subject to some degree of deflection. Deflection may lead to settlement behind the retaining structure. A major factor in controlling deflections is the degree to which a bracing system is prestressed when it is installed. One important function of prestressing is that it tends to close up any spaces between braces, wales, sheeting, and soil, which could otherwise result in significant movements during or after construction. The main function of prestressing, however, is to stress the retaining system to the level that would be expected after excavation, so that there is very little deflection remaining to occur. Proper prestressing of the bracing system can greatly reduce vertical and lateral deflections outside an excavation. Prestressing of an internal bracing system involves jacking or wedging a preload into the bracing system and then securing that preload into the system with welded shim plates or wedges. Tieback anchors effectively prestress the retaining system when the anchors are tested and locked off against the retaining structure to some percentage of the anchor design load.

For situations where deflections are particularly critical or where the adequacy of the bracing system is in question, deflections should be monitored with strain gauges, slope indicators, survey monitoring, or some other instrumentation. Continuing significant movements are a warning that corrective measures need to be taken.

Wales

Bracing systems using tieback anchors or internal braces usually employ continuous or intermittent, horizontal members called *wales*. The wales are in contact with the earth-retaining structure and transfer the loads from the retaining structure to the tiebacks or braces. The main purpose of the wales is to permit the tiebacks or braces to be placed far enough apart that they can be designed for high loads, minimize interference with the construction operations, and permit work to be performed economically.

When designing with wales, the span between points of support should be governed by the deflection of the wale as well as by the difference in cost of heavier wales as compared to additional tiebacks or braces.

Prior to the early 1970s, most deep excavations employed internal steel bracing systems. Braced wales are generally HP or WF sections. When using tieback anchors to brace a retaining structure, the wales are usually double channel sections or, less frequently, single HP or WF sections or double WF sections. When channels are used for tieback anchor wales, the channels are paired together with their webs parallel to each other and spaced apart sufficiently to allow the tieback anchor tendon, and sometimes the tieback anchor drill bit or casing, to pass freely between the channel webs. For braced wales, it is good practice to choose as a wale a member that has a minimum 8 in (20 cm) wide flange. This minimum flange width provides a large bearing area for connection to the raker or cross braces. Great care must be given to details of the connections between wales and tiebacks or braces. Most failures in bracing systems occur at connections that are not properly designed or constructed. Welded steel wale stiffeners should be installed where required to prevent crippling of the wale web and to resist wale torsion. Wale brackets, commonly referred to as *roll chocks*, should be placed to resist the vertical component of inclined tieback anchors and raker braces. Generally, tieback anchors are installed with the strong axis of their wales oriented parallel with the tieback tendons. Therefore, roll chocks are not needed, but wale brackets are needed to properly align the tiebacks and to connect the tieback wales to the retaining wall's steel sheet piling or soldier piles. Refer to Fig. 7.20 for typical wale details.

Tiepoint/Thru-Tie Connections

Another common excavation bracing technique is a system in which the individual HP or WF soldier piles in a pile and lagging wall are each retained by one or more tieback anchors without using wales. These tieback anchor connections are commonly referred to as *tiepoints* or *thru-tie connections* and are illustrated in Fig. 7.21. The advantages compared to a wale-braced system are the smaller clearance required between the temporary retaining wall and the proposed permanent structure, fewer obstructions in the work space, and less steel used in the bracing. Perhaps the biggest advantage with a tiepoint or thru-tie connection system is that the proposed permanent structure can be constructed immediately against the earth-retaining structure. This "wall line" excavation support method reduces the required amount of structure excavation, eliminates backfilling of the foundation wall, eliminates the need to remove tieback anchor wales when backfilling, and often allows the permanent structure to be located closer to the property line. One possible disadvantage is that a larger number of tieback anchors may be required, although in areas where high-capacity tieback anchors cannot be installed because of ground conditions, the cost differential may be small.

Tiepoint or thru-tie connections, however, are somewhat sensitive to proper installation of their components, as each soldier pile is essentially an independent member. The tieback connection is positioned either off-center of the HP or WF soldier pile and next to the pile web, or is installed through the pile web. This eccentric tiepoint connection

(a)

(b)

FIGURE 7.20   (a) Braced wales with roll chocks; (b) double channel tieback wales. (*Photographs courtesy of Moretrench American Corporation*)

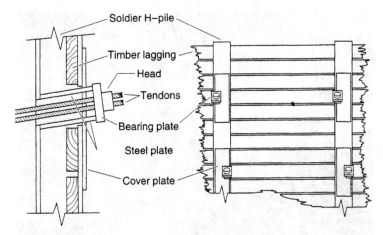

FIGURE 7.21 Tiepoint/thru-tie connection to soldier pile.

induces twisting stresses in the soldier pile. Therefore, multiple tieback anchors in a single pile must be installed on alternating sides of the pile in order to prevent pile rotation. The concentric thru-tie connection does not induce twisting. However, both the tiepoint and thru-tie connections require cutting and removing a portion of the soldier pile at each tieback anchor and, therefore, significantly reduce the strength of the soldier pile. The amount of strength reduction is a function of the size of the tieback anchor drill hole or casing pipe that needs to pass through the pile. For this reason, installation of these connections usually includes a fabricated replacement for some or all of the pile's lost strength through the addition of welded flange plates, web plates, or a pipe sleeve. Soldier piles that rotate or wander when driven require careful evaluation to decide how to connect the tiebacks and whether added structural stiffening is needed.

When it is not possible or desirable to install tieback anchors through an HP or WF soldier pile or when there is concern about drivability of the pile, it is usually preferable to use a drilled-in-place soldier pile fabricated from a pair of steel channels or WF sections. These members are paired with a space between them that allows the tieback anchors to pass concentrically between the members without the need to reduce their combined strength. Each tieback anchor is then connected to the front flanges of the double-member pile with a welded, properly angled, seat connection often consisting of a heavy steel angle or a pair of parallel wedge plates.

Internal Bracing

Internal braces are usually HP, WF, HSS (tube), or pipe sections. For high loads or long braces, pipe sections offer the advantage of requiring less lacing (intermediate bracing) than the other sections, but the higher cost of the pipe material may exceed the savings involved. When installing cross-lot braces (see Fig. 7.2), one end is usually welded to the wale and the other end is often stressed with plates and wedges when it is important to minimize wall movement. When using inclined braces (see Fig. 7.22), the end bearing against the wale is usually welded to the wale and the other end is either cast into a temporary footing called a *heel block* (for light brace loads) or the load is transferred to the heel block with

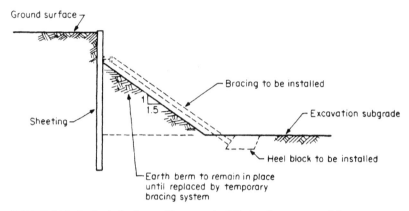

FIGURE 7.22   Inclined raker brace with concrete heel block and temporary earth berm.

FIGURE 7.23   Inclined raker braces with roll chocks and offset wales for wall line sheeting. (Concrete foundation has already been poured against the sheeting wall.) (*Photograph courtesy of Moretrench American Corporation*)

bearing plates and steel wedges (see Fig. 7.23). If deflections of the sheeting system are to be kept to a minimum, the bracing system can be prestressed by means of jacks. However, wedging both cross-lot and inclined braces can also impart significant preload to the braces and is more commonly done than jacking.

Long, internal braces should be designed for combined stresses due to the axial load from the lateral earth pressures and all expected surcharges and from the bending moment due to the brace dead load. It is also important, but too frequently overlooked, to design the bracing for any superimposed surcharges, such as accumulated excavation spoils along the braces and wales and any loads from any personnel, materials, equipment, or utilities that may be supported on the bracing system. No surcharge loads should be supported on the bracing system unless the system and its connections have been designed for those specific loads.

Tieback Anchor Systems

Tieback anchors are structural elements that act in tension and receive their support in earth or rock. The tieback anchor system consists of the earth or rock that provides the ultimate support for the system; a tension member (tendon) that transfers the load from the earth-retaining structure to stable earth or rock behind the retaining structure; a transfer agent, such as cement grout or a bearing plate, that transfers the load from the tendon to the soil or rock; and a stressing unit, such as a bearing plate with an anchor nut or wedges that engage the tendon, permit the tendon to be stressed, and allow the load to be maintained in the tendon (see Fig. 7.24).

Earth anchors are usually installed at an angle of 15° to 45° below horizontal. This angle aids in placement of the anchor and grout, allows the anchors to pass beneath underground utilities, and provides sufficient overburden for developing the anchor's required tension capacity. If acceptable soil is not encountered at a lower-degree angle, it is necessary to increase the angle to engage the proper soil stratum. A relatively flat angle has the advantage of not introducing a large, vertical load into the earth-retaining structure that happens when placing a rock anchor at an angle of 45°. At this angle, the vertical load introduced at the bottom of the earth-retaining structure is equal to the lateral load being supported.

In situations where excavations extend through soil into rock, with tieback anchors restraining the sheeting wall and the excavation extending below the toe of the vertical members, great care must be exercised in protecting and bracing the excavated rock face under the vertical members (see Fig. 7.25). Good construction practice would be to keep the vertical members a safe distance behind the face of the rock excavation and to line-drill the rock face. Line drilling often consists of drilling 2- to 4-in (5- to 10-cm)-diameter, vertical holes at approximately 6 to 12 in (15 to 30 cm) on center along the edge of the proposed rock cut below the earth-retaining structure. The setback, or rock bench, used is typically 3 to 5 ft (91 to 152 cm) in front of the retaining structure but could be less in hard, competent rock; the decision should be made by someone experienced in rock mechanics. Excavation of the rock should be made in small lifts, with the excavated rock face being bolted or pinned, as required, as each lift is blasted and removed.

Tieback Anchor Tendons

Tieback anchor tendons generally are medium- to high-strength continuously threaded bars [75 or 150 ksi (517 or 1034 MPa)] or multiple, high-strength strands [270 ksi (1862 MPa)]. This permits the use of a high-capacity tension member in a small-diameter drill hole. Either multistrand tendons or threaded bar tendons may be used. Because tieback anchors are tension members, the tendons should not be subjected to bending stresses. Therefore, if tendons are to be used in a situation where settlement of the ground around the tendons could cause bending of the tendons, appropriate measures should be taken to prevent tendon bending.

Suppliers of tieback anchor materials furnish special anchor heads or wedging assemblies to connect the anchor tendons to the retaining structure. These assemblies fall into two categories: wedge connections and threaded connections. The wedge connection is used with strand tendons while the threaded connection is used with threaded bar tendons.

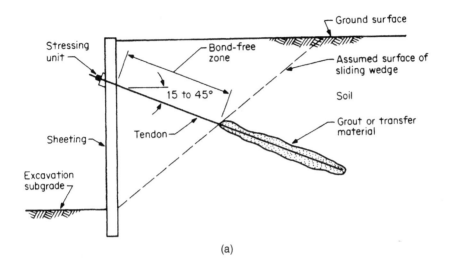

(a)

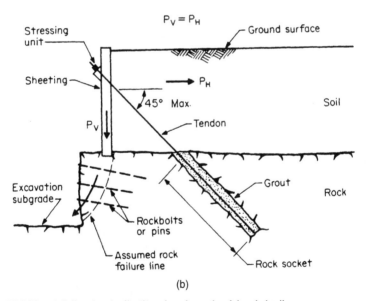

(b)

FIGURE 7.24 (a) Soil anchor details; (b) rock anchor and rock bench details.

The maximum working or service load of a tieback anchor is 60 percent of the tendon's guaranteed ultimate strength (GUTS) for high-strength threaded bar tendons [150 ksi (1034 MPa)] and strand tendons [270 ksi (1862 MPa)]. For lower- to medium-strength threaded bar tendons where the yield stress is either 60 ksi (414 MPa) or 75 ksi (517 MPa), the maximum working or service load is 60 percent of the tendon's yield strength (Fy). All tieback anchors should be test-loaded to a percentage greater than their service design load.

FIGURE 7.25   Line-drilled, rock-bolted rock bench below tiedback wall. (*Photograph courtesy of Peirce Engineering, Inc.*)

This test load is usually between 120 and 133 percent of the design load for proof tests and is usually 133 percent of the design load for performance tests. Tieback anchors are then secured, or locked off, to the retaining structure at approximately 75 to 100 percent of their design load. A hollow-ram, hydraulic jack with calibrated pressure gauges permits the tendons to be stressed to a predetermined test load and then be locked off at the desired load. The contractor should be advised that jacks for higher capacity tieback anchors are extremely heavy and require equipment to handle and place them. Some tieback anchor tendons, especially permanent, corrosion protected tiedback anchors, require off-site fabrication of the tendons. This can cause delay in placing the tendons if the tendon length is subject to field conditions. If the drill holes or rock sockets are unstable, the time lag could be critical and the drill hole may collapse while waiting for the tendon. It may be to the contractor's advantage to use a system that is flexible and permits timely fabrication or adjustment of the tendons at the jobsite. This will permit placing and grouting the tendon as soon as the drilling of the hole is complete.

Laboratory tests and field experience indicate that when using strand or threaded bar tendons, the bond between the tendons and the grout is not as critical as that between the grout and the soil or rock, except possibly for rock anchors with short, but larger-diameter, bonded lengths.

Drilling for Tieback Anchors

The types of equipment and techniques used to install tieback anchors are usually determined by the type of soil and/or rock to be drilled, the depth to the supporting stratum, and whether groundwater will be encountered. When tieback anchors first came into use, they were installed in rock, and eventually in soil, using percussion drills and, when necessary, driven temporary casing pipe to prevent the drill holes from collapsing. Although not very common today, rock drills or air-tracks were often used to install both soil and rock tieback anchors. This equipment permits installation of a continuous casing, where necessary, which can be withdrawn after the tendon has been grouted in place. Ordinary drills of this type can install casings up to 4.5 in (11.4 cm) in diameter, while larger ones can handle casings up to 8 in (20 cm) in diameter. When used to install soil anchors, the casing can be driven with a point or plug in the bottom, eliminating the necessity for cleaning the pipe. When withdrawing the casing, the plug can be knocked out or the point left in the ground. When used to place rock ties, the usual procedure is to drive the casing to the top of rock, attempting to seat the casing in the rock. Seating the casing in rock is quite important to creating a seal that prevents soil from entering the hole as the rock socket is being drilled. After the rock socket is drilled, it is flushed and cleaned and the tendon is placed and grouted. It is common to have large quantities of groundwater entering the drill holes through seams in the rock. Should this occur, the holes are filled with water and the grout placed by tremie methods. This type of drilling is less effective when it is necessary to drill through obstructions or boulders and back into soil again. Pressure-grouting fractured bedrock and then redrilling the hole are sometimes performed to control inflow of water prior to insertion and final grouting of the tendon.

During the early 1980s, if tieback anchors were to be installed in a cohesive soil, a large-diameter, uncased shaft could be drilled with a truck- or crane-mounted drill. The bottom of the hole was then sometimes belled out, the tendon and concrete placed, and the tieback was stressed. More often, the bell was eliminated and the length of shaft was increased to develop the load in friction rather than by bearing. Starting in the later 1980s, it became more common to install tieback anchors in both soil and rock using smaller drilling equipment such as track-mounted, rotary drills and eventually rotary-percussive drills (see Fig. 7.26).

FIGURE 7.26 Rotary-percussive tieback drill, casing pipes, and grout mixer. (*Photograph courtesy of Moretrench American Corporation*)

These drills usually drilled smaller-diameter holes into which the tieback tendons would be placed and be grouted. For cohesive soils, regroutable tieback tendons were then developed for use in smaller-diameter drill holes. The increased capacity resulting from regrouting an anchor more than compensated for the smaller drill hole diameter. During the course of drilling, if layers of soil that will not remain open without being supported are encountered, temporary casings are used to keep the holes open. This protection through the unstable layer increases the anchor cost but still permits the use of drilling equipment that is locally available. Use of rotary and rotary-percussive drills makes it possible to drill in various types of soil and rock, while casing the drill holes as needed. Today, rotary and rotary-percussive drills are the most often used drilling equipment for tieback anchor contractors.

Another common method used to install ties through unstable soil is to drill with a continuous flight, hollow-stem auger. This equipment permits rapid drilling and, with the hollow pipe stem, allows the tendon to be placed and the grout installed through a completely protected hole as the auger is being withdrawn. This method loses its advantage if boulders or very hard layers of soil are encountered. The drilling equipment used is very large, so it is not adaptable for operations in limited space. Holes can also be drilled and maintained open using drilling mud or slurry. The concrete or grout can then be placed by tremie methods, which are acceptable and provide the desired results.

When installing tieback tendons into the drill holes, centering devices called *centralizers* are installed along the tendons at approximately 10 ft (3.0 m) on center. These keep the tendon approximately in the center of the hole, permitting complete encasement of the tendon with the concrete or grout.

Grouted Tieback Anchors

The most common materials used to transfer the loads from the tendon to the soil or rock are concrete, grout, and epoxy. Concrete is most often used where large-diameter holes have been drilled, say 12 to 18 in (30 to 45 cm) in diameter. Where small-diameter holes are drilled, grout or epoxy is used. For small-diameter holes in rock, most contractors use neat cement grout, use no additives or expanding agents, and use a grout with low water-cement ratio. Although lower-speed paddle mixers are commonly and satisfactorily used to mix grout, faster mixing and better mix results are obtained when using a high-speed, colloidal or shear mixer that ensures a thorough wetting of the cement. These high-speed mixers usually provide a discharge pressure that is normally sufficient to place the grout without the use of another grout pump. A grout pump is used to tremie grout the drill hole from the bottom up to the face of the retaining wall. If pressure grouting is required or if the grout must travel a long distance from place of mixing to point of placement, a higher-pressure grout pump should be used.

Pressure grouting introduces larger quantities of grout into the hole and, when used with earth anchors, will probably permit higher anchor capacities than if pressure grouting were not used. This technique requires special equipment and procedures, such as the use of packers that seal the hole and help maintain the pressure. More often, earth anchors are pressure-grouted by using re-grout tubes and then post-grouting the anchor's bond length. A re-grout tube is a small-diameter tube that has holes or grout ports located along the anchor bond length at approximately 5 ft (1.5 m) on center. These grout ports are covered with a tight-fitting, rubber sleeve. When high-pressure grout is pumped through the re-grout tube, the sleeved port opens and the high-pressure grout cracks the initial tremie grout, consolidates the soil in the bond length, and increases the bond of the grout to the soil. After re-grouting an anchor, the re-grout tube is washed out so that the anchor can be re-grouted again if necessary. After the final re-grout, the re-grout tube is left in place alongside the tieback tendon. As an alternative to using packers or re-grout tubes, pressure grouting can be done by attaching the grout hose to the top of the temporary drill casing immediately with a pressure cap after the tendon has been inserted and tremie grouted. As the casing pipe is being withdrawn, grout is pumped through the casing pipe with high pressure. High-pressure grouting can be done when anchors are being installed in granular soil. Anchors in cohesive soils cannot be pressure-grouted but can be post-grouted. Rock anchors are pressure-grouted only when necessary to seal rock fractures and control loss of grout due to groundwater flow but not to increase effective bond diameter. Capacities of the earth anchors installed with pressure grouting have reached several hundred kips with great consistency and reliability. Grout pressure should be measured in the grout hose near the tieback anchor or at the casing pipe. Measuring grout pressure at the grout pump may not be a true indication of the grout pressure down in the drill hole. Grouting is considered to be high-pressure grouting when the grout is pumped at a minimum of 50 psi (345 kPa) and, preferably, at least 150 psi (1034 kPa).

The cost of pressure-grouted or post-grouted grouted anchors is higher when compared with that of tremie-grouted anchors. However, this increased cost may be offset by the ability to use fewer tieback anchors having higher capacities.

Epoxy grouts come in the form of cartridges (sausages) that are placed in the holes after drilling has been completed. The steel rod, which is to be used as the tendon, is then connected to the drill rig, inserted into the hole, and turned with the drill rig. This procedure penetrates the cartridge casing and mixes the components of the epoxy. The advantage of epoxy grout is that it sets up faster than any other type of grout, permitting testing of the anchors in a relatively short time, and that it can be used successfully under water.

Possible problems can occur in weathered or decomposed rock that fractures and breaks up during drilling and cleaning (blowing) operations. Poor-quality rock may not bond properly with the epoxy. If the drill holes become enlarged during drilling, the cartridges have room to move and spin without the casing breaking. Under these conditions, the cartridge manufacturers recommend grouting the holes, then redrilling through the grout and using the epoxy. Here it is probably to the contractor's advantage to use the neat cement grout as the transfer agent and use epoxy grout in more competent rock.

All tieback anchors should have an "unbonded" length, an area in which no transfer of load between the soil and anchor is desired during the stressing operations. The length of this bond-free zone varies according to the design criteria. However, the methods used to establish this zone are independent of the type of tendon or method of installation. For temporary applications, threaded bar tendons have an unbonded length that is usually sheathed in a smooth, plastic tube. Temporary strand tendons usually have an unbounded length that is coated with corrosion resistant grease and sheathed in an extruded, smooth, plastic tube. The drill hole along the unbounded length is usually grouted along with the anchor's bonded length (Fig. 7.24). A tendon's smooth, plastic sheath prevents the unbounded length from bonding to the soil or rock.

There are many hypotheses that have been advanced for tieback anchor design, some primarily theoretical, some empirical; all require the application of judgment factors. Experience has shown that the only real proof of anchor capacity is by full-scale field tests.

It is standard practice to test every anchor to a load greater than the design load before incorporating it into the bracing system (see Fig. 7.27). For all tieback anchors, an analysis should be made of the creep characteristics of the soil or rock in the anchor's bonded length to ensure that significant movements will not occur during the time when the excavation is to remain open.

Mechanical Tieback Anchors

Several types of mechanical anchors for soil are commercially available, such as helical anchors (Fig. 7.28) and toggle-plate anchors. The main advantages of these types of

FIGURE 7.27   Tieback anchor testing setup. (*Photograph courtesy of Peirce Engineering, Inc.*)

FIGURE 7.28   Helical anchor tendons with multiple helices. (*Photograph courtesy of Peirce Engineering, Inc.*)

anchors over grouted anchors are that mechanical anchors can often be installed quickly with smaller equipment, usually do not require removal of soil, and can be tested immediately after installation. The disadvantages of mechanical anchors are that they may be difficult to install in dense or rocky soils, cannot easily penetrate underground obstructions, may sometimes require predrilling, and often require larger-diameter penetrations through the retaining structure. Most importantly, in our experience with these anchors, we have not been successful in obtaining working loads greater than 25 to 30 tons with great consistency or economy.

Berms

The vertical spacing between tiers of bracing is critical to the stability of the entire retention system. As excavation proceeds, the earth just below an excavation level provides the lateral support until that bracing level is installed. If the earth below the brace level is excavated on a slope with the toe of the slope at a level lower than the brace level, the wedge of earth that remains to support the sheeting is called a *berm* (see Fig. 7.22).

An analysis should be made of the entire system at each stage of excavation. The stability of the berm at each bracing level, before the braces are installed, may be critical and should be checked. The toe, or embedded portion of the earth retaining system below the excavation subgrade, should be checked at completion of excavation and bracing installation to verify stability of the toe. The slope of the berm should be analyzed to ensure its stability. In good soils slopes of 1 vertical to 1.5 horizontal, or flatter, are recommended. Poor soils might require slopes of 1:2 or 1:2.5. To maintain excavation stability under adverse conditions, it may also be necessary to install the sloped braces one unit at a time, excavating trenches for brace installation before the general excavation. It is important to consider the effect of groundwater when checking the stability of both toe embedment and berms. Buoyant weight of soils due to high groundwater can greatly and adversely affect the design of toes and berms.

Tiedback systems usually do not have the same problem with berm stability as internal brace systems. Tiebacks anchors are generally installed with the excavation being performed in reasonably horizontal lifts, eliminating the need for sloped berms. A lift is usually located approximately 1 to 2 ft (30 to 60 cm) below the tieback anchor level and temporarily supports the retaining wall until that level's tieback anchors are installed. The lift also serves as a stable work bench for the tieback anchor drilling equipment. The width of a lift, perpendicular to the wall, is usually between 20 and 60 ft (6 to 18 m), depending on the size of the drilling equipment.

FREEZING AND GROUTING

Ground freezing and grouting are not typically the first choices for straightforward earth support applications. However, there are circumstances in which these geotechnical methods can offer certain advantages. Temporary ground freezing is highly effective in controlling groundwater as well as providing a stabilized soil face that can be excavated without additional shoring and can be accomplished in virtually all soils, including those containing cobbles, boulders, and other obstructions. Jet grouting creates a homogeneous, low-permeability mass that can fulfill both an excavation support and underpinning function. Both methods will impart a high compressive strength, improved ground product. Frozen walls and jet grout walls can both be installed to significant depths, although ground freezing is generally more suitable at depths of 100 ft (30 m) or more. When the earth support is required to address both water cutoff and earth support, both freezing and jet grouting can be attractive options.

Ground Freezing

In the ground freezing process, in situ pore water is converted into ice through the circulation of a chilled liquid (typically brine) through a system of closed-end, small-diameter pipes (see Fig. 7.29). The frozen water bonds the soil or rock particles together, creating a frozen mass of significantly improved compressive strength and impermeability. Long-term creep strengths will vary between 200 and 600 psi (1.4 and 4.1 MPa) for fine-grained and granular soils, respectively. Significantly higher short-term strengths are achievable.

FIGURE 7.29 Freeze pit and refrigeration system. (*Photograph courtesy of Moretrench American Corporation*)

Brine ground freezing is primarily used to provide a perimeter wall of stabilized earth for deep shaft excavation and is also used for other tunnel-related excavations such as launch and retrieval blocks and connector tunnels. Because ground freezing is achieved by thermal process rather than displacement of the ground, this method is particularly applicable to difficult, disturbed, or sensitive ground where other methods of ground improvement are problematic (Schmall, 2006). Where a shallow excavation is required to remain open for a short period of time, for a small underground connection or structural tie-in for example, rapid ground freezing with liquid nitrogen has been used.

Advantages and Limitations. Ground freezing offers a number of advantages for earth support:

- Both excavation support and groundwater cutoff are provided in one operation.
- Freezing can be accomplished in virtually any subsurface condition.
- Ground freezing does not displace, modify, or disturb the greater soil mass, requiring only minimal penetration for pipe installation.
- For deep shaft construction, the freeze can be implemented perfectly through the difficult geology of the soil/rock interface.
- The ground typically returns to original condition after freezing is discontinued.
- Ground freezing installation is virtually vibration-free.
- The work can be performed with mobile refrigeration units.

Drawbacks to ground freezing include:

- The time for a brine freeze to achieve closure in the target soils so that excavation can begin is typically measured in weeks or months. This method is therefore only suitable for larger, long-term projects.
- Specialized equipment, such as roadheaders, is often required for excavation of frozen ground where it may intrude into the excavation.
- The method is sensitive to moving groundwater.
- Given the higher costs associated with liquid nitrogen consumption, this is typically limited to smaller, short-term projects or emergency situations where achieving a rapid freeze is critical.

Freezing Method. For brine freezing, the portable refrigeration plant consists of a compressor, condenser, chiller, and cooling tower. Plants range in capacity from 60 to 150+ tons (200 to 525 kW) of refrigeration. The brine is cooled to temperatures typically between 5°F and −13°F (−15°C and −25°C), pumps down a drop tube to the bottom of the freeze pipe, and flows up the annulus, withdrawing heat from the soil. Typically, the freeze pipes are hooked up in series parallel. The brine is returned to the refrigeration plant where ammonia or other refrigerant, contained within the freeze plant itself, is used to chill the recirculated brine.

For nitrogen freezes, the liquid nitrogen is stored in an insulated pressure vessel that is periodically refilled from special over-the-road tank trucks. During the freezing process, withdrawal of heat from the soils results in boiling of the liquid. Exhaust vapor is vented to the atmosphere.

Quality Control. Essential quality-control elements for brine freezes include

- Surveying of individual freeze pipes to confirm freeze pipe alignment
- Instrumentation to confirm adequate freeze propagation
- Installation of piezometers to measure groundwater gradients
- Monitoring of brine temperatures

Additionally, for liquid nitrogen freezing in confined spaces, care must be taken to ensure that exhaust vapor is piped to a safe disposal point. The extremely low temperatures reached during liquid nitrogen freezing also require special freeze pipes, fittings, and thermal insulation.

Jet Grouting

Jet grouting is a specialized technique whereby high-pressure, high-velocity jets hydraulically erode, mix, and partially replace the in situ soil or weak rock with cementitious grout slurry to create an engineered soil-cement product of high strength and low permeability. This product is sometimes referred to as *soilcrete*.

Jet grouting has a number of construction-related applications, including structural underpinning, excavation support, groundwater control or cutoff, utility support, temporary or permanent soft soil improvement, slope stabilization, and hazardous waste containment. However, in the United States, underpinning and excavation support are the most common applications.

Jet grouting is typically performed vertically or inclined and above or below the water table. The three basic systems in general use are single-fluid (grout only), double-fluid (grout/air), and triple-fluid (grout/air/water). Selection of the most appropriate system is dependent on the in situ soil characteristics and the application. A number of different geometries can be achieved by the jet grouting process including full and half columns, wedge shapes, discs, and panels. For excavation support and structural underpinning, full columns are typically used.

Advantages and Limitations.   The jet grouting system offers several advantages over other excavation support methods:

- Both underpinning and excavation support can be provided in one operation.
- Jet grouting can facilitate ground improvement where direct vertical access to the target soils is limited by underground or surface obstructions.
- The work can be accomplished in most subsurface stratigraphies from cohesionless soils to clays.
- Low-headroom, maneuverable equipment is available, allowing the work to be accomplished under restricted access conditions.

However, there are potential drawbacks that must be considered:

- The specialized equipment, material storage, and handling typically necessitate a large staging area.
- Obstructions such as cobbles, boulders, and timber piles may result in reduced penetration of the jets, resulting in incomplete soil-cement geometries.
- A considerable amount of spoil is generated, particularly in the double- and triple-fluid processes, which must either be contained on site if space permits or removed by vacuum trucks for off-site disposal.

Equipment.   This highly specialized technique demands sophisticated equipment built for the process. A typical basic system will include a track-mounted drill rig equipped with a jet grout monitor that delivers the fluid(s) through the jet grout nozzles (Fig. 7.30) to the target soils; high-volume mixing and batching equipment to ensure a constant grout supply; high-performance, heavy-duty piston pumps to generate the pressure required to achieve the soil erosion velocities needed; air compressors if double- or triple-fluid jet grouting is performed; and an automated data acquisition system to monitor and record operational parameters.

Grout Mix Design.   Grouts for strength are typically composed of cement and additives, as appropriate. The water/cement ratio is selected taking into consideration the

FIGURE 7.30   Dual jet grouting nozzles. (*Photograph courtesy of Moretrench American Corporation*)

soil grain-size composition, the water content of the soil, and the average quantity of grout per unit volume of treated soil. The water/cement ratio typically ranges from 0.6 to 1.2 by weight.

The achievable strength of the soil-cement mass is dependent on a number of factors. These include the in situ soil type, water-to-cement ratio of the grout, jet grouting installation method, and jetting energy applied. The strength of the final product will therefore vary from site to site, and this consideration should be factored into the grout mix design. The highest strength product, with unconfined compressive strengths of up to around 3000 psi (20.7 MPa), is achieved in clean sands and gravels, with that of clays most often around 100 to 300 psi (0.7 to 2 MPa).

Installation Method.   A jet grouted wall is composed of a series of overlapping integrated columns. If the wall is to provide underpinning in addition to excavation support, coring of existing foundations may be necessary. A borehole is advanced to design depth and jetting is initiated from the bottom of the hole and progresses upward. It may be necessary to use a stabilizing agent such as bentonite to maintain borehole stability during drilling. Uniform rotation and controlled extraction of the jetting tool create a continuous column to near surface. During column construction, spoil is expelled from the annulus between the jetting tool and the borehole wall. Column construction is sequenced to allow primary columns to cure before installation of the overlapping secondary elements.

Quality Control.   Prior to production work, a test section should be constructed, exhumed for visual examination if possible, and tested to establish that the grouting parameters selected have achieved the required product strength, continuity, and column diameter. If exhumation is not possible, core samples at the column interstices should be taken and laboratory-tested.

During production grouting, a number of quality-control measures should be in place, including

- Laboratory testing of retrieved wet samples for unconfined compressive strength
- Monitoring of the batching operation for grout consistency
- Real-time data acquisition to monitor and record drill rotation and withdrawal rate, fluid pressures, and flow rate
- Post-grouting verification holes
- Instrumentation to monitor for surface heave and any movement of nearby structures

EVALUATION OF SITE CONDITIONS

The first step in planning for and designing a sheeting and bracing system is to evaluate conditions at the site that will affect construction and performance of the system. Items to consider are topography, soil conditions, structures adjacent to or near the construction area, underground and overhead utilities, groundwater, and construction procedures and requirements.

Topography

Topography has obvious effects on construction activities. Flat or near-flat areas are relatively easy to work, while steep slopes along a retaining structure alignment or even moderate slopes across alignment may require special consideration in planning site work. There are two situations in which topography may have a significant effect on design. The first situation is where there is a significant difference in surface elevation on the opposite sides of an excavation: it may be awkward to slope braces so as to have reasonably equal earth loads on both sides, or, in cases with large differences, there may not be sufficient resistance readily available on the low side to restrain the loads from the high side. In the latter case, some auxiliary reactions must be developed such as reaction blocks, piles, or tiebacks.

The second situation is where there is a sharp rise in ground level, a short distance beyond the excavation area proper. Digging a trench or large excavation on or at the base of a high slope may cause instability of the slope, resulting in extremely high pressures on a retaining structure. Such situations require a thorough investigation and study of conditions by a geotechnical engineer.

Soil and Groundwater

Soil and groundwater conditions are obviously matters to be considered, but all too often are not properly evaluated. Sands of different densities, soft and stiff clays, all have different characteristics regarding difficulty of excavation and field measures needed to handle them. Add groundwater, alternating stratifications, and rock, and conditions can become quite difficult. Knowing what is there ahead of time permits preparing for it, but one must recognize that at least a few small surprises are likely no matter how carefully an area has been studied.

Obtaining adequate knowledge of subsurface conditions is critical to design, and, if sufficient information is not made available along with plans and specifications for a project, it must be obtained (by borings, test pits, seismic lines, or other means) and properly analyzed by a geotechnical engineer in order to plan a suitable sheeting and bracing system. Engineering analyses must evaluate not only lateral earth pressures but also the stability of the bottom of the excavation, all as affected by soil strengths and groundwater pressures. Finally, not the least important is the matter of groundwater, both in terms of pumpage and in terms of how it may affect adjacent areas. Lowering a groundwater table for construction can sometimes cause settlements in surrounding areas, due to either migration of fines (locally) or increased overburden loads (at some distance), or may even affect nearby wells. Avoiding these potentially adverse effects required planning.

Rock

Excavations into rock may or may not require retention, depending on the nature of the rock and the joints and fractures present. Typically, retention for rock excavation involves rock bolts but rarely sheeting (except possibly skeleton sheeting in soft or broken rock). The presence of rock at depth may hinder the placement of sheeting and bracing down to that level if the rock is hard enough to prevent keying or embedment of the toe of the sheeting into the rock. Where blasting is used, its effect on any sheeting and bracing already in place (as well as on nearby structures) must be considered.

Adjacent Structures and Utilities

Surface structures—buildings, streets, or whatever—can be observed simply by a visit to an excavation site or along a pipe alignment. Where they are within the area to be excavated, the easiest method of removal is about the only consideration. Where alongside the excavation, a determination must be made of the type of foundation existing, whether it can withstand some limited deflection or whether underpinning is needed. A compilation has been made of typical deflections near excavations relating to the type of retaining system, depth of excavation, distance to structure, etc. (see Figs. 7.31 and 7.32); from this, an estimate can be made as to whether it will be necessary to consider underpinning. Where underpinning is needed, an experienced engineer should be called on to design and execute it.

Subsurface structures at a site may consist of shallow underground utilities, sewer and water mains, culverts and tunnels, or other buried structures. Some may simply be removed; some must be left in place; others may require temporary or permanent rerouting. Again, knowing what to expect minimizes unnecessary damage and can speed up the job if appropriate equipment is on hand to cope with the things encountered.

From the design aspect, subsurface pipe and structures that must remain in place can control the type of retaining structure to be used. For instance, vertical steel sheeting could not be used in such a situation unless soil conditions permitted, without adverse effects, leaving a gap in the sheeting that could be left open or blocked during excavation. Generally, pipes or conduits alongside an excavation will not be a problem and do not require special support, although deflections should be evaluated (see Figs. 7.31 and 7.32) to make certain that predicted movements will not cause damage. A tunnel or other large structure wall alongside may make conditions easier to deal with.

An important design consideration in sheeting and bracing design is surcharge loads (loads applied at or below the surface) alongside trenches. Such loads could be due to buildings or other structures immediately alongside the trench, construction or other vehicles and equipment, excavated material, or other stockpiled material. Designing for these loads results in heavier members and increased costs, so it is often desirable, where practicable,

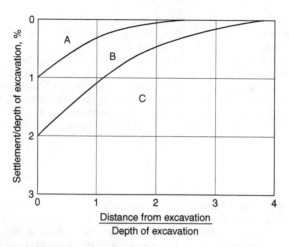

Relevant soil conditions:

Zone A—sand or firm to hard clay [s > 500 lb/ft² (24 kN/m²)]

Zone B—very soft to firm clay [s < 500 lb/ft² (24 kN/m²)] where there is a limited depth of clay below the base of the excavation, or where $\gamma H/s < 5$

Zone C—very soft to firm clay [s < 500 lb/ft² (24 kN/m²)] where there is a significant depth of clay below the base of the excavation and where $\gamma H/s > 5$

Note: Good workmanship assumed. Poor workmanship (such as unfilled voids behind sheeting) will increase deflection.

FIGURE 7.31   Settlement versus depth of excavation. (*From Peck, 1969.*)

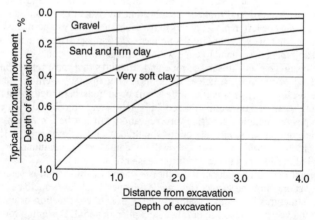

FIGURE 7.32 Lateral displacement versus depth of excavation. (*From Crofts et al., 1977.*)

to keep such loads away. However, if these loads are applied without the retaining structure being designed for them, excessive deflections or failure can be expected to result.

Where structures alongside an excavation are underpinned, there is of course no surcharge load to be considered. Often, it is possible and desirable to design the underpinning to serve as the earth-retaining structure as well as the building foundation.

DESIGN

The proper evaluation of lateral earth pressure acting on sheeting is critical to safe design. It is generally desirable, and essential in situations involving deep excavations or important structures adjacent to excavations, to have conditions analyzed by an experienced and qualified geotechnical engineer. This section starts with examples of sheeting and bracing design using basic theory, which can be the preferred approach, particularly with weaker soils. This is followed by simpler timesaving empirical solutions that can be used where conditions are less critical. It is assumed that the reader has a basic knowledge of static structural analysis and strength of materials.

Earth-Pressure Theories

As related to sheeting and bracing, the underlying concept of earth-pressure theory is that there are limiting conditions that develop in soil at the verge of failure (1) when it is allowed to expand laterally behind a deflecting vertical wall and (2) when it resists pressure applied to a vertical wall (see Fig. 7.33). One simplifying assumption used is that a plane surface represents the limit of the failing zone; analysis of the mechanics shows that these surfaces will slope at $45° \pm \varphi/2$, as shown in Fig. 7.33 (φ = internal friction angle of the soil).

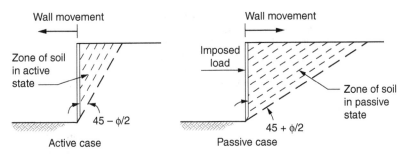

FIGURE 7.33   Limiting soil conditions.

With further assumptions (e.g., that soil pressures act parallel of the soil surface and that the presence of the wall does not affect shear stresses at the soil-wall interface), Rankine developed expressions for the active and passive earth pressures, respectively, for soils with both cohesion and friction.

$$P_a = \gamma Z K_a - 2c\sqrt{K_a}$$
$$P_p = \gamma Z K_p + 2c\sqrt{K_a}$$

P_a and P_p are active and passive earth pressures, respectively, at depth Z below the ground surface; γ is the effective unit weight of the soil; c is the cohesive strength of the soil; K_a and K_p are functions of the internal friction angle of the soil and the slope of the ground surface. For the typical case of level backfill, the expressions for these factors are

$$K_a = \tan\left(45 - \frac{\phi}{2}\right)^2$$

$$K_p = \tan\left(45 + \frac{\phi}{2}\right)^2$$

where $(45 \pm \phi/2)$ is the angle of the failure plane from vertical.

Coulomb thought it important to include the effect of soil friction (δ) on the wall. This friction is likely to develop a downward force on the wall as the soil slides down in the active case and an upward force as the soil in the passive condition tries to slide upward. With these considerations, his evaluation resulted in

$$K_a = \frac{\cos(\phi)^2}{\cos(\delta) \cdot \left(1 + \sqrt{\dfrac{\sin(\phi+\delta) \cdot \sin(\phi-\beta)}{\cos(\delta) \cdot \cos(\beta)}}\right)^2}$$

$$K_p = \frac{\cos(\phi)^2}{\cos(\delta) \cdot \left(1 - \sqrt{\dfrac{\sin(\phi+\delta) \cdot \sin(\phi+\beta)}{\cos(\delta) \cdot \cos(\beta)}}\right)^2}$$

which give apparently more realistic values to the coefficients. With wall friction δ and the slope of the ground surface β equal to zero, these expressions reduce to those given above for Rankine.

The inclination of the failure plane for the Coulomb assumptions is determined from another complex expression.

Further evaluation and analysis have shown that the assumption of a plane surface for the failure plane is not realistic; it is really curved (approximating a log spiral). The error for the active coefficient is generally quite small, but it becomes large for the passive coefficient with a wall friction angle more than one-third the soil friction angle. Caquot and Kerisel developed expressions for the log-spiral values; the chart (Fig. 7.34) developed from those expression provides a realistic basis for selecting active and passive coefficients in the ranges of normally encountered values for various parameters.

Thus, for simple cases of horizontal ground surface and horizontal stratification, active and passive earth pressures may be calculated from the Coulomb expression or from the log-spiral method (Fig. 7.34). The passive-pressure coefficient may be calculated somewhat conservatively from the Coulomb equation if δ is set equal to zero; however, this Coulomb coefficient will still be less conservative than a Rankine passive-pressure coefficient.

For complex stratifications, sloping strata, and other variations, reference should be made to texts that treat the topic in greater detail (*Navy Design Manual 7.2*, Lambe and Whitman, Fang, Leonards, and others noted in the Bibliography). While cantilevered earth-retaining structures and some anchored or braced structures are designed for triangular lateral earth pressures, most anchored or braced structures are designed using one of many empirical, rectangular, or trapezoidal earth-pressure distributions. These commonly used, empirical, earth-pressure distributions are discussed later in this chapter.

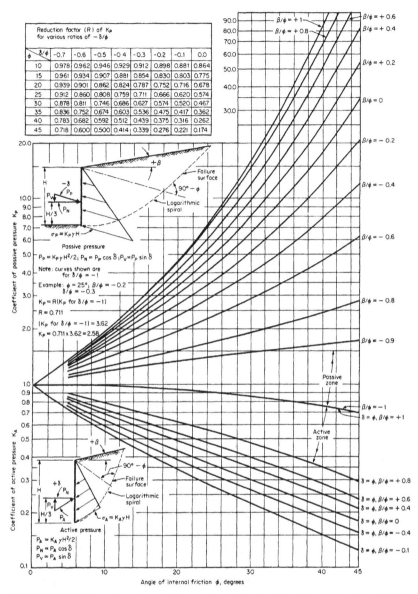

FIGURE 7.34 Active and passive coefficients with wall friction. (*From Navy Design Manual 7.2.*)

Cantilever Wall Design

For soldier pile, sheet pile, secant pile, and similar earth-retaining structures with exposed wall heights up to approximately 12 to 14 ft (3.6 to 4.3 m), the vertical sheeting member usually can stand unsupported except for its embedment below the excavation subgrade,

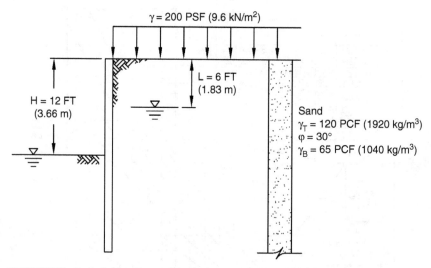

FIGURE 7.35   Example: assumed conditions.

provided the soil is not very loose or soft, the structure is not supporting significant sur-charge loads, or controlling wall deflections is not critical.

To demonstrate how these principles are applied in practice, along with the effects of water, and other items, a step-by-step example of the calculations for a cantilever wall is given, starting with Fig. 7.35, for all-sand soil conditions. This figure shows the basic conditions: a 12-ft (3.6-m) cut, 200 lb/ft² (9575 N/m²) surcharge load up to the edge of the cut, sand having the properties shown, groundwater outside the cut 6 ft (1.8 m) below the surface, and at ground level inside the cut. For a wall of this height retaining a cohesive soil, a total stress analysis includes cohesion in the design and often shows that the soil can stand unsupported, at least temporarily, without the need for a retaining structure. In this case, it is recommended to design the retaining structure using a drained or effective stress analysis where there is no cohesion ($c = 0$), but an effective soil friction angle is used ($\varphi > 0$). For the design of a cantilevered retaining structure for cohesive soil, a drained condition will be similar to the design of a structure in sand.

In the following example, a steel sheet-pile wall is assumed. The selection of representa-tive groundwater levels for design should be done carefully. With tight sheeting, using the natural water level is reasonable or slightly conservative; with pervious sheeting, although the seepage line into the excavation may be about at the bottom of the excavation, to account for seepage forces, the assumed external level should be a little above this. A fully dewatered site will have no water showing, but, if the water level is at shallow depth below the excavation, it may still have a strong effect on passive and active pressures. When designing an earth-retaining structure for a fully dewatered condition, consideration must be given to the possibility that the dewatering system may shut down either expectedly or unexpectedly. If this happens, the critical design case will be when the groundwater level rises to the excavation subgrade in front of and behind the wall. This condition will cut the passive resistance approximately in half as the resisting soil's unit weight changes from moist to buoyant.

Figure 7.36 shows the basis for calculating the lateral earth, water, and surcharge pres-sures. The pressures must be separated into (effective) horizontal soil pressures, hydrostatic

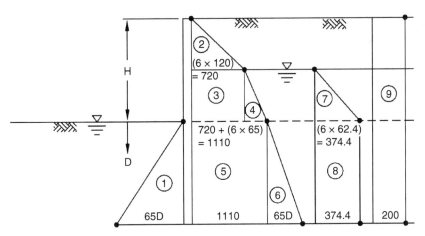

FIGURE 7.36   Vertical pressures, lb/ft².

pressures, and surcharge pressures. Also given are the expressions by which the lateral pressures can be calculated, at the respective levels in the diagram. Although most reference texts, including earlier editions of this text, and design manuals advocate combining these lateral pressures into a net or resultant diagram, maintaining separate diagrams simplifies the mathematics involved in designing the wall and also allows the designer to more clearly understand the individual loads on the retaining structure.

The active, passive, hydrostatic, and surcharge pressures that occur on each side of the wall are shown in Fig. 7.37. The total lateral pressure to be considered is the sum of the vertical soil pressures times the appropriate active and passive coefficients of lateral pressure, plus the vertical surcharge pressure times the appropriate coefficient of lateral pressure, plus the hydrostatic pressure. Going to the Caquot-Kerisel charts (Fig. 7.34) with $\varphi = 30°$ and β (ground slope angle) $= 0°$, $K_a = 0.31$ when δ (friction angle of soil on the wall) $= 0.3\varphi$. For $\delta = 0$, K_a would be 0.33, a very small difference. At this value, $K_p = 6.5(0.627) = 4.08$. Then, following the basic relationship described above, the numeric values of the lateral pressures can be calculated for various depths, as shown in Fig. 7.37, and expressions are developed that express the slopes of the pressure diagram below the bottom of the excavation, which are used in the required sheeting penetration calculation.

Once the pressure diagrams are developed, use them to sum the moments about point 0. Setting the sum of the moments to 0, solve for the depth D of the sheet piling's penetration below subgrade where moment equilibrium will occur. This process is illustrated in Fig. 7.38. The maximum moment in the sheet piling will occur at the point of zero shear. This point can be found by taking the derivative of the moment equation, setting it to 0 and then solving for the depth at which the shear is 0. Once this depth is known, enter it into the moment equation to calculate the maximum bending moment in the sheet piling. Because passive resistance is a function of the embedment depth squared, to obtain a factor of safety of approximately 1.5 or 2.0, increase the calculated depth of penetration D to $1.2D$ or $1.4D$.

Many texts, including earlier versions of this text, and design manuals discuss summation of horizontal forces when designing cantilevered retaining walls. This is unnecessary. In order to provide moment equilibrium, the passive pressure will always be greater than

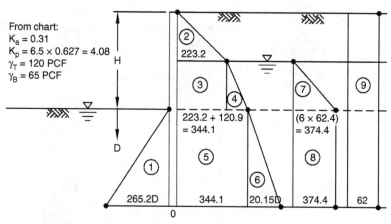

From chart:
$K_a = 0.31$
$K_p = 6.5 \times 0.627 = 4.08$
$\gamma_T = 120$ PCF
$\gamma_B = 65$ PCF

Area:
1   $65 \times 4.08 \times D = 265.2D$ LB/FT^2
2   $6 \times 120 \times 0.31 = 223.2$ LB/FT^2
3   223.2 LB/FT^2
4   $6 \times 65 \times 0.31 = 120.9$ LB/FT^2
5   $223.2 + 120.9 = 344.1$ LB/FT^2
6   $65 \times 0.31 \times D = 20.15D$ LB/FT^2
7   $6 \times 62.4 = 374.4$ LB/FT^2
8   374.4 LB/FT^2
9   $200 \times 0.31 = 62$ LB/FT^2

FIGURE 7.37 Horizontal pressures, lb/ft^2, and calculations.

$\Sigma M_0 = 0 = M_1 + M_2 + M_3 + M_4 + M_5 + M_6 + M_7 + M_8 + M_9$

$M_1 =$	$+1/2 \times D \times D/3 \times 265.2 \times D$	$=$	$+44.2D^3$		
$M_2 =$	$-1/2 \times 6 \times 223.2 \times (D + 6 + 6/3) =$			$-669.6D$	$-5,356.8$
$M_3 =$	$-6 \times 223.2 \times (D + 6/2)$	$=$		$-1339.2D$	$-4,017.6$
$M_4 =$	$-1/2 \times 6 \times 120.9 \times (D + 6/3)$	$=$		$-362.7D$	-725.4
$M_5 =$	$-344.1 \times D \times D/2$	$=$	$-172.05D^2$		
$M_6 =$	$-1/2 \times D \times 20.15 \times D \times D/3$	$=$	$-3.36D^3$		
$M_7 =$	$-1/2 \times 6 \times 374.4 \times (D + 6/3)$	$=$		$-1123.2D$	$-2,246.4$
$M_8 =$	$-374.4 \times D \times D/2$	$=$	$-187.2D^2$		
$M_9 =$	$-1/2 \times 62 \times (D + 12) \times (D + 12) =$		$-31D^2$	$-744D$	$-4,464$

$\Sigma M_0 = +40.84D^3 \quad -390.25D^2 \quad -4238.7D \quad -16,810.2$

$\Sigma M_0 = 0 @ D = 17.06$ ft

Add 20% minimum to calculated depth.

Use 17.06 ft $\times$ 1.2 = 20.5 ft minimum.

$M' = V = 122.52D^2 - 780.5D - 4,238.7$

$V = 0 @ D = 9.8741$ ft

$M_{max} = M_{9.8741} = 57,395$ ft-lbs/lin.ft of wall = 688,740 in-lbs/lin.ft of wall

For sheet piling, assume Fy = 50,000 psi.

Required section modulus = $S = M_{max}/ (0.66 \times F_y) = 688,740/(0.66 \times 50,000) = 20.9$ in^3/lin.ft of wall

PZC 13 steel sheet piling has a section modulus = $S = 24.2$ in^3/lin.ft of wall. OK, > 20.9 in^3/lin.ft of wall

Use PZC 13 sheet piling, or equal, with a sheet length of 20.5 ft minimum.

FIGURE 7.38 Cantilevered steel sheet piling design calculations.

the sum of the lateral earth, surcharge, and hydrostatic pressures. However, even though the passive pressure is greater, it is impossible for the wall to be pushed backward toward the retained soils. For a properly designed cantilevered sheeting wall, the passive pressure is a mobilized resistance force, not an applied load.

The passive resistance indicated in this example assumes a static water condition. If there is an upflow into the bottom of the excavation, the effective weight of the soil must be reduced by the seepage pressure. Some of the calculations can be short-cut by a method suggested by Teng. The *Steel Sheet Piling Design Manual* from United States Steel and its updated replacement, *Sheet Pile Design by Pile Buck* from Pile Buck International, give charts that, if used carefully, can save considerable effort for certain soil conditions (see Fig. 7.48 in "Base Stability"). These charts may be used to determine the required sheet-pile embedment needed to prevent piping or boiling, a quick condition caused by unbalanced hydrostatic pressure. For higher retaining structures, the flexibility of the structural members can result in lower pressures and consequently lower moments in the members.

Anchored/Braced Wall Design

If the retained soils or the soils in the embedment area are weak, or if applied surcharge pressures or hydrostatic pressures are significantly high, the retaining structure may not be able to perform satisfactorily as a cantilevered wall. In this case, the structure will need to be anchored or be braced to provide the required lateral support.

Calculations for braced or tieback sheeting in cohesive soils are illustrated, starting with the conditions shown in Fig. 7.39. The same basic principles given under earth-pressure theories apply, but their application involves different procedures because different soil characteristics are involved.

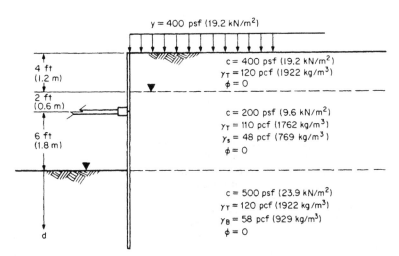

FIGURE 7.39 Conditions: restrained sheeting.

In cohesive soils, unless there is clear-cut evidence otherwise, the water table should be assumed at the soil surface and total stresses are used in calculations, with $\varphi = 0$ and $c > 0$. This applies to short-term excavations, without standing water. An excavation open for an extended period can mean that other assumptions are needed regarding

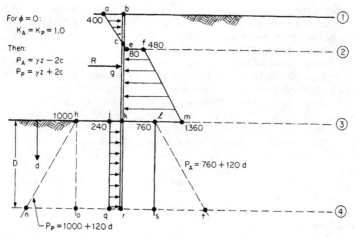

Left side:

$$P_{P3} = \gamma_z + 2c = 0 + 2 \times 500 = 1000$$
$$P_{P4} = 10 \times 120 + 1000 = 2200$$
$$\text{Net } P_P = (1000 + 120d) - (760 + 120d) = 240$$

Right side:

$$P_{A1} = \gamma_z - 2_c = 0 + 400 - 2 \times 400 = -400$$
$$P_{A21} = 4 \times 120 + 400 - 2 \times 400 = 80$$
$$P_{A22} = 4 \times 120 + 400 - 2 \times 200 = 480$$
$$P_{A31} = 400 + 4 \times 120 + 8 \times 110 - 2 \times 200 = 1360$$
$$P_{A32} = 400 + 4 \times 120 + 8 \times 110 - 2 \times 500 = 760$$
$$P_{A4} = 760 + 10 \times 120 = 1960$$

FIGURE 7.40   Horizontal earth pressures, lb/ft²—restrained sheeting.

water conditions, resulting in important differences in methods of computations, and consequently in results.

Figure 7.40 shows the lateral pressure profile developed from conditions given in Fig. 7.39. From the expression $K = \tan^2 (45 \pm \varphi/2)$, with $\varphi = 0$, $K_a = K_p = 1$; then $P_a = \gamma z - 2c$ and $P_p = \gamma z + 2c$. Because of cohesion, a net negative soil pressure (tension) is calculated at the ground surface. For the purpose of these calculations, all tensile stresses are taken as zero; it is assumed the soil may develop a crack. At the lower levels, the lateral soil pressure changes abruptly depending on soil cohesion. Below the excavation line, the net pressure (line jq) is the difference between the total passive pressure (line hn) on the excavated side and the total active pressure (line lt) on the unexcavated side.

Calculations for the sheeting and bracing of a one-tier earth-retaining structure start by taking the moments of the pressure diagram about point g, the level of the brace. The depth of sheeting penetration d is entered as an unknown. In this design example, tensile stresses above line ef are not included, and the very small positive pressure above line ef is ignored for simplicity. After developing and solving the cubic moment equilibrium equation, the calculated sheeting penetration is then 6.2 ft (1.9 m). Next, the horizontal forces are summed to calculate the bracing reaction on the wall per foot (or meter) of wall. Then, the required sheeting penetration, or vertical member embedment, below subgrade is found by adding between 20 and 40 percent to the calculated penetration length as a factor of safety.

An addition of 20 percent is very common in practice, while an addition of 40 percent is, not surprisingly, common in design literature produced by various steel manufacturers.

The sheeting section required is found by calculating the maximum moment developed, which will be at the point of zero shear at or below the brace level (see Fig. 7.41). In this

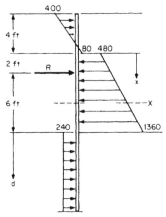

Sum of moments about g

$$\Sigma M_g \;(+)\; 8 \times 480 \times 2 + (\frac{1360 - 480}{2}) \frac{8}{2} (6 - \frac{8}{3}) - 240 \, d \, (6 + \frac{d}{2}) = 0$$

$$d = 6.2 \text{ ft}$$

Sum of horizontal forces:

$$R - 8 \times 480 - (\frac{1360 - 480}{2}) \frac{8}{2} + 240 \,(6.2) = 0 \quad R = 4112 \text{ lb/ft}$$

Add 30% to d for safety factor, required penetration = 8.1 ft

To calculate maximum moment in sheeting, find point of zero shear at or below R by sum of horizontal forces:

$$R - x \,(480 + 110 \, x) = 4112 - 480x - 110x^2 = 0 \quad x = 4.31 \text{ ft}$$

Sum of moments of forces above X:

$$M = (4.31 - 2.00)\, 4112 - 480 \frac{(4.31)^2}{2} - 110 \frac{(4.31)^3}{6} = 3573 \text{ ft lb}$$

Selection of sheeting by section modulus:

$$\text{Required section modulus } S = \frac{\text{Moment developed}}{\text{Allowable stress}} = \frac{M}{f}$$

$$S = \frac{3573 \times 12}{1000} = 42.9 \text{ in}^3 \text{ for wood}$$

$$\text{Section modulus for rectangular section} = \frac{bh^3}{12}$$

$$\text{calculate h: } 42.9 = \frac{12h^3}{12} \qquad h = 3.5 \text{ in}$$

4 in nominal thickness planking will serve satisfactorily.

(a)

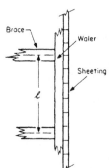

To select waler size: Assume brace spacing ℓ = 8 ft

Sheeting load R = 4112 lb/ft

$$\text{Maximum moment in waler } M = \frac{P\ell^2}{8} = \frac{4112 \times 8^2}{8} = 32896 \text{ ft-lb}$$

Required section modulus for waler (assume wood)

$$S = \frac{M}{f} = \frac{32,896 \times 12}{1000} = 395 \text{ in}^3$$

For a square member $S = \dfrac{d^3}{6} = 395 \quad d = 13.3$ in.

Try ℓ = 7 ft then M = 25,186 ft-lb S = 302 d = 12.2 in

Try ℓ = 6 ft then M = 18,504 ft-lb S = 222 d = 11.0 in

Use 12 in × 12 in timber, braces at 6 ft

Brace load = 6 × 4112 = 24,672 lb compressive force

At 1000 psi, need 25 in²

Nominal 6 × 6 has 30.5 in. OK, but must check allowable stress vs. ℓ/d for member.

ℓ = unsupported span length

d = least dimension of member

See American Institute of Timber Construction Design Manual for allowable stresses

(b)

FIGURE 7.41 Calculations for sheeting and bracing, lb/ft².

calculation, x is the distance below the top of the pressure diagram [4 ft (1.2 m) below top of the wall] and, because the lateral pressure varies with depth, x is determined by a quadratic equation. If the lateral earth pressure were constant at the point of zero shear, x would be determined by a linear equation. The moment in the sheeting is then determined arithmetically about the point of zero shear. Selection of a suitable section of sheeting depends first on the material to be used. Once the maximum bending moment and shear are known, the appropriate sized sheeting member (soldier pile, sheet pile, secant pile, vertical wood, etc.) can be selected. For this example, 4 in (10 cm) nominal thickness, vertical, timber sheeting could be used. If this is considered not feasible, another vertical member, such as a PS 27.5, flat, steel sheet-pile section could be used.

The sizing of wales and braces follows simple static and structural analysis. In the previous example, it is found that an 8-ft (2.4-m) spacing for braces puts too high a bending moment into the wale if timber is to be used. However, a 6-ft (1.8-m) brace spacing is structurally acceptable. The allowable, timber cross brace size is then quite small, 6 × 6 in (15 × 15 cm). Practicality often dictates that the braces be the same size as the wale for simplicity of construction and ability to withstand accidental impact during trench work. If a stronger (high-section-modulus) wale were used, perhaps a steel wide flange or H-pile section, the spacing between braces could be extended. Wider brace spacing is often desirable because it allows for easier excavation and placement of materials, such as utility pipes or precast manholes, into the shored excavation.

It should be noted that in actual construction, during excavation to install the upper or top brace, the sheeting will be acting in cantilevered manner and should be checked to see that the moment developed is not excessive and that sufficient embedment is in place. In this case, it can be seen by inspection that there will be no significant cantilevered loads.

For an anchored or braced wall retaining a cohesionless soil, triangular, lateral earth pressures can be determined using the method previously shown for a cantilevered wall retaining sand, or one of the empirical earth-pressure distributions can be used. If triangular, lateral earth pressures are calculated, it is common practice to calculate the total earth pressure per unit length of wall and then increase the total load by 30 percent. This increased total load is then reshaped into a rectangular or trapezoidal shape that produces the same total load per unit length of wall. Calculation of maximum bending moments, maximum shears, anchor or brace loads, and sizes for wales and braces then can be performed as described in the previous example.

The analysis given above for restrained sheeting can be expanded on to cover situations with more than one brace level. To illustrate this, consider an excavation planned for three levels of bracing (Fig 7.42*a*). Although a triangular earth-pressure distribution is shown, the design method is the same when using an empirical earth-pressure distribution. The first point requiring a computational check is the cantilever condition existing when the excavation for the first brace level has been completed (Fig. 7.42*b*). If we assume that sheeting has already been driven to the required depth, there is usually no need to calculate the required penetration, but the required section modulus must be checked as described in the previous section on cantilever design.

Next, a calculation is made with the first brace in place and the excavation completed for the second brace (Fig. 7.42*c*). This calculation follows the steps given above for restrained sheeting design. The following stage of excavation, with two braces in place and excavation made for the third level (Fig. 7.42*d*), requires either that the sheeting be considered a beam and appropriate reactions be calculated or, much more simply, that a hinge be assumed at the second brace level (Fig. 7.42*e*). The latter permits the calculation to follow the pattern of restrained sheeting design.

The final step, with three braces in place and excavation to final grade, can be solved in the same relatively simple manner by assuming an additional hinge at the third brace (Fig. 7.42*f*). It is also commonly accepted to assume a hinge at the ground surface in

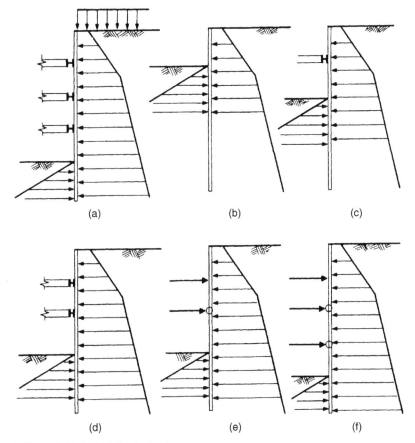

FIGURE 7.42 Multiple bracing levels.

front of the retaining structure at each stage of the design except for the first cantilevered stage. Once a hinged ground reaction is found, it can be used to calculate the required wall penetration or embedment for each restrained stage as necessary.

Although intermediate design stages are usually not more critical than the final stage, moments in the vertical sheeting member can be calculated for each level of bracing to verify that the sheeting has adequate section modulus to carry the applied pressures. Efficient design would have approximately equal moments between each brace, that moment being the maximum which can reasonably be carried by the section modulus of the sheeting selected. Alternatively, the structure can be designed to more equally balance the loads to the wales and braces or tieback anchors. More experienced designers often try to reasonably balance both the moments and the bracing loads.

Soldier Pile Design

Calculations for soldier piles and lagging (see Figs. 7.4 and 7.5), or any other sheeting system with full height, discrete, vertical members and horizontal sheeting, use the same principles as the calculations described in the preceding sections, but with a few added considerations.

Earth-retaining structures with continuous structural members, such as sheet piling and diaphragm walls, are designed based on a unit length of wall—either per linear foot or meter of wall. However, soldier piles are designed based on the horizontal spacing between individual piles. This horizontal spacing usually is between 7 and 10 ft (2.1 and 3.1 m); the actual spacing depends on factors such as total wall length, available pile material or strength, soil strength, tolerable wall deflections, or spacing of bearing piles or drilled shafts required for the proposed structure. Where driven soldier piles are most often installed at a spacing of 8 ft (2.4 m) or less, drilled-in soldier piles backfilled with lean-mix concrete are often spaced at 10 ft (3.1 m) or less.

When designing soldier piles, the tributary load width above subgrade is the soldier piles' horizontal spacing. Below subgrade, the tributary load width is a function of the pile flange width for driven piles or the diameter of the concreted drill hole for drilled-in piles. The passive resistance for a driven pile is usually calculated for an effective width of 3 times the flange width. For drilled-in piles backfilled with low-strength, lean-mix concrete or flowable fill, the passive resistance is usually calculated for an effective width of 3 times the concreted drill hole diameter. This effective width factor would not be appropriate in very soft clays, although a higher factor could safely be used in very dense granular soils. This effective width factor should be applied only to normal sizes of piles and drill holes; very wide, drilled shafts, etc., should be evaluated as independent pile units (see NAVFAC DM-7.2). Additionally, the effective width of the soldier pile embedment should not be more than the soldier pile spacing.

Because passive earth-pressure coefficients are usually approximately 10 times (or more) greater than active earth-pressure coefficients, and because the passive effective width is approximately 3 times greater than the active width below subgrade, the available passive resistance for the embedded soldier pile can easily be 30 times (or more) greater than the active force that is, therefore, relatively insignificant. For this reason, in design of temporary soldier pile walls, the active earth pressure acting on the back of the soldier pile flange or drill hole below subgrade is often ignored.

Cantilevered soldier piles are designed using 100 percent of their triangular earth-pressure distribution. However, for braced or tiedback soldier piles, empirical, rectangular, or trapezoidal earth-pressure distributions are most often used. These empirical distributions evolved from the work of Terzaghi and Peck who, using actual field measurements, developed their pressure distributions for calculating the maximum anticipated brace loads for sheeting walls. Therefore, it is common to design soldier piles, or other vertical wall members, for less load than is used to calculate the brace or tieback loads. Various early references recommend designing braced or tiedback soldier beams and sheet piling for as little as 67 to 80 percent of the empirical earth-pressure distribution.

Because soldier piles are usually fully supported in the plane of the wall (i.e., confined along their weak axis) by a combination of lagging, soil, and/or lean-mix concrete or flowable fill; there is no "unbraced length" and the soldier beams can be designed for their maximum allowable bending stress. Where soldier piles are required to support additional vertical loads (e.g., from supported utilities or roadway decking), or where the wall is supported by steeply inclined tieback anchors, the soldier beams should be designed for the combined axial and bending stresses caused by the applied, vertical loads and the lateral earth and surcharge pressures.

Lagging Design

Soldier pile walls utilize horizontal sheeting boards (lagging) that span between the vertical members (soldier piles) and pick up the lateral soil and surcharge pressures and carry them to the vertical members. Experience has shown that, in general, a given lagging size can restrain soil where the typical pressure distributions used in retaining structure calculations indicate the lagging should be highly overstressed. The explanation is that the soil mass arches between the soldier piles; the lagging deflects sufficiently to greatly reduce the pressure on it; this pressure is just enough to hold the soil that is acting as the arch in position. A possible version of the resultant pressure distribution is shown in Fig. 7.43a, where line 1 represents the average soil pressure (p) to be expected at the level being evaluated, line 2 is the idealized representation of the total pressures in the soil immediately behind the sheeting, and line 3 approximates the pressure actually reaching the lagging. No "correct" lagging calculation method has been established. For this reason, lagging is usually not designed, but its size is based on past local practice on similar projects. However, when they must be provided, calculations for the required size of lagging are sometimes based on the pressure distribution shown in Fig. 7.43b. Figure 7.43c might be considered closer to fact, but the two figures result in the same maximum moment ($pl^2/12$), so this is probably a reasonable value. A similar method for designing wood lagging is recommended in the New York City Transit Authority's *Field Design Standards* which allow increasing by 50 percent the allowable bending stress for the wood lagging. The FHWA states in *Lateral Support Systems & Underpinning, Vol. 1* that the lateral pressure used to design lagging boards may be decreased by 50 percent ($M = 0.5pl^2/8 = pl^2/16$) or the allowable flexural stress of the lagging may be increased by 50 percent ($M = pl^2/8/1.5 = pl^2/12$).

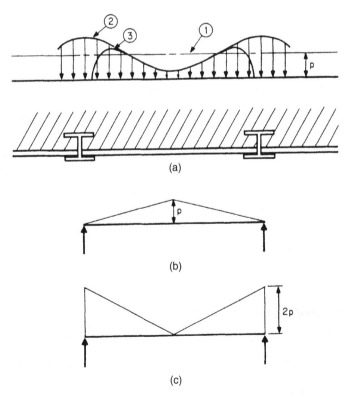

FIGURE 7.43 Pressure on lagging.

As an alternative to calculating the required lagging thickness, the FHWA has published a table of Recommended Thicknesses of Wood Lagging (see Fig. 7.6). The FHWA also states in its *Design Manual for Permanent Ground Anchor Walls* that the lagging thicknesses in their table "were developed for applications where displacements needed to be limited. Limiting displacements is an unnecessary requirement for most permanent ground anchor walls for highway applications. Therefore, thinner lagging boards than those recommended …. can be used for most highway walls." The FHWA also states in *Geotechnical Engineering Circular No. 4, Ground Anchors and Anchored Systems*, that "For temporary SOE walls, contractors may use other lagging thicknesses provided that they can demonstrate good performance of the lagging thickness for walls constructed in similar ground."

It should be noted that arching can be assumed in granular soils and stiff-to-hard cohesive soils. In soft cohesive soil, however, where the soil overburden pressure approaches 4 times the cohesive strength of the soil, arching will not develop and the full active pressure of the soil will act on the lagging. This may also occur where an excavation in firm cohesive soil is open for an extended period.

Steel Sheet Piling Design

Calculations for steel sheet piling, or any other sheeting system with full height, continuous, vertical members, use the same principles as the calculations described in the preceding sections, but with a few added considerations.

Earth-retaining structures with continuous structural members, such as sheet piling and diaphragm walls, are designed based on a unit length of wall—either per linear foot or meter of wall—for both active and passive earth pressures and applied surcharges, both above and below subgrade. Active or at-rest earth pressures will be applied to the back of the retaining wall from the top of the sheet piling to its tip. Passive earth pressures will be applied from excavation subgrade, or dredge line, to the tip of the sheet piling.

Cantilevered sheet piling is designed using 100 percent of the triangular earth-pressure distribution. However, for braced or tiedback sheet piling, empirical, rectangular, or trapezoidal earth-pressure distributions are most often used. Therefore, the sheet piling may be designed for as little as 67 to 80 percent of the empirical earth-pressure distribution used to calculate the brace or tieback loads. When the sheet-pile wall is supporting very soft or very loose soils or when the wall bracing or tieback anchors are located unusually high on the wall, as is common with waterfront bulkheads, it may be more appropriate to design the wall with a triangular earth-pressure distribution rather than with an empirical, rectangular, or trapezoidal pressure distribution.

Because the steel sheet piling is continuous and fully supported in the plane of the wall, there is no "unbraced length" and the sheet piling can be designed for its maximum allowable bending stress. Where the sheet piling is required to support additional vertical loads (e.g., from supported utilities or roadway decking), or where the wall is supported by steeply inclined tieback anchors, the sheet piling should be designed for the combined axial and bending stresses caused by the applied, vertical loads and the lateral earth and surcharge pressures.

Commonly Used Pressure Distributions

The difficulties and complications of developing soil-pressure diagrams for different situations, and the extended calculations resulting from them, have led to the use of simplified pressure distributions. These distributions have been developed and revised over a period of many years and are based to a large extent on actual field measurements as well as experience.

Diagrams illustrating the more commonly used versions of these pressure distributions are shown in Fig. 7.44. The diagram for sand, Fig. 7.44a, covers all common situations: loose to dense sands and both braced and tied-back excavations. It is important to use an angle of internal friction representative of the in situ material; a loose sand will not have high-friction angle; a dense sand will.

For clays, a greater variability must be taken into account, so different distributions are needed, as illustrated in Fig. 7.44b through 7.44e.

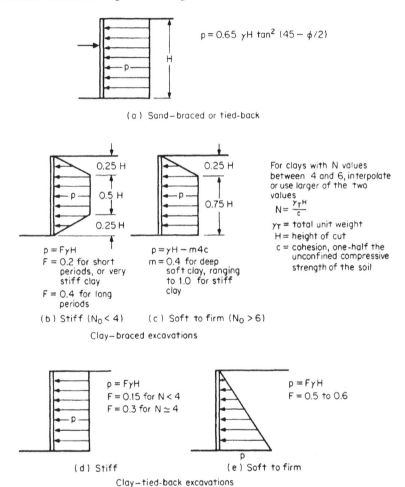

$$p = 0.65 \, \gamma H \tan^2 \, (45 - \phi/2)$$

(a) Sand—braced or tied-back

0.25 H

0.5 H

0.25 H

$p = F\gamma H$
F = 0.2 for short periods, or very stiff clay

F = 0.4 for long periods

(b) Stiff ($N_0 < 4$)

0.25 H

0.75 H

$p = \gamma H - m4c$
m = 0.4 for deep soft clay, ranging to 1.0 for stiff clay

(c) Soft to firm ($N_0 > 6$)

Clay—braced excavations

For clays with N values between 4 and 6, interpolate or use larger of the two values

$$N = \frac{\gamma_T H}{c}$$

γ_T = total unit weight
H = height of cut
c = cohesion, one-half the unconfined compressive strength of the soil

$p = F\gamma H$
F = 0.15 for N < 4
F = 0.3 for N ≈ 4

(d) Stiff

$p = F\gamma H$
F = 0.5 to 0.6

(e) Soft to firm

Clay—tied-back excavations

FIGURE 7.44 Simplified lateral earth-pressure distributions.

Because of newer research sponsored by the FHWA, the American Association of State Highway and Transportation Officials (AASHTO) and State Departments of Transportation currently recommend the use of nonsymmetrical, trapezoidal earth-pressure distributions for earth-retaining structures utilizing both one level and multiple levels of bracing and tieback anchors. The dimensions of the trapezoids are dependent on the location(s) of the tieback anchor(s), as shown in Figs. 7.45 and 7.46.

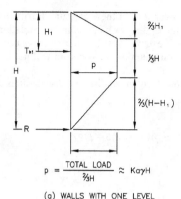
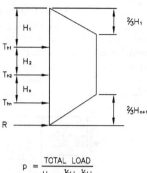

$$p = \frac{\text{TOTAL LOAD}}{\tfrac{2}{3}H} \approx K_a\gamma H$$

$$p = \frac{\text{TOTAL LOAD}}{H - \tfrac{1}{3}H - \tfrac{1}{3}H_{n+1}}$$

(a) WALLS WITH ONE LEVEL OF GROUND ANCHORS

(b) WALLS WITH MULTIPLE LEVELS OF GROUND ANCHORS

H_1 = DISTANCE FROM GROUND SURFACE TO UPPERMOST GROUND ANCHOR
H_{n+1} = DISTANCE FROM BASE OF EXCAVATION TO LOWERMOST GROUND ANCHOR
T_{hi} = HORIZONTAL LOAD IN GROUND ANCHOR i
R = REACTION FORCE TO BE RESISTED BY SUBGRADE (i.e., BELOW BASE OF EXCAVATION)
p = MAXIMUM ORDINATE OF DIAGRAM

TOTAL LOAD = $0.65 K_a\gamma H^2$

FIGURE 7.45   FHWA-recommended apparent earth-pressure diagrams for sands.

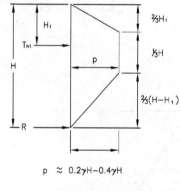
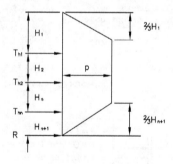

$p \approx 0.2\gamma H - 0.4\gamma H$

$p \approx 0.2\gamma H - 0.4\gamma H$

(a) WALLS WITH ONE LEVEL OF GROUND ANCHORS

(b) WALLS WITH MULTIPLE LEVELS OF GROUND ANCHORS

H_1 = DISTANCE FROM GROUND SURFACE TO UPPERMOST GROUND ANCHOR
H_{n+1} = DISTANCE FROM BASE OF EXCAVATION TO LOWERMOST GROUND ANCHOR
T_{hi} = HORIZONTAL LOAD IN GROUND ANCHOR i
R = REACTION FORCE TO BE RESISTED BY SUBGRADE (i.e., BELOW BASE OF EXCAVATION)
p = MAXIMUM ORDINATE OF DIAGRAM

TOTAL LOAD (kN/m/METER OF WALL) = $3H^2 - 6H^2$ (H IN METERS)

FIGURE 7.46   FHWA-recommended apparent earth-pressure diagrams for stiff to hard clays. (For soft to medium clays, refer to Fig. 23c and Section 5.2.6 in FHWA's *Geotechnical Engineering Circular No. 4, Ground Anchors and Anchored Systems.*)

Except for the FHWA pressure distributions, these empirical pressure distributions were developed before the use of computers became common. In order to simplify the hand calculations required to design a wall, a hinge or pinned connection was assumed at subgrade and at each brace or tieback elevation except for the top elevation. The FHWA pressure distributions, therefore, make this same assumption. The brace or tieback anchor reactions may be calculated assuming that each support carries the load developed over half the distance to the next reaction; this is essentially the same as assuming hinges at each reaction (except the top). The required soldier pile or sheet-pile embedment is then calculated using the calculated, pinned, subgrade reaction. Today, the use of computers makes it possible to easily design these walls without the simplifying pin or hinge assumption.

When there is surcharge load alongside a trench, it should be taken into account as an added, horizontal, rectangular pressure, pK_a, or as an added, horizontal, Boussinesq surcharge pressure. The effect of a groundwater table within the cut depth cannot readily be added to these pressure distributions except where water is at the ground surface and buoyant unit weights are used in calculating earth pressures. In cases where groundwater is present at an intermediate depth, it would be appropriate to go to full analysis. It should also be noted that, for a temporary earth-retaining structure, the water table to be considered is that which will exist during construction, which may be different from that originally or finally existing in the ground.

It must be emphasized that there is no "correct" or "incorrect" pressure distribution. Over many, many years, innumerable retaining walls have been successfully built using all of these empirical earth-pressure distributions. It is more important to design with reasonably correct soil properties and to build with good workmanship than to worry about which pressure distribution is "correct." Additionally, walls have been successfully designed and built using both simplified, hand calculations and sophisticated, computer analyses. Either method of analysis should be suitable for most wall designs.

Base Stability

A factor important to check for all excavations is bottom stability. In clays, it is basically related to shear strength; in sands, except for unusually low-friction angles, the problem is primarily related to groundwater. Failure to take these factors into account is a primary cause of difficulties with braced and tieback excavations.

For cohesionless soils, a simple verification of base/friction angle stability should be made as shown in Fig. 7.47, to see if it is permissible to terminate the sheeting at the bottom of the excavation. In general, any material with an internal friction angle over 25° will be stable in this configuration regarding strength. Groundwater, however, can still have important influence. Whenever groundwater is above the bottom of the proposed excavation in granular soils, there is a real potential for bottom heave, piping, or similar instability of the bottom. These problems can be averted by extending steel sheet piling below the excavation and/or by some degree of dewatering (lowering the hydrostatic head) so as to meet the criteria described as follows and depicted in Figs. 7.48 and 7.49.

When considering sheeted excavations in pervious soil, the most basic case is that with an "infinite" extent of uniform sand (see the upper diagram in Fig. 7.48). To qualify for this, the uniform deposit must extend a distance below the bottom of the cutoff wall at least equal to the width of the excavation. The chart defines the depth of penetration of sheet piling required (regarding the head of water above the bottom of the excavation) for various safety factors in loose and dense sands. An average safety factor of 1.0 is not adequate; it means that in some areas of the trench the bottom will become quick.

For locations where an impervious layer is at moderate depth below the bottom of the excavation, use the lower diagram of Fig. 7.48. In these figures, a layer with a permeability

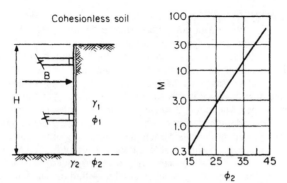

Factor of safety for stability of bottom of cut:

$$F_s = M(\gamma_2/\gamma_1)$$

γ_2 and φ_2 apply to material to a depth of $H/3$ or B below bottom of cut, whichever is less. For water at base of cut, use buoyant unit weight for γ_2. If sheeting stops at bottom of cut, groundwater must not be above that level or piping will occur at the bottom corners. With higher groundwater, deeper sheeting is required (see Fig. 7.48), and/or groundwater level must be lowered.

FIGURE 7.47   Bottom stability—cuts in sand.

of less than one-tenth that of an adjacent material should be considered impervious. For impervious layers near the bottom of the excavation, see Fig. 7.49. For layers with limited differences in permeability, see Fig. 7.50.

For excavations in cohesive soils, bottom stability is affected by the depth of cut, cohesive strength of the soil, and the thickness of clay below the bottom of the excavation. Charts and formulas are given in Figs. 7.51 through 7.54. Additionally, groundwater conditions below the excavation must be checked if there is a pervious layer present. For large excavations, the total weight of the soil between the bottom of the excavation and the top of the previous layer must be greater than the total hydrostatic head at the latter point. For narrower excavations, where the width is less than 1.5 times the thickness of the impervious material, advantage may be taken of the strength the clay; total forces (excavation-wide, for a unit length along the trench) may be considered, then including vertical shear in line with the sides of the excavations. The pertinent relationships are shown in Fig. 7.55.

For wide excavations in deep cohesive soils, where sheeting restraint and/or passive reactions are within the same deposit, the overall stability of the excavation must be checked. This may be accomplished with stability charts, or may require a circular arc or wedge-type analysis (refer to *Navy Design Manual 7.2*, Taylor, or Lambe and Whitman in the Bibliography). In certain cases, it may be appropriate to assume a failure surface that passes through a retaining member (such as tieback), although, more frequently, the critical surface will encompass the entire wall and its restraining members.

Moment Reduction

When a retaining system consisting of steel sheet piling, restrained by one tie level, and penetration into the subgrade is installed in a medium-dense or dense granular deposit, it has been found that the pressure on the sheeting and consequently the moment in the sheet

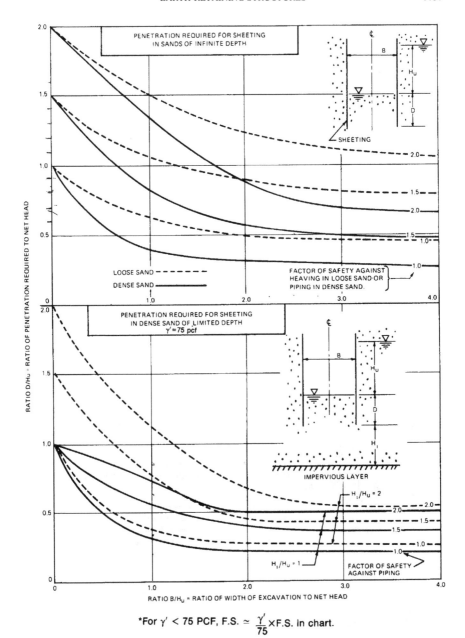

$$\text{*For } \gamma' < 75 \text{ PCF, F.S.} \simeq \frac{\gamma'}{75} \times \text{F.S. in chart.}$$

FIGURE 7.48 Sheeting penetration versus piping in isotropic sand: (*a*) penetration required for cutoff wall in sand of infinite depth; (*b*) penetration required for cutoff wall in dense sand of limited depth. (*From Navy Design Manual 7.1.*)

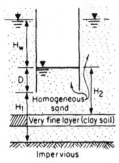

Fine layer in homogeneous sand stratum: If the top of fine layer is at a depth greater than width of excavation below cut-off wall bottom, safety factors of Fig. 7.48 apply, assuming impervious base at top of fine layer.

If top of fine layer is at a depth less than width of excavation below cut-off wall tips, pressure relief is required so that unbalanced head below fine layer does not exceed height of soil above base of layer.

If fine layer lies above subgrade of excavation, final condition is safer than homogeneous case, but dangerous condition may arise during excavation above the fine layer and pressure relief is required as in the preceding case.

To avoid bottom heave, $\gamma_T \times H_3$ should be greater than $\gamma_w \times H$

γ_T = total unit weight of the soil

γ_w = unit weight of water

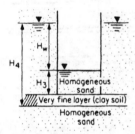

FIGURE 7.49   Sheeting penetration versus piping in stratified sand with impervious layer. (*From Navy Design Manual 7.2.*)

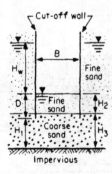

Coarse sand underlying fine sand: Presence of coarse layer makes flow in fine material more nearly vertical and generally increases seepage gradients in the fine layer compared to the homogeneous cross section of Fig. 7.48.

If top of coarse layer is at a depth below cut-off wall bottom greater than width of excavation, safety factors of Fig. 7.48 for infinite depth apply.

If top of coarse layer is a depth below cut-off wall bottom less than width of excavation, the uplift pressures are greater than for the homogeneous cross section. If permeability of coarse layer is more than than 10 times that of fine layer, failure head H_w = thickness of fine layer H_2.

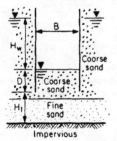

Fine sand underlying coarse sand: Presence of fine layer constricts flow beneath cut-off wall and generally decreases seepage gradients in the coarse layer.

If top of fine sand layer lies below cut-off wall bottom, safety factors are intermediate between those for an impermeable boundary at top or bottom of the fine layer using Fig. 7.48.

If top of the fine layer lies above cut-off wall bottom, the safety factors of Fig. 7.48 are somewhat conservative for penetration required.

FIGURE 7.50   Sheeting penetration versus piping in stratified sand with differing permeabilities. (*From Navy Design Manual 7.2.*)

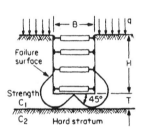

Cut in clay, depth of clay unlimited (T > 0.78)
If sheeting terminates at base of cut:

Safety factor $F_s = \dfrac{N_c\,C}{\gamma_T\,H + q}$

N_c = bearing-capacity factor, Fig. 7.52 which depends on dimensions
of the excavation: B, L and H
C = undrained shear strength of clay in failure zone beneath and
surrounding base of cut
q = surface surcharge

If safety factor is less than 1. , sheeting must be carried below base of
cut to insure stability.
Force on buried length:

If $H_1 > \dfrac{2}{3}\dfrac{B}{\sqrt{2}}$, $P_H = 0.7\,(\gamma_T\,HB - 1.4\,CH - \pi\,CB)$

If $H_1 > \dfrac{2}{3}\dfrac{B}{\sqrt{2}}$, $P_H = 1.5 H_1\left(\gamma_T\,H - \dfrac{1.4 CH}{B} - \pi\,C\right)$

Cut in clay, depth of clay limited by hard stratum (T < 0.7)

Sheeting terminates at base of cut. Safety factor:

Continuous excavation $F_s = N_{CD}\,\dfrac{C_1}{\gamma_T\,\dfrac{H + q}{C_1}}$

Rectangular excavation $F_s = N_{CR}\,\dfrac{C_1}{\gamma_T\,H + q}$

N_{CD} and N_{CR} = bearing capacity factors from Figs. 7.53 and
7.54, which depend on dimensions of the excavation: B, L and H

Note: In each case friction and adhesion on back of sheeting is disregarded.
Clay is assumed to have a uniform shear strength = C throughout
failure zone.

FIGURE 7.51   Stability of base for braced cut. (*From Navy Design Manual 7.2.*)

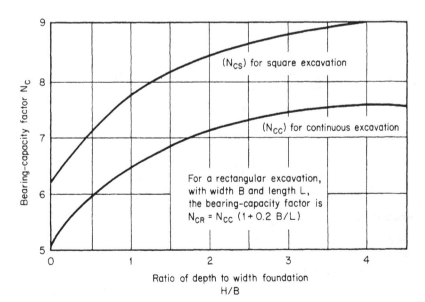

FIGURE 7.52   Bearing capacity factors for deep uniform clays. (*From Navy Design Manual 7.2.*)

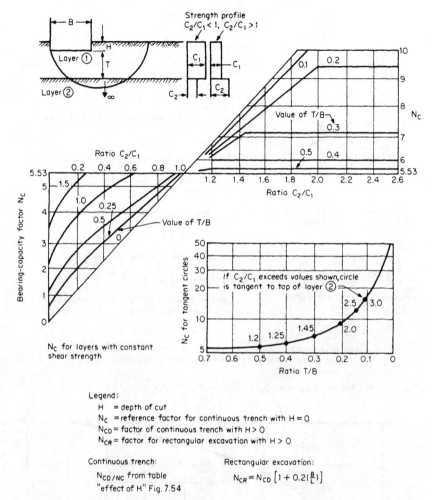

FIGURE. 7.53 Bearing capacity factors for two-layer cohesive soils ($\varphi = 0$). (*From Navy Design Manual 7.2.*)

piling (or other vertical support members), can be reduced in relationship to the flexibility of the sheeting. Using the chart in Fig. 7.56, an allowable reduction factor may be obtained for a given flexibility number (calculated per foot of wall). The flexibility number derives from the moment of inertia of the sheeting.

In summary, the steps are as follows:

1. Determine the maximum moment (per linear foot) by standard calculations for assumed retaining structure configuration.

2. Select an appropriate maximum allowable unit stress in the sheeting material (f_s), note its modulus of elasticity (E), assume a type of sheeting, and identify its moment of inertia (I).

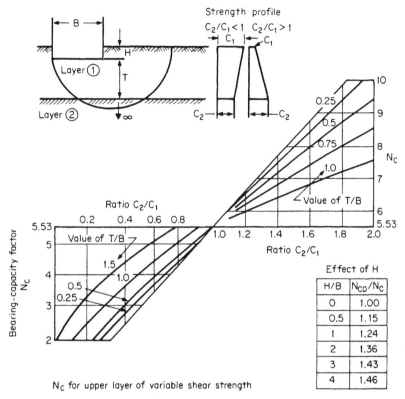

FIGURE. 7.54 Bearing capacity factors for two-layer cohesive soils ($\varphi = 0$).
(*From Navy Design Manual 7.2.*)

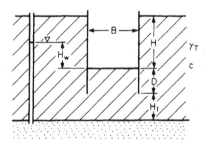

FIGURE 7.55 Excavation in clay with pervious layer below.

3. With the depth of the excavation (H) and the sheeting penetration below the excavation (D), calculate the flexibility number (ρ); then determine the appropriate reduction factor from the diagram.

4. Apply the factor to the normally calculated maximum moment; then calculate the design stress for the sheet piling. If it is not near the maximum allowable stress, select a different size of sheet piling and recalculate until the most efficient sheet piling section is found.

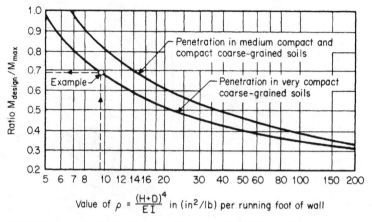

Value of $\rho = \dfrac{(H+D)^4}{E\,I}$ in (in^2/lb) per running foot of wall

FIGURE 7.56   Moment reduction factor. (*From Navy Design Manual 7.2.*)

Allowable Overstress

In designing temporary earth retention structures, it is common to size discrete or continuous, vertical, sheeting members (soldier piles or sheet piles), wales, braces, etc. for a higher bending or shear stress than would be allowed for a permanent structure. This allowable overstress is often 10 to 33 percent of the usual allowable stress, with 20 percent being most commonly used. Where deflections of the support system are critical, use of a temporary allowable overstress may not be desirable, as excessive deflections could result. A temporary overstress may be more appropriate for projects with relatively low, specified design stresses from agencies such as railroads and highway departments. Often, applying a reasonable overstress to their specified, permanent, allowable design stresses will still result in temporary design stresses that are still less than those allowed by the American Institute of Steel Construction (AISC) for similar, permanent structures.

In design of temporary, earth-retaining structures, there are several commonly used methods for reducing the size of structural members. Some of these methods include designing for 80 percent of the earth pressure, using moment reduction, allowing a temporary overstress, and considering soil arching. While each of these methods is referenced in the literature and may be appropriate for a particular design, it could be inappropriate, and possibly unsafe, to simultaneously implement more than one. Therefore, design and construction experience, seasoned with a strong dose of common sense, is critical for safe but economical design and construction of earth-retaining structures.

Factors of Safety

The selection of approximate factors of safety is a critically important matter. With too low a factor, failure, or at least large movements, could occur. If so, replacement of the retaining structure or repair of adjacent buildings or other structures could be required— at considerable financial cost! Using too high a safety factor means an uneconomical, earth-retaining structure that could be the reason for losing a bid or having an unprofitable project. Just what factor of safety is right depends on a number of factors, and experience is the best guide.

As a gross generalization, support for permanent structures might be designed with a safety factor of 2.0; temporary structures might require a safety factor of about 1.5. There are specific situations where lower factors of safety are acceptable, but the selection of a typical factor of safety must rest in part on consideration of the following:

1. Accuracy to which the soil characteristics are known. With a thorough investigation—frequent borings and laboratory testing of representative samples—it may be reasonable to reduce the indicated factors of safety by 10 to 20 percent. More commonly, boring information is limited and there are few, if any, tests. In this case, knowledge of an area would be needed to hold to moderate safety factors. It is hard to pick a high enough factor of safety for situations where soil conditions are unknown or their reporting is unreliable. The only reasonable approach here is to go out and get enough information to allow the use of a moderate factor of safety. It is always possible to use very-low-strength parameters, but this forces an ultraconservative (and high-cost) design.

2. The consequences of failure. Once identified as a factor, this is inherently obvious. So long as the safety of personnel is properly attended to, a relatively high risk factor is allowable on a trench across an open field. But an excavation near existing structures needs a conservative factor of safety to prevent failure or detrimental movement. The responsibility for and cost of repairing a structure, if it should become damaged, should be weighed against the upfront cost of installing an extra strong retaining structure.

3. Extreme soil conditions. Very soft clay should be considered a liquid; its low strength should not be relied upon. If soft to firm clay is stressed to its limit, it will start to yield and then, due to remolding, lose strength. This can occur whenever horizontal pressures exceed 4 times the cohesive strength of the soil. Very loose sands will not "set up" or arch; thus full active pressures must be taken into account at all points. Very dense, granular soils may act as if they have a higher-friction angle than is shown by laboratory test. Note that, with deflection, sand may change density (increase or decrease), and it will then have somewhat different characteristics than those with which it started. Soil parameters and/or the factor of safety must be adjusted accordingly.

4. Need to restrict deflection. A low factor of safety will result in at least some deflections of the soil (and/or stratum) behind the retaining wall. To restrict deflections to very low values, a factor of safety on the order of 2.0 or higher is needed, along with prestressing of the retaining structure and its supports.

5. The conservatism of the design criteria. The commonly used pressure diagrams (Fig. 7.44) were based on the maximum, anticipated, bracing values as determined by actual field measurements. Hence, the inherit factor of safety (usually 1.2 to 1.3 compared to the theoretical earth pressure) for these diagrams may seem to be relatively low. However, when coupled with the factors of safety (i.e., allowable stresses), used with the individual wall components (i.e., approximately 1.5 to 2.0 for soldier piles, sheet piling, braces, wales, tieback anchors, anchor bond stresses, etc.), sufficient total factor of safety is available and it is not necessary to further increase the empirical earth pressure.

Load and Resistance Factor Design

Traditionally, earth-retaining structures have been designed using the *allowable stress design* (ASD) method, also referred to as the *service load design* (SLD) method, which has been previously described in this chapter. For most private projects and non-highway structures, this is still true. More recently, however, most, if not all, state transportation departments have been requiring the use of the *load and resistance factor design* (LRFD) method for the design of both permanent and temporary, bridge and highway structures, including earth-retaining structures. The switch from ASD to LRFD for earth-retaining structures still has not been received well by engineers who learned structural design prior to AISC's 1986 introduction of its LRFD steel design specification. Additionally, while methods for concrete design moved toward strength design LRFD in the 1970s, timber design methods have only recently been updated to include LRFD when the *National Design Specification for Wood Construction* (*NDS*) was updated in 2005.

As a simplified comparison of ASD and LRFD, consider the following: The ASD method uses actual, service loads and allowable stresses for analyzing the structural members, soil, and rock. The allowable stresses are the ultimate or yield stresses reduced by a safety factor. All of the service loads—earth, wind, hydrostatic, live load and dead load surcharges, etc.—are combined and applied to the structural member using one safety factor. The LRFD method uses various load factors (each > 1) and member resistance factors (each < 1). Each different type of load is "factored up" (increased) by a load factor that depends on the accuracy or certainty to which the load is known. The ultimate or yield strength of each structural member—steel, timber, concrete, soil, rock, etc.—is "factored down" (reduced) by a resistance factor that depends on the consistency or certainty of the member material properties and surrounding ground conditions.

In developing the LRFD method for earth-retaining structures, the FHWA attempted to determine load and resistance factors that would result in structure designs that were similar to those designed using ASD, while at the same time, considering the variability or certainty of the loads, materials, and ground conditions. These factors were then adopted, or modified, by AASHTO and individual state transportation departments. For more detailed information on LRFD design methods and load and resistance factors, refer to the FHWA *Geotechnical Engineering Circular No. 4*, Publication No. FHWA-IF-99-015, to the current AASHTO LRFD *Bridge Design Specifications*, and to the appropriate state transportation department's structure design manuals that may be available or downloadable online at the department's publications web site.

BIBLIOGRAPHY

Bowles, J. E.: *Foundation Analysis and Design*, 4th ed., McGraw-Hill, New York, 1988.

Canadian Foundation Engineering Manual, Foundations Committee, Canadian Geotechnical Society, Montreal, March 1978.

Chassie R. G. "Soil Nailing Overview" and "Soil Nail Wall Facing Design and Current Developments," presented at the Tenth Annual International Bridge Conference, Pittsburgh, PA., 1993.

Crofts, J. E., B. K. Menzies, and A. R. Tarzi: "Lateral Displacements of Shallow Buried Pipelines due to Adjacent Deep Trench Excavations," *Geotechnique*, vol. 27, no. 2, 1977, p. 165.

Das, B. M.: *Principles of Geotechnical Engineering*, 3d ed., PWS Publishing Co., Boston, MA., 1994.

Design Manual 7.1: Soil Mechanics, Department of the Navy, Naval Facilities Engineering Command, May 1982.

Design Manual 7.2: Foundations and Earth Structures, Department of the Navy, Naval Facilities Engineering Command, May 1982.

Diaphragm Walls and Anchorages, conference proceedings, Institution of Civil Engineers, London, 1975.

Dunnicliff, J.: *Geotechnical Instrumentation for Monitoring Field Performance*, NCHRP 89, Transportation Research Board, National Research Council, Washington, D.C., 1982.

"Earthquake Resistant Design for Civil Engineering Structures," *Earth Structures and Foundations in Japan*, The Japan Society of Civil Engineers, 1977.

"Earth Reinforcement—New Methods and Uses," *Civil Engineering*, vol. 49, no. 1, January 1979.

Fang, H. Y.: *Foundation Engineering Handbook*, 2d ed., Van Nostrand-Reinhold, New York, 1991.

Field Design Standards, New York City Transit Authority, Engineering and Construction Department, New York, January, 1988.

Goldberg, D. T., W. E. Jaworski, and M. D. Gordon: *Lateral Support Systems and Underpinning*, vols. I, II, III, Federal Highway Administration Report no. FHWA-RD-75-130, Washington, D.C., 1976.

Ground Engineering, conference proceedings, Institution of Civil Engineers, London, 1970.

Lambe, T. W. and R. V. Whitman: *Soil Mechanics*, Wiley, New York, 1969.

Leonards, G. A.: *Foundation Engineering*, McGraw-Hill, New York, 1962.

Lindahl, H. A. and D. C. Warrington: *Sheet Pile Design by Pile Buck*, Pile Buck International, Inc., Vero Beach, 2007.

"New York City Building Code," International Code Council, 2008.

"Occupational Safety and Health Standards—Excavation and Trenching; Title 29 of the Code of Federal Regulation (CFR), Part 1926.650," Occupational Safety and Health Administration, Department of Labor, Washington, D.C., 2002.

Open Cut Construction, Temporary Retaining Structures, seminar proceedings, Metropolitan Section ASCE, New York, 1975.

Peck, R. B.: "Deep Excavations and Tunneling in Soft Ground," International Conference on Soil Mechanics and Foundation Engineering, Mexico City, 1969, pp. 225–290.

Powers, J. P., A. B. Corwin, and P. C. Schmall: *Construction Dewatering and Groundwater Control*, Wiley, New York, 2007.

Prentis, E. and L. White: *Underpinning*, 2d ed., Columbia University Press, New York, 1950.

Recommendations for Prestressed Rock and Soil Anchors, Post-Tensioning Institute, Phoenix, 2004.

Sabatini, P. J., D. G. Pass, and R. C. Bachus: *Geotechnical Engineering Circular No. 4, Ground Anchors and Anchored Systems*, Federal Highway Administration Report no. FHWA-SA-99-015, Washington, D.C., June 1999.

Schmall, P. C., D. K. Mueller and, K. E. Wigg: "Ground Freezing in Adverse Geology & Difficult Ground Conditions." Proceedings of the North American Tunneling Conference (NATC), Chicago, IL, 2006.

Schnabel, H. and H. W. Schnabel: *Tiebacks in Foundation Engineering and Construction*, 2d ed., McGraw-Hill, New York, 2002.

Schuster, J. A.: "Controlled Freezing for Temporary Ground Support," *Proceedings*, Rapid Excavation and Tunneling Conference, Chicago, 1972.

Steel Sheet Piling Design Manual, United States Steel, Pittsburgh, updated and reprinted by the U.S. Department of Transportation/FHWA, Washington, D.C., July, 1984.

Taylor, D. W.: *Fundamentals of Soil Mechanics*, New York, 1948.

Teng, W. C.: *Foundation Design*, Prentice-Hall, Englewood Cliffs, New Jersey, 1962.

Terzaghi, K. and R. B. Peck: *Soil Mechanics in Engineering Practice*, Wiley, New York, 1948.

Tuozzolo, T. J.: "Soil Nailing: Where, When and Why—A Practical Guide." Proceedings, 20th Central Pennsylvania Geotechnical Conference, Hershey, PA 2003.

Weatherby, D. E.: *Design Manual for Permanent Ground Anchor Walls*, Federal Highway Administration Publication no. FHWA-RD-97-130, Washington, D.C., September, 1998.

CHAPTER 8
DIAPHRAGM/SLURRY WALLS*

GianCarlo Santarelli, Civil Engineer

---

*This chapter is a revision and update of *Diaphragm/Slurry Walls* from the second edition of this book, where it was co-authored by GianCarlo Santarelli and Robert T. Ratay.

GENERAL

Definition

The term *diaphragm* or *slurry wall* refers to a reinforced concrete wall constructed below ground level, utilizing the slurry method of trench stabilization. A trench with the width of the intended thickness of the concrete wall is excavated. As the digging progresses, the excavated material is immediately and continually replaced by a bentonite slurry mix. The lateral pressure created by the slurry prevents the sides of the trench from collapsing. After the trench is excavated to its final elevation, caged rein-forcing steel is lowered through the slurry. Concrete, placed by the tremie method, fills the trench from the bottom up, displacing the lighter-weight bentonite slurry, which is pumped away (see Fig. 8.1). Once the concrete cures and the wall attains its intended strength, the soil is excavated from one side and lateral bracing is installed, making the wall a temporary, and then later a permanent, retaining or foundation wall.

History

Slurry wall construction is a fairly new technology, developed in the early 1950s. Originally it was called cast-in-place wall or continuous diaphragm wall construction. The technique was developed practically at the same time in 1952–1953 by two engineers in Italy.

Bentonite slurry (drilling mud) has been employed for rotary drilling in the oil-drilling industry since 1901. Consequently it was adapted to the foundation industry for drilled pile installation. Through the experiences acquired by the observations that the bentonite slurry would support circular-shaped excavation, the next step was to apply the same technique to secant piles installation with the concrete tremie pipe method (placing concrete through a pipe from the bottom of the trench up). The most important development of the technique was the assertion that the bentonite slurry would support not only circular but rectangular-shaped excavation as well. Originally the length of a rectangular excavation (panel) was in the 6-ft (2-m) range. Gradually with the advances of experience and technology, the lengths of excavation have been extended up to 20 and then 30 ft (6 and then to 9 m) lengths.

The first principal tool for the rectangular excavation of the trench was the free-hanging mechanical clamshell, a current example of which is shown in Fig. 8.2. Another system of trench excavation was developed, mainly for difficult soil conditions and deep excavations, called "reverse circulation." It is described under that heading later in this chapter.

It was soon realized that the excavation by the bucket was more economical, and with the advent of better and more advanced clamshells this method of excavation became the preferred one in the industry. Today the majority of the slurry walls throughout the world are still excavated by the use of buckets. Nevertheless, during the last decade there has been an increase in the use of the reverse circulation method for projects with unique characteristics.

After the first applications in Italy and then through Europe in the early 1950s, the slurry wall technique was introduced to the North American continent first in 1956 in Canada and then in 1963 in the United States. For the first 15 years there was reluctance from the North American engineering community to use this technique, except in extreme situations, owing to the limited knowledge of it and the lack of experience in local projects. In addition, because of the absence of specialized slurry wall contractors, the cost for diaphragm wall installation was very high. With the establishment of specialized local contractors and with the experiences and good results from the early 1970s, the diaphragm wall technique became more and more accepted and since the early 1980s has become very widely used in the United States.

As real estate, particularly in congested metropolitan areas, is becoming scarce, and as environmental considerations assume an increasingly important role, the need for extensive

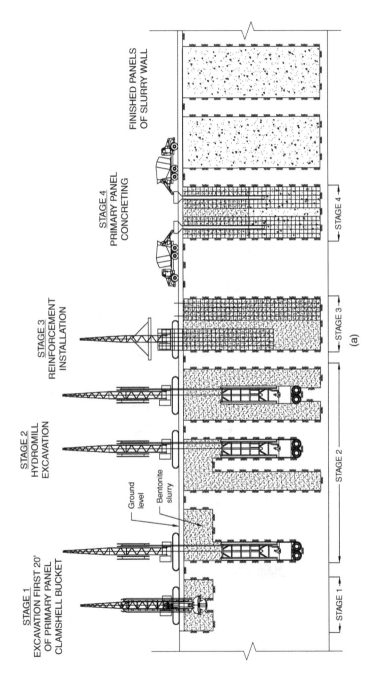

FIGURE 8.1 Typical construction sequences of slurry walls shown in four stages: (*a*) primary panels.

8.3

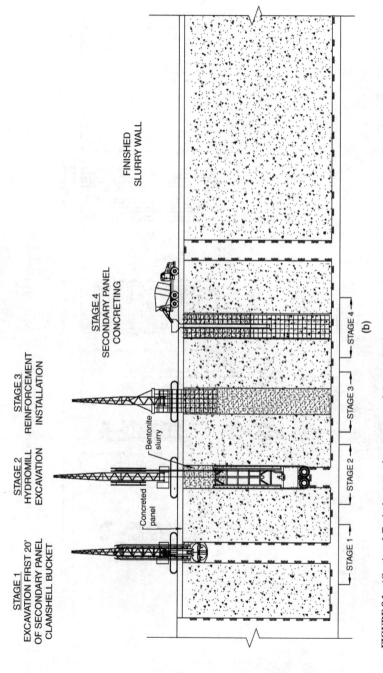

FIGURE 8.1 *(Continued)* Typical construction sequences of slurry walls shown in four stages: *(b)* secondary/closing panels.

(a)

(b)

FIGURE 8.2 Typical high-performance tools for slurry wall excavation: (*a*) free-hanging mechanical clamshell. (*b*) Trench cutter (hydromill). (*From BENCOR Corp.*)

8.5

deep construction in cities is on the rise. There is a trend in urban areas to build garages, mass transit, and highways underground in order to reduce congestion and pollution. The construction of these underground facilities frequently has to take place in the immediate vicinity of existing buildings and utilities, which results not only in tight construction sites but also in stringent constraints to limit wall movements of both the temporary and permanent earth retainage systems.

Uses and Advantages

The slurry wall technique allows the execution of projects that may not be feasible without this technology. Diaphragm walls are used for the construction of continuous vertical concrete walls cast in place in a trench excavation, thus forming a vertical diaphragm in the natural or filled soil. If properly planned and executed, it results in a watertight structural retaining wall with a minimum soil decompression and movement. It also enhances the construction of below-ground walls through layers of dense soil where the use of sheet piling is not feasible. The diaphragm wall method has numerous applications, with both reinforced and unreinforced concrete.

Reinforced diaphragm walls are used to form permanent or temporary structural retaining walls and structural bearing elements. The range of applications includes subways, depressed highways, quays, basements, retaining walls for excavations, cofferdams, and shafts. In most cases, partial or nearly total impermeability is achieved in addition to the load-bearing functions. Unreinforced diaphragm walls are used to form impermeable cutoff walls below dams, levees, and other hydraulic structures.

Diaphragm walls often serve as temporary earth retainage structures, but they possess a unique attribute compared to other commonly used earth retainage systems (e.g., soldier piles and lagging, sheet piling) in that they can be an integral part of the permanent structure. This attribute of diaphragm walls, perhaps more than any other, contributes to their efficiency, and in fact has resulted in the development of some new techniques such as "top-down" construction.

Some other useful characteristics of diaphragm walls are as follows:

1. They are rigid.
2. They can be made watertight.
3. They have the ability to support substantial vertical and lateral loads.
4. They are installed virtually without any vibration.
5. They can be constructed in low-headroom areas.

Through the years, a variety of different procedures and applications has evolved around the diaphragm wall technique to allow the safe and economical execution of projects, which would have been impossible or extremely expensive with traditional techniques. (A detailed discussion is presented under "Applications and Uses.")

INSTALLATION

Jobsite Planning

The on-site operation for a diaphragm wall installation is equipment-intensive; it is necessary to plan the logistic and operational phases well ahead of construction. The planning evolves around the diaphragm wall shop drawings that have to show panel sizes, their sequence of installation, top of concrete elevation, type of reinforcing, etc., taking into account the installation time, the removal of the excavated material and waste bentonite, the fabrication and placement of resteel, and the delivery and placement of concrete.

For major or difficult projects it is wise to produce a site-utilization plan locating storage, fabricating, and mixing areas, as well as a plan detailing the phasing of the work in accordance with the overall project schedule. It is important to keep track of the preparatory work. For example, if not enough guide walls are installed ahead of the trenching operation, there is a risk of delays and/or stoppage of the excavation work. It is also important to plan for the possible modification of the installation sequence and panel sizes in the event of encountering excavation problems, bentonite losses, obstructions removal, utilities relocation, etc. In the overall picture, one needs to be aware that a diaphragm wall jobsite is generally muddy, which can create problems regarding vehicular site accessibility.

Use of Excavation Fluids

A bentonite water slurry is used to stabilize the trench and support the sides. The bentonite suspension has thixotropic properties that, in pervious soils such as sand and gravel, hold in place the particles of in situ soil and form a filter cake or membrane on the sides of the trench. It gels when left undisturbed and becomes fluid when agitated. The slurry that has penetrated and become entrapped in the surrounding soil forms an impervious barrier against the ingress of water. When the excavation tool penetrates the fluid to continue excavating the trench bottom, it does not affect the membrane already formed on the sides of the trench. By carefully maintaining sufficient bentonite suspension (density) in the trench, an excess hydrostatic pressure is exerted against the two side membranes during both excavation and the subsequent concreting operation to support the sides of the excavation.

Properties and Preparation of Bentonite Slurry

Bentonite slurry is essentially a mix of water and a special clay called *bentonite* (Fig. 8.3). The name bentonite derives from the mining area of Fort Benton, Wyoming. The Wyoming bentonite, or sodium montmorillonite, is the most commonly used clay to obtain desired viscosity and fluid loss control. It has the ability to become hydrated with water many times its own weight, thus forming a suspension occupying a volume several times that of the original solid. It also has the property of forming a gel or semi-plastic solid when allowed to stand quiescent for a short period of time. Agitation will quickly destroy this gel, restoring the fluid state. This property is reversible and is known as thixotropy.

Generally good bentonite slurry is obtained by mixing water with 5 to 6 percent by weight of bentonite powder. In some instances, owing to the water quality or the soil conditions, it is necessary to modify the water-bentonite mix ratio and to add different types of additives. The same applies when working in the presence of seawater where more sophisticated additives are needed to maintain the slurry properties. (In the oil industry the mixing, control, and treatment of the drilling mud is more sophisticated and elaborate in order to handle various difficulties of the deep drilling range and geological diversity.) Because of the high volume of slurry employed in diaphragm wall construction, it is not economically feasible to use expensive additives and sophisticated treatment techniques.

A basic difference between oil drilling and diaphragm wall construction is that the latter includes not only the action of drilling a hole but also the activity of pouring a concrete wall that needs to be watertight. In order to keep the trench from caving in, it is desirable

FIGURE 8.3 Automatic bentonite mixing plant. (*From BENCOR Corp.*)

to employ a thick rich slurry. However, the use of this thick slurry during the concrete pouring operation carries the inherent risks of poor concrete flow and thus the possibility of trapping heavy bentonite pockets within the concrete, a poor jointing system between panels, and other finished product defects. For these reasons, the optimum consistency of the bentonite slurry during concrete pouring should be very lean, looking like dirty water. In order to achieve a happy medium, it is necessary to solve these contradictions on each project based on experience.

To help in the control of bentonite slurry, four basic testing procedures that are borrowed from the oil-drilling industry are employed:

- Control of density by the *mud balance*
- Control of viscosity by the *marsh funnel*
- Sand control by the *sand content tube*
- Filtration property and filter cake property control by the *filter press*

All these testing procedures are well described in specialized publications, some of which are listed in "Bibliography" to this chapter. Particularly useful is the *Recommended*

Practice and the Specification for Drilling-Fluid Materials of the American Petroleum Institute.

The bentonite and water slurry is mixed in a high-speed mixer. As water is introduced into the mixer tank under pressure, the bentonite powder is added to the water and the bentonite-water slurry is recirculated by a high-speed pressure pump until it reaches homogeneity. Once the homogeneity of the mix has been attained, the slurry is sent to storage tanks for full hydration and storage.

During the excavation operation the bentonite slurry acquires suspended solids, in most cases sand. It is then necessary to circulate the bentonite slurry through a shaker screen and/or centrifugal cyclones in order to reduce the suspended solids to the specified maximum allowed quantity. Shaker screens (see Fig. 8.4) are used for the removal of coarser material and hydrocyclones for the removal of finer material. Another simple method to reduce the solids in suspension is to let them settle at the bottom of a used bentonite slurry settlement pond. This cannot be achieved if the bentonite slurry is too thick, thus impeding the settlement of the solids. It is to be noted that the disposal of contaminated bentonite is costlier than producing fresh slurry.

FIGURE 8.4   Shaker screen for coarse solids removal. (*From BENCOR Corp.*)

The bentonite slurry treatment operation is more complex during the reverse circulation process where the soil (solids) and the slurry are fully mixed together, as discussed later in this chapter. The need for extensive slurry treatment also occurs during the bucket excavation procedure, if it is accompanied by an extensive amount of percussion excavation by chisels, because soil and rock fragments become blended with the bentonite slurry.

Guide Walls

A pair of concrete guide walls are constructed to establish the location of the diaphragm wall and to prevent the top of the trench from falling in. A trench is excavated along the

(a)

(b)

FIGURE 8.5   Concrete guide walls: (*a*) Guide wall with temporary bracing, (*b*) Using styrofoam as a concrete form for circular shafts. (*From BENCOR Corp.*)

alignment of the diaphragm wall to a width that will accommodate two lightly reinforced concrete guide walls (see Fig. 8.5) with a slight clearance to allow passage of the excavating tools.

Where possible, the trench for the guide walls is excavated to undisturbed solid ground. The guide walls themselves are formed on their inner face, and in the majority of cases are cast directly against the ground on their outer faces. Where the ground is very unstable or recently filled, it is necessary to form the walls on their outer faces also, as is seen in Fig. 8.6. Once the concrete forms are removed, it is necessary to brace, as in Figs. 8.5 and 8.6, or backfill the inside of the guide walls in order to prevent their lateral movement. There are several configurations of guide wall construction, a few of which are shown in Fig. 8.6. Some common problems and their remedies are illustrated in Fig. 8.7.

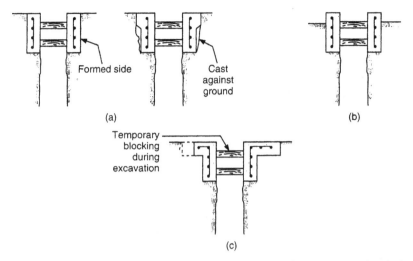

FIGURE 8.6   Concrete guide wall details: (*a*) Simple configurations in stable ground. (*b*) Raised above grade for higher-than-ground slurry level. (*c*) Shape in unstable ground or adjacent to heavy surcharge loads. (*After Xanthakos. Slurry Walls as Structural Systems, 2d ed., McGraw-Hill.*)

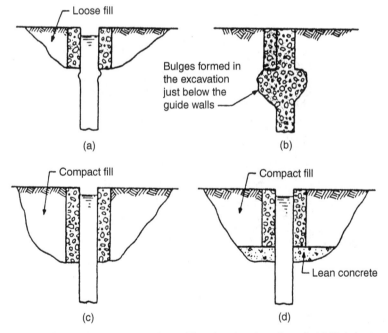

FIGURE 8.7   Guide-wall construction problems in soft or in caving soil: (*a*) Variation in slurry level induces flow in outside soft soil and creates cavities. (*b*) Shallow guide walls and loose fill behind high water table cause the formation of bulges. (*c*) Preventive action provided by deep guide walls and compact fill behind. (*d*) Preventive action provided by lean concrete at the base of guide walls and compact fill. (*From Xanthakos: Slurry Walls as Structural Systems, 2d ed., McGraw-Hill.*)

Excavation Tools

Buckets.   The first slurry wall clamshell buckets were free-hanging two-cable buckets operated by two drum winches. There are round jaws and square jaws. The excavation by the round jaws results in a trench with round ends, which facilitates the placement of the panel-to-panel stop-end joint pipes. The square jaws are more appropriate when square panel-to-panel joints are used in the wall construction.

The jaws can be fitted with different teeth so as to facilitate the excavation process in soils of different densities. For easy and soft soil types, that is, clay, the employment of teeth on the bucket jaws is not needed; generally a serrated-type jaw will suffice.

For the free-hanging excavation bucket, a system of guiding is needed to minimize the rotation caused by the holding cable of the bucket. To ensure a vertical excavation, the bucket is equipped with a top guide that is of the same thickness as the jaws (see Fig. 8.8). This also adds weight, which improves the bucket performance.

FIGURE 8.8   Free-hanging clamshell with long top guides. (*From BENCOR Corp.*)

Another type of excavating bucket is attached to a Kelly bar (see Fig. 8.9). The Kelly can be tubular, telescopic, or simply a large H-beam. The mechanism is operated from a standard crane by means of a specially designed attachment, including a Kelly bar and a Kelly bar guide. The use of the Kelly bar ensures an easier operation for the insertion and removal of the bucket into and from the guide walls, thus increasing the cycles of excavation per unit of time. In addition, the weight of the Kelly helps the excavating bucket's penetration into the soil. One of the drawbacks of the Kelly equipment is that for difficult soil conditions it is necessary to use percussion tools (see Fig. 8.10) to remove cobbles or

FIGURE 8.9 Excavating bucket attached to telescoping Kelly bar. (*From BENCOR Corp.*)

FIGURE 8.10 Percussion chisels to break up boulders and penetrate rock. (*From BENCOR Corp.*)

boulders to penetrate the rock. In those conditions another crane is necessary to employ the percussion tools.

A good slurry wall bucket needs, above all, to be simple and strong. It needs to be of sturdy construction in order to be able to act as a chisel with open jaws in dense soil. Some of the buckets at the site need to be equipped with different types of teeth to deal with the different types of soil.

For the past 40 years the main tools to excavate deep diaphragm walls have been cable-suspended or Kelly guide clamshells and percussion chisels based on the direct circulation system, but new and progressive techniques have been developed in Japan and Europe.

Reverse Circulation System. The difference between direct and reverse circulation consists of the following: By direct circulation the soil is excavated with a clamshell and is disposed of, as the bentonite slurry is pumped into the trench to replace the excavated material. By reverse circulation the excavation is performed with the aid of a slurry trench cutter (hydromill), and the excavated soil and cuttings are conveyed to the surface through the drill stem with the aid of direct suction or air lift. There is continuous recirculation of the drilling fluid that is brought up through reverse lines, and then the soil cuttings and the bentonite slurry are separated by means of a shaker screen and desander, and the drilling fluid is returned to the trench through the supply lines.

One of the systems developed in Japan in the early 1970s is the BW system, the originator of rectangular reverse circulation excavating technique. The BW system is the excavation method using the BW long wall drill developed by the Tone Boring Co., Ltd., in Tokyo, Japan. The system uses a submersible excavator with multiple rotary drill bits. Reverse circulation is employed for the continuous removal of trench cuttings, and a suction pump is generally used to remove the spoils. One of the drawbacks of the BW system of excavation has been the inability to excavate bouldery soils and rock.

Another, further development from the BW system is the slurry trench cutter developed by several European companies. This equipment consists of a frame with two hydraulic drives and cutter wheels attached to its base (see Figs. 8.11 and 8.12). The drives rotate on horizontal axles in opposing directions. The cutter wheels equipped with different teeth continuously loosen and break up the soil material that is mixed with the bentonite slurry (see Fig. 8.13). The spoil-rich slurry is pumped through hoses to a bentonite treatment plant for removing the spoil in suspension, regenerated, and eventually returned to the trench. The set of equipment including the cutter, desander, and bentonite treatment facilities at a jobsite is shown in Fig. 8.14.

Jointing of Panels

After the excavation of the trench for a panel is completed, joint equipment is installed in the unconcreted end of the trench to provide for connecting to the next panel. One of the most widely applied methods is the use of the stop-end pipe (see Fig. 8.15). This system consists of installing a pipe of slightly smaller diameter than the width of the excavated trench (approximately 2 in (5 cm) smaller). The length of the joint pipe is determined by the depth of the panel and site restrictions. Joint pipes vary in lengths of 5 to 50 ft. (1.5 to 15 m). The connection between ends of sections of pipe should be strong enough to allow for later extraction by hydraulic jacks or other means. Crawler cranes or hydraulic jacks are used for the gradual pulling up of the joint pipes. Once the joint pipe has been extracted, a concave joint surface remains for the jointing of the next neighboring panel. Water leakage at the panel to panel joint can be prevented by appropriate scraping of the concrete in the semicircular concave face of the installed panel so that contaminating material does not remain sandwiched between adjoining panels.

Other types of panel joints have also been developed, for example, wide-flange beams left in place (see Fig. 8.16) or custom stop-end element shaped joints (see Figs. 8.17 and 8.18).

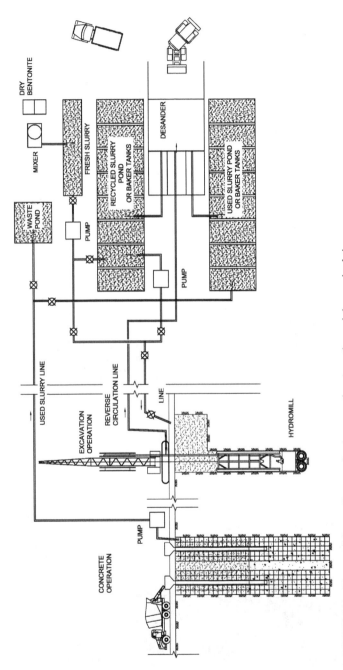

FIGURE 8.11 Diagram of slurry wall system showing excavation, concreting, and slurry recirculation.

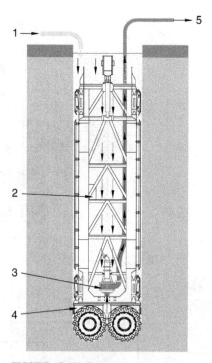

FIGURE 8.12 Hydromill cutters with guide plates. (*From BENCOR Corp.*)

FIGURE 8.13 Hydromill flow diagram. (*From BENCOR Corp.*) 1. Clean bentonite 2. Descended control position 3. Centrifugal pump 4. Hydromill Head 5. Bentonite and cuttings to treatment plant

FIGURE 8.14 Hydromill with supported equipment. (*From BENCOR Corp.*)

FIGURE 8.15 Stop-end pipes extended up from slurry trench. Note heavy steel framing to hold pipes in place. (*From BENCOR Corp.*)

FIGURE 8.16 Wide-flange beam left in place as both panel-to-panel joint and vertical reinforcement. (*From BENCOR Corp.*)

FIGURE 8.17 Stop-end element to make shaped panel to panel joints. (*From BENCOR Corp.*)

FIGURE 8.18 Panels with locking joint. (*From BENCOR Corp.*)

The latest improvement for slurry walls joint connection is technique that allows the connection of well-defined joints with the installation of water stop (see Fig. 8.19).

When the hydromill excavation equipment is used, it will excavate a joint directly into the concrete of two adjacent panels ensuring a tight and clean connection (see Fig. 8.20).

Reinforcement

Reinforcement is made up usually on site and formed into three-dimensional cages for each panel. The cages are picked up by cranes (see Fig. 8. 21) and lowered into the trench through the bentonite slurry. One of the most asked questions is whether the presence of

(a)

(b)

FIGURE 8.19 Water stopper joints: (*a*) stop-end element with water stop, (*b*) stop-end element in place. (*From BENCOR Corp.*)

(c)

FIGURE 8.19   (*Continued*) Water stopper joints: (*c*) joint shape. (*From BENCOR Corp.*)

the bentonite slurry reduces the bonding of the concrete to the resteel bars. Based on many years of diaphragm wall installation and after several on-site studies, it has been widely accepted and demonstrated that the film of bentonite mud on the resteel is automatically removed by mechanical contact with the concrete, that is, by the drag of the flow of the concrete around the bars.

Since reinforcing steel cages are lowered into the trench by a crane, they need to be of certain rigidity. Large and heavy cages are usually spot-welded at bar crossings to maintain their shape during handling and placing. They can also be attached to bulkheads that provide stability to the cages and at the same time serve as panel-to-panel jointing material (see Fig. 8.22). (Note the temporary styrofoam blocking placed against the web of the wide flange. Its purpose is to keep soil away from the steel surfaces for easier cleaning before concreting.) In cases of deep diaphragm walls the cages are fabricated in several sections and then assembled over the trench. The first section is partially lowered into the trench, suspended by the guide walls. The next section is positioned on the top of the already installed section and is coupled to it by means of mechanical couplers or by overlapping and tying together with cable clips.

In order to center the steel cages in the trench and to ensure the required concrete cover in both faces of the wall, spacers are attached to the cages. Either concrete block roller spacers or steel plate spacers are used for this purpose.

Another type of slurry wall reinforcement is in the so-called soldier-pile-tremie concrete (SPTC) walls. This type of wall was first installed in San Francisco in the early 1960s for the construction of several subway stations for the Bay Area Rapid Transit System. The diaphragm wall is made by first installing vertical steel beams at several-foot spacings in the excavated slurry trench. The beams (soldier piles) will serve both as vertical reinforcing and as panel joints. Concrete is placed by the tremie method into the trench between the

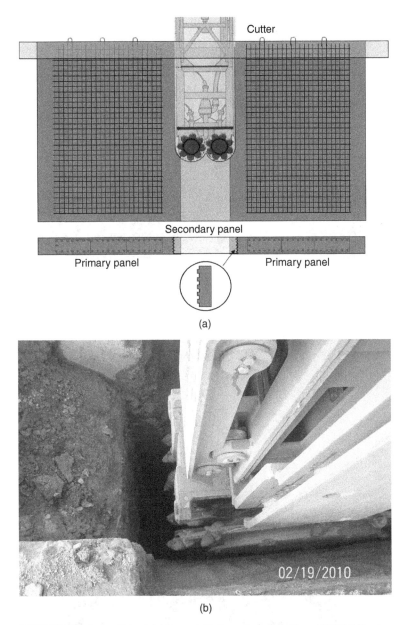

FIGURE 8.20 Hydromill joint. (*a*) Concept. (*b*) Excavation panel. (*From BENCOR Corp.*)

FIGURE 8.21 Reinforcing steel cage being lifted by crane. (*From BENCOR Corp.*)

FIGURE 8.22 Reinforcing steel cage attached to wide-flange beam bulkheads. (*From BENCOR Corp.*)

FIGURE 8.23 "Soldier pile tremie concrete" (SPTC) composite wall. (*From BENCOR Corp.*)

vertical steel beams (see Figs. 8.17 and 8.23). This type of wall is well suited where high vertical bending strength is required.

Tremie Concreting

Placement of the concrete for a diaphragm wall is by the tremie pipe method. Concrete is discharged continuously through a tremie pipe that reaches from the ground surface down to near the bottom of the trench (see Fig. 8.1). The concrete mix is designed to flow easily and without segregation. The slump needs to be a minimum of 8 to 9 in (20 to 22.5 cm) to allow adequate fluidity. The rate of concreting is usually 30 to 90 yd^3/h (22 to 66 m^3/h).

Concrete placement should begin as soon as feasible after the reinforcing steel has been installed. This will enable better removal of the bentonite film from the reinforcement, thus achieving a good-quality bond between the concrete and the reinforcing steel.

Flush-joint-type tremie pipe columns of 8 to 10 in (20 to 25 cm) diameter are lowered to the bottom of the slurry trench. The pipes are equipped with a quick-disconnecting jointing system in order to facilitate the removal of the pipe sections. Based on the horizontal length of the panel, more than one tremie pipe is placed. The ideal horizontal length of a panel is between 6 and 10 ft (1.8 to 3 m), but there have been cases where up to five tremie pipes have been placed in one long panel (see Fig. 8.24).

Concrete is introduced into the tremie pipes through hoppers (see Figs. 8.17 and 8.24). The concrete travels from the hoppers to the bottom of the trench through the pipes by gravity flow. Since the concrete's density is greater than that of the bentonite, the bentonite is displaced at the top of the trench as the concrete rises.

Care should be taken to keep the bottom of the tremie pipes 5 to 10 ft (1.5 to 3 m) below the top of the concrete. However, the embedment of the tremie pipe into the concrete should not be excessive in order to avoid poor flowing of the concrete.

FIGURE 8.24   Concrete being placed through five tremie pipes simultaneously. (*From BENCOR Corp.*)

The concrete placement is completed once the concrete has reached the design elevation in the trench. It is preferable to have the final top of concrete elevation right at the top of the guide walls in order to better control the top finish, but it is possible to stop the concrete pour several inches or feet below the guide wall level, if so required by the design.

It needs to be noted that owing to the contamination of the top of the concrete with the bentonite slurry, the top layer of the wall is of inferior concrete quality and needs to be trimmed off until good concrete is reached [usually 10 to 20 in (25 to 50 cm) below the top].

Wall Bracing Systems

During excavation of the work area the diaphragm wall needs to be supported against the lateral earth pressure, hydrostatic pressure, and surcharges stemming from nearby structures. The selection of the bracing system has to take into account the shape and form of the diaphragm wall, the forces to be resisted, allowable lateral movements, site constraints, economical availability of bracing material, etc.

The following is a list and brief summary with illustrations of the different commonly used bracing systems.

Tieback Anchors.   The tieback bracing system is widely used and offers the great advantage of a clear and unobstructed area of excavation (see Fig. 8.25). One possible problem in the use of the tieback system is the requirement of an easement by adjacent property owners if the tiebacks extend into other private properties.

Cross-Lot Bracing.   In the case where the construction consists of two parallel walls and the distance between them is not excessive, such as subway or below-ground roadway tunnels, the cross-lot bracing support system is feasible and most likely economical. The braces can be steel wide-flange beams (Fig. 8.26), unbraced steel pipes, or other shapes.

FIGURE 8.25 Diaphragm walls with tiebacks for a deep excavation. (*From BENCOR Corp.*)

FIGURE 8.26 Wide-flange beams used as cross-lot wall bracing. Brace is point-bearing against diaphragm wall. (*From BENCOR Corp.*)

FIGURE 8.27   Cross-lot bracing with steel pipes for an underground transportation tunnel. (*From BENCOR Corp.*)

They can bear directly against the concrete wall or against waler beams (Fig. 8.27). In order to employ this system, it is necessary that the lateral earth pressures on the two opposite walls be similar in distribution and magnitude.

Combined Supports.   Dimensions, soil conditions, or jobsite constraints sometimes require the use of more than one type of wall support on the same project. A combination of tiebacks in the long sidewalls and corner struts stabilizing the short end wall against the two sidewalls of a subway station or building foundation projects is shown in Fig. 8.28. Particular attention must be paid to the compatibility, especially the relative stiffnesses, of the different types of supports against one wall.

Interior Inclined Bracing.   In a case where there is a relatively shallow cut and the soil in the bottom of the excavation is of firm and relatively incompressible nature, the use of the inclined bracing system is feasible and probably economical (see Fig. 8.29).

Berm and Permanent Floor Bracing.   Another method of bracing is the one utilizing components of the permanent structure, such as floor slabs. The diaphragm wall is partially excavated; then an appropriate berm is left while completing the excavation of a smaller interior area to the final grade. The center section of the structure is then built. The permanent floor slab is extended to the diaphragm wall and used as the final bracing at that level. As the slabs, one below the other, are extended out to the diaphragm wall, the berm is gradually removed and the next level slab is installed until the final bottom of the excavation of the entire area within the walls is reached. One of the drawbacks of this method is that part of the final excavation is done underneath the installed slab, which creates limitations on the equipment usage and production rates.

Top-Down Construction.   Although this is an overall system of multilevel basement construction, it is outlined here because of its efficiency with diaphragm walls. It involves the construction of successive basement levels from the ground down to the lowermost subbasement (see Figs. 8.30 and 8.31). Construction begins with the installation of the

FIGURE 8.28   Mixed tieback and corner strut bracing. (*From BENCOR Corp.*)

FIGURE 8.29   Inclined bracing of a diaphragm wall for a relatively shallow excavation. (*From BENCOR Corp.*)

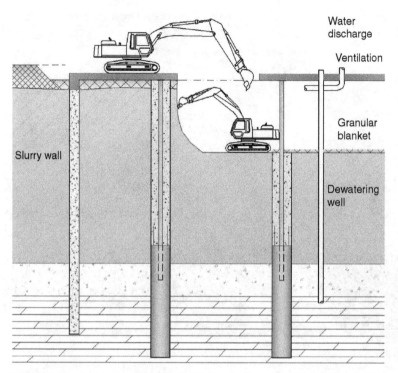

FIGURE 8.30 Berm and permanent floor bracing of a diaphragm wall. (*After Xanthakos: Slurry Walls as Structural Systems, 2d ed., McGraw-Hill.*)

perimeter foundation and basement walls by the slurry trench method. Interior column foundations are drilled in from the ground level to the bearing stratum using either conventional drilled shaft or slurry trench techniques. The shafts are concreted up to the lowermost basement level to form foundation piers. Cast-in-place concrete or structural steel columns are installed in the open shafts to rise from the foundation piers to the ground-floor level. The ground-floor slab is cast either on a mud mat over unexcavated soil or on a drop-down form system anchored to the columns. After the slab is cast and cured, its diaphragm action makes it a continuous lateral brace against the perimeter walls. The soil is excavated by an operation similar to low-headroom horizontal mining below the previously completed structural slab. Access openings have to be left in the overhead slab for vertical soil removal and for supply of equipment and materials. The next lower level basement slab is poured, becoming the next lower level wall bracing, and the process is repeated down from level to level. Thus each floor acts as both temporary and permanent bracing for the perimeter walls, eliminating the need for temporary bracing or tiebacks, as well as the need for separate retaining walls. An additional great advantage of this technique is the ability of the system to minimize wall and soil movements, and consequently minimize or even prevent settlements of adjacent structures.

An extension of the top-down system is the up-down construction of buildings, whereby erection of the superstructure begins as soon as the perimeter walls, the interior columns below the ground floor, and the ground-floor slab are completed. Thence the constructions of the subbasement floors and of the superstructure proceed simultaneously, saving significant project construction time.

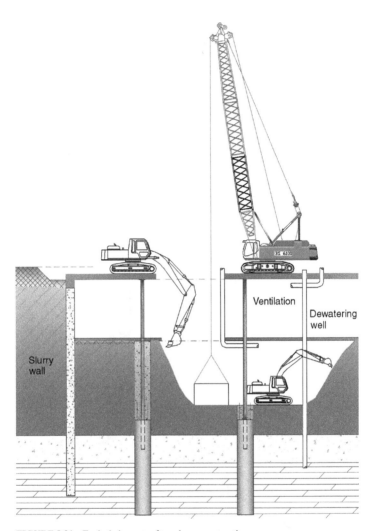

FIGURE 8.31 Typical elements of top-down construction.

CONSTRUCTION PROBLEMS

There are practical limitations, problems, and risks involved in diaphragm wall construction. The most apparent of these are (1) quality of appearance of the exposed wall surface, (2) misalignment of the wall, (3) permeability of the wall, (4) cave-in of trench excavation, and (5) difficulties in concreting.

Surface Appearance.   The roughness of the exposed face of the diaphragm wall reflects the type of the excavated soil. Underground structures and abandoned utilities that are encountered in the trenching will also show their marks on the wall surface. Depending on the final use of the space along the wall, the surface may have to be finished or clad.

Misalignment. Vertical misalignment is caused by underground boulders or other obstructions, inclined layers of hard soil strata, and alternating layers of soft and hard soils (see Fig. 8.32). The excavation tool has the tendency to deviate from the vertical alignment when encountering these conditions. Accumulation of the deviation can be easily corrected, if detected early.

During excavation panel verticality is assured by continuous real-time readout of excavation in the operator's cabin (Jean Lutz system), where adjustments are made immediately

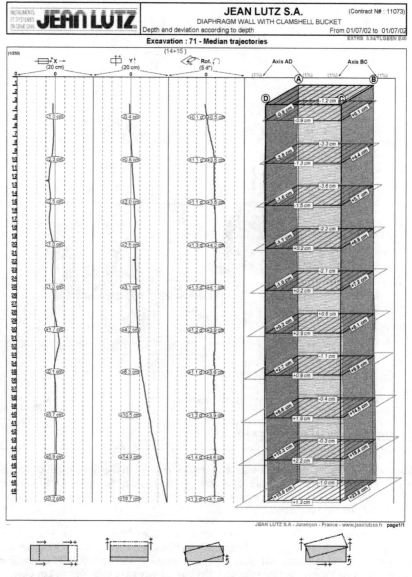

FIGURE 8.32 Typical Jean Lutz output. (*From BENCOR Corp.*)

to the clam bucket and the hydromill. In case there is an evidence of deviation, the operator should immediately rectify the panel.

Both during and after the panel excavation is concluded, another check of the verticality of the panel can be performed by means of a Koden device. The Koden is an ultrasonic device that determines depth, width, and the verticality of the excavated trench walls (see Fig. 8.33).

FIGURE 8.33   Koden. (*From BENCOR Corp.*)

The Koden test performs at the joints of each panel (primary and secondary) that in addition of the Jean Lutz output produces an additional check of verticality and alignment of the excavated panels.

The Koden output reading shows the actual final verticality of the fully excavated panel. The Koden test can be witnessed by the project QA/QC representatives (see Fig. 8.34).

Permeability.   It must be realized that normal cast-in-place concrete is not totally watertight. It is possible to experience moisture patches or sweat areas even on perfectly constructed diaphragm walls. These should be acceptable. What should not be acceptable is any amount of running water through the wall.

Excavation Cave-in.   One of the main and most frequent accidents in diaphragm wall construction is the loss of bentonite with the consequent loss of stability of the excavated trench. A natural cause of the loss of bentonite is very permeable soil conditions. To counteract this situation, it is necessary to add sandy, silty materials, or sawdust, to the bentonite.

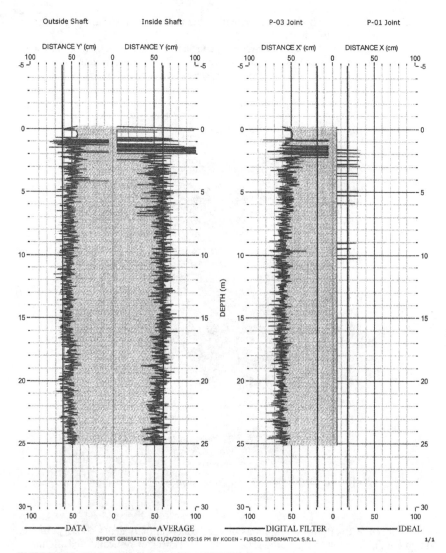

FIGURE 8.34   Typical Koden reading. (*From BENCOR Corp.*)

It is to be noted that generally there will be more bentonite loss at the beginning of the excavation of the diaphragm wall trench than later on. The loss will diminish after the excavation of the first few panels as more bentonite seeps into the adjacent ground.

The most dangerous situation is the sudden loss of bentonite from the trench to unknown old pipes, underground cavities, etc. Sudden loss of bentonite can lower its level in the trench within seconds with the result of almost immediate trench collapse. It will require quick action of backfilling the trench, researching the cause of the bentonite losses, and correcting the situation. Generally, the remedial trench backfill is done with the excavated material or with a lean mix of sand-cement backfill.

Another cause of caving of the trench is the increase in the elevation of the water table. The bentonite cake formed on the surfaces of the trench generally prevents groundwater leakage into the trench. However, in the cases where the water pressure is great because of high water table, or the groundwater overflows the ground surface, the density of the bentonite solution needs to be increased. In addition, the guide walls and/or working platform elevation need to be raised in order for the bentonite slurry to exert a higher head differential against the water level. In extreme cases a wellpoint system may be needed to lower the water table to effectively attenuate this situation.

Concreting Problems. The majority of concreting problems are caused by poor delivery schedule of the concrete to the trench. During a long wait for the next delivery, the concrete already in place may stiffen to the degree that it will not rise properly in the trench as fresh concrete is supplied through the tremie pipe.

It needs to be noted that bentonite slurry in contact with concrete has a tendency to chemically react and flocculate. This results in a jellying of the bentonite slurry which, in some cases, can reach the thicker consistency of fresh concrete. A layer of this affected material is sometimes found at the contact of the concrete and bentonite. In some instances, if the tremie pipe is not embedded well into the good concrete, there is a strong possibility of creating a cold joint with contaminated concrete sandwiched in (see Fig. 8.35*a*). This is one of the most serious defects in slurry wall construction, and it can result in a dangerous lack of strength along a seam in the wall.

Lack of tightness in panel joints is generally caused by contaminated bentonite trapped between the two adjacent panels during the pouring of the second panel (see Fig. 8.35*b* and 8.35*c*). The cause of this problem is either the poor quality of the bentonite slurry or the low plasticity of the concrete or a combination of the two.

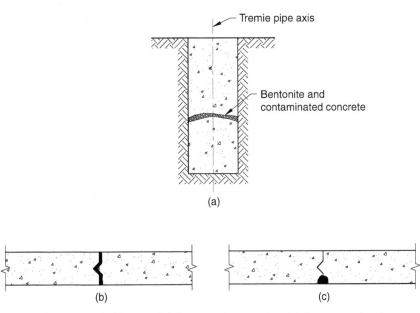

FIGURE 8.35 Discontinuities caused during concreting operation: (*a*) Slurry-contaminated concrete entrapped in wall creates a horizontal "cold joint." (*b*) Slurry or soil not completely removed from the end of the hardened concrete panel creates a defective vertical joint. (*c*) Soil entrapped in the surface of the wall creates a cavity.

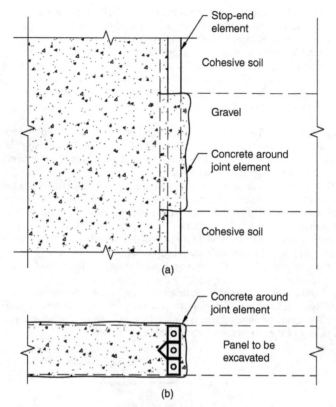

FIGURE 8.36 Penetration of fresh concrete beyond the stop-end element:
(a) Elevation. (b) Horizontal section through gravel layer. (*Xanthakos: Slurry Walls as Structural Systems, 2d ed., McGraw-Hill.*)

If the temporary stop-end joint element is not tight in the slurry trench, because the trench excavation is wider than specified or the side of the trench crumbles in a particular soil layer, the fresh concrete penetrates around the joint element into the adjacent unexcavated panel space (see Fig. 8.36).

A potential hidden defect is discontinuous concrete coverage of the reinforcing steel. This can be the result of poor-quality bentonite slurry entrapped in the concrete along a rebar, and/or of inadequate flow of concrete to surround the rebar completely. Heavily reinforced areas of the wall are most prone to this defect.

The lack of specified thickness of concrete cover over the reinforcing steel in the exposed side of the wall is another hidden defect. It is generally caused by poor and/or contaminated slurry that builds up a thick cake on the side of the trench, thus reducing the trench width and the available space for concrete cover.

The upward drag of the rising concrete in the trench can create a large uplift force on the reinforcing steel cages, which can cause the reinforcing steel cages to uplift during the concrete pouring operation. This is influenced by the shape and fabrication of the resteel cages, and exacerbated by low plasticity of the concrete or improper concrete-pouring operation. One solution to attenuate this situation is to shorten the tremie pipes, thus reducing the depth of their embedment in the fresh concrete, so that a shallower depth of the resteel is subjected to the upward drag. The other is to bend some of the vertical bars at their deep ends in order to act as anchors in the concrete. In cases where the reinforcing steel cages are not installed all the way down to the bottom of the trench, it is prudent to extend at least some of the vertical steel of the cages to the bottom of the trench to act as anchorage for the upper resteel cage assembly.

A problem that generally occurs at the end of the concrete pour is the near impossibility of removing the joint pipes. Generally, the difficulty increases with the depth of the diaphragm wall. This problem can be caused by several factors: late removal of the joint pipe, early setting of the concrete, or early removal of the pipe with consequent siphoning of concrete inside the joint pipe, causing difficulties in its extraction.

Another construction problem related to the concrete-pouring operation is the finishing of the top of the diaphragm wall. In some cases, the top of the trench is higher than that of the finished concrete elevation of the diaphragm wall, so it is necessary to stop the concrete pour at a certain distance below the top of the trench. This is not a simple task. Generally, the good concrete at the top of the wall is covered by a layer of contaminated bentonite and concrete, the thickness of which is difficult to determine. (Depth sounding during concrete pouring is achieved by a heavy plumb bob weight attached to a graduated cable.) Because of the heavily contaminated bentonite at the top of the wall, it is likely that the true top elevation of the good concrete will not be determined accurately, and it will end up too low or too high.

Owing to the nature of the tremie pouring technique, in the case when the top of the concrete is stopped below the top of the trench, it is common that the top surface of the concrete ends up being not flat but convex. This is not to be considered a defect of the concrete installation; however, provisions need to be made in the contract documents for level finishing of the top of the diaphragm wall concrete.

PERFORMANCE

The performance of diaphragm walls is generally evaluated from two main aspects: watertightness and structural performance, the latter including both strength and lateral movement.

Watertightness. The wall's watertightness depends upon the nature of the soil, the type of joints in the slurry wall panels, and the construction technique. With good construction practice, a filter cake of bentonite forms in the soil in the side of the trench against the outer face of the slurry wall that minimizes leakage through the wall. Usually only minor moist spots appear at some locations of the joints. With poor construction techniques, considerable problems of honeycombing and leakage in the walls have been noted. In such cases extensive and expensive repairs, including reconstruction, patching, and grouting or combinations thereof, have to be made to control and minimize groundwater infiltration. Another cause of seepage is created when temporary prestressed braces are released and the load is transferred to the finished building floor slabs. This may result in some inward deflection of the diaphragm wall. If these deflections are unequal in adjacent panels, the joints can open up and lead to excessive or unusual seepage even if no seepage was noticed while the wall was supported by the temporary braces.

Structural Performance. The structural performance of a diaphragm wall, in terms of its movements and strength, is usually much better than that of a concrete building wall constructed within a separately retained excavation. The diaphragm wall is usually thicker, so its stiffness is greater than that of a common foundation wall of a building basement. Furthermore, the design spans of a diaphragm wall during construction are usually greater than those in the completed permanent structure; hence it has more reserve capacity than a conventionally built permanent structure wall.

An important aspect of diaphragm wall performance is lateral movement during the excavation of the enclosed work area. In this regard, the performance of the diaphragm wall is superior to that of common retention systems owing to its greater stiffness. Bending of the wall between support locations is minimal, indeed often negligible, because of its great thickness. Displacement of the support locations is effectively controlled by the stiffness of the bracing (especially cross-lot bracing) and/or tiebacks.

In order to achieve substantially greater wall stiffness than that obtained with flat panels, diaphragm walls can be constructed in T shapes in plan. The trench is excavated so as to follow the outline of the T shape, the reinforcement cage is fabricated to fill the T shape, and the web and flange of each wall panel are cast monolithically. The increased strength and stiffness of the wall allows much greater vertical spacing of braces and tiebacks than with conventional flat walls. Posttensioned T-panel diaphragm walls have been built successfully with braces and tiebacks at vertical spacing of as much as 30 ft (9 m).

APPLICATIONS AND USES

Diaphragm walls have been employed for a variety of civil engineering projects. Their use is gaining more acceptance from the engineering community in underground construction projects with the potential of groundwater problems, where adjacent structures need to be protected against settlement, where sheet piling meets with installation difficulties and driving noise restrictions, where the excavation has to extend through difficult soils, and where the temporary retention system during construction has to remain as part of the permanent underground structure. The use of diaphragm walls is particularly favored where sheet piling would be required as a temporary retention system and the soil and groundwater conditions would necessitate dewatering to lower the water table with the consequent possibility of nearby structure settlement. Another great advantage of diaphragm walls over other types of earth retention is their inherent vertical load carrying ability. The following examples illustrate some of the applications of diaphragm walls.

Basement Walls.   Diaphragm walls have been extensively used for basement walls for earth retention, load-bearing purposes, and groundwater seepage control, particularly in urban areas where sloping sides of excavation are restricted by space and dewatering is risky (see Fig. 8.37).

Subway Construction.   Diaphragm walls have been used for deep excavation support in subway construction. More recently, diaphragm walls have also been utilized in the "Milan Metro system method" as part of the permanent subway tunnel or station structure. With this method, the diaphragm walls are installed on the two sides of the future subway line, a temporary deck is constructed at ground level on unexcavated ground between the two walls, the soil is excavated below the temporary deck, the permanent roof slab of the tunnel is poured on forms and falsework, and finally the base slab of the tunnel is poured (see Fig. 8.38). Construction of the tunnel is achieved with minimal interruption of surface traffic. The same procedure has also been used in highway underpass construction.

Circular Shafts.   During the past 10 years advances in large and deep diameter diaphragm/slurry walls shafts have grown in North America. This is due to the improvements of design, construction method, and excavation equipment capabilities.

Circular diaphragm/slurry walls are built using the same basic technique used to build strait walls. The length of the walls (panels) is limited in order to take advantage of length of the excavating tool and the geometry of the structure (see Fig. 8.39).

The use of diaphragm/slurry walls for deep shafts is a very cost-effective method for support of excavation for circular shafts. The basic procedure as described are used on a larger scale and routinely in major projects in North America especially to focus on

FIGURE 8.37 Diaphragm walls with tieback for building basement construction. (*From BENCOR Corp.*)

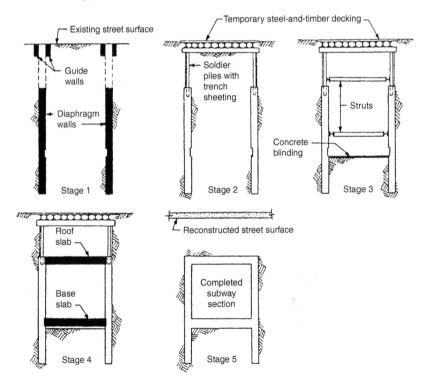

FIGURE 8.38 The "Milan Metro system method" illustrated in five construction stages. (*From Xanthakos: Slurry Walls as Structural Systems, 2d ed., McGraw-Hill.*)

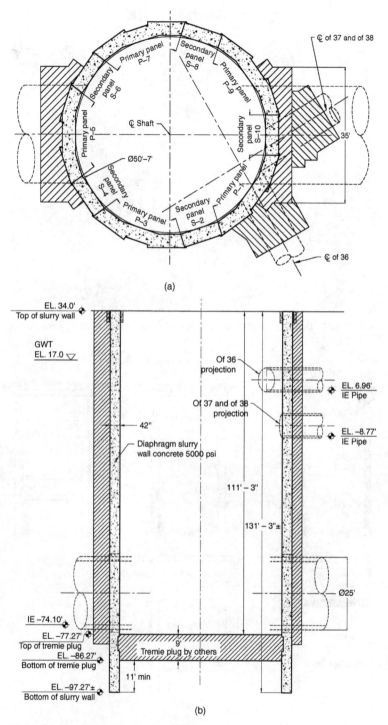

FIGURE 8.39 Typical circular shaft with break-in/out: (*a*) Plan view. (*b*) Typical section. (*From BENCOR Corp.*)

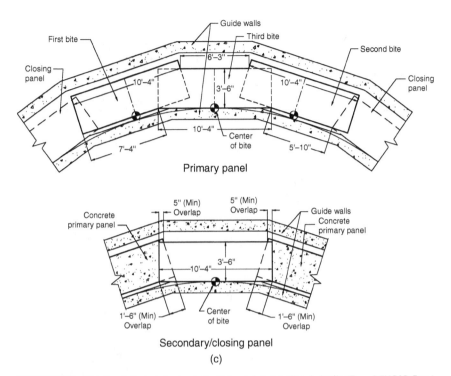

FIGURE 8.39 (*Continued*) Typical circular shaft with break-in/out: (*c*) Panels details. (*From BENCOR Corp.*)

Environmental Protection Agency (EPA)-mandated combined sewer overflow (CSO) structures built recently nationwide.

The advantage of circular diaphragm/slurry walls is the elimination of support systems (struts or tiebacks). The mass excavation within the shaft can be mined without excavation sequences in addition due to the circular shape the diaphragm/slurry wall achieves a great wall stability.

As described previously in view of the advances of construction techniques and equipment (hydromill) is possible to install the walls with very small vertical tolerance. This control is done through on-board and real-time management instrumentation that allows to correct deviation (see Fig. 8.40).

Typical specified tolerances are in the order of 1 percent of wall height, but tolerances of 0.5 percent have been commonly achieved in practice.

Retaining Walls and Slope Stabilization. Diaphragm wall systems have been used for retaining walls and slope stabilization along deep excavations for buildings, highways, tunnels, and projects with tight working conditions (see Fig. 8.41). Avoidance of the noise and vibration of a sheet-pile driving operation and the use of the diaphragm wall as part of the permanent retaining wall structure favor the use of the slurry wall technique.

Where unusually high strength and stability of a retaining wall are needed, one efficient solution is to install T-shaped elements to create a buttress-type wall, reducing the external bracing requirements.

Underpinning Alternative. Diaphragm walls have been used effectively as an alternative to the underpinning of structure foundations adjacent to deep excavations. Instead of constructing conventional underpinning, a diaphragm wall with internal bracing

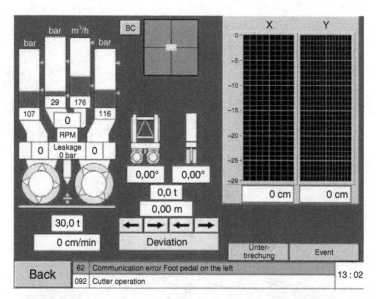

FIGURE 8.40 Trench cutter/hydromill control system monitor. (*From Bauer.*)

FIGURE 8.41 Slurry wall used for slope stabilization and earth retaining. Note the concrete-bearing pads at the tieback stressing locations. (*From BENCOR Corp.*)

(Fig. 8.42) or with tiebacks (Fig. 8.43) is installed along the outside of the existing foundation to retain the subsoil. Ground movements during a bentonite-supported excavation are very small compared with those during a conventional underpinning operation; and the rigidity of the diaphragm wall ensures limited elastic deformations during the excavation and bracing operations. Thus the existing adjacent foundation is effectively restrained from settlement.

FIGURE 8.42 Slurry walls with cross-lot bracing used to obviate the need for underpinning the adjacent building foundations during subway construction. (*From BENCOR Corp.*)

FIGURE 8.43 Slurry walls with tiebacks used to underpin a building adjacent to a very deep excavation. (*From BENCOR Corp.*)

ANALYSIS AND DESIGN

Reliable determination of the stresses and deformations in a diaphragm wall and in its supports requires analyses of the interaction between the wall and the soil during the successive temporary stages of excavation, as well as after completion of the permanent facility.

Usually, the design documents for the permanent facility indicate the location and dimensions of the wall, and its performance requirements under the temporary construction and permanent service life conditions. The detailed design of the wall and its supports is usually performed by the diaphragm wall contractor's own engineers or by specialized consultants. Both geotechnical and structural engineering expertise are needed to produce a safe and economical design. An earth-retaining wall installed by the slurry method may perform the same function and may be supported in the same general manner as steel sheet piling or soldier piles with lagging. Therefore, it may appear logical to use the same methods of analysis. However, very different movements and earth pressures are encountered with a stiff diaphragm wall, which necessitate different approaches of analysis. Several publications deal with the methods of analysis and design of diaphragm walls (Refs. 1 to 4), but there is no definitive recommendation or requirement in any national design code for the methods). Only a brief overview is presented in this chapter.

Lateral Earth Pressures

The design of braced flexible walls, such as sheet piling, and soldier pile and lagging usually involves the use of apparent earth pressures.[5,6,10] For diaphragm walls, however, these may underpredict the applied pressures, especially near the bottom of deep excavations.[7]

There are two important characteristics of diaphragm walls that influence the earth pressures: (1) Concrete walls constructed with the slurry method are relatively thick and therefore much stiffer than other typical temporary walls, and (2) the bracing of diaphragm walls is often more rigid both during staged excavation (e.g., cross-lot bracing) and in their permanent states (e.g., internal floor structure). These stiffnesses restrain the movement of the wall and of the retained soil, giving rise to different distributions and greater magnitudes of earth pressures than those behind the conventionally constructed more flexible temporary earth-retaining walls.

It is suggested that staged excavation analyses for diaphragm walls be performed using triangular earth-pressure distributions.[8,9]

The earth-pressure coefficient to be used depends on the expected wall displacement; it will start from the "at-rest" pressure coefficient before excavation begins and may ultimately decrease to the value of the "active" pressure condition. It is generally accepted that active pressure will act when there is a lateral movement of approximately 0.05 percent of the wall height in dense sand, 0.2 percent in loose sand, 1.0 percent in stiff clay, and 2.0 percent in soft clay. (See Chap. 3, Sec. 2, Fig. 1 of NA VFAC DM-7.2, Ref. 10.) Field measurements of actual diaphragm wall installations have shown that in many cases movements of 0.1 percent induce active pressures in most soils.

Where it is apparent that active earth pressures can develop, the analyses presented in Chap. 7 of this handbook may be used.

Some designers use an earth-pressure coefficient corresponding to the average of the "at-rest" and "active" values. Where deflections are somewhat restricted, some designers use the "at-rest" pressures with coefficients in the neighborhood of 0.5 for most soils. Even higher earth-pressure coefficients may be appropriate in clays and in some unusual situations.

For cohesive soils (clays) the earth pressure used during short-duration temporary staged excavation analysis, as opposed to the permanent state, should represent the undrained state; that is, the active and passive pressures should be functions of the soil's undrained strength.

Since earth pressures are calculated using effective stresses derived from the soil's unit weight, hydrostatic pressures have to be considered separately in addition to the calculated earth pressures by the soil only.

Staged excavation and bracing cause arching of the soil near the upper levels of bracing. This results in localized soil pressures that are greater than those predicted by a triangular pressure distribution. Therefore, it is good practice to set at least a moderate value of applied soil pressure in the analysis near the top of the wall.

Prestressing and preloading of the bracing system can significantly increase the earth pressures behind the wall. However, this increase will usually be relieved in time as a result of wall movement.

Both the distribution and magnitude of earth pressure will, of course, change with the successive stages of excavation of the work area, and therefore have to be updated in each stage of the analysis.

For the structure's permanent condition, it is customary to apply at-rest earth pressures in order to account for the adjustment of soil stresses during the life of the structure.

It is clear that a thorough awareness of jobsite conditions, understanding of soil mechanics, experience, and good engineering judgment are important for establishing the proper design earth pressures.

Structural Analysis and Design

A diaphragm wall has to be analyzed and designed for each successive stage of excavation and bracing, as well as for the permanent condition after completion of construction. Five main methods of structural analysis are used for diaphragm walls:

Continuous beam on rigid supports

Continuous beam on nonrigid supports

Beam on elastic foundation

Finite elements

Limit analysis

These methods require different levels of sophistication of the engineer, and offer different degrees of reliability and accuracy. The choice of the analysis method should depend on whether only the wall strength or both wall strength and ground movement are of concern, on the needed reliability and accuracy of the results, on the availability of the soil parameters, and last but not least on the ability of the designer. (Perhaps the complete explanation of some of the analysis methods can be found in Ref. 9's Chap. 6 and App. B.)

Continuous Beam on Rigid Support. Often referred to as the "rigid" method, this is the simplest but least accurate analysis. (In the sheet-piling industry this is also known as the "free" or "fixed earth support" analysis method.) A vertical strip of the wall is treated as a multi-span beam on rigid supports that are located at the bracing (or tieback) points. The lowest support location is assumed to be below the bottom of excavation at the point of zero net earth pressure. The portion of the wall below this point is neglected (see Fig. 8.44). The two pressure diagrams are the design earth pressure against the outside and the passive earth pressure against the inside face of the wall. The former extends from the top of ground to the actual bottom of the wall, the latter from the bottom of excavation to the actual bottom of the wall. This model can even be analyzed by hand by such methods as the Hardy-Cross moment distribution or the slope-deflection equations. A new earth-pressure diagram is constructed and a new analysis is performed for each stage of excavation when a new level of bracing is introduced.

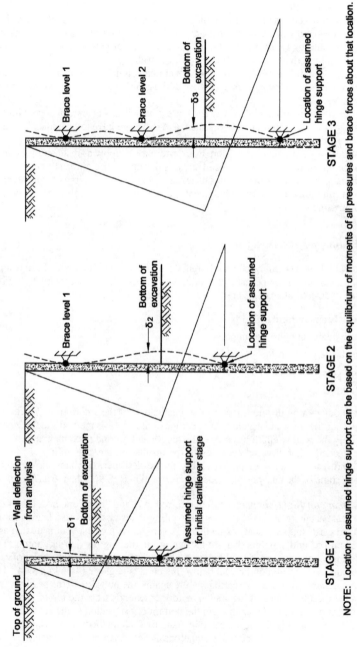

NOTE: Location of assumed hinge support can be based on the equilibrium of moments of all pressures and brace forces about that location.

FIGURE 8.44 Modeling for staging analysis by continuous beam on rigid support method.

8.44

This analysis gives conservative results of wall moments and bracing reactions, and provides no information on wall and ground movements. Support points displacements, especially near the bottom, may result in moment reversals in the wall that is not provided by this analysis.

Continuous Beam on Nonrigid Supports. This approach is identical to the continuous beam on rigid supports with two significant exceptions: (1) The braces (or tiebacks) are modeled not as rigid reactions to the wall but as axially loaded elastic members (see Fig. 8.45). (2) To account for the wall deflection that takes place at the level of the next lower brace (or tieback) before it is installed, an initial support displacement is applied at the fixed end of that next brace (or tieback). This initial displacement δ is equal to the wall deflection δ computed at that level immediately prior to the installation of the brace (or tieback).

This is a simple, logical method that provides an approximation of the amounts of wall movements and their influence on structural stresses. It may give smaller wall stresses and smaller forces in the lower braces (or tiebacks) than the rigid support method. The simplest of structural analysis computer software should be adequate to handle this model.

Beam on Elastic Foundation. The entire depth of the wall is modeled as a continuous beam, the braces are represented by spring supports, and the passive resistance of the soil below the bottom of excavation is modeled by a series of parallel springs with initial loads in them that equal the at-rest soil pressures against the wall (see Fig. 8.46.). (An alternative approach is to apply the at-rest soil pressure directly against the inside face of the wall and use the soil springs to provide only the additional resistance up to the limiting passive pressure.[3]

The stiffnesses of the brace spring supports are derived from the axial stiffnesses of the braces or tiebacks. The passive soil springs are usually linear elastic or bilinear elastoplastic; their stiffnesses represent the soil's coefficient of subgrade reaction and their plastic limits represent the ultimate passive soil pressures calculated from classic earth-pressure theory. (Note that the ultimate passive pressure for each soil spring has to be recalculated for each analysis run, based on its distance below the bottom of the current stage of excavation.) The soil's modulus of subgrade reaction is dependent both on the soil and on the structural bearing element, and its values are often subject to uncertainty.

By introducing the loads in the passive soil springs at each stage of analysis, a number of considerations have to be included, such as (1) the initial spring loads in the at-rest condition before excavation began, (2) unloading due to earlier stages of excavation, and (3) accumulated displacements and stresses due to earlier stages of excavation.

The applied earth pressure against the outside face of the wall is usually kept the same through the successive stages of an analysis, but sophisticated soil mechanics are sometimes used to approximate the changes in the applied pressure from stage to stage of construction.

Wall deflection δ at a brace level prior to the installation of that brace can be represented either by a fictitious load $k\delta$ applied against the reaction spring at that level or by imposing the δ displacement on that spring. This $k\delta$ force will have to be subtracted from the final calculated spring reactions in order to get the real brace forces.

Prestressing, if any, in a brace or tieback can be modeled simply by applying an appropriate concentrated force at its location against the wall.

This method provides more insight into the behavior of the wall and the soil, but its reliability is much dependent on the judgment calls of the designer. It can be performed with the aid of most general-purpose structural analysis computer software.

Finite-Element Methods. These are the most advanced methods of analysis and can predict reasonably well the movements of the wall and of the soils behind it, including ground settlements. However, even these methods do not always provide results that correspond to field observations. Despite the sophistication of the numerical analysis, uncertainties are introduced by the assumptions made for the soil-structure model (particularly the material model used for the soil) and by the unpredictability of the in situ properties.

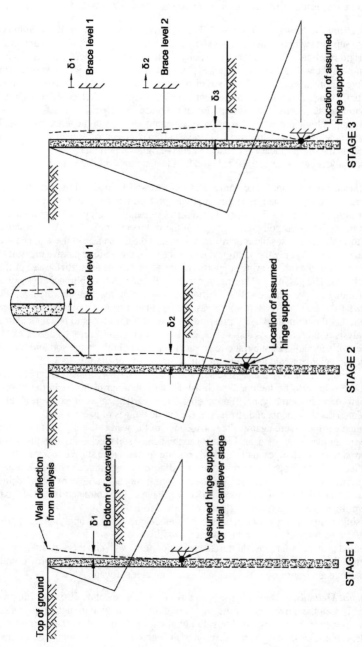

NOTE: Location of assumed hinge support can be based on the equilibrium of moments of all pressures and brace forces about that location.

FIGURE 8.45 Modeling for staging analysis by continuous beam on nonrigid support method.

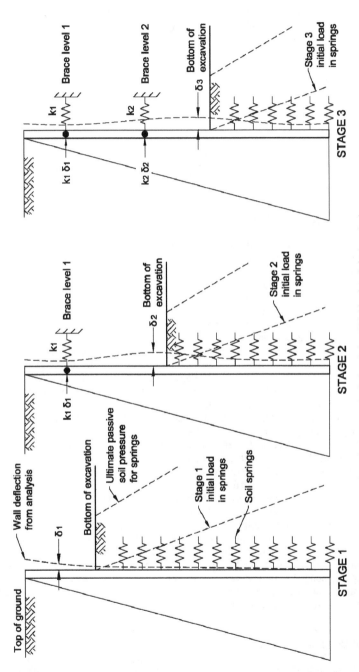

FIGURE 8.46 Modeling for staging analysis by the beam on elastic foundation (Winkler foundation) method.

A two- or three-dimensional finite-element model is developed; the wall is represented by beam or plate elements and the soil by plane strain or brick elements. Material properties of the wall and soil are introduced through definitions of their unit weights, elastic moduli, plastic limits, creep coefficients, etc. The pressures on the wall are generated during the calculations from the self-weight of the soil and from surcharge or superimposed ground surface loads. Excavation is simulated by the removal of layers of finite elements on one side of the wall. A rather advanced structural analysis computer software is necessary for efficient analyses.

Limit Analysis.   Limit analysis principles are generally not used in foundation engineering work but do have some advantages in diaphragm wall design. In a limit analysis the bending moments in the wall at the brace locations are assumed to be the yield or ultimate moments of the cross section. The model so obtained is statically determinate and can be easily analyzed with either rigid or nonrigid supports. The analysis gives little or no estimation of the ground movements behind the wall. More development work and comparisons with field data are needed to establish the reliability of this method.

Estimation of Wall and Ground Movements

One of the frequent reasons for the use of diaphragm walls is the need to control surrounding ground movement. Therefore, knowledge of the lateral displacement of the wall is essential as it affects both the bracing system and, indirectly, the settlements of structures and utilities near the excavation.

Three of the five general methods of analysis described above can provide displacement results. Their reliability depends on the sophistication of the method, particularly on the degree to which the soil and structure can interact in the analysis. The *continuous beam on rigid supports* and the *limit analysis* when coupled with rigid supports do not predict wall and ground movements. The *continuous beam on nonrigid supports* gives the least reliable displacement results, because the effect of the soil on the structure is prescribed as fixed earth-pressure values. The *beam on elastic foundation* method gives more reliable displacements, since not only the braces above the bottom of excavation but the soil on the passive side of the diaphragm wall below the bottom of excavation is also modeled as springs with attendant deformations. The accuracy of the calculated displacements is strongly dependent on the soil's modulus of subgrade reaction that is used in the analysis. The most accurate displacement predictions come from the *finite-element analyses*, since the soil is permitted to exert both its driving and resisting pressures on the wall according to its deformed configuration. Furthermore, soil deformation in the analysis occurs in two or three dimensions; thus the shear deformations of the soil, which is the principal mechanism controlling soil behavior, are captured in the model. However, the calculated deformations are highly influenced by the assumed soil model and soil parameters.

Another approach to estimating wall and ground movements is the so-called semiempirical methods. These methods are calibrated from actual case histories and theoretical analyses. The maximum lateral wall movements may be obtained from the geometry and dimensions of the excavation, and the stiffness of the bracing system.[11] The vertical settlements of the ground at various distances from the excavation are, in turn, obtained from the maximum lateral wall movements. The effects of vertical settlements on structures and utilities adjacent to the excavation may then be estimated.[12]

Practical Considerations in Design

In the absence of a national design code or specification for diaphragm/slurry walls at the time of this writing, engineers rely on their experience and judgment, and follow the provisions of ACI 318, for both reinforced and unreinforced concrete where applicable.[13] This code has been developed primarily for concrete structures above ground where their exposed portions can be inspected. Therefore, conservatism is advised in the design and detailing when applying the ACI code provisions to slurry walls that are constructed in the blind. Tolerances in dimensions, inaccuracies in the placement of reinforcing steel, defective patches of concrete, and other imperfections in the wall may be unnoticed.

The load factors and load combinations in ACI 318 were developed for the permanent conditions of structures. When using them for a staged construction analysis, it may be prudent to apply some special load combinations and to increase or decrease some load factors to account for the degree of certainty of the magnitudes of the loads, for the expected short duration of exposure to the loads, as well as for the temporary performance criteria of the diaphragm wall.

In addition to providing the required dimensions, strength, and stiffness, the detailing of the reinforcement and other features of the wall must be such as to fit the site constraints and allow for the efficient use of the contractor's equipment.

Significant economy of diaphragm walls can be realized by the omission of reinforcing steel cages in the trench. This can be accomplished by the SPTC wall system. If the spacing of the embedded soldier piles is small in relation to the thickness of the wall so that the horizontal flexural and shear stresses between the piles are less than those allowed by ACI 318 for plain concrete, this code permits the use of unreinforced concrete between the piles. (Without lowering a reinforcing cage into the trench, local collapses or misalignments in the slurry-filled trench may go unnoticed, and localized weak pockets in the concrete will not be bridged over by reinforcement.)

Composite action between the soldier piles and the concrete in vertical bending can be a great additional economic advantage but must be investigated carefully. It is not established at the time of this writing how much bond can be developed between the concrete and the steel I-section surfaces that had been covered with slurry prior to concreting. It may be reasonable to count on composite action when estimating deformations under service loads but neglecting it in strength calculations under ultimate load conditions. Welded studs or other projecting attachments can, of course, enhance composite action, but they may be cumbersome during the lowering of the pile into the trench and during the concreting operation.

Circular slurry walls are used more and more frequently. They are particularly effective when the conditions are such that the circumferential stresses are pure compression under all possible load combinations, obviating the need for reinforcing steel. Their analyses, however, demand more than the routine methods for plane walls.

REFERENCES

1. Kerr, William C. and George J. Tamaro: "Diaphragm Walls—Update on Design and Performance," *Proceedings ASCE Conference on Design and Performance of Earth Retaining Structures*, Cornell University, Ithaca, N.Y., 1990.

2. Bechara, Camille. H: "Tips for Slurry Wall Structural Design," *Civil Engineering Practice*, Fall/Winter 1994.

3. Tamaro, G. J. and J. P. Gould: "Analysis and Design of Cast In-Situ Walls (Diaphragm Walls)," *Retaining Structures*, Proceedings of the July 20–23, 1992 Conference of the Institution of Civil Engineers, London, edited by C.R.I. Clayton, Thomas Telford Publishing.

4. Slurry Walls: *Design, Construction, and Quality Control*, STP 1129, American Society for Testing and Materials, Philadelphia, Pa., 1992.

5. Peck, R. B.: "Deep Excavations and Tunneling in Soft Ground," *Seventh International Conference on Soil Mechanics and Foundation Engineering*, 1968.

6. Tschebotarioff, G. P.: *Foundations, Retaining and Earth Structures*, 2d ed., McGraw-Hill, New York, 1973.

7. Goldberg, D. T., W. E. Jaworski, and M. D. Gordon: "Lateral Support Systems and Underpinning," Report FHWA-RD-75-128, vol. I, Federal Highway Administration, Washington, D.C., 1976.

8. Shields, D. R.: "Comments for Panel Discussion No. 1." *Boston Society of Civil Engineers Geotechnical Seminar on Design, Construction and Performance of Deep Excavations in Urban Areas*, Boston, Mass., 1988.

9. *Final Geotechnical Engineering Report, Design Section DO17A, 1-93lCentral Artery-Congress Street to North Street, Central Artery (I-93)/Tunnel (I-90) Project, Boston, Mass.*, Prepared for the Massachusetts Highway Department, by GEI Consultants, Winchester, Mass., October 1992.

10. *Design Manual 7.2, Foundations and Earth Structures*, NAVFAC DM-7.2, Department of the Navy, Naval Facilities Engineering Command, Alexandria, Va., 1982.

11. Clough, G. W. and T. D. O'Rourke: "Construction Induced Movements of In-situ Walls," *Proceedings ASCE Conference on Design and Performance of Earth Retaining Structures*, Cornell University, Ithaca, N.Y., 1990.

12. Boscarding M. D. and E. J. Cording: "Building Response to Excavation-Induced Settlement," *Journal of Geotechnical Engineering*, ASCE, vol. 115, no. 1, 1988.

13. *Building Code Requirements for Structured Concrete*, ACI 318-95, American Concrete Institute, Detroit, Mich., 1995.

BIBLIOGRAPHY

A Review of Diaphragm Walls, A Discussion of "Diaphragm Walls and Anchorages," Institution of Civil Engineers, London, 1977.

Boyes, R. G. H.: *Structural and Cut-Off Diaphragm Walls*, Wiley, N.Y., 1975.

BW System Diaphragm Walling, Bulletin No. BW 102B, Tone Boring Co., Ltd. Tokyo.

Catalano, N., M. Kirmani, and G. Aristorenas: *Post-Tensioned Diaphragm Wall T-Panel for Large Unbraced Excavation Spans*, Proceedings of the 19th Annual Conference and Meeting of the Deep Foundation Institute, Boston, Mass., October 3–5, 1994.

Design and Performance of Earth Retaining Structures, Special Publication 25, Proceedings of an ASCE Conference, Cornell University, Ithaca, N.Y., June 1990.

Diaphragm Walls and Anchorages, Proceedings of the 1974 Conference of the Institution of Civil Engineers, London, 1975.

Drilling Mud Data Book, Baroid Division of NL Industries, Houston. Tex., 1975.

Gill, Safdar A.: *Applications of Slurry Walls in Civil Engineering Projects*, Preprint 3355, ASCE Convention, Chicago, Ill., October 1978.

Hanna, Thomas H.: *Foundations in Tension-Ground Anchors*, Trans Tech Publications and McGraw-Hill, New York, 1982.

Mud Engineering, Dresser Magcobar Division of Dresser Industries, Inc., Houston, Tex., 1968.

Proceedings from the Symposium on Design and Construction of Slurry Walls as Part of Permanent Structures, Federal Highway Administration, Washington, D.C., March 1980.

Recommended Practice—Standard Procedure for Field Testing Oil-Based Drilling Fluids, API 13B-2, American Petroleum Institute, Washington, D.C., 1990.

Rogers. Walter F.: *Composition and Properties of Oil Well Drilling Fluids*, Gulf Publishing Co., Houston. Tex., 1958.

Seminar on Diaphragm Walls and Anchorages, Institution of Civil Engineers, London, 1976.

Slurry Wall Construction for BART Subway Stations, Preprints of the ASCE National Meeting on Structural Engineering, Pittsburgh, Pa., 1968.

Specification for Drilling—Fluid Materials, API Specification 13A, American Petroleum Institute, Washington, D.C., 1990.

Xanthakos, Petros P.: *Slurry Walls as Structural Systems*, 2d ed., McGraw-Hill, New York, 1994.

CHAPTER 9

CONSTRUCTION DEWATERING AND GROUNDWATER CONTROL

Arthur B. Corwin, P.E.
Paul C. Schmall, P.E., D.GE
Robert G. Lenz, P.E.

INTRODUCTION

Construction dewatering has as its purpose control of the subsurface hydrologic environment in such a way as to permit the structure to be constructed safely and "in the dry." Although construction dewatering is a key element of the construction program, often on the critical path, it is important to remember that the overall purpose is to build a structure, not to dewater the excavation. The entire dewatering program and all its components should therefore be oriented toward allowing the construction operations to be conducted in an environment where the groundwater and surface water problems are under control, and the methods of achieving that are not an impediment to the construction process.

Dewatering involves the removal of groundwater from pores or other open spaces in soil or rock formations. This stabilizes the material to facilitate construction activities. This leads into concepts like predrainage of soil, control of groundwater and surface water, and even the improvement of physical properties of soils and the relief of loads on cofferdam structures, all of which are included in the general subject of dewatering. These go far beyond the idea of pumping water out of a hole, which can be termed "unwatering." Pumping is the easy part of dewatering. The important part is the collection of groundwater and of surface water to achieve the purpose of the dewatering program.

It is assumed that the reader has a background in construction or engineering or both and is familiar with some of the more elementary concepts of soil mechanics and hydrology. The purpose of this chapter is to give the reader an overall picture of the issues that are involved in construction dewatering: a bibliography is included for those who wish to go beyond this scope.

It is not intended to enable readers to become dewatering experts. Rather, it provides an overview of a construction specialty that invariably is on the critical path and is frequently one of the more important determining factors in the success or failure of the project as a whole. The dewatering program conducted at Lock & Dam 26 Replacement near Alton, IL, on the Mississippi River is a case in point (Fig. 9.1). This landmark project involved construction of two navigation locks and seven very large tainter gate bays. The work was done in three

FIGURE 9.1   Lock & Dam 26, Phase III. (*Courtesy Moretrench.*)

separate cofferdams that spanned the Mississippi River. The project took in excess of 12 years to complete the three separate phases and remains the largest, deepest predrainage project ever constructed in the United States. Each of the dewatering systems pumped in the vicinity of 100,000 gal/min (378,500 L/min) during high-water stages, which involved handling 84 ft (25.6 m) of differential head between the river levels and the deepest subgrades.

HISTORY OF DEWATERING

Modern dewatering dates back to soil stability problems in connection with construction of the Kilsby Railroad Tunnel in England in the 1830s. In that project, a pocket of "quicksand" 1200 ft (366 m) long was encountered and the material was stabilized by pumping from a series of shafts and boreholes. That project, one of the first recorded engineering efforts to understand the movement of water in soils and to analyze methods of accomplishing the dewatering results, was described by Robert Stephenson, the engineer in charge, in a report dated 1841:

> As the pumping progressed, the most careful measurements were taken of the level at which the water stood in the various shafts and boreholes; and I was much surprised to find how slightly the depression of the water level in the one shaft influenced that of the other, notwith-standing a free communication existed between them through the medium of the sand, which was coarse and open. It then occurred to me that the resistance which the water encountered in its passage through the sand to the pumps would be accurately measured by the inclination which the surface of the water assumed toward the pumps and that is would be unnecessary to draw the whole of the water off from the quicksand, but to preserve in pumping only in the precise level of the tunnel allowing the surface of the water flowing through the sand to assume that inclination which was due to its resistance.
>
> The simple result, therefore, of all the pumping was merely to establish and maintain a channel of comparatively dry sand in the immediate line of the intended tunnel, leaving the water heaped up on each side by the resistance which the sand offered to its descent to that line on which the pumps and shafts were situated.

These careful observations and deductions eventually led to the discovery of the parabolically shaped "cone of depression" and to the concept that a construction excavation may be predrained by maintaining a balanced relationship between pumping rates and drawdowns.

From these beginnings, the use of shafts and boreholes (today called sumps and wells) spread throughout the construction industry. Also developing at this time was the art of handling unstable soils with only the crudest of tools, with excellent results. Gravel, salt hay, french drains, sheeting, sandbags and steam-powered pumps were the tools of the trade until the first practical wellpoint was developed by Thomas F. Moore in the mid-1920s. Moore, at that time a designer and manufacturer of conveying excavator equipment for the trenching industry, developed his wellpoint system in response to problems encountered in Hackensack, N.J., by the lessee of one of his machines. Notoriously unstable saturated silts along the trench alignment had slowed excavation and pipe laying to a crawl. Although earlier wellpoints were in existence, they had only been tested in clean sands and quickly clogged on the Hackensack site. Moore's wellpoints, however, proved to be very successful in drawing down the groundwater, marking the beginning of a new direction for the dewatering industry.

During the 1930s, dewatering was largely an art. The physical sciences were not incorporated into the procedures. In the late 1930s and certainly in the 1940s and 1950s, the various disciplines of science, including soil mechanics and hydrology, were integrated into the art of dewatering.

Today, the dewatering practitioner can, and does, draw regularly from all the engineering disciplines so as to produce the best possible result. There have been changes in dewatering practice in recent years, for the most part in the technology available to engineer system hydrogeological design. However, dewatering operations still require people with the practical sense and ability to visualize the movement of water in the ground; the art has not been entirely replaced by science. People are still needed who know and understand water and soils and have the practical experience to work with the materials at hand to accomplish the end result.

FACTORS INVOLVED IN DEWATERING

Defining the Dewatering Objective

To fully understand the problems that might be associated with building a structure below the groundwater table, all relevant facts and conditions must be gathered and understood. These include the physical characteristics of the structures to be built; their length, width, and depth; the types of cofferdams and/or excavation support that may be considered (earthen dikes, steel sheet piling, H-beams and lagging); the location of deeper parts of the structure, including sumps, foundation piles, potential overexcavation, and backfilling (to remove unsuitable materials); and whether the work will be conducted as a single excavation or the particular structure is one of several that may be done in sequence or as a group.

With these facts in hand, together with a good understanding of the subsurface soil and rock conditions, the nature and scope of the dewatering problem can be appraised. For example, is the dewatering problem simply a question of lowering the water level in a free-draining granular aquifer of substantial depth with no particular concern for pressure relief in the underlying aquifer or the stability of questionable soils? Or, to the contrary, are there underlying aquifers that will require independent or supplemental pressure relief? Are there subsurface soils present which, in their natural state, will present construction stability problems, and should the dewatering program address their improvement?

A plan and sectional drawing should be employed to summarize the surface and sub-surface conditions. On occasion, contours of subsurface layers can be revealing and on large projects a peripheral profile along the line of a dewatering system can reveal potential trouble areas.

In short, for a dewatering operation to properly meet its objectives, a thorough under-standing of the prevailing site conditions is essential. These include soil (and rock) charac-teristics, geology, geography, hydrology, construction methodology, and potential impacts to existing structures.

Soils and Rock Characteristics

The nature and extent or continuity of the soils involved in an excavation, and below, is the single most important determining factor with respect to how much water must be pumped and what techniques should be used. To that end, the available soils information can be assembled and a determination made if the scope of the project requires additional information.

With the availability of a series of borings with soil classifications that are understood, one can construct a soil profile, paying particular attention to the stratification in the under-lying soils. The continuity of all layers is important because that has a profound effect on groundwater movement. The densities of various soil strata are also important. Information concerning density can be obtained from standard penetration testing (blow counts), cone penetrometer testing (CPT), or other information contained in the borings. Occasionally, soils can be described in colloquial terms that can convey significant meaning to the experienced analyst: bull's liver, sugar sand, and gumbo are terms that are encountered in local areas and each of them has a specific meaning and a characteristic that can be very important. Other descriptors such as cohesive, loose, dense, soft, hard, cemented, and varved can be significant.

Permeability information should be available that can be used reliably in the dewatering analysis. The grain-size distribution and fines content is directly related to the soil perme-ability. In most soils, the horizontal movement of water is frequently governed by the most pervious layer, whereas the vertical movement of water is governed by the least pervious layer. Physical samples of the soils are often necessary to the evaluation of the continuity of layers. The color and other characteristics of the material can give the analyst a good idea of continuity between borings.

In addition to the soils, the underlying rock and its location should be known and under-stood. The permeability of the rock will determine if it is an aquiclude or perhaps an aqui-fer, such as some of the limestone or basalt formations existing in various parts of the country. If the piles for the structure will extend to a lower stratum, it must be confirmed that penetration of otherwise not dewatered or pressure-relieved soil strata will not cause a vertical flow along the pile, leading to a boil or a blow in the excavation. Quite frequently, the specific location of rock is not known in dewatering investigations, but again the engi-neer must have a concept as to where the rock is because the thickness of the overlying aquifers will determine to a great extent the amount of dewatering effort required.

Geology, Geography, and Meteorology

As indicated previously, the soils information from borings and other sources should be developed into a geological profile that will provide a greater understanding of the potential movement of groundwater in the ground. The geotechnical engineer should be aware of and evaluate the geological history of the site. The deposition of the soils should be indicated, particularly if they are of alluvial or sedimentary origin. Other characteristics of the soils should be indicated, such as the degree of under- or overconsolidation. Has there been preloading through glacial action or previous structures at the site? If the site is in a river valley, what are the deposition characteristics of the river? What has been its flooding history? How have scour and redeposition affected the dewatering problem? Will river scour be a problem during a dewatering program?

These are all questions that can never be fully or positively answered, but an understanding of the site and its geological history can frequently help in avoiding problems. A site's location can dictate much about a structure's design, and procedures involved in excavation, dewatering, and cofferdam construction will be different in rural and urban settings. Maintaining traffic and existing utilities may be a factor impinging on dewatering. Are there regulations regarding pumping volumes and water quality that will affect dewatering methods?

To fully understand the dewatering problems and to develop a good design for their solutions, one must be aware of all the environmental factors involved. The climate can be rather important and a major factor in selecting dewatering methods; the Arctic has its problems and the Tropics have theirs.

Hydrology

Having assimilated the information concerning the nature of the excavation and the soils in which it is to be made, it is important to understand where the water levels are and the source of water when the dewatering system is functioning.

The hydrological characteristics of the site can be evaluated from simple soils and geological analyses, assuming certain characteristics of the soils involved, or they may be measured directly by in situ and aquifer pumping tests. Occasionally, it is important to measure the permeability and transmissivity of a particular stratum rather than the whole profile. Should this be the case, pumping tests must be designed accordingly. Frequently, a pumping test at one location can be augmented by test pumping of piezometers at other locations. Sometimes this procedure can be used to substantiate the geological inferences that are frequently drawn from pumping tests. It is important to understand the sources of water that might be encountered in pumping tests and, if necessary, make provisions to grout holes caused by drilling which are connections between aquifers to avoid problems during the actual dewatering program. As a case in point, it is common practice in deeper excavations in Florida limestone to be sure that all borings are sealed and/or taken outboard of the excavation so that vertical flows from untapped aquifers will not present problems during excavation.

Adjacent Structures

Adjacent structures can dictate dewatering techniques and methods and may even require artificial recharging of groundwater beneath them. The effect of the dewatering to the adjacent structure is sometimes very significant too. Was that structure built with dewatering methods that involved underdrains and sumps that can create local dewatering problems that must be recognized and faced? Some of the most difficult dewatering problems occur when a new structure goes deeper than an older existing structure and is directly adjacent to it. Previous dewatering experience can frequently be helpful with respect to anticipating problems.

DEWATERING METHODS

It may be useful to consider different dewatering methods in the same manner as one considers different tools. A given method or tool may be technically the most appropriate for a particular problem, but sometimes other approaches may be more advantageous, with all issues considered. Awareness of the different tools or methods of dewatering can help one decide on the most appropriate procedure or, if circumstances warrant, combination of procedures.

Open Pumping Methods: Sumps, Ditches, Trenches

Sumps (Fig. 9.2) are important to any dewatering job in collecting surface water, storm water, and, perhaps, seepage that does not readily find its way to other predrainage devices. In almost every major construction job, sumps are necessary because of the limited ability of soils to accept major quantities of storm water and construction water. Water falling within the excavation limits will accumulate at the deep places and should be removed quickly to facilitate the work.

Occasionally, sumping can be the primary dewatering method. In favorable situations, sumps may be established and pumped to depress the general water table prior to excavation. In various parts of the country, techniques have been developed to address local soil conditions and sometimes sumping is practiced to a very sophisticated degree. For example, in the New York area years ago, a common method of excavation below the water table in highly unstable bull's liver soils was referred to as a "peel the onion" technique. Under this procedure, very shallow slices would be taken out of an excavation and, after each slice, additional dewatering would be accomplished by lowering the sumps within the excavated area. In the south, "rim ditch and sump" is an expression that also describes a method, under certain conditions, to conduct a satisfactory sumping operation.

Occasionally, groundwater can be controlled at the toe of a slope, or at and below subgrade, by means of ditches in which either gravel or a drainage pipe is placed for the purpose of collecting the water and conveying it to a sump (Fig. 9.2). When this procedure is used to intercept seepage and stabilize the subgrade, it is frequently referred to as "French drain." In circumstances where an impervious layer exists near or above subgrade, French drains are frequently necessary even when the major dewatering work is accomplished with deep well systems or even wellpoints. A toe drain installed, say, at the toe of a dam embankment is functionally very similar to a French drain, but constructed with proper filtration criteria so the movement of soil or "piping" will not occur.

Sumps, ditches, and trenches are typically relied upon for handling surface and construction water as well. A successful program must also include provisions for handling

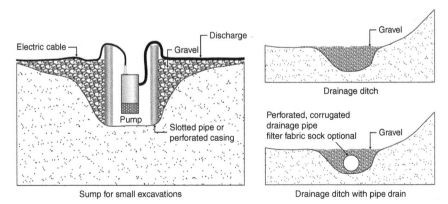

FIGURE 9.2   Open pumping methods.

surface water for a severe storm event. Tidal variations, river-level fluctuations, and localized rainy seasons are all factors that should be considered in the overall dewatering plan. The excavation should be graded so that construction water and surface water may flow freely to the ditches and sumps. This is particularly important in low-permeability soils that are more adversely affected by water infiltration at the surface. Temporary sumps are necessary until permanent facilities can be established.

Wellpoints

Small pipes, up to 2.5 in (6.3 cm) in diameter, connected to screens at the bottom and to a vacuum header pipe at the surface constitute a wellpoint system (Fig. 9.3). A few to several thousand individual wellpoints may be connected to the vacuum header pipe and typically to a single wellpoint pump (combination centrifugal and vacuum) that separates and disposes of air and water so that the system can be constantly and effectively primed and the water removed. The effective suction lift of a wellpoint system is governed by the atmospheric pressure and vacuum that can be generated and by friction losses in the system.

Wellpoints are highly effective in low-permeability soils. In low-flow conditions, wellpoints will apply the system vacuum to the soils and enhance the drainage of low-permeability soil that may not drain well under gravity flow (i.e., no vacuum) conditions.

Effective lifts of 15 ft (4.5 m) are quite common at sea level and, under certain circumstances, lifts can be increased to as much as 20 ft (6 m). Wellpoints typically have capacities ranging from a fraction of a gallon per minute to more than 100 gal/min (400 L/min) and they may be used in single stages or in multiple stages (Fig. 9.4) to accomplish deep dewatering. Each wellpoint is equipped with a valve for regulating flow and limiting air intrusion. The wellpoints are typically PVC or steel, with the headers made of quick coupled PVC or aluminum, and the swing connections made with flexible suction hose. Wellpoint pumps are typically skid-mounted units equipped with several basic components; a centrifugal water pump, a vacuum pump, and some type of air/water separator. Special modifications of the pumping equipment are required for very-low-flow conditions.

Wellpoints are often the right tool where dewatering is shallow (i.e., within a single stage of wellpoints), the soil is of low permeability and the vacuum application is beneficial to the drainage, or where a geological interface warrants close spacing of the dewatering devices.

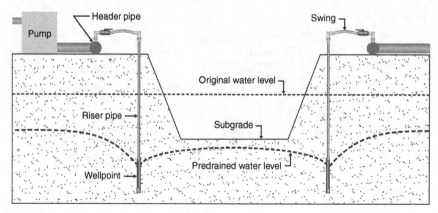

FIGURE 9.3   A basic single-stage wellpoint system.

FIGURE 9.4   Two-stage wellpoint system. (*Courtesy Moretrench.*)

Wellpoints are appropriate for high-flow conditions also. The wellpoints can be constructed with larger diameters, higher-capacity intake screens and be connected to larger-diameter vacuum header piping and to higher-capacity pumping stations.

Deep Wells

A deep well can be considered as a dewatering device equipped with a submersible pump (Fig. 9.5). Deep wells may be as shallow as 25 ft (7 m) and as deep as several hundred feet. They can pump as little as a fraction of a gallon and as much as several thousand gal/min. They may be powered with high-volume line-shaft turbine pumps, either electrically or engine-driven, or by electric submersible pumps capable of fractions of gallon per minute as well as flows up to several thousand gal/min. Figure 9.6 shows a peripheral deep well system installed to dewater a prolific alluvial aquifer to a depth of approximately 85 ft (26 m) immediately adjacent to the Mississippi River. Dewatering was accomplished with a series of high-volume, diesel-driven, deep wells that pumped as much as 48,000 gal/min (182,000 L/min). Wells may be vented to the atmosphere or may be operated under vacuum to enhance the drainage of low-permeability soil. Dewatering wells are typically constructed with a filter pack between the wellscreen and the natural formation, although unfiltered wells may be utilized where the natural soils are not susceptible to continued ground loss with pumping. They may employ various types of screens from simple slots or louvers to more sophisticated trapezoidal wire and mesh configurations. Wells can be as small as 3 in (75 mm) and as large as 48 in (1.2 m) in diameter. Developing procedures are generally employed to bring wells up to their maximum potential. Each well may have a single discharge or multiple wells may have a common discharge system.

Electrical distribution can be a major cost component of the dewatering system if a large number of units are to be electrically powered. This may also entail consideration of standby power sources, generally provided by generating sets.

Wells are most often the preferred dewatering technique because they are the most cost-effective tool on a per-foot of perimeter basis where the ground conditions are favorable for their use. Those favorable conditions are where the soil will drain by gravity and permeable soil underlies the excavation, with no difficult geological interface within the excavation or near subgrade.

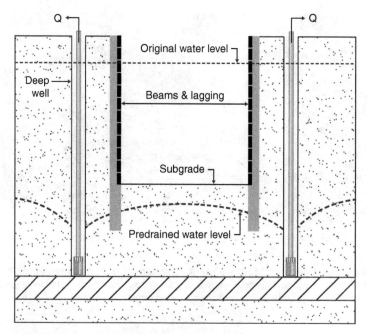

FIGURE 9.5 A deep well system.

FIGURE 9.6 A high-capacity deep well system installed in Mississippi River alluvium. (*Courtesy Moretrench.*)

Ejectors

The ejector (sometimes called an eductor) is a down-well pumping device that pumps by vacuum or suction but overcomes the ordinary limitations of suction lift by employing a nozzle and venturi located within the ejector body (Fig. 9.7). The ejector bodies may be installed at the bottom of wells or within wellpoint riser pipes above a wellpoint screen. The in-well ejector apparatus is powered by high-pressure water and requires separate supply and return risers that may be concentric or parallel. Separate supply and return headers are utilized at the surface and water under substantial pressure flows from the supply header down the supply riser and through a nozzle and venturi, creating a vacuum within the ejector body. Passages within the body permit water flow from the wellpoint screen, generally through a foot valve. This water then joins the supply water and flows through the return piping to a tank. Generally, the supply water is drawn off the bottom of the tank and the overflow of the tank becomes the net yield of the system. The ejector mechanism can create very high vacuums, which makes ejectors a rather effective tool in fine-grained soils.

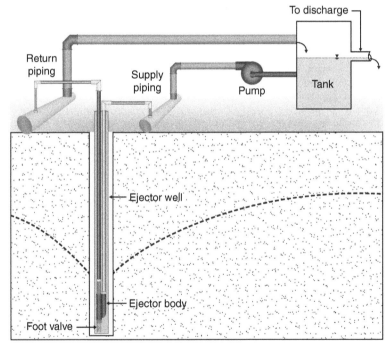

FIGURE 9.7   Ejector system with recirculating supply and return piping.

Ejectors are commonly used to lift water as much as 100 ft (30 m). They are not generally used for higher lifts because the efficiency of an ejector system is rather poor—generally on the order of 5 to 15 percent on a water horsepower basis. Other dewatering systems may approach 75 percent efficiency. Ejector systems are frequently operated when the volumes of water to be pumped are typically less than 5 gal/min (19 L/min) per well. In most ejector systems, it is particularly important that the piping, nozzles, and venturi be sized to suit the circumstances because of the inherent low efficiency of the method.

Horizontal Dewatering Installations

Horizontal dewatering systems are often referred to as trench drains or sock drains. Systems are installed with specially designed trenching machines. A horizontal dewatering installation consists of a flexible, corrugated, perforated polyethylene pipe encased in a geotextile filter sock. Where this technology is used for dewatering work, installations are typically less than 15 ft (4.5 m) deep with a 4-in (100-mm)-diameter drainage pipe. Deeper and larger-diameter installations are possible. The individual lengths of drainage pipe are connected to wellpoint pumps. These systems can be very cost-effective on projects where shallow dewatering is required from a relatively thin water-bearing formation of moderately permeable soils with few obstructions to trenching. Horizontal drains are occasionally used to drain unstable soils such as highway cuts and spoil piles.

Horizontally oriented dewatering devices can also be installed by drilling techniques, including horizontal directional drilling. While drilling techniques readily permit the insertion of horizontal pipes into predrilled holes, the installation of proper filters and screens and the techniques for developing the hole are very limited and drilled-in horizontal drains therefore have limited applicability in construction dewatering.

Wick Drains

The fabric wick drain, driven into soft compressible soil utilizing a thin-blade crane-mounted mandrel, is most commonly used to provide accelerated drainage and consolidation from a compressible soil. The wick collects soil moisture that is transmitted to a surface drainage blanket.

Wicks can be installed quickly and on very close spacing to provide a short drainage path for soil consolidation. In this procedure, compressible soils may be consolidated during a construction program by improving the vertical drainage and causing a pressure differential to exist between the pores of the compressible soil and the wick drain so that water will flow from the soil. This differential pressure may be caused by common surcharge or sometimes by supplemental pumping operations, with or without vacuum effects. Consolidation projects require an integrated design involving the evaluation of soil consolidation characteristics, drainage devices, spacing and capacity, surcharge time, and installation techniques.

Recharging

Groundwater recharging may be done for a number of reasons, including

1. To control the migration of saltwater or contamination within an aquifer
2. The return of potable water to groundwater storage
3. To maintain water levels to prevent or avoid settlement and consolidation of compressible soils

The latter reason is the most common in connection with temporary dewatering work, although major recharging programs have been established to maintain water supply in a number of different localities.

Water can be returned to the ground through a number of different procedures. Recharging basins can provide a means of adding storm water to the groundwater table over a long period of time simply by preventing runoff into adjacent bodies of water. Similarly, ditches can be effective in returning the water to the ground along the periphery of a temporary dewatering project.

More frequently, though, groundwater is injected into the ground by means of wellpoints or wells. Water can be injected into a recharging system under pressure, although the pressure is generally limited by the amount of cover above the recharge zone. The mechanical design of the system is very important to avoid air entrainment.

Perhaps the single most important factor in a recharging program (other than the permeability of the soils through which the water is being injected) is the character of the water itself. Most sources of water contain dissolved and suspended solids. It is necessary to treat even potable municipal water supplies for recharging purposes. As a well is pumped, the water migrates through progressively larger pore spaces. In the recharging process, the procedure is reversed and the water flows through progressively decreasing pore space, thus the increased tendency for plugging of the devices.

Whenever the temperature or pressure of the water is changed, as happens in pumping operations, the characteristics of the water with respect to materials in solution can change and result in significant well plugging. The control of sediments and precipitation of solutionized minerals is the key to successful recharge. It is not unusual to see a recharging system's efficiency deteriorate in several weeks because of problems in the water quality. Treating and redevelopment of wells may restore or improve the effectiveness of a system under such conditions.

DEWATERING DESIGN AND DESIGN CONSIDERATIONS

Dewatering design by analytical methods requires basically two steps: determining total pumping quantities and determining the type of dewatering device and the quantity needed. An understanding of the aquifer type and behavior is necessary.

A number of parameters will determine pumping quantities and aquifer behavior. They may be determined by full-scale pump tests, by results from previous relevant experience and/or by careful analysis of soils and hydrology and extrapolation of their characteristics into design values. The authors hesitate to treat this subject in great detail because proper understanding of the limitations of various approaches is necessary to avoid expensive mistakes. The following discussion is intended only to give the reader an overview of the various factors that determine pumping volumes and aquifer characteristics.

Relevant Hydrogeological Parameters

For any given dewatering scenario, the dewatering design will be determined by (1) the structural considerations, that is, the dimensions of the excavation, and (2) the hydrogeological conditions. The hydrogeological conditions include the amount of groundwater lowering required, the permeability of the soil and aquifer thickness, and the effective proximity of recharge to the aquifer. The structural considerations are straightforward, but the hydrogeological parameters require experience to evaluate.

The permeability of the ground is a measure of the soil's ability to transmit water. More specifically, the permeability of a given soil is the amount of water that will pass in a unit of time through a square unit of area, under a unit of groundwater pressure. The aquifer thickness is also very significant in determining the dewatering requirements. *Transmissivity* is defined as the product of the permeability and aquifer thickness and is a useful term that can be used to describe the potential of an aquifer to produce water.

The radius of influence of pumping is a hydrogeological parameter that takes into consideration the proximity of recharge. The presence of sources of recharge as well as aquifer barriers or discontinuities is taken into consideration with the parameter of radius of influence. The radius of influence of a pumping system can be very large, particularly if recharge boundaries such as rivers or other bodies of water are not present. Barrier boundaries created by discontinuities in the aquifer will affect pumping volumes. Often in construction dewatering, the radius of influence will be different in different directions. This, too, must be considered when calculating pumping volumes. The pumping quantity will be very much affected by the radius of influence that, of course, determines the gradient (i.e., slope of the water table) that is produced when the water flows from its source of recharge to the dewatering system. In a given aquifer, the steeper the gradient, the higher the flow. It is not unusual to conduct dewatering operations in alluvial plains where the radius of influence can sometimes be many thousands of feet and the gradients very flat. By way of contrast, dewatering for, say, a subaqueous tunnel approach on an island in a bay surrounded by water may have a radius of influence not much greater than the distance to open water.

Aquifer Types and Their Behaviors

The design of a dewatering system must take into consideration if the water bearing formation(s) is an unconfined (or "water table") aquifer or a confined aquifer (Fig. 9.8). The two will respond differently with pumping.

The unconfined or water table aquifer differs from the confined aquifer in that it has a phreatic surface, or "water table," that rises and falls with changes in recharge or in pumping. The upper confining bed is missing. This is the more common aquifer condition encountered with dewatering work. The pumping quantity is actually composed of two primary components. The first is the steady-state flow to the pumping system that will continue after drawdown has been achieved. The second is the quantity of water that must be pumped to effect storage depletion, which is essentially pumping down of the cone of depression. The release of stored water in the aquifer is a time dependent task. The unconfined aquifer that has a water table that falls during pumping releases stored water slowly by gravity drainage. Similarly, there is a time associated with rebound of an aquifer once pumping ceases. Even in the case of aquifers with high transmissivities, storage depletion can be the dominant factor influencing system design, particularly when the time available

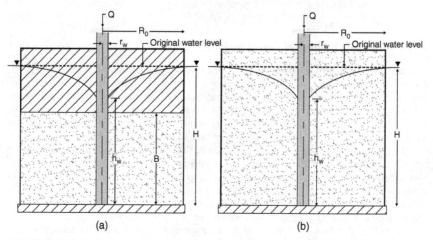

FIGURE 9.8   Confined aquifer (*a*) and water table aquifer (*b*).

for dewatering is limited. This, incidentally, sometimes determines the dewatering method. For example, a peripheral deep well system can be activated and pumped prior to the beginning of excavation and storage depletion can be caused over a long period of time. In the typical wellpoint application, it is important that the drawdown be achieved quickly so as not to delay the excavation. Under those circumstances, sufficient capacity to cause storage depletion must be provided. It is not unusual for construction dewatering systems to pump only one-third to one-half of the initial pumping quantity after drawdown and after the bulk of the storage has been depleted. Again, mathematical relationships can be used to calculate storage volumes and depletion quantities.

Where the water level in a confined aquifer is never drawn down below the top of the aquifer, one would more accurately say that formation is *pressure relieved* rather than *dewatered*. In such a situation, there is no physical draining of the soil that would occur with dewatering of a water table aquifer. The response to pumping of a confined aquifer is an immediate pressure reaction and occurs instantaneously. When pumping ceases, the pressure or water level rebounds instantaneously as well.

A dewatering program involving only pressure relief theoretically has a linear relationship between the pumping quantity and the depth to which the pressure is relieved. In theory, the pumping quantity remains constant with time, perhaps being influenced by factors such as water temperature and seasonal changes in the normal static head caused by variations in rainfall, river stages, etc. Occasionally a construction dewatering job will have these characteristics. If that is the case, the job will be sensitive to interruptions of pumping. The water levels observed in piezometers can go up and down drastically on an almost instantaneous basis as pumping is started and stopped.

Sometimes in the course of a dewatering program, an aquifer will act as a confined aquifer until the water is drawn down below the top of the layer, at which point it will react as a water table aquifer. Mathematical analyses must be adapted to this situation when applicable.

In a confined aquifer, transmissivity is a constant. In a water table aquifer, it decreases as the water level is drawn down and the saturated zone through which the water flows decreases. As such, the design of a dewatering system for a confined aquifer is much more straightforward than for a water-table aquifer.

Calculating Total System Flowrate

The basic mathematical relationships concerning flow in soils assume that the aquifer is an ideal aquifer. This means that it extends horizontally in all directions beyond the area of interest without encountering recharge or barrier boundaries. The thickness is uniform throughout. It is isotropic; that is, its permeability in the horizon and vertical directions is the same. Water is released from storage, the head is reduced instantaneously and the pumping well is frictionless, very small in diameter, and fully penetrates the aquifer. In the ideal water table aquifer, the additional assumption is that the phreatic surface will rise and fall with pumping operations. The basic flow relationships for ideal confined aquifers and water table aquifers, respectively, as illustrated in Fig. 9.8, are

$$Q = \frac{2\pi K B (H - h)}{\ln R_0 / r_w} \tag{9.1}$$

$$Q = \frac{\pi K (H^2 - h^2)}{\ln R_0 / r_w} \tag{9.2}$$

where Q = pumping rate and K = permeability.

As may be seen from the above formulas, pumping quantities are determined by dimensional factors and the hydrogeological characteristics of the soil. Obviously, the more one lowers the water level, the more water one must pump. The larger the dewatered area is, the higher the pumping volume will be. Pumping volumes will also vary directly with the permeability of the soil. Empirical relationships that take these factors into account have been developed, and an estimate of permeability may be made from grain-size curves of most soils, making allowances for particle shape and density. Seldom is an aquifer isotropic. Invariably, the horizontal permeability is many times the vertical permeability. Suitable recognition of this must be made where appropriate.

It should be noted that basic mathematical relationships have been developed for fully penetrating pumping systems. The well fully penetrates the aquifer and draws from it evenly over its entire depth. Frequently, in practice this is not the case and a suitable mathematical adjustment must be made in the design.

Construction dewatering jobs seldom qualify as ideal aquifers; therefore, a number of experience-based modifications of the mathematical relationship must be made for use in calculating flows in actual situations. A very good presentation of this subject is given in *Construction Dewatering and Groundwater Control: New Methods and Applications* (see "Bibliography").

Several different methods may be used to determine the pumping volume for any particular project. Those based on mathematical relationship must be carefully adapted to the job at hand. The experienced designer has the added ability to correlate mathematical results with a practical, experienced concept of pumping quantities. It enables the designer to at least know that the decimal point is in the right place. Without experience-based judgment on how a particular situation may deviate from the ideal aquifer assumptions, a dewatering design may be off by an order of magnitude.

The experienced designer may use several techniques in evaluating dewatering flows. With one of the simplest analytical models, the designer could calculate the flow to a dewatering system that might have a circular configuration by simply assuming that the radius of the effective pumping well is the radius of the dewatering periphery.

The designer can calculate a dewatering flow by utilizing a cumulative drawdown method wherein the drawdown at any particular point on the project is the cumulative drawdown from each of the pumping wells. In this case, appropriate assumptions are made with respect to the pumping quantity for each well and a sufficient number of wells are included in the design to give the required drawdown. The latter procedure is particularly valuable when the well array is irregular and the well system cannot be analyzed simply as a large-diameter equivalent well.

Determining the Number and Spacing of Dewatering Devices

The first step in dewatering design is determining a total pumping flowrate required for the given aquifer type. The second step is then determining how many dewatering devices are required to achieve the total pumping flowrate. Here, of course, one must be aware of the dewatering tools available and the advantages and limitations of each dewatering approach. The method of collecting the water can influence the pumping volume. For example, fully penetrating wells will pump more water per well than partially penetrating wells, but they will also be more expensive and require more water to be pumped to achieve a given result. However, a well whose capacity and efficiency deteriorate significantly when the water levels are lowered can be a very expensive well. The designer must have a feel for costs when drawing conclusions as to how to design the dewatering systems.

The number of wells required is determined by the ability of each well to pump water from the ground and by a number of other considerations. In well systems, a larger number of wells provide more redundancy in the system as a whole. Keep in mind that the pump in each well must lower the water level sufficiently in the well to achieve the design purposes. With factors such as storage depletion entering into the process, the steady-state pumping quantities are invariably less than the pump's capacity.

In many dewatering situations, the pumping quantity is not particularly critical. The key factor may be the spacing of dewatering devices so as to achieve drawdown close to an impervious layer. Here, designs do not lend themselves to hydrogeological calculation but must rely upon experience and judgment of what can be achieved with different dewatering devices. In such a situation, closely spaced wellpoints or ejector wells may be more advantageous than wells.

Aquifer Pumping Tests in Design

Pumping tests are sometimes conducted when the success of dewatering programs is critical. The test, which generally involves the installation of a well or a series of wells with piezometers radiating in several directions, is conducted by pumping and observing the shape of the resulting cone of depression. It is appropriate to perform a pumping test to evaluate total flowrate, radius of influence, the effective proximity of a suspected source of recharge (man-made or natural), or to provide pumping test data for reliable system design where partial penetration is to be considered. Mathematical relationships are used to determine aquifer characteristics as a result of the observed drawdowns. Sometimes the extrapolation of pumping test results beyond the area of good quality data can lead to costly errors. For example, this may be the case where an inadequate array of piezometers is installed and a large radius of influence is projected beyond the area instrumented with piezometers.

The specific capacity of a well or even a piezometer that may be test-pumped can be valuable in terms of confirming assumptions about the continuity of the aquifer and its characteristics. Frequently, the recovery characteristics of a well when pumping is stopped can be indicative not only of the well's efficiency and performance but also of aquifer characteristics. Plots of drawdown and recovery can be made and used to classify an aquifer as confined or water table, to indicate the presence of recharge and barrier boundaries and, in general, to convey to the designer a good understanding of the behavior of the aquifer. Instrumentation can identify well losses.

Groundwater Modeling

The authors have employed groundwater modeling (Fig. 9.9) as a predictive tool prior to construction and as a valuable aid in design and optimization of dewatering systems during construction.

Analytical models do not take into consideration aquifer heterogeneity, anisotropy, or other complexities. There are certain aquifer conditions and geometry where the use of an analytical model either becomes too complicated or the simplifying assumptions involved differ greatly from reality.

Groundwater modeling with computers can be employed to reach more accurate solutions for dewatering situations that heretofore defied analysis. Computer models describe the groundwater flow system in detail, with both spatial and temporal variations in aquifer properties, boundaries and applied stresses defined for specific regions in the model. Numerical (computer) models can therefore accommodate aquifer heterogeneity, anisotropy, complex and irregular boundary conditions, partial aquifer penetration, and transient and steady-state flow simulations. Two dimensional (2D) or three dimensional (3D),

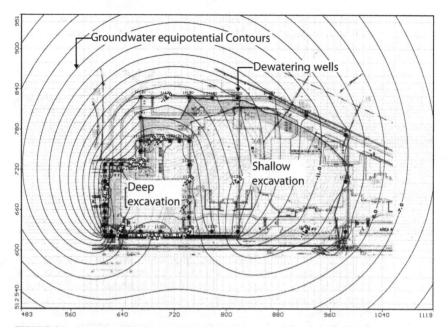

FIGURE 9.9   Groundwater model output showing steady-state groundwater equipotential contours.

transient or steady-state, confined or unconfined models can consider both vertical and horizontal components of flow.

A groundwater model is only as good as the information on which it is based. Calibration is required for confidence in the model. Calibration of the model involves adjustment of aquifer parameters until a reasonable match is achieved between model predictions and a known state of aquifer stress (usually hydraulic heads) measured in the field. The ultimate test of a groundwater model is to calibrate it to actual field data.

The skill and experience of the person doing the modeling is closely related to the quality of the model results. Any model is only an approximation of the real groundwater system. It should be noted that

- The model is not right until there is a correlation with field data.
- A calibrated model is only one of a number of possible solutions to the given data.
- A model is no substitute for the practical experience and judgment derived from the analysis of dewatering systems and subsequent observation of their performance in the field.

Factor of Safety

Every dewatering design should include a factor of safety for a number of different reasons. Some of them have been enumerated, but perhaps the most compelling reason is that one is dealing with ground conditions that were determined by nature and are interpreted through investigations and experience. Designs cannot be completely accurate. To the extent that the subsurface investigation suggests potential variability in the ground conditions, the dewatering system design should have flexibility so that while the system is being installed, the variations that occur in nature can be appraised and modifications can be made if necessary. For example, one might change wellpoint spacing to more nearly suit probable worst case soil conditions or to be sure that some screens are within a particular stratum where they are likely to exist. Well design may contemplate the need for a larger pump in the well.

Redundancy elements should be built into a dewatering system where warranted. Sometimes it can be done at very little cost. Valving on a discharge system may be included so as to provide alternative flow routes in the event that pipes are damaged. The experienced designer will consider redundancy, keeping in mind the potential variability of the ground, duration of the dewatering job, the susceptibility of the structure or the excavation to damage in the event of dewatering failure, and any number of other factors that determine the value of adding redundant features to the dewater system.

Design Considerations

Instrumentation. Dewatering systems should be instrumented to verify the result expected to be obtained. Instrumentation has the value of appraising the characteristics of the system not only as an installation proceeds, but can be a means of monitoring the continuing adequacy of the dewatering system after the job has been dewatered.

Piezometers should be installed where necessary. Pumps should be equipped so that pressure gauges may be utilized to monitor operating conditions and performance. Vacuum gauges may be utilized to monitor operating conditions and performance for wellpoint systems. Vacuum gauges should be spaced along the wellpoint header to monitor the characteristics of the whole system. Flowmeters or other measurement devices should be included in discharge lines so that an evaluation can be made of the performance of a system.

Other methods of monitoring may be used in lieu of direct reading devices such as piezometers and flowmeters. These include recovery tests in active pumping wells, test pits, and flow measurements through individual well bypass valves.

Potential for System Fouling. Corrosion and incrustation are always of concern in major construction dewatering projects, particularly those of extended duration. Occasionally, groundwater and some surface waters are extremely corrosive to the degree that even short-term dewatering programs must entail the use of specially designed equipment and materials. The majority of projects will never have a problem, but, when a corrosion or incrustation problem does occur, it tends to be severe, often with rapid deterioration of the system.

The gas and mineral content of the water to be pumped should be evaluated and its effect on the program known. Hydrogen sulfide, free CO_2, chlorides, miscellaneous salts, dissolved oxygen, iron, oxides, manganese, and alkaline hardness are some of the more common constituents of water that can have a great bearing on corrosion and incrustation. If the water to be pumped from the ground is to be used for recirculating purposes or groundwater recharging purposes, it becomes critical to understand the corrosion and incrustation problems so that suitable chemical control or alternative procedures may be adopted. Corrosive groundwater conditions, should they exist, must be clearly indicated as such in the project geotechnical report(s). If surface waters or wastewater from industrial processes leach into the ground or are to be pumped by segments of the dewatering system, the component materials of the dewatering system must be selected accordingly. Corrosion of wellscreens can have several detrimental effects, including opening of the well slots and the subsequent pumping of sand and loss of structural strength that leads to collapse. Plastics are inert and therefore resistant to most corrosive groundwater and where feasible they should be used. Piping systems are frequently made of polyvinyl chloride (PVC) because of its ready availability, reasonable cost, and ease of installation. Pumps in the smaller sizes are available with critical parts manufactured from plastic.

Mineral incrustation usually results in the formation of a deposit or a scale that causes a reduction in effective pipe diameter or bridges across wellscreen openings, reducing the open area available for water to flow through the screen and ultimately reducing water flow through that soil zone. While there are many reasons for mineral incrustation to form in water well systems, the primary cause is a change in the water chemistry equilibrium brought on by pumping. Incrustation is influenced by pressure drops and by temperature, and occasionally dewatering systems must be designed to minimize pressure changes so as to minimize incrustation.

Frequently, incrustation is also linked to biological activity, particularly when the effects are observed within a short period of time. The most prevalent form of biological incrustation in dewatering work is due to iron bacteria (also referred to as iron fixing or iron oxidizing bacteria). The iron bacteria form stalks or sheaths of iron and are the most debilitating biological incrusting agent due to the direct blockage of well screens and gravel packs. There are ways to construct and operate a dewatering system to mitigate the impact of iron incrustation but at a significant additional cost.

Groundwater conditions where there is a potential for incrustation must be clearly indicated as such in the project geotechnical report(s).

Potential Adverse Effects of Dewatering.   The most common cause of settlements associated with construction dewatering is not the lowering of groundwater tables under an adjacent structure and the subsequent increase in load caused thereby. Rather, most frequently, the water being pumped from an excavation contains suspended solids and fines because of inadequate sumping procedures or dewatering systems that have not been designed and installed properly. If a pumping operation is moving fines and if it is to continue for a substantial length of time, it is not difficult to compute how much material will be removed from the ground by the pumping operation. This can create voids in the ground and cause settlements. On the other hand, lowering the groundwater table does remove the buoyancy from soil particles and the effective weights of those soils are increased with respect to the loads on lower compressible layers. If a layer is compressible under the increment of loading, difficulties could ensue. Groundwater recharging can be a means of minimizing these problems.

Cost Considerations.   All through the decision-making procedures previously described, the designer must have an awareness of costs. This is an involved subject and can only be discussed here in general terms. Some very basic factors must be considered. The use of two different dewatering systems may present duplicate operating labor costs because of local union contract agreements. Should this be the case, a more economical solution might be to deal with one system or the other, sacrificing some of the results of the program or substantially decreasing the efficiency and increasing the installation cost. The designer should be aware when the size and complexity of a particular well is such that two more simple wells could achieve better results at a lower cost. Similarly with a wellpoint system, cost factors involved in very close wellpoint spacing are not necessarily linear because of the reduced installation cost on a per-unit basis. The system discharge location may be a considerable cost factor; permit and per-gallon discharge fees may apply. A storm-water-handling component of a system may have a separate piping system so as to forego disposal and treatment costs that may be applied to the groundwater component. The factors involved in any particular job determine the most cost-effective measures, but the designer must be aware of the costs when considering the alternatives.

Materials of Construction.   *Piping design* must be made in accordance with standard practices for evaluating flowrate, frictional considerations, nature of the pipe, and the precision with which it might be laid. *Pump selection* is based on the design requirements for the pump, of course. However, frequently pumps are selected because of their cost or availability or perhaps because of their favorable characteristics under low net positive suction head (NPSH) conditions. The total dynamic head under which the pump must operate should be evaluated and an allowance made for the fact that the pump, even if it is new when first utilized, will wear and its ability to pump the water from the deepest point in the excavation must be evaluated. The total dynamic head under which the pump must function is composed of the static head, the friction head, the velocity head, and the net positive suction head. *Component materials* such as screens, piping, and pumps must be selected with regard to routine maintenance or corrosion and incrustation characteristics of the water to be pumped. Most commonly, systems are constructed of PVC. PVC is lightweight and resistant to corrosion but susceptible to construction damage. Wellpoint pumps commonly have cast-iron, bronze, and stainless steel components. Well screens can be made of PVC, galvanized steel, plain steel, or stainless steel. Well pumps can be cast iron or bronze or, in the smaller sizes, PVC with stainless-steel fittings. Special corrosion-resistant materials have been developed for pump construction such as zincless bronze and stainless steel.

SYSTEM INSTALLATION, OPERATION, AND REMOVAL

Jetting and Drilling—Installation of Wells and Wellpoints

Dewatering devices are installed utilizing many different methods. At one time or another, every method of advancing a hole in the ground has been employed in the installation of dewatering systems, sometimes with very poor results. Methods that cause compression of fine-grained soils, remolding of cohesive soils, smear, and other negative characteristics that inhibit the flow of water from the soil should be avoided. Some methods will entail a mandatory developing procedure to remove borehole smear and drilling additives.

Wellpoints, ejectors, and even deep wells are frequently jetted into place through the use of a self-jetting device built into the tip of a screen section. Internal valves regulate flow direction for jetting or for pumping. Continuity during filter sand installation is important in order to minimize segregation of filter sand, particularly if it is well graded.

There are other methods of installing wellpoints and wells, including jetting pipes and holepunchers. A holepuncher is a 6- to 24-in (15- to 60-cm)-diameter, heavy-duty pipe equipped with air and water connections (Fig. 9.10). It may be equipped with a removable head so that the dewatering device may be inserted and then the holepuncher withdrawn. A holepuncher may be used with an external sanding casing so as to facilitate the placement of the device and its filter. Casings 24 in (60 cm) in diameter and 100 ft (30 m) deep have been used for the installation of wells. Methods employing the jetting processes result in a better installation, particularly in fine-grained soils, than those involving drilling and driving simply because smear, remolding, and compression of the soils are avoided.

Many types of drilling equipment have been utilized for well installations. A number of these machines have been adapted from the foundation drilling industry. These include bucket auger-type rigs for large-diameter, shallow boreholes, and direct rotary-type drill rigs that are used for smaller-diameter and/or deeper holes. Duplex drilling involves a casing advanced with the drill steel to prevent collapse of the borehole or to isolate a particular aquifer. Dual rotary drilling can be used, which is similar to duplex drilling except both the inner rod and the casing can be advanced independently. Percussion drills are used in dense soil conditions or when penetrating rock. Today, even sonic-type rigs are being utilized. On occasion, reverse rotary equipment, which is typically used for water supply wells, is used for high-capacity, large-diameter deep wells. The choice of the type of screen used on the dewatering device is sometimes affected by the installation method. The need to develop

FIGURE 9.10 Holepuncher and casing installing 60-ft-deep wells. (*Courtesy Moretrench.*)

the well will influence the choice of the screen also. Drilling additives such as bentonite and biodegradable Revert are typically used to advance the borehole but must be removed by the developing process. It is important that the choice of screen and filter, installation method, and developing technique all be coordinated so as to produce an effective dewatering device.

Piping Systems and Pumping Stations

Dewatering piping systems most frequently utilize aluminum or plastic piping coupled with both flexible and rigid couplings. Aluminum is often chosen for reusable components and PVC chosen for disposable components. There is an endless variety of connectors that have been developed for special purposes in connection with dewatering system applications. In connection with pipe laying, it is important to remember that the line frequently must handle air as well as water, and the lines should be sloped so as to minimize air pockets. In the case of wellpoint headers, devices such as air/water separators and suction manifolds are frequently used to improve vacuum and hydraulic conditions. A typical well piping system should include a check valve so that water cannot flow back into the well when the pump is shut off. A throttling valve should be used so that the flow may be regulated without causing severe oscillation of water level within the well, and it should be equipped with connections for pressure gauges and flow measurement so that the performance and condition of the well might be monitored throughout the life of the job. Sampling taps are typically installed to check water quality.

On major jobs, standby pumps are essential. Generally with dewatering systems, pumps run 24 hours/day, 7 days/week. This creates the need for standby facilities and preventive maintenance. The pump station should be located in a convenient place for access and fueling. Discharge lines that go through earth cofferdams should be equipped with suitable water stops around the pipe and, if there is a possibility of a river rise, the discharge lines should be equipped with check valves to prevent siphoning and to permit pump relocation at high river stages.

Turbine pumping units (Fig. 9.11) are employed in situations where large quantities of water must be handled. They may also be employed where multiple stages of header can be handled with one pumping station, as is the case in some deep cofferdams. Characteristically, a turbine pumping unit will have superior air-water separation capability and its hydraulic operation characteristics are excellent.

FIGURE 9.11   Vertical turbine wellpoint pumps. (*Courtesy Moretrench.*)

Powering a Dewatering System

The mode of power for a dewatering system must be considered throughout the design process.

Electric power sources for pumping operations should be investigated from a cost and reliability standpoint. Invariably, major dewatering jobs will require 100 percent power standby. This may consist of generator sets or sometimes standby diesel-operated pumping equipment. Large, high-capacity deep wells can be adapted to both electric and diesel operation through a well head, which will permit the mounting of the motor in line with the pump shaft on the top of the device and a diesel engine on a right-angle drive.

The electrical distribution of power to a large number of operating submersible electric pumps in deep wells around the perimeter of a large excavation is an important and costly component of a dewatering system. Not only must the power be distributed with regard to amperage draw and voltage drops, but the job may be divided into multiple circuits, permitting disconnection of individual circuits should repair or relocation be required. In this connection, redundancy concepts should be explored. For example, if an accident knocks out the power substation and adjacent generating facilities, what would this do to the dewatering program? In some cases, the dewatering system should have power drops at two locations and the standby facilities subdivided into two components.

Where commercial electric power is used, it is generally advisable to provide standby generating capacity in the event of power failure. Generators are diesel-powered. If a well system has individual diesel power on each well, it may not be necessary to provide any standby power but simply to be in a position to repair or replace a malfunctioning diesel engine.

System Operation

Anticipation of potential operational issues and careful monitoring during the progress of the work are essential to timely recognition and resolution of operation and maintenance issues. Some jobs are sensitive in that a malfunction of the pumping system can produce trouble in minutes; other jobs may take hours or even days for trouble to show. This difference is often the difference between pressure relief of a confined aquifer and dewatering at a water table aquifer. Standby pumps should be provided and set up for quick use. Engine and pump maintenance procedures should be set up so as to minimize downtime. Piezometer readings, vacuum and pressure readings, fuel consumption, flow measurements, and sand content tests should be taken periodically and recorded. The variations or trends in the various measurements can suggest deterioration of pumps due to wear, clogging of well screens due to chemical incrustation, leaks in the system due to corrosion, or a multitude of other conditions that, if detected early, may simply constitute maintenance and pump replacement. If, however, the eventual progression of a deteriorating condition is such as to cause system failure, the results can be catastrophic.

Occasionally on a dewatering job, the quantity of water to be pumped will vary, for example with tide levels, river stages, or precipitation. Depending on the type of dewatering system, this can be a very difficult problem to cope with. It could manifest itself, for example, in inadequate capacity at high tide and vacuum losses at low tide in the case of a wellpoint system whose wellpoints were tuned for a high tide condition. Occasionally, the severity of such a condition can be influential in choosing the dewatering method or particular system components.

System Removal

Frequently, perhaps most of the time, there is nothing complex about removing dewatering systems, the primary considerations being the disassembly and cleaning of expensive equipment so that it may be used again another time. Occasionally, however, dewatering system removal can be very complex. Sometimes the installation sequence is such that a withdrawal sequence from a deep excavation must be prepared as well. This is particularly the case with multistage wellpoint systems. One must decide whether backfilling and hydrostatic uplift pressure considerations will permit the removal of the lower portions of the system and the shifting of operations to the upper portions, or whether the lower portions must be continually operated and ultimately abandoned in the backfill or within the completed structure. Should this be the case, suitable thought should be given to the grouting of piping and screens and perhaps even filter packs.

Sometimes it is possible to relocate components of a dewatering system simply to facilitate the backing-out procedures. When it is necessary to bury equipment and grout the piping, the program should be prepared carefully and executed properly. Sometimes control valves are extended up through the backfill so that proper operating control of the system may be retained. When backfill is placed over an operating segment of a dewatering system, care should be taken to avoid breakage. The system itself should be designed and installed so as to function under this circumstance, and the backfilling should be done carefully with proper materials.

Deep well systems that are installed around the periphery of an excavation present fewer removal issues. The removal of pumps and piping typically will not affect backfilling operations and, upon backfilling above the water table, can readily come out. Sometimes it is necessary to grout the filter pack in addition to the well to avoid a permanent connection between different aquifers. Occasionally regulatory authorities will be concerned about contamination of one aquifer by water from another, particularly if the one aquifer is a water-supply aquifer and the other is subject to contamination from surface sources.

In the case of dewatering programs associated with structures that have continuing additions planned (e.g., where subsequent units of a generating station will be built either directly adjacent to or close by the existing units), the removal of a dewatering system should be given very serious thought. Occasionally, the dewatering system can be designed and constructed so as to be useful for dewatering subsequent units. At other times this is not the case and unless properly removed, sealed, or otherwise contained, old buried piping and gravel trenches can be the source of significant problems in dewatering a new adjacent excavation. As noted previously, unanticipated consequences of system removal are not common, but they are sometimes serious and should be given proper consideration.

It is important to remember the relationship of the dewatering systems to the project as a whole. The dewatering program is intended to improve the conditions under which the structure is to be built. The installation and removal procedures should not monopolize job facilities. Ramps should be kept open. In the case of sequential installations, dewatering work should be coordinated with other operations at the site so as to minimize interference and waiting. Jetting water should be controlled so as not to flood out other operations. Similar procedures must be followed when removing the system so as not to impact the ongoing work.

COST ELEMENTS OF DEWATERING

Economic considerations play a large part in the choice of dewatering methods. Frequently, more than one type of dewatering approach is possible and it is necessary to make an appraisal of the cost and relative merits of more than one system. The characteristics and limitations of each dewatering method must be correlated with the other construction procedures to be employed on the project and the decisions made on the basis of many different factors. The following items have very direct effects on the cost of a dewatering system and should be determined before beginning a cost estimate:

1. Type of dewatering system
2. Size of dewatering system and components
3. Pumping volumes and drawdown required

4. Duration of pumping
5. Depth of installations
6. Weather
7. Availability, quality, and cost of local labor
8. Availability, quality, and cost of local materials
9. Union work rules
10. Surface and subsurface conditions
11. Specification requirements

Estimating dewatering costs should consider many different elements. For example:

1. Mobilization—equipment requirements and availability
2. Storage and staging areas
3. Accessibility of the work area
4. Installation challenges associated with drilling or jetting
5. Source of power (electric or diesel)
6. Schedule limits and sequencing of work
7. Continuity of operations
8. Coordination with other activities (cofferdam construction, excavation support, maintenance, backfill procedures, and sequence)
9. Standby requirements—consequences of pump failure
10. Stormwater and surface water handling requirements
11. Removal and cleanup—demobilization
12. Testing, evaluating, monitoring
13. Potential impacts (plus or minus) on other construction activities
14. Potential need for supplemental efforts and the cost of remobilization
15. Contingency allowances

Detailed consideration of each of these items (and possibly others) in any given project should be made by competent and experienced people who can make knowledgeable evaluations so that the dewatering estimate is completely and accurately finalized.

Mobilization, Coordination, and Responsibilities

Mobilizing a dewatering job involves the planning and execution of logistics, designs, and procedures necessary for the work. Long in advance of the actual work, the various logistical problems must be addressed. Equipment pertinent to the system should be secured and scheduled for delivery. The designs upon which the programs have been based should be updated in the light of new information and should be clearly understood by those responsible for the installation and operation of the system so that if field conditions differ materially from design assumptions, suitable changes may be made. The construction equipment and labor necessary for the installation must be arranged for and programmed so as to be in the right place at the right time. Perhaps additional borings or even a pumping test should be done prior to making actual commitments to installation procedures or to pumping equipment.

Another ingredient of the program concerns itself with the standards that are to be followed in conducting the dewatering program. Most frequently, job specifications will contain only performance requirements with respect to the dewatering results. This is actually a prudent procedure on the part of most design engineers because the actual techniques of conducting the dewatering program should properly be within the province of the general contractor or a specialty dewatering subcontractor. This will allow maximum flexibility with respect to the available equipment, methods, techniques, etc. The owner's engineer should be properly concerned with the result but not necessarily with the means of creating the result. There are exceptions to this, generally relating to situations where temporary construction dewatering systems are incorporated into the permanent function of a structure, such as might be the case in a drydock. Occasionally, to expedite major construction work, contracts covering different phases of the work are let in sequence, in which case the dewatering involved in the initial contract might extend into later contracts. Under these and, perhaps, other circumstances, the owner has a direct interest in the type and quality of the dewatering system.

However, in the typical case of a performance specification, it would be well to draw up performance and operational standards for one's own reference with respect to the conduct of the work. Having predetermined the objectives of the dewatering system and the design assumptions and criteria, one should know what constitutes a satisfactory yield from a dewatering device and what constitutes satisfactory results from each component of a dewatering system. Should initial results not sustain these requirements, installation procedures could be re-examined and perhaps even repeated, ground or aquifer conditions could be reassessed, or fundamental design assumptions could be challenged. It is important that each component of the system be installed properly and functions in the manner in which it was intended to function. Liaison with other activities at the site is extremely important. Coordination of dewatering and excavating can be helpful to both the dewatering contractor and the excavating contractor. If they do not cooperate, each can be hindered by the other.

Typically, the owner will require a submittal explaining the dewatering program, even if the method of accomplishing the dewatering is a contract option. The dewatering plan should always follow the dewatering objectives and should provide for some reasonable flexibility. Soil conditions are rarely as uniform as they are depicted in soil profiles. The system must cope with reasonable variations in soils. However, if unanticipated conditions develop, they should be recognized as soon as possible and suitable changes made in the dewatering program to address these conditions. It is during the mobilization phase and perhaps the installation phase that contingency considerations are reevaluated and investigated so as to ensure the success of the dewatering program.

Labor and Materials

The quality of local labor, materials, and equipment for a dewatering program will affect the performance of the resulting system. Where are the sources of sand and gravel, of cranes, drill rigs, loaders, and other construction equipment? Owners require the contractor to provide a fully trained and safety-conscientious workforce. This is also necessary to meet the Occupational Safety and Health Administration (OSHA) requirements. In both the union and non-union environment, it may be necessary for the specialty contractor to utilize an experienced workforce to perform the work safely. This can add to the overall cost of the dewatering program. All these factors must be understood in order to develop a practical program that can be executed well.

Sequence and Duration of Work

Can the dewatering system be installed external to the excavation? Must portions of it be installed internal to the excavation? Must portions of the dewatering system be relocated as construction proceeds? What are the requirements for pressure relief as the structure is being built? Is uplift a consideration even after the foundation and the backfill are completed? Must the superstructure be in place prior to discontinuing pumping? These are some of the questions that have to be answered, and in order to do so excavation, fine grading, pile driving, mud mats, base pours, wall pours, backfilling, machinery installation, and superstructure programs should be understood.

Operational Considerations

There are a number of direct and indirect costs associated with the operation of a dewatering system. These may include union manning, equipment housing, standby power sources, safe water handling, and provision for controlled flooding of an excavation adjacent to a large body of water to avoid overtopping.

The number of pumping units that can be covered by one man is sometimes determined by local union regulations. In some cases, the distance between units or the size of the discharge pipe will determine the union manning required. A pump operator around the clock, 7 days/week, represents a substantial element of the total dewatering cost. If working rules impose limitations on the number or location of pumping units, this can influence the design of a dewatering system and the choice of methods.

Housing a pump station is dependent on weather and the duration of the job. If internal combustion engines are used, they will generally provide the required heat in the pump house during the winter. If electric pumps are used in very cold weather, heat must be provided separately, particularly for vacuum piping systems and inactive water lines. Sometimes insulation of piping systems is necessary. Dewatering programs have operated under weather conditions in all parts of the world. Systems can be operated dependably at −40°F (−40°C), although this obviously entails special precautions and procedures.

Generating facilities used for standby purposes should have a source of heat in extremely cold weather. Damp conditions should be avoided. All standby units should be checked out and exercised at regular intervals. This testing should be done under load to ensure the adequacy of the equipment in an emergency. In today's environment, when diesel engines are used to drive pumping units or electric generators, fuel storage on jobsites can require special permits as well as dual containment tanks or facilities to prevent a spill should the tank become damaged or fail. The quantity of fuel stored on site can also be limited by the permit. Therefore it is necessary to use a reputable supplier with the ability to deliver fuel on weekends if necessary.

Commonly, storm water and other forms of surface water can present significant challenges, particularly when the site is initially excavated. Site drainage is often not complete and the conditions in the hole are geared to excavating efficiencies, not to drainage. A good contractor should anticipate what would happen if it rained heavily overnight and, within reason, do the small amount of preparatory work that might be necessary to channel rainwater into areas not critical to the excavation or into ditches and sumps provided for such purposes.

In conducting an excavation program in a large area, it is generally wise to take the perimeter down deeper than the middle so that water drains off and can be collected and pumped. Pumping of the water is not the difficult part; collecting it is. Frequently, excavations can become bogged down for days after a rain if the water is allowed to accumulate and soak into previously dried fine-grained soils. It might take a long time to extract that water again, and it is generally advisable to try to collect the water before it seeps into the ground.

Overall site drainage should be considered as well. Water that might tend to flow into the excavation should be kept outside by a dike around the top of the slope. Where ramps enter into an excavation, it is a good idea to put in an artificial hump at the top of the ground so that traffic on the ramp cannot wear grooves that would let the ramp drain the adjacent site area. Thought should be given to slope erosion. Frequently on long-term jobs, it is entirely practical to seed slopes so that the vegetation can provide resistance to erosion. Occasionally, storm water is handled on slopes by means of collecting ditches, culverts under roadways, berms, and drop pipes to pumping facilities.

On some projects, particularly cofferdams adjacent to large bodies of water, it is necessary to provide controlled flooding facilities to avoid the overtopping of the cofferdam or to permit flooding in the event of dewatering-system failure. Occasionally, jobs on rivers will be designed so as to flood the cofferdam at high water and interrupt construction operations until the river recedes. In this event, the dewatering system must incorporate features that permit its submergence and the rapid flooding and subsequent unwatering of the cofferdam. More often than not, it is very difficult to convey large quantities of water into an excavation quickly. Typical procedures involve the use of large-diameter pipe siphons. Pipes are occasionally installed through the cofferdam, with suitable valving to prevent accidental flooding. The difficulty with procedures of this nature concerns itself with the inability to test-operate the facility. Occasionally, an earth dike can be breached so as to cause controlled flooding. Unless the character of the excavation lends itself to the flow of large quantities of water to the bottom, suitable erosion protection such as riprap should be installed along the probable flow path.

Removal of Dewatering Systems

As previously discussed, the removal of a dewatering system can become a critical feature of the program. Sometimes it is not feasible to remove dewatering systems that are buried under structures and consideration should be given to grouting of piping to prevent voids in the future and possible detrimental effects to subsequent construction work. Appropriate costs for ultimate removal or abandoning in place must be included in the dewatering estimate.

ENVIRONMENTAL APPLICATIONS OF DEWATERING

Environmental concerns play a significant role in the application of the groundwater control process, particularly when working in an urban environment. There are numerous instances where the installation of a construction dewatering system must deal with contaminated groundwater. The quality of the groundwater not only impacts the design of the groundwater control system but will require a special effort in dealing with the discharge from a pumping system. These efforts typically involve on-site treatment or, rarely, off-site disposal.

In choosing the groundwater control or dewatering method at a contaminated site, one's initial reaction is to use a passive or non-pumping method of groundwater control such as sheeting, a slurry trench cutoff or ground freezing. However, with mobile, on-site temporary treatment units, it can be more cost-effective to pump and treat groundwater than to install a cutoff. Years ago, the treatment aspect of the project was handled by a separate firm. Today, the dewatering subcontractor handles all aspects of the project. If the required pumping volume is high, perhaps in excess of a thousand gal/min, on-site treatment is typically not economically feasible and a cutoff should be considered.

A significant concern when dealing with discharge of treated groundwater is obtaining the necessary discharge permits. These vary from location to location and can be time-consuming to obtain. The authors strongly recommend that the owner start this process during the project design phase and not wait until a dewatering subcontractor is chosen. This can delay the entire construction process. In applying for the permit, the owner should discuss with knowledgeable people the potential maximum flow that might be expected from the dewatering system. There have been several instances where the permits have been obtained but the allowed flowrate was way too low, which seriously impaired the construction schedule.

The following sections discuss the impact that contaminated groundwater has on the choice of the dewatering method, materials of construction, and on-site treatment options that are available.

Materials

When designing a groundwater pumping system at a contaminated site, the designer must understand the contaminants that are present and the impact they may have on the materials of which the groundwater system is constructed. One must also take into account the length of time that the system will be in operation. The major components of any groundwater pumping system are screens, risers, the collection piping (Fig. 9.12), and the pumping unit.

The most commonly used type of wellscreen is slotted PVC. It is inexpensive and relatively easy to handle. It is chemically resistant; however, there are certain compounds, such as xylene, which react with PVC, causing it to deform. Therefore it is important to understand the specific contaminants in the groundwater. PVC is also the most commonly used material in wellpoint applications. In longer-term applications (greater than 1 year), the use of stainless steel wellscreen can be considered. In short-term dewatering applications, such as sewer construction where wellpoints are in place for less than 1 month, reusable steel, self-jetting wellpoints are more often used.

FIGURE 9.12   PVC dewatering system piping. (*Courtesy Moretrench.*)

With regard to the collector pipe, again PVC is most commonly used because it is inexpensive, lightweight, and easy to handle. The swing from deep well to the collector line is also typically constructed of PVC, including the valve. In wellpoint applications the connecting swing is typically hose. In a contaminated situation, the designer must make certain that the connecting hose is chemically resistant or consider using rigid PVC piping.

PVC pipe is typically joined or connected utilizing glued couplings. The pipe is cleaned with primer then glued. The glue and primer contain solvents that can leach into the discharge water. At certain sites where there are strict requirements to monitor the quality of the discharge, solvent-glued PVC is not permitted, in which case it is necessary to utilize threaded and coupled PVC piping for well materials and mechanical couplings for discharge pipe.

The most important component of the pumping system, naturally, is the pump, whether it is a submersible pump for a well system or a wellpoint pump. In deep well applications, inexpensive stainless steel pumps that are resistant to most contaminants are readily available. The designer must be careful that the seals in the pumps are also chemically resistant. The same is true of wellpoint pumps. Again, time can be the overriding factor. In short-term applications, bronze pump impellers are acceptable; but in longer-term applications, it may be necessary to utilize zincless bronze impellers.

Earlier in this chapter there was discussion regarding what is involved during the removal of a dewatering system. When working at a hazardous or contaminated site, there is an additional factor that must be addressed. The reusable dewatering equipment must be decontaminated after it has been removed from the ground but prior to being shipped off site. This usually involves steam cleaning all the equipment at a temporary decontamination pad. The decontamination pad must be a lined, contained area where the equipment can be steam-cleaned and the residual water confined and treated. This does require some effort, not only because of the time it takes to steam-clean but the equipment also has to be rehandled prior to being loaded on trucks for shipment elsewhere. These decontamination procedures also apply to construction equipment, including drill rigs. These costs must be considered when preparing the dewatering budget. Owing to the costs associated with removal and decontamination, there are times, with the exception of the pumps, when all the dewatering equipment including wells and collection piping are either disposed of or abandoned in place.

Temporary Treatment

In urban environments, it is quite common to encounter groundwater that is contaminated. Typically, contaminants are either hydrocarbons from leaking underground storage tanks, volatile organic compounds, or solvents that have been utilized in manufacturing or even dry-cleaning applications. Typically, on-site treatment (Fig. 9.13) prior to discharging from the pumping system can be accomplished with a reasonable effort and cost.

The most common first step in the treatment process is either a large settling tank or a series of bag filters to remove the suspended solids. Both types of units require routine maintenance; either the bags changed or the tanks cleaned out. Depending on the type and level of contamination, these wastes must be properly disposed of, which represents another cost to the project.

When dealing with volatile organic compounds, the most cost-effective method of treatment is air stripping. Air stripping units are available in sizes to handle from a few to over 1000 gal/min (4000 L/min). The discharge from the dewatering system is directed to the air-stripping unit. Fresh air is blown through the unit, which releases the volatile organics into the atmosphere or to a vapor phase carbon filter. It is not uncommon for air-stripper towers to have efficiencies of 95 to 99 percent removal of volatile organic compounds.

FIGURE 9.13   Temporary trailer-mounted groundwater treatment system. (*Courtesy Ground/ Water Treatment & Technology, Inc.*)

The water discharged from the base of the unit can be directed to either a storm sewer or a surface stream. In addition to the costs of providing and installing an air-stripping tower is the power costs associated with the blower.

Another method for dealing with volatile organic compounds is the use of granulated activated carbon or carbon adsorption. The discharge from the dewatering system is directed to a vessel that is filled with carbon. The volatile organic compounds attach themselves to the carbon particles and therefore the discharge from the vessel is no longer contaminated. It is not uncommon for carbon systems to have 100 percent removal efficiencies. They are very easy to install. There is a connection from the pumping system to the inlet of the carbon unit and then from the outlet pipe to the discharge point. However, use of carbon can be very expensive. Typically, 100 lb (45 kg) of carbon can collect 1 lb (450 g) of compounds. Given a 100 gal/min (378 L/min) pumping system with 5 ppm (parts per million) of compounds to be removed, over 500 lb (225 kg) of carbon per day would be consumed. Unfortunately, carbon removes all compounds whether they are contaminants or inert, such as calcium or hardness. Consuming 500 lb (225 kg) of carbon per day for any significant period of time can become quite expensive. Therefore, carbon is commonly used on low-volume, low-concentration applications. There are times when carbon is used in conjunction with air stripping. The air stripping tower removes the bulk of the contaminants and the carbon is then used to polish the water, removing the balance of the compounds that remain after air stripping. In addition to the initial cost of purchase, the carbon must be properly disposed of after is has been used. As stated previously, it is not uncommon to utilize thousands of pounds of carbon. The costs to dispose of the carbon can be as much as 1.5 times the original purchase cost.

In many locations, environmental regulations require an air discharge permit when utilizing an air-stripping tower. These permits are not always readily available in the time frame during which the contractor is required to complete the work. Since carbon adsorption units are self-contained with no emissions, no air discharge permit is required. When dealing with hydrocarbons and free-phase product, such as gasoline or heating oil, which have leaked from underground storage tanks, an oil-water separator can be used to remove the free product. Basically this consists of a large settling tank in which a series of baffles is installed, and coalescing media that cause the oil to separate from the water. After the separation takes place, the water is released to the discharge point and the free product is skimmed from the top of the oil-water separator, collected, and disposed of off-site. One of the difficulties in utilizing an oil-water separator in construction dewatering applications is that the pumping unit, whether it be a deep well submersible pump or a wellpoint pump, mixes or emulsifies the product and the groundwater and it is therefore necessary to allow sufficient time for the oil and water to separate again. This may require oversizing of the oil-water separator. Another option in deep well applications is to utilize two pumping units in each well: a deep well pump for lowering of the groundwater and an upper skimming pump for removal of the free product. In this case, the oil and water are not emulsified and the product collected from the skimming pumps is collected and disposed of offsite.

The pH of water to be discharged may be a concern. Many temporary treatment systems include pH adjustment as the final step. The discharge water is directed to a tank with mixers. Monitors measure the pH and either caustic soda or acid is added through a chemical metering system to achieve the proper pH.

Applications of Construction Dewatering in Environmental Cleanup

Construction dewatering techniques are applicable to the cleanup of environmentally contaminated sites. Perimeter wellpoint systems surrounding the contaminated groundwater plume are effective in containing and collecting the plume. The wellpoints can be installed on wide spacing, as much as 20 to 25 ft (6 to 8 m). The overall objective of the system is to maintain a small gradient toward the plume to prohibit contaminants from migrating off site. In this type of application, wellpoints may be the preferred method owing to their relatively lower installation cost (when compared to deep wells, for example), but more importantly they have the ability to be finely tuned to minimize the amount of water pumped to maintain the gradient (as compared to deep wells on larger spacing). Because of the costs associated with treatment of the discharge, it is advantageous to minimize the amount of water pumped while still maintaining the necessary gradients.

It is common to utilize cutoff methods in conjunction with dewatering tools to minimize the amount of groundwater drawn from clean areas that would have to be subsequently treated. This may be a cutoff wall installed just outside of a contaminated area where a wellpoint system is used to contain and capture a plume. Other projects might use combinations of deep wells, slurry trenches, and interceptor trenches.

GROUNDWATER CUTOFF AND EXCLUSION METHODS

Steel Sheet Piling

Interlocked steel sheet piling is frequently used in an effort to cut off or limit the flow of water. When steel sheet piling is installed in open water, such as for a river cofferdam, it is frequently necessary to seal the interlocks so that the flow through them can be reduced. When installed in soils, continuous interlocked steel sheet piling can sometimes eliminate dewatering problems. The sheeting typically needs to make a positive cutoff with an impervious layer of soil or rock capable of resisting the differential pressures and the interlock leakage should be small. In fine or well-graded soils, interlock leakage is generally quite low. In clean permeable soils, it could be quite large unless interlock tension pulls the interlock tight and seals water passages. Frequently, the use of steel sheet piling does not eliminate dewatering but simply minimizes the scope of a dewatering program. For example, it may be necessary to cause pressure relief below subgrade and below an impervious cutoff even though the walls of an excavation may be constructed of interlocked steel sheet piling. On occasion, it is desirable to reduce the loads on bracing or tieback systems by lowering the water level outside the cofferdam.

Slurry Trenches

The use of a slurry trench backfilled with an impervious material to control groundwater seepage is very common. A narrow trench is excavated to cutoff, typically using a backhoe (Fig. 9.14). During excavation, the trench walls are stabilized by bentonite slurry that provides a positive head and prevents groundwater from entering the excavated trench and causing sloughing. As the trench is excavated, a selected, low-permeability backfill is placed. Generally, the backfill is mixed with the slurry prior to placement, and the backfill placement techniques can be extremely critical in ensuring that voids will not exist in the wall. The procedure results in a trench backfilled with an impervious material that is somewhat flexible.

FIGURE 9.14   Slurry trench installation. (*Courtesy Moretrench.*)

Structural Slurry Walls

Structural slurry (diaphragm) walls provide both groundwater control and ground support, typically serving as either the cofferdam during the construction period or, as is often the case, part of the permanent construction. Structural slurry walls will provide a relatively rigid excavation support system.

A specially designed clamshell or a hydromill is used to excavate panels. For groundwater control purposes, the panels will typically extend into an underlying cutoff stratum, typically rock or impermeable soil. One of the most challenging parts of the work can be providing a positive seal at a soil/rock interface in a transitional geology with cobbles and boulders or a steeply sloping rock surface.

Diaphragm walls are constructed by alternating primary/secondary panel excavation and construction sequence. The stability of the ground during excavation is maintained by bentonite slurry. The joints between panels can be formed with beams or temporary end stops. Reinforcement, with rebar cages or beams, is installed for the full depth of the excavation and concrete is tremie-placed to displace the slurry and complete the panel.

Diaphragm walls offer increased strength and stiffness compared to other cutoff wall methods and can be installed in a wide range of ground conditions and to appreciable depths. The installation method generates less noise and vibration than driven steel sheet piling, which may be an advantage in builtup environments. However, a large work area is required for equipment and reinforcing cage fabrication. Panel joints may be an increased source of leakage potential compared to, say, a slurry trench. The cost of diaphragm wall construction is also high relative to other cutoff methods, although this may be offset if the wall is designed to be incorporated into the completed structure.

Ground Freezing

Ground freezing provides both water cutoff and excavation support. To create a frozen earth cofferdam, or frozen soil mass, closed-end freeze pipes, consisting of an external casing and an internal drop pipe, are inserted into drilled holes in a pattern consistent with the shape of the area to be stabilized and the required thickness of the wall or mass. A coolant, typically calcium chloride brine, is delivered via a supply header to the drop tube and flows up the annulus, extracting heat from the soil and causing it to freeze. An insulated return header delivers the circulated brine to the refrigeration plant where it is re-chilled and recirculated through the freeze system. A typical freeze layout is shown schematically in Fig. 9.15.

Among the most important conditions that can adversely affect the ground freezing operation are natural or pumping-induced groundwater velocities, which have been known to prevent the effective formation of a frozen wall; the possibility of subsequent movement of water

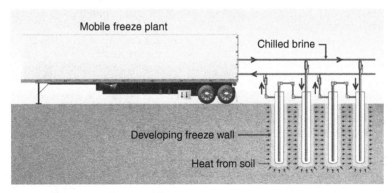

FIGURE 9.15   Freeze plant schematic.

around the toe of a freeze wall; and the possible presence of extraordinary heat sources such as pipelines adjacent to or penetrating a frozen wall. When soils to be frozen have substantial clay layers, design consideration must be given to "creep" of the frozen clay. In some applications, where freezing is performed above the water table, temporary artificial raising of soil moisture content must be considered in order to provide an effective result.

Temperature instrumentation by means of supplemental monitor holes is a very important part of ground freezing program. This information provides data to establish rates of closure and a mechanism by which the success of the program can be evaluated and, if necessary, supplemental efforts made prior to excavation. The formation of the frozen wall requires substantial energy and significant time. However, once formed, the wall is not subject to rapid deterioration unless a significant heat source is concentrated at one point or unless moving water erodes the frozen soil. The use of frozen walls in construction sometimes has a decided advantage in terms of their lack of sensitivity to interruption that sometimes causes difficulties in dewatering projects.

Ground freezing applications are common in connection with small, deep structures such as shafts or tunnels. Many frozen shafts have been constructed in New York City, some to depths in excess of 250 ft (76 m), for construction of the 60-mile (96-km) long City Water Tunnel No. 3. In recent years, several projects involving "mass freezing" have been accomplished, typically as part of the tunneling process. The largest and most innovative project accomplished in the United States to date involved the jacking of three massive box tunnels through frozen ground under South Street Station as part of the Boston "Big Dig."

Jet Grouting

Like ground freezing, jet grouting (Fig. 9.16) can provide both water cutoff and excavation support and an improved ground product of high compressive strength and low permeability. However, unlike ground freezing, which is a temporary application where the ground reverts to its original state after completion, jet grouting permanently changes the structure of the treated soil.

In the jet grouting process, the in situ soil is hydraulically eroded, mixed, and partially replaced with cementitious grout slurry, injected at high velocity, to create an engineered soil-cement product. Jet grouting can be performed above or below the water table and in virtually all soils from cohesionless soils to clays. It can also be targeted to a specific vertical soil zone that may be particularly important for groundwater control or cutoff. The three basic systems of jet grouting in general use in the United States today are single-fluid (slurry grout only), double-fluid (slurry grout sheathed in a compressed air collar), and triple-fluid (separate water/air and grout nozzles). Each offers a differing degree of in situ soil improvement/replacement.

Depending on the application, the jet grouting system can be designed to create a number of different overlapping or interlocking soil-cement geometries. *Columns*, which can be oriented vertically or battered, are typically used for structural underpinning/excavation support. For groundwater control, columns are used to infill around utilities or other obstructions that create "gaps" in otherwise continuous barrier walls, such as sheet pile walls, slurry walls, or secant pile walls. They are also typically used as the main groundwater barrier in areas where more conventional methods cannot be installed and are effective in providing permanent encapsulation for environmental purposes. *Wedge shapes* allow the concentration of grout where it is needed, as in sealing behind gaps in sheetpiling, and minimizes waste thus providing a more economical application. *Discs*, or relatively thin columns, constructed in an overlapping pattern, have been used to create horizontal groundwater barriers, often referred to as a "bottom seal" or "plug." *Panels* can form a very effective, economical groundwater barrier.

FIGURE 9.16 Jet grouting equipment: drill rig, high-capacity grout mixer, high-pressure grout pump, and cement silo. (*Courtesy Moretrench.*)

The overall effectiveness (permeability) of a jet grout system can only be obtained with a pump test of the composite jet grout structure. A full-scale pump test and/or recharge test should be performed of the jet grouted cofferdam or test cell utilizing piezometers installed both inside and outside of the jet-grouted cell.

Permeation Grouting

Permeation grouting is the flow of grout into the pores of granular soil without displacing or changing the soil structure, resulting in modification of the characteristics of the ground with the hardening or gelling of the grout. Permeation grouting may be used to increase soil strength and cohesion or to decrease permeability.

For groundwater control purposes, permeation grouting can be used as a site perimeter water cutoff measure, to exclude contamination on environmental projects, to lower the quantity of water pumped from excavations in urban areas, to close "windows" or gaps in "bathtub" excavations, and to seal off high-permeability backfill that may be encountered in or adjacent to an excavation.

Permeation grouts fall under two categories: chemical grouts (true chemical solutions or colloidal suspensions) and particulate grouts. The most commonly used chemical permeation grouts for groundwater control are sodium silicates due to their safety and environmental compatibility. Particulate grouts consist at least partly of cementitious materials. Particulate or suspension grouts typically contain ordinary Portland cement as the "active ingredient" but will more often than not contain other particulate materials such as bentonite, fly ash, slag, or other components to improve the characteristics of the grout. The particulate grouts may vary from the very viscous bentonite-cement grouts, which can only penetrate highly permeable ground such as clean gravel, to ultrafine cement that can penetrate sand.

The penetrability, or groutability, of the soils is the single most significant factor in the selection of a grout material and grouting technique, particularly for the purpose of groundwater control. The amount of permeability reduction achievable in a particular situation depends on several factors such as the permeability of the ground, viscosity of the grout, surface tension, grout pipe spacing, injection sequencing, quality control, stability of the cured grout, and so forth. Fines content is a significant determinant in the permeability and groutability of soils.

Permeation grouting in soil is typically performed with multiple injections of fixed, predetermined volumes of grout to create a composite grout mass of overlapping grouted injections. Tube a manchette (TAM), or "sleeve port," pipes provide the greatest control over grout placement. Driven pipes or "grout needles" can be utilized for shallow grouting applications.

Permeation grouting in rock can also be referred to as rock curtain grouting. It involves the filling of fractures and fissures, typically using Portland cement grout materials, to reduce permeability and in some cases strengthen or stabilize the rock. Many times, this method is used to control seepage beneath a cutoff wall extending into rock to ensure the requirement for a "bathtub" excavation is met.

BIBLIOGRAPHY

Powers, J. P., A. B. Corwin, P. C. Schmall, and W. E. Kaeck: *Construction Dewatering and Groundwater Control: New Methods and Applications*, 3d ed., John Wiley & Sons, New York, N.Y., 2007.

Pyne, R. D. G.: *Groundwater Recharge and Wells: A Guide to Aquifer Storage Recovery*, CRC Press, Boca Raton, FL, 1995.

Schneiders, J. H.: *Chemical Cleaning, Disinfection and Decontamination of Water Wells*, Johnson Filtration Systems Inc., St. Paul, MN, 2003.

Sterrett, R. J. (ed): *Groundwater and Wells*, 3d ed., Johnson Filtration Systems, Inc., St. Paul, MN, 2007.

CHAPTER 10

UNDERGROUND/TUNNELING SUPPORTS

Vincent Tirolo, Jr., P.E.

INTRODUCTION

There have been a number of significant recent developments in tunneling technology and in the understanding of interactive relationships between the excavated tunnel opening and its supports.

Many of the significant developments in tunneling over the last decade involved the extensive use of technology. This tunneling technology includes the common use of tunnel boring machines (TBMs) in both rock-mixed face and soft ground earth-pressure balance machines in lieu of compressed air in soft ground, sequential excavation method [SEM, previously known as new Austrian tunneling method (NATM)], remote-controlled microtunneling machines for larger-diameter tunnels, and direction drilling. The technology of shaft construction also includes the more extensive use of slurry walls, ground freezing, and jet grouting.

The impact of the common use of these new technologies on the supports of tunnel openings, particularly in rock, is only beginning to be understood. Engineers now understand what miners and sandhogs have understood for many years: that the loads depend on "what (selection of support), when, where, and how" the initial liner elements are installed.

GENERAL INFLUENCES ON TUNNEL SUPPORT LOADING

Contractors generally utilize an initial support system to control ground movement during the excavation phase of tunnel and shaft construction. Although this system is sometimes referred to as "temporary support," in almost all cases the initial support remains in place. Contractors and miners have often considered the initial supports to be the major load-carrying system within the tunnel, even after the final supports are installed. Hence initial supports have also been called "primary supports."

Increasingly, tunnel designers incorporate the initial liner into the design of the final support, usually a concrete lining. The final support, including a water proofing system, is usually installed after the completion of all excavation.

Examples of this approach are

1. Deep-shaft construction, where, in many parts of the world, the final concrete lining is traditionally advanced so as to remain within a few diameters from the bottom of the shaft excavation. This is done to minimize the amount of temporary support required.

2. Deep shafts constructed using slurry walls that act as both the initial and final shaft lining.

3. Tunnels utilizing precast concrete, fabricated steel, or extruded concrete liners that are installed within the tail of a shield or tunneling boring machine (TBM) and act as both the initial and the final tunnel lining.

4. SEM (previously NATM) in rock and soft ground where an initial support medium, usually shotcrete, used alone or in conjunction with rockbolts, is later incorporated into the final lining that may be either cast-in-place concrete or a sprayed concrete line.

Historically, the presence of an initial support system was disregarded by the engineer designing the final support. It was often said that initial supports, which included timber lagging and blocking or thin-gauge liner plates, would deteriorate with time and thus reduce their structural capacity. Also, initial supports were often struck and damaged by mining equipment during construction. Thus disregarding the presence of initial supports by the engineer provided an extra margin of final liner reliability. Final liner reliability was also enhanced by ignoring concrete between the initial supports and the rock surface. In drill-and-blast rock tunnels where 12 to 24 in (0.3 to 0.6 m) of overbreak is common, the thickness of cast-in-place concrete between the inside face of steel rib initial supports and the rock surface can be substantial.

In an attempt to curb the cost of tunnel construction, designers now consider the load-sharing capability of the initial and final linings. Often designers will specify a number of different initial support options that change with the ground conditions. In rockbolt-supported excavations, the thickness of the permanent concrete lining can be reduced by considering the additional support offered by the owner-specified "temporary" tensioned rockbolts and/or untensioned rock dowels that remain in place after concreting.

This text will assist the underground engineer in the design of initial supports when the owner does not specify them. By necessity, each topic is limited to a brief discussion, but each method or theory is referenced well enough for the serious reader to follow it up in much greater detail than space allows in this chapter. The discussion of field practice for the installation of each type of support is, again, brief.

Ground Characterization

The "ground" is the both the medium through which the tunnel must be mined and, with groundwater, also the source of primary source of the load on the tunnel liner. Our understanding of the ground has changed and evolved. Initially, the loads on underground mined structures were considered the same as the load on underground structures constructed by cut-and-cover methods. Designers used full overburden pressures whether the mined tunnel was in soft ground or rock. Today we consider the ground to be a partner in the support system that spans a tunnel bore or a major mined cavern. We also know that the load on a liner is directly related to ground movements and thus the stiffness of the liner.

In the following sections we discuss the evolution of both empirical and analytical design methods for calculating loads on mined underground openings. However, before we begin this discussion, it is worthwhile to review the general material characteristics of "the ground."

Influence of Scale—CHILE versus DIANA

Hudson and Harrison[1] define two useful acronyms (*CHILE* and *DIANA*) for "the ground" for material modeling purposes. These acronyms are useful to understand the complexities inherent in determining ground loadings. They also illustrate the influence of scale in ground modeling.

Ground that can be classified as *CHILE* is continuous, homogeneous, isotropic, and linearly elastic. Ground that can be classified as CHILE requires that the ground components, soil particles, or rock blocks are small compared to the dimensions of the underground structure. Generally soils, residual soil, and highly fractured rock can be considered CHILE materials. Massive rock without discontinuities is also considered to be CHILE material. In general, ground is classified as a CHILE material if the sizes of the soil or rock components are at least two or three orders of magnitude smaller than the smallest dimension of the underground structure. Of course, scale effects also have a profound impact on other engineering parameters such as permeability. Hudson and Harrison[1] illustrate this by referring to figures provided by Long.[2]

Ground that is classified as *DIANA* is discontinuous, inhomogeneous, anisotropic, and nonelastic. In general rock is considered a DIANA material. Priest[3] states "Rock masses usually contain such features as bedding planes, faults, fissures, fractures, joints and other mechanical defects which, although formed from a wide range of geological processes, possess the common characteristics of low shear strength, negligible tensile strength and high fluid conductivity...." It is these characteristics that make most rock masses DIANA materials. Their behavior is governed by these discontinuities and, except for highly fractured or residual rock, these discontinuities define rock blocks that are within one order of magnitude of the dimensions of typical underground structures.

Time, Deformation, and Loading

Tunnel ground deformation in reaction to excavation and the support loading that results from these deformations do not occur instantaneously. The in situ existing stress field around the tunnel is altered as the tunnel is excavated. Indeed, the stress field is altered as the tunnel heading approaches and passes a specific tunnel face location. The critical area for tunnel support analysis is at the tunnel face and the ground immediately behind the face. Sometimes geotechnical investigations and/or probing reveal that supports are required forward of the tunnel face. These support methods include spiling (forepoling), grouting, pre-drainage, and even, in extreme cases, ground freezing.[4] Examples of the relationship of radial pressure, radial convergence (deformation) of the tunnel opening, and support application are best illustrated by Wood[5] and Lunardi.[6]

Groundwater

Groundwater plays a major role in tunneling. In all tunnel projects, groundwater must be controlled both during construction and permanently. One method of controlling groundwater during construction is dewatering. Historically, dewatering has been associated with soft-ground rather than rock tunneling.

Dewatering is not always possible, particularly in urban areas where groundwater drawdown can result in increased overburden stresses and cause settlement of structures along the tunnel alignment. On the other hand, dewatering can reduce settlements due to loss of ground at the tunnel face. This is particularly true when tunneling through non-cohesive sand and nonplastic silts. Tunnels act as a drain, and the groundwater seepage gradient toward the tunnel face can cause these soils to run and cause the tunnel face to collapse.

Dewatering for tunneling is generally a 24-h, 7-day-a-week operation and can be very expensive. Also, the volume of water pumped can be very large and have environmental impacts. Temporary supports installed in tunnels mined through dewatered ground are normally not designed for hydrostatic pressures. These temporary linings, whether ribs and lagging or precast segments, are not watertight. Permanent tunnel lining, the subject of this chapter, would be designed for full hydrostatic pressure.

Alternatively, groundwater in soft-ground tunnels can be controlled by selecting a mining system, tunnel-boring machine (TBM), which controls groundwater at the tunnel face while installing a watertight tunneling lining within the tail of the TBM. These TBMs are discussed in the next section.

Groundwater is also an issue in rock tunnels. However, because groundwater movement is usually associated with flows through discontinuities, volumes of water infiltration into the tunnel are significantly lower than those in soft-ground tunnels. Often, groundwater can be controlled by pre-excavation fissure grouting at the shaft and perimeter wall contact and rock mass grouting and fissure grouting through probe holes drilled ahead of the tunnel face during tunneling.

Groundwater lubricates the discontinuities in the rock mass, particularly if these discontinuities are filled with soft materials such as clay gouge. Groundwater pressures also reduce the shear strength between rock blocks. Groundwater acting as either a lubricant or a medium that reduces the effective stress between rock blocks can have a major impact on rock mass stability.

Methods of Construction

Construction Sequence. Successful in-tunnel support systems are a function of many factors. Wood[4] lists a number of factors to account for when developing a support system:

- Characteristics of the ground, including variability
- Degree of tolerance to water
- Tunnel geometry
- Time-dependent ground behavior
- Impacts on adjacent structures, for example settlement of building
- Observation and monitoring
- Scheme of construction

The last factor often has the greatest impact on the cost of completing a tunnel and therefore is of primary concern to tunnel contractors. A tunnel is a linear structural element. All activities revolve about advancing the tunnel face systematically and as rapidly as possible. It is the "scheme of construction," or construction sequence developed by the tunneling contractor during the bid process, that is critical to the project success.

The construction sequence selected is the tunnel contractor's methodology in executing the construction. This methodology is in part based on whether the tunnel can be excavated full face, by a top heading and bench, or finally by multiple top headings with one or more benches. For example, as stated by the American Association of State Highway and Transportation Officials (AASHTO),[7] "The Sequential Excavation Method (SEM), also referred to as the New Austrian Tunneling Method (NATM), is a concept that is based on the understanding of the behavior of ground as it reacts to the creation of an underground opening." Sometimes the tunnel geometry dictates these choices. Other times a contractor's choices are dictated by the ground characteristic curve or some of the factors described by Wood[4] in the previous list.

Drill and Blast. Drill-and-blast tunneling is construction method in which the tunnel is advanced through rock using a repetitive sequence of drilling and packing drill holes with explosives, blasting, ventilation, mucking, and support. Rock removed from a tunnel is called "muck." This term also includes materials that are by-products of the tunneling operations such as waste concrete, timbers, steel, debris, etc.

Drill-and-blast tunneling is a versatile construction method and can be used in a wide range of rock geology. Drill and blast can also be used in conjunction with soft-ground tunneling methods in mixed-face tunnels. Mixed-face tunnels are tunnels where the tunnel face is partially in soil and partially in rock. Drill and blast is also an economic option in short tunnels, complex tunnel geometries, and in tunnel caverns. Drill-and-blast methods can be used with SEM (or NATM) tunneling. Drill-and-blast methods are also used in conjunction with TBM tunneling to provide a tunnel portal/starter tunnel for the TBM cutterhead and grippers.

The primary disadvantage of drill-and blast-tunneling from the viewpoint of tunnel support is the damage done to the rock mass surrounding the tunnel opening resulting from blasting vibrations. Disturbance of the rock can extend a number of feet beyond the tunnel perimeter. Blasting can also loosen potentially unstable rock blocks. Figure 10.1 shows disturbed rock surrounding a tunnel portal excavated using drill-and blast-methods. A number of blasting techniques can be used to mitigate this disturbance, but it cannot be eliminated. Typical supports used in drill-and-blast tunnels include rockbolts, straps, wire mesh, shotcrete, lattice girders, and steel ribs.

FIGURE 10.1   Drill and Blast Tunnel Portal.

Rock TBM.   The first rock TBMs were developed in the late 1950s, but they did not have a major impact on tunneling until the 1970s. It was in the 1970s when progress began on soft-ground TBMs.

The first successful hard rock TBM was built in 1956 by Robbins for a 3-mile-long, 11-ft-diameter sewer tunnel in Toronto, Canada. The TBM mined through sandstone, shale, and limestone with UCS that ranges from 750 to 27,000 psi. The machine was originally designed with drag bits and disc cutters. The drag bits were removed as an experiment that turned out so successfully that most rock TBMs now have only disc cutters on a single rotational head.

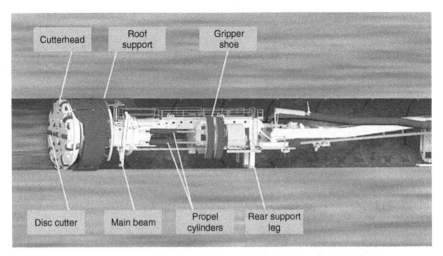

FIGURE 10.2 Illustration of an Open Gripper Main Beam TBM. (*Robbins Company.*)

TBMs are full-face machines that excavate by rotating a cutterhead. The cutterhead is the front end of the TBM. It has a rotational speed up to 8 to 12 r/min (revolutions per minute). TBMs are currently powered at 3500 to 4000 HP.

Various types of rock TBMs are available depending on site requirements: Open Gripper Main Beam TBMs, Single Shield TBMs, and Double Shield TBMs. Open Gripper Main Beam TBMs are best suited for slightly fractured rock, Single Shield TBMs in fractured rock, and Double Shield TBMs over the entire range of rock mass conditions. Both the Single- and Double- Shield TBMs facilitate the erection of precast segmental tunnel liners within the tail of the TBM. Single- and Double- Shield TBMs also can propel the TBM off the erected tunnel liner if the rock does not allow the use of side grippers. Figure 10.2 shows an illustration of a Robbins Open Gripper Main Beam TBM.

Disk-shaped cutters are mounted on a cutterhead. Thrust on cutters is as high as 70,000 lb each. Cutter diameters are up to 19 in. Typically cutterheads for large-diameter tunnels have 40 to 50 disc cutters. The useful life of a disc cutter depends on its location on the cutterhead, abrasiveness and hardness of the rock, and other factors. A disc cutter's useful life, before refurbishment, is measured as 125 to 250 yd^3 per cutter. For a 20-ft-diameter tunnel, that life would be 11 to 22 lin ft of tunnel. Figure 10.3 shows the cutterhead of a partially assembled Main Beam TBM. Figure 10.4 shows a typical disc cutter.

The following tests are usually used to estimate Rock TBM Parameters.

- Brittleness: To determine Brittleness Value S_{20}, an impact device is used that compares percent of materials passing through sieves before and after 20 impacts, for example Quartzite $S_{20} = 42$ while for Limestone $S_{20} = 52$.
- Surface Hardness: Sievers' J-value (SJ) gives a measure of rock's surface hardness using a miniature drilling device. SJ is the mean value of penetration of drilled hole, measured in tenths of a mm, after 200 revolutions, for example Quartzite SJ = 0.6, Limestone SJ = 70.4.
- Wear Capacity: the Abrasion Value (AV) and the Abrasion Value Cutter Steel (AVS) measure the rock's wear capacity on tungsten carbide and cutter ring steel, respectively.

FIGURE 10.3 Partially Assembled Main Beam TBM. (*Photograph provided by Christopher Kohr.*)

FIGURE 10.4 Typical Disk Cutter. (*Photograph provided by Christopher Kohr.*)

AV and *AVS* are defined as the weight loss of the tungsten carbide and the cutter ring steel pieces in milligrams after 5 min and 1 min, respectively, for example Quartzite AV = 68.2, AVS = 42.2; Limestone AV = 1, AVS = 1.

S_{20}, SJ, AV, and AVS are used in various empirical formulas to determine the following indices used to measure TBM bid parameters:

* Drilling Rate Index (DRI), for example Quartzite DRI = 25 (very low) and in Limestone DRI = 62 (very high)
* Bit Wear Index (BWI), for example Quartzite BWI = 91 (extremely high) and Limestone BMI = 15 (very low)
* Cutter Life Index (CLI), for example Quartzite CLI = 3 (extremely low) and Limestone CLI = 72 (very high)

Figure 10.5 shows side grippers on a typical Open Gripper Main Beam TBM.

Support requirements in TBM-mined tunnels include rockbolts, straps, wire mesh, and steel rings. In highly fractured rock, precast segmental concrete liners are sometimes required. The major advantage of rock TBMs is that the rotational cutting minimizes disturbance of the rock mass surrounding the tunnel opening.

FIGURE 10.5   Open Gripper Main Beam TBM side grippers and thrust cylinders.

Soft-Ground Tunneling.   Soft-ground tunneling is a construction method in which the tunnel is advanced through soils. Soft-ground tunneling has existed since ancient civilization. However, we have only been able to mine soft-ground tunnels below groundwater since the mid-1800s. At that time, Marc Brunel and Thomas Cochrane invented the "tunnel shield" as a means of advancing a tunnel under the Thames River in London. Brunel's original shield was square with a compartmental opening for 36 miners. The tunnel was mined through impervious soils and rock under the Thames River. Shield tunneling became the standard mining method in soft ground in the early 1900s with the introduction of compressed-air shield tunneling. By balancing the hydrostatic forces acting on the tunnel face when mined below groundwater level, compressed air allows tunneling through soils that otherwise would have been entirely unstable and the work too hazardous to attempt.

The use of compressed air has its limitations. Miners working in compressed air are subject to caisson disease and therefore must go through long periods of decompression in an air lock when exiting a compressed-air tunnel. They also must spend time in an air lock before entering the tunnel. Work time is also limited and decreased with increased pressures. During some subaqueous crossing of the Hudson River between New York and New Jersey, pressures exceeded 40 psi above atmospheric pressure and miners worked only 1 hr per 8-hr shift. Since compressed-air tunneling is a 24-hr, 7-day-a-week operation, the costs associated with this work are very high. However, compressed air is still used on a limited basis. It is often used as a means of egress when attempting to repair or remove an obstruction in front of the cutterhead of soft-ground TBMS.

Tunnel supports in soft-ground shield tunnels are entirely dependent on groundwater conditions. If dewatering in soft-ground tunnels is feasible and permitted, the initial lining can be ribs and lagging or sometimes liner plates. Liners are installed within the tail of the shield and the shield jacked forward by reacting against the previously installed liner. When mining is done below groundwater in compressed air, a watertight segmental liner must be installed. In the early 1900s these liners were almost exclusively made of cast iron. Today the material of choice for segmental liners is precast concrete.

Soft-Ground TBMs.   The two types of soft-ground TBMs have almost entirely replaced traditional shield tunneling with compressed air. These TBMs are the Earth Pressure Balance (EPB) and Slurry Face Shield (SFM). Both types of machine are similar in that

- They have a revolving cutter wheel.
- They have both an internal bulkhead that traps cut soil against the face (hence, they are called closed face) and that maintains the combined effective soil and water pressure and thereby stabilizes the face.
- No workers are at the face; mechanization and computerization control all functions except segment erection in the machines' tail.
- Precast concrete segments erected in the shield tail, with the machine advanced by shoving off those segments.
- Both machines have advantages and disadvantages but generally work best when mining through certain ground conditions. EPBs are better in clay, clayey and silty soils, and sand. SFMs are better in sand and finer gravels. Slurry pipes have a tendency to clog in clay soils. Very coarse gravels may cause face collapse. SFMs are better in high-water-pressure environments.
- It is problematic to use an SFM in soils with fines content exceeding 20 percent. However, treatment of the muck with soil conditioning foams and polymers may mitigate clogging. Soil conditioners may also aid in muck removal in EPMs.
- Because EPBs and SFMs are always used below groundwater, the liners associated with these are always watertight. In general these liners are precast segmental concrete.

Sequential Excavation Method (SEM). The first rock tunnel excavation using SEM (then known as NATM) occurred in the early 1960s. NATM began to be used for soft-ground tunneling in the late 1960s or early 1970s. In the United States the first NATM tunnel was the Mount Lebanon Tunnel in Pittsburgh, Pennsylvania in the 1980s. NATM tunnels on the Red Line and Wheaton Station were also completed in Washington DC as part of the WMATA system.

The basis of NATM philosophy was

1. Preserve the ground's inherent strength.
2. Control ground deformation.
3. Support using rockbolts and shotcrete.
4. Timing of installation is crucial and varies with ground.
5. Initial support partially or completely represents final support.
6. Tunnel unsupported length as short as possible.
7. Installation of a structural invert closure, when necessary, will stiffen ground arch by creating a closed tube.
8. Excavation sequence is critical to overall stability. Excavate full face if possible but use drifts in crown, bench, and invert sequence as ground conditions require.
9. Maintain rounded shapes of all openings. See Fig. 10.6.
10. Stakeholders must adopt a cooperative attitude toward decision making.

FIGURE 10.6 Sequential Excavation Method (SEM)—maintaining oval excavations.

NATM was associated with "shotcrete"; over time shotcrete has been replaced by the term Sprayed Concrete Lining or SCL. Gradually the term *NATM* was abandoned and replaced by *SEM*.

SEM can be used in both soft ground and rock. SEM has maximum geometric flexibility. It is flexible in its ability to handle various unanticipated ground conditions. However, SEM

 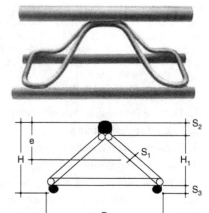

FIGURE 10.7   Lattice Girders.

is not a forgiving system. Careful adherence to NATM's basic principles is necessary. These principles were given by Muller and Fecker,[8] and listed and explained by Chapman et al.[9]

The ground support tools normally associated with SEM are shotcrete, steel mesh, rockbolts, and lattice girders (Fig. 10.7). However, the SEM ground support toolbox also includes modifying the excavation geometry through the use of top heading and invert, sidewall drift, and dual sidewall options. These tools reduce the amount of tunnel face excavated without initial support at any one time. Pre-support measures such as spiling and pipe roof supports can be installed to add to stand-up time prior to excavation. Stabilization of the face enhanced using face stabilization measures such as face bolts, additional shotcrete and wire, drainage, and pressure relief.

CALCULATING ROCK OR SOIL LOADS ON UNDERGROUND OPENINGS

The first and most difficult step in the design of an underground support system is the determination of the magnitude of the load acting upon the support system. There is no universally accepted method to determine the loads on a tunnel liner, nor is there an accepted method to design the tunnel liner once these loads are determined. Design methods are based on either an analytical or an empirical model. In an attempt to account for uncertainty in methodology, commonly combinations of empirical and analytic methods are employed.

The ground through which a tunnel is mined is three-dimensional, nonhomogeneous, anisotropic, nonlinear, and sometimes discontinuous. Often the tunnel is mined below groundwater. Peck[10] stated that tunnel liner loading not only is a function of the characteristics of the ground but is also dependent on the relative stiffness of the liner when compared to the ground. The loading on the tunnel liner is in fact four-dimensional, with its dependence on the time of installation and time-dependent ground characteristics.

Currently the most widely accepted tunnel design methods contain elements of both analytic and empirical models. These methods have evolved over the last 100 years or so. Paradoxically, as ability to solve complex analytic problems by sophisticated computer modeling has improved, the confidence of tunnel designers in these models varies. Kuesel[11] has recommended that these sophisticated models be used primarily for ground and liner parameter studies rather than in the final design of tunnel liners. Empirical methods are becoming increasingly popular. Schwartz and Einstein[12] have recommended combining empirical and analytic methods that are tested by field observational data.

Time Available to Install Supports

An important factor that must be considered in the design of temporary supports is the time available to install them after the ground has been excavated. This time is referred to as the *stand-up time*, and it indicates the time during which the ground remains stable without support. (In his earlier papers on tunnel load analysis, Terzaghi referred to this time as the bridge-action period.) A short stand-up time may dictate the type of support regardless of any structural design analysis to the contrary. For example, if a designer, through rock-loading and support-design analysis, determines that steel sets on 4-ft (1.2-m) centers are adequate to support a given rock load, the design must also determine whether the ground has adequate stand-up time to excavate 4 or 5 ft (1.2 or 1.5 m) ahead of the lead steel set and install the new support before rock fallout starts to occur.

A qualitative definition of stand-up time was put forth by Lauffer[13] in 1958, who defined the active span to be supported as the smaller of either the tunnel diameter or the distance between the new face and the last-installed support. Seven types of rock masses were charted (see Fig. 10.8), with type A the best and type G the worst.

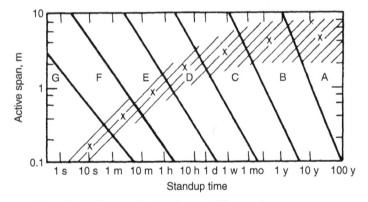

FIGURE 10.8 Lauffer's stand-up time for seven different rock types.

Stand-up time is important to the tunnel or shaft contractor because the longer the stand-up time, the easier it is for the work crew to install the supports in an organized manner, since they can be farther away from the working face and its inherent congestion. In drill-and-blast heading with steel-rib supports, the farther the ribs are from the working face, the easier it is to position the machines used to drill the perimeter blast holes, and the result will usually be a tunnel with less overexcavation, or "overbreak." Contractors will thus be rewarded by removing less, and if a final lining is required, they can also reduce their overall concrete costs.

In summary, contractors take good advantage of the stand-up time available to increase their rates of advance and decrease their overbreak by more effectively utilizing crews and equipment.

The support-design methods often do not consider the stand-up time allowed to install the supports. The two Commercial Shearing Inc., publications,[14,15] as well as Szechy's text,[16] outline various techniques, such as compressed-air, shield-driven tunneling, well-point pumping, and the installation of spiling that are used to reduce ground movement in poor ground until the supports can be installed. It will be up to the judgment of the support designer to determine if the ground conditions will permit the support to be installed safely before the occurrence of excessive ground movement.

Empirical Methods in Rock

Steiner and Einstein[17] have identified four different types of empirical methods in tunneling. Over the last 10 years, the use of empirical methods has tended increasingly to move from indirect or qualitative rock-load methods (type A) toward multiple parameter quantitative direct methods (type D2).

Type A empirical methods, or qualitative indirect or rock-load methods, calculate rock loads using a qualitative assessment of rock conditions (rock class). Rock class is based on both general rock-mass attributes, that is, geologic classification, age, origin, and geomorphology, and on-site specific rock structures such as joint spacing, bedding, and formation orientation, all considered with respect to the tunnel alignment. Once the rock load is determined, standard techniques of structural analysis are used to design the initial liner.

Type B empirical methods, or qualitative direct methods, qualitatively characterize the rock mass in more detail than type A methods. A particular construction procedure and initial liner are associated with each rock-mass characterization.

Type C empirical methods first classify the rock mass quantitatively, and then a rock load is determined based on the tunnel dimensions, unit weight of rock, and various constants determined for each rock classification. Water level and/or construction method may also be considered in the determination of rock load. Once the rock load is determined, the tunnel liner is designed using conventional structural analysis.

Type D1 empirical methods quantitatively classify the rock mass using a single parameter, and *type D2 empirical methods* quantitatively classify the rock mass using multiple parameters. Both type D1 and D2 then directly recommend support elements, for example, size and spacing of the initial liner.

Various geotechnical engineers have tabulated guidelines for estimating rock loads on tunnels. A widely used rock-load estimation method in North America is Terzaghi's[18] table of loads for nine different rock conditions. Other rock and soils engineers, such as Stini[19] and Bierbaumer,[20] have also published guidelines for rock-load estimating. These tables and guidelines are type A empirical methods. Deere et al.[21] published a combined table of five separate rock-load classifications, comparing each to Deere's rock-quality designation (RQD) and fracture-spacing index. The method used to log rock cores to determine RQD is shown in Fig. 10.9. Deere and Deere[22] described a low RQD value as a "red flag" indicator of poor ground that would require additional investigations. The combined table includes Terzaghi's original load table but adds a tenth condition (for sand and gravel) and modifies the loads for circular tunnels. The combined tabulation as shown in Table 10.1 is an example of a type C empirical method Deere[23] described data, originally developed in 1970, correlating data from Table 10.1 with the support of 20- to 40-ft-diameter tunnels. These correlations are shown in Table 10.2. In Fig. 10.10, Merritt[24] showed correlations between RQD and initial support elements for various size tunnels. Table 10.2 and Fig. 10.10 are examples of type D1 empirical methods.

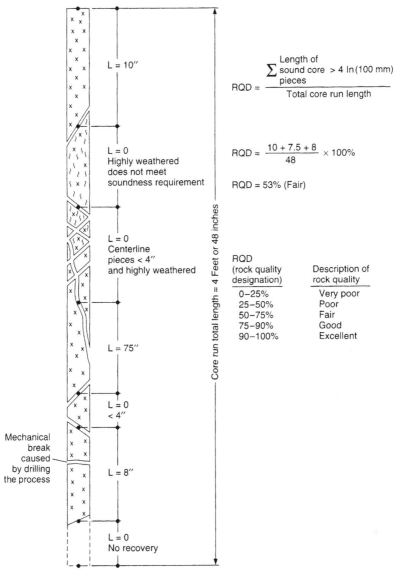

FIGURE 10.9 RQD logging. [*From Deere, "Rock Quality Designation (RQD) After Twenty Years,"* USAEWES, 1989.]

It is important to note that the empirical methods described above were based on data from drill-and-blast tunnels. Support requirements will generally be less for tunnels excavated with either a full-face rotating cutter head or roadheader TBM.

The questions that initially confront the user of these tables are how to identify the correct rock conditions and whether to use the high or low end of the rock-load range for the rock condition once it is identified. The greatest error in support design would most

TABLE 10.1 Deere's Combined Table of Rock Loads and Classifications

Fracture spacing, cm	RQD		Rock load H_p, m Initial	Rock load H_p, m Final	Remarks		H_p, m	Remarks
100	98	Coherent — 1. Hard and intact	0	0	Generally no side pressure — Lining only if spalling or popping	1—Stable	0 to 0.5	
50		Hard stratified or schistose — 2.	0	0.25B	Spalling common	2—Nearly stable	0.5 to 1	Few rock falls from loosening with time
	95	3. Massive moderately jointed	0	0.5B	Side pressure if strata is inclined, some spalling	3—Lightly broken	1 to 2	Loosening with time
	90							
20	75	4. Moderately blocky and seamy	0	0.25B to 0.35C	Ertic load changes from point to point	4—Medium broken	2 to 4	Immediately stable: breakup after few months
10	50	5. Very blocky, seamy, and shattered	0 to 0.6C	0.25C to 1.1C	Little or no side pressure	5—Broken	4 to 10	Immediately fairly stable; later rapid breakup
5	25	6. Completely crushed		1.1C	Considerable side pressure. If seepage, continuous support required	6—Very broken	10 to 15	Loosens during excavations; local roof falls
	10							
	2	Weak and Coherent — 7. Gravel and sand	0.54C to 1.2C	0.62C to 1.38C	Dense — Side pressure			
			0.96C to 1.2C	1.08C to 1.38C	$\sigma h = 0.3\gamma\,(0.5H_t + H_p)$ — Loose			
2		8. Squeezing moderate depth		1.1C to 2.1C	Heavy side pressure	7—Lightly squeezing	15 to 25	High pressures
		9. Squeezing great depth		2.1C to 4.5C	Continuous support required	8—Moderately squeezing	25 to 40	
		10. Swelling		up to 80 m	Use circular support. In extreme cases use yielding support	9—Heavy squeezing	40 to 60	Very high pressures

Terzaghi [18]

1. For rock classes 4, 5, 6, 7 when above groundwater levels, reduce loads by 50%
2. For sands (7), $H_{p\,min}$ is for small movements (−0.01C to 0.02C); $H_{p\,max}$ for large movements (0.15C)
3. B is tunnel width, $C = B + H_t$ = width + height of tunnel. For a circular tunnel, $H_t = 0$*

Stini [19]

Note: Loads are for 5-m-wide tunnel
For L meter wide tunnel
$H_p = H_{p5m}\,(0.5 \text{ to } 0.1L)$

*The reader should be cautioned that by Setting $H_p = 0$ for a circular tunnel, can arbitrarily cut the design rock load in half. This may be true for a homogeneous soil that may distribute its load more uniformly on a circular opening (as compared to a horseshoe shaped opening). If however, very blocky ground is encountered, the circular opening may offer no better load distribution advantages over those of a noncircular opening.

TABLE 10.1 Deere's Combined Table of Rock Loads and Classifications (*Continued*)

	Rock load H_p, m Initial/Final	Side pressure, m Initial/Final	Invert pressure, m		
		Little loosening		A—Stable	Sound
				B—Unstable after long time	
Slightly broken	0/3 to 4	0/0	0	C—Unstable after short time	Sound stratified or schistose (some fissures?)
				D—Broken	Strongly fissured
Very broken	3/11 to 13	Loosening with time 0/1	1–2	E—Very broken	Fully mechanically disturbed
Extremely broken	5 to 10/11 to 15	Roof falls, loosening at time of excavation 2 to 4/2 to 6	4		Gravel and sand
				F—Squeezing	Pseudo-sound rock (properties change with time)
Soft, squeezing, or flowing; moderate depth	10 to 13/15 to 25	4/4	6		Some squeezing (genuine rock pressures), small overburden
Soft, heavy, squeezing; great depth	15 to 25/47 to 75	8/6	12	G—Heavy squeezing	Heavy squeezing; large overburden Swelling Silt. clay
Bierbaumer[20] and others				Lauffer[13] Note: This classification is correlated with stand-up time	Rabcewicz[51] Note: This classification has been used for evaluating feasibility of rockbolt types

likely result from improper assumption of rock load, rather than from an approximation of a particular stress-distribution method. It is therefore important that the designer of tunnel supports acquire all available geotechnical information about the project.

Many civil works projects are designed after an extensive boring, rock and soil testing program have been accomplished by a consultant engineering firm. These geotechnical reports, called the Geotechnical Data reports (GDR) and the geotechnical interpretative report (GIR), describing the basis of initial and final lining design, are available to all contractors bidding the work. Many of these reports contain a narrative section that may discuss similar projects performed in the area, highlighting the successes (or failures) of the construction procedures and of the support design. References to other available reports are also usually listed. In many cases, empirical field data from nearby tunnel projects, or projects in similar type rock or soil, will help the tunnel constructor in determining the support requirements. Some bid documents also include a geotechnical baseline report (GBR). The GBR establishes specific ground design parameters that contractors can consider a baseline during bid preparation.

RSR Method.   Wickham and Tiedemann[25] developed an empirical method for determining tunnel rock loads and support requirements, utilizing support data from 53 tunnel projects (primarily located in the western United States) and correlating the load and support requirements of each project with the characteristics of the rock being supported. Using three parameters, the characteristics of the rock being supported are graded to provide a rock structure rating (RSR). RSR methods are limited to steel rib sets. The RSR

TABLE 10.2   Guidelines for Selection of Primary Support for 20- to 40-ft Tunnels in Rock[22]

| Rock quality | Construction method | Steel sets | | |
		Rock load (B = tunnel width)	Weight of sets	Spacing[‡]
Excellent RQD > 90	Boring machine	(0.0–0.2) B	Light	None to occasional
	Drilling and blasting	(0.0–0.3) B	Light	None to occasional
Good RQD = 75–90	Boring machine	(0.0–0.4) B	Light	Occasional to 5–6 ft
	Drilling and blasting	(0.3–0.6) B	Light	5–6 ft
Fair RQD = 50–75	Boring machine	(0.4–1.0) B	Light of medium	5–6 ft
	Drilling and blasting	(0.6–1.3) B	Light of medium	4–5 ft
Poor RQD = 25–50	Boring machine	(1.0–1.6) B	Medium circular	3–4 ft
	Drilling and blasting	(1.3–2.0) B	Medium to heavy circular	2–4 ft
Very poor RQD < 25	Boring machine	(1.6–2.2) B	Medium to heavy circular	2 ft
(excluding squeezing and swelling ground)	Drilling and blasting	(2.0–2.8) B	Heavy circular	2 ft
Very poor, squeezing or swelling ground	Both methods	Up to 250 ft	Very heavy	2 ft

TABLE 10.2 Guidelines for Selection of Primary Support for 20- to 40-ft Tunnels in Rock[22] (*Continued*)

Rockbolts* (conditional use in poor and very poor rock)		Shotcrete† (conditional use in poor and very poor rock)		
Spacing of pattern bolts	Additional requirements and anchorage limitations*	Total thickness		Additional support†
		Crown	Sides	
None to occasional	Rare	None to occasional local application	None	None
None to occasional	Rare	None to occasional local application 2–3 in	None	None
Occasional to 5–6 ft	Occasional mesh and straps	Local application 2–3 in	None	None
5–6 ft	Occasional mesh and straps	Local application 2–3 in	None	None
4–6 ft	Mesh and straps as required	2–4 in	None	Provide for rockbolts
3–5 ft	Mesh and straps as required	4 in or more	4 in or more	Provide for rockbolts
3–5 ft	Anchorage may be hard to obtain. Considerable mesh and straps required	4–6 in	4–6 in	Rockbolts as required (approx. 4–6 ft cc)
2–4 ft	Anchorage may be hard to obtain. Considerable mesh and straps required	6 in or more	6 in or more	Rockbolts as required (approx. 4–6 ft cc)
2–4 ft	Anchorage may be impossible. 100% mesh and straps as required	6 in or more on whole section		Medium sets as required
3 ft	Anchorage may be impossible. 100% mesh and straps required	6 in or more on whole section		Medium to heavy sets as required
2–3 ft	Anchorage may be impossible. 100% mesh and straps required	6 in or more on whole section§		Heavy sets as required

NOTE: Table reflects 1969 technology in the United States. Groundwater conditions and the details of jointing and weathering should be considered in conjunction with these guidelines, particularly in poor-quality rock. See Deere et al.[21] for discussion of use and limitations of the guidelines for specific situations.

*Bolt diameter = 1 in length − 1/3 to ¼ tunnel width. It may be difficult or impossible to obtain anchorage with mechanically anchored rockbolts in poor and very poor rock. Grouted anchors may also be unsatisfactory in very wet tunnels.

†Because shotcrete experience is limited, only general guidelines are given for support in the poorer-quality rock.

‡Lagging requirements for steel sets will usually be minimal in excellent rock and will range from up to 25 percent in good rock to 100 percent in very poor rock.

§In good and excellent quality rock, the support requirements will in general be minimal but will be dependent.

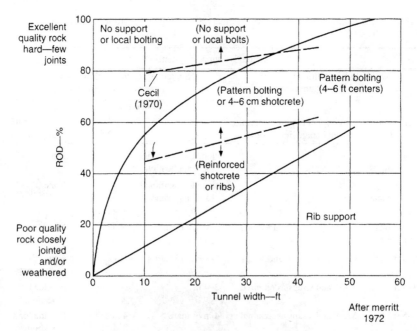

NOTE: Support data from igneous and metamorphic rocks where rock
pressures or swelling/squeezing ground did not exist

FIGURE 10.10   Rock quality and support requirements for tunnels of varying dimensions. [*From Deere and Deere, "Rock Quality Designation (RQD) After Twenty Years." USAEWES, 1989.*]

case histories include steel sets designed using the upper-limit loading determined by the Terzaghi method. The RSR method is a multiple-parameter type D2 empirical system. A type D2 system quantitatively classifies the rock mass and then directly recommends a tunnel liner.

The method first determines the value of parameters A and B and then, using the sum of A and B, determines the value of parameter C.

Parameter A considers the general area geology; its values are tabulated in Table 10.3. Parameter B considers the joint direction and spacing in relation to the direction the tunnel is being driven; its values are tabulated in Table 10.4. Parameter C considers the effect of groundwater inflow and joint condition, along with the sum of parameters A and B; its values are tabulated in Table 10.5.

The RSR is the sum of these three parameters: A, B, and C. RSR has a maximum value of 100. After calculating the RSR, the user can determine the rock load on the tunnel arch by referring to Table 10.6, or can utilize the rib ratio-RSR relationship to arrive directly at a tunnel rib size and spacing. The rib ratio is a ratio of the desired rib size and spacing, to a datum rib size and spacing. The datum rib is designed for a tunnel of the same diameter as the desired rib but subjected to a Terzaghi's condition 7 load of 1.38 $(B+H)$ below the water table. Theoretical spacing for this datum condition can be calculated by using Proctor and White's[14] table of load capacities for various rib sizes and tunnel diameters, assuming a maximum bending stress of 27,000 lb/in² (18,620 N/cm²) in the rib flange. This datum spacing is shown for some typical sizes of ribs and tunnel diameters in Table 10.7.

TABLE 10.3 Rock Structure Rating, Parameter A, General Area Geology
Maximum value = 30

Basic rock type*	Massive	Geological structure		
		Slightly faulted or folded	Moderately faulted or folded	Intensely faulted or folded
1	30	22	15	9
2	27	20	13	8
3	24	18	12	7
4	19	15	10	6

*The basic rock types are

	Hard	Medium	Soft	Decomposed
Igneous	1	2	3	4
Metamorphic	1	2	3	4
Sedimentary	2	3	4	4

The primary equation in the RSR method is the determination of rib ratio. Terzaghi's empirical formula for the maximum roof load for a tunnel mined through a loose, cohesionless sand below groundwater is

$$P_1 = [1.38(B + H_t)](B)(w_t)$$

where P_1 = vertical load on steel sets, lb/ft of tunnel
B = tunnel width, ft
H_t = tunnel height, ft
w_t = unit weight of sand (assumed as 120 lb/ft^3)

For tunnels of circular or horseshoe cross section $H_t = B = $ diameter D. Therefore, the equation for P_1 reduces to

$$P_1 = 165.6 D(D + D)$$

or

$$P_1 = 331 D^2$$

Using the load table from Proctor and White,[5]

$$P_t = (P_r)(D)$$

where P_t = theoretical maximum (allowable) load on steel sets
P_r = Proctor and White allowable load per foot of tunnel width for rib size and tunnel diameter selected

To find the theoretical rib spacing S_d for the datum condition:

$$S_d = \frac{P_t}{P_1}$$

$$S_d = \frac{(P_r)(D)}{331 D^2}$$

$$S_d = \frac{P_r}{331 D}$$

TABLE 10.4 Rock Structure Rating, Parameter B Joint Pattern—Direction of Drive
Maximum value = 45

	Strike perpendicular to axis					Strike parallel to axis		
	Both	Direction of drive				Direction or drive		
		With dip		Against dip		Both		
		Dip of prominent joints				Dip of prominent joints		
	Flat	Dipping	Vertical	Dipping	Vertical	Flat	Dipping	Vertical
1. Very closely jointed < 2 in (50 mm)	9	11	13	10	12	9	9	7
2. Closely jointed 2–6 in (50–150 mm)	13	16	19	15	17	14	14	11
3. Moderately jointed 6–12 in (150–300 mm)	23	24	28	19	22	23	23	19
4. Moderate to blocky 1–2 ft (0.3–0.6 m)	30	32	36	25	28	30	28	24
5. Blocky to massive 2–4 ft (0.6–1.2 m)	36	38	40	33	35	36	34	28
6. Massive > 4 ft (1.2 m)	40	43	45	37	40	40	38	34

NOTES: Flat, 0–20°: dipping, 20–50°: vertical, 50–90°.

TABLE 10.5 Rock Structure Rating, Parameter C, Groundwater Joint Condition
Maximum value = 25

Anticipated water inflow, gal/min/1000 ft	Sum of parameters $A + B$					
	13–44			45–75		
	Joint condition					
	Good	Fair	Poor	Good	Fair	Poor
None	22	18	12	25	22	18
Slight (< 200 gal/min)	19	15	9	23	19	14
Moderate (200–1000 gal/min)	15	11	7	21	16	12
Heavy (> 1000 gal/min)	10	8	6	18	14	10

Joint condition: Good = tight or cemented. Fair = slightly weathered or altered. Poor = severely weathered, altered, or open.

TABLE 10.6 Correlation of Rock Structure Rating to Rock Load and Tunnel Diameter

Tunnel diameter D, ft	Rock load on tunnel arch W_r, kips/ft²											
	0.5	1.0	1.5	2.0	3.0	4.0	5.0	6.0	7.0	8.0	9.0	10.0
	Corresponding values of rock structure ratings (RSR)											
10	62.5	49.9	40.2	32.7	21.6	13.8						
12	65.0	53.7	44.7	37.5	26.6	18.7						
14	66.9	56.6	48.3	41.4	30.8	22.9	16.8					
16	68.3	59.0	51.2	44.7	34.4	26.6	20.4	15.5				
18	69.5	61.0	53.7	47.6	37.6	29.9	18.8					
20	70.4	62.5	55.7	49.9	40.2	32.7	26.6	21.6	17.4			
22	71.3	63.9	57.5	51.9	42.7	35.3	29.3	24.3	20.1	16.4		
24	72.0	65.0	59.0	53.7	44.7	37.5	31.5	26.6	22.3	18.7		
26	72.6	66.1	60.3	55.3	46.7	39.6	33.8	28.8	24.6	20.9	17.7	
28	73.0	66.9	61.5	56.6	48.3	41.4	35.7	30.8	26.6	22.9	19.7	16.8
30	73.4	67.7	62.4	57.8	49.8	43.1	37.4	32.6	28.4	24.7	21.5	18.6

TABLE 10.7 Theoretical Spacing S_d, ft, of Typical Rib Sizes for Datum Condition

Rib size	Tunnel diameter, ft										
	10	12	14	16	18	20	22	24	26	28	30
4I7.7	1.16										
4H13.0	2.01	1.51	1.16	0.92							
6H15.5	3.19	2.37	1.81	1.42	1.14						
6H20		3.02	2.32	1.82	1.46	1.20					
6H25			2.86	2.25	1.81	1.48	1.23	1.04			
8W-31				3.24	2.61	2.14	1.78	1.51	1.29	1.11	
8W-40					3.37	2.76	2.30	1.95	1.67	1.44	1.25
8W-48						3.34	2.78	2.35	2.01	1.74	1.51
10W-49								2.59	2.22	1.91	1.67
12W-53										2.19	1.91
12W-65											2.35

The *RR* is the measure of the actual tunnel support compared to the datum and is expressed by

$$RR = \frac{S_d}{S_a} \times 100$$

where $S_d = \dfrac{Pt}{P_1}$ is the datum rib spacing and S_a is the actual rib spacing.

Wickman and Tiedemann[25] developed the following empirical relationships:

$$(RR + 80)(RSR + 30) = 8800$$

and

$$W_r = \frac{D}{302}\left(\frac{8800}{RSR + 30} - 80\right)$$

where W_r is the theoretical potential rock load.

Geomechanics or RMR Method.   The Rock-Mass Rating (RMR) method, sometimes referred to as the Geomechanics classification system, was developed by Bieniawski[26,27]. The RMR method is similar to the RSR but contains additional rock-classification parameters, including RQD, intact rock strength, and more detailed descriptions of the rock joints. Bieniawski reintroduced the Terzaghi method concept of "stand-up time" into his definition of rock-mass class.

The general classification of the RMR system is similar to that of the RSR system. The final result is a "rock class" that can be directly correlated to support requirements developed from over 268 case histories. The RMR system is a multiple-approach type D2 empirical system.

The RMR design parameters are partitioned into five categories:

1. Uniaxial compressive strength (or point load strength index) of intact rock; maximum value 15
2. RQD; maximum value 20
3. Spacing of discontinuities; maximum value 20
4. Condition of discontinuities; maximum value 30
5. Groundwater; maximum value 15

Additional parameters used to determine the final rock-mass class are strike, and dip. The RMR primary design parameters and rating adjustments for strike and dip in tunneling are given in Table 10.9.

Each classification parameter and adjustment factor is rated based on the judgment of Bieniawski as to their relative importance in rock mass during tunneling.

The procedure used in the application of the RMR system is as follows:

1. Identify regions along the tunnel alignment that have uniform geologic features. Each identified region should be independently classified. Identification of these regions is subjective and as such is based on experience and the availability of geologic data.
2. Each region is then evaluated using the RMR system. The sum of the rating for the five rock-mass evaluation parameters is then adjusted for strike and dip orientation of joints with respect to the tunnel alignment. The final sum is the RMR.

3. Tunnel support load can be determined from the RMR system as proposed by Unal:[28]

$$P = \left[\frac{100 - RMR}{100}\right]\gamma B$$

where P = support load, kN
$\quad$ B = tunnel width, m
$\quad$ γ = rock density, kg/m^3

4. Stand-up time and recommended tunnel supports can be obtained from Tables 10.8 through 10.10 and Fig. 10.11.

Rock Mass Quality Rating or Q-System Method. *The Rock-mass Quality Rating Method* (Q system) was developed by Barton et al.[29] for the Norwegian Geotechnical Institute. Barton[30] described the Q system as a numerical assessment of rock-mass quality using RQD and five additional parameters. Their aim was to provide a numerical value of rock-mass behavior that included several important factors missing in the Terzaghi and RQD methods. In a manner similar to the RSR method, these parameters are further subdivided into additional variables that are prioritized using a parameter weighing system. The Q system was based on the analysis of 212 European case histories, particularly from Scandinavia. Steiner and Einstein describe the Q system as a multiple-parameter type D2 empirical system.

The Q system contains the following six general parameters that describe rock-mass behavior, calculate rock load, and provide a basis for recommendations of tunnel liners:

1. Rock-quality designation (RQD) partitioned into the five ranges from very poor to excellent.

2. Joint set number J_n partitioned into nine categories from massive (no joints) with $J_n =$ 0.5 to crushed soil-like rock with $J_n = 20$.

3. Joint roughness number J_r partitioned into nine categories from a discontinuous joint set $J_r = 4$ to slickensided joint $J_r = 0.5$.

4. Joint alteration number J_a partitioned into 16 categories from tightly healed joint $J_a =$ 0.75 to joints filled with thick continuous clay $J_a = 20$.

5. Joint water-reduction factor J_w partitioned into six categories of flow from dry joint $J_w =$ 1.0 to a joint with exceptional high flows $J_w = 0.1$.

6. Stress-Reduction Factor (SRF) partitioned into 16 categories from weak zones containing clay to chemically disintegrated rock.

These six parameters are combined into the following three crude measures of rock-mass behavior:

1. Relative block size RQD/J_n
2. Interblock shear strength J_i/J_a
3. Active stress J_w/SRF

Detailed descriptions of the parameters used in the Q system are contained in Table 10.11. The general methodology when using the Q system is

1. Identify regions along the tunnel alignment that have uniform geologic features. Each identified region should be independently classified. Identification of these regions is subjective and as such is based on experience and the availability of geologic data.

TABLE 10.8 Geomechanics Classification of Jointed Rock Masses[27]

	Parameter		Ranges of values						
1	Strength of intact rock material	Point-load strength index	> 10 MPa	4–10 MPa	2–4 MPa	1–2 MPa	For this low range —uniaxial compressive test is preferred		
		Uniaxial compressive strength	> 250 MPa	100–250 MPa	50–100 MPa	25–50 MPa	5–25 MPa	1–5 MPa	< 1 MPa
	Rating		15	12	7	4	2	1	0
2	Drill core quality RQD		90–100%	75–90%	50–75%	25–50%	< 25%		
	Rating		20	17	13	8	3		
3	Spacing of discontinuities		> 2m	0.6–2 m	200–600 mm	60–200 mm	< 60 mm		
	Rating		20	15	10	8	5		
4	Condition of discontinuities		Very rough surfaces Not continuous No separation Unweathered wall rock	Slightly rough surfaces. Separation < 1 mm Slightly weathered walls	Slightly rough surfaces. Separation < 1 mm Highly weathered walls	Slickensided surfaces or Gouge < 5 mm thick or Separation 1–5 mm Continuous	Soft gouge > 5 mm thick or Separation > 5 mm Continuous		
	Rating		30	25	20	10	0		
4	Ground-water	Inflow per 10 m tunnel length	None	< 10 L/min	10–25 L/min	25–125 L/min	> 125		
			or	or	or	or	or		
		Ratio: joint water pressure / major principal stress	0	0.0–0.1	0.1–0.2	0.2–0.5	> 0.5		
			or	or	or	or	or		
		General conditions	Completely dry	Damp	Wet	Dripping	Flowing		
	Rating		15	10	7	4	0		

Rating adjustment for joint orientations

Strike and dip orientations of joints		Very favorable	Favorable	Fair	Unfavorable	Very unfavorable
Ratings	Tunnels	0	-2	-5	-10	-12
	Foundations	0	-2	-7	-15	-25
	Slopes	0	-5	-25	-50	-60

Rock-mass classes determined from total ratings

Rating	100←81	80←61	60←41	40←21	<20
Class No.	I	II	III	IV	V
Description	Very good rock	Good rock	Fair rock	Poor rock	Very poor rock

Meaning of rock-mass classes

Class No.	I	II	III	IV	V
Average stand-up time	10 years for 15-m span	6 months for 8-m span	1 week for 5-m span	10 h for 2.5-m span	30 min for 1-m span
Cohesion of the rock mass	> 400 kPa	300–400 kPa	200–300 kPa	100–200 kPa	< 100 kPa
Friction angle of the rock mass	> 45°	35–45°	25–35°	15–25°	< 15°

TABLE 10.9   Effects of Discontinuity Strike and Dip Orientation in Tunneling

Strike perpendicular to tunnel axis			
Drive with dip		Drive against dip	
Dip 45–90°	Dip 20–45°	Dip 45–90°	Dip 20–45°
Very favorable	Favorable	Fair	Unfavorable

Strike parallel to tunnel axis		
Drive with dip		Irrespective of Strike
Dip 20–45°	Dip 45–90°	Dip 0–20°
Fair	Very unfavorable	Fair

2. For tunnels the ground region from the tunnel invert to $2B$ above the tunnel crown is evaluated using the Q system. The sum of the ratings for the six classifications is then placed in the following equation:

$$\text{Rock-mass quality } Q = \frac{\text{RQD}}{J_n} \frac{J_r}{J_a} \frac{J_w}{\text{SRF}}$$

     The range of possible values of Q is from 0.001 (heavy squeezing ground) to 1000 (sound unjointed rock).

3. Barton et al. introduced a parameter that relates the purpose of the excavation and the degree of safety required. This parameter is called the excavation support ratio (ESR), shown in Table 10.12. ESR has a range from 0.8 for critical facilities and 1.0 for permanent facilities to 5.0 for temporary mine openings.

4. Once Q is calculated and a value of ESR established, the type of tunnel support can be determined using the Q-system tables for support categories I through 38 contained in Table 10.13 and Fig. 10.12. It is, however, first necessary to determine the "Equivalent Dimension," which is the tunnel width (or span) B divided by ESR. Figure 10.12 correlates values of Q and SPAN/ESR with the support systems contained in the 212 case histories used to develop the Q system included in Table 10.13.

Empirical Methods in Soft Ground

The empirical methods previously discussed are all primarily used for the design of liners in rock tunnels. By assuming rock conditions to be highly fractured, that is, rock condition 6 or RQD less than 25 percent, these methods are adaptable to determining design loads for liners in soft ground. However, that was not the original intent of these methods.

The American Method.   The American method is an empirical approach specifically used for the design of tunnel liners in soft ground or highly fractured rock. The American method originated with Peck.[10]

     Kuesel,[31,32] in response to a world survey questionnaire of tunnel design models, argued that questions concerning methods of structural analysis, statistical models, modulus of

TABLE 10.10 Geomechanics Classification Guide for Excavation and Support in Rock Tunnels (Horseshoe Shape. 10 m Wide. Vertical Stress < 25 MPa. Drilling-and-Blasting Construction)[19]

Rock-mass class	Excavation	Support		
		Rockbolts (20 mm dia., fully bonded)	Shotcrete	Steel sets
1. Very good rock RMR: 81–100	Full face: 3 m advance	Generally no Support required except for occasional spot bolting		
2. Good rock RMR: 61–80	Full face: 1.0–1.5 m advance, complete support 20 m from face	Locally bolts in crown 3 m long, spaced 2.5 m with occasional wire mesh	50 mm in crown where required	None
3. Fair rock RMR: 41–60	Top heading and bench: 1.5–3 m advance in top heading; commerce support after each blast; complete support 10 m from face	Systematic bolts 4 m long, spaced 1.5–2 m in crown and walls with wire mesh in crown	50–100 mm in crown and 30 mm in sides	None
4. Poor rock RMR: 21–40	Top heading and bench: 1.0–1.5 m advance in top heading; install support concurrently with excavation—10 m from face	Systematic bolts 4–5 m long, spaced 1–1.5 m in crown and walls with wire mesh	100–150 mm in crown and 100 mm in sides	Light ribs spaced 1.5 m where required
5. Very poor rock RMR: < 20	Multiple drifts: 0.5–1.5 m advance in top heading; install support concurrently with excavation; shotcrete as soon as possible after blasting	Systematic bolts 5–6 m long, spaced 1–1.5 m in crown and walls with wire mesh. Bolt invert	150–200 mm in crown and 150 mm in sides and 50 mm on face	Medium-to-heavy ribs spaced 0.75 m with steel lagging and forepoling if required. Close invert

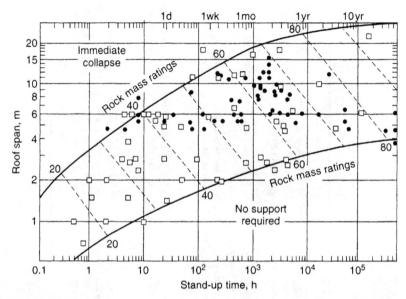

FIGURE 10.11 Geomechanics classification of rock masses output for mining and tunneling.[27]

TABLE 10.11 Ratings for the Six Q-System Parameters[30]

1. Rock quality designation (RQD)	
A. Very poor	0–25
B. Poor	25–50
C. Fair	50–75
D. Good	75–90
E. Excellent	90–100

NOTE: Where RQD is reported or measured ≤ 10 (including 0), a nominal value of 10 is used to evaluate Q. RQD intervals of 5, i.e., 100, 95, 90, etc., are sufficiently accurate.

2. Joint set number J_n	
A. Massive, no or few joints	0.5–1.0
B. One joint set	2
C. One joint set plus random	3
D. Two joint sets	4
E. Two joint sets plus random	6
F. Three joint sets	9
G. Three joint sets plus random	12
H. Four or more joint sets, random, heavily jointed, "sugar cube," etc.	15
J. Crushed rock, earthlike	20

NOTE: For intersections use $(3.0 \times J_n)$. For portals use $(2.0 J_n)$.

TABLE 10.11   Ratings for the Six Q-System Parameters[30] (*Continued*)

3. Joint roughness number J_r	

(*a*) Rock wall contact and
(*b*) Rock wall contact before 10 cm shear:

A. Discontinuous joints	4
B. Rough or irregular, undulating	3
C. Smooth, undulating	2
D. Slickensided, undulating	1.5
E. Rough or irregular, planar	1.5
F. Smooth, planar	1.0
G. Slickensided, planar	0.5

NOTE: Descriptions refer to small-scale features and intermediate-scale features, in that order.

(*c*) No rock wall contact when sheared

H. Zone containing clay minerals thick enough to prevent rock wall contact	1.0
J. Sandy, gravelly, or crushed zone thick enough to prevent rock wall contact	1.0

NOTE: Add 1.0 if the mean spacing of the relevant joint set is greater than 3 m. J_r = 0.5 be used for planar slickensided joints having lineations, provided the lineations are orientated for minimum strength.

4. Joint alteration number		
	J_a	ϕ_r (approx.)

(*a*) Rock wall contac

A. Tightly healed, hard, nonsoftening, impermeable, filling, i.e., quartz or epidote	0.75	(–)
B. Unaltered joint walls, surface staining only	1.0	(25–35°)
C. Slightly altered joint walls. Nonsoftening mineral coatings, sandy particles, clay-free disintegrated rock, etc.	2.0	(25–30°)
D. Silty- or sandy-clay coatings, small clay fraction (nonsoftening)	3.0	(20–25°)
E. Softening or low-friction clay mineral coatings, i.e., kaolinite or mica. Also, chlorite, talc, gypsum, graphite, etc., and small quantities of swelling clays	4.0	(8–16°)

(*b*) Rock wall contact before 10 cm shear

F. Sandy particles, clay-free disintegrated rock, etc.	4.0	(25–30°)
G. Strongly overconsolidated nonsoftening clay mineral fillings (continuous, but < 5 mm thickness)	6.0	(16–24°)
H. Medium or low overconsolidation, softening, clay mineral fillings (continuous but < 5 mm thickness)	8.0	(12–16°)
J. Swelling clay fillings, i.e., montmorillonite (continuous, but < 5 mm thickness). Value of J_a depends on percent of swelling clay-size particles, and access to water, etc.	8–12	(6–12°)

(*c*) No rock wall contact when sheared

K. Zones or bands of disintegrated or crushed rock and clay (see G, H, J for description of clay condition)	6, 8, or 8–12	(6–24°)
L. Zones or bands of silty or sandy-clay, small clay fraction (nonsoftening)	5.0	(-)
M. Thick, continuous zones or bands of clay (see G, H, J for description of clay condition)	10, 13, or 13–20	(6–24°)

TABLE 10.11 Ratings for the Six Q-System Parameters[30] (*Continued*)

5. Joint water-reduction factor		
	J_w	Approx. water pressure, kg/cm^2
A. Dry excavations or minor inflow, i.e., < 5 liters/min locally	1.0	< 1
B. Medium inflow or pressure, occasional outwash of joint fillings	0.66	1–2.5
C. Large inflow or high pressure in competent rock with unfilled joints	0.5	2.5–10
D. Large inflow or high pressure, considerable outwash of joint fillings	0.33	2.5–10
E. Exceptionally high inflow or water pressure at blasting, decaying with time	0.2–0.1	> 10
F. Exceptionally high inflow or water pressure continuing without noticeable decay	0.1–0.05	> 10

NOTE: Factors C to F are crude estimates. Increase J_w if drainage measures are installed, Special problems caused by ice formation are not considered.

6. Stress-reduction factor SRF

(*a*) Weakness zones intersecting excavation, which may cause loosening of rock mass when tunnel is excavated	
A. Multiple occurrences of weakness zones containing clay or chemically disintegrated rock, very loose surrounding rock (any depth)	10
B. Single weakness zones containing clay or chemically disintegrated rock (depth of excavation ≤ 50 m)	5
C. Single weakness zones containing clay or chemically disintegrated rock (depth of excavation > 50 m)	2.5
D. Multiple shear zones in competent rock (clay-free), loose surrounding rock (any depth)	7.5
E. Single shear zones in competent rock (clay-free) (depth of excavation ≤ 50 m)	5.0
F. Single shear zones in competent rock (clay-free) (depth of excavation > 50 m)	2.5
G. Loose open joints, heavily jointed or "sugar cube," etc. (any depth)	5.0

NOTE: Reduce these values of SRF by 25 to 50 percent if the relevant shear zones only influence but do not intersect the excavation.

	σ_c/σ_1	σ_t/σ_1	SRF
(*b*) Competent rock, rock stress problems			
H. Low stress, near surface	> 200	> 13	2.5
J. Medium stress	200–10	13–0.66	1.0
K. High stress, very tight structure (usually favorable to stability, may be unfavorable for wall stability)	10–5	0.66–0.33	0.5–2
L. Mild rock burst (massive rock)	5–2.5	0.33–0.16	5–10
M. Heavy rock burst (massive rock)	< 2.5	< 0.16	10–20

NOTE: For strongly anisotropic virgin stress field (if measured): when $5 \le \sigma_1/\sigma_3 \le 10$, reduce σ_c and σ_t to $0.8\sigma_c$ and $0.8\sigma_t$. When $\sigma_1/\sigma_3 > 10$, reduce σ_c and σ_t to $0.6\sigma_c$ and $0.6\sigma_t$, where σ_c = unconfined compression strength and σ_t = tensile strength (point load), and σ_1 and σ_3 arc the major and minor principal stresses.

Few case records available where depth of crown below surface is less than span width. Suggest SRF increase from 2.5 to 5 for such cases (see H).

TABLE 10.11 Ratings for the Six Q-System Parameters[30] (*Continued*)

6. Stress-reduction factor SRF (*Continued*)	
SRF	
(*c*) Squeezing rock: plastic flow of incompetent rock under the influence of high rock pressure	
N. Mild squeezing rock pressure	5–10
O. Heavy squeezing rock pressure	10–20
(*d*) Swelling rock: chemical swelling activity depending on presence of water	
P. Mild swelling rock pressure	5–10
R. Heavy swelling rock pressure	10–15

subgrade reaction, nonlinear behavior, etc., are not relevant in current U.S. tunnel design practice. He stated that generally the design details of tunnel liners are governed by non-mathematical considerations which include, but are not limited to, ease of fabrication, erection, and watertightness. Schmidt[33] carries Kuesel's argument even further by stating that the stresses in real tunnels are affected by many factors often ignored in analytic design methods, including

1. The vertical load will partially arch over the tunnel.

2. Horizontal pressure will be altered owing to vertical load arching.

3. Inelastic behavior of the soil.

4. Consolidation of cohesive soils in plastic zone around the liner could increase stresses with time.

5. Anisotrophy and variability of ground parameters, that is, ground modulus, overconsolidation ratio, plastic index, etc.

6. Ground relaxation prior to liner installation.

The basic assumptions of the American method are

1. If the liner is flexible and is permitted to deform from a circle to an ellipse during loading, external pressures against the liner will approach a uniform distribution and bending moments will approach zero.

TABLE 10.12 Excavation Support Ratio (ESR) for a Variety of Underground Excavations[30]

Type of excavation	ESR	Number of cases
A. Temporary mine openings, etc.	ca. 3–5?	2
B. Permanent mine openings, water tunnels for hydro power (excluding high-pressure penstocks), pilot tunnels, drifts, and headings for large openings	1.6	83
C. Storage caverns, water-treatment plants, minor road and railway tunnels, surge chambers, access tunnels, etc.	1.3	25
D. Power stations, major road and railway tunnels, civil defense chambers, portals, intersections	1.0	79
E. Underground unclear power stations, railway stations, sports and public facilities, factories	ca. 0.8?	2

TABLE 10.13   Support Recommendations for the 38 Categories Shown in Fig. 10.12[21]

Support category	Conditional factors			Type of support	Notes
	RQD/J_n	J_r/J_a	SPAN/ESR		
1*				sb(utg)	
2*				sb(utg)	
3*				sb(utg)	
4*				sb(utg)	
5*				sb(utg)	
6*				sb(ulg)	
7*				sb(utg)	
8*				sb(utg)	

NOTE: The type of support to be used in categories 1 to 8 will depend on the blasting technique. Smooth wall blasting and thorough barring-down may remove the need for support. Rough-wall blasting may result in the need for single application of shotcrete, especially where the excavation height is > 25 m. Future case records should differentiate categories 1 to 8.

Support category	RQD/J_n	J_r/J_a	SPAN/ESR	Type of support	Notes
9	≥ 20			sb(utg)	
	< 20			B(utg) 2.5–3 m	
10	≥ 30			B(utg) 2–3 m	
	< 30			B(utg) 1.5–2 m + clm	
11*	≥ 30			B(tg) 2–3 m	
	< 30			B(tg) 1.5–2 m + clm	
12*	≥ 30			B(tg) 2–3 m	
	< 30			B(tg) 1.5–2 m + clm	
13	≥ 10	≥ 1.5		sb(utg)	I
	≥ 10	< 1.5		B(utg) 1.5–2 m	I
	< 10	≥ 1.5		B(utg) 1.5–2 m	I
	< 10	< 1.5		B(utg) 1.5–2 m + S 2–3 cm	I
14	≥ 10		≥ 15	B(tg) 1.5–2 m + clm	I, II
	< 10		≥ 15	B(tg) 1.5–2 m + S(mr) 5–10 cm	I, II
			< 15	B(utg) 1.5–2 m + clm	I, III
15	> 10			B(utg) 1.5–2 m + clm	I, II, IV
	≤ 10			B(tg) 1.5–2 m + S(mr) 5–10 cm	I, II, IV
16* See note XII	> 15			B(tg) 1.5–2 m + clm	I, V, VI
	≤ 15			B(tg) 1.5–2 m + S(mr) 10–15 cm	I, V, VI

TABLE 10.13 Support Recommendations for the 38 Categories Shown in Fig. 10.12[30] *(Continued)*

Support category	Conditional factors			Type of support	Notes
	RQD/J_n	J_r/J_a	SPAN/ESR		
17	> 30			sb(utg)	I
	$\left(\begin{array}{c}\geq 10 \\ \leq 30\end{array}\right)$			B(utg) 1–1.5 m	I
	< 10		≥ 6 m	B(utg) 1–1.5 m + S 2–3 cm	I
	< 10		< 6 m	S 2–3 cm	I
18	> 5		≥ 10 m	B(tg) 1–1.5 m + clm	I, III
	> 5		< 10 m	B(utg) 1–1.5 m + clm	I
	≤ 5		≥ 10 m	B(tg) 1–1.5 m + S 2–3 cm	I, III
	≤ 5		> 10 m	B(utg) 1–1.5 m + S 2–3 cm	I
19			≥ 20 m	B(tg) 1–2 m + S(mr) 10–15 cm	I, II, IV
			< 20 m	B(tg) 5–1.5 m + S(mr) 1–10 cm	I, II
20* See note XII			≤ 35 m	B(tg) 1–2 m + S(mr) 20–25 cm	I, V, VI
			< 35 m	B(tg) 1–2 m + S(mr) 10–20 cm	I, II, IV
21	≥ 12.5	≤ 0.75		B(utg) 1 m + S 2–3 cm	I
	< 12.5	≤ 0.75		S 2.5–5 cm	I
		> 0.75		B(utg) 1 m	I
22	$\left(\begin{array}{c}> 10 \\ < 30\end{array}\right)$	> 1.0		B(utg) 1 m + clm	I
	≤ 10	> 1.0		S 2.5–7.5 cm	I
	< 30	≤ 1.0		B(utg) 1 m + S(mr) 2.5–5 cm	I
	≥ 30			B(utg) 1 m	I
23			≥ 15 m	B(tg) 1–1.5 m + S(mr) 10–15 cm	I, II, IV, VII
			< 15 m	B(utg) 1–1.5 m + S(mr) 5–10 cm	I
24* See note XII			≥ 30 m	B(tg) 1–1.5 m + S(mr) 15–30 cm	I, V, VI
			< 30 m	B(tg) 1–1.5 m + S(mr) 10–15 cm	I, II, IV

TABLE 10.13 Support Recommendations for the 38 Categories Shown in Fig. 10.12[30] (*Continued*)

Support category	Conditional factors			Type of support	Notes
	RQD/J_n	J_r/J_a	SPAN/ESR		
	> 10	> 0.5		B(utg) 1 m + mr or clm	I
	≤ 10	> 0.5		B(utg) 1 m + S(mr) 5 cm	I
25		≤ 0.5		B(tg) 1 m + S(mr) 5 cm	I
26				B(tg) 1 m + S(mr) 5–7.5 cm	VIII, X, XI
				B(utg) 1 m + S 2.5–5 cm	I, IX
			≥ 12 m	B(tg) 1 m + S(mr) 7.5–10 cm	I, IX
27			< 12 m	B(utg) 1 m + S (mr) 5–7.5 cm	I, IX
			> 12 m	CCA 20–40 cm + B(tg) 1 m	VIII, X, XI
			< 12 m	S(mr) 10–20 cm + B(tg) 1 m	VIII, X. XI
28* See note XII			≥ 30 m	B(tg) 1 m + S(mr) 30–40 cm	I, IV, V, IX
			(≥ 20 / < 30 m)	B(tg) 1 m + S(mr) 20–30 cm	I, II, IV, IX
			< 20 m	B(tg) 1 m + S(mr) 15–20 cm	I, II, IX IV, VII, I
				CCA(sr) 30–100 cm + B(tg) 1 m	X, XI
	> 5	> 0.25		B(utg) 1 m + S 2–3 cm	
29*	≤ 5	> 0.25		B(utg) 1 m + S(mr) 5 cm	
		≤ 0.25		B(tg) 1 m + S(mr) 5 cm	
	≥ 5			B(tg) 1 m + S 2.5–5 cm	IX
30	< 5			S(mr) 5–7.5 cm	IX
				B(tg) 1 m + S(mr) 5–7.5 cm	VIII, X, XI
	> 4			B(tg) 1 m + S(mr) 5–12.5 cm	IX
31	≤ 4 ≥ 1.5 < 1.5			S(mr) 7.5–25 cm	IX
				CCA 20–40 cm + B(tg) 1 m	IX, XI
				CCA(sr) 30–50 cm + B(tg) 1 m	VIII, X, XI

TABLE 10.13 Support Recommendations for the 38 Categories Shown in Fig. 10.12[30] (*Continued*)

Support category	Conditional factors			Type of support	Notes
	RQD/J_n	J_r/J_a	SPAN/ESR		
32 See note XII			≥ 20 m	B(tg) 1 m + S(mr) 40–60 cm	II, IV, IX, XI
			< 20 m	B(tg)l m + S(mr) 2–40 cm CCA(sr) 40–120 cm + B(lg) 1 m	III, IV, XI IX IV, VIII, X, XI
33*	≥ 2			B(tg) 1 m + S(mr) 2.5–5 cm	IX
	< 2			S(mr)5–10 cm S(mr) 7.5–15 cm	IX VIII, X
34	≥ 2	≥ 0.25		B(tg) 1 m + S(mr) 5–7.5 cm	IX
	< 2	≥ 0.25		S(mr) 7.5–15 cm	IX
		< 0.25		S(mr) 15–25 cm CCA(sr) 20–60 cm + B(ig) 1 m	IX VIII, X, XI
35 See note XII			> 15 m	B(tg) 1 m + S(mr) 30–100 cm	II, IX, XI
			≥ 15 m	CCA(sr) 60–200 cm + B(tg) 1 m	VIII, X, XI, II
			≥ 15 m	B(tg) 1 m + S(mr) 20–75 cm	IX, III, XI
			< 15 m	CCA(sr) 40–150 cm + B(tg) 1 m	VIII, X, XI, III
36*				S(mr) 10–20 cm S(mr) 10–20 cm + B(tg) 0.5–1.0 m	IX VIII, X, XI
37				S(mr) 20–60 cm S(mr) 20–60 cm + B(tg) 0.5–1.0 m	IX VIII, X, XI
38 See note XIII			≥ 10 m	CCA(sr) 100–300 cm	IX
			≥ 10 m	CCA(sr) 100–300 cm + B(tg) 1 m	VIII, X, II, XI
			< 10 m	S(mr) 70–200 cm	IX
			< 10 m	S(mr) 70–200 cm + B(ig) 1 m	VIII, X, III, XI

*Authors' estimates of support. Insufficient case records available for reliable estimation of support requirements.
Key:
  sb = spot bolting
  B = systematic bolting
  (utg) = untensioned, grouted
  (tg) = tensioned (expanding shell type for competent rock masses, grouted posttensioned in very poor quality rock masses; see note XI)
  S = shotcrete

TABLE 10.13　Support Recommendations for the 38 Categories Shown in Fig. 10.12[30] (*Continued*)

(mr) = mesh reinforced
clm = chain-link mesh
CCA = Cast concrete arch
(sr) = steel reinforced

NOTES:

I. For cases of heavy rock bursting or "popping," tensioned bolts with enlarged bearing plates are often used, with spacing of about 1 m (occasionally down to 0.8 m). Final support when "popping" activity ceases.

II. Several bolt lengths are often used in the same excavation, i.e., 3, 5, and 7 m.

III. Several bolt lengths are often used in the same excavation, i.e., 2, 3, and 4 m.

IV. Tensioned cable anchors are often used to supplement bolt support pressures. Typical spacing 2–4 m.

V. Several bolt lengths are used in some excavations, i.e., 6, 8, and 10 m.

VI. Tensioned cable anchors often used to supplement bolt support pressures. Typical spacing 4–6 m.

VII. Several older-generation power stations in this category employ systematic or spot bolting with areas of chain-link mesh, and a free span concrete arch roof (25–40 cm) as permanent support.

VIII. Cases involving swelling, for instance, montmorillonite clay (with access of water). Room for expansion behind the support is used in cases of heavy swelling. Drainage measures are used where possible.

IX. Cases not involving swelling clay or squeezing rock.

X. Cases involving squeezing rock. Heavy rigid support is generally used as permanent support.

XI. According to Barton's[21] experience, in cases of swelling or squeezing, the temporary support required before concrete (or shotcrete) arches are formed may consist of bolting (tensioned shell-expansion type) if the value of RQD/J_n is sufficiently high (i.e., > 1.5), possibly combined with shotcrete. If the rock mass is very heavily jointed or crushed (i.e., RQD/J_n < 1.5, e.g., a "sugar cube" shear zone in quartzite), then the temporary support may consist of up to several applications of shotcrete. Systematic bolting (tensioned) may be added after casting the concrete (or shotcrete) arch to reduce the uneven loading on the concrete, but it may not be effective when RQD/J_n < 1.5, or when a lot of clay is present, unless the bolts are grouted before tensioning. A sufficient length of anchored bolt might also be obtained using quick-setting resin anchors in these extremely poor quality rock masses. Serious occurrences of swelling and/or squeezing rock may require that the concrete arches are taken right up to the face, possibly using a shield as temporary shuttering. Temporary support of the working face may also be required in these cases.

XII. For reasons of safety the multiple drift method will often be needed during excavation and supporting of roof arch. Categories 16, 20, 24, 28, 32, 35 (SPAN/ESR > 15 m only).

XIII. Multiple drift method usually needed during excavation and support of arch, walls, and floor in cases of heavy squeezing. Category 38 (SPAN/ESR > 10 m only).

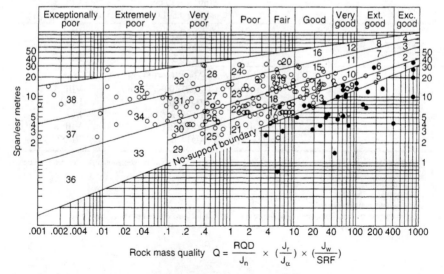

FIGURE 10.12　Rock support category given by box numbers 1 to 38.[30]

2. The amount of liner deformation required to achieve a uniform external pressure distribution is a function of the stiffness of the surrounding ground, its time-dependent stress-strain characteristics, and the dimensions and depth of the tunnel.

3. Liner stresses under a nearly uniform external pressure will be limited to axial (hoop) stresses.

The design of the tunnel liner can be accomplished by

1. Designing the liner to accommodate hoop stresses. Hoop loads for shallow tunnels can be assumed to be equal to the full overburden pressure.

2. Designing the liner to accommodate distortion (increase in horizontal diameter with a decrease in vertical diameter).

3. Checking local transverse buckling stresses and construction loadings, jacking pressures, watertightness.

Table 10.14, from Schmidt,[33] proposes the following ultimate distortions:

TABLE 10.14 Ultimate Distortion of Tunnels $\Delta R/R$, versus Soil Type [33]

Soil type	$\Delta R/R,\%$
Stiff to hard clays, overload factor[2] < 2.5–3	0.15–0.40
Soft clays or silts, overload factor > 2.5–3	0.25–0.75
Dense or cohesive sands, most residual soil	0.05–0.25
Loose sands	0.10–0.35

Add 0.1 to 0.3 percent for tunnels in compressed air.
Add appropriate distortion for external effects such as adjacent construction.

The moment is based on the assumed deformation:

$$M = \frac{3\,EI}{R_m}\left(\frac{\Delta R}{R}\right)$$

The thrust will equal:

$$T = \gamma_T zR$$

To check for buckling:

$$P_{cr} = \frac{3\,EI}{R^3}$$

where γ_T = total unit weight
 z = depth to tunnel centerline (springline)
 R_m = average liner radius
 R = radius of excavated opening
 ΔR = change in liner radius
 I = moment of inertia of tunnel liner per linear foot at tunnel
 E = Young's modulus of tunnel liner

Analytical Methods

Beam Spring Model.   Recent publications by the British Tunnel Society (BTS),[34] American Society of Civil Engineers (ASCE),[35] and the American Association of State Highway and Transportation Officials (AASHTO)[36] describe a number of analytical methods for tunnel design. Among the analytic methods discussed is the beam spring model as shown in Fig. 10.13. In the beam spring model, the tunnel liner is modeled as a series of beam elements and radial and tangential springs. The structure is loaded with gravity loads, either soil or rock, and allowed to deform. Any springs that go into tension are eliminated (made inactive) since soil or rock cannot support tensile loads. After each trial, more springs are eliminated until only springs that are in compression remain. The model is comparable to the Winkler Beam solution for a beam on an elastic foundation.

The model can be considered either no-slip or slip interface conditions between the ground and the liner. Spring constants are based on contributory area and an estimate of the ground deformation modulus.

The Schwartz and Einstein Method.   Schwartz and Einstein,[37] after a review of tunnel design methods, proposed a simplified approach to ground structure interaction that incorporated many of the significant factors from previous empirical and analytic methods. These significant factors included:

1. Relative stiffness of the tunnel liner and the ground

2. Lag time in installation of liner

3. Yielding ground mass as its shear strength is exceeded

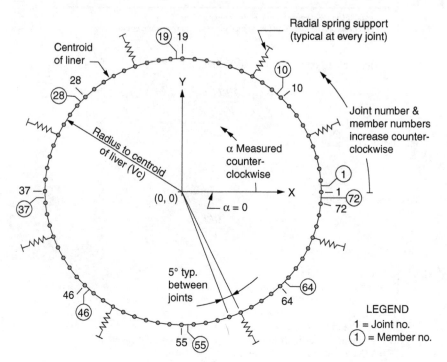

FIGURE 10.13   Beam-spring model. (*AASHTO*).[7]

Schwartz and Einstein used a simple closed-form plane strain model supplemented with the results of finite-element parameter studies. These parameter studies were used to develop correction factors for the elastic plane strain variables used in the initial model. Two of these correction factors were incorporated into the final closed-form model.

Their Simplified Analysis Method (SAM) calculated stresses in the tunnel liner, not in the ground mass surrounding the opening. SAM is based on the concept of excavation unloading which assumes that the tunnel is constructed in ground that is initially stressed. Schwartz and Einstein confirmed their model using data obtained from five tunnels in which liner loads were measured during construction.

Schwartz and Einstein's assumptions, used to develop the simplified analysis method, were

1. Plane strain conditions.

2. The tunnel cross section is circular.

3. The tunnel is deep enough so that (a) the influence of the ground surface on tunnel behavior is insignificant; (b) the variations in the ground stresses over the height (diameter) of the tunnel are negligibly small relative to the magnitude of the total stress.

4. The ground is an isotropic homogeneous continuum.

5. Neither the support nor the ground mass exhibits time-dependent behavior. (Time-dependent behavior includes squeezing or swelling ground, and creep in the liner or rock support.)

6. The strength of the ground can be modeled using an elastoplastic failure criterion.

7. The tunnel liner is made of materials that are linearly elastic.

8. The tunnel liner is continuous around the entire tunnel perimeter.

9. The excavation is full-face without compressed air.

10. The tunnel liner is watertight and the hydrostatic pressure can be considered as a separate applied load.

The simplified analysis method has three major steps:
Step 1. Determine relative stiffness of the liner compared to the ground.
Step 2. Adjust for delay in liner installation.
Step 3. Consider yielding of the ground.

Step 1—Determine Relative Stiffness of Liner and Ground. Two dimensionless variables are used as measures of relative stiffness, the compressibility ratio $C*$ and the flexibility ratio $F.*$ The liner is considered as an elastic circular shell embedded in an elastic ground medium that is loaded with a surface overpressure P and a lateral pressure KP.

The compressibility ratio is a measure of the relative circumferential stiffness of the ground to the liner, that is, compressive strength on both vertical and horizontal planes. $C*$ is defined as

$$C* = \frac{ER(1 - v_L^2)}{E_L A_L (1 - v^2)}$$

The flexibility ratio $F*$ is a measure of the relative "flexural" stiffness of the ground to the flexural stiffness of the liner under an asymmetric loading, that is, horizontal stress equal but opposite to the vertical stress. $F*$ is defined as

$$F* = \frac{ER^3(1 - v_L^2)}{E_L I_L (1 - v^2)}$$

(Liner properties are identified using the subscript L.)

The circumferential liner thrust is primarily related to $C,*$ and the bending moments are primarily related to $F*$.

The two limiting conditions are "full-slip" and "no-slip." No-slip conditions mean there is no relative displacement between the liner and the surrounding ground. No-slip conditions result in variable shear stress transfer between the ground and the liner circumference. Full-slip conditions assume some relative movement. Relative movement results in a uniform shear stress between the liner and the ground. For full-slip, excavation unloading conditions:

$$\frac{T}{PR} = \frac{1}{2}(1+K)(1-a_0*) + \frac{1}{2}(1-k)(1-2a_2*)\cos^2\theta$$

and

$$\frac{M}{PR^2} = \frac{1}{2}(1-k)(1-2a_2*)\cos^2\theta$$

$$\frac{u_L E}{PR(1+v)} = \frac{1}{2}(1+K)a_0* - (1-K)[(5-6v)a_2* - (1-v)]\cos^2\theta$$

$$\frac{v_L E}{PR(1+v)} = \frac{1}{2}(1-K)[(5-6v)a_2* - (1-v)]\sin^2\theta$$

a_0*, a_2* = coefficients, having the expressions

$$a_0* = \frac{C*F*(1-v)}{C*+F*+C*F*(1-v)}$$

$$a_2* = \frac{(F*+6)(1-v)}{2F*(1-v)+6(5-6v)}$$

For no-slip, excavation unloading conditions:

$$\frac{T}{PR} = \frac{1}{2}(1+K)(1-a_0*) + \frac{1}{2}(1-K)(1+2a_2*)\cos^2\theta$$

$$\frac{M}{PR^2} = \frac{1}{4}(1-K)(1-2a_2*+2b_2*)\cos^2\theta$$

$$\frac{u_L E}{PR(1-v)} = \frac{1}{2}(1+K)a_0* + \frac{1}{2}(1-K)[4(1-v)b_2* - 2a_2*]\cos^2\theta$$

$$\frac{v_L E}{PR(1-v)} = -(1-K)[a_2* + (1-2v)b_2*]\sin^2\theta$$

and

$$b_2* = \frac{C*(1-v)}{2[C*(1-v)+4v-6[cf]\hat{b}-3[cf]\hat{b}C*(1-v)]}$$

$$\hat{b} = \frac{(6+F*)C*(1-v)+2F*v}{3F*+3C*+2C*F*(1-v)}$$

$$a_2{}^* = \hat{b}b_2{}^*$$

where T = thrust in tunnel liner
P = in situ ground stress
R = radius of tunnel measured to centerline of liner
K = lateral in situ stress coefficient
E = modulus of elasticity of ground
θ = angular coordinate measured from tunnel springline
v = Poisson's ratio for the ground
u_L, v_L are, respectively, the radial and tangential displacements of the tunnel liner

Step 2—Adjust for Support Delay.   The reduction in the liner load due to a delay in the installation of the liner can be represented by a delay factor:

$$\lambda_d = \frac{T'}{T}$$

where T is the liner thrust ignoring the effect of support delay calculated from the equations in step 1. T' is the support thrust reduced because of the delay in installing the tunnel liner.
The multiplication factor λ_d is used to modify the liner forces calculated in step 1:

$$T_2 = \lambda_d T_1$$
$$M_2 = \lambda_d M_1$$

T_2 and M_2 are the thrust and moment modified because of the delay in installation of the liner. The liner installation delay factor λ_d is a function of the ratio of the delay length L_d, defined as the distance between the face of the tunnel and the midpoint of the leading edge of the liner support, and the radius of the tunnel R.
Based on axisymmetric finite-element analysis, Schwartz and Einstein determined the following:

$$\lambda_d = 0.98 - 0.57\left(\frac{L_d}{R}\right)$$

If there are additional ground movements other than those associated with liner installation, they must be included in the delay factor λ_d. Defining u_0' as these additional preliner installation ground movements and u_f as the total elastic radial displacement of the wall of the unlined tunnel, the complete liner installation delay factor is

$$\lambda_d = 0.98 - 0.57\left(\frac{L_d}{R}\right) - \frac{u_0'}{u_f}$$

Step 3—Consideration of Ground Yielding.   The ground yield factor λ_y was derived similarly to λ_d to represent the effects of ground yielding:

$$\lambda_y = \frac{p_s{}^*}{P_s'} = \frac{P_s{}^*}{\lambda_d P_s}$$

A detailed explanation of the calculation of λ_d can be found in Schwartz and Einstein.[37]

The final thrust in the liner can be obtained by correcting the results of step 1 by the results of steps 2 and 3:

$$T^* = \lambda_y \lambda_d T_1$$

Schwartz and Einstein compared the results of these equations with stresses obtained from the in situ monitoring of five actual tunnel projects. The results of these comparisons indicated that the errors were within ±30 percent, with an average error of 15 percent. However, the errors in the predicted values of liner thrust ranged between extremes of −79 and +62 percent. Schwartz and Einstein concluded that the method was accurate in a wide range of tunnel projects. They considered the high standard deviation of their results "inherent in real tunneling situations."

Steel Sets.   One of the more widely used methods of designing a steel tunnel support is suggested by Proctor and White in Chap. 11 of *Rock Tunneling with Steel Supports*.[14] Some assumptions used by Proctor and White are outlined below:

1. The rock bearing load is assumed to be an inverted U-shaped load bearing on an inverted U-shaped tunnel crown. In order to simplify the calculations, only one-half of the arch is considered, as in Fig. 10.14.
2. Blocking points are spaced in a uniform radial pattern from arch springline to crown. The uniform radial spacing results in a vertical load W at each blocking point, which is not uniform but depends on the horizontal thickness of each rock block.
3. Vertical loads are resolved graphically into radial and tangential components, as illustrated in Fig. 10.15. Owing to friction, the tangential components can never have a direction of more than 25° from the horizontal.
4. Moments in the curved rib are calculated by multiplying the axial force or thrust T in a particular section of the rib by h, or the distance between the chord drawn between two adjacent blocking points and the neutral axis of the rib section, as illustrated in Fig. 10.16.
5. Since the arch approximates a curved, continuous beam, actual moments are somewhat less than $T \times h$, and reduction factors are given for the two-piece rib set as follows: (a) Maximum moment, which is normally located near the arch splice, may be reduced by using a multiplier of 0.86. (b) Moments at blocking points may be reduced by using a multiplier of 0.67. (This is the multiplier used to check moment in the straight portion, or leg, of the rib set.)

Circular Tunnel Supports.   Many analyses have been carried out for designs of permanent circular tunnel linings, Szechy[16] reviews about 10 such analyses, but the only one that considers the hinge effect of bolted segments is the Hewett-Johannesson method,[38] originally published in 1922. This analysis took into account the weight of the segments, the vertical and horizontal pressures acting on the tunnel due to soil load, and the effect of compressed air if it is driven as a pressure tunnel. The analysis assumed inner segments bolted rigidly in the longitudinal direction, and assumed as a tour-hinged circle (as in Fig. 10.17a) in the transverse direction. None of these analyses addressed the condition faced by the tunnel contractor installing sectional steel ribs in a circular tunnel in soil.

Contemporary rib-and-lagged primary tunnel supports used in conjunction with shield or boring-machine excavation more closely approximate the sections shown in Figs. 10.17b and c, and are not rigidly connected to each other in the longitudinal direction. Shortly after the ribs emerge from the shield, they are generally jacked apart and shimmed with spacers, or "dutchmen," at jacking points so as to minimize voids (and earth movement) at the crown as soon as possible. The use of timber lagging does not allow the grouting of voids outside the primary support until after the final (concrete) lining is placed. Thus some

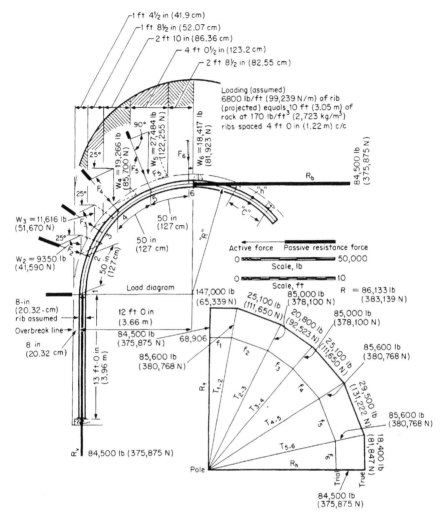

FIGURE 10.14 Proctor and White's steel rib loading and force diagram. (*Commercial Shearing, Inc.*)

complex nonuniform loads may act on the ribs and lagging during the excavation phase, especially in soils where large boulders or voids may be encountered. Since the wide-flange rib is of uniform section throughout, the worst-case moment conditions would be used to design the entire rib set.

The following *expedient analysis* assumes an uneven distribution of soil load (in this case, equal to the entire load as a function of tunnel diameter and soil pressure acting on only five-eighths of the length of the rib's chord distance) upon a four-piece rib set (Fig.10.18*a*) and a three-piece rib set (Fig. 10.18*b*). The direction of the load is assumed perpendicular to the chord of the rib arch. No movement of the rib sections or their hinged ends is assumed; nor is passive soil reaction assumed as a result of rib movements or deformations; and the rib configuration is still assumed as circular, in spite of the jacking

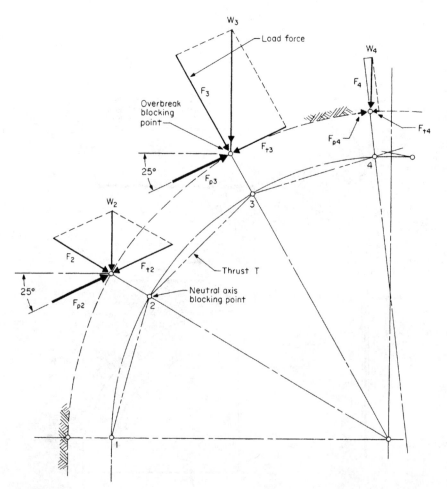

FIGURE 10.15 Proctor and White's resolution of forces at overbreak blocking points. (*Commercial Shearing, Inc*)

movement and the insertion of a spacer. The ribs are also considered slender enough so that moment and compression failure will occur before shear failure. The longitudinal force due to thrust is not considered in this analysis.

Leontovich's equations[39] for two-hinged, constant-section circular bridge arches subjected to uniform loads over only a portion of their span arc used in analysis. The total soil load is calculated by using the soil pressure multiplied by the full tunnel width (see Fig. 10.18). Each arch rib is then subjected to one-half the total load, shown as a uniform load acting on only five-eighths of the arch. An earth-pressure coefficient K is then used to reduce the load, since it is no longer acting in a vertical direction. This coefficient was derived from Coulomb's original earth-pressure theory (ca. 1776) used to reduce active soil loads acting on sloping retaining walls. (Figure 10.20 indicates values of K for various soil friction angles.)

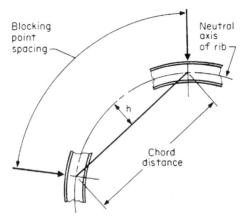

FIGURE 10.16 Chord distance h between two blocking points on Proctor and White's steel rib.

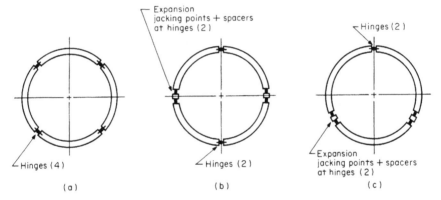

FIGURE 10.17 Three methods of bolting circular supports (a) Hewett-Johannesson's four-piece bolted segments. (b) Four-piece rib set with expansion jacking points at springline. (c) Three-piece rib set with expansion jacking points 30° below springline.

A loading diagram for each type of rib set, plus a moment and axial load plot for each rib (flattened for case of plotting), are shown in Fig.10.19.

A commentary is offered on the foregoing analysis. By inspection of Fig. 10.20, note that as the value of ϕ decreases, the values of K for the three-piece arch approach those of the four-piece arch—both being close to unity. Thus the maximum fiber stresses will be greater for the three-piece arch than for the four-piece arch, because the moment multiplier at maximum negative moment location from Fig. 10.19b is approximately one-third greater than from Fig.10.19a. This would appear to be incorrect, that a three-piece arch would be "weaker" than a four-piece arch, but remember the assumption of no transmission of moment across the bolted butt splices, that is, a true hinged connection. It should also be pointed out that as the number of hinged joints increases, the maximum moments in each segment will decrease, but the ability of the structure as a whole to retain its circularity will be diminished.

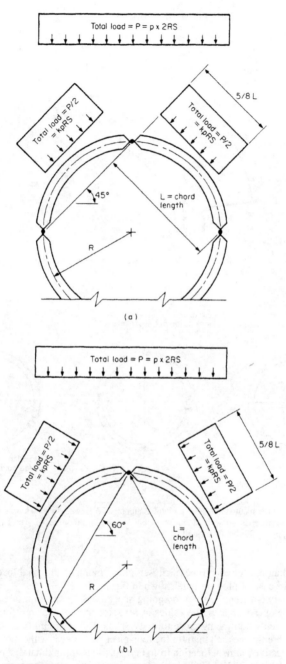

FIGURE 10.18 Assumed load distribution for four-piece and three-piece rib sets: (*a*) Four-piece rib set at 90° per piece. Total load *kpRS*, perpendicular to chord of arch piece and acting over five-eighths of chord length. (*b*) Three-piece rib set at 120° per piece. Total load *kpRS*, perpendicular to chord of arch piece and acting over five-eighths of chord length. (*P* = total load, *p* = soil pressure, *S* = rib spacing, *R* = tunnel radius, *K* = earth-pressure coefficient.)

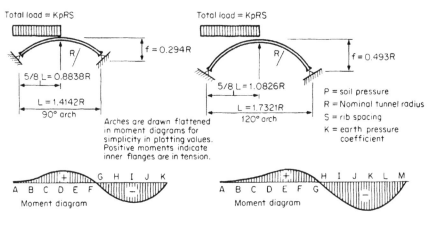

Location	Moment	Axial Compression	Location	Moment	Axial Compression
A	0	0.9557 KpRS	A	0	0.6378 KpRS
B	+0.0053 KpR²S	0.8636 KpRS	B	Negligible	0.7647 KpRS
C	+0.0139 KpR²S	0.7766 KpRS	C	+0.0046 KpR²S	0.6796 KpRS
D*	+0.0229 KpR²S	0.7252 KpRS	D	+0.0155 KpR²S	0.5940 KpRS
E	+0.0209 KpR²S	0.6667 KpRS	E*	+0.0233 KpR²S	0.5254 KpRS
F	+0.0079 KpR²S	0.6663 KpRS	F	+0.0212 KpR²S	0.4857 KpRS
G	-0.0147 KpR²S	0.7018 KpRS	G	+0.0097 KpR²S	0.4848 KpRS
H	-0.0369 KpR²S	0.7283 KpRS	H	-0.0178 KpR²S	0.5248 KpRS
I*	-0.0443 KpR²S	0.7338 KpRS	I	-0.0465 KpR²S	0.5624 KpRS
J	-0.0310 KpR²S	0.7212 KpRS	J*	-0.0599 KpR²S	0.5761 KpRS
K	0	0.6906 KpRS	K	-0.0564 KpR²S	0.5722 KpRS
			L	-0.0389 KpR²S	0.5510 KpRS
			M	0	0.5138 KpRS

*Denotes maximum moment.

(a)

*Denotes maximum moment.

(b)

FIGURE 10.19 Value of moment and axial compression for four-piece and three-piece arch rib: (*a*) values for four-piece arch rib; (*b*) values for three-piece arch rib.

INSTALLATION OF SUPPORTS

Liner Plates

In soft ground tunnels, the plates are generally installed close to the working face by the same crew that excavates the tunnel. The erection and bolting take place in a crowded and dirty environment using only the miners' muscles, aligning pins, and spud wrenches. At most, the only mechanical aid may be an occasional air-operated impact wrench. Flange holes are generally oversized to allow for alignment errors and the bolts and nuts have coarse, quick-acting threads, so they can be tightened even if coated with grime. The longitudinal joints or seams are alternately staggered to minimize planes of weakness in that direction. The plates are 37-11/16 in, or π ft (95.7 cm), in length; therefore, a 10-ft (3-m)-diameter tunnel lining will consist of 10 plates around its perimeter.

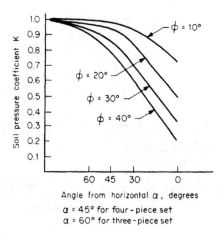

Angle from horizontal α , degrees

α = 45° for four - piece set

α = 60° for three - piece set

Soil friction angle φ, degrees	Loading 45° from horizontal (as in the case of a four-piece arch rib set)	Loading 60° from horizontal (as in the case of three-piece arch rib set)	Loading 0° from horizontal (horizontal load)*
10	1.00	0.94	0.71
20	0.94	0.84	0.49
30	0.87	0.72	0.33
40	0.80	0.60	0.22

*Values of the horizontal load factor compare closely with

$$k = \frac{1 - \sin \phi}{1 + \sin \phi}$$

also known as the active earth-pressure coefficient.

FIGURE 10.20   Soil-pressure coefficient K versus angle from horizontal α for selected values of soil friction angle ϕ.

   Figure 10.21 illustrates the installation of four-flanged liner plates on a California tunnel project. The second row of plates has a plate at springline containing a threaded grout hole fitted with a pipe plug. Plates should be backpacked with pea gravel and then grouted to limit deflection to within 3 percent of nominal tunnel diameter. A certain percentage of plates should therefore be purchased with grout fittings, Backpacking normally starts at the invert, so that the completed ring does not sag or deflect under loads from the mucking or haulage equipment. It is difficult to backpack the sidewalks and arch until enough liner plates are installed so that the gravel or grout does not run toward the "free" end of the lining at the working face. In stiff soils it is sometimes advantageous to pack and grout in stages as the tunnel advances, and straw or excelsior is sometimes used to pack the free end of the rings to minimize grout losses.

   Poor ground can sometimes be controlled by installing arch plates one ring ahead of the remaining plates and by maintaining a sloping work face during the advance. If conditions

FIGURE 10.21   Installation of pressed-steel liner plates. (*Commercial Shearing, Inc.*)

warrant further arch support, commercially manufactured hydraulic poling plates, as illustrated in Figs. 10.22 and 10.23, may be utilized to contain the crown until plates can be installed. Some contractors utilize job-fabricated devices to control the crown in a similar fashion.

The engineer on a tunnel project utilizing liner-plate supports is responsible for certain functions to ensure a safe and orderly advance as well as the efficient utilization of workers and materials.

1. The engineer should maintain an accurate inventory of liner plates delivered to the job-site and update orders to compensate for loss or damage to the plates. This is especially important for elliptically shaped tunnels or for tunnels with curves in their alignment,

FIGURE 10.22   Schematic diagram of a hydraulic cylinder shoring and policy plate forward from a liner-plate ring. (*Commercial Shearing Inc.*)

FIGURE 10.23 Using hydraulically actuated poling plates on a French tunnel project. (*Commercial Shearing Inc.*)

which will require liner plates of more than one size and shape. The engineer should also maintain an adequate supply of nuts and bolts to compensate for normal losses.

2. Plates should be plainly marked and staged on flat cars prior to being sent to the heading. This minimizes confusion and delays by the erection crew. The proper number of plates with grout plugs should be marked and supplied when needed. Plates with grout plugs are more expensive than regular plates and should be used according to the engineer's or superintendent's schedule, not the miners'.

3. Lasers or other alignment aids should be checked frequently and advanced with the heading.

4. Accurate measurements of horizontal and vertical inside dimensions of the installed plates should be taken and recorded. Measurements within a few diameters of the face should be taken daily, farther back, at less frequent intervals or as required by field conditions.

5. A record of each quadrant grouted, along with its tunnel station, should be maintained to ensure that proper backpacking is being carried out.

Steel Ribs in Rock

Steel ribs are usually installed by the heading crew after the shotrock (muck) has been removed from the heading. Stand-up time determines the length of the round to be shot and the maximum distance from the new working face to which the new rib sets should be erected. After scaling off loose rock, the miners prepare for rib erection by first digging to firm rock where the posts will stand. In order to avoid standing a lopsided set, foot blocks are installed to ensure that the foot plates at the bottom of both posts are at the same elevation. Figure 10.24 illustrates a typical two-piece rib set, with an optional invert strut, for a small-diameter tunnel; Fig. 10.25 illustrates three methods of foot blocking. Once the foot blocks are leveled up, the miners will stand the posts. In smaller tunnels with light-steel ribs,

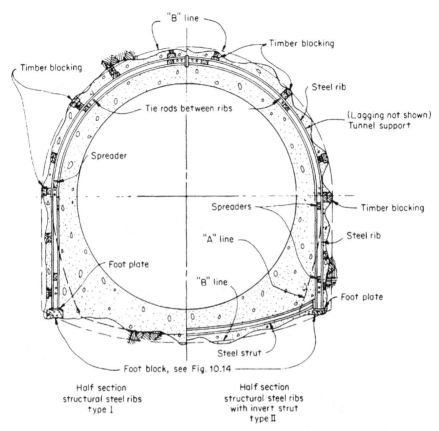

FIGURE 10.24 A typical two-piece rib set for a small-diameter rock tunnel. The right half of the drawing illustrates an optional invert strut, to be utilized when side loads require it. (*Bureau of Reclamation, U.S. Department of Interior.*)

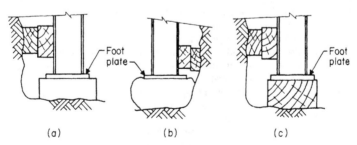

FIGURE 10.25 Three methods of foot blocking: (*a*) precast concrete; (*b*) sacked concrete; (c) timber. (*Bureau of Reclamation, U.S. Department of Interior.*)

the set is usually a two-piece unit with a single splice at the arch. If each half weighs less than 300 lb (136 kg), the miners can usually muscle the halves into place. For heavier sections, four-piece rib sets are commonly utilized, and the miners generally use the hydraulic booms on their drill jumbo, or a specialized erector to hold all the pieces up, until the splices can be bolted together. During this maneuver, it is important to insert the longitudinal tie-rods into their appropriate holes. In Fig. 10.26, Washington, D.C., miners are shown muscling an arch piece into position while inserting the tie-rod into its hole in the web.

FIGURE 10.26   Assembling a rib set in a Washington, D.C., subway tunnel. (*Commercial Shearing, Inc.*)

Tie-rods are a necessary part of a rib-support system because they minimize longitudinal movement. Tie-rods are used in conjunction with spreaders between the rib sets. Spreaders are generally made of cut lumber, but they may also be made of steel pipe (with the tie-rod inside the pipe) or of angle iron. Timber spreaders are generally the most economical, and they also have the advantage that they can be trimmed to fit at the heading when adjustments to rib alignment and plumbness are necessary. A disadvantage is that they are subject to rot, or that they create voids in concrete, if the tunnel is to have a secondary lining. For these reasons, some owners specify that all or some of the timber spreaders must be removed prior to placing the secondary lining. The cost of removing the timber may, at times, offset its initial economy.

After the erection and assembly of the set, it should be blocked firmly against the rock surface according to the design requirements. The miners should be provided with precut timber of random thicknesses and lengths and an adequate supply of hardwood wedges to accomplish the blocking. Lagging (timber or steel channels) may also be installed at this time, if required to support the rock in the area between the rib sets. In order to speed up the tunnel drive, a portion of the blocking and lagging may be completed at the rear of the jumbo during the drill cycle. This option, of course, depends on the stand-up time available to the tunneler. Repairs to rib sets damaged by previous blasts should be carried out at this time, as well as reblocking or rewedging of any sets that have been loosened by the blasting or mucking operations.

Figure 10.27 shows steel rib arches with tie-rods and timber spreaders between the ribs, and also shows steel channel lagging placed outside the rib flanges to contain loose rock between the ribs. Notice also the horizontal wall plate beams between the arch ribs and the vertical posts. Many large-diameter tunnels are driven "top-heading-and-bench"—that is, the top half is first driven horizontally and the lower half is later "benched" as in a conventional quarry operation. In this type of excavation it is sometimes difficult to mate a post exactly to its respective rib as the bench is removed. The wall plate in this case distributes arch loads evenly to the posts. Installed during the top heading drive, it also bridges the gap between bench excavation and post installation.

FIGURE 10.27   Steel ribs with wall plates and vertical posts in Chicago water tunnel. (*Commercial Shearing Inc.*)

In ground with poor arch rock or a very short stand-up time, it sometimes is necessary to cantilever crown bars or spilling ahead of the leading rib set to control the arch.

Figure 10.28 shows crown bars which are shoved forward as the rib sets are advanced along with the tunnel heading. Figure 10.29 shows spiling extended ahead from the leading rib set and later trimmed to provide the proper clearance for the secondary lining. Spiling may also be drilled and inserted above and ahead of the working face. Figure 10.30 shows spiling being installed in a rock tunnel.

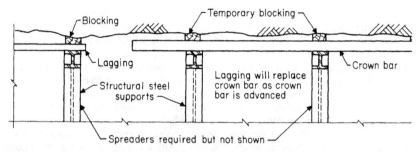

FIGURE 10.28 Longitudinal view of crown bars used in a steel-rib-supported tunnel. (*Bureau of Reclamation, U.S. Department of Interior.*)

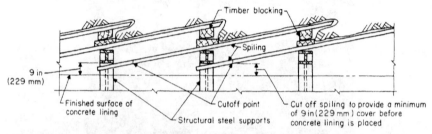

FIGURE 10.29 Longitudinal view of spiling used in a steel-rib-supported tunnel. In this case the owner requires that the tail end of the spiling be cut off prior to placing the secondary lining, to provide a minimum cover of 229 mm (9 in). (*Bureau of Reclamation, U.S. Department of Interior.*)

FIGURE 10.30 Spiling (forepoling) in a rock tunnel.

The dues of the tunnel engineer in a steel-rib-supported tunnel are generally the same as outlined for a liner-plate tunnel—that is, to provide adequate materials to the work crews, to assist them in proper tunnel alignment, and to monitor settlements or rib movements.

Steel Ribs in Soil

Steel ribs in shield-driven tunnels are generally installed at the rear, or "tail," of the shield by hand or with the aid of a mechanical erector. Figure 10.31 shows a mechanical erecting device

FIGURE 10.31   Main beam tunnel-boring machine. (*Robbins Compnay.*)

on a tunnel-boring machine to be used on a Washington, D.C., subway project. A worker is shown near the transporter, which picks the support (in this case a segmental liner plate that will be utilized as a permanent support) off a flatcar and moves it forward into the tail-shield area (flatcar not shown). A rotating erector can be seen in the tail shield. The erector picks up each segment and rotates it into position, where it is bolted together. Generally a full circle of timber lagging is installed between each rib set to counteract the thrust used to propel the shield forward. Since the diameter of the rib set is slightly smaller than the hole bored by the shield, the ribs must be jacked at selected splice points to expand the rib set so it is in intimate contact with the soil. Figure 10.32 shows a miner inserting a spacer at springline after completion of jacking. Figure 10.33 shows a completed circular tunnel. When curves are negotiated, care must be taken in cutting and installing the proper length of lagging in its particular position around the ring.

Optimally the contractor would prefer to install ring beams behind a shieldless tunnel-boring machine and avoid the inherent congestion near the work face, but stand-up time does not always allow this option. When ribs are installed close to the cutterhead, rib-erection equipment is generally supplied by the machine manufacturer as an integral part of the machine.

FIGURE 10.32 Expanded circular ribs on a California tunnel project. The hydraulic jack pushes out on reaction pads welded to the inside of the ribs. (*Commercial Shearing, Inc.*)

FIGURE 10.33 A circular rib-supported tunnel in Ohio. The four-piece rib sets are fully lagged and spacers are inserted at springline on both sides of the tunnel. (*Commercial Shearing, Inc.*)

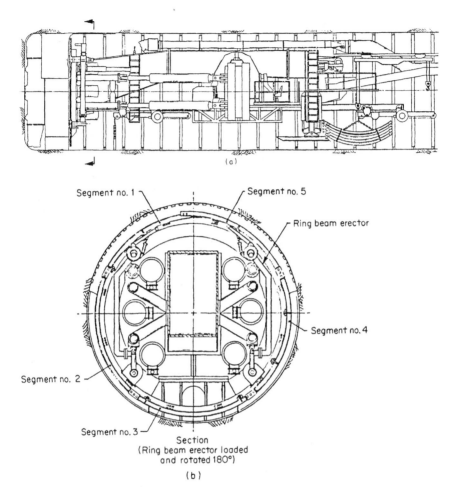

FIGURE 10.34 (*a*) Longitudinal section through a nonshielded tunnel-boring machine with steel-rib installation close behind the cutterhead. (*Robbins Company.*) (*b*) Section through the erector. (*Robbins Company.*)

Figure 10.34*a* shows a longitudinal section through a large-diameter tunneling machine. A hoist and monorail picks an unassembled rib set off a flatcar and transports it under the main body of the machine to a rotating erector located just behind the cutterhead. Figure 10.34*b* illustrates how the erector clamps onto the respective parts of the five-piece set and rotates them into position. A hydraulic expander, in this instance, is located at the crown splice.

TUNNEL SUPPORTS USING ROCKBOLTS

A rockbolt consists of a bar, usually a high-strength steel, with one end anchored to the rock at the bottom of a drill hole and the other end restrained by a bearing plate at the collar of the hole on the rock surface. The bolt's anchor may be of the mechanical type (i.e., a steel shell that is expanded and forced against the sides of the drill hole), or it may be of the epoxy (resin) type that glues the rockbolt to the sides of the drill hole. By tensioning the bolt, the rock surrounding the bolt hole is considered an *active* support. Inversely, if the bolt is not tensioned when it is installed, it will later become tensioned as a result of rock relaxation and movement toward the shaft or tunnel, and this type of bolt is considered a *passive* support.

Bolt Design.   Several approaches to rockbolt design are presented here. One approach uses the bolt-induced compression in the rock, which locks fractures and shears in the rock together and creates an arch of rock that protects the openings beneath it. Other approaches examine the weight of the rock mass, or of specific rock blocks within the mass, and then design a bolt pattern to resist the movement of the mass or the blocks. Regardless of the design approach, geotechnical engineers generally agree that a contractor must develop an installation and inspection program to ensure that the bolts and anchorages are capable of providing the design tension force. Installation programs include a hydraulically jacked pull test, or a torque-tension test, to check the integrity of the bolt and its anchorage. Inspection programs include selective testing of bolts installed in the past. If a regular inspection and corrective maintenance program cannot be carried out owing to operation constraints of the construction procedure (i.e., the bolts can no longer be reached or are covered over), then the contractor should grout the bolts after their initial testing with a cement grout or resin.

Flat Roofs in Laminated Rock.   This condition is commonly encountered in coal mining, where the rock overlying the coal seam is usually shale or sandstone. When the layer of rock adjacent to the roof of the opening is in a shale or other weak formation, it is usually bolted to an overlying layer of competent cock, such as sandstone. Obert and Duvall[40] consider reinforcement by suspension from a thick body of competent overlying rock (see Fig. 10.35*a*), where load per bolt *W* is

$$W = \frac{tL\gamma}{(n_1 + 1)(n_2 + 1)}$$

where n_1 = number of rows of bolts in a roof
n_2 = number of bolts per row
γ = unit weight of underlying rock seam, lb/ft^3 (kg/m^3)
t = thickness of underlying rock seam, ft (m)
B = width of opening, ft (m)
L = length of opening, ft (m)

This equation was based on a mine roof of infinite width and breadth, that is, with no sidewalls to carry part of the lower seam's weight. However, the authors also examined the same seam with side and end support, and found, for a thick, self-supporting upper seam, that the weight per bolt approaches that for the infinite seam condition; that is, the above equation is valid in both cases. To effectively transmit the bolt load from its anchor to the competent rock, the length of the bolt should at least equal the sum of the underlying rock thickness, plus one-half the bolt-spacing distance (see Fig. 10.35*b*). In most cases, bolt spacing should not exceed 5 or 6 ft (1.5 or 1.8 m).

The roof-support designer should check with the bolt manufacturers for their recommended drill-hole size, type of anchorage, and bolt-tensioning procedures. It is also important to ensure that the bolting crew is consistently developing the recommended bolt tension, and if not, to determine why not. (For example, the rock may be too soft for the type of mechanical anchor used, requiring a longer anchor to perhaps the use of resin anchorages.)

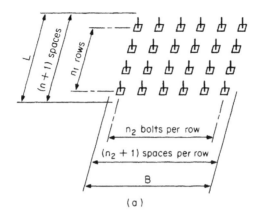

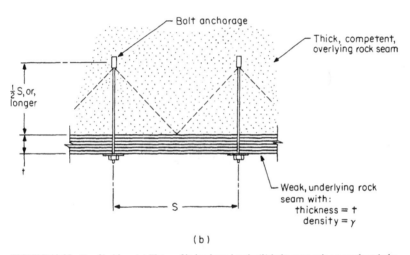

FIGURE 10.35 Roof bolting: (*a*) Flat roof in laminated rock; (*b*) bolts supporting a weak underlying rock in a flat roof.[40]

Panek and McCormick[41] give a rough estimate of the amount of torque required to develop tension in a bolt by the equation (in U.S. customary units).

$$\text{Tension (lb)} = C \times \text{torque (ft-lb)}$$

where C = 50 for 5/8-in bolts
　　　 = 40 for ¾-in bolts
　　　 = 30 for cone-neck, or self-centering, headed bolts
　　　 = 60 for using hardened-steel washers between bearing plate and bolt head of nut

The recommended tensions for various grades and diameters of bolts are given in Table 10.15.

TABLE 10.15 Tension Loads for Various Diameters and Grades of Steel Rockbolts

Bolt diameter, in	Stressed area at thread, in^2	Grade 30 bar yield load, lb*	Grade 55 bar yield load, lb*	Grade 75 bar yield load, lb*
5/8	0.226	—	12,400	17,000
¾	0.334	10,000	18,400	25,100
7/8	0.462	13,900	25,400	24,700
1	0.606	18,200	33,300	45,500
1¼	0.969	—	53,300	—
1-3/8	1.555	—	63,500	—
1½	1.405	—	77,300	—

*It is customary to use 60 percent of the yield loads as the design working load of the bolt (Panek and McCormick[41]).

Source: ASTM F432-76, copyright, ASTM 1916 Race Street, Philadelphia, Pa. Reprinted with permission.

Rockbolting Circular Openings.   This configuration is usually encountered in arched-roof, civil works construction.

Talobre's[42] approach utilizes a method of calculating radial stress around a circular opening to determine the load on an array of rockbolts. Figure 10.36a illustrates a somewhat circular opening in a rock formation that has relaxed and moved slightly toward the opening. Talobre assumes that a hydrostatic condition exists in the general area of the opening, that is, where horizontal and vertical components of the residual stresses are equal. The shaded area represents an arch or vault, of radially compressed, consolidated rock. In the general case, the length of the rockbolt will be equal to the thickness of the arch t plus the rockbolt spacing s. The radius to the arch midpoint r is equal to the radius of the tunnel a, plus one-half the length of the rockbolt, l, or

$$r = a + \frac{1}{2} = a + \frac{t+s}{2}$$

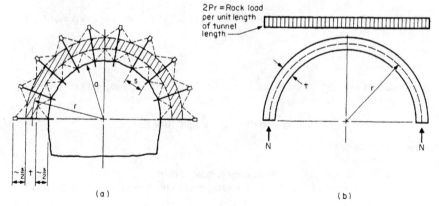

FIGURE 10.36 Support a circular opening in rock with steel rockbolts: (*a*) Typical rockbolt pattern in arch; (*b*) free-body diagram of the supporting arch.[42]

Figure 10.36*b* shows a free-body diagram of the top half of a 1 -ft- (30-cm)-long section of the arch. Thrust N at the springline would equal

$$\frac{P \times 2r}{2} = P \times r$$

where P is the rock pressure around the opening.

Talobre then assumes that the circumferential or tangential unit stress at springline equals thrust divided by the thickness of the arch, or

$$t = \frac{N}{t}$$

(For a fuller understanding of tangential and radial stresses around a hole, see the article "Stresses in Rock about Cavities" by K. Terzaghi and F. Richart, Jr., in the vol. 3, 1952–1953 issue of *Geotechnique,* and see the appendix to that article by Richart.)

Talobre further assumes that this tangential stress is the major stress in a Coulomb = Mohr's circle, where from inspection

$$\frac{\sigma_r}{\sigma_t} = \frac{\sigma_{min}}{\sigma_{max}} = \frac{1 - \sin\phi}{1 + \sin\phi} = \tan^2\left(45 - \frac{\phi}{2}\right)$$

where ϕ = angle of friction of rock
σ_r = radial stress in arch
σ_t = tangential stress in arch

He then slates that the load on the rockbolt is equal to the area by each bolt, multiplied by the radial stress. The bolt is then tensioned to twice the amount for a safety factor of 2. The approximate friction angles for various types of rock are shown in Table 10.16.[43] Since the hydrostatic condition assumes that the rock around the

TABLE 10.16   Friction Angle ϕ for Various Rocks

Rock	$\phi°$ (intact rock)	$\phi°$ (residual)
Andesite	45	28–30
Basalt	48–50	
Chalk		
Diorite	53–55	
Granite	50–65	31–33
Greywacke	45–50	
Limestone	30–60	33–37
Monzonite	48–65	28–32
Porphyry		30–34
Quartzite	64	26–34
Sandstone	45–50	25–34
Schist	26–70	
Shale	45–64	27–32
Siltstone	50	
Slate	45–60	25–34

Source: E. Hoek, "Estimating the Stability of Slopes in Opencast Mines," *Trans. Institution of Mining & Metallurgy,* London, 1970.

opening has moved slightly toward the opening, the residual values of ϕ should be used in the calculations.

Jaeger[44] states that rockbolt systems designed to cause average pressures on the surface of the rock cavity of between 10 and 20 lb/in² (7 and 14 N/cm²) are usually sufficient for a sustaining arch. Cording and Maher[45] point out where support pressures as high as 20 lb/in² were required to stabilize a 120-ft- (86-m)-high cavern sidewall in rock with an excellent RQD, but also show where rock pressures of only t lb/in² were required in 20-ft- (6.1-m)-wide openings in rock with a poorer RQD, and suggest that support pressures may be related to the size of the rock opening. A survey of a few underground caves in North America would reveal that the average compression in the three major underground arches at the Churchill Falls Hydro Project[46] was specified at 12.5 lb/in² (8.62 N/cm²); the three chamber arches at the NORAD Expansion Project[47] were compressed to an average of 8.5 lb/in² (5.9 N/cm²); and a subway arch constructed in a major U.S. city is compressed to about 8 lb/in² (5.5 N/cm²). In our sample problem we compressed 16 ft² (1.5 m²) to 17,856 lb (79.420 N), resulting in an average pressure of 7.75 lb/in² (5.34 N/cm²).

Bolting across Joints.   Haas [48] ran a series of tests on rockbolts acting across a shear plane and found that rockbolts are most effective across a shear plane when they are positioned so that they tend to elongate as the shear plane is subject to movement, least effective when shear plane movement tends to compress the bolts axially. In order to avoid trouble, the safest action that a tunneler should take in blocky ground is to try to avoid subjecting rockbolts to shearing action and to position the bolts so they do their work in tension. If this cannot be avoided, an analysis of each block and how it acts in relation to the rock opening should be carried out by the tunneler or geotechnical consultant.

Figure 10.37 illustrates our rock arch again but this time considers the effect of a wedge of rock in the crown of the opening. The wedge of rock is estimated to be 13 ft (4 m) high by 12 ft (3.6 m) wide with a weight per linear foot of tunnel equal to

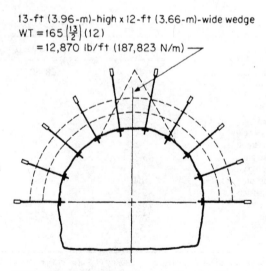

13-ft (3.96-m)-high x 12-ft (3.66-m)-wide wedge

$$WT = 165 \left(\frac{13}{2}\right) (12)$$
$$= 12,870 \text{ lb/ft } (187,823 \text{ N/m})$$

FIGURE 10.37   Rockbolted arch in blocky ground subjected to a wedge failure in the crown.

12,870 lb/ft (19,150 kg/m), acting in such a manner as to subject the bolts in the crown to simple tension. The weight of the wedge per 4-ft (1.22-m) bolt spacing would then be 12,870 × 4 = 51,480 lb (23,350 kg). By inspection, if the two bolts in the crown of the arch were lengthened [to approximately 14 ft (4.3 m) in this case], they would be able to hold up a 51,480-lb (23,350-kg) wedge prior to failure, because the yield strength of each bolt was 33,300 lb (148,120 N).

Grouted Rockbolts. Bolts may be grouted after tensioning either by Portland cement grout or, in the case of resin bolts, by using a fast-setting resin in the anchoring portion of the drill hole and a slower-setting resin in the remainder of the hole to allow time for tensioning the bolt in the interim between each resin set. In examining the advantages of a fully grouted rockbolt system versus the same system without grout, we have arbitrarily chosen an 8-ft- (2.44-m)-long bolt [less 1 ft (30 cm) for anchorage length] traversing a joint in the rock, which is dilating owing to ground movement after bolt installation (see Fig. 10.38).

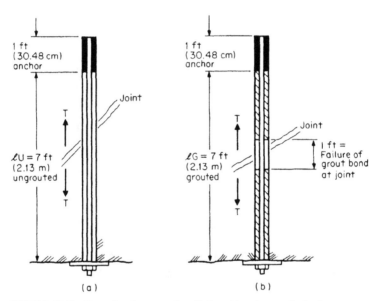

FIGURE 10.38 Grouted and ungrouted rockbolts subjected to tensile loading across a joint: (*a*) type U (grouted); (*b*) type G (grouted).

It can be seen that the extension of the ungrouted bolt will be 7 times that of the grouted bolt for the same force *T*. Thus many geotechnical engineers consider rockbolt grouting a way to limit small rock movements after bolt installation, movement that can slowly lead to an unstable condition in a shaft or tunnel.

Alternative active support devices that eliminate the need for grouting are the Split Set (trademark of Ingersoll Rand Co.) friction rock stabilizer (see Fig. 10.39) and Swellex bolts (trademark of Atlas Copco) (see Fig. 10.40). Both methods are particularly well suited for TBM and NATM where bolts are generally installed immediately behind the tunnel face.

This Split Set is a longitudinally split steel tube that is forced into a predrilled hole in the rock. The driving force, usually supplied by the rock drill used to drill the hole, loads the rock in compression. In addition to hard-rock support, it has been utilized successfully

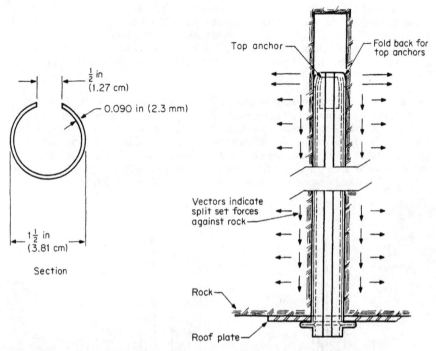

FIGURE 10.39   Split Set friction rock stabilizer.

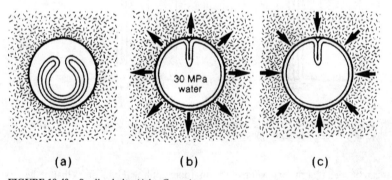

FIGURE 10.40   Swellex bolts. (*Atlas Copco.*)

in rock too soft for mechanical anchors and in certain shales that cannot be reliably grouted with cement or resin. According to Scott,[49] testing has shown that it can develop an anchoring force of from ¾ to 2 tons/ft (22 to 58 kN/m) of borehole. An additional advantage is that if it is overloaded, it will slip slightly and still adhere to the borehole after the rock has completed its movement. Mechanical anchors, on the other hand, usually break or lower their load-carrying capacity once they are overloaded.

Passive Supports Related to Rockbolting. Rockbolts require tension to do their work. Passive supports, on the other hand, are installed untensioned and do their work by taking up tension as the rock relaxes and the joints dilate. Two common types of passive supports are

1. Perfo bolts (see Fig.10.41*a*) are steel reinforcing bars grouted to the rock. The grout is inserted and held in the drill hole by a perforated sleeve until the reinforcing bar is inserted. For the most part, the perfo bolt has been supplanted by the untensioned rebar resin spiling.

2. Untensioned rebar resin spiles (see Fig. 10.41*b*) are steel reinforcing bars inserted in drill holes and are completely encapsulated with resin. The bars are in many cases driven in a fan-shaped pattern, pointing ahead of the tunnel or shaft working face. Manufacturers will recommend compatible drill-hole and bar sizes in order to install them efficiently by minimizing the amount of resin required.

When using passive devices of this nature, they should be installed as close to the working face as possible, and as soon as possible, to minimize initial joint dilation. Some geotechnical engineers claim that these devices, when installed at the working face as fast as possible, can be as effective as a tension-bolt system installed many hours after the ground is opened up. On many civil works projects, a combination of nontensioned bolting at the face, followed by tensioned bolting within one or two diameters of the face, is used effectively to limit ground movement.

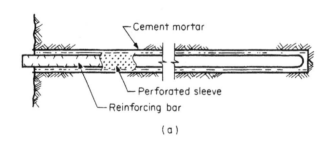

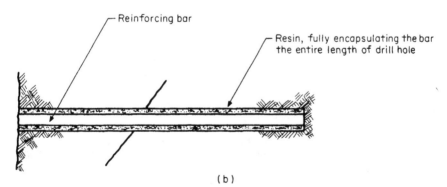

FIGURE 10.41 Passive support devices related to rockbolt: (*a*) Perfo bolt; (*b*) Untensioned rebar resin spile.

Supplemental Materials

1. Bearing plates are used to transmit the reaction of the bolt's tension to the rock surface. As the rock becomes weaker, larger plates are used. Bearing plates should be thick enough to minimize "dishing," or deforming to a concave configuration, when the bolt is fully tensioned.

2. Roof mates, sometimes called "pans" or rock straps, are made from 14- or 16-gauge sheet metal and are utilized to support weaker rock between the bolts.

3. Shotcrete is used on many occasions to support rock surfaces in between the bolt pattern. It can be utilized as plain shotcrete or can be reinforced with welded wire mesh.

4. Chain-link fabric is utilized, not as supplemental support, but to protect workers from loose pieces of rock that may fall from the arch.

Installation of Rock or Resin Bolts, Spiles, Friction Rock Stabilizers.   The first step the miner must take is to thoroughly scale loose rock from the new surface. In addition to providing a safe working environment for the driller, the scaling also uncovers sound rock with a surface flat enough to collar a drill hole, and later, flat enough to provide a good solid surface for the bearing plate.

Drilling a straight hole with the proper diameter at the far end (anchorage end) of the hole is important to the speed of the operation and the ability of the anchor to seat according to the manufacturer's recommendation. If handheld jackleg drills are used, the miner usually is equipped with a set of drill steels in 2-ft (60-cm) increments. For example, an 8-ft- (2.4-m)-deep hole would require a 2-ft (60-cm) starter steel to allow the miner good stability and leverage while collaring the hole, followed by longer drill steels, each with its own bits, to deepen the hole to 8 ft (2.4 m). The longest drill steel must have a bit diameter equal to the bolt manufacturer's recommended diameter, and the shorter drill steels must therefore have bit diameters slightly larger to allow clearance for the final bit. If, for example, three drill-size changes are required for the miner to arrive at a final hole diameter of 1-5/8 in (41 mm), the first or shortest steel would have a bit diameter of 1¾ in (45 mm), the second steel would have a bit diameter of 1-11/16 in (43 mm), and the final, or longest, drill steel would have the 1-5/8-in (41-mm) bit. It is therefore important to have the appropriate drill steels on hand, clearly marked to avoid delay.

The contractor may choose to utilize a drill jumbo to drill the bolt holes. The jumbo is generally equipped with power-fed drills having a higher energy output than jackleg drills, and it can usually drill holes 8 to 12 ft (2.4 to 3.6 m) long without a change of drill steel. The power feed is usually 2 to 4 ft (0.6 to 1.2 m) longer than the length of hole it can drill in a single pass; therefore, a power-fed drill designed to drill an 8-ft (2.4 m) hole in a single pass would have an overall length of 10 to 12 ft (3 to 3.6 m). Thus we can see that the 8-ft- (2.4 m)-long bolt in our sample problem would require a jumbo drill that would just barely fit in our tunnel with a 12-ft (3.6-m) diameter. A platform would also be required for the miner to torque or pull-test the bolts in the upper portion of the arch.

Figure 10.42 shows a miner driving a Split Set into the hole he has just drilled with his handheld jackleg drill. Note how the blocky rock has spalled from under the bearing plates on the left side of the photo. Ungrouted rockbolts with mechanical anchors would be ineffective if the bearing plates were to become loose, whereas Split Sets or grouted bolts would still be effective in this situation. Note also how the chain-link fabric keeps loose rock from falling from the arch.

Figure 10.43 shows a drill jumbo drilling and installing Split Sets, while Fig. 10.44 shows a rockbolting drill mounted on a tunnel-boring machine. Mechanized rockbolting devices have become more commonplace in recent years as the demand has increased for

FIGURE10.42 Installation of Split Sets with a hand-held jackleg drill in blocky ground. (*J. J. Scott and Ingersoll-Rand Co.*)

FIGURE 10.43 A specialized jumbo for drilling (left side) and installing (right side) Split Sets. (*J. J. Scott and Ingersoll-Rand Co.*)

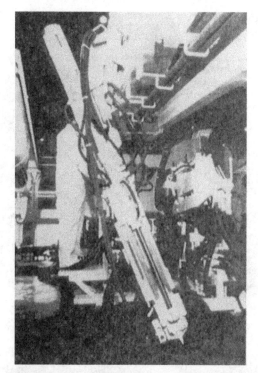

FIGURE 10.44   Rockbolter on a boring machine. (*Robbins Company.*)

safer and more efficient ground-support methods in high-speed tunneling projects as well as in coal, uranium, and other energy-related mining projects.

Resin bolting has become very commonplace in civil works construction since the early 1970s. Prepackaged resin cartridges can be stored at the jobsite for relatively long periods of time until they are required at the heading. This convenience, as compared with the older method of preparing Portland cement grout mixtures at the heading when required, has added to the popularity of resin grouts. Figure 10.45 shows the steps necessary to install and grout bolts using prepackaged resin cartridges. Resin manufacturers provide assistance in determining the proper amount of cartridges to fully encapsulate the bolt with grout. Care must be taken in fissured or seamy rock to insert extra resin to allow for losses to the fissures. Figure 10.46 shows how a hydraulic jack equipped with a center hole to accommodate the bolt, or a bolt-extension coupler, is used to provide the proper tension. Torque wrenches can tension the bolt faster than the jacking device, but a certain percentage of torqued bolts should be checked with the hydraulic ram to ensure proper quality control.

Considerations for Construction Expedience in Rockbolting

1. Does the rockbolt length exceed the diameter of the shaft or tunnel, requiring a coupled connection (for both the rockbolt and the drill steel used to drill the hole)? Does the diameter of either coupling require a drill-hole diameter larger than the manufacturer's recommended hole size at the anchor? This will require one to step down the size of the drill hole.

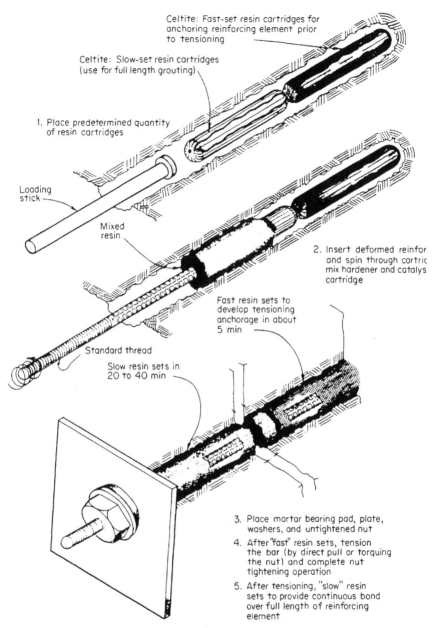

Celtite: Fast-set resin cartridges for anchoring reinforcing element prior to tensioning

Celtite: Slow-set resin cartridges (use for full length grouting)

1. Place predetermined quantity of resin cartridges

Loading stick

Mixed resin

2. Insert deformed reinfor and spin through cartric mix hardener and catalys cartridge

Fast resin sets to develop tensioning anchorage in about 5 min

Standard thread

Slow resin sets in 20 to 40 min

3. Place mortar bearing pad, plate, washers, and untightened nut

4. After "fast" resin sets, tension the bar (by direct pull or torquing the nut) and complete nut tightening operation

5. After tensioning, "slow" resin sets to provide continuous bond over full length of reinforcing element

FIGURE 10.45 Step-by-step method of installing resin-grouted, tension rockbolts. For vertical bolting, the manufacturer will supply parachute-shaped keepers to hold the resin cartridges in position until a bolt is inserted in the drill hole. (*Celtite, Inc.*)

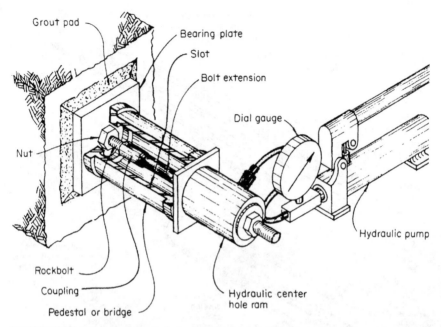

Grout pad

Bearing plate

Slot

Bolt extension

Dial gauge

Nut

Hydraulic pump

Rockbolt

Coupling

Pedestal or bridge

Hydraulic center hole ram

FIGURE 10.46   A portable hydraulic device for tensioning bolts. (*Celtite, Inc.*)

2. Can workers reach the bolts to torque them and later to test them (also by torquing)? A worker on a ladder can effectively apply only 75 to 100 ft · lb (100 to 135 m · N) of torque to a rockbolt. Two workers on a sturdy platform can attain a maximum torque of 400 to 500 ft · lb (540 to 675 m · N) with off-the-shelf torque wrenches. Air-operated impact wrenches tend to "wind and unwind" very long bolts, without transmitting the proper torque to the anchor, thus requiring the workers to use either a hand-operated wrench or resin bolts with the thread on the "outside end" of the bolt.

3. Can changing the bolt-spacing length of required tension eliminate, or at least minimize, any of the aforementioned problems? Will the change require more rockbolts, thus off-setting the cost advantage? Will the change allow one to assume drilling and blasting sooner, or will it tie up the working face for a longer period of time?

TUNNEL SUPPORTS USING SHOTCRETE

Shotcrete design is not much different from concrete design. The engineer in both cases assumes that through reasonable quality control in the selection, batching, and placement of materials, it is possible to ensure a reliable design strength on which to base calculations. The shotcrete designer, however, can vary the early strength of the mix and can use this early strength to advantage in dealing with progressive rock deterioration in the tunnel or shaft. The authors will briefly discuss methods of ensuring reasonable early and final design strengths of a shotcrete mix, and then focus on a few contemporary approaches in using shotcrete to support an underground opening in rock.

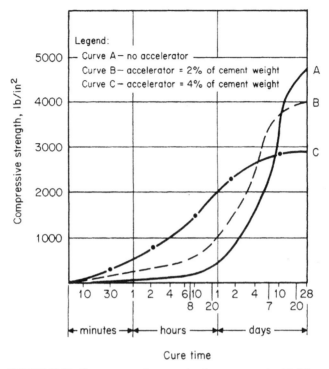

FIGURE 10.47 Shotcrete strength vs. cure time for a common mix with differing quantities of accelerator added.

Mix Design. Prior to reviewing shotcrete design techniques, it is important to examine the strength-versus-time characteristics of coarse aggregate shotcrete mixes using commercial, fast-setting agents, commonly referred to as accelerators. Figure 10.47 shows strength versus time for three such mixes, with the proportion of each of the three mixes being the same, except for the amount of accelerator introduced to the mixes, expressed as a percentage by weight of the cement in the mix.

Note that for a constant cement content, the best 28-day strength is attained by using no accelerator in the mix. This, however, limits the early strength (4 to 8 h) of the in-place shotcrete, which the tunneler usually needs to prevent rock loosening in the blocky ground. Note also that a large percentage (4 percent in our case) of accelerator may initially yield a high early strength, but at a sacrifice of 28-day strength.

A suggested mix, which is readily obtainable under normal field conditions, would have the following characteristics:

1. Aggregate ratio
   a. Sand [¼ in (6 mm) minus]: 60 percent by weight
   b. Coarse aggregate [¾ in 19 maximum size]: 40 percent by weight
2. Cement and accelerator
   a. Cement: 650 to 700 lb/yd³ (385 to 415 kg/m³)
   b. Accelerator: 1 to 2 percent by weight of cement quantity
3. Water
   a. Water-cement ratio of 0.38 to 0.45 by weight

4. Design properties

    a. Compressive strength at 8 h = 500 lb/in^2 (345 N/cm^2)

    b. Compressive strength of 7 days = 2500 lb/in^2 (1725 N/cm^2)

    c. Compressive strength of 28 days = 4000 lb/in^2 (2750 N/cm^2)

It should be mentioned that regulated set (reg-set) cements are being developed that promise to yield much higher early strengths with very little sacrifice of final strength. But for the most part, contractors are still using commercially produced Portland cements and accelerators at this writing.

Design of Shotcrete Support Systems.   Shotcrete can provide primary support in a rock tunnel or shaft, by itself or in conjunction with rockbolts or steel ribs. In many cases, it provides both the primary and permanent support to an underground opening. Rabcewicz[50,51] examined the hydrostatic forces acting on a shotcrete support opening according to the elastoplastic stress distribution of Kastner.[52] He then used Fenner's[53] and Talobre's equations to express the following relationship:

$$p_i = -C \cot\phi + [C \cot\phi + p_0(1-\sin\phi)]\left(\frac{r}{R}\right)^{(2\sin\phi)/(1-\sin\phi)}$$

where p_i = required lining resistance

    C = cohesion

    ϕ = angle of friction

    p_0 = earth pressure, or γH, where H is overburden

    r = radius of cavity

    R = radius of protective zone

Assuming negligible cohesion, the equation is simplified to

$$p_i = p_0(1-\sin\phi)\left(\frac{r}{R}\right)^{(2\sin\phi)/(1-\sin\phi)} = p_0(n)$$

where n is shown as a function of ϕ in Fig. 10.48

When the radius of the protective zone equals the radius of the opening,

$$p_i = p_v(1-\sin\phi)$$

and the cavity attains equilibrium without deformation. Examination of Fig. 10.48 indicates that using $r/R = 1.0$ would give the user the maximum p, or the thickness lining, while r/R values lower than 1.0 would give low support requirements. The fact that one can calculate support requirements without regard to the size of the opening in rock has led Hopper et al.[54] to introduce another term, $\gamma(R - r)$, to the Fenner-Talobre equation as follows:

$$p_i = p_0(1-\sin\phi)\left(\frac{r}{R}\right)^{(2\sin\phi)/(1-\sin\phi)} + \gamma(R-r)$$

where γ = unit weight of the ground, and all other terms remain the same.

By inspection, it can be seen that as the ratio r/R decreases, the value p decreases to some minimum value, then increases at a rapid rate. Some designers feel that if R is more

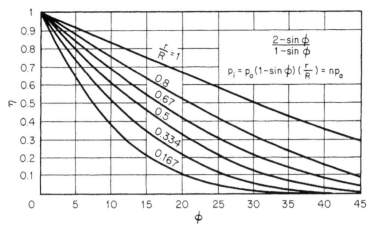

FIGURE 10.48 Skin resistance p_i required to establish equilibrium of a cavity as a function of ϕ, angle of internal friction, and $p_0 = \gamma H$. (*From L. V. Rabcewicz, Water Power, November/ December 1964 and January 1965.*)

than twice the value of r, the value of p may be overly conservative. This equation was utilized to explain the differences in support requirements between the small pilot drift for a large-diameter highway tunnel and the main tunnel itself. Without the $\gamma(R - r)$ factor, the value of p would have been calculated as the same for either tunnel size. The authors claim that the value for R was compatible with the extensometer readings of the pilot bore, but in no way imply that this will always be the case. Mahar et al.[55] use the calculated value of p to determine the thickness t of the support system by the equation

$$t = \frac{p_i \times r}{f'_c \times FS}$$

where t = thickness of support system (shotcrete in our case)
      p_i = calculated lining resistive pressure
      f'_c = compressive strength shotcrete
      FS = factor of safety (of 2 to 3)

Subsequent studies by the University of Illinois revealed the need for a more conservative design; thus the equation should be modified to

$$t = \frac{2p_i \times r}{f'_c} \times FS = \frac{4p_i \times r}{f'_c}$$

where the factor of safety is 2.

Deere et al. examined the progressive failure of rock supported solely by shotcrete, owing to fallout of key rock blocks, after the shotcrete has been installed and has reached partial strength. They suggest that the diagonal tension properties of the shotcrete multiplied by the perimeter of the rock block at the intersection with the shotcrete surface should equal or exceed the weight of the block. The progressive steps to failure are shown in Fig. 10.49. It should be pointed out that shotcrete can sometimes hold up far more weight than the diagonal tensile strength at the rock face would indicate through design computations, because in many cases the geometry of the rock blocks can key in a much greater

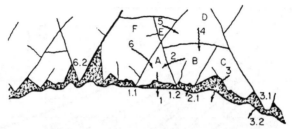

Step 1. Block A drops down, shearing through shotcrete along 1.1 and 1.2 and at each end of the block.

Step 2. Block B rotates counterclockwise and drops out, falling shotcrete in tension at 2.1.

Step 3. Block C rotates counterclockwise and drops out breaking rock-shotcrete bond at 3.1.

Step 4. Block D drops out followed by block E.

Step 5. Block F rotates clockwise and drops out, breaking poor bond between shotcrete and clay that was along weathered joints at 6.1 and 6.2.

FIGURE 10.49   Progressive failure in shotcrete-supported rock.

weight. Of course, if the rock-block geometry is of poor structural configuration, shotcrete may not be effective.

Mahar et al. suggest the following rules of thumb for using a thin membrane of shotcrete as the sole support in loosening ground: (1) the diameter of the opening should be less than 9 m (30 ft); (2) rock joints should be rough and clean; and (3) rock blocks should be less than 1.5 m (5 ft) in size, with no large wedges apparent.

We have isolated one such block and show it in Fig. 10.50 as trying to rotate counterclockwise into a tunnel around an axis *A-B*. The only resistive force is provided by a

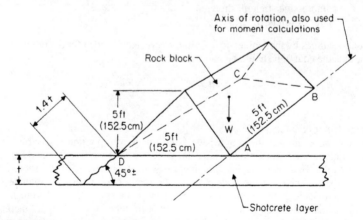

FIGURE 10.50   Shotcrete layer preventing a block from falling out due to rotation around axis *A-B*, based on Cecil's[59] theory.

layer of shotcrete with thickness t and a certain diagonal tension strength f_v (commonly referred to as "shear strength"), acting at 45° to the plane of the shotcrete over a diagonal thickness of $1.41\ t$. Deere uses $fv = 4\sqrt{f'c}$ as the diagonal tension strength of the concrete or shotcrete.

Heuer[56] examines the long-term requirement of shotcrete to support the slowly increasing swelling-and-squeezing pressure on a tunnel lining. He suggests that

$$t_e = \frac{LF}{0.85\phi}\left(\frac{p_i r}{f'_c}\right)$$

where $\quad t_e$ = effective thickness of shotcrete lining

$LF/0.85\ \phi$ = load-reduction and capacity-reduction factor for which a value of 2 is recommended for temporary shotcrete support and 2.5 to 3.0 for a final lining

r = radius of tunnel

p_i = average radial pressure

f'_c = normal unconfined compressive strength of shotcrete

The effective thickness t_v is somewhat less than the nominal thickness of the applied coating of shotcrete. Heuer suggests that t_c is 2 in (5 cm) less than the nominal thickness for machine-bored tunnels, or 4 in (10 cm) less in the case of drilled and blasted tunnels.

Therefore, for temporary shotcrete support in a drilled and blasted tunnel, with slow swelling pressures anticipated, the nominal amount of shotcrete required would be

$$t_{\text{nominal}} = 2\frac{p_r}{f'_c} + 4 \text{ in } (10 \text{ cm})$$

Fernandez-Delgado et al.[60] found, through both field testing and large-scale laboratory studies, that the bond between the shotcrete and rock is quite significant in determining its effectiveness as a support in tunnels with rock blocks that may in time loosen and bear against the lining. In rock with poor adhesion characteristics, additional shotcrete may have no significant effect after a certain thickness is reached.

For the most part, shotcrete design is based on empirical data obtained from previous successes (or failures) in the area of shotcrete-supported tunnels. Here we again refer to Lauffer's chart (Fig. 10.51) used by Linder[57] to compare rock quality types A and G with the shotcrete thickness required to support a tunnel alone or in conjunction with some other support device. However, Muller[58] has concerns about the use of shotcrete based on empirical data only.

So, we can choose from a variety of solutions for our requirement to temporarily support a tunnel opening solely by shotcrete. It appears that 3 or 4 in (7.6 or 10 cm) of shotcrete would be sufficient for early protection of the tunnel from deterioration due to progressive block failure, while 6 in (15 cm) of shotcrete would provide more long-term protection from moderate squeezing and swelling. Lauffer's short for condition F/E would recommend approximately 6 in (15 cm) of shotcrete, plus marginal use of rockbolts.

Empirical data from nearby projects could also provide significantly more accurate design basis than this approach. In many civil works tunnels, rock chambers, or shaft projects, the owner or engineer will specify the type of shotcrete support required, with particular emphasis on early and 28-day strengths and the lining thickness. Temporary bolts or steel ribs may also be specified to be installed in conjunction with the shotcrete.

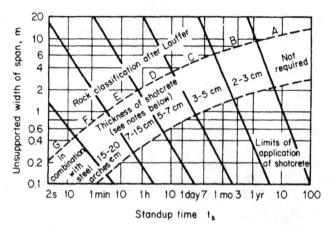

Standup time t_s

NOTES:

B. Alternatively, rockbolts on 1.5- to 2.0-m spacing with wire net; occasionally reinforcement needed only in arch.

C. Alternatively, rockbolts on 1.0- to 1.5-m spacing with wire net; occasionally reinforcement needed only in arch.

D. Shotcrete with wire net; alternatively, rockbolts on 0.7 to 1.0-m spacing with wire net and 3-cm shotcrete.

E. Shotcrete with wire net; rockbolts on 0.5- to 1.2-m spacing, with 3- to 5-cm shotcrete sometimes suitable; alternatively, steel arches with lagging.

F. Shotcrete with wire net and steel arches; alternatively, strutted steel arches with lagging and subsequent shotcrete.

G. Shotcrete and strutted steel arches with lagging.

FIGURE 10.51 Lauffer's chart according to Linden.[57]

Wire Reinforcement. Wire mesh greatly assists in providing the tensile strength necessary in blocky ground that the shotcrete cannot provide on its own. Opponents of using the wire mesh would argue that to expect any shotcrete (whether reinforced or not) to be reasonably effective, we must have an intimate bond between the shotcrete and the rock surface, and that the wire mesh reduces this bond by causing voids in the shotcrete behind the mesh. An uneven rock surface could conceivably create voids between the wire and rock, and even greater voids would result if the field crew were only a little bit careless in the mesh installation and the shotcrete application. The opponents would also argue that in order to be effective, shotcrete should be applied as soon as possible after the ground is opened up. The time required to attach the mesh closely to the rock surface (usually by using many short rockbolts) would detract from the shotcrete's effectiveness, plus exposing the workers to unnecessary risk while installing wire mesh.

Some designers arrive at a halfway point in this argument by first installing a coat of nonreinforced shotcrete immediately after opening up the rock, then installing wire mesh and a second coat of shotcrete. This provides a safer working environment for the meshing crew and a smoother surface upon which to install the mesh. This method has been successfully utilized on many projects.

A recent development is to introduce steel wires, or "fibers," into the shotcrete mix to increase the strength of the coating. Although there are still some field problems associated with plugging of the mixing and application machinery and excessive rebound of the fibers during application, some dramatic increases have been observed in the strength of the coating. Alberts,[61] using 1.3 to 1.4 percent (by volume of mix), reports a 50 percent increase in tensile strength and 180 percent increase in flexural strength as compared with a plain shotcrete mix.

Further testing by Moran[62] with shotcrete mixes containing 0.5 to 1.5 percent fibers by volume reveals that although the compressive strength may not increase substantially over nonfiber mixes, the toughness index, or the ability of the fiber mix to carry a load after cracking and deformation, is dramatically increased over the toughness of a nonfiber mix.

Field Practice. Shotcrete may be applied by the wet or dry process. In the wet process, the mix is delivered as a low-slump concrete and is generally pumped to a nozzle. At the nozzle, compressed air is injected into the stream of flowing concrete and propels it toward the rock surface. Wet shotcrete may also be applied by loading the mix into a sealable chamber and then forcing the mix through the hose and nozzle by pressurizing the chamber with compressed air. In the wet process, the accelerator will be in liquid form and will normally be introduced to the mix at the nozzle.

The dry process generally utilized a mixture of dry cement and sand and aggregate, which is introduced into rotary guns. The guns propel the damp mix through a hose to the nozzle where the proper amount of water is introduced into the stream of material. Accelerator may be introduced to the mix either as a liquid at the nozzle or as a dry powder as the mix is loaded into the hopper of the rotary gun. The nozzleman is the key to an efficient and high-quality shotcrete program. Some owners require that the contractor's nozzleman must have creditable prior experience on similar shotcrete projects and may also require a qualified backup nozzleman.

Figure 10.52 illustrates a nozzleman applying a wet shotcrete coating to a rock surface. Prior to shotcreting, he should clean the rock surface of all dust and dirt from blasting operations to ensure a good bond between the rock and the coating. He is shown positioning the nozzle so that the stream of shotcrete hits the rock in a perpendicular direction. He also positions the nozzle so that it is 3 to 5 ft (0.9 to 1.5 m) from the rock surface. Some contractors mount the nozzle on a remote-controlled boom equipped with a swivel head. In this case, the nozzleman would operate the boom and water controls from a position at the rear of the nozzle, where he would be exposed to less shotcrete rebound and possible rockfalls. The contractor should provide the nozzleman and other members of the shotcrete crew with face-mask filters, eye shields, and protective skin ointment to protect them from the caustic effects of the cement and accelerator.

As in any concrete operation, care in supplying the sand, aggregate, cement, and additives under strict quality-control standards as well as good placing procedures will result in a high-quality shotcrete. A cautionary note: Deficient design or field application can prove more dangerous than no shotcrete at all, because the coating will cosmetically hide potentially treacherous rock conditions that would otherwise be detected.

FIGURE 10.52   Application of wet shotcrete over wire mesh reinforcement in a large-diameter rock shaft. Proper manipulation of the nozzle will minimize rebound and will result in a dense, high-quality coating.

Sprayed Concrete-Lined (SCL) Tunnels.   All the elements described previously come together in the Sequential Excavation Method (SEM). In SEM the elements are combined through a sprayed concrete liner (SCL) for tunnels. The components of a sprayed concrete liner are sprayed concrete applied with a pneumatic hose or pipe (commonly called shotcrete); mesh or bar reinforcing; or alternatively, fiber reinforcement, either metallic or nonmetallic. Lattice girders are not used for reinforcement themselves but rather as a control measure for tunnel shape and also to support mesh or other reinforcement during sprayed concrete applications. Thomas[63] is an excellent source of detailed information on SCL design and applications. An SCL can also be used at the final liner in lieu of the traditional cast-in-place liner in TBM-mined rock tunnels where the initial support is rockbolts, mesh and/or steel sets, or even precast segmental liners when those segmental liners are part of a two-pass system.

SCL liners are not a watertight tunnel liner. Usually a prep (smoothly) layer of shotcrete is applied to facilitate the waterproofing membrane. The smoothness tolerance of these membranes is specified by their manufactures. Recently, sprayed-on membranes are being used in tunnel areas with complex geometries. These membranes also require careful surface preparation.

SHAFT SUPPORTS

The shaft contractor uses the same materials as the tunnel contractor (i.e., plates, ribs, bolts, etc.) to provide temporary support while sinking the shaft. For deeper shaft excavated below ground water level, ground freezing offers an effective means of ground-water control and temporary ground support.

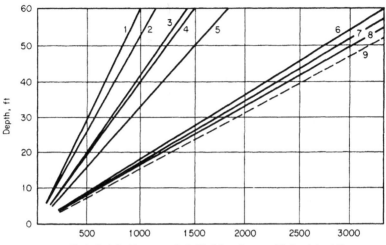

Equivalent liquid pressure P, lb/ft of diameter, per 12-in plate width

1. **Clay: Lumpy and dry**
   Earth: Loose and either dry
   or slightly moist
2. **Earth: Fairly most and packed**
3. **Earth: Perfectly dry and packed**
4. **Clay, sand, and gravel mixture**

5. **Drained river sand**
6. **Earth: Soft flowing mud**
7. **Clay: Damp and platic**
8. **Earth: Soft, packed mud**
9. **Hydrostatic pressure of water**

FIGURE 10.53 Equivalent fluid pressure for caisson construction. (*American Iron and Steel Institute, Handbook of Steel Drainage and Highway Construction Products, 2d ed.*)

Design. Liner plates may be utilized as the sole support in small-diameter earth shafts. The American Iron and Steel Institute provides a method (Fig. 10.53) to determine the equivalent fluid pressure acting on a liner-plate shaft. As in tunnel liner-plate design, thrust on the plates, is determined by the equation $T = PD/2$, where P is the radial pressure and D is the shaft diameter. The thickness of the plates may be increased in stages as the shaft deepens, to meet the increasing thrust on the seam.

Earth pressures on shafts may also be calculated using triangular, trapezoidal, or rectangular load distributions as in soldier-pile or cofferdam construction, covered elsewhere in this book.

The designer should consider the additional load due to groundwater pressure when utilizing liner plates. However, if ribs and timber lagging are used, the water will drain through the support, and its effect on support loading should not be considered.

Design of ground freezing is better left to specialty subcontractors. Ground freezing requires a refrigeration plant and a distribution system for the circulation of coolant to the ground. Design issues include frost heave during construction of the frozen ground and subsequent thaw settlement following construction. Lacy and Floess[64] provide a detailed explanation of the minimum requirements for temporary support with frozen ground. Figure 10.54 shows the start of shaft excavation after the ground has been frozen.

Field Practice. The contractor usually builds a collar or bearing set to support the liner plates or ring beams in a vertical direction and prevent them from settling as the shaft is deepened.

FIGURE 10.54   Frozen ground shaft supports. (*Freezewall Corp.*)

In Fig. 10.55 a contractor is shown constructing a collar for the first 12 ft (36 m) of a steel-rib and liner-plate shaft that will extend an additional 70 ft (21 m) to sound rock. Concrete will be placed between the outside of the liner plate and the surrounding soil. The concrete collar will provide a bearing surface to counteract the weight of the ribs and plates as the shaft is deepened. Note the tie-rods and pipe spreaders on the left, as well as the hardwood wedges between the ribs and plates. The outside diameter of the liner-plate rings is 36 ft (11 m). Pouring a 2-ft (60-cm)-thick permanent concrete lining outside the plates will result in a 32-ft (9.75-m)-inside-diameter shaft.

Some contractors hand the temporary earth support from a steel bearing set. Figure 10.56a and 10.56b illustrates a rectangular bearing set, resting on concrete bearing piles, used to support a large ring beam and timber-lagged earth shaft. The bearing piles may be drilled to rock or to a competent earth stratum. Note also in Fig. 10.56a that the timber lagging is between the rib flanges but is wedged together against the outside flange to minimize voids.

Contractors sinking deep mine shafts in rock generally advance a permanent concrete lining along the excavation, keeping it within a few diameters of the working Face. Figure 10.57 illustrates a deep rock shaft under construction. The camera is pointing upward from the bottom of the shaft. Split Sets and chain-link fabric provide temporary ground support until the concrete lining is placed. The forms are immediately above the temporary support. Note also the bottom of the multiple deck staging with its wells or openings for the muck-removal buckets. The staging allows miners to install temporary and permanent supports concurrent with the excavation. The staging is equipped with its own independent hoisting system to allow more flexibility to the concurrent operations.

Shafts are generally sunk to provide access for constructing a horizontal tunnel. At the shaft bottom, the contractor will cut a hole through the temporary shaft support in

FIGURE 10.55 Collaring a 32-ft-diameter mine shaft in Colorado, using steel ribs with 36-ft-diameter liner plates. (*Harrison Western Corp.*)

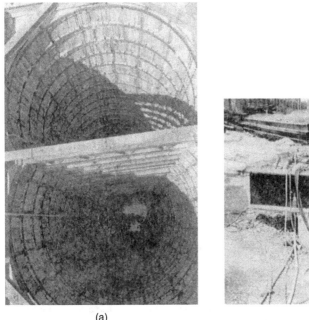

(a)

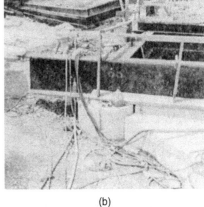

(b)

FIGURE 10.56 A 90-ft-deep center studded shaft in Chicago overburden: (*a*) and (*b*) Circular ring beams and lagging are supported by tie-rods hung from a rectangular structural steel bearing set. (Corner of the bearing set is supported by drilled and concrete-filled bearing pile.) (*S. A. Healy Co.*)

FIGURE 10.57 View from shaft bottom of a deep shaft under construction in Idaho. (*J. S. Redpath Corp.*)

order to provide a tunnel portal. Bracing must be provided to redistribute the vertical and tangential shaft loads around the tunnel portal. Figure 10.58 illustrates a tunnel portal at the bottom of a shallow ring beam and lagged earth shaft. Vertical posts will redistribute the vertical loads acting on the shaft lining, while the tangential loads may be redistributed by a lintel beam or by doubling up on the shaft ring beams at the elevation of the portal. Also note that owing to the shallowness of the shaft, the contractor chose to drive vertical timber lagging outside the ring beams as the shaft was deepened.

RISK MANAGEMENT OF UNDERGROUND AND TUNNEL SUPPORTS

The contractor is required to estimate, during a competitive bidding period, the location, type, and quantity of tunnel supports that are required in tunnels, caverns, and other underground structure. Tunneling methods—Drill and Blast, SEM, and TBM—are intrinsically risky. There is little room for error. When tunneling operations cease at the tunnel face, there are few, if any, concurrent operations to absorb the crews and equipment costs that continue to escalate. In addition, tunnels are usually on a tight schedule. Competition is also fierce. These factors, together with the inherent uncertainties associated with the ground, make this work inherently risky. In the past this responsibility and risk were left

FIGURE 10.58 A shallow Chicago earth shaft with tunnel portal at its bottom. (*J. Lattyak, Metro Sanitary District of Greater Chicago.*)

entirely to the contractor. This was an extremely inefficient and costly approach to the design of supports.

In recent years, the risk associated with the estimate of tunnel support requirements is beginning to be shared, at least in part, by the owner. Tunnel design and construction methodology are also being reexamined in order to identify potential risk areas early in the design of the project before construction begins.

One methodology to reduce risks has been proposed by the International Tunneling Insurance Group (ITIG).[65] The ITIG Code of Practice states as its goal:

The objective of this Code is to *promote and secure best practice* for the *minimization and management of risks* associated with the design and construction of tunnels, caverns, shafts and associated underground structures including the renovation of existing underground structures, referred to hereafter as Tunnel Works. It sets out practice for the *identification of risks, their allocation* between the parties to a contract and Contract Insurers, *and the management and control of risks through the use of Risk Assessments and Risk Registers.*

Risk Registers identify hazards and risks, and propose mitigation and contingency actions. Most important, the Risk Registers also assign responsibilities for these actions within a specified time of completion. The decision-making process involves managing these risk considerations of political, social, and economic issues, as well as strictly technical issues.

Some basic definitions associated with risk management approach are

- *Hazard* is defined as an *event* that has the potential to adversely impact the project, for example excessive ground subsidence adjacent along tunnel alignment during mining.
- *Consequence Analysis* represents an undesired outcome or impact from a hazard. What can happen, for example building collapse, building tilt, and utility settlement, can be expressed in terms of monetary losses, life safety, public relations, etc.
- *Likelihood* is defined as the probability of an undesirable consequence occurring (subjective probability).
- *Risk* is defined as the likelihood of an undesirable consequence occurring and the significance of this consequence occurring. It is expressed as $R = L \times C$.

The ITIG Code of Practice has its limitations and there are concerns within the tunneling community about its implementation. Some of these concerns are because the code was developed by one stakeholder in the tunneling industry, Insurance Companies, with a narrow self-interest focus.

Another means of addressing risk has been suggested by ASCE and the AIME.[66] This approach addresses risks by advocating best practices in the tunneling industry. These practices include having a Geotechnical Design Summary Report (GDSR) to be included in the contract documents during the bidding process. The GDSR has evolved over time into the Geotechnical Baseline Report (GBR) that includes geotechnical baselines for identification of differing site conditions. Also included as part of best practices are Geotechnical Data Reports (GDR) and a Geotechnical Interpretative Report (GIR). The latter report usually provides for information only and is not considered a "contract document." ASCE and AIME also recommend the use of escrow documents to document the successful bidders' estimate to assist the owner in evaluating the cost of a differing site condition. Finally, it is recommended that large underground projects include a Disputes Review Board (DBR). This three-member board, one picked by the owner, one by the contractor, and the chair picked by the first two, hears all major disputes and makes recommendations to resolve them. In essence, each party is given an opportunity to present their case to a group of relatively impartial experts who will make nonbinding recommendations to resolve the dispute. DBRs have been very successful in reducing the adversarial climate that can sometimes exist between the owner, the engineer and the contractor on complex, difficult underground projects.

TABLE 10.17 Typical Hazard Identification Matrix

No.	Hazard	Consequence
1.	Unknown obstructions encountered during tunneling	Reduced production Removal of interventions
2.	Problems tunneling through rock-soil interface (mixed face)	Reduced production Loss of face pressure Alignment control issues
3.	Replacement/repair of TBM main bearing	Reduced production Work stoppage Hyperbaric intervention
4.	Delays in TBM procurement or manufacturing process	TBM not available on earliest starting day
5.	Final design errors and omissions liability	Improperly constructed facilities Problems with third-party stakeholders

REFERENCES

1. Hudson, John A. and John P. Harrison: *Engineering Rock Mechanics: An Introduction to the Principles*, Elsevier Science Ltd., Oxford, U.K., 1997.

2. Long, J. C. S: Investigations of Equivalent Porous Medium Permeability in Networks of Discontinuous Fractures, Ph.D. Dissertation, University of California, Berkeley, California, 1982.

3. Priest, Stephen D.: *Discontinuity Analysis for Rock Engineering*, Chapman & Hall, London, U.K., 1993.

4. Sopko, Joseph A. and Michele R. Norman: "Ground Freezing for Tunnel Support No. 7 Line Subway Extensions New York City," Proceeding Rapid Excavations and Traveling Conference, San Francisco, California, 2011.

5. Wood, Alon Muir: Tunnelling Management by Design, E & FN Spon, London, 2000.

6. Lowardi, Pietro: Design and Construction of Tunnels—Analysis of Controlled Deformations in Rocks and Soils (ADECO-RS), Springer Verlag, Berlin Heidelberg, 2008.

7. American Association of State Highway and Transportation Officials (AASHTO): *Technical Manual for Design and Construction of Road Tunnels—Civil Elements*, Washington, DC, 2011.

8. Muller, L. and E. Fecker: *"Grundgedanken und Groundsatze Neuen Osterrekhischen Tunnel bauweise,"* Trans Tech Publications 247–2, Zurich, Switzerland, 1978.

9. Chapman, David, Nicole Metje, and Alfred Stark: *Introductions to Tunnel Constructions*, Spon Press, London, U.K., 2010.

10. Peck, R. B.: "Deep Excavations and Tunnelling in Soft Ground," *7th International Conference on Soil Mechanics and Foundation Engineering*, Mexico City, State of the Art Volume, pp. 225–290, 1969.

11. Keusel, T. R.: "The Structural Behavior of Tunnel Linings," *Proceedings, Seminar on Tunneling and Underground Construction,* ASCE, Metropolitan Section, New York, 1983.

12. Schwartz, C., and H. Einstein: *Improved Design of Tunnel Supports*, vol. 1, *Simplified Analysis for Ground-Structure Interaction in Tunneling*, Report UMTA-MA-06-0100-80-4, U.S. DOT, UMTA, 1980.

13. Lauffer, H.: "Gebirgsklassifikation für den Stollenbau," *Geologie und Bauwesen*, vol. 24, no. 1, pp. 46–51, 1958.

14. Proctor, R. V. and T. L. White: *Rock Tunneling with Steel Supports*, Commercial Shearing and Stamping Company, Youngstown Printing Company, Youngstown, Ohio, Revised 1968.

15. Proctor, R. V. and T. L. White: *Earth Tunneling with Steel Supports*, Commercial Shearing and Stamping Company, Youngstown, Ohio, Revised 1977.

16. Szechy, K.: *The Art of Tunnelling*, Akademiai Kiado, Budapest, Hungry, 1970.

17. Steiner, W. and H. Einstein: *Improved Design of Tunnel Supports*, vol. 5, *Empirical Methods in Rock Tunneling, Review and Recommendations*, Report DOT-TSC-UMTA-MA-80-27.V, 1980.

18. Terzaghi, K.: Section I—Rock Defects and Loads on Tunnel Supports, in *Rock Tunneling with Steel Supports*, R. V. Proctor and T. L. White, The Commercial Shearing and Stamping Company, Youngstown, Ohio, 1946, Revised 1968.

19. Stini, J.: *Tunnelbaugeologie*, Springer, Vienna, Austria, 1950.

20. Bierbaumer, A.: *Die Dimensionierung des Tunnel-bauerworks*, Engleman, Liepzig, 1913.

21. Deere. D., R. B. Peck, J. E. Monsees, and B. Schmidt: "Design of Tunnel Liners and Support Systems," DOT Contract 3-0152, University of Illinois, February 1969.

22. Deere, D. U. and D. W. Deere: "The Rock Quality Designation (RQD) Index in Practice," *Symposium on Rock Classification Systems for Engineering Purposes*, ASTM, STP 984, 1988.

23. Deere, D. U.: "Rock Quality Designation (RQD) after Twenty Years," USAEWES Geotechnical Laboratory, Contract Report GL-89-1, 1989.

24. Merritt, A. H.: "Geologic Predictions for Underground Excavations," *Proceedings North American Rapid Excavation and Tunneling Conference*, vol. 1, 1972.

25. Wickham, G. and H. R. Tiedemann: "Ground Support Prediction Model (R.S.R. Concept)," Tech. Rep. 125, Jacobs Associates, for Advanced Research Projects Agency of the U.S. Department of Defense, Washington, D.C., 1974.

26. Bieniawski, Z. T.: *Tunnel Design by Rock Mass Classifications*, Technical Report GL-79-19, Report of Office, Chief of Engineers, U.S. Army, Washington, D.C., 1979.

27. Bieniawski, Z. T.: "The Rock Mass Rating (RMR) System (Geomechanics Classification) in Engineering Practice," ASTM STP 984, *Rock Classification Systems for Engineering Purposes*, 1988.

28. Unal, E.: Design Guidelines and Roof Control Standards for Coal Mine Roofs, Ph.D. dissertation, Pennsylvania State University, 1983.

29. Barton, N., R. Lien, and J. Lunde: *Analysis of Rock Mass Quality and Support Practice in Tunneling and Guide for Estimating Support Requirements*, NGI Internal Report 54206, 1974.

30. Barton, N.: "Rock Mass Classification and Tunnel Reinforcement Selection Using the Q-System," *Rock Classification Systems for Engineering Purposes*, ASTM STP 984, 1988.

31. International Tunnelling Association: *Views, on Structural Design Models for Tunnelling, Synopsis of Answers to a Questionnaire*, H. Duddeck, ed., 1981.

32. Kuesel, T. R.: "The Structural Behavior of Tunnel Linings," *Proceedings, Seminar on Tunneling and Underground Construction*, ASCE, Metropolitan Section, New York, 1983.

33. Schmidt, B.: "Tunnel Lining Design—Do the Theories Work?" *4th Australian/New Zealand Geomechanics Conference*, Perth, 1984.

34. The British Tunnelling Society and The institution of Civil Engineers: Tunnel Lining Design Guide, Though as Telford, London, 2004.

35. O'Rourke, T. D.: Guidelines for Tunnel Lining Design, Technical Committee on Tunnel Lining Design, UTRC, ASCE, 1984.

36. Duddock, H. : "Guidelines for the Design of Tunnels," ITA Working Group, Tunnelling and Underground Space Technology, vol. 3, no. 3, pp. 237–249, 1988.

37. Schwartz, C., and H. Einstein: *Improved Design of Tunnel Supports*, vol. I, *Simplified Analysis for Ground-Structure Interaction in Tunneling*, Report UMTA-MA-06-0100-80-4, U.S. DOT, UMTA, 1980.

38. Hewett, B. H. M. and S. Johannesson: *Shield and Compressed Air Tunneling*, McGraw-Hill, New York, 1922.

39. Leontovich, V.: *Frames and Arches*, McGraw-Hill, New York, 1959.

40. Obert, L. and W. Duvall: *Rock Mechanics and the Design of Structures*, Wiley, New York, 1967.

41. Panek, L. and J. McCormick: Chap. 13 in Panek and Williams, *Mining Engineering Handbook*, Society of Mining Engineers, Englewood, Colorado, 1973.

42. Talobre, J. A.: *La Mecanique des Roches*, 2d ed., Dunod, Paris, France, 1967.

43. Hoek, E.: "Estimating the Stability of Slopes in Opencast Mines," *Trans. Institution of Mining & Metallurgy*, London, 1970.

44. Jaeger, C.: *Rock Mechanics and Engineering*, Cambridge University Press, London, U.K., 1972.

45. Cording, E., and J. Maher: "Index Properties and Observations for Design of Chambers in Rock," *Engineering Geology*, Elsevier, Amsterdam, The Netherlands, 1978.

46. Benson, R., R. Conlon, A. Merritt, P. Joli-Coeur, and D. Deere: "Rock Mechanics at Churchill Falls," *Symposium Trans. American Society of Civil Engineers*, New York, January 1971.

47. Provost, A. and G. Griswold: "Excavation and Support of the NORAD Expansion Project," Paper 11, British Tunnelling Society, *Tunnelling, '79*, London, March 1979.

48. Haas, C.: "Rock Bolting to Prevent Shear Movement," *Symposium Proceedings*, M&E Series 79-08, SMA—American Institute of Mining Engineers, New York, February 1979.

49. Scott, J.: "Testing of Friction Rock Stabilizers," *Proc. SME-American Institute of Mining Engineers*, New York, March 1977. ("Split Set" is a patented trade name of the Ingersoll-Rand Company and James J. Scott.)

50. Rabcewicz, L.: "New Australian Tunnelling Method," Parts I and II, *Water Power Magazine*, November, December 1964, January 1965.

51. Rabcewicz, L.: "Dr Ankerung inn Tunnel bar ersetzt bisher gebraüchliche Einbaumethoden," Schweiz, Bauztg, vol. 75, 1957.

52. Kastner, H.: *Statik Dei Tunnel und Stollenbaues*, Springer, Berlin, Germany, 1965.

53. Fenner, R.: *Untersuchungen zur Erkenntis die Gebirgsdrucks*, Gluckauf, 1937.

54. Hopper, R., T. Lang, and A. Mathews: "Construction of Straight Creek Tunnel, Colorado," *Proc. North American Rapid Excavation and Tunneling Conference*, New York, 1972.

55. Mahar, J. H. Parker, and W. Wuellner: "Shotcrete Practice in Underground Construction," U.S. Department of Transportation, Rep. U1LU-ENG-75-2018, Urbana, Ill., 1975.

56. Heuer, R.: "Selection/Design of Shotcrete for Temporary Support," Publication SP45, *Use of Shotcrete for Underground Structural Support*, American Society of Civil Engineers—American Concrete Institute, pp. 160–174, 1973.

57. Linder, R.: "Spritzbeton inn Felshohlraumbau," *Die Bautechnik*, October 1963.

58. Muller, L.: "Removing Misconceptions on the New Austrian Tunnelling Method," *Tunnels and Tunnelling Magazine*, London, October 1978.

59. Cecil, O.: *Correlations of Rockbolt Sholcrete Support and Rock Quality Parameters in Scandinavian Tunnels*, University of Illinois Thesis, Urbana, Ill., 1970.

60. Fernandez-Delgado, G. E. Cording, J. Mahar, and M. VanSintjan: "Thin Shotcrete Linings in Loosening Rock," *Proc. Rapid Excavation and Tunneling Conference*, Atlanta, June 1979.

61. Alberts, C.: "Steel Fiber Shotcrete," Report for Stabilator AB, Stockholm, on its part in the North Research Project, 1979.

62. Moran, D.: "Steel Fibre Shotcrete—A Laboratory Study," *Concrete International*, January 1981.

63. Thomas, Alun: *Sprayed Concrete Lined Tunnels*, Taylor & Francis, London, U.K., 2009

64. Lacy, H. S. and C. H. Floess: "Minimum Requirements for Temporary Support with Artificially Frozen Ground," *Transportation Research Record*, 1190, 1988.

65. The international Tunneling Insurance Group: A code of Practice for Risk Management of Tunnel Works, 2006.

66. Technical Committee on Contracting Practices: Avoiding and Resolving Disputes in Underground Construction, UTRC of ASCE and AIME, 1989.

CHAPTER 11
UNDERPINNING*

Francis J. Arland, P.E.
Rudi van Leeuwen, P.E.

*This chapter is an update, revision, and expansion of "Underpinning" by Rudi van Leeuwen, P.E. and Fred N. Severud, P.E. in the Second Edition of the *Handbook of Temporary Structures in Construction*.

INTRODUCTION

Underpinning is the installation of temporary or permanent support to an existing foundation to provide either additional depth or an increase in bearing capacity. There are several existing conditions that may lead to the need for underpinning, such as

- Construction of a new project with a deeper foundation adjacent to an existing building
- Settlement of an existing structure
- Change in use of a structure
- Addition of a basement below an existing structure

When the foundation for a new structure is deeper than that of an adjacent existing structure, the installation of some type of protection for the existing one is usually required. Underpinning may be needed even though the new excavation is not directly adjacent to an existing structure.

The location of the influence line of the new excavation in relation to a nearby structure has to be considered, as well as the soil conditions, the water table, the type of existing construction, and any temporary retaining structures, such as soldier piles and lagging or steel sheet piling, installed between the excavation and the existing structure. Installation of a soldier pile or steel pile sheeting system is generally not considered an acceptable substitution for underpinning. In some cases, underpinning of interior columns or wall footings may not be required, if the exterior wall footing is underpinned and provides protection for the interior foundations.

Settlement of existing structures in many cases is caused by the lowering of the water table due to tidal fluctuations, wells for a water district, or the construction of deep foundations with long-duration round-the-clock dewatering. This lowering of the water table can cause the tops of timber piles to decay over time and will require remedial underpinning as illustrated in Fig. 11.1.

With certain soil profiles, rising of the water table can effect a decrease in the bearing capacity of the soil, causing settlement and requiring underpinning to save the building.

Another frequent reason for settlement is the construction of structures on unsuitable bearing material or over a compressible layer (peat, organic silt, or a poorly compacted backfill or sanitary fill). Such a structure may settle, either during or after construction. When existing friction piles for a structure that is to remain are exposed during excavation, underpinning may be required to support the weakened foundation and prevent settlement.

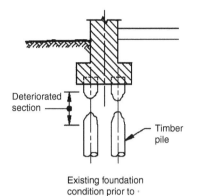

Existing foundation
condition prior to ·
underpinning

Step 1. Shore existing construction, excavate
approach pit, dewater as needed,
and expose existing timber piles.
Remove top portion of piles and cut
piles at new cut-off elevation.

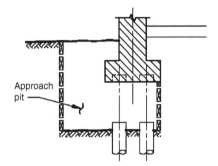

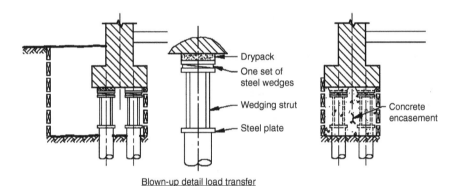

Blown-up detail load transfer

Step 2. Install steel plates, drypack,
and wedging strut.
Transfer load into pile by
means of steel wedges.

Step 3. Placement of concrete
encasement, backfill approach
pit.

FIGURE 11.1   Underpinning timber-pile foundations.

To prevent any unexpected settlement of a new structure, it is of the utmost importance
that proper field investigations be performed, including test borings, test pits, and laboratory
soil tests. The cost of these tests is minor compared to the expenses of field investigation and
survey of structural damage, design of remedial work, and the actual costs of the underpinning and repairs after the building shows signs of distress.

Underpinning may also be necessary if the requirements of a structure have changed, for
example, when heavier floor loads are imposed or additional floors are added to an existing
building.

Another use for underpinning (especially during the last 60 years) has been to provide extra basement space under existing buildings. In such applications exterior walls are
underpinned and the interior column or wall loads are transferred to new columns, resting
on new footings placed at the lower elevation. This creates new space without altering the
height above the curb, which is often limited by zoning laws.

DETERMINING THE NEED FOR UNDERPINNING

When a structure starts showing signs of settlement or distress, it is of the utmost importance to establish level readings and offset readings. These should be taken by a professional on a daily, weekly, or monthly basis, depending on the severity of the movements. Plotting these readings will indicate if the movements are decreasing or increasing, and by analyzing the results, a decision can be made whether or not underpinning or other measures are required to safeguard the structure and its occupants.

Before the start of excavation for a new structure, it is advisable to have a professional engineer examine all structures in close proximity to the construction site, to determine whether or not underpinning is necessary. A report will be produced, consisting of a description of the structure with all its preexisting structural defects, backed up by photographs and/or video recordings. Many insurance carriers and city agencies today insist on this type of report, which must be submitted to them before the start of demolition or excavation operations.

DESIGN CONSIDERATIONS

Building Code and Agency Requirements

After the need for underpinning is established, the current local building code and building department along with other local agencies having jurisdiction should be consulted for design and construction requirements, and for submittals requiring their approval. Obtaining these approvals during the design phase or before construction can eliminate substantial delays or work stoppage during construction, if the building department is called to the site by an adjacent owner alleging property damage. Some jurisdictions have specific requirements pertaining to the information that shall be shown on the submittals. For example, the New York City Building Department has requirements for preparing site-specific plans and sections, and has minimum requirements for underpinning design.

Some building codes require a preconstruction condition survey of the existing structures be performed within the zone impacted by construction. This should be performed regardless of whether it is required by the building department. Performing a preconstruction condition survey of structures to be underpinned is essential to managing construction risk and assessing the impact of underpinning and new construction on the adjacent structures.

Subsurface Investigations

A geotechnical investigation, typically including borings and test pits, should be performed to determine subsurface conditions in the area of the structure to be underpinned. The investigation should determine subsurface stratigraphy; soil types and properties; the location of groundwater; and type, geometry, and elevation of the foundation to be underpinned. This information is used to establish the depth, extent, and method of underpinning and the need for dewatering. In many applications, underpinning is also designed to provide excavation support below the exterior wall of the structure being underpinned. Lateral soil pressures and the lateral surcharge loads from adjacent interior foundations, if bearing at higher elevations than the planned underpinning, need to be considered.

Influence Lines

Influence lines are generally used to evaluate the need for underpinning. The loads in the existing structure must be safely transmitted to a lower soil stratum, capable of supporting these loads. The bottom elevation of the underpinning must be at a sufficient depth so that the general excavation for individual column footings or pile caps will not endanger the adjacent structure and deep enough so that the underpinning will not impose surcharge loads onto the future structure. A thorough investigation of the relationship between future subgrade, the bottom elevation of the foundation of the existing structure, and the distance between the existing and new structure are needed to establish the zone of influence of the existing structure. This evaluation must also consider the local soil conditions, elevation of groundwater, and the impacts of construction dewatering, if needed.

To avoid undermining of installed underpinning during excavation, the bottom of the underpinning must extend at least 6 in (15 cm) below future subgrade (see Fig. 11.2). The use of influence lines and slope lines is illustrated in Figs. 11.3 through 11.6. The angle of the slope line depends on the soil or rock conditions.

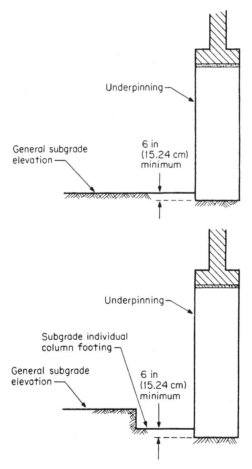

FIGURE 11.2 Minimum depths of underpinning.

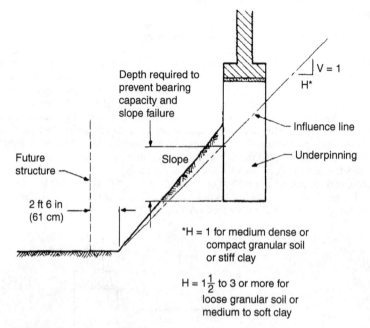

FIGURE 11.3   Influence lines in soil.

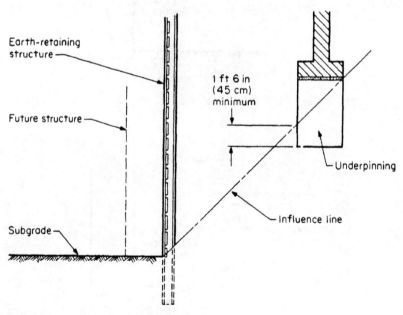

FIGURE 11.4   Influence line behind sheeting.

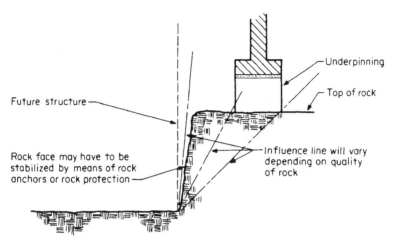

FIGURE 11.5 Influence lines in rocks.

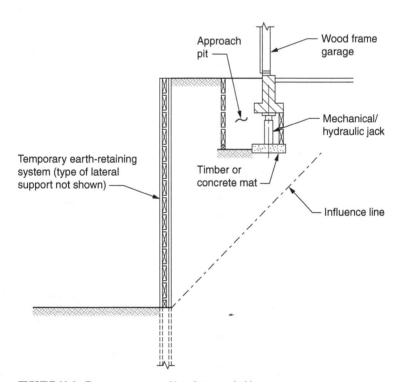

FIGURE 11.6 Temporary support with maintenance jacking.

Figure 11.4 shows soldier pile sheeting installed in front of an adjacent structure. Underpinning should be extended below the influence line to the depth required to provide competent bearing and lateral support of the structure being underpinned.

When a structure rests on rock or the underpinning is seated on rock, influence lines can normally be ignored. However, if the rock is of poor quality or is decomposed, the underpinning must be carried to sound rock or to an elevation determined by an influence line.

Figure 11.5 shows underpinning seated on rock. The slope of the influence line depends on the soundness of the rock formation, which may vary from vertical in sound hard rock, to 1:1 in soft, highly weathered and decomposed rock. If subgrade is below the top of the rock, the quality of the rock face must be closely examined during rock removal. In case faulty rock is discovered, measures must be taken to prevent loss of bearing and provide rock face stability, by means of rock bolting and/or rock protection. Rock anchors or rock face protection may be required locally or for the full extent of the rock face, depending on the rock condition.

PRECAUTIONS

The design of underpinning is far from an exact science. The conditions of the existing structure may be known at the time of the design, but the bearing capacity of the soil at subgrade may change from one location to another. It is very important to design and prepare working drawings before the start of the underpinning operation. These plans will indicate what one hopes to find, but in many cases, especially with old structures without as-built record drawings, one must be prepared to make drastic changes to the design to cope with conditions encountered in the field.

During and after the installation of any type of underpinning, the following precautions should be taken:

1. During any excavation and sheeting installation, loss of ground is almost impossible to prevent, but it should be kept to a minimum by backfilling behind the boards where and when possible. Loss of ground will create voids behind the sheeting, which may instigate settlements of floor slabs and interior footings.

2. Upon completion, the building load must be properly transferred into the underpinning system by means of drypack or with steel plates and wedges. If the transfer is not performed properly, additional settlement will occur.

3. The installation of an adequate lateral bracing system will prevent movement in the structure and in the new underpinning. A poorly designed bracing or tieback system will add to the total and final settlement of the structure.

SETTLEMENTS CAUSED BY UNDERPINNING

The most cautiously installed underpinning will be accompanied by some minor settlement of the structure. The settlement will vary with the type of underpinning system installed, the soil conditions encountered, the type of structure, and, above all, the quality of workmanship and supervision on the job. Most settlements for this reason can be held to within ¼ to ½ in (6 to 12 mm). It is very important to keep the settlement of a building uniform, since the difference in settlement from one point to another may cause structural damage. Normally, the excavation adjacent to a structure will be carried to within 6 in (15 cm) above the bottom of the existing column or wall footings to facilitate the start of approach pits. This may cause unloading of the bearing soil, creating minor settlement.

The installation of a dewatering system in the new excavation may displace fines under the wall and column footings, creating a cause for additional settlements. The installation of piles and caissons adjacent to existing structures may be the cause of movement and settlement, depending on the soil condition and the method of installation used.

The installation of any type of bearing pile in sandy material, especially when driven with a vibratory hammer, adjacent to or in the vicinity of an existing structure requires precise monitoring of the structure before and during the pile installation. Considerable settlement and consequently damage to the adjacent structure can take place due to this type of operation. The type and method of installing of lateral support for the structure itself and for the underpinning are very important. Any lateral movement of the structure will add loading to the bearing strata on which the underpinning is founded and thereby will increase the total settlement. The lateral support can consist of cross-lot bracing, inclined bracing, or tiebacks in earth or rock. The vertical component of the tiebacks must be added to the building load in each underpinning unit to determine the minimum bearing area. Substitution of soldier piles or steel sheet-pile sheeting for underpinning does not prevent horizontal and vertical movement of the structure. In case of light structures, maintenance jacking may be used to prevent differential settlement, with the actual installation of underpinning. Whatever method of underpinning is employed, it is always necessary to correct structural damage to walls and footing before the start of the underpinning, especially since the first step in an underpinning operation is to excavate under a footing. By doing so, the bearing capacity of the foundation will be decreased. In certain instances the use of temporary shores or needles is required. In the case of a loose rubble foundation wall that must be underpinned, it may be necessary to remove the rubble wall in short sections and replace it with a concrete wall. Alternatively, the rubble wall may be reinforced by repointing the joints with mortar or grout or by placing a concrete wall against the rubble wall and tying the walls together through the bonding action of the freshly placed concrete.

MINIMIZING DISTURBANCE DUE TO UNDERPINNING

In most instances, underpinning can be installed without any inconvenience to the occupants of the structure, especially if only exterior walls and footings are affected.

Some underpinning may be installed from within the basement of a structure when the locations of the existing boiler plant or other utilities do not obstruct the sinking of the approach pits. However, in many instances, structures have been underpinned without entering the building, with a minimum disturbance to pedestrians on the sidewalks. Performing the work entirely below the sidewalk and covering the areas of underpinning immediately on completion help keep the disturbance to the public to a minimum.

TEMPORARY SUPPORT WITH MAINTENANCE JACKING

Light structures (e.g., wood-frame garages) that fall within the influence line of an adjacent excavation and that do not warrant the expense of an underpinning installation may be supported on timber or concrete mats. If settlement occurs, the structure will be kept at the same level by means of mechanical or hydraulic jacks (see Fig. 11.6). At completion of the work in the adjacent lot, the jacks are replaced with short steel columns, as in Fig. 11.1, and the void is filled with concrete.

UNDERPINNING PIERS

This method is the one most frequently used in the foundation industry. If the water table is above the level to which the new underpinning has to be extended, a dewatering system must be installed before the start of the underpinning operation, so that all the pits can be sunk in the dry. If the excavation must be performed partially in the wet, this not only makes the work more costly but also makes loss of ground unavoidable. The piers may be continuous for the full length of an exterior wall, or an intermittent system of piers may be used. The building loads are transferred onto the piers by means of "drypack," a mixture of sand and Portland cement with very little water that is rammed into the space between the existing wall or column footing and the hardened concrete underpinning pier, placed from within the approach pit. This method is more completely described subsequently.

Procedures

When it is decided to sink underpinning piers under a structure, the following procedures should be followed as illustrated in Fig. 11.7:

1. General excavation to within 6 in (15 cm) above the bottom of the existing wall or column footings.

2. An approach pit is excavated and supported by 2- by 8-in (5- by 20-cm) dressed lumber. Minimum size of the approach pit is 3 by 4 ft (90 by 120 cm) to allow a laborer room to excavate and install the pit boards.

3. The normal depth of an approach pit is 4 ft (120 cm) except where clearance requirements warrant deeper pits.

4. After the approach pit is completed, the excavation is extended under the wall or column footing while horizontal wood sheeting is placed.

5. The underpinning pit can now be extended downward to its proper depth by carefully excavating and sheeting of the shaft to the proper elevation.

6. Inserts are installed for an inclined bracing or tieback system.

7. All loose masonry and dirt are removed from the bottom of the existing wall or column footing.

8. Formwork is installed in the approach pit to facilitate the placement of the concrete up to within 3 in (7.5 cm) of the bottom of the existing footing.

9. Drypack, a damp mixture of Portland cement and mason sand, is installed between the bottom of the existing footing and top of the new concrete, not less than 12 h after the concrete is placed. Due to the rather dry mixture and the total overall thickness of the drypack layer (3 to 5 in), no noticeable shrinkage will take place during the curing process. The use of nonshrink cement in the mix is unnecessary.

10. Alternative to drypack. Instead of leaving a gap for drypacking between the underpinning concrete and the footing to be underpinned, the concrete may be placed with a "lip" [carried to an elevation of 2 ft (60 cm), higher than the footing bottom], which places underpinning concrete under sufficient pressure to eliminate the need for drypack. This requires careful design of the concrete mix, using superplasticizer to make the mix flowable, and minimizing curing and drying shrinkage of the concrete. The lip should be removed the day after placement, while the concrete is still green and has not reached strength that will make removal difficult.

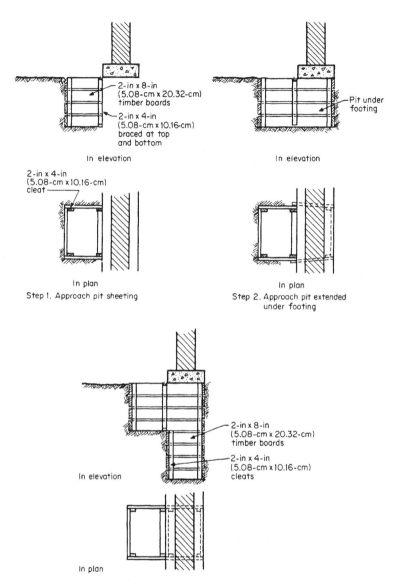

FIGURE 11.7   Underpinning piers.

The excavated material is normally removed from the pit by means of buckets, using a well wheel and hand power. However, if the piers are deep, mechanical equipment such as hoists is used.

If the soil is fairly cohesive, the shaft excavation can progress in 2- to 3-ft (80- to 90-cm) lifts before the pit boards are installed; and in the case of very hard materials, only skeleton boards [one ring of 2- by 8-in (5- by 10-cm) boards at 16 in (40 cm)] vertically will be required.

However, in very fine, dry sand, because it runs freely when the excavation progresses, before the next board can be placed, steps must be taken, such as pregrouting of soils to improve stand-up time, thorough wetting of the soil and/or the use of 2- by 4-in (5- by 10-cm) lumber, closely spaced. It is advisable to use louvered boards at all times in order to enable backfilling behind the boards where necessary. (Louvered boards are boards placed with spaces left between them, which can be packed with sand and salt hay in case the sand starts running after it dries out.) Untreated lumber is generally used for underpinning piers and sheeting. While the lumber will eventually deteriorate, it typically does not lose its volume and will therefore not cause additional settlement. The only valid reason for using treated lumber is where termites are prevalent; otherwise it is not cost-effective.

Sequence

The sequence of the installation of underpinning pits under bearing walls depends on the type and condition of the structure and on the soil conditions, but in general no pits closer than 12 ft (3.6 m) on center should be opened at the same time. The installation of underpinning pits under column footings should not be done with more than one pit opened at a time under the same column and pits under adjacent column footings should not be opened simultaneously.

In case an exterior wall footing has to be underpinned, but continuous underpinning is not required, owing to the light loads in the structure, intermittent underpinning piers may be installed, as shown in Fig. 11.8. However, it must be recognized that the resulting bearing pressures under the intermittent piers will likely be greater than the bearing pressure below the larger footing area of the continuous wall. This may limit the addition of future load to the underpinned building wall if adding additional height to the building is later considered.

The sequencing for continuous underpinning also applies for intermittent piers. The spacing between the piers depends on the loading in the structure and the soil conditions. The interpier sheeting, as indicated in Fig. 11.8, has to support the soil behind the underpinning system. The thickness of the boards depends on the span between the concrete underpinning piers: 2 in (5 cm) boards up to 4 ft (1.2 m) and 3 in (7.5 cm) boards thereafter.

In case the total amount of settlement has to be kept to a minimum or pile driving will occur in the vicinity of the structure, it may be advantageous to install jacking pockets in each individual underpinning pier. Then, if additional settlement occurs, mechanical or hydraulic jacks can be installed and the structure can be maintained.

Lintels

Some municipal building codes require continuous support of any exterior wall footing, and in some instances the structural integrity of the exterior wall footing does not allow intermittent underpinning piers. In these cases the installation of lintels between underpinnings piers is a workable alternative, as illustrated in Fig. 11.9.

There are many types of lintel construction that can be used, but, in case of an emergency, a steel beam is the fastest solution. After the steel lintel is in place, drypack can be installed immediately between the steel beam and the bottom of the wall footing; permanent concrete encasement can be placed later. If time is not of the essence, either of the two methods in Fig. 11.9 can be used.

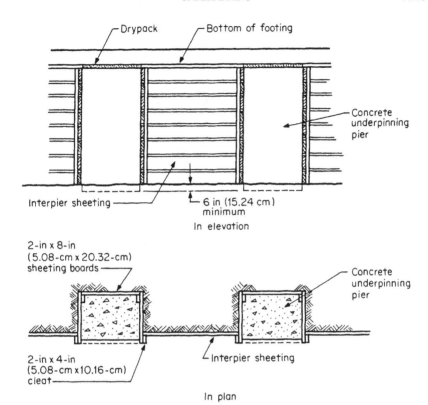

FIGURE 11.8 Intermittent underpinning piers.

Exposed Underpinning

In certain instances it is necessary to incorporate the underpinning into the new construction, especially where space is very tight, as in construction of new tunnels and corridors. The underpinning should be installed in such a manner that the exposed face is neat and clean.

Figure 11.10 indicates how the eventually exposed underpinning piers and the formwork are installed. After all the underpinning piers are fully completed, the general excavation is lowered while the interior sheeting is being installed as indicated. When the excavation is completed, formwork is set between the underpinning piers and the concrete for the fill-in wall is placed.

Overlapping Structures

Sometimes design criteria or lack of space make it necessary for a new column footing to project under the existing wall footing of an adjacent structure. When the adjacent structure requires concrete underpinning, a special method must be utilized to facilitate this condition. The procedure of underpinning is identical to that of a normal installation, except for the following measures: The bottom of the excavation has to be lowered at least 12 to 18 in (30 to 45 cm) below the bottom of the future column footing beneath the existing wall, as shown in Fig. 11.11.

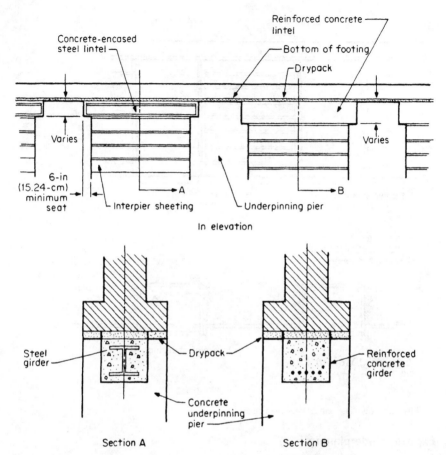

FIGURE 11.9 Lintels.

Concrete is placed up to the level of the bottom of the future column footing. Steel posts are set on this concrete mat after the concrete has set. Sand is placed and compacted up to the top of the future column footing. The underpinning pier is constructed from the compacted sand up to within 3 in (7.5 cm) of the bottom of the wall footing, and drypack is installed. After the general excavation has reached subgrade elevation, the sheeting within the area of the future column footing has to be stripped and the sand previously placed in the underpinning piers removed. The existing building load is now transferred through the previously placed steel posts into the soil. Formwork, reinforcing steel, and concrete for the new column footing can now be placed continuously under the adjacent structure.

Underpinning of Column Footings

The underpinning of column footings, individual or combined with wall footings, demands a great deal of care because of the heavy concentrated loads that must be extended to a

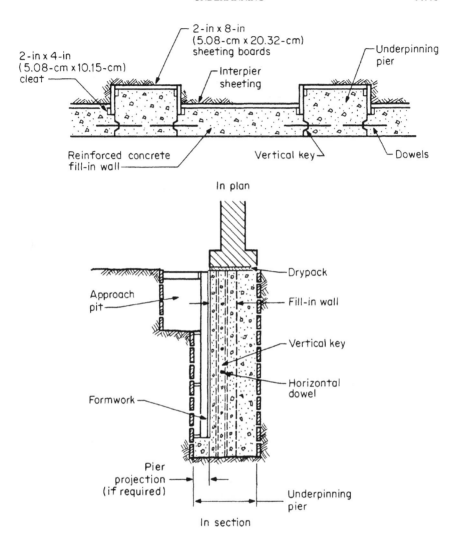

2-in x 8-in
(5.08-cm x 20.32-cm)
sheeting boards

2-in x 4-in
(5.08-cm x10.15-cm)
cleat

Interpier
sheeting

Underpinning
pier

Reinforced concrete
fill-in wall

Vertical key

Dowels

In plan

Drypack

Approach
pit

Fill-in wall

Vertical key

Horizontal
dowel

Formwork

Pier
projection
(if required)

Underpinning
pier

In section

FIGURE 11.10   Exposed underpinning.

lower elevation. The most difficult situation is the underpinning of a small individual column footing. In case the existing column footing is 5 by 5 ft (1.5 by 1.5 m) or smaller, the
only way to underpin the footing with concrete piers is to temporarily shore the column by
means of inclined shores or needle beams supported on temporary mats, before the installation of the one or two underpinning piers required.*

Figure 11.12 indicates the underpinning of a 6- by 6-ft (1.8- by 1.8-m) individual column
footing with two alternatives. Any loss of ground during the installation of the underpinning

*Almost all underpinning operations must be executed by an experienced specialty contractor, but in this
instance it is a requirement.

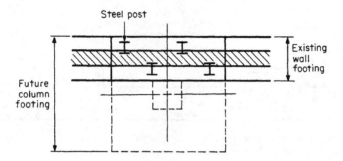

Plan—future column footing projecting underneath adjacent wall footing

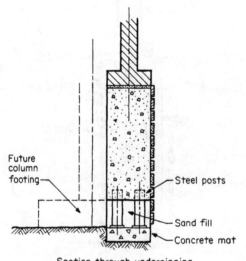

Section through underpinning

FIGURE 11.11   Overlapping structures.

piers will cause immediate settlement. The construction sequence of the scheme with the three piers is 1, 2, and finally 3.

The only time that the alternative, with only two piers, can be used is when the soil condition improves with depth and less bearing area is required for the underpinning piers than for the original column footing. The installation of the lintel is a necessity to prevent overstressing of the existing column footing.

Figure 11.13 indicates the underpinning of a large column footing incorporated into the wall footing on both sides. To avoid entering the basement of the structure or living quarters of tenants, all piers are approached from the street side of the building. The procedure is as follows:

1. Install approach for pier 1.

2. Excavate underpinning pit 1.

3. Place concrete in pit 1 to bottom of approach pit.

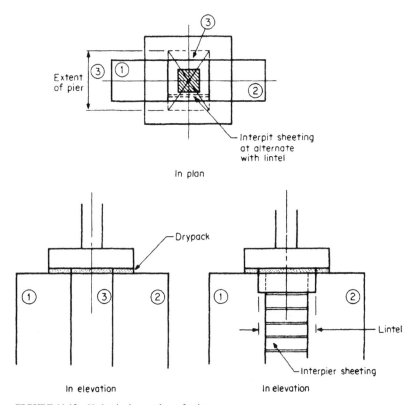

FIGURE 11.12   Underpinning a column footing.

4. Place steel posts as indicated, transferring part of the column load into the posts by means of steel plates and wedges.
5. Excavate underpinning pit 2.
6. Place concrete in pit 2 and drypack pit 2.
7. Install formwork in approach pit to facilitate the placement of the concrete for pier 1
8. Place concrete in remainder of pit 1 and drypack. The same procedure is used for the installation of piers 3 and 4.

Belled Piers

If ground conditions permit, it is possible to enlarge the bearing area of underpinning piers by belling out the bottom section of the shaft. Belling out in dry, loose, runny silt and sand or in water-bearing sandy materials is very risky and will expose the structure to additional settlement due to loss of ground. However, in cohesive soils the belling out of underpinning piers is feasible and will diminish the cost of the project due to lesser shaft excavation, especially when the underpinning piers are deep.

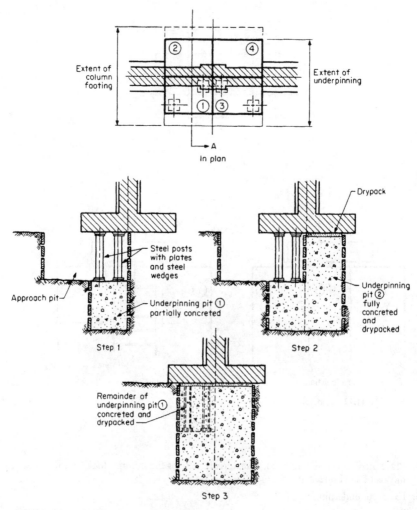

FIGURE 11.13 Underpinning a column footing.

Figure 11.14 indicates two different methods of using belling in underpinning an individual column footing. The actual column footing dimensions for the example are 8 by 10 ft (2.4 by 3 m), and four piers of 4 by 5 ft (1.2 by 1.5 m) each will accomplish the job. Total volume of pit excavation and concrete, assuming the depth of underpinning to be 15 ft (4.5 m), is 44.4 cy (34 m³). The first method shown indicates four piers, each with a 3- by 4-ft (90- by 120-cm) shaft and a 1-ft (30-cm) bell in two directions. Total area in bearing is 80 ft² (7.4 m²) with a total volume of pit excavation and concrete of 28.8 cy (22 m³).

Because of soil conditions, it may be very difficult to install bells on four sides in an excavated shaft, and the second alternative shown in Fig. 11.14 may be the answer to the problem: four piers, each with 4- by 4-ft (120- by 120-cm) shaft and 1-ft (30-cm) bells on two of the four sides only.

Total area in bearing is 80 ft² (7.4 m²) with a total volume of pit excavation and concrete of 36.1 cy (27.6 m³). If soil conditions allow, a third alternative (not shown, but similar to

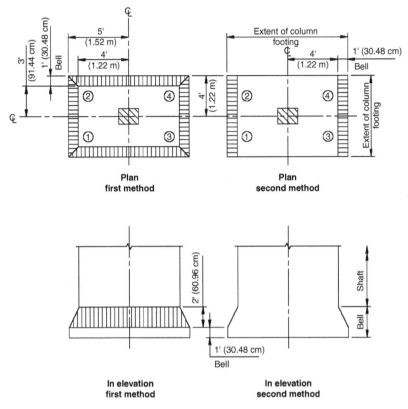

FIGURE 11.14 Belled piers.

the second method) could be the four piers, each with 3- by 4-ft (90- by 120-cm) shaft and 2-ft (60-cm)-wide by 4-ft (120-cm)-deep bells on two sides only. Total volume of excavation is 29.0 cy (22 m³).

Belling of individual pits or continuing pits under bearing walls can reduce the total cost of underpinning. However, care must be taken that the belled-out portion of the underpinning does not interfere with future permanent foundations on the adjacent site, and it must be kept in mind that the soil condition in the area of belling must be very hard or stiff to allow belling without loss of ground.

PRETEST PILES

On many occasions, the pretest pile method is the most viable solution for underpinning a structure: for example, in case the water table cannot be lowered to facilitate hand-dug pit underpinning, or when the bearing capacity of the soil at future subgrade is not sufficient to carry heavy concentrated loads and the tip elevation of underpinning must be lowered to bear on deeply seated strata of hardpan or rock.

Pretest piles consist of steel pipe 10 to 24 in (25 to 60 cm) in diameter with a minimum wall thickness of 0.25 in (6 mm), up to 100 ft (30 m) in length, installed in 4- to 6-ft (1.2- to 1.8-m) sections. In most instances the pipe is cleaned and backfilled with structural concrete. The load is transferred by means of hydraulic jacks, steel plates, wedging struts, and wedges. A more detailed description of this method is presented subsequently.

Procedures

The starting procedure for pretest piles is the same as the one used for the installation of concrete underpinning piers. The approach pit, however, will be deeper to facilitate the placement of steel pipe sections into the jacking pit. The length of the pipe sections will vary from project to project depending on field conditions. In case the water table is high and cannot be lowered by pumping, this condition will determine the depth of the jacking pit and the length of the pipe sections, which may vary from 3 to 7 ft (0.9 to 2.1 m). In case boulders are suspected in the underlying strata, it may be advisable to install a cutting edge on the bottom section of pipe to prevent damaging the tip of the pile. After completion of the jacking pit, a steel bearing plate is placed against the bottom of the footing, with drypack as a leveling device to compensate for the uneven bottom surface of the footing, as shown in Fig. 11.15. The first section of pipe is placed in the jacking pit. After the pipe is set plumb, a temporary steel plate is set on top of the pipe. Two double-acting hydraulic jacks are placed on this plate. By applying pressure to the jacks, the pipe is forced downward. It is of the utmost importance to monitor the structure during the jacking operation to avoid any uplift or damage to the structure, especially when the jacking progress is slow and high pressures are being applied to the jacks. After the jacks are extended to 75 percent of their stroke capacity, the pressure in the jacks is reversed and jacking dice are placed on top.

This procedure is repeated until the next section of pipe can be installed. The splice between two pipe sections can be accomplished by welding or with the use of external pipe sleeves. If the pretest pile is exposed only to vertical loading, a single weld pass will suffice. Splicing of the pipe section is very time consuming due to the lack of adequate work space and the awkward location at the bottom of the pit where the weld has to be applied. To speed up the splicing operation, an external tightfitting sleeve may be used. The sleeve is normally placed at the top of the pipe before lowering it into the jacking pit. (If alignment, watertightness, or bending is a problem, the splice must have a full-penetration weld.) The advantage of the sleeve is that most of the welding will be performed outside the confinement of the jacking pit, and welding between the sleeve and the next section of pipe is a down-hand weld, which is easier to make than a full-penetration weld.

Before the next section of casing is placed, the usual procedure is to remove the soil from inside the casing to reduce friction, thereby keeping the loads applied to the jacks to a minimum. The soil is typically removed by using a pancake auger or an orange-peel bucket.

If the soil is very soft, it may be advisable to add the next section and continue jacking without cleaning the pipe. This procedure is repeated until the pipe is seated on the proper bearing stratum or until the jacking loads become too high for the safety and integrity of the structure above.

Jacking through Soft Clay

Where piles are to be jacked through a stratum of soft clay, as in the Chicago Loop area, it is good practice to place a compacted sand plug at the bottom of the first pipe section before the start of jacking. The sand plug will be forced upward slightly during the jacking of the pile to the hardpan layer, but it will prevent the soft clay from filling the pipe, and thus little or no cleaning of the pile is required. This procedure can save time and reduce costs.

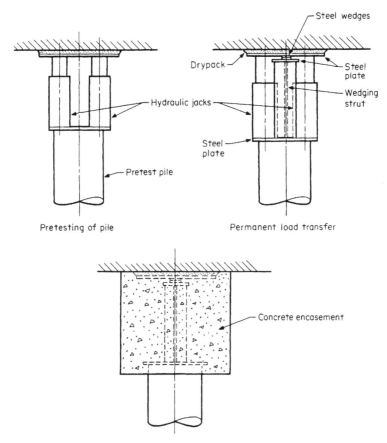

FIGURE 11.15   Pretest piles.

Overcoming Difficulties in Jacking

During the course of jacking the pretest pipe, hard layers of soil mixed with gravel, cobbles, or even boulders may be encountered. Jacking pressures will get too high for safety, and conventional methods of removing the soil do not work. In this case, it is necessary to drill or chop out the obstructions with special equipment, such as Bucyrus-Erie churn drills, to remove the obstacles.

In some instances, when the high frictional forces on the outside face of the pipe cause it to hang up, it may be necessary to drive the pipe with an internally placed impact or vibratory hammer, to advance the casing, if space allows. This, however, increases the risk of inducing settlement.

Cleaning out Piles

In most cases, jacked piles must be cleaned out and filled with concrete. There are different methods of cleaning them out, depending on the soil conditions and length of the piles. Most methods are tedious and time consuming, especially with long piles. The most common cleaning methods are orange-peel buckets, pancake augers, short flight augers, and air lifting.

It is of the utmost importance in areas with a high water table to leave a plug of soil inside the tip of the pile at all times. Overexcavation inside the pile (especially past the end of the pipe) may result in loss of ground, which will cause additional settlement.

Where the load on the pile is very light, corrosion of the pipe is not a problem, and the steel pipe alone is capable of carrying the load, the soil does not need to be removed from inside the pipe, saving considerable time and expense.

Pretesting

When the pile reaches a certain bearing stratum or the tip is at a predetermined elevation, the capacity of the pile has to be tested. Typically, 150 percent of the working load is applied as a test load to the empty shell. If no substantial movement occurs, the pile is ready for concreting. However, if the pile moves considerably [½ in (12 mm) to 1 in (25 mm)] under the 150 percent load over a period of 15 min, the jacking of the pile must be continued. When the pile passes this test, it is filled with structural concrete, and a permanent plate (bearing firmly on the steel pipe and the concrete) is placed on top of the pile. Two hydraulic jacks are set on top of the bearing plate in a manner that a wedging strut can be placed between them.

The pretest load is now applied to the pile (between 150 and 200 percent) of the working load, depending on code requirements). The structure is carefully monitored for vertical movement and signs of distress in the areas of the pretest pile while the load is applied. If no substantial vertical movement occurs while the pressure in the jacks is maintained (taking into account the elastic shortening that will take place), a steel wedging strut with steel plates and one set of steel wedges is placed between the two jacks. The steel wedges are driven home until the pressure in the jacks drops. The pile is now completed, except for the concrete encasement around the top of the pile and the wedging strut.

By preloading the pile in this manner, the rebound of the pile after removing the jacks will be diminished from as much as ¾ in (19 mm) to between 1/16 in (1.5 mm) and 1/32 in (0.8 mm).

Bearing Capacity

The bearing capacity of pretest piles depends on the size of pile being used and the bearing material on which the tip of the pile is resting. The capacity may vary from 80 tons (73 t) for piles sitting on a sand and gravel stratum to 120 tons (110 t) and higher for piles bearing on sound rock.

It is very important to have a sufficient number of borings taken in close vicinity of the structure to be underpinned with pretest piles. If the bearing stratum is fairly thin, it is possible to punch through it by overjacking or overexcavating the piles. The amount of additional pile length to be installed to reach the next bearing stratum, if present at all, could be very costly.

BRACKET PILE UNDERPINNING

Installation

When both the existing and future structures belong to the same owner, the use of bracket piles is very economical. (Most municipal building codes do not allow a building to be supported on a foundation that is located on someone else's property.) The steel bracket piles are driven or placed adjacent to the future structure in pre-augered holes that are then backfilled with a lean sand-cement mix. The load is transferred from the structure into the pile through a steel bracket welded to the backside of the pile. A combination of steel plates, wedges, and drypack is installed to ensure a tight fit between the structure and the bracket, as shown in Fig. 11.16.

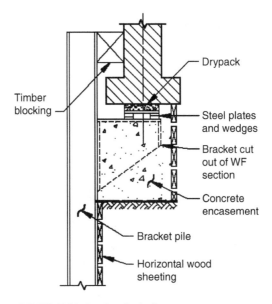

FIGURE 11.16   Bracket pile detail.

This type of underpinning can be utilized for multiple story buildings, depending on the structural and bearing capacity of the bracket pile. The toe penetration of the piles is determined by the passive resistance below subgrade and by the vertical load distribution of the bracket pile. The spacing of the piles depends on the load distribution in the existing structure and also on the ability of the soil to arch from pile to pile for horizontal sheeting installation. The maximum spacing should be determined based upon the acceptable span length of the wall or footing being underpinned, and generally should not exceed 8 ft (2.4 m).

MICROPILE UNDERPINNING

Micropiles, another type of pile used for underpinning, are small diameter non-displacement piles, typically less than 12 in. in diameter, which are generally installed using rotary drilling equipment. The piles can be installed in low headroom environments and therefore have been used frequently to underpin buildings or other structures where access and/or headroom is limited. The use of high-strength steel casing and large-diameter reinforcing bars have permitted underpinning design loads in excess of 200 tons to be placed on piles of diameters less than 10 in. The pile develops its capacity in friction between the grout and surrounding soil along the pile length primarily within the bearing stratum commonly referred to as the pile bond zone. Placement of the grout under pressure within the bond zone, similar to that of a ground anchor, enhances the load-carrying capacity between the pile and soil. Micropiles can also be socketed into rock, which significantly increases their capacity. Micropiles can be installed through existing concrete foundations and directly bonded to the foundation by side friction, connected to a footing by bracket or be used to support needle beams or other types of underpinning framing members.

Figure 11.17 shows the use of micropiles to extend a column footing to create additional basement space below the bearing level of an existing spread foundation. This technique is used where the magnitude of column loads and distance between columns make the

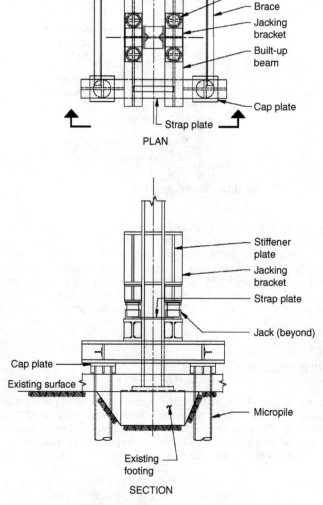

FIGURE 11.17 Underpinning of a column using micropiles.

method of supporting jacking beams on cribbing bearing on the existing basement floor between columns followed by excavation of a sheeted pit below the column impractical. Micropiles are installed by low headroom equipment around the existing foundation to be lowered. The micropiles support a jacking frame comprising structural steel members that temporarily support the column load. A jacking bracket attached to the existing column and temporary jacks are used to make the load transfer to the micropile underpinning.

Once the column load is temporarily supported by the micropiles, the existing column footing can be removed, the basement excavated, the column extended, and a new foundation constructed at the lower bearing level. The column load is then transferred back to the extended column and lowered foundation. The jacking bracket and frame are removed, and the micropiles cutoff below the new floor slab. The advantage of this method is that all of the basement columns to be extended can be temporarily supported on micropile underpinning at one time to permit excavation of the basement in one stage. This is typically advantageous to schedule and construction staging.

Figure 11.18 illustrates the use of micropile underpinning to support a subway tunnel to permit excavation for a new tunnel that crosses below it. The piles were installed through the existing subway structure from the ground surface above it to minimize the out-of-service time needed to perform the work. A track framing system supported on the micropiles was installed during weekend outages. The micropiles were braced as the excavation was made below the subway structure as shown in Fig. 11.19.

Pile Installation

The micropile is typically installed using hydraulic drill rigs. The size of the drill rig can vary significantly depending upon the application such as operating in a low overhead

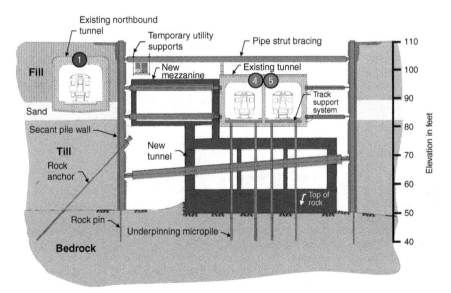

FIGURE 11.18 Underpinning of an existing subway tunnel using micropiles to permit construction of a crossing tunnel below it. (*Courtesy of Mueser Rutledge.*)

FIGURE 11.19 Excavation below an existing subway tunnel temporarily underpinned on micropiles. (*Courtesy of Mueser Rutledge.*)

basement or an area where access and overhead clearance are not limited. The micropile is generally installed by advancing casing using rotary methods. Drill cuttings within the casing are removed using external or internal flushing techniques. Once the casing is advanced to the tip elevation of the pile, the pile is cleaned of drill cuttings, pile reinforcing installed, and the pile filled with grout using tremie methods. For piles in soil, the bond zone is typically grouted by pressurizing the grout column either after or during extraction of the casing through the bond zone. The method of grouting is usually determined based upon soil type and the desired pile capacity. Re-grout tubes can also be installed in the pile for post-grouting of the bond zone. In most piles, the casing is usually left in place above the bond zone or extended 3 to 5 ft into it. The pile reinforcing and/or casing is used to make the connection to the structure.

Load Transfer

Figures 11.20 and 11.21 illustrate the use of micropiles to underpin the column foundations of a building to permit mining of a tunnel beneath it. Micropiles were installed within the 7-ft high basement along the centerline and the perimeter of the future tunnel. Low headroom drill rigs using 3 ft lengths of casing installed the micropiles. The 200-ton design capacity micropiles support a framing system of permanent girders and beams that span over the binocular tunnel as shown in Fig. 11.20.

Multiple columns are supported on the same girders that span over the tunnel. Column loads were temporarily transferred to the underpinning frame using jacking beams and manifolded locknut jacks to permit the column base plate to be disconnected from the pile cap and the granite stone pile cap removed. Permanent support beams were placed below the column base plate as shown in Fig. 11.21. Transfusion-type flat jacks filled with oil

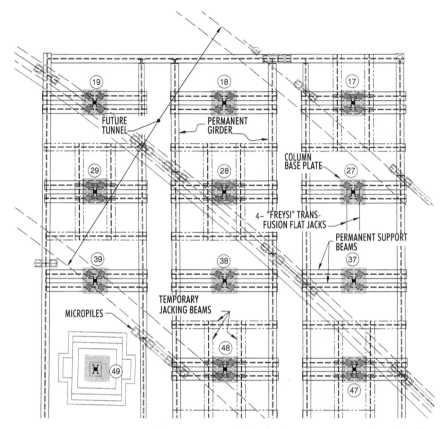

FIGURE 11.20 Permanent building column underpinning framing system supported on micropiles. (*Courtesy of Mueser Rutledge.*)

and connected to a manifold were placed between the base plate and support beams to provide a means for making load adjustments as additional columns were added to the same permanent girder. After final adjustments were made, the flat jacks were filled with epoxy grout and the remaining space between the beams and plate filled with nonshrink grout.

HELICAL PILE UNDERPINNING

A helical pile is a manufactured, steel-segmented pile comprising a central shaft with one or more helical-shaped bearing plates welded to the lower portion of the shaft. The pile is screwed into the ground by applying torsion to the shaft. Multiple shafts are coupled together to extend the pile to the desired depth into the bearing stratum. The depth of penetration is dependent upon soil type and density. These type piles are generally used with pre-engineered angle and plate brackets to underpin light commercial and residential structures as illustrated in Fig. 11.22. Design compression loads are typically limited to about 30 tons although some manufacturers are producing larger-diameter piles to support higher design

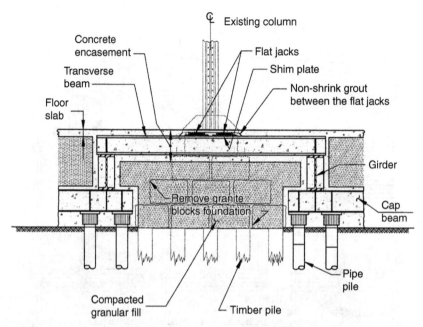

FIGURE 11.21 Permanent underpinning of building column above subway tunnel. (*Courtesy of Mueser Rutledge.*)

loads. The advantages of this pile type are that it can be installed in limited access and low headroom areas, generates minimal vibrations during installation, and can be installed relatively quickly compared to the other pile types. The disadvantages are that obstructions are not easily penetrated and that the small-diameter shaft may buckle in soft or loose soils or if lateral soil support is removed by excavation adjacent to the underpinned structure. Installation of the pile also requires that the shaft of the pile be offset from the existing structure for clearance of the helical bearing plates. This creates a bending moment at the connection between the pile and the existing structure limiting the capacity of the pile.

Helical Pile Installation

The helical pile is typically installed using a hydraulic auger drive unit or torque motor. These drive units can be mounted on a wide variety of equipment of differing sizes permitting piles to be installed in hard to reach places or areas with limited access or clearance. The pile is installed by attaching the lead section containing the helical plates to the torque motor and rotating it into the soil using downward pressure (crowd) from the drilling equipment. The rotational speed and crowd should be selected to screw the pile into the ground minimizing soil return to the surface and strain relief of the soil below the existing foundation to be underpinned that can result in settlement. The number of helical plates and spacing between them is determined considering soil type and density and the desired capacity of the pile. After the lead section is advanced, additional shaft sections are mechanically coupled to the installed end of the shaft and the pile advanced until it reaches the desired tip elevation or the desired torque. Relationships between torque and capacity developed by installers of this pile type or

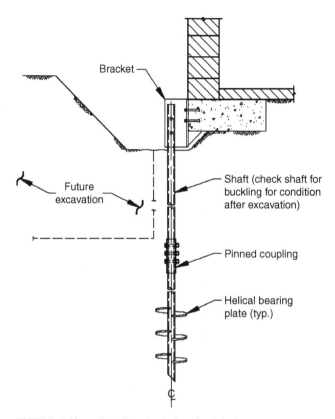

Bracket

Future excavation

Shaft (check shaft for buckling for condition after excavation)

Pinned coupling

Helical bearing plate (typ.)

FIGURE 11.22   Helical pile underpinning of wall footing.

the manufacturers are commonly used to estimate pile capacity. In the absence of this data or local experience with this pile type, pile load tests should be performed to confirm capacity. There are various methods to attach the pile to the existing foundation of the structure. Some of these methods include attaching a pre-engineered plate supplied by the manufacturer to the pile and anchoring it into the side of a foundation wall footing supporting the structure. Alternatively, a pre-engineered angle bracket can be attached to the pile and positioned to provide support at the bottom edge of the existing footing.

SLANT-AUGERED PILE UNDERPINNING

This method resembles that of the bracket pile underpinning and is most frequently used on the west coast for light one- and two-story structures. One advantage of this method is that the support system is located fully under the existing structure without infringing on the adjacent property, as shown in Fig. 11.23. Also, this prevents eccentric loads causing additional lateral forces in the earth support system.

Augered holes at a slight angle (slant) are drilled under the existing foundation at intervals of 8 ft (2.4 m) maximum. The top portion of the hole is enlarged with a special tool

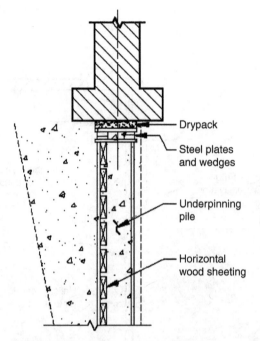

Drypack

Steel plates
and wedges

Underpinning
pile

Horizontal
wood sheeting

FIGURE 11.23   Slant-augered pile detail.

to facilitate the placement of a steel beam in the hole, which will be set plumb. The pile is encased in a lean sand-cement mix. The building load is transferred by means of steel plates and wedges between top of pile and underside of foundation. In case the pre-augered holes do not stand open because of groundwater or soil conditions, a bentonite slurry mix may be necessary to allow the setting of the steel beams and placing the sand-cement mix without serious loss of ground under the existing structure.

SLURRY AND SECANT PILE WALLS

The slurry and secant pile methods of support installed adjacent to an existing structure can only take place when there is ample room between the old and proposed building or when the slurry or secant pile wall can be incorporated within the proposed structure. It is used instead of installing a support system under a structure. Both of these methods offer the added benefit of providing a groundwater cutoff if the excavation extends below the groundwater and soils beneath the structure being protected could consolidate, resulting in building settlement from lowering the groundwater beneath it.

Slurry Wall

If a slurry wall is used in lieu of underpinning, and depending on the type of structure to be supported, the panel length should be reduced to 4 to 8 ft (1.2 to 2.4 m). The lateral support system must take into account the surcharge loads introduced into the slurry wall by the structure sitting directly behind it. The design must consider soil type and bearing capacity, proximity to the existing foundations, and whether panel excavation below the adjacent foundation will cause a local bearing failure of the trench wall or strain relief of the soil supporting the foundation load that could result in building settlement.

Secant Pile Wall

A secant pile wall consists of a row of drilled piles overlapped (secant to each other) to provide a continuous wall. The row of piles typically comprises primary and secondary piles. A guide wall is constructed to provide a template for secant pile locations. The primary piles are initially installed. The secondary piles, which overlap the primary piles and form the continuous wall, are installed a day or so following primary pile installation. An H-pile or wide flange section is typically installed in each of the secondary piles to reinforce the wall for installation of tiebacks or rakers during excavation. The piles are typically 30 to 42 in. in diameter and are filled with concrete. The advantage of this method is that the pile is advanced by drilling casing to the desired tip elevation while cleaning spoil materials from within the casing. Therefore, the drill hole is supported by casing at all times during installation compared to slurry support for the slurry trench wall. Additionally, each secant pile is a much smaller sequential excavation than one panel for a slurry wall excavation. These advantages of the secant pile wall are desirable in applications where the adjacent structure and soils beneath it are sensitive to disturbance and/or movement.

JET GROUTING

Jet grouting is a method of underpinning where a high-pressure jet is used to cut and erode the natural soil in order to mix and partially replace it with grout (a cement and water slurry). This method creates a soil-cement column that can be designed to support foundation loads and lateral pressures and to provide groundwater control during excavation (Fig. 11.24).

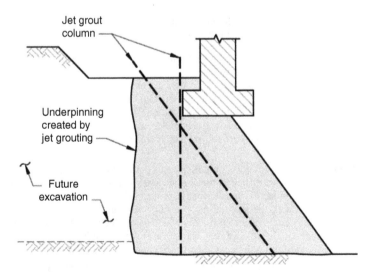

FIGURE 11.24   Jet grout underpinning of foundation wall.

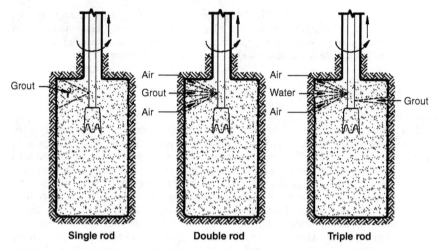

FIGURE 11.25   Jet grouting systems.

Jet grouting is most effective in cohesionless soils but can be used in all types of soils, including silts, clays, and in some instances decomposed to highly weathered rock. For underpinning, jet grout columns are constructed sequentially at primary, secondary, and tertiary locations similar to the sequencing performed for pit underpinning. Columns can be overlapped to provide a continuous wall. A single reinforcing bar, bundled bars, or structural steel shapes depending upon soil-cement column diameter and fluidity can be placed through the fluid soil-cement to increase load-carrying capacity or to resist lateral pressure. There are typically three generalized methods used for jet grouting as illustrated by Fig. 11.25. Other methods of jet grouting such as using intersecting jets and double cutting to create larger-diameter columns are not discussed herein as these methods are not typically used in underpinning applications:

1. **Single fluid.** A single tubular rod with a cutting bit is drilled into the soil to the specified depth. Cement grout is then pumped under high pressure through the rod into the soil as the rod is withdrawn. Since the cutting bit is of larger diameter than the rod, this leaves an annular space through which some of the soil-grout mixture is pushed upward by the pressure of the grout. The grout nozzle is positioned along the side of the rod, above the cutting bit. This method is the least desirable for underpinning because of the high grout injection pressures used with this method and the subsequent potential for ground heave.

2. **Double fluid.** The jet grout rod is configured to permit the separate supply of high-pressure air and grout to different concentric nozzles, which discharges a shroud of air around the grout injection nozzle that is positioned above the cutting bit. The air enhances the erosion of the soil during grout injection and mixing of the soil. Soil-cement column diameters of 3 to 6 ft can be achieved in cohesionless soils using this method. As in the single-fluid system, the grout pressure displaces some of the soil-grout mixture upward through the annular space created by the cutting bit.

3. **Triple fluid.** The jet grout rod contains three tubes to provide a separate source of grout, air, and water. High-pressure water shielded in a cone of air is used to cut and erode

the soil. Cement grout is injected under lower velocity than the other methods into the soil just below the air/water nozzle. This method generally results in a better quality soil cement column compared to the other methods and is the preferred method for underpinning. Soil-cement columns ranging from 3 ft to more than 5 ft in diameter can be achieved using this method.

For all three systems, the drill rod with jet grout tool and cutting bit is advanced to the design depth and rotated while grout is injected and the drill rod withdrawn to create a soil-cement column. The diameter and strength of the soil-cement columns is a function of soil type; the method of jet grouting; the operating pressure; and flow rate of air, water, and grout, along with rotation and withdrawal speed of the system. The methods employed are illustrated in Fig. 11.26.

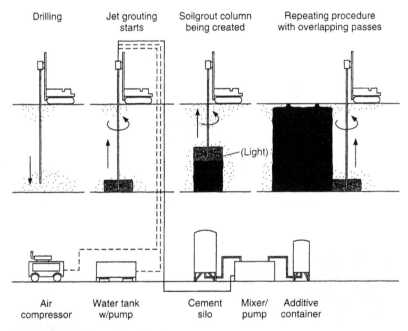

FIGURE 11.26   Jet grouting plant setup.

Jet grouting requires a highly experienced contractor, with the special equipment and knowledge to engineer and produce the expected results. Many of the jet grout rigs have the capability to monitor the drilling and jetting parameters during installation such as rotation speed; withdraw rate; pressure; and flow rate of water, air, and grout. Monitoring of these parameters during installation along with performing specific gravity testing of the grout mix and compressive strength tests on representative samples recovered from the installed columns provides valuable information during installation for assessing column uniformity, geometry, and strength.

A thorough geotechnical investigation program is essential, with detailed, reliable soil descriptions, strata identification, laboratory gradation and strength data where appropriate, and location of groundwater levels. A jet grout test program should be performed at the proposed location to determine the pressures and rate of withdrawal to be used to produce the strength and diameter of the soil-cement columns required by the design. The test program is also needed for the contractor to demonstrate that the method of jet grouting will not heave or damage the structure being underpinned. With these caveats, jet grouting for underpinning can be an effective and economical solution, especially in areas where difficult access precludes the use of other types of underpinning methods.

GROUND FREEZING

Ground freezing is another method that has been successfully used to temporarily underpin structures until the permanent support can be constructed. Although not routinely used for underpinning, it can be advantageous where difficult ground conditions, high groundwater levels and complicated below-ground excavations are required below or adjacent to existing structures. Ground freezing involves the installation of a closed circuit of piping that includes freeze pipes spaced throughout the mass of soil to be frozen. A brine solution chilled by an onsite refrigeration plant is circulated through the network of piping to freeze the ground. Typical brine circulation temperatures are on the order of −20°F, which results in frozen ground temperatures in the range of 5°F to 15°F.

This technique was used for a tunnel project in Boston, Massachusetts, which required that a historic 7-story building be temporarily underpinned to permit a binocular transitway tunnel to be constructed approximately 10 to 15 ft below its foundations. The tunnel alignment passed below a number of interior column foundations supported on timber piles and the building façade. Groundwater levels were tidal and within a few feet of the lower floor slab. Ground freezing was selected because it could stabilize the soft compressible soils at the crown of the tunnel, provide a groundwater cutoff during tunneling, and stabilize the soils above the tunnel that would permit temporary support of the building columns on the frozen soil. Figure 11.27 shows the typical freeze pipe layout between columns to be temporarily supported.

FIGURE 11.27 Ground freezing piping and freeze pipes between existing columns being temporarily underpinned. (*Courtesy of Mueser Rutledge.*)

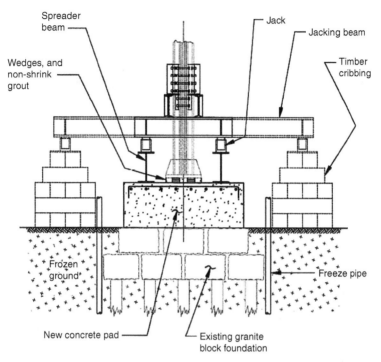

FIGURE 11.28 Load transfer from temporary underpinning supported on frozen ground to timber piles re-supported on roof of new tunnel. (*Courtesy of Mueser Rutledge.*)

A jacking bracket and related beams were installed to temporarily support the 500- to 700-kip column loads on the frozen ground and permit replacement of the upper granite block foundation with a reinforced concrete jacking pad as shown in Fig. 11.28. Jacks were placed between the jacking beams and timber cribbing to make adjustments for heave or settlement while the column was temporarily supported on the frozen ground. After the tunnel was mined below an individual column, the timber piles of that column foundation were cutoff and re-supported on the tunnel roof lining. Jacks were placed on the jacking pad to re-level the column as load was transferred to the tunnel roof.

SAFETY PRECAUTIONS

Since, in most underpinning operations, workers are working under an existing building, often in pits dug into the founding layer, safety precautions are very important. Most municipalities include provisions regarding stability of trenches and pits. In most cases, sheeting of pit sides is required. In deep pits or trenches, bracing must be provided.

Some of the factors affecting stability of excavation sides or side slopes are type and density of soil, elevation of groundwater table, methods of excavation, nearby construction operations, vibrations, and surcharge loadings. The designer must always examine the applicable codes to be sure that the method chosen meets all code and life-safety requirements.

Type and Density of Soils

Loose sands and soft clays present the greatest design difficulties, since lateral pressures developed are the greatest in these soils. On the other hand, dense sand-gravel mixtures and stiff clays may be self-supporting at near vertical slopes for a considerable time, leading to limited loads on sheeting and bracing. The presence of soft organic soil layers may lead to sliding failures of slopes and extremely high lateral pressures.

Elevation of Groundwater

This factor is especially important in sandy and silty soils. Excavation below the groundwater table should be avoided, if possible, by dewatering to draw down the water table before the start of excavation. If the excavation is in a tidal area, the fluctuation in the groundwater table may seriously affect dewatering operations and another type of underpinning may have to be chosen.

Methods of Excavation

Owing to high labor costs, some contractors have replaced some of the hand excavation for approach pits and shallow underpinning pits with machine excavation, using a small backhoe. This can only be done if the following conditions are met:

1. The soil on the sides of the cut must be capable of standing vertically without support.
2. The Occupational Safety and Health Administration (OSHA) requirements must be met.
3. The structure (foundations, walls) must be sound enough to span 6 ft (180 cm) without soil support.
4. Competent machine operators and laborers are available, along with capable and knowledgeable supervision.
5. Proper backfilling around pit boards is performed without delay.
6. Maximum depth of underpinning does not exceed 5 ft (150 cm).

Procedure of installing underpinning using some machine excavation

1. Make prefabricated timber rings, consisting of 2 by 8 in (5 by 20 cm) lumber.
2. Excavate opening under wall 5 or 6 ft (150 or 180 cm) wide, to proper elevation [not more than 5 ft (150 cm) below bottom of footing].
3. Place rings in excavation, interconnected with vertical 2 by 4 in (5 by 10 cm) lumber at internal corners, to 6 in (15 cm) below footing bottom, to allow placement of concrete inside the sheeted area.
4. Backfill around rings.
5. Place concrete.
6. Install drypack not less than 12 h after the concrete is placed or place concrete with a lip.

Nearby Construction Operations

Some operations that may lead to failures of slopes or high lateral forces on sheeting and bracing are vibrations and loads caused by pile driving, heavy equipment, dewatering, blasting, and tunneling.

Surcharge Loads

The presence of slopes and/or surcharges adjacent to the excavation will also add considerable lateral load. The effects of all these loadings must be anticipated in proper design of excavation support.

Vibrations

The impact from vibrations should be considered. Sources include subways, railroad lines, highway traffic, and seismic loads. Construction-induced vibrations can also be generated by pile driving if a driven pile foundation is selected for the new structure.

CONSTRUCTION DOCUMENTS

Drawings for an underpinning operation are in the form of shop drawings. They must be very specific, indicating all the dimensions and materials and a detailed description of methods to be used. Specifications are often given directly on the drawings as notes.

Usually the engineer of record will review drawings and specifications but will not assume the responsibility of approving or disapproving them. In most jurisdictions, the drawings require the seal of a professional engineer and must be submitted to building officials for review. They will likewise not assume responsibility for the methods and materials used.

RESPONSIBILITY

In most projects, underpinning is a "design-build" operation. This means that the contractor who performs the work either has a professional engineer on staff or engages a professional engineer to do the design. In either case, the contractor assumes the risk of damage to existing structures or utilities. To minimize this risk, the professional engineer needs complete information regarding the existing construction and the on-site subsurface conditions. This must include a sufficient number of new and existing test borings and/or test pits, construction documents (including, if possible, as-built drawings) of the affected buildings, construction sequences, and any other information that could affect the underpinning operation.

Before beginning the underpinning operation, a complete survey of existing structures (with photos, sketches, and reference measurements) is absolutely necessary. If this precaution is neglected, the team doing the underpinning may be held responsible for preexisting conditions.

When underpinning is indicated on structural drawings, the matter of responsibility should be carefully spelled out, and unless the structural engineer completely designs and inspects the underpinning, the method should not be shown on the structural drawings. The drawings should only indicate the extent of underpinning required.

INSPECTION

The contractor's professional engineer has the responsibility for the design of the underpinning and therefore this professional engineer should inspect the work to be sure that it is being installed according to the design and to make sure that actual field conditions are as assumed. In some cases, changes to the design may be necessary, when soil conditions and/or the existing construction are different from what was assumed.

The responsible professional engineer who personally inspects the operation can spot these differences before they lead to problems and can make necessary changes expeditiously. This will prevent failures and prevent unnecessary delays to the project. The professional engineer may delegate the daily supervision and field inspections to a qualified and experienced representative who must be in daily contact with the professional engineer.

BIBLIOGRAPHY

Chellis, Robert D.: *Pile Foundations*, 2d ed., McGraw-Hill, New York, NY, 1961.

Holtz, Robert D., Kovacs, William D., and Sheahan, Thomas C.: *An Introduction to Geotechnical Engineering*, 2d ed., Prentice-Hall, Upper Saddle River, NJ, 2010.

Prentis, Edmund A. and White, Lazarus: *Underpinning*, Columbia University Press, New York, NY, 1950.

Tschebotarioff, Gregory P.: *Foundations, Retaining and Earth Structures*, 2d ed., McGraw-Hill, New York, NY, 1973.

CHAPTER 12
ROADWAY DECKING

Bernard Monahan, Sr., P.E.

INTRODUCTION

Roadway Decking is defined as a framing system that is designed to carry the heaviest trucks and equipment that are utilizing these roads. Contractors learn their profession by applying theoretical concepts to the real world of urban construction. This is particularly true of temporary structures. This subject matter is vulnerable to the often-heard criticism that public agencies and responsible professionals are immune to change.

Methods for decking over streets are fixed in the subconscious of the experienced engineer. Experience is a marvelous teacher; just as the homebuilder who erects a foundation at the ocean edge learns the meaning of tides and wave action, so too the construction engineer who builds a deck that spans a major urban intersection learns the meaning of safety. The variables are great and often defy rational analysis, not because of our inability to define the forces and resulting stresses, but more often because of our limitations we are unable to weigh all of the possibilities in a manner that is safe for the vehicle operator or pedestrian using the surface and also the construction worker who is performing his or her duties underneath. The surface decking system must be integrated into the general support system that is required for overall stability. (See Figs. 12.1 to 12.3.)

Bridging may be fabricated out of timber, concrete, or metal. Future generations may develop the use of fiberglass or other lightweight products to the point where they may be substituted for traditional materials. The use of fiberglass sections or plates to span street

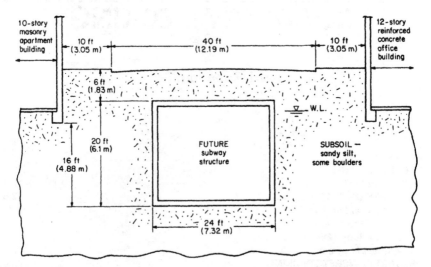

FIGURE 12.1   Typical cross section, simplified in order to clarify the basic requirements.

FIGURE 12.2   Roadway decking must be designed to support heavy construction cranes and concrete trucks.

FIGURE 12.3 A precast concrete deck in New York City, which spans a proposed subway extension. Concrete decking systems must be designed for construction equipment as well as for public use.

openings has not yet been recognized as a medium, which would result in savings of life-cycle costs for subway or other subsurface construction. Perhaps future students and practitioners will recognize fiberglass as a viable alternative. For the present, however, this chapter provides the details that an engineer must evaluate regardless of the element used to span the opening.

Another chapter in this handbook deals with the spanning of a simple trench; this chapter analyzes the decking of an entire street in an urban metropolis.

The construction engineer must generalize the design criteria and then address each segment of the design as a special situation. The need is simply that the engineer is to design a floor-framing system that is capable of supporting vehicles that are normally found traversing urban crossways, heavy construction equipment, and pedestrians while providing a safe work area for the craftsmen underneath. (See Fig. 12.4.) The engineer, given this responsible charge, must come up with a practical design that will accomplish this and withstand the forces of nature as well. Decking systems must be capable of withstanding impact and vibration loads resulting from various construction methods such as blasting, drilling, and pile driving.

FIGURE 12.4   Concrete panels (5 × 10 ft) utilized as decking and supported by steel framing.

DECKING DESIGN

A decking or flooring system might be envisioned. The beams and the decking itself are considered to act as simply supported beams. They are supported by girders, which in turn are supported by columns that transmit the loads to the foundation. A proper analysis starts with the imposition of a design load, which should include an allowance for impact and a safety factor to cover unknown variables in the construction procedure and in the makeup of the structural members.

A typical live load analysis might assume a loaded concrete delivery truck with its wheels positioned at the mid span or a combination of worst conditions. (See Fig. 12.5.) Two trucks possibly are positioned side by side while they await discharge instructions. Usually the live load is computed by assuming a uniform load of 200 to 300 lb/ft^2 (9500 to 14,400 N/m^2) or by positioning an anticipated worst condition loading.

DECK SUPPORTS

Each column or soldier beam will support one-half of the designated loads for the segment of roadway decking that is located between column supports. (See Fig. 12.6.)

A settlement or load analysis must be performed on the pile sections to determine the validity of assuming that the pile can support the design load. The analysis can consist of a load test or judgments based on past experience, or on accepted criteria such as the *Engineering News Record formula.*

Each case is different; however, usually a high driving resistance will convince most engineers that sufficient penetration has been achieved. An exceptional condition is the case

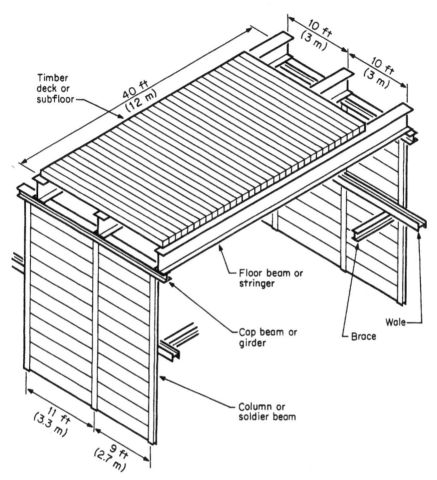

FIGURE 12.5 Concrete or timber decking with cap beams.

in which a soldier beam is driven and intercepts a boulder, which prevents its further penetration. This soldier pile, acting as a column, is obviously unsatisfactory; it must be freed; however, the choices are limited and almost always time consuming. If the engineer is sure that the obstacle is natural and not a manufactured utility such as an electrical duct bank or a water main or some other in-service utility that is not shown on the available plans, a drill steel might be used to sound the depth of the obstacle. It might be that the borings failed to locate a natural ridge in the substrata rock formation. However, more likely it is a sizable boulder of 2 ft (60 cm) or more in diameter. A pit could be excavated alongside the soldier pile, and hand drilling and chipping techniques could be used to free the tip of the pile. Frequent obstructed piles will cause a contractor to predrill the anticipated pile locations. This may consist of drilling to subgrade with a 2- to 3-in (5- to 7.5-cm) auger or boring tool and then redrilling holes that are known to contain obstacles with a well drillers rock bit that has a larger diameter than the steel soldier pile that is needed to support the sheeting and decking.

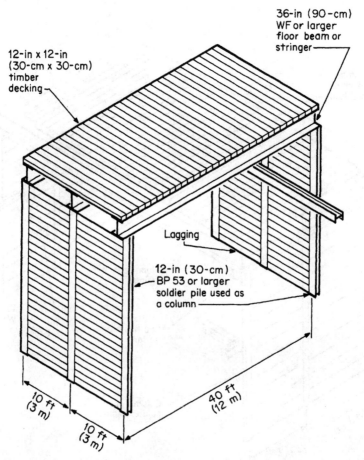

FIGURE 12.6   Timber decking supported by stringer beams.

Blasting techniques are rarely used in these circumstances because of the proximity to surface utilities and the need for heavy charges when no relieving face exists. A more common approach is to drive the soldier to the top of the obstruction and then to underpin the pile during excavation. This case requires that the decking system be completed and the short pile be resupported, usually by means of welding a spliced section to the bottom of the obstructed pile. The method, as well as the use of temporary supports, must be judged on the basis of conditions at the time when re-support is necessary. Usually the waler support can be welded to the soldier pile and thus act as a needle-beam support to ensure a bridging of the soldier pile during underpinning procedures.

CONCRETE OR TIMBER DECKING

At this point, let us assume that the pile columns have been installed satisfactorily and that the contractor is now ready for the actual decking sequence. The use of the girder or cap beam is voluntary; if the pile column spacing is the same as the decking—beam spacing, say 10 ft (3 m), each cross beam could sit directly on a column or a soldier pile. Remember that the lateral soil pressure mandates the column-support spacing, and if timber lagging is to be installed, 10 ft (3 m) will probably result. The omission of a cap or girder beam requires that the pile column or soldier beam have a fairly close tolerance, say within 6 in (15 cm) of design location. Those soldier piles, which fall outside of the tolerances, will then require a cap beam that spans a minimum of two bays or three soldiers. (See Fig. 12.7.)

FIGURE 12.7 Concrete truck and crane positioned on roadway decking.

After the question of cap or girder beam is answered, the engineer must now choose a decking or floor beam that will support the given loads and whose deflections will not be excessive.

A typical segment, in this example, considers that bending, shear, and deflection calculations dictate the use of a W33 × 240 stringer beam in this case that spans 40 ft (12 m). Deflection of support stringer beams is a factor in assessing hanger requirements. Excessive movements may cause leaks in waterlines and shorts in electric ducts that are hung from these roof beams.

The important point to remember in analyzing a specific decking requirement is that in a heavily trafficked urban thoroughfare assumptions should always be made conservatively. Concrete or timber decking systems must be designed for the support of construction equipment and for public use. (See Figs. 12.8 to 12.11.)

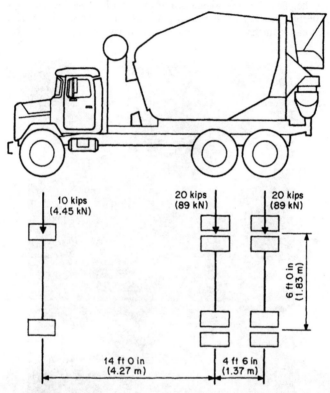

FIGURE 12.8 Wheel loading for a rear discharge concrete truck with a 15-yd³ capacity. [Wheel load = 10 kips (44.5 kN) per wheel; resultant of two rear wheels = 20 kips (89 kN); front axle load = 20 kips (89 kN); rear axle load = 40 kips (179 kN) each.]

The resulting deck shows the optimum case of the deck and floor beams sitting directly on the evenly spaced column supports. In this circumstance the girder or cap beam can be deleted, while the various other members of the decking support, such as braces and tie rods, are required.

The connections between columns and floor beams are normally bolted to prevent a gradual "walking" of the floor beam off the head of the column. The tops of the columns and the ends of the floor beams should be secured against rotation or independent movement due to live loads or vibration. Normally a spacer or tie beam placed on each end and at the third points will be sufficient to prevent rotation on spans of less than 50 ft (15 m).

The floor beam must be checked for buckling and stiffener plates added at vital areas. Usually stiffener plates are required at the supports, where the pile column or soldier, because of the heavy load concentrations, supports the floor beam.

Having designed the floor or decking beam, the engineer may wish to develop a plan to install this 33-in (84-cm) beam without disrupting the normal traffic pattern. These projects

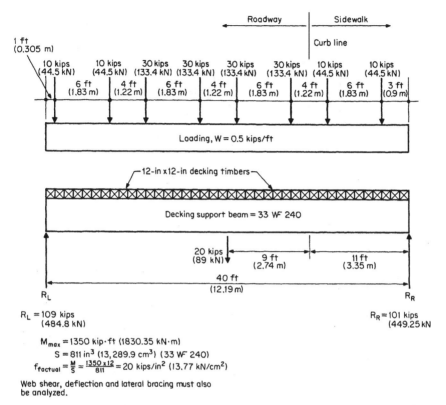

FIGURE 12.9   Decking support beam analysis.

create a great deal of inconvenience and following a standard procedure can minimize the hardship to the public. A typical sequence is as follows:

1. Install soldier column piles.
2. Reroute gas lines and other utilities that either obstruct the floor beam installation or would cause a hazard to the construction crews.
3. Excavate and vehicular-plate the cross or slot trenches that are to receive the floor beams, which are to span the entire street below the surface.
4. Close street; reroute traffic, preferably during off-hours; and install floor beams.
5. Reinstall vehicular plates and reopen roadway.

Specific periods or hours are usually prescribed for installation of floor beams or any operation that is expected to interrupt the normal flow of traffic. In certain business districts, it might require a 10 P.M. to 6 A.M. operation with noises muffled to minimize complaints.

Having completed the design and installation of the floor-beam support system, one must consider the decking or surface design. The design must be simple, for the deck sections will be subjected to all of the known design variables and a safety factor is necessary to provide for some others that were not envisioned. The decking must be durable, for it

FIGURE 12.10   Steel beams supporting street plates at a busy intersection.

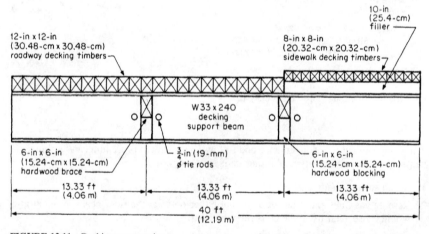

FIGURE 12.11   Decking cross section

will be in service for at least 3 or more years. The decking must be portable, for it must be remembered that the decking is a cover or roof that enables a contractor to install the actual structure below while the city continues its daily routine. Finally, the decking must be simple and safe in order to avoid errors in loading and installation. (See Fig. 12.12.)

One system that fits all these requirements is a timber-decking sequence. A structural grade of 12- × 12-in (30- × 30-cm) timber is a simple beam that is durable and portable. It is relatively light and can often be moved into position by three workers. The timber beams are standard and may be interchanged; the friction factor of wood is relatively good, when

FIGURE 12.12   Concrete decking panels of 5 by 10 ft placed parallel to traffic flow.

compared with steel or other metal, for braking auto vehicles. Complaints due to slipping and falling are usually minor and are not greater than those of a normal busy roadway. Disadvantages are that wood is a natural material that is often subject to the discontinuities of nature. A timber beam may develop a fault line through an internal knot or flaw after having been installed, and a heavy loading may cause cracking or actual failure. Repetitive wetting and drying cycles may weaken the resiliency of the section and may lead to failure. All things considered, timber has proved itself to be a dependable material for decking. Timber beams tend to redistribute stresses to adjacent members when overloaded. Timber is not fragile or brittle; it is rather a flexible material that seems to always meet our expectations even under adverse conditions.

The following definitions and design procedures for a timber or concrete deck are now introduced:

Structural Lumber.   This term refers to material that has been graded for strength in accordance with nationally recognized standards and to which specific values have been assigned for the various strength properties.

Design Procedures.   Lumber or concrete used structurally in temporary buildings, bridges, and other facilities is subject to the same design procedures and basic formulas that are applied for other kinds of structural materials like steel. Lumber is not an isotropic material and this condition must be recognized in proper design.

Use of Actual Dimension.   Calculations to determine required sizes of timber members must be based on actual or net dimensions.

Allowable Stresses.   Typical allowable bending stresses as determined for single-member uses are increased 15 percent. Allowable unit stresses for normal loading duration may be increased, as follows, when duration of the full design load does not exceed the periods indicated: 15 percent for 2 months, as for snow; 25 percent for 7 days, as for loads during construction; 33 percent for wind or earthquakes; 100 percent for impact.

Precautions must be taken to prevent the timbers from gradually walking off of the floor beam flanges. A welded seat angle or metal runner is often used to prevent this type of serious failure.

A 10-ft (3-m) timber deck beam weighs approximately 500 lb (225 kg), and if it falls to the floor of an excavation, it becomes a deadly missile. Carelessness of this nature is negligence that may result in criminal penalties as well as crews refusing to continue to work and in stop-work orders by governing agencies. Contractors realize that safety and good workmanship pay high dividends, and for this reason especially skilled workers are normally used when placing decking beams or precast concrete slabs in crowded urban centers. Daily follow-up inspections are necessary to ensure continued safety to the workers below. (See Figs. 12.13 to 12.15.)

This discussion has pointed toward the 10-ft (3-m) span that is commonly found in the northeastern metropolitan area. They have a laissez-faire attitude toward trucks and their allowable wheel loads. The streets of the northeastern metropolis's are among the busiest in the world, and they are often found to contain vehicles that punish temporary structures beyond our proposed safety factors. They care for their street and property, but they realize that business seems to prosper under a hands-off approach. Overloaded vehicles are common in the city and trucks are rarely weighed. The result of this knowledge is that a 10-ft (3-m) span is considered a maximum for a concrete or timber deck in the northeast cities. In other areas, this is not the case; truckloads are more reasonable and enforcement is such that a 12-ft (3.7-m) spacing is allowable and considered a safe practice. Concrete trucks are limited to 10 yd³ (7.65 m³) in many other areas, rather than the 15 yd³ (11.5 m³) allowed in New York and other northeastern cities. (See Fig. 12.16.)

All the previously mentioned design considerations are pertinent to the installation of either a 10- or 12-ft (3- or 3.7-m) decking slab or beam span. The engineer must always determine what type and level of loading can be expected under normal conditions and what the maximum anticipated loading is.

FIGURE 12.13   Concrete decking system installed along a jersey-type safety barrier.

FIGURE 12.14 Ten-foot roadway decking timbers that span between two support beams are utilized to provide access hatches for construction operations. (*Thomas Crimmins Contracting Co.—Photo Collection.*)

FIGURE 12.15 Excavation progresses by removing decking. Spacer timbers prevent lateral movements in decking support beams.

FIGURE 12.16   Precast concrete panels over new subway construction. System is designed to support maximum allowable loadings.

PRECAST CONCRETE DECKING

The precast concrete slab, used as temporary decking on a busy urban thoroughfare, is becoming more popular. A 5- by 10-ft (1.5- by 3-m) section is prefabricated, under controlled conditions, and shipped to the jobsite. The concrete slab, when positioned, is more durable than the timber 12 × 12 in by 10 ft (30 × 30 cm by 3 m) as a decking beam system; however, it is more costly. The concrete slab weighs three times the weight of wood, and the design normally requires a heavier support system. Larger-support floor beams require deeper steel sections and result in a lowering of additional utilities that might be hung from the underside of the installed floor girder system.

The minimum thickness of a precast concrete decking panel is usually 9 in (22 cm) with the preferred thickness being 12 in (30 cm).

"Do not argue with success." Subway contractors and designers compare alternative systems on each job and choose either concrete or timber based on their past experience. Many contractors believe that the most reliable and cost-effective system remains the timber-beam design. Owners that prefer the precast concrete deck are challenging it; however, the added weight of the concrete system poses a formidable extra cost every time a labor supervisor is instructed to lift a slab at an access opening. The 12- × 12-in by 10-ft (30- × 30-cm by 3-m) timber is easier for hand labor to remove and replace, whereas the concrete slab requires the use of a mechanized crane for its removal and replacement.

UTILITIES

It is worthwhile to consider the standard conditions, if indeed; they exist on the proposed project under study by the subway contractor. (See Fig. 12.17.)

The intersection contains utilities running in all four directions. Some of these utilities are close to the surface and interfere with the proposed decking system. Other utilities, such

FIGURE 12.17 City street where electric and telephone ducts are temporarily supported.

as gas, are automatically rerouted outside of the subway excavation, for it is feared that an undiscovered leak or break might cause a serious explosion.

Sewer and steam take a high priority, for a gravity sewer must be maintained at grade and a steam line requires condensate pump pits at low points in the line. Electric, telephone, and water are usually routed in a manner that allows the normal construction operations to proceed. That is, the duct bank usually requires the removal of the concrete duct encasement, while the waterline might be made satisfactory by utilizing hangers and anchors that secure the line to the underside of the floor beam system.

FUTURE TRENDS

What are some of the innovations that are on the horizon? What might future engineers expect in the twenty-first century? The answers to these or similar questions fall into a hypothetical area; however, certain generalizations might be logically presumed.

Tunneling methods will probably become so advanced and cost-effective that the massive open-cut type of excavation will become obsolete and be replaced by a tunnel that is installed at a level below that of the utilities. This would enable the contractor to concentrate on the debris removal entrances rather than the open-cut type of problem.

The central city might become less populated and the need for constant access along and on top of subway structures might become less desirable. This would enable the contractor to fence in the entire excavation, while providing pedestrian access along sidewalks, during the construction period. This open-cut type of excavation is presently used to build foundations and basements for large office buildings.

Neither of these two descriptions is really innovative; both are practiced today where the conditions are suitable. However, where a new subway is to be placed in a metropolitan setting, the old but tested methods will be slow to give way to innovation.

The use of aluminum mats and fiberglass panels in place of concrete or timber decks is a likely possibility. Temporary bridge panels may gain in popularity and be used between intersections.

DESIGN CONSIDERATIONS FOR BEAMS AND COLUMNS

Design of Wood Beams.   Investigation of the strength and stiffness requirements of a wood beam under transverse loading should take into consideration the following factors: bending moment induced by the load, deflection caused by the load, horizontal shear at supports, and bearing on supporting members. Any one of these four factors may control the design, although deflection of a temporary beam is usually not a matter of safety and a limit is applied only where appearance or comfort is important.

Design of Simple Solid Columns.   The most common form of wood column is a simple solid member of rectangular cross section. A critical factor in the design of such columns (or braces) is the slenderness ratio, or l/d for rectangular sections and l/r for round and other sections. In determining the slenderness ratio, the laterally unsupported length is measured along the longitudinal axis and is the distance between supports, which prevent lateral movement in the direction of the least dimension d or least radius of gyration r. The l/d ratio for a simple solid rectangular column should not exceed 50.

For short solid columns where the l/d ratio is in the order of 10 to 12 (and sometimes higher depending on the species and grade of lumber), the allowable load is governed by the axial compressive strength of the timber without reduction for slenderness.

Treatment of Timber.   For pressure-impregnated preservative-treated wood, no reduction in allowable stresses is necessary. For wood pressure impregnated with a fire-retardant chemical treatment, a 10 percent reduction in the allowable stresses is often recommended.

REGULATIONS AND RESPONSIBILITIES

The installation must conform to the requirements of the Occupational Safety and Health Act (OSHA) of the federal regulations, and the code requirements of the various states and municipalities. The local building code usually contains many of the minimum standards that are considered acceptable and is beneficial in guiding the engineer in design requirements. Safety to the public is paramount in determining the minimum limits for installing decking or other structures that are designed and installed to protect the public during construction operations.

The inspection of construction installations is the responsibility of the governing agency. On some construction sites, a multitude of agencies will inspect and issue a variety of violations that must be corrected. The basic requirement for on-site inspection remains with the owner who often delegates the work to an inspection agency.

The contractor is often required to propose temporary construction details that have been prepared by a consulting professional engineer in the contractor's employ; the design engineer of record then reviews these details. The project specifications often state that temporary structures are the responsibility of the contractor and that the examination of these proposed details by the design engineer does not relieve the contractor of this responsibility. In practice, accidents usually result in damages that are shared by all parties. The legal system seeks out the deepest pockets to assess blame and often find those that have the highest insurance limits as partially responsible.

DRAWINGS AND SPECIFICATIONS

Drawings for temporary decking or sheeting should be equal in detail and completeness to the contract drawings. Specifications for temporary facilities should be provided to the contractor and should clearly outline the minimum requirements for acceptable installations.

COSTS

Estimating the cost of various decking installations is dependent on the contractor's experience regarding previous installations. There are two methods for estimating these costs. The first is to work up the basic ingredients of cost for labor, materials, equipment, and insurance on a self-perform basis. Labor and equipment costs are dependent on time and productivity factors, whereas materials are simply the number of items multiplied by their unit cost. The second approach is to separate the work into packages and to solicit estimates from subcontractors for each group of tasks. In practice contractors usually combine the two methods to arrive at their estimate based on the values that they feel most confident in.

A standard approach is to estimate the productivity of a work crew during an average week on the assumption that the crew will complete a segment of decking during that pay period. For analysis purposes, let us assume that the crew will complete 200 linear ft (60 m) of decking 50 ft (15 m) in width during a 5-day period. The night shift will install four decking beams per night, or 20 per week. The day shift will install concrete or timber decking on the previously installed beams. The cost of this operation may be estimated, based on prevailing wages and equipment rates; figuring on a straight-time, three-shift basis; or allowing for overtime, if weekend work is desired. A slower productivity rate, or a higher crew cost, would increase the labor and equipment factor. Cost estimates vary with union restrictions and traffic patterns in urban centers.

CHAPTER 13

CONSTRUCTION RAMPS, RUNWAYS, AND PLATFORMS

Bernard Monahan, Sr., P.E.

INTRODUCTION

The planning of access is basic to the completion of every construction project. City traffic must coexist with construction endeavors, engineers must plan to cover trenches during peak traffic periods. Trucks should be provided with ramps or runways to load and unload materials. Cranes require platforms within the construction project in order to maintain clear streets for the delivery of supplies and to provide a safe turning radius for the crane.

The purpose of this chapter is to present some of the pertinent considerations that must be evaluated when designing a support system for construction access.

DEFINITIONS

A ramp is a sloped surface that joins different levels. It is utilized on a construction project to provide access between a foundation or basement and the street level.

A runway is a strip of level road that forms a track for wheeled vehicles. It is a narrow extension out into the construction site.

A platform is a raised horizontal surface of wood or steel used by cranes and vehicles to bring materials into construction projects.

DESIGN AND RESPONSIBILITIES

The designs of ramps, runways, and platforms are often assigned to the contractor. The engineer of record will often require that plans for temporary structures be submitted by a professional engineer retained by the contractors, the design is then examined by the owner's designer. The responsibility for the design of the temporary structure remains with the contractor, the logic being that only the contractor could be aware of the problems and needs of the construction crews. Owners and government authorities often believe that access and supplies during construction operations are best planned for by the contractor. The contractor as a major participant in the building team is often willing to accept this charge and will be required to retain or employ professional engineers to plan for access to the construction project.

In order to properly plan for access, the constructor must segregate the job into building blocks and realistically grasp the magnitude of the problem. A high-volume traffic access might be approached on a different level from a rarely used secondary emergency approach. However, under any circumstances the design loading should not be reduced.

The physical conditions of ramps, runways, and platforms might be considered similar in that each structure must be designed for the most stressful or worst condition; however, a single-usage design must be supplemented by specific controls, such as a flagman, whereas a highly trafficked approach might require wider access and higher safety factors.

The engineer responsible for the design of temporary ramps, runways, and platforms must understand the physical conditions that could lead to failure of the structure. Only by having a clear understanding of the weakest link in the support chain can the professional plan to overcome the condition that could cause collapse.

The cost of construction is increasing; the cost of a construction failure is excessive and avoidable. Good training and quality control can reduce the cost of failures in terms of human life and monetary considerations. It is up to the construction industry team to provide the expertise required to meet the challenge for safe and efficient access passageways.

The basic human factor in the construction industry is the civil engineer, employed or retained by contractors, design consultants, and by government. The civil engineer's responsibilities include the design and supervision of the installation of ramps, runways, and platforms. Too often a superintendent will delegate the responsibility for important structures to construction craftsmen rather then to trained professionals. This leads to dangerous practices that may cause failures.

VEHICULAR RUNWAY PLATES

The most common type of ramp or runway covering is the steel plate that is used to support vehicular traffic where trenching is required. There are many intersections where utility relocations require extensive plating for temporary runway access. Vehicular steel plates must also be in place during nonworking periods to allow safe passage of traffic (see Figs. 13.1 and 13.2).

Temporary runways are often required to expedite the handling of materials during construction. These projects require the contractor to install shoring to support cranes and delivery trucks. In many cases the trucks delivering concrete to the site are the determining factor in designing a roadway support system. In other cases it is a special crane handling situation that determines the design values for the runway support system. In either case, rolling loads can be dangerous, for the typical ready mix concrete truck may weigh 50 tons (45 t) when fully loaded. The problem becomes even more serious if consideration is given to the fact that the loading is repetitive and the friction loadings in a horizontal direction can cause serious failures. The usual method of securing street plates to roadways is to drive a series of 6-in (15-cm) spikes into the existing pavement surface. This method is effective

FIGURE 13.1   Street plates are utilized to temporarily cover excavations.

FIGURE 13.2   Street excavations sometimes require that the entire street be covered with vehicular plates that are normally 1 in by 5 × 10 ft.

in low-traffic areas where the trench plating is left in place for short periods. However, in high-traffic areas where the runway plate is repeatedly removed and replaced, the spikes create a failure plane that may lead to a serious vehicular accident.

The use of steel plates supported on timber or steel supports is commonly seen at intersections where trenches tend to branch out in order to follow various utilities that might require relocation. Typically, utilities such as sewers and steam lines require the right of way, for they must be installed to grades suitable for drainage. Other utilities must be diverted around these priority installations. The trenching system at a busy intersection may lead to a platform or covering of most of the right of way. A safe method of covering an intersection is imperative and is mandated by law (see Figs. 13.3 and 13.4).

Sheeting and bracing of the sides of the trenches serve a twofold purpose in that they support the pavement and earth and provide vertical support for the street plating system; 6-in spikes are usually used as fasteners to prevent the lateral shifting of plates off of

FIGURE 13.3 Vehicular plates must be secured to prevent movement due to the vibrations caused in the plate from the flow of traffic.

FIGURE 13.4 Horizontal movements of vehicular street plates may be prevented by the proper use of bolts, spikes, wedges, and asphalt edging.

their supports. The support headers are generally oversized 8- by 8-in (20- by 20-cm) or 12- by 12-in (30- by 30-cm) timber because of its resistance to rolling and its structural characteristics.

Runway support systems have long been relegated to a relatively minor role, with the philosophy that the system was beyond the planning control of the design engineer and belonged to the artisan for the skilled employment of materials that were capable of supporting vehicular traffic. This viewpoint is no longer practical and it is now up to the engineering profession to provide design standards for temporary street coverings or runways and other vehicular traffic hazards. Standards require professional controls of quality and performance, and this can only be accomplished through additional restrictions imposing professional inspections of field installations.

TIMBER TRESTLE BRIDGES

The timber trestle bridge is one of the simplest types of bridges. Spans are usually limited to 25 ft (7.5 m) using timber stringers. The timber stringers rest on trestle bents and abutments (see Fig. 13.5).

The first priority is to classify the loading limitations that are expected. An error in judgment regarding weight capacity can be avoided by clearly marking the safe capacity of the bridge. Many construction engineers overlook this standard requirement; they design and construct a bridging system with a certain crane loading as the design model. A new or rented crane requires a structural support revision that may not have been carried out. This omission may result in a bridging failure that leads to loss of life and serious injury. The design is based on a single lane approach by a concrete truck or crane. The driver or operator must be limited as to the rate of speed while on the vehicular bridge and must be cautioned regarding sudden stops. A major hazard regarding the construction of temporary wooden bridging is the tendency of spiked timbers to loosen. A daily engineering inspection

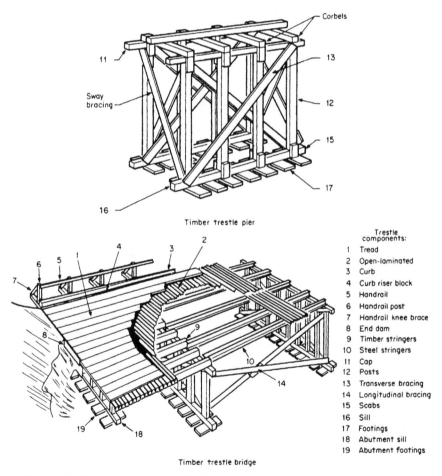

Trestle components:

1 Tread
2 Open-laminated
3 Curb
4 Curb riser block
5 Handrail
6 Handrail post
7 Handrail knee brace
8 End dam
9 Timber stringers
10 Steel stringers
11 Cap
12 Posts
13 Transverse bracing
14 Longitudinal bracing
15 Scabs
16 Sill
17 Footings
18 Abutment sill
19 Abutment footings

Timber trestle bridge

FIGURE 13.5 Timber trestle pier and bridge.

of a heavily trafficked bridge is recommended. Loose spikes, timbers, and main supporting stringers must be tightened and secured.

Main supporting stringers must be reviewed regarding bearing surfaces. Stringers have a tendency to move with the load and may creep off the cap support. Overlapping stringers are designed to provide adequate support, but a bolted or spiked connection will often be preferable. A substantial end dam or heel is needed to prevent this tendency for horizontal shifting. The bridge thread and decking planks are often combined. A 4- by 12-in (10- by 30-cm) timber laid on the flat will usually provide the required strength for construction equipment. (See Table 13.1)

Bridge footings require special attention; on permanent structures they are often driven to rock in order to provide a secure support. On temporary structures the tendency is to overlook the need for adequate bearing for the footing timbers. The 4- by 12-in (10- by 30-cm) planks used as footings are really only leveling sills, and, if a site investigation leads the engineer to suspect the bearing capacity of the underlying soils during rains or other adverse weather conditions, a separate soils analysis is in order. Investigations of this nature must be based on borings and specific load testing data.

Usually a temporary structure will not have the benefit of a detailed soil analysis and the construction engineer will visually judge a safe bearing capacity. Sands might fall in the 3 ton/ft² (30 t/m²) capacity area while a silty clay or organic meadow mat might be incapable of providing any meaningful support. Piles may have to be driven or a raft type of floating support system may have to be designed. These are judgments that must be made on site observations and investigations.

In many cases the stringer design is the most critical member of the bridge; however, a check must be made of the capacity of the posts and the cap beams. A 12- by 12-in (30- by 30-cm) post is not normally designed to carry a load in excess of 36 tons (32 t). The maximum unsupported length of the column post is critical and must be limited to a safe span. Cross bracing of bents is important, for it prevents a side sway force from causing a collapse (see Figs. 13.6 and 13.7).

TABLE 13.1   Bridge Components of Timber Trestle Bridge

No.	Bridge components	Common sizes for reference
1	Tread	2-in planking
2	Open-laminated deck	4-in planking
3	Curb	6 by 6 in
4	Curb riser block	6 by 10 in
5	Handrail	2 by 4 in
6	Handrail post	4 by 4 in
7	Handrail knee brace	2 by 4 in
8	End dam	4 by 12 in
9	Timber stringers	12 by 12 in
10	Steel stringers	12 BP 53
11	Cap	12 by 12 in
12	Posts	12 by 12 in
13	Transverse bracing	4 by 12 in
14	Longitudinal bracing	4 by 12 in
15	Scabs	4 by 12 in
16	Sill	12 by 12 in
17	Footings	4 by 12 in
18	Abutment sill	12 by 12 in
19	Abutment footings	4 by 12 in

Note: 1 inch = 2.54 cm.

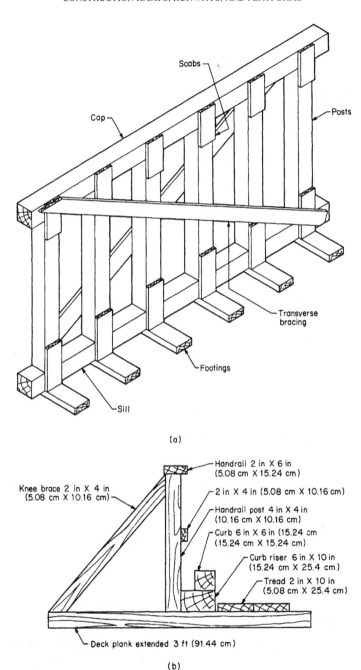

FIGURE 13.6 (*a*) Typical timber trestle bent and (*b*) curb handrail system.

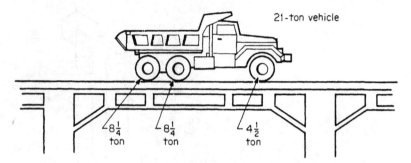

FIGURE 13.7   Effect of vehicle on bridge depends on (1) gross weight of vehicle; (2) weight distribution to axles; (3) speed at which vehicle crosses bridge.

PREFABRICATED BRIDGES

The Bailey-type bridge panel provides a rapid means of temporarily bridging streams and excavations for various types of construction equipment. It can be assembled in different ways for various spans and loadings. It is a through-truss bridge supported by two main trusses formed from 10-ft (3-m) steel sections called panels. The roadway decking can be either timber or steel. The reliability of this type of bridge is unequaled. It requires a minimum of maintenance, and the standard components may be reused under many conditions. Acrow supplies a similar type of system (see Figs. 13.8 through 13.13).

A standard panel-bridging system has various single- and double-lane widths and includes several types of decking for different traffic loading requirements. Timber or steel decking can be provided; the former for temporary use and the latter for semipermanent or permanent use. A special feature of a panel bridge is that foot walks may be cantilevered outside the main trusses and are therefore considered safe from roadway

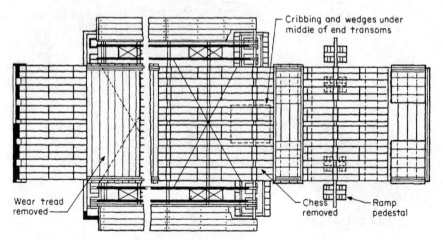

FIGURE 13.8   Steel panel-fixed bridge, Bailey type (plan).

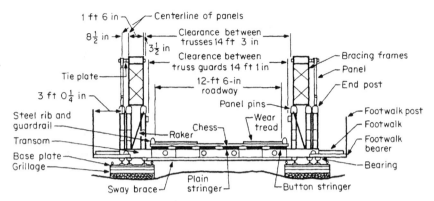

FIGURE 13.9   Steel panel-fixed bridge, Bailey type (end view).

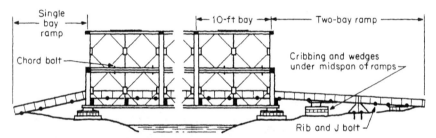

FIGURE 13.10   Steel panel-fixed bridge, Bailey type (elevation).

FIGURE 13.11   Acrow panel bridges are suitable for various spans.

FIGURE 13.12 Pedestrian panel bridges over busy urban traffic.

FIGURE 13.13 Pedestrian bridge spanning roadway.

traffic. Foot walks of 3 to 4 ft (0.9 to 1.2 m) widths can be provided, capable of accepting a live load of up to 120 lb/ft^2 (5750 N/m^2). The foot walks may be steel, chequer plate, or timber. Sloping bridges or sections should not be steeper than 1 on 15 for normal vehicular traffic. Typical widths of bridges are about 14 ft. Prefabricated bridges are available and engineers and contractors are advised to consider their use whenever bridging a gap is necessary.

FOUNDATION RAMPS

Office and apartment building foundations often require the use of temporary ramps for access during construction.

Shallow basements (one and two levels) are normally approached by means of earth fill in the form of a ramp. Deeper excavations may require timber or steel ramps. The basic rule is that ramps must not exceed a 10 percent slope for a fully loaded 10-wheel trailer to climb to street level to exit the construction site. Errors in the slope can be very costly with an expensive installation that requires winches to assist trucks in and out of the cut or foundation excavation. This defeats the beneficial advantages of a construction ramp (see Figs. 13.14 and 13.15).

Timber and steel ramps are preceded by earth or fill ramps. The excavating machines dig their own ramp as the work progresses; it is only after the excavation has reached subgrade that a steel or timber ramp is constructed.

One-Basement Structure

In this case the excavation is about 15 ft (4.5 m) or less and the earth or fill ramps are normally excavated by a front-end loader or shovel as it progresses toward subgrade. Earth ramps may be totally unsatisfactory from a traction point of view, and for this reason they are often topped with loose bricks. The rough edges of the brick form high spots for the truck tires to grab on when utilizing the ramp access. The brick topping should form a blanket at least 12 in (30 cm) in depth to minimize repairs. The topping must be maintained in order to preclude the forming of potholes; a ready supply of brick and stone aggregate is necessary to ensure a continuous operation. The existence of fine silts or clays on the ramp will lead to a loss of traction for the trucks; these soils, when wet, are slow to dry and are known to form slipping planes.

The key ingredient of a good earth ramp is drainage. The surface must be drained with materials that are heavy enough to resist erosion during heavy rains. Rutted roads make for poor transportation access and an unusable ramp can cause serious construction delays or deadly accidents.

Heavy snows or freezing rains will prevent the utilization of a ramp, as is the case with all roads. A ramp is limited in length and most projects will find it advantageous to maintain an ice- or snow-free ramp. This is accomplished by removing snow accumulations with snow plow-type equipment, and by blowing the remaining snow off to the side of the ramp using compressed-air blow pipes. The more common are 6-ft (1.8-m)-long and ¾-in (19-mm) pipes hooked to the air lines and having a hand-operated shutoff valve. A ready supply of salt or calcium chloride will often provide the assurance needed to determine whether a foundation project operates during icy conditions.

Two-Basement Excavations

These deep excavations of 30 ft (9 m) or more are approached in a similar fashion to one-basement excavations. In addition, the need for prior planning of access is more urgent; an error in judgment will require considerable effort to revise the access ramp's slope.

Some of the typical features of a timber access ramp include steel stringers and a surface decking that consists of 4- by 12-in (10- by 30-cm) rough planks. The decking rests on a series of steel stringers interwoven with 12- by 12-in (30- by 30-cm) timbers. For uniformity and simplicity, the steel and timbers are 12 in (30 cm) in depth and the typical ramping detail is similar in style and appearance to temporary bridging.

Construction

General

Acrow Panel Bridge is supplied in four roadway widths.

Standard ⎫
Extra wide ⎬ Single lane
Ultra wide ⎭
Double wide Two lane

Roadway widths and clearances are shown below.

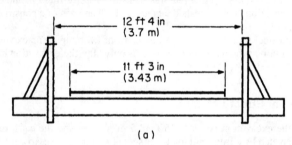

(a)

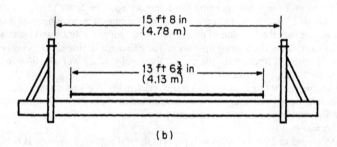

(b)

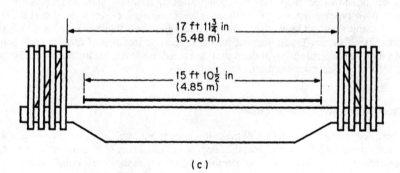

(c)

FIGURE 13.14   Acrow bridge dimensions: (*a*) standard; (*b*) extrawide; (*c*) ultrawide.

FIGURE 13.15 Construction ramp for a large building foundation. (*Thomas Crimmins Contracting Co.—Photo Collection.*)

The steel stringers are utilized to support the actual wheel loads of the vehicles that are expected to use the ramp. Care must be exercised to route the vehicles in a manner that is safe for the designed trespass. Timber 8- by 8-in (20- by 20-cm) guide blocks or curbs are utilized to maintain the truck access in a path that is fully supported by the stringers. Walkways are usually positioned along the edge of the ramp, and they are provided with secure rails, all in accordance with the Occupational Safety and Health Administration (OSHA) requirements (see Figs. 13.16 and 13.17).

A ramp is a sturdy bridging system that must meet stringent quality controls. The ramp must be planned simply and it must be constructed in a secure fashion in order to meet the requirements of daily use. Like all bridges, it must be secured in place; the likelihood of horizontal movement is great due to the laterally induced forces. The standard method of resisting these thrusts is to secure the toe and the cap of the ramp against movement (see Fig. 13.18).

The design of the cap footing depends on the bearing value of the surface materials; they must be carefully examined. It may be necessary to continue the cap footing until a firm bearing is established, possibly even to rock or to increase the cap footing pad in order to lower the required bearing value (see Figs. 13.19 through 13.24).

RUNWAYS AND PLATFORMS

Construction runways are normally utilized when the construction is to proceed in a horizontal rather than a vertical direction. The runway is built on top of existing or new construction, utilizing the walls and grade beams as supports (see Figs. 13.25 through 13.27).

FIGURE 13.16   Construction ramps may be fabricated using a combination of steel and timber. (*Thomas Crimmins Contracting Co.—Photo Collection.*)

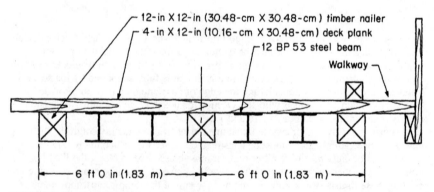

FIGURE 13.17   Typical vehicular ramp cross section.

Important details, regarding the construction of a crane runway platform over a newly constructed sewage channel or over concrete girders, include the position for lifting that must be carefully shored. It is when the crane is swinging over a single outrigger that the maximum stresses are induced into the support members. Reinforced turnoffs or additional platforms may be added to provide for special lifts.

Working platforms are sometimes required to provide access for structural steel erection. The placement of timber mats and shores may be utilized to strengthen existing or new construction (see Figs. 13.28 through 13.30).

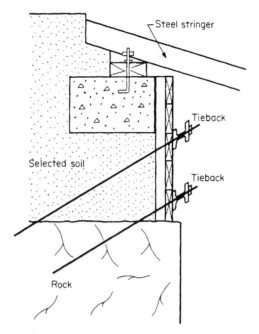

FIGURE 13.18 Typical cap detail.

FIGURE 13.19 Ramps—stage 1. (*Thomas Crimmins Contracting Co.—Photo Collection.*)

FIGURE 13.20   Ramps—stage 2. (*Thomas Crimmins Contracting Co.—Photo Collection.*)

FIGURE 13.21   Ramps—stage 3. (*Thomas Crimmins Contracting Co.—Photo Collection.*)

FIGURE 13.22 Ramps—stage 4. (*Thomas Crimmins Contracting Co.—Photo Collection.*)

FIGURE 13.23 Ramps—stage 5. (*Thomas Crimmins Contracting Co.—Photo Collection.*)

FIGURE 13.24   Ramps—stage 6. (*Thomas Crimmins Contracting Co.—Photo Collection.*)

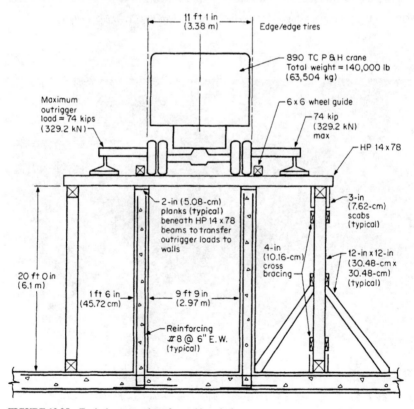

FIGURE 13.25   Typical cross section of a working platform.

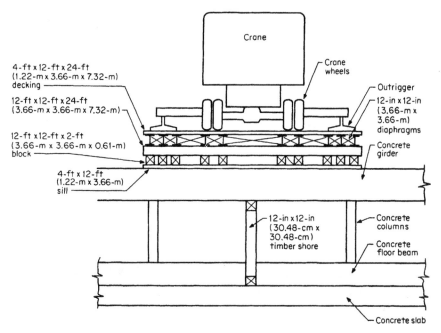

FIGURE 13.26 Typical crane platforms.

FIGURE 13.27 Construction runway provides access for cranes and trucks. (*Thomas Crimmins Contracting Co.—Photo Collection.*)

The width of the support system is determined on the basis of the widest vehicle usage, with its outriggers extended. A crane runway may be over 20 ft (6 m) wide enough for the placement of outriggers, while a truck runway need only be the standard 12.5 ft (3.8 m) clear. The length of a ramp or runway is determined by the physical characteristics of a project, such as its depth and size, as well as the lifting capacity of the site crane (see Figs. 13.31 and 13.32).

FIGURE 13.28 Timber runways are required for reinforcing and for concrete deliveries. (*Thomas Crimmins Contracting Co.—Photo Collection.*)

FIGURE 13.29 Concrete grade beams may be strengthened by means of intermediate shores. (*Thomas Crimmins Contracting Co.—Photo Collection.*)

FIGURE 13.30 Timber mats being placed to provide crane access for the erection of steel. (*Thomas Crimmins Contracting Co.—Photo Collection.*)

FIGURE 13.31 Shoring of concrete deck in order to provide support for a crane platform. (*Thomas Crimmins Contracting Co.—Photo Collection.*)

FIGURE 13.32   Timber mats placed on sills. (*Thomas Crimmins Contracting Co.—Photo Collection.*)

The city of New York requires that a crane permit be obtained from the city before establishing a large crane within the city's limits. The firm seeking the crane permit must have a professional engineer's certification that the crane's intended position has been examined and found it to be suitable for supporting the crane during its intended use. The engineer's report must include an evaluation of the bearing value of the subsoil, and also a statement regarding the possible presence of sidewalk vaults under the bearing pads, and calculations showing the intended use and safe loadings. The bearing value of the soil is not to be assumed greater than 1500 lb/ft^2 (71,820 N/m^2) unless the engineer has specific knowledge regarding the nature of the subsurface soils. Borings and soil testing programs are not usually carried out for standard temporary crane installations; most engineers rely on a visual inspection of the soil-bearing value and assign a value of 1500 lb/ft^2 (71820 N/m^2) unless the subsoil appears to be incapable of providing even this minimum level of support.

A standard rubber tired crane has four outrigger pads whose bearing surfaces are 4 ft^2 (0.37 m^2) each for a total of 16 ft^2 (1.5 m^2). A loading of 60,000 lb (27,180 kg) per outrigger is common; a footing must be provided to distribute this loading over a larger surface area. Typically each outrigger will be placed on a timber crib that may consist of 12- by 12-in (30- by 30-cm) timbers bolted together to form a mat of approximately 40 ft^2 (4 m^2).

The question of subsurface vaults or pockets is more difficult to solve; urban construction sites may have had a history of four or five earlier buildings whose drawings are not known and whose construction may have included a sidewalk vault for storage or a manhole for utilities. These unknown earlier vaults or boxes are usually brick and may provide poor bearing for surface cranes. In the interest of public safety, it is important that a responsible engineer investigates and determines the safe bearing values before a crane is permitted to function in a metropolitan area.

ROADS AND DRIVEWAYS

Access to a construction site is usually not a serious problem. Trucks and other vehicles can operate on the natural ground with a minimum of difficulty. Dust can be alleviated by sprinkling with water, a temporary expedient that requires frequent repetition, or by the application of an asphalt oil palliative. Heavy and frequent traffic requires more extensive measures.

Drainage is of primary importance. Water must be removed from the roadway before it accumulates into large puddles; this is accomplished by crowning the surface, sloping, and side drainage ditches. A stabilized soil is not a suitable wearing surface and should be protected. A common method is the road mix method in which asphalt emulsion or cutback asphalt is mixed with the surface material and the layer compacted by traffic. This provides a good wearing surface that is easy to construct, sheds water, can be repaired easily, and can be removed simply with a minimum of effort when no longer needed. Road mix asphalt can be applied to almost any soil except clay.

Another method is the application of a seal coat that consists of the spray application of liquid asphalt to the compacted and graded roadway surface, usually reinforced by the application of a thin layer of sand or fine gravel. The asphalt may be cutback liquid asphalt, asphalt emulsion, or hot asphalt cement.

Access roads that are to become part of the permanent road installation must be constructed in accordance with the project specifications.

RAILROADS

Rail facilities are necessary in tunnel construction as well as for moving bulk and heavy materials that are needed for the project. Panelized track units are used for sidings and industry spurs, one advantage being the speed of construction. Another advantage for the contractor is their high salvage value upon completion of their temporary duty or their rental costs. Several subcontractors have centralized fabricating plants in which they build their own track panels. One development is a prefabricated standard gauge steel tie and track section 39 ft (12 m) in length. The sections are shipped to the site by rail or truck and can be handled by a small crane. They are easily laid and joined with a minimum of skilled labor, and can be salvaged to be reused upon completion of their service on this project (see Fig. 13.33).

Rails are secured to the steel ties by pairs of steel jaws inserted in holes drilled in the ties on each side of the rail seats. The joint is secured by a tapered steel key or wedge. Similar panels may also be fabricated of rail and wood ties. Load-carrying capacity depends on the weight of the rail used and the spacing of the ties. The steel tie units recommended for infrequent light rail traffic include 70-lb (32-kg) or 90-lb (40-kg) relay rail with the steel ties on 36-in (90-cm) centers. A stronger unit, designed for frequent heavier traffic, would consist of 100-lb (45-kg) or heavier relay or new rail with ties spaced 24 in (60 cm) or possibly 30 in (75 cm). A 39-ft (12-m) panel consisting of 90-lb (40-kg) rail with ties spaced at 30 in (75 cm) weighs about 4700 lb (2130 kg). Panels are laid on the prepared subgrade, ballasted, raised to grade, and the ballast is then tamped and compacted by hand or power tools.

FIGURE 13.33 Panelized railroad track units are used to span major urban roadways.

Tunnel construction requires temporary rail lines to remove muck from the heading and most importantly to move supplies into the tunnel. Tunnel tracks are often narrow gauge, some as small as 24 in (60 cm). Panel tracks are used in underground work because of their convenience. Special steel and wood composite ties are available for use in machine made circular cross-sectional tunnels. When they are fitted with wood end blocks, the ties are designed to fit the contour of the tunnel. The blocks can be reused and modified to fit tunnels with different diameters.

CHAPTER 14
SCAFFOLDING

Michael S. D'Alessio, P.E.

Scaffolding has been used for 5000 years to provide access areas for building and decorating structures taller than the people who worked on them. The word "scaffolding" refers to any raised platform or ramp used for ingress and egress for pedestrian movement and/or the passage of building materials. The word "staging" is often used synonymously with scaffolding since the Shakespearean era when plays were staged on a raised platform. A scaffold also means a raised platform used for executions. In broad context, the scaffolding in this case is a "work place" for those who do the hanging or beheading.

It can be surmised that the first crude forms of scaffolding were developed when humans moved from single story huts at ground level to higher structures; most certainly they used scaffolds made of tree trunks and limbs lashed together with fibrous vines.

By the Middle Ages, people became more adept at devising and using more complex mechanical devices: raw trees became smoothed poles held together with longer-lasting ropes, twisted or braided from hemp fibers. In the Far East, light hollow bamboo poles are still used extensively. Joints are made with bamboo bark strips that are wet and then tied. They dry out to become a considerably rigid joint binding.

It was not until the mid-1920s that the concept of using steel pipes fastened together with metal-formed or -cast clamps (clamps) instead of poles and ropes was introduced. This advance lowered erection costs and provided more predictability for the strength of the finished scaffold. Aluminum alloy pipes and clamps were developed for their lighter weight and speedier construction. Aluminum alloy is only two-thirds as strong as steel, but it is only one-third to one-half its weight. Because of higher initial cost, aluminum is restricted mostly to building maintenance scaffolds and suspended platforms where lightweight easy maintenance and appearance are important.

Although pipe was the original basic structural component, thinner-wall structural tubing has become popular in the United States because it has similar strength with lower weight. Although the words "pipe" and "tube" are often used synonymously (especially in Europe), they differ substantially in standard sizes, strengths, construction, and nomenclature. Pipe has fairly thick walls to enable thread cutting at the ends with minimum strength loss. Standard rolled pipes, pressure- or butt-welded, are now generally of ASTM designation

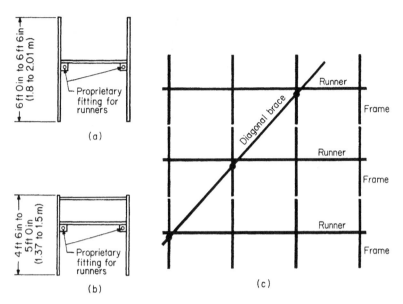

FIGURE 14.1 Pipe scaffold: (*a*) section; (*b*) section; and (*c*) side view.

A-50 steel. The strength properties of tubes used for scaffolding vary considerably with proprietary manufacturers, as covered later.

Most US scaffolding is made from thin-wall, high-strength structural steel tubing similar to ASTM A-500 grade B. In recent times, much of the scaffold tubing and finished scaffold components are imported from China with the steel designation Q235. This is deemed to be equivalent to the ANSI 1050 grade B.

In the late 1930s and 1940s the concept of welding pieces of pipe and tube into prefabricated "frames" or "panels" enabled simplification of tube and clamp scaffolds by sectionalization. Each frame replaced at least three pieces of tube and at least four clamps, thereby speeding erection times. Some early frames incorporated clamp devices built into them so that they could be joined longitudinally with standard scaffolding pipe runners. In Europe, frames using this principle are still in use. (See Fig. 14.1.) However, in the United States, the construction industry quickly took advantage of the convenience of frames, and by the end of the 1950s frames had become the most common type of scaffolding in use. They evolved to have quick longitudinal attachment of the frames with pivoted diagonal cross bracing made in various lengths to give *fixed* spacings between frames, the most popular being between 5 and 10 ft (1.5 and 3 m). Quick-acting mechanical devices were developed to attach the braces to the frames. Threaded studs with wing nuts progressed to sliding or hinging devices, as shown in Fig. 14.2. Today the original wing-nut type of fastening has been almost completely outmoded by the quicker-acting devices, all resulting in shorter erection and dismantling times and requiring no special wrenches or other tools as are needed for tube and clamp scaffolds.

Continuing sophistication in design of scaffolding components is commonplace, and most manufacturers carry a large line of special-purpose accessories. The two basic types of scaffolding, sectional frames and tube and clamp scaffolds, including small sections equipped with wheels and known as rolling towers, are commonly referred to as "built-up" scaffolds. This is to differentiate between these and other specialized types of scaffolding, such as those consisting of scaffold platforms suspended by cables from overhead structural components and known as "hanging" or "suspended" scaffolds of which there are a number of types. Technical advances in both built-up and hanging scaffold designs now include power-operated components: either

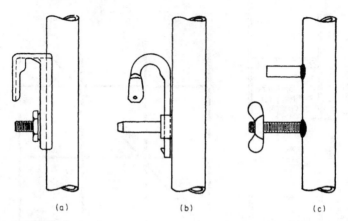

FIGURE 14.2 Bracing attachments on sectional frame scaffolds: (*a*) slide lock; (*b*) slide lock; and (*c*) stud and wing nut.

air-powered, electromechanical, or electrohydraulic. These "motorized" scaffolds generally allow for fast relocation of the scaffold work platforms and are of two main types:

1. For vertical movement of a suspended scaffold platform
2. For vertical and/or lateral movement of a platform supported from a wheeled base

GENERAL DESIGN CONSIDERATIONS

Commonly, all types of scaffold have incorporated in their designs a minimum safety factor of 4. This criterion is generally stated as: "Scaffolds and their components shall be capable of supporting without failure at least 4 times the maximum intended load." An exception to this is suspension wire rope that, when used to support scaffolds, must have a safety factor of 6.

There are three ways of complying with this requirement:

1. Use AISC load and resistance factor design procedure with the live load factor set at 4 and all other factors set at one (1).
2. Construct samples or prototypes of the proposed design and subject them to test loads equivalent to 4 times the design load applied to the specimens in a manner that simulates actual field usage of assembled components and method of loading. This is the method most used in the industry.
3. Base a design for a specific purpose on data from applicable results of a similar scaffold configuration obtained from method 2.

The analysis of scaffold structures can be very complex. Often, representative segments of a scaffold system are tested to destruction and the results used to verify the analysis. This system of analysis is then applied to the particular scaffold to ensure that the structure, as a whole, is sound and efficiently designed.

This level of analysis is usually reserved for complex scaffolds that are to be subjected to unusual loadings. Most scaffolds can be properly built using the safety rules and instructions provided by the manufacturers.

The industry standard factor of safety for scaffolding is 4, the rationale for which is not known. The safety factor is applied to the maximum intended load (mil) that is designated

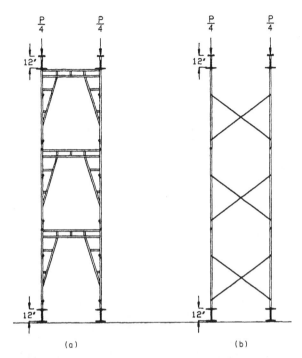

FIGURE 14.3 Load testing: (*a*) front elevation; (*b*) side elevation.

as "the total load of all persons, equipment, tools, materials, transmitted loads, and other loads reasonably anticipated to be applied to a scaffold or scaffold component at any one time." Most scaffolds are designed against failure, a method that lends itself very nicely to load and resistance factor design.

Figure 14.3 shows the Scaffolding, Shoring, and Forming Institute, Inc., standard test for scaffolding from which the basic load-bearing capacity of a scaffold module is determined. It is the fundamental test from which we verify our design assumptions. The scaffold is loaded at a rate of 1000 lb/min until load refusal.

CODES OF STANDARD PRACTICE

It is law that all scaffolding be designed, constructed, and used in compliance with the pertinent regulations and standards of the Department of Labor Occupational Safety and Health Administration in the *Federal Register* as follows:

Standards 29 CFR part 1926. Safety and Health Regulations for Construction

Subpart L—Scaffolds (1926.450, 451, 452)

Subpart M—Fall protection 1926.500 (a)(2)(vii), (a)(3)(i)

Subpart N—Helicopters, hoists, elevators, and conveyers

Standards 29 CFR part 1910.28, .29

Subpart D—Walking and working surfaces

Subpart F—Powered Platforms, Manlifts and Vehicle-Mounted Work Platforms (1910.66, .67).

Both regulations and standards are required to be applied in conjunction with all other applicable provisions of their other accompanying subparts containing various general provisions such as those for personal protective equipment and fire protection. Some general standards are also applied to construction work (1910.29).

A number of states have their own federal OSHA approved regulations such as CAL/OSHA in California and WA/OSHA in Washington. In the event of conflict between federal and state OSHA regulations, the more (apparently) stringent and/or more specific provisions prevail. Some states have dual-jurisdiction status with both federal and state inspection machinery in effect.

A new Subpart M 1926.500 was issued in the October 1994 revision that deals specifically with fall protection. This section does *not* apply to scaffolds. Section 1926.500 (a) (i) directs the reader to Subpart L—*Scaffolding* where in section 1926.451(g)(2) it states "Effective September 2, 1997, the employer shall have a competent person determine the feasibility and safety of providing fall protection for employees erecting or dismantling supported scaffolds. Employers are required to provide fall protection for employees erecting or dismantling supported scaffolds where the installation and use of such protection is feasible and does not create a greater hazard." Standard OSHA Interpolation of (12/04/1997) speaks to the development of an Appendix B to subpart L that "will provide a non-mandatory set of guidelines that a competent person would take into account when evaluating access and fall protection options for erectors and dismantlers of supported scaffolds."

As of this writing, the reserved Appendix B location remains blank.

The following is an example of a fall protection plan for scaffold erectors and dismantlers that has been submitted by the competent person in charge of the erecting process:

We propose the following fall protection plan as developed by the competent person in charge of the scaffold erection crew:

- All scaffold erectors working 6 ft or more above a level will be equipped with a full-body harness attached to a 6-ft-long shock-absorbing lanyard.

- All scaffold erectors, when in a stationary position and located within the scaffold, will tie off to a horizontal scaffold member located directly above their heads, provided that that horizontal scaffold component is part of a completed portion of the scaffold.

- The competent person has determined that the handling and positioning of scaffold plank is the most arduous part of the erecting process. The labor to install two planks is equivalent to the labor to install one section of the scaffold. Therefore we will limit the erector's deck to two planks wide. Planking the erecting deck fully would more than double the time to erect the scaffold and hence significantly increase the time each erector is exposed to the hazards.

- The leading edge erector will also be equipped with full-body harness and, 6-ft shock-absorbing lanyard; however, it has been determined that there is no suitable place for this erector to tie off. This erector is specially selected and trained to deal with this hazard and need not tie off while in motion. This erector will tie off when stationary and when a suitable place, as determined by this erector, exists.

- The support erectors are required to stay within the braced portion of the scaffold, that is, within a completed module. The leading edge erector is the only person that may assemble or dismantle the unbraced portion of the scaffold.

- We anticipate that the scaffold components will be passed by hand from level to level by support erectors located within the cross-braced scaffold. The scaffold components, however, may be passed along the outside of the scaffold, in particular the scaffold frames and trusses.

• The scaffold will be accessed via adjacent floors, scaffold stair units, or scaffold ladders.

• All scaffold erectors will be fully trained in the erection and dismantlement of scaffolds and will present scaffolder-training cards to attest thereto.

There exists, however, at least one OSHA interpretation that exempts scaffold erectors from the use of personal fall protection equipment, if their work calls for them to be in motion. Some form of fall protection or prevention must be provided for stationary scaffold erectors.

Subpart M does address fall protection for employees on the face of formwork requiring:

1926.501(b)(5)
Formwork and reinforcing steel. Each employee on the face of formwork or reinforcing steel shall be protected from falling 6 ft (1.8 m) or more to lower levels by personal fall arrest systems, safety net systems, or positioning device systems.

1. Wire Mesh. Wire mesh midrails, screens, mesh, intermediate vertical members, solid panels, and equivalent structural members of a guardrail system shall be capable of withstanding, without failure, a force applied in any downward or horizontal outward direction at any point along the midrail or other member of at least 75 lb (333 N) or guardrail systems with a minimum 100-lb toprail capacity, and at least 150 lb (666 N) for guardrail systems with a minimum 200-lb toprail capacity.

2. Planking. All plank used for scaffolding purposes must be graded as "Scaffold Grade Plank." For solid sawn planks the more common species used are Douglas fir, with an allowable bending stress of 2200 psi, and southern yellow pine that meets the grading rules for scaffold plank. Other species can be used if they meet the grading rules of a recognized agency for the species of wood being used. Scaffold plank is usually called for in their nominal or finished dimensions. A 2-in-thick plank measures 1½ in. in thickness and 10-in-wide plank measures 9½ in. in width unless the term *rough cut* or *full thickness* is used in reference to the cross section.

Manufactured scaffold plank has provided the industry with a ready source of scaffold plank. Laminated veneer lumber (LVL), manufactured in the customary thicknesses, is widely used for planking scaffolds. The manufacturers claim these planks are stronger, lighter, more reliable, and will last longer than solid sawn lumber. The use of LVL is still new in the industry, and these claims are yet to be proved. In any case, manufactured scaffold plank is a ready and reliable source of decking.

3. Wire Ropes. These are to have a minimum safety factor of 6 when used for suspending scaffolds, instead of the normal 4 for all other materials.

4. Open Sides and Ends. (These are required to have guardrails, midrails, and toeboards at outsides of all scaffolding platforms.) The term *open sides and ends* means the edges of a platform that are more than 14 in (36 cm) away horizontally from a sturdy, continuous, vertical surface (such as a building wall) or a sturdy, continuous, horizontal surface (such as a floor), or a point of access. Exception: For plastering and lathing operations the maximum horizontal threshold distance is 18 in (46 cm).

5. Rolling Scaffolds. Under the OSHA construction regulations, subpart 1926.425(w), persons are allowed to ride on moving, manually propelled mobile scaffolds (rolling towers) under certain listed conditions. We recommend, however, that this practice be discouraged unless extra precautions are taken to ensure the safety of the people riding the scaffold.

It is important to emphasize that compliance with OSHA regulations is necessary by law. It is impossible to provide specific circumstances that lead to scaffold-related accidents in the field; only the broadest caution—using only safe procedures and being alert at all times—can prevent accidents. Therefore, it is recommended that appropriate ANSI standards for type and use of equipment always be followed whether or not their current versions are promulgated into OSHA regulations. Further, almost all scaffolding manufacturers publish and make freely available safety rules and instructions for their various products. Following these publications can further promote safety and reduce hazards. An industry association that publishes safety rules and other free safety-enhancement information is Scaffold industry Association Inc 400 Admral Blvd. Kansas City, Missouri 64106.

6. List of ANSI Standards.   A list of scaffold-related ANSI standard publications follows:

A 10.4—2007 *Personnel Hoists*

A 10.5—2006 *Material Hoists*

A 10.8—2001 *Scaffolding*

A 10.11—2010 *Safety Nets*

A 10.13—2001 *Steel Erection*

A 10.18—2007 *Temporary Floor and Wall Openings, etc.*

A 10.22—2007 *Workmens' Hoists, Rope-Guided and Non-Guided*

Z359.1—2007 Fall Arrest Systems

A 120.1—2006 *Powered Platforms*

A 92.2—2009 *Vehicle Mounted Elevating and Rotating Aerial Devices*

A 14.1—2007 *Portable Wood Ladders*

A 14.2—2007 *Portable Metal Ladders*

A 14.4—2009 *Job Made Ladders*

A 14.5—2007 *Portable Reinforced Plastic Ladders*

A 92.3—2006 *Manually Propelled Elevating Work Platforms*

A number of judgments in product liability lawsuits have established that over and above regulatory compliance, *voluntary* compliance with national consensus codes such as ANSI is a positive defense and indicative of maintaining pace with current state-of-the-art manufacturing practices.

DESIGN LOADS

In accordance with OSHA and ANSI criteria and common practice for many years, design load ratings for *scaffold platforms* are as follows:

Light-Duty Loading.   25 lb/ft^2 (1200 N/m^2) maximum working load for support of people and tools (no equipment or material storage on the platform).

Medium-Duty Loading.   50 lb/ft^2 (2400 N/m^2) maximum working load for people and material not to exceed this rating, often described as applying to bricklayers' and plasterers' work, but not confined thereto.

Heavy-Duty Loading. 75 lb/ft^2 (3600 N/m^2) maximum working load for people and stored material often described as applying to stone masonry work, but not restricted thereto.

These ratings imply uniform load distribution. They are intended to communicate the type of use that the scaffold is intended to accommodate.

A light-duty scaffold is designed to support personnel and light hand tools only. Material may not be stored, or be allowed to accumulate on the scaffold. The rating "light duty" (25 lb/ft^2) is adequate, for personnel, in that it allows one person rated at 250 lb in every 10 ft^2 of scaffold work surface. Workers can stand about 3 ft from each other (i.e., arm's length). A 50-ft-long by 5-ft-wide scaffold can support 25 people—much more than is usually assigned to that short a length of wall. It would take 100 people on this scaffold, at one time, to overload it. There is little danger of overloading a scaffold with personnel. If the scaffold is to be used as a walkway bridge, or a viewing platform, or a grand stand, even though it is meant to carry only people, there is a very real danger that it will be overloaded. These are *not* scaffolds, they are structures in their own right and they should be designed to comply with applicable building codes and/or fire codes, not scaffolding codes.

The rating "medium duty" (50 lb/ft^2) is usually associated with masonry work, and allows for the storage of brick, block, and mortar on the scaffold. A half-pallet of brick can weigh 1000 lb. A 5- by 8-ft bay of scaffold rated at 50 lb/ft^2 is designed for a load of 5 ft × 8 ft × 25 lb/ft^2 = 2000 lb. A tub of mortar weighs about 100 lb/ft^3. Therefore a bay of scaffold can support half-pallet of brick, one tub of mortar, and three or more people—just about enough capacity. If the scaffold is used for storage of materials, however, it does *not* have enough capacity. This scaffold's ultimate capacity is 50 lb/ft^2 × 8 ft × 5 ft × 4 safety factor = 8000 lb; 5 bays = 40,000 lb—*one truckload* of material. Contractors who do not have a place to unload their shipment of masonry have been known to load it on the scaffold; the masonry is going to be put there eventually, anyway. Most of the time the scaffold holds the load; sometimes it doesn't.

The rating "heavy duty" (75 lb/ft^2) is usually associated with stone setting. If LVL planks are used, we can still use 8-ft-long bays. Solid sawn plank can only span 6 ft. One could theoretically load about 10 ft^3 of stone on the scaffold, that is, about 2000 lb, and still have 1000 lb left for personnel and mortar. The only time I have ever seen stone handled this way was in the building of a gothic cathedral in Manhattan. They did so because they lacked the funds for a crane. Stone setting applications usually employ alternative means to handle the stone. The scaffold is usually designed for medium duty. Heavy-duty scaffolds are predominantly used for refacing of buildings and for demolition work. The broken fascia, usually brick, marble, or granite, is wheeled around the scaffold in carts to a place where it can be lowered to the ground. When the fascia is removed, and subsequently is to be put back, the scaffold user may want to store the fascia material on the scaffold, and so eliminate moving it to the ground and then loading it back on to the scaffold. Unfortunately, it is rarely possible to build a scaffold with enough capacity to carry all the wall material. Sometimes scaffolds, adjacent to the main scaffold, are erected to provide areas where the material can be laid awaiting reuse. This does not prove cost-effective, in most cases, because it doubles the scaffold cost but only allows for one more loaded level per working level, that is, 5 ft × 8 ft × 75 lb/ft^2 = 3000 lb/2 = 1500 lb; live load on scaffold leg from one level 1500 lb × 4 SF = 6000 lb; ultimate capacity of welded tubular frame scaffold leg = 8000 lb. Add the self-weight of the scaffold say 1000 lb × 23/12 = 1917 lb; 6000 + 1917 lb = 7917 lb. OK—just enough. The designer must always be aware of the end user's expectations and design the scaffold accordingly. We address the need for communication between the scaffold designer and the scaffold user later in the chapter. Figure 14.4 illustrates how scaffolds can be modified to increase its load rating when the rating is governed by bearer or plank capacity.

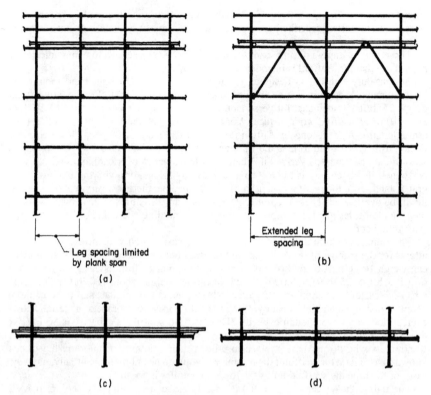

FIGURE 14.4   Scaffold leg spacings. (*a*) Spacing limited by allowable span of planks, (*b*) Additional bearer in center of bay, runner reinforced with diagonal bracing, (*c*) Use of double level of planking, (*d*) Use of 2½-in (6.45-cm)-OD bearers instead of 2-in (5.08-cm)-OD bearers.

TUBE AND CLAMP SCAFFOLDS

Tube and clamp scaffolds are assembled from three basic structural elements: the uprights, or posts, which rise from the ground or other solid support; the bearers, which support the working platforms and/or provide transverse horizontal connections between the posts; and the runners, which attach to the posts directly below the bearers and provide longitudinal connections along the length of the scaffold. These three elements are usually connected with standard or fixed clamps that provide a 90° connection in two places. (The word "standard" derives from the European terminology of naming the upright posts as standards.) These three elements form the basic structure shown in Fig. 14.5*a* and are repeated in the horizontal and vertical planes to build the scaffold to its desired size and egg-crate-type configuration with the components and fittings shown in Fig. 14.5*b* to 14.5*g*.

Diagonal bracing is used to stiffen the structure as necessary—most importantly in the longitudinal direction. Bracing is generally connected to the posts with "adjustable" or "swivel" clamps that have the facility of adjusting a full 360° so that two members in parallel planes are connectable at any angle. Diagonal bracing should always be attached to the posts as closely as practical to the "node" points formed by the runner-bearer connections. The use of lateral diagonal bracing in the transverse direction (i.e., in the plane of the bearers) is used extensively in Europe but to a much lesser extent in the United States, as discussed later.

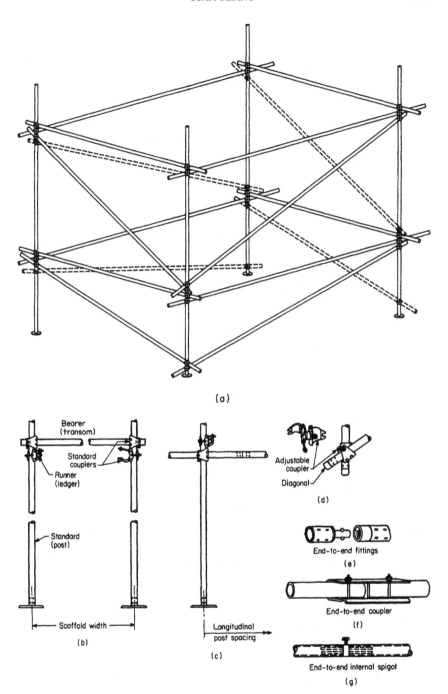

(a)

Bearer
(transom)

Standard
couplers

Runner
(ledger)

Standard
(post)

Scaffold width

(b)

Longitudinal
post spacing

(c)

Adjustable
coupler

Diagonal

(d)

End-to-end fittings

(e)

End-to-end coupler

(f)

End-to-end internal spigot

(g)

FIGURE 14.5 The basic assembly and components of tube and clamp scaffolds.

Another important structural element is the building tie that connects the scaffold to the wall or structure and is needed to provide rigidity and anchorage of the scaffold in the transverse direction. Figure 14.6 shows a number of methods of accomplishing this. The scaffold in the transverse (narrow) direction forms a repeated series of braced, columnlike frames connected by runners along the length of the scaffold; these column frames or bents need to be laterally supported; otherwise they are unstable because of their height-to-width ratio and have low strength to resist wind and other lateral forces.

Application

Tube and clamp scaffolds can be assembled in numerous ways because of the flexibility of their assembly dimensions in the horizontal and vertical planes (within their structural limitations). Unlike sectional frame scaffolds they are not restricted by frame width in the transverse direction, by brace length in the longitudinal direction, or by frame height in the vertical direction. Consequently, they are preferred for access to workplaces having irregular dimensions and contours, for example, churches, old auditoriums, and chemical processing structures.

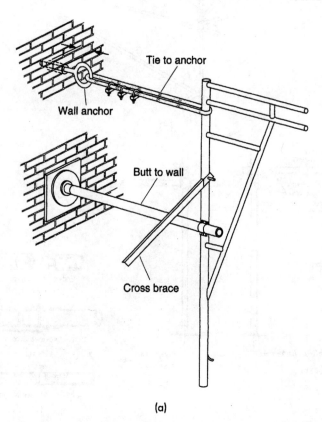

(a)

FIGURE 14.6   Methods of stabilizing against a building: (*a*) wall tie and anchorages.

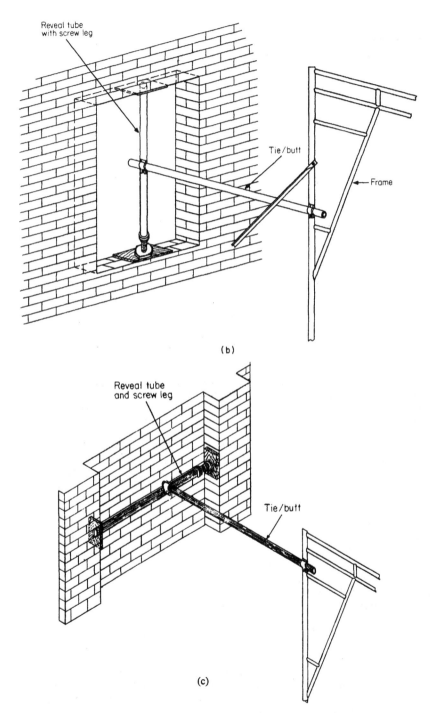

FIGURE 14.6 (*Continued*) Methods of stabilizing against a building: (*b*) window reveal tube, (*c*) reveal between pilasters.

Additionally, the chemical and petrochemical industries often have structures or vessels with multiple protrusions, such as walking platforms, curved stairs, or piping, which tend to make construction of scaffolds of sectional frames difficult and hazardous. Such protrusions are often added *after* the scaffold is in place and often pass *through* the scaffold; frequently, cutting sectional frames is required for removal. This is a wasteful process obviated by using tubes and clamps. Church interiors and exteriors and other old buildings with flying buttresses, multiple decorative corbeils, and cornices are similarly more safely and easily scaffolded with tubes and clamps. Scaffolding with tubes and clamps is the work of experienced specialists because it is not always possible to preplan a scaffold design for a church exterior (for instance) because accurate drawings of the structure are not always available and dimensions of embellishments, spires, and other obstacles frequently cannot be ascertained prior to the erection and must be done "on the job" as these various obstacles are reached and become measurable.

Basic Configurations

The basic configurations are as follows:

1. Double Pole.   Also called "independent" wall scaffolds, these are used for access to vertical surfaces for construction, alteration, or surface finishing and repair. They consist of repetitive *pairs* of posts along the length, connected by bearers and runners.

2. Single Pole.   Also called "putlog" wall scaffolds, these are used for construction of masonry walls. They consist of *single* posts 3 to 5 ft (90 to 150 cm) away from the wall surface spaced at regular or varying intervals along the wall. The different feature of this type of scaffold is that the inside ends of the bearers are supported at joints or courses in the wall being built instead of by inside posts. Flat plates are often attached to the bearers to seat in the mortar joints; when used this way, the bearers are called "putlogs." The putlogs are removed from the mortar courses and the holes thus left are grouted as the scaffold is dismantled. This method has little use nowadays because it is slow and cumbersome compared with other available methods, but is mentioned because it is included in most regulations and codes. Originally, it was devised as an inexpensive means of providing scaffold access for a very few workers who constantly relocate around the perimeter of the building.

3. Tower Scaffolds.   These consist of one or a few bays in either horizontal plane, constructed to a required height for access to ceilings or for specialized load support requirements not conveniently achievable with sectional frames. They may be mounted on casters and become mobile scaffolds or *rolling towers* as shown in Fig. 14.7. A specialized use of tower scaffolds is to provide stair access to unusual structures such as cooling towers. The posts can be angled and offset so as to conform reasonably closely to the curved outer surface. Such applications as this exist when it is not practical or feasible to use a mechanized personnel elevator.

4. Wide-Area Scaffolds (Birdcage Scaffolds).   These are used for building interiors for construction and renovation of walls and ceilings. Because of their flexibility in dimensional spacings, tube and clamp scaffolds are easily adaptable to irregular wall and column shapes, domed or sloping ceilings, sloping floors, positioning between seat rows, etc. While modern church and auditorium interiors tend to be simple and plain-surfaced and easily scaffolded with sectional frames, those with special acoustic-dependent surfaces and older buildings needing renovation are usually scaffolded better with tubes and clamps.

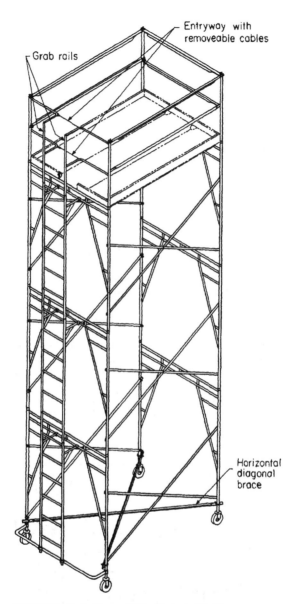

FIGURE 14.7 Rolling scaffold with stand-off ladder.

Tube and Clamp Materials

The most popular size pipe or tube used throughout the world is approximately 2 in (50 mm) OD. Standard-sized pipe and its structural tubing equivalent are compared in Table 14.1. *Pipe* is generally classified by its *internal* diameter: ID in the United States and NB (nominal bore) in Europe. Both measure approximately 1-29/32 in (48 mm) OD.

TABLE 14.1 Tube and Clamp Scaffolding Terms

Item	U.S. nomenclature	U.K. nomenclature
Tube connection	Clamp	Fitting
Structural members	Tube or pipe	Tube
Vertical supports	Posts (uprights, poles, legs)	Standards
Lateral pairs of posts	Bent	Frame
Spacing between bents	Bay	Bay
Longitudinal members	Runners	Ledgers
Plank support members	Bearers	Transoms
Vertical structural intervals	Lifts	Lifts
Diagonal bracing	Same	Same
Platform members	Planks	Boards

Tubing is generally classified by its *outside* diameter, which for scaffolding purposes is manufactured to 1-29/32 in (48 mm) OD and often designated as "pipe-size" tubing. Although there is also exact 2-in (50.8-mm)-OD tubing available, the pipe-size tubing is often referred to as 2-in (50-mm) nominal OD pipe or tube.

European pipe of 1½ in (38 mm) NB conveniently equals 48 mm OD in metric measurement. It is still common European practice to refer to pipe as tubing. The materials used to make tube and clamp (T&C) scaffolding vary extensively all over the world. Much of it in the United States has been made from 1½ standard pipe or 1.9 OD tubing with a 0.097 wall depending on the manufacturer. In modern times the material is 50,000-psi yield, welded tubing having specifications tailored to the manufacturer's requirements. As of this writing most new scaffolding is made of Q345 material, the Chinese equivalent of A-500 grade B tubing.

Tube and Clamp wall scaffolds are still widely used in Great Britain. The U.S. construction of a T&C wall scaffold, as shown in Fig. 14.8, has not been used since "welded tubular frame scaffolding" (WTFS) came on the market. WTFS is faster and easier to erect, and it can carry more load. Figure 14.9 shows the construction of a typical British wall scaffold as can be seen all over London today. It is advised that the U.S. configuration shown in Fig. 14.8 require the use of 4-in wide-bodied ridged 90° clamps. Wide-bodied clamps have given way to the 2-in narrow-bodied clamps. These clamps do not provide a moment connection. Without moment connections the T&C wall scaffold defaults to the British construction method as shown in Fig. 14.9.

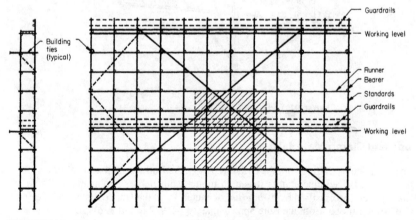

FIGURE 14.8 U.S. construction of wall scaffold (rarely used in the United States).

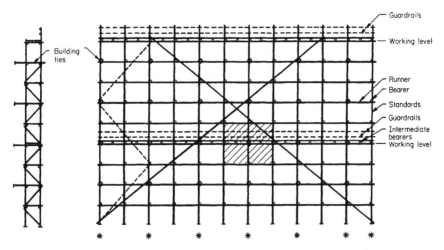

FIGURE 14.9 British construction of wall scaffold (*indicates location of walkway bracing).

Design Loads for Tube and Clamp Scaffolds

Tube and Clamp scaffolding is manufactured all over the world, to widely varying specifications and standards. For this reason we cannot provide finite loading information here. T&C scaffolding must be designed for each application utilizing the information available from each manufacturer and supplier. The designer should be aware that the scaffold posts (legs) are loaded eccentrically. (See Fig. 14.10.) Figure 14.11 shows the historical methods for construction tube and clamp scaffolds for the three generally accepted load ratings. Please note that method (d) generally requires nominal two and one half inch bearers (2-3/8 od × 0.154 wall 50 ksi yield tubing or pipe).

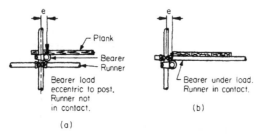

FIGURE 14.10 Eccentricity e of loading on post.

SYSTEM SCAFFOLDS

Components

System scaffolding is the most popular type of scaffolding used in the industrial maintenance industry. It lies between sectional frame scaffolding and tube and clamp scaffolding in flexibility and ease of erection. It is well suited to scaffolding around irregular surfaces,

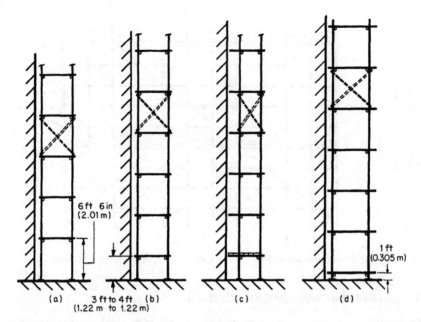

FIGURE 14.11　Suggested construction methods for duty-rated tube and clamp scaffolds: (a) light-duty scaffold, (b) medium-duty scaffold, (c) alternative medium-duty scaffold, and (d) heavy-duty scaffold.

such as cracking towers, digesters, etc., the kind of structures found in oil refineries, power plants, paper mills, chemical plants, etc.

Posts (Standards).　Posts have locking rings, or cups, welded to them in height increments of 19½ to 21 in (49.5 to 53.3 cm). These rings are used for the attachment of horizontal bearers, runners, and diagonal bracing.

Horizontals.　Bearers, runners, and guardrails constitute the range of fixed-length members that have a fitting built into each end; these fittings are usually wedge-type, although other types exist.

Diagonals.　These are also fixed-length members of varying sizes for the appropriate horizontal members of the scaffold. With these components, flexibility of application approaching that of tube and clamp scaffolds is achieved, but it is accomplished with the simplicity of connection of sectional-frame scaffolds. It is believed that labor savings of up to 50 percent are achieved over the labor costs of installing similar tube and clamp scaffolds.

In Figs. 14.12 and 14.13 an example of this type of scaffold is shown; it is the QES (QUICK ERECT SCAFFOLD) manufactured by Harsco Infrastructure, Inc., Division of Harsco Corporation. As many as 10 competitive and similar designs are available in the United States.

Application

System scaffolds offer much of the flexibility of Tube and Clamp scaffolds but are much faster and easier to erect and dismantle. The two systems are so similar that the design

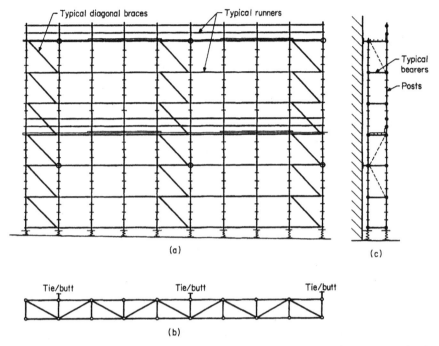

FIGURE 14.12 QUICK ERECT modular scaffold system: (*a*) elevation, (*b*) plan view at anchorage level, and (*c*) vertical section.

rules for them can be used interchangeably, provided full-moment clamps are used on the tube and clamp scaffold. System scaffolds offer a host of handy components that go along with their systems. Some of which are shown in Fig 14.13. When used as façade scaffolds (wall scaffolds), system scaffolds are tied to the building at vertical intervals of 4 times the minimum base width, or less, and every third or fourth lift thereafter—depending on the manufacturer's recommendations—and at horizontal intervals not exceeding 30 ft. All scaffolds must be tied to the building a maximum of one lift below the top. Diagonal braces are installed on both the inner and outer faces, at both ends, and in every fifth bay in between.

Many manufacturers recommend that horizontal braces be installed at the butt and tie levels to ensure full-leg-load capacity of the scaffolds (see Table 14.2). Horizontal diagonal braces interfere with hooked plank and deck platforms. These braces must be moved before the planks or platforms are laid down. Hooked plank or deck platforms can be used at the butt and tie levels in place of the horizontal diagonals. They act as membranes and will provide the desired lateral support.

Most manufacturers rate facade scaffolds at 2000 lb per leg, (or post) when braced properly and when butt and tied to every 4th lift of the building. Some brands increase the leg load to as many as 4000 lb, if the scaffold is tied to the building at every second lift.

Closing the horizontal tie spacing and/or increasing the frequency of diagonal bracing will only increase leg capacity by a few hundred pounds. Reducing the vertical tie spacing and horizontal distance between ties is the best way to increase the scaffold's load-bearing capacity.

Scaffold ratings differ from manufacturer to manufacturer; consult your scaffold's manufacturer or distributor for more information.

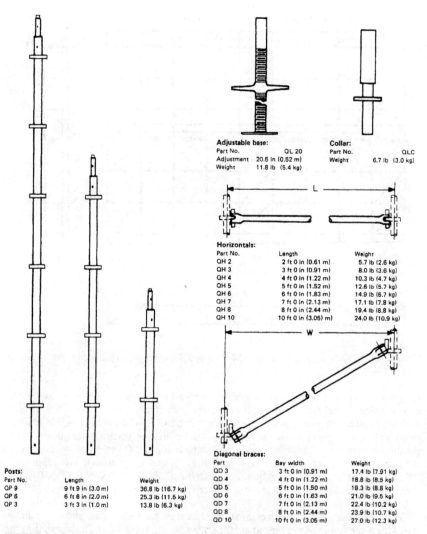

Adjustable base:

Part No.		QL 20
Adjustment	20.5 in (0.52 m)	
Weight	11.8 lb (5.4 kg)	

Collar:

Part No.		QLC
Weight	6.7 lb (3.0 kg)	

Horizontals:

Part No.	Length	Weight
QH 2	2 ft 0 in (0.61 m)	5.7 lb (2.6 kg)
QH 3	3 ft 0 in (0.91 m)	8.0 lb (3.6 kg)
QH 4	4 ft 0 in (1.22 m)	10.3 lb (4.7 kg)
QH 5	5 ft 0 in (1.52 m)	12.6 lb (5.7 kg)
QH 6	6 ft 0 in (1.83 m)	14.9 lb (6.7 kg)
QH 7	7 ft 0 in (2.13 m)	17.1 lb (7.8 kg)
QH 8	8 ft 0 in (2.44 m)	19.4 lb (8.8 kg)
QH 10	10 ft 0 in (3.05) m)	24.0 lb (10.9 kg)

Diagonal braces:

Part	Bay width	Weight
QD 3	3 ft 0 in (0.91 m)	17.4 lb (7.91 kg)
QD 4	4 ft 0 in (1.22 m)	18.8 lb (8.5 kg)
QD 5	5 ft 0 in (1.50 m)	19.3 lb (8.8 kg)
QD 6	6 ft 0 in (1.63 m)	21.0 lb (9.5 kg)
QD 7	7 ft 0 in (2.13 m)	22.4 lb (10.2 kg)
QD 8	8 ft 0 in (2.44 m)	23.9 lb (10.7 kg)
QD 10	10 ft 0 in (3.05 m)	27.0 lb (12.3 kg)

Posts:

Part No.	Length	Weight
QP 9	9 ft 9 in (3.0 m)	36.8 lb (16.7 kg)
QP 6	6 ft 6 in (2.0 m)	25.3 lb (11.5 kg)
QP 3	3 ft 3 in (1.0 m)	13.8 lb (6.3 kg)

FIGURE 14.13 Basic components of QUICK ERECT SCAFFOLD.

TABLE 14.2 Calculated Leg-Load Capacities of Various Bay Widths

Bay width ft	ASD exterior lb	LRFD exterior lb	Bay width ft	ASD interior lb	LRFD interior lb
2	6200	5420	6	5875	5080
3	5875	5080	7	5735	4940
4	5600	4800	8	5600	4810
5	5380	4600	10	5380	4600

The reader is cautioned not to extrapolate from rated capacity to ultimate capacity. The load ratings are arrived at through testing of specific scaffold configurations. Only the manufacturer or testing agency can provide accurate ultimate load capacity for any particular application.

Wide-Area Scaffolds (Birdcage Scaffolds)

"Birdcage scaffold" is a term of art that refers to a scaffold that fills a volume, for instance a scaffold that fills the interior of an auditorium.

System scaffolds are particularly well suited to this application. In most products the wedged connection between the leg and horizontal is tight enough to form a moment connection. In some system scaffolds, the connection is stiffer than it would be if the leg were welded to the horizontal. When diagonal bracing is added, side sway of the frame is inhibited, the effective unbraced length of the column is reduced, and the load-bearing capacity of the scaffold leg is increased significantly.

Design Loads for System Scaffolds

Manufacturers rate the leg capacity of scaffolds so configured, at about 5000 lb, depending on the length of horizontals used to form the "cube," and the number of diagonal braces in the plane of the leg.

Interior legs, that is, legs that have two horizontals framed into them (at the top and bottom of the leg in the same plane) have more load-carrying capacity than the legs at the periphery of the scaffold that have only one horizontal framing into them (at the top and bottom). The difference can be as much as 1000 lb, depending on the length of the horizontals.

Historically, the failure load of scaffold legs has been calculated using the Allowable Stress Design, ASD Eqs. (E2-1) and (E2-2) less the safety factor divisor.

$$\frac{5}{3} + \frac{3(Kl/r)}{8\,Cc} - \frac{(Kl/r)^3}{8\,Cc^3}, \text{ or the multiplier } 12/23 \text{ in Eq. (E2-2)}.$$

The steel's mechanical properties, straightness, and consistency of the shape were controlled and these equations most closely matched our test results.

Now that scaffolding is being manufactured all over the world. The variation in steels and the consistency of shapes are not as reliable as we would like. Therefore, in the absence of specific load ratings that are backed by the manufacturer, we prefer Load and Resistance Factor Design, LRFD formulas E3-2 and E3-3 for the determination of the critical buckling load of the leg. (See Table 14.2.)

Design Loading of Wide-Area (Birdcage) Scaffolds

A "birdcage scaffold" is usually used to gain access to the walls and ceiling of the volume it fills. Therefore, the outer periphery and the top level of these scaffolds are where the work decks are usually placed. The outer bays are made smaller than the interior bays. They are built just wide enough to provide workspace. The bearers (horizontal members on which the planks rest are often shorter and therefore stiffer than interior horizontals. Their shorter length, and therefore greater stiffness, is enough to bring the leg-load capacity of the outermost legs up, nearer to that of the interior legs (see Table 14.2). As you can see, the aforementioned 5000-lb leg capacity fits the ASD calculations well.

The diagonal braces that inhibit side sway are commonly placed in every fourth bay. The load capacity of the diagonals for these systems is usually about 2000 lb and is not entirely dependent on the strength of the diagonal tube itself. The connection between the diagonal and the leg fixture is outside the plane of horizontals and the diagonals are attached to the legs with a wedged connection that can slip around the attachment point under load. This limits both the strength and stiffness of the connection, hence the diagonal load capacity is usually limited to 1500 to 2000 lb.

Figure 14.14 shows a typical bracing scheme for a bird cage scaffold. Please note that every plane in every lift and every bay is diagonally braced (lean-on bracing).

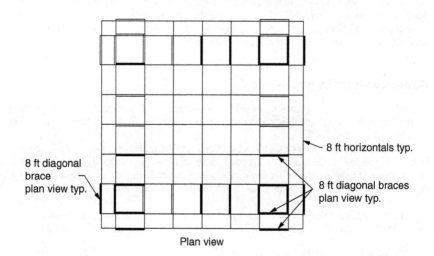

Plan view

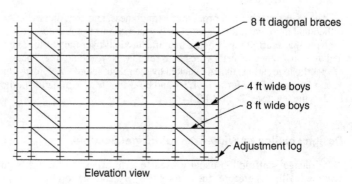

Elevation view

Typical bracing of a birdcage scaffold

FIGURE14.14 Typical bracing of a birdcage scaffold.

SECTIONAL SCAFFOLDING (WELDED TUBULAR FRAME SCAFFOLDING)

The construction principle of sectional scaffolding is shown in Fig. 14.15. The most common material used in the fabrication of steel frames is 1-5/8-in (41.3-mm)-OD tubing with a wall thickness between 0.086 and 0.105 in (2.18 and 2.67 mm). The most common grade of steel used for this purpose is AISI designation A1050, a high-carbon alloy having a minimum yield stress of 50,000 lb/in^2 (34,475 N/cm^2) with a corresponding ultimate stress of over 75,000 lb/in^2 (51,712 N/cm^2). Some manufacturers use lower-carbon steels of A1020 to A1025. The higher-carbon steel is generally preferred because its lower ductility and greater rigidity make it more resistant to damaging and bending of the members and because it has greater strength. Much of the scaffolding manufactured in China uses the steel designation Q235. This is deemed to be equivalent to the ANSI 1050. Many frame designs utilize smaller tubes for *internal* frame members, generally 1 or 1-1/16 in (25 or 27 mm) OD, of various wall thicknesses. These smaller secondary members generally perform bracing and rigidity functions. The welded joints are of two types: coped to conform to the contour of the tubes to which they abut or flattened at the ends as shown in Fig. 14.16. Generally, the coped joint is stronger.

The availability of frame configurations is almost limitless, depending on designs for specific uses and user preferences. Most frames are available in widths of 2, 3, and 5 ft

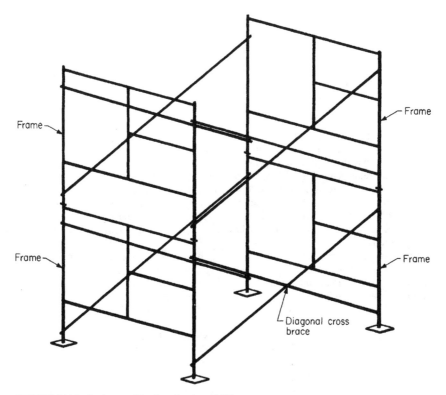

FIGURE 14.15 Basic assembly of sectional scaffolding.

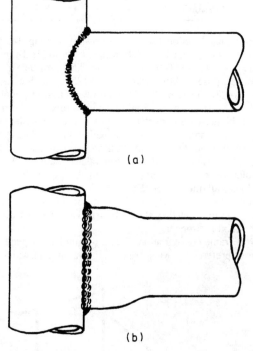

FIGURE 14.16 Welded frame joints: (*a*) coped and circumferentially welded joint; (*b*) flattened and welded joint.

(60, 90, and 150 cm); some special-purpose frames are available in 4-ft (120-cm) and 6-ft (183-cm) widths, the latter generally used in sidewalk canopies where the greater width is more convenient for the passage of pedestrians.

Standard frame heights are 3, 4, 5, and 6 ft, and 6 ft 6 in (90, 120, 150, 180, and 200 cm) high, with some areas using 4-ft 6-in (140-cm) high frames for masonry requirements. Special-purpose frames for sidewalk canopies are also available in heights of 7 ft 6 in, 8 ft, and 10 ft (230, 240, and 300 cm). Some typical representative frame designs are shown in Fig. 14.17, although each manufacturer has its own variants for internal dimensions and configurations.

Pivoted diagonal cross bracing, or X bracing, is generally made of approximately 1-in (25-mm)-diameter tubing or 1¼-in (32-mm) angle, these being flattened at the ends that have holes punched in them for attachment to fastening devices on the frames. The X configuration is achieved by riveting or bolting two pieces together at their geometric centers; they therefore have the facility of triangulation when in use and the ability to be folded flat when stored or transported.

Cross braces are dimensioned so as to give whole-foot incremental spacing between the frames they were designed to be used with. The most popular spacings are 5, 6, 7, 8, and 10 ft (150, 180, 210, 240, and 300 cm). The distance between brace connections on frames varies somewhat with the manufacturer. A specific brace size when connected to frames of different size having different brace stud spacing dimensions will result in the frames being spaced at some fractional-foot distance which is sometimes convenient

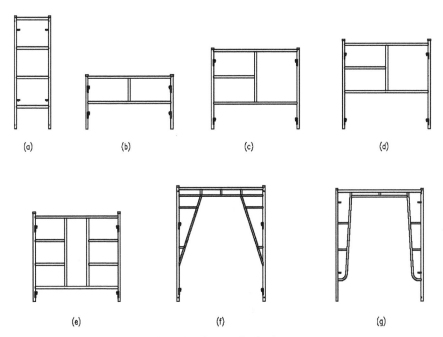

FIGURE 14.17 Representative designs of sectional scaffolding frames.

in enabling a scaffold to fit more closely to physical limitations of the building. The odd distance can be easily calculated, once the brace connection dimensions are known, by "squaring the triangles." For example, a brace designed to be used with 48-in (122-cm) brace connections at 84-in (213-cm) spacing is required to be connected to a frame having a 24-in (61-cm) connection spacing. To find the new span, or frame space, proceed as follows:

$$\text{New spacing} = \sqrt{48^2 + 84^2 - 24^2} = 93.72 \text{ in } (238 \text{ cm})$$

Other basic components necessary for the modular construction of sectional frame scaffolds are the brace-attaching devices built into the frames. The design of these is as varied as the imagination of manufacturers, and it is impossible to show them all. Three types are shown in Fig. 14.2 as being representative of the choices available. With some minor exceptions, the braces were originally attached by means of threaded studs welded to the frame legs, securing being achieved by wing nuts, which necessitated their removal and rethreading every time a brace was attached or removed. This was undesirable because it was so time-consuming, spurring the design of the various fast-acting, non-thread-reliant devices now available.

The last basic component (the coupling pin Fig. 14.18) is used to provide vertical alignment (and frequently positive connection) between frames when vertically stacked. These devices are also known as sprockets, connectors, joint pins, etc. Most are available with holes drilled in the upper and lower portions of the coupling pin that correspond to alignment holes in the frame legs through which solid or formed connecting pins can be installed to prevent the coupling pins from falling out and being lost—which at times can be a major expense. These connecting pins are available in various configurations such as hinge

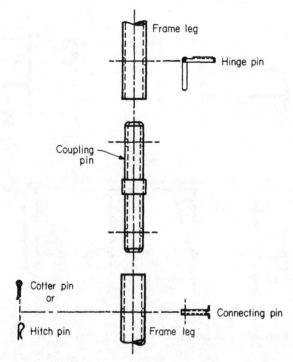

FIGURE 14.18 Typical coupling pin or stacking pin coupling sprocket.

pins and those retained with cotters. Connecting pins also provide a means of preventing uplift of a frame from the frame below it, and in most types of manufacture they are utilized extensively to positively attach various accessories used in conjunction with the basic frames and braces. Some manufacturers weld coupling pins in the frames. This has both advantages and disadvantages.

The frame and brace module gives the flexibility of unlimited horizontal and vertical extension of a scaffold within the limitations of (1) available modules of frame size and brace spacing and (2) allowable height of the scaffold (see design section for limitations).

Standard Accessories

There is a major difference in the design concept of sectional scaffolding from that of tube and clamp scaffolding. With the latter, any element, such as inset cantilevers for plank support as used by masons and finishers, guardrailing, and other formed configurations, are achieved by suitable assembly of the three basic components. Sectional scaffolding requires that special-design accessories must be used for each individual purpose and, more importantly, *available* at the site; this entails substantially greater preplanning and scheduling. There are two types of accessories: (1) standard, such as base plates, casters, adjustable screw legs, vertical guardrail supports, and horizontal guardrails (and midrails) and (2) special purpose, such as sidewall brackets for inset planks of various types, trusses for spanning over openings or obstacles, attachment accessories, and accessories for vertical irregularities. Most manufacturers have special accessories to overcome the majority of problem conditions experienced in the field.

Codes of Standard Practice for Sectional Scaffolding

There are some special OSHA and ANSI provisions applicable to sectional scaffolding. Frames must be erected plumb and square with each other. Because frame legs are not adjustable for minor ground irregularities as are tube and clamp posts, the use of adjustable screw legs at the base is required except on known level flat surfaces, where plain base plates should be used. The use of connecting pins or other locking devices in conjunction with coupling pins is required "where uplift may occur." Drawings and specifications for all frame scaffolds over 125 ft (38 m) in height above the base plates are to be designed by a licensed professional engineer, per ANSI A10.8 1988. Most manufacturers have suitable printed drawings available for this purpose for standard construction; for special construction and loadings, individual design should be made for the application. In the OSHA "General Requirements" section, construction regulations subpart 1926.451(c)(2)(ii), it is stated that "unstable objects shall not be used to support scaffolds or platform units." This requirement is generally more directed against sectional scaffolding, which uses adjustable screw legs, than against tube and clamp scaffolding, which, within commonsense safe practices, can be self-compensating, as mentioned earlier. The use of such blocking is most prevalent with subtrades and is a most dangerous practice.

Requirements for guardrailing, as for all scaffolds, are stated in terms of installation of wood 2 × 4s (5 × 10 cm) for guardrails and 1 × 6s (2.5 × 15 cm) for midrails. In general guardrails are designed to support a load of 200 lb (90 N) applied in an outward or downward direction.

OSHA section 1926.451(e) requires access means be provided for all employees, including employees erecting or dismantling scaffolds (1926.451(e)(9). This section allows the erectors to climb end frames or ladder frames that incorporate ladders within their design (Fig. 14.17*f* and 14.17*g*), and is written such that mason frames (Fig. 14.17*a* and 14.17*b*) would qualify as climbable. This section does not directly address frames shown in Fig. 14.17*g*. These frames have horizontal footholds that erectors can and do use for climbing. The industry assumes that these frames are suitable for erectors to climb.

Essentially, manufacturers of sectional scaffolding provides three means to climb scaffolds: (1) climbing the frames, provided they are of the "mason" type having multiple and relatively regularly spaced rungs and having been designed for climbing. (2) Climbing hook-on ladder accessories that attach to the frame. Older types have the rungs immediately adjacent to frame legs and newer ones have a "standoff provision so that *clearance* is provided between the rungs and the frame legs. (3) Climbing internal stair sections positioned within the confines of the 5-ft (150-cm) frame widths and generally the 7-ft (210-cm) bracing length. Alternatively, properly installed job-made "cleated" ladders and portable wood or metal ladders can be used for scaffold access. Figure 14.19 shows the previously described methods.

Design Loads for Sectional Scaffolding

The design load statements made earlier in this chapter apply to sectional scaffolding. In general and with the exception of some cities' regulations the pound per square foot (newton per square meter) load ratings are not aligned to any particular type, configuration, or allowable spacing of scaffolding frames for use by specific building trades. Allowable plank loading vs. span is usually the determinant of the frame spacing. Because the majority of scaffolding work involves work platforms that are progressively relocated, the economies

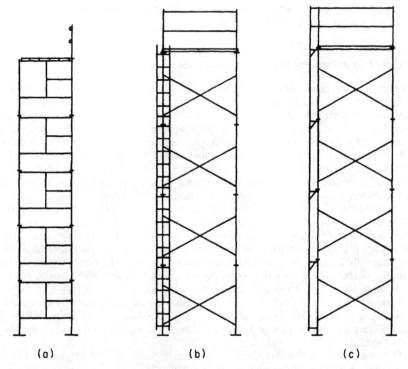

FIGURE 14.19 Climbing scaffolds: (*a*) frame rungs, (*b*) hook-on ladders, and (*c*) stand-off attachable ladders.

achieved by larger spacings between the scaffolding members are not applicable. It is *common practice* in most areas to use 7-ft (210-cm) frame spacing for masonry work and 8- or 10-ft (240- or 300-cm) frame spacing for painting and similar renovation work.

The stated safety factor of 4 is applicable with a caution. Platform loads at levels of application are transmitted to frame legs by means of bearers, generally called "head bars," and consequently apply a bending movement to the frame legs, which can result in their having lower capacity than when concentrically loaded through the legs from upper sections of the scaffold. There are three determinants of frame capacity: (1) allowable load on the frame head bar or other load transfer member, which is usually less than the allowable concentric load of *one* frame leg [depending on design and method of loading, 5-ft (150-cm)-wide frame head bars can typically support loads between 2000 and 3000 lb (9000 and 13,500 N), uniformly distributed; point loads applied close to the frame legs plus other intermediate point loads can give higher values, such a typical application being the use of 19-in (48-cm) or so prefabricated planking that applies loads through hooks or siderails at the edges of the planks (Fig. 14.20)]; (2) allowable leg loads with loading transferred through the head bars; (3) allowable leg loads with loading applied

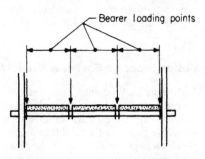

FIGURE 14.20 Prefabricated planking.

concentrically. Typically, the maximum leg load condition in a given scaffold will be a combination of (2) and (3) when a loaded platform is located close to the bottom of a high scaffold.

It is advised that the default leg load for 6-ft 6-in high walk through sectional frames when used, as scaffolding is 2000 lb per leg. Higher leg loads are achievable when the bottom frame is horizontally braced through the walkway or a mason frame is used at the bottom of the stack-up. The load-bearing capacity of the frames is determined by test. When leg loads above 2000 lb are required, consult the manufacturer or distributor. When used as shoring, the frame capacity can be increased because of the difference in the safety factor used for the two different applications. Scaffolding requires a safety factor of 4 whereas shoring generally is designed with a safety factor of 2.5. For shoring then we can increase the leg loads by $4/2.5 \times 2000$ lb. = 3200 lb. The same is true for the increased capacity attained by the special stack-ups mentioned above.

Another frequently overlooked loading application is that used for caster-wheel attachments. Casters should be allowable load-rated for two conditions: static loaded and rolling, or mobile loaded. The static load rating is simply how much the caster will safely carry. The mobile or rolling load is the load the caster can carry and still roll when a 5 percent horizontal force is applied to the rolling tower. For steel-wheeled casters the two loads are close, for tiered casters it is a function of how much load can the tier carry before it is flattened to the point of inefficiency or overstress.

Mobile scaffolds with cantilevered loads, such as a rolling benchwork for a suspended scaffold (Fig. 14.21), must be restrained from overturning by counterweighting or physical restraints. Safe practice requires that the counterweighting or restraining forces required be calculated using 4 times the maximum applied load to determine the overturning moment. From this can be deducted the restraint moment of the inboard self-weight of the scaffolding equipment *before* determining the amount of counterweight or restraint force. A common miscalculation is to multiply the *total out of balance* overturning moment by 4,

SCAFFOLDING

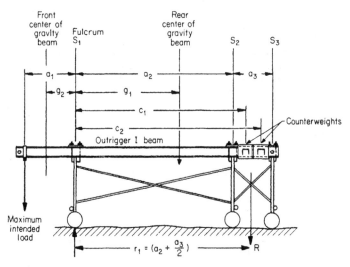

FIGURE 14.21 Rolling scaffold used to suspend a two-point swinging scaffold (swing stage).

having included the self-weight restraint moment in this calculation. This method is wrong because, by so doing, the self-weight moment restraint has been multiplied by 4, which understates the external restraint force required.

Safe allowable loads on all structural scaffolding components should always be determined from the equipment manufacturer(s) or their authorized agents. Because of the great variance in materials, design, and allowable loads, it is most hazardous to connect intermingled components of different manufacturers to each other, although many are apparently physically compatible; even then, differences in tolerances and fit can cause incalculable strains between them, resulting in unsafe loading conditions.

Figure 14.10 shows how the loads applied to posts from platform-supporting bearers are eccentric in nature.

SCAFFOLD PLANK

Wood Scaffold Plank

The strength of wood used for scaffolding planks is possibly the least comprehensible of all scaffolding elements. Wood is organic and affected by many factors, such as region of growth, rate of growth, moisture condition at time of sawing and grading, and, naturally, conditions of use. These factors do not permit use of exact stress design data similar to those used for relatively homogeneous metals. The complexity is compounded further by the standard commercial practice to cut a batch of trees from a "stand," that is, a certain small area that may contain a number of trees of similar but different species from the major one. These different species generally will have varying strength qualities.

Essentially, published wood strengths are derived from procedures published in ASTM Designation D2555, *Standard Methods for Establishing Clear Wood Strength Values.* Various statistical "weighting" factors are applied to the results of testing small "clear" samples of woods from a given stand or area using complicated statistical procedures. A clear sample is a piece of visually perfect wood.

The cut wood is then further graded by an official grading association authority using visual grading criteria. Subsequently, in accordance with ASTM Designation D245, additional downgrading factors are applied to the wood-strength values to arrive at published "working" stresses for various grades of that wood. Such factors as species preference or preponderance, species "grouping," commercial marketing practices, design techniques, safety, and wood graded for specific-product end uses all contribute to the complexity of determining end-strength values that can only be averages.

The final visual grading of the woods is relatively arbitrary, and therefore published working stresses should be used as a guide rather than finite values. In general, they have a ratio of approximately 1:4 below the clear strength values for the species, but not always.

Finally, these published values need to be factored by the end purchaser or user in accordance with conditions of use such as moisture content *during* use and use with the wider dimension on the "flat" such as for planking. Unless otherwise specified, strength values are for woods used as beams with the largest dimension vertical. Wood in use is considerably weaker when green or wet than when it is dryer because the wood fibers are more easily able to slide along one another and this reduces bending strengths and moduli of elasticity (stiffness). The most expensive woods are those that have been slowly kiln-dried to a very low moisture content and that exhibit the highest and most consistent strengths for a certain wood species and grade. Next are woods that have been gradually air-dried or seasoned. Both are generally too expensive for use as scaffolding planks and are mostly used for furniture and other ornamental purposes.

The one exception is lumber that has been mechanically (or machine) stress rated (MSR). This is a process in which each piece of lumber has its moisture content measured and is nondestructively tested and stamped by the machine with its allowable fiber stress F_b and modulus of elasticity E. Typically, the grading stamp will include the letters MSR followed by values for these two stresses such as "1200f-1.3E." The 1200f value of F_b given is that for "single member" loading, explained later. Each grade authority book of rules will give tables showing other design stress values assigned to the grade mark. This MSR lumber also has to pass some additional visual grading inspection and is quite expensive in comparison with visually graded lumber.

Most scaffold plank material is cut green, or unseasoned, and, if stacked properly during shipping and in yards before use, may season somewhat. The degree of moisture content is not easily assessed or determined by the user. Even though its sawn density may include less than 19 percent moisture content, or even 15 percent, it can absorb additional moisture from humidity of the air and during its placement on scaffolds when subjected to rain and snow. A recent change in grading practices now permits wood having a moisture content of less than 19 percent to be classified as "dry." Green or unseasoned wood is assumed to have more than 19 percent moisture content by density.

The importance of yard stacking under cover, with good airflow between the planks after receipt from the mill source in order for the wood to season slowly, cannot be overemphasized. This allows the wood to dry out slowly with the least shape distortion such as crooking, warping, bowing, and end splitting. Consequent exposure to climatic moisture will subsequently have the least effect on its strength in use. Putting green wood immediately into use, especially in hot sunlight, will usually distort much of it too severely for further use and at best may drastically shorten its service lifetime because of end-cracking and "cupping."

Selection of wood for scaffolding plank is made from select structural or dense select structural grades by an appropriate grading authority for the species in accordance with individual visual criteria that vary somewhat in knot sizes allowed, slope of grain, and other factors; they are not given here but are available from the various grading authorities listed. Administrative and marketing considerations often make it expedient to combine different species having relatively similar physical properties into a single marketing combination. Published data of allowable stresses generally give two values for maximum fiber stress in bending (F_b): single and repetitive. Single values assume that any single piece of wood may contain maximum strength-*reducing* characteristics allowed in the grade, on the basis of each individual member carrying its *full* design load permanently. Repetitive values assume that members are used in a system with common supports or joining surfaces (such as plywood) so that applied loads are distributed over a number of members. Scaffolding plank can be considered to fall into the repetitive category, because large, single point loads are virtually never applied to only one plank. An exception to this might be the point load from a wheel of a wheelbarrow or buggy, but even then the loaded total weight of the vehicle will be distributed over one or more wheels or between a single wheel and the driver's foot when the foot is supporting the carrying handles. Also, such loads are seldom static and are usually transitory in nature. Pallets of bricks or blocks generally apply their loads over two to three planks, and the duration of full loads is short, maybe 2 to 4 days.

Wood has the unique ability to withstand high overloads for short periods of time in accordance with Table 14.3. All published wood strength data give working stresses for permanent (normal) duration during loading, a category in which scaffold planking does not belong. Therefore, the criteria assumed hereafter are quite conservative because no increase is assumed in the limiting stresses resulting from short-term loading. Thereby, an additional safety factor, which can compensate for planks in less than perfect condition, is built in.

The supplement gives stress values for all major species in one reference source; it gives the grading authority from which the stresses are obtained and lists the various modification factors applied by the grading authorities. It does not give stresses for "special" end-use grades

TABLE 14.3   Working Strength Increase for Duration of Loading

Maximum load duration	Type of loading	Allowable increase of working stresses, %
1 s	Impact	100
1 min		72 (interpolated)
1 h		47 (interpolated)
1 day	Wind/earthquake	33
7 days	Concrete formwork	25
1 month	Snow	17
2 months	Snow	15
1 year		7 (interpolated)
10 years	Normal	0
50 years	Permanent	−10

such as scaffolding plank. The latest supplement no longer publishes repetitive values separately, but rather gives an adjustment factor of 1.15 to be applied to the tabulated value for F_b.

Without having the assurance afforded by buying (expensive) machine stress rated lumber, the lowest possible evaluation of published strengths must be made taking into account all the potential strength-affecting factors.

Table 14.4, together with the appropriate modifying factors and end-use stresses after modification. The limiting stress values used have considered the likelihood of such species groupings and intermingling and are extremely conservative as a measure of commonsense practicality. *Caveat emptor* is the guiding principle that should be applied because the final purchaser usually has little knowledge of the real origins of the wood purchased.

TABLE 14.4   Recommended Design Stresses for Scaffold Planking

	Basic design stresses		Factored values for use		
	F_b repetitive, lb/in^2	$E \times 10^{-6}$ lb/in^2	F_b factors	F_b repetitive, lb/in^2	$E \times 10^{-6}$ lb/in^2
Douglas fir	1500	1.4	$\times 1.2 \times 0.85$	1550	1.3
Southern pine	1600	1.6	$\times 1.2 \times 0.85$	1650	1.4
Spruces	1425	1.3	$\times 1.2 \times 0.85$	1450	1.2

Using factored load values and maximum deflections of $L/60$, a few allowable span versus load relationships have been calculated. The calculated allowable values for the nationally prescribed loads of 25, 50, and 75 lb/ft^2 (1200, 2400, and 3600 N/M^2) are shown in Tables 14.5 and 14.6.

TABLE 14.5   Allowable Standard Load Ratings for Various Spans

Span	Condition*	7 ft	8 ft	9 ft	10 ft
Douglas fir	R	75	75	75	50
	D	75	75	50	25
Southern pine	R	75	75	75	75
	D	75	75	50	25
Spruces	R	75	75	75	50
	D	75	75	50	25

*R = 1.875 in (4.75 cm), rough; D = 1.5 in (3.81 cm), dressed.
NOTE: 1 ft = 0.305 m.

TABLE 14.6   Allowable Spans for Standard Load Ratings, ft

Species	Condition*	Loading, lb/ft²		
		25	50	75
Douglas fir	R	10	10	9
	D	10	9	3
Southern pine	R	10	10	10
	D	10	9	8
Spruces	R	10	10	9
	D	10	9	7

*R = rough; D = dressed.
NOTE: 1 ft = 0.305 m; 1 lb/ft² = 47.9 N/m².

Prefabricated Plank

Also available are a number of types of manufactured planking, including the following major types:

1. Laminated veneer lumber (plywood-type plies on flat)
2. Steel fabricated
3. Aluminum frame with plywood deck
4. Aluminum frame with aluminum deck

Laminated veneer lumber (LVL) scaffold plank accounts for about one third of the plank used in North America. It has become popular because it has greater strength, is more readily available, and its load capacity, when new, is backed by its manufacturer. Solid sawn scaffold plank, however, still accounts for approximately 65 percent of the total scaffold plank produced and used in North America. Their structure is akin to 1½-in (38-mm)-thick plywood, available in all desired widths and lengths. High bending strength and longer life are claimed for this material.

The fabricated metal or metal-plywood combination planking is available from many proprietary manufacturers in fixed, 1-ft (30-cm) incremental lengths, the most popular being 7, 8, and 10 ft (210, 240, and 300 cm). They are equipped with hooks at each end to seat on scaffold supporting members. Depending on manufacturer, they are rated at various load capacities up to and including 75 lb/ft² (3590 N/m²). The ones with metal walking surfaces generally incorporate some slip- or skid-resisting provisions, and the aluminum-plywood deck types are available or treatable with skid-resistant paint. The planks with aluminum frames are, of course, susceptible to damage from rough handling and are generally more favored by subtrades for rolling scaffold applications than for general-duty construction scaffolding, where the solid or laminated wood planking is preferred for more arduous usage and flexibility.

Testing and Care of Wood Scaffold Plank.   There is some controversy as to the desirability of testing new, solid wood planking before use. Advantages are that testing may reveal hidden flaws in strength which are not visually apparent. Wood is one of the few materials whose strength is related to its stiffness. This relationship allows for an easy way to nondestructively measure the strength of scaffold plank.

Scaffold grade plank should have a modulus of elasticity of 1,800,000 psi and an allowable stress of about 2200 psi. Because of this unique relationship between the stiffness and strength, we can test the strength of scaffold plank by measuring its deflection under load.

Before any testing, scaffold plank must be inspected for defects that would compromise its strength. The SIA recommends that any nondestructive testing of scaffold plank should be done only after mandatory visual inspections.

Planks, with saw cuts, saw kerfs, splits, holes, dry rot, gauges, dents, discoloration, softness, face cracks, edge cracks, or frangible areas, should be discarded or cut down to a shorter lengths to eliminate the defect.

It is important to eliminate localized defects because they may not affect the plank's overall stiffness, and so may not show up as weak plank in the bending stiffness test.

Any plank that has dry rot, shows chemical discoloration, or was used in a corrosive atmosphere like a paper mill or a chemical plant should be cut up and discarded immediately. The long-term effects of chemical contamination make these planks too risky to reuse.

The oldest, and most often used, though not the best, method of testing scaffold plank is known colloquially as the "Gorilla test." We have it on good authority that this test is going to be dropped from OSHA, but as of this writing we have nothing to replace it with. The principal objection to this method is that the personnel who do the testing are exposed to risk. They are the test weights. This test could be conducted using test weights instead of people as some scaffold companies do.

In this test two blocks, 2-3/8 thick, are placed on level ground, 7 ft apart. The plank is laid on the blocks. Two men whose combined weight is at least 325 lb stand in the center of the plank. If the plank touches the ground beneath them or sounds of cracking or splitting are heard, or if the plank does not fully return to its unloaded deflection, it is rejected; if not, it is accepted for use.

Planks that have failed inspection and testing should be immediately cut into short lengths and discarded, or used for nonstructural purposes. The men must never jump on the plank, for this could actually weaken it.

The philosophy behind the Gorilla test is that, if two men can stand in the middle of a single plank, and it doesn't break or sound like it's about to break, and it is not permanently deformed by the load, the plank is good.

Testing plank is a complicated process that is far beyond the scope of this book. In general the process is as follows:

1. Inspect the plank for damage—saw cuts, rotting, and visible weakening. The Scaffold Industry Association *Solid Sawn Field Pocket Handbook* is industry-standard reference for this procedure. For additional reference, one can follow the recommendations of the Scaffold Industry Association *Solid Sawn Field Pocket Handbook*.

2. Measure the planks cross section. Calculate the moment of inertia.

3. Load the plank in bending; up to its working stress or as permitted for the test span by the manufacturer.

4. Measure the deflection and compare it to the theoretical deflection at that load for the plank species, using the plank cross section previously measured.

5. The plank passes if the deflection is not more than 25 percent above the theoretical deflection. The 25 percent allows for variations due to moisture content, dimensional variation, and species stiffness variations. For LVL plank we recommend that the deflection be matched to the span chart provided by the manufacturer.

Testing plank using the previous outlined procedure is delicate, and time-consuming, but it is the only procedure recommended by the makers and distributors of manufactured wooden scaffold plank.

Some of the makers of manufactured wood plank recommend a more scientific method. They recommend that when the plank be placed on the supports, a small load be applied at the center of the plank to settle it in place and then to set the measurement apparatus to zero. The test load is then applied and the deflection measured. They recommend loading the plank to slightly more than the working stress and accurately measuring the resulting deflection. Users of LVL plank should consult their supplies for the proper testing of that brand of plank.

Plank testing machines are available on the market they make testing less complex, faster, and easier.

Short scaffold planks (10 ft or less) are sometimes cleated (Fig. 14.22), particularly when they are used for erection or dismantling. These planks should always be tested cleats down in the same orientation they are used. Mostly, however, planks are lapped (Fig. 14.23) so that the top of the plank is not readily discernible. They can be marked to indicate the top or, in some cases where the owner's or manufacturer's name is stamped into the side, they should be used and tested so that the lettering is right side up.

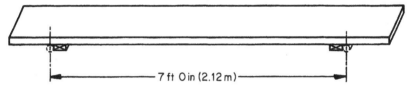

FIGURE 14.22   Cleated single-span planks.

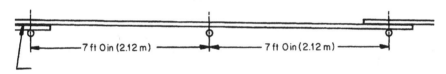

FIGURE 14.23   Overlapped planks.

For longevity, the ends of planks should always be protected from propagation of splits by means of a number of end-protective devices available, such as spiked, formed steel strips which are hammered into the top and/or bottom surfaces and annular ring nails. (See Fig. 14.24.)

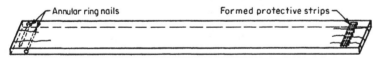

FIGURE 14.24   End-protected plank.

Stacked planks should be intermediately separated by dunnage or firring strips in accordance with handling methods or machinery, ensuring that all such strips are placed vertically over each other—not offset so as to cause local bending. (See Fig. 14.25.) Where possible, keep planks dry, protect them from rain, or arrange them so that they will drain rapidly. The top level of planks in a pile will receive most rain exposure, so be extra cautious in their use.

Wood Deterioration.   Aging of wood alone does not cause it to lose strength, and with care an old well-used dry plank is possibly more trustworthy than an untried, new one. Moist conditions over a period of time can reactivate growth of incipient decay fungi

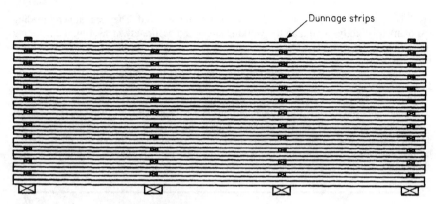

FIGURE 14.25 Stacked scaffolding planks.

already present; moisture does not initiate decay and wood does not itself decay. Decay is a parasitic fungus often present in wood grown in moist, warm conditions, which after drying or seasoning becomes dormant. Prolonged reexposure to moisture can cause these fungi to reactivate, and they do so by eating through the cell walls to eat the contents of the cells. This is why decayed wood is lighter than comparable wood of the same species, so planks that appear lighter than their associated ones should be suspected of being decayed. ("Light" planks may also be of a lighter density, weaker subspecies as previously discussed.) Nevertheless, the lightness indicates one or the other condition. Once decay has started, it is impossible to stop.

GENERAL APPLICATIONS OF SCAFFOLDS

Wall Scaffolds.   The use of horizontal diagonal bracing is recommended on high scaffolds; it should normally be positioned close to the same height increments and levels as the building ties. Although not always used, horizontal bracing does impart additional rigidity to the scaffold. To facilitate installation, braces should be positioned also at the lowest frame level in order to ensure squareness of frames in relation to each other; all brace connections have sufficient tolerance of fit to allow a nonsquare scaffold.

Figure 14.26 shows a scaffold using 5-ft (1.5-m)-high mason-type frames, which, because of their inability to provide intermediate platform egress, are generally used for work at the top platform level only, which is progressively raised as frames are added to the height and planks relocated upward. These frames are generally 4 ft 6 in, 5 ft, and 6 ft 6 in (140, 150, and 200 cm) high.

Figure 14.27 shows an adaptation of 6-ft 6-in (7.98-m)-high walk-through-type frame construction adapted for use by masons. This method involves the use of inset sidewall brackets (also known as outriggers—wrongly), extensions, and lookouts. Used for masonry, they give the facility of providing alternating work-level positions at intermediate heights of the frames, allowing easier reach to the work and more convenient height locations than offered by frame support bars. For masons' work above 18 ft (5.5 m) or so, the use of the 6-ft 6-in (200-cm)-high frame with brackets provides substantial labor economy over top-platform-level 4-ft 6-in or 5-ft (140- or 150-cm)-high frames. For any given height, there is a 25 percent or more savings of the total quantities of frames and braces required and

FIGURE 14.26 Masons' scaffold, using 5-ft (1.5-m)-wide and 5-ft (1.5-m)-high mason frames, working from the upper west level. *(Courtesy of Patent Construction Systems.)*

25 percent fewer plank relocations; the additional expense involved in the brackets and planks for them is minimal.

Figure 14.28 shows a sidewalk scaffold for exterior renovation or similar work. Such scaffolds can be constructed in accordance with manufacturers' instructions to provide appreciable pound per square foot (newton per square meter) loadings on the overhead planked level in addition to support of the platform load(s) from the upper working level(s) and the weight of the superimposed scaffold.

Rolling Towers (Mobile Scaffolds). With suitable bracing and accessories, these are adaptable to almost any work requirement. Ranging from the single width, single bay tower, they can be expanded in both frame width and brace length directions. (See Fig. 14.7.)

Large rolling towers should always use ties connecting the lower legs, so that there is multiplied resistance against excessive deflection caused by one leg or caster hitting a hole or obstruction. All frames in rolling towers should be connected to each other in the vertical tiers by means of positive-locking connection pins, as mentioned previously, because uplift is always a hazard more prevalent to this type of construction than most others. Casters should always be pinned or otherwise securely connected to the bottom frames.

Horizontal diagonal bracing should always be installed on rolling towers to avoid racking; OSHA regulations require that it be positioned at levels (1) as close to the rolling surface as possible and (2) at least at successive height intervals of not more than 20 ft (6 m).

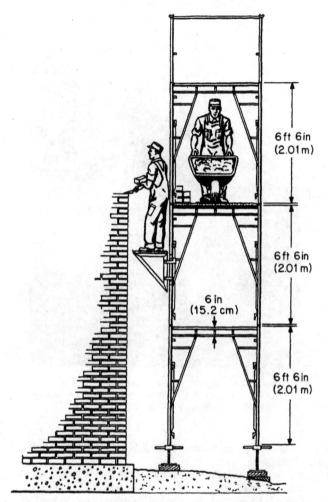

FIGURE 14.27   Walk-through-type frames 6-ft 6-in (1.98 m)-high used by masons.

Free-standing towers must not be used when the top working platform height exceeds 4 times the minimum base dimension—which in most instances is the frame width—*unless* they are securely tied off to a building structure and/or the base is widened by adding more frame and brace elements to widen the base dimensions. Some manufacturers have special frame-type accessories, often called "outriggers," to assist in accomplishing this. Some states may have more restrictive dimensional requirements than the mentioned 4:1 height-to-base ratio (3:1 or 3½:1). Rolling towers frequently need to be equipped with adjustable screw legs to compensate for slightly sloping floors and for irregularities in the rolling surface; few surfaces are truly flat or level and a minor 1-in (25-mm) deviation could cause a caster to lose contact with the surface transferring the weight of the scaffold leg it supports to the interconnecting braces. This could result in the overload of casters adjacent to it. Frequent readjustment of screw legs can overcome this problem.

FIGURE 14.28 Sidewalk scaffold. (*Courtesy of Patent Construction Systems.*)

To refer back to the OSHA construction regulations, subsection 1926.451(f)(5) and 1926.452(w) allows personnel to ride scaffolds while they are being moved, provided the floor is within 3° of level and is free from pits, holes, or obstructions; provided the scaffold has a height-to-base ratio of 2:1 (instead of the normal 4:1); and provided all tools and materials are secured or moved from the platform. As a safe practice, riding of towers is not recommended because obstructions may not be obvious. Columns may not be noticed if the scaffold rolls out of line, and what may appear to be an obstruction-free surface could suddenly become obstructed by an unrelated worker at the job placing some material in the path of the scaffold. The resulting shock as a caster hits such an obstruction could cause a rider to fall on the platform. When moving all towers, whether using outriggers or other means of lateral support or not, the scaffold should preferably be moved in the direction of its longest base dimension and the motive force applied as close to the base as possible. Casters must always be locked when the scaffold is in use or being climbed.

Interior Scaffolds. The application of large-area interior scaffolds is limitless; access to almost any wall or ceiling surface can be achieved using ingenuity *and* preplanning the use of the various correct accessories. Typical scaffolds often use spanning trusses (to minimize the amount of vertical support components), cantilevered short trusses, brackets, and tube and clamps to reach the various surfaces. All open sides of such scaffolds require guardrails to be installed in accordance with the criteria mentioned earlier. Availability of variously sized cross braces facilitates dimensional adaptability to existing seat row spacings and similar obstructions. However, depending on the availability of special accessories and the experience of available erectors, such interiors may well be just as conveniently scaffolded using tube and clamp scaffolds. Sometimes building configurations are such that, without expert help to lay out the work, the time needed to make sectional scaffolding fit is greater than the normal time saved when using sectional scaffolds instead of simple scaffolds; thereby both methods may be equally feasible.

There are occasions when economies are achievable by spanning prefabricated platforms between stationary rows of sectional scaffold. Such platforms as used for two-point suspended swinging scaffolds are sometimes used for this purpose. They are normally available in sizes up to 28 in (71 cm) wide by 32 to 36 ft (9.75 to 11.0 m) long. If use of this method is proposed, extra precautions must be taken. It is not safe to rest the end of such a platform (usually aluminum) on one or two wood planks running atop the sectional scaffold; the point load reactions are excessive, and safely securing the platforms is difficult. This method is favored when only a portion of the area needs to be worked on at one time, necessitating frequent relocation of the (heavy) aluminum platforms. Depending on the work (i.e., spot work such as electrical or decorative work requiring a reasonably large contiguous work area), guardrails must be used along the sides of the platforms when the platforms are not directly adjacent to one another. A major difference between using this method and using rows of scaffolds connected by trusses is that the latter provides bracing rigidity between the scaffold rows; the U-bolted connections used with the platforms do not provide this feature so that on scaffolds over 20 ft (6 m) high, alternative, separate means of stabilizing the scaffold rows is required.

Outrigger Scaffolds.   These consist of a series of outriggers, or thrust-outs, projecting from the face of a structure. The outboard cantilevered ends support a planked platform or a built-up scaffold. The inboard end must be anchored by securely strutting it to a floor slab or beam above it or restrained from movement by a positive means such as U bolts passing over the outrigger and (1) passing through the floor or (2) connecting to looped steel inserts previously set in a concrete floor. The outriggers must be securely braced against lateral overturning, and the point of bearing at the fulcrum point of the beam is required (by OSHA) "to extend at least 6 inches in each horizontal dimension." Although this type of scaffold originated with wood beam outriggers, current practice is to use steel beams or sectional steel trusses. From the steel outriggers, built-up sectional scaffold can be erected on the cantilevered end or, when necessary, built down, or hung, from the outrigger. Figure 14.29 shows a scaffold such as used for surface finishing (such as thin stripes of brick veneer, special mortar course pointing, flashing, caulking) where the work is occasional and not sufficiently continuous to require a two-point suspension scaffold. Figure 14.30 shows a scaffold used to reach *down* from the outrigger level using a truss as an outrigger. In calculation of uplift restraint forces at the inboard end, the previously mentioned safety factor of 4 is applicable using 4 times the live and dead loads of the workers, material, and equipment supported by the cantilevered part of the beam.

Shipbuilding Scaffolds (Steel).   These are generally specialized semifabricated steel components, having evolved over the years from a combination of tube and clamp scaffolding and sectional frame scaffolding. Special OSHA maritime regulations cover shipbuilding and repairs: subparts 1915, 1916, 1917, and 1918.

Wood Built-up Scaffolds.   It is not intended to describe wood scaffolds in any detail. The OSHA and ANSI regulations give more complete and comprehensive guidelines for wood scaffolds than for any other type. One seldom sees wood scaffolds in any quantity, their use being generally restricted to low-rise housing. Many states and cities still give extensive construction details in regulatory bylaws. In principle, the construction of wood scaffolds is

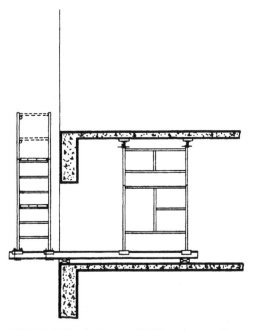

FIGURE 14.29 Outrigger scaffold braced against floors inside the building.

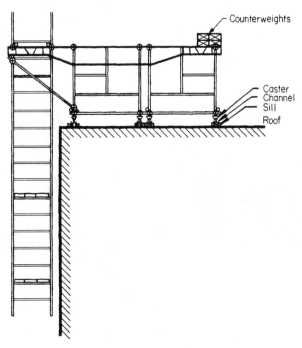

FIGURE 14.30 Outrigger rolling scaffold for work below the roofline.

essentially the same as that of tube and clamp scaffolds; however, because the common fastening is nails, which have little moment rigidity, wood scaffolds require substantial amounts of diagonal bracing in both transverse and longitudinal planes. Construction labor costs are high in comparison with those for metal scaffolds. Carpenters' bracket scaffolds, bricklayers' square scaffolds, and horse scaffolds are prefabricated wood modules used essentially on small housing projects. They fall into the same categories as built-up wood scaffolds.

There are also a number of wood and steel bracket scaffolds with restricted use for formwork platforms, steelworkers platforms, and the like. These are quite specialized in nature and of little value for generalized use.

SUSPENDED (HANGING) SCAFFOLDS

Multi-Point Suspension Scaffolds. The first mechanization of equipment specifically designed for use as scaffolding occurred between 1870 and 1880. The concept involved simple, manually operated, exposed-gear winches, used in pairs at regular intervals along the length of a building face and developed for high-rise masonry work. Each pair of winches supported a steel "putlog" which served as a bearer to support longitudinal planking. Each winch had a toothed drum upon which ½-in (12.7-mm) steel wire rope was wound by a simple ratchet lever mechanism. The wire ropes were suspended in pairs from I-beam outriggers cantilevered from a floor or roof of the building. The inboard ends of the outriggers were anchored by U bolts to the concrete slab or structural steel beams of the building. Cranking the ratchet lever of each pair of winches simultaneously wound the wire rope on the drums and raised the putlog at that position. Successive cranking of each pair of drums was repeated along the length of the scaffold in approximately 9- to 12-in (23- to 30-cm) increments so that the whole scaffold platform was constantly raised to keep the masons at an optimum working height in relation to the top course of masonry.

This scaffolding continues to be popular 100 years later, with some minor refinements over the years (Fig. 14.31). The winches have been replaced with mechanical rope climbing "traction hoists" that grip the wire rope. The wire rope no longer winds around a drum

FIGURE 14.31   Heavy-duty suspended scaffolding for masonry work around the building. (*Courtesy of Patent Construction Systems.*)

but hangs freely below the stage (scaffold). Usually it is coiled up and pulled onto a floor below. Two lifting mechanisms, plus their connecting putlog (bearer), plus the outrigger I beam and its anchorage bolts were collectively known as a "machine" and continue to be termed thus. Originally, the platforms were 5 ft (150 cm) wide; 6-ft 6-in and 8-ft (200- and 240-cm) platforms became available, so as to give more access room for the passage of materials. Putlogs were developed which gave one-plank or two-plank insets between the inside scaffold winch and the masonry wall so that masons would work uninterrupted by passage of materials. Overhead canopy attachments were added to provide the masons with some protection from hazards caused by other work being done overhead, and this is now an OSHA requirement.

The inset plank methods were later refined to be positioned 16 to 20 in (40 to 50 cm) below the material platform. This lessened fatigue resulting from the masons needing to constantly bend over to reach their brick and mortar. The wider 6-ft 6-in and 8-ft (200-and 240-cm) width platforms required stacked pairs of outriggers to obtain the necessary cantilever strength. The outriggers used were once a standard size popularly used as intermediate joists between larger structural beams, being 7-in, 15.3-lb (18-cm by 7-kg) beams, 15 ft (4.5 m) long. This is no longer a popular size, but they are still available in large quantities from suppliers of this equipment.

Multiply stacked 7-in (18-cm) beams become logically disfavored because of inconvenience. In some areas of the country 8-in (20-cm), 20-ft (6-m) long steel and aluminum beams are used.

For stone masonry, methods were developed to use the 7-in, 15.3-lb (18-cm, 7-kg) beams as monorails to facilitate movement of stone blocks along the scaffold from an end lifting location, using 1- or 2-ton trolleys and chain hoists. The use of this type of equipment on even taller skyscrapers involved a special procedure known as "jumping." Economy dictates the use of finite lengths of wire rope. These segments are spliced together to form lengths long enough to cover the height of the building. When the stages reach the splice, they are tied off to the upper length of wire rope, the lower section is removed, and the upper section is threaded through the machine so that it is free to climb the next segment. This is repeated until the scaffold can reach all the way to the outriggers. The determination whether to use extension cables or outrigger "jumps" is generally based upon the rate of rise of the building frame and the timing of the beginning of the masonry construction.

As with all other scaffolding components, the universal safety factor of 4 is used except for the steel wire ropes, which OSHA requires to have a safety factor of 6. The labor required to install and remove suspended scaffolds of this type is high. Steel frame scaffolds are cheaper overall, but costs rise geometrically with height so that for any given project, there is a break-even height above which the suspended scaffold is more economical overall for masonry. Approximately, this break-even height varies between 80 and 120 ft (24 and 36 m). Naturally, high buildings tend to use masonry less and less because of the availability of a multiplicity of other exterior cladding materials, often installed from the building floors and requiring no external scaffolding. Much institutional work using federal and state funding continues to specify brick exteriors, which are preferred for their superior insulation qualities.

Two-Point Suspension Scaffolds (Swing Stages). Originally conceived for shipbuilding and repairs and later adapted for construction work, this type of scaffold began as a planked platform supported at both ends by ropes suspended from an overhead anchorage. To obtain mechanical advantage, multipart blocks were used at the platform ends to make it easier for people to lift their own weight (see Fig. 14.32). Such platforms are still used on ships, by some light-work building trades, and for some window cleaning and painting operations.

Platforms. Early platforms were merely planks spanning between the winches, which limited platform length. These evolved into use of planked ladders lying flat, which were

FIGURE 14.32 Typical two-point swinging scaffold. This type of scaffold consists of a platform suspended by two manually operated Gold Medal Junior machines. Manually operated scaffolds such as this are ideal for tuck-pointing, caulking, and similar light-duty work that does not require the higher speed obtainable with electric or air-powered machines. (*Courtesy of Patent Construction Systems.*)

relatively unsafe, to extend the span and give increased lateral working area for efficiency. Eventually, the ladder type platform was developed using stronger siderails and supporting light planks or wooden slats on top of the rungs.

The demands by industry and construction trades for lighter, easier handling equipment spurred the development of aluminum platforms, or "picks," generally designed with tubular or extruded I sections or C sections for siderails and with aluminum rungs to support either plywood or extruded aluminum slats; the latter is the most popular form.

Metal platforms are available up to 40 ft (12 m) long. However, the norm is limited to approximately 32 ft (10 m) for handling reasons. The design of power hoists became more sophisticated, embodying safety devices to sense and prevent (1) too high a rate of speed when lowering, (2) slippage of wire rope, (3) rope slippage when power is off, and (4) overloading. Powered hoists are of two major types: drum wound and traction. "Drum wound" is self-explanatory; the "traction" type is designed to have the wire rope reeved around grooved drums for between three and four wraps, one of the drums being driven by the electric or air motor. The rope passes through the winch and exits at the bottom; thus the winch and supported platform effectively *climb* the rope. The traction type has replaced the drum type, because the latter is hampered by the weight of the wire rope wound and stored on the drum. It limits practical application to low heights compared to the traction type that can be operated for any height over which the wire rope is long enough to extend. Heights of 1200 ft (365 m) and greater have been scaffolded with these machines. The wire rope, of course, needs to be a single length; it is not practical or safe to use with tying-off "jumping" procedures as for multiple-point suspended scaffolds.

Metal platforms for swinging scaffolds are required by OSHA to be "approved" by Underwriters Laboratories in accordance with "rated" safe load capacity, which must be indelibly marked on the platform and/or have a load rating affixed to it in a manner and format prescribed by UL. The generally used load ratings are for: (1) two workers and hand tools not exceeding 500 lb (226 kg) and (2) three workers and hand tools not exceeding 750 lb (340 kg). These loads are applied within 3 ft (90 cm) of the center of the platform span. The maximum rated load is multiplied by 4 for testing in accordance with the standard requirement for a safety factor of 4.

Similarly, all winches for swinging scaffolds must be of a design tested and listed by Underwriters Laboratories or Factory Mutual Engineering Corporation.

Rigging. Anchorage of wire ropes for scaffolding winches is accomplished by three basic methods: (1) positive attachment to solid and secure structural elements of the building, (2) attachment to "cornice hooks," also known as "S hooks" used over a building cornice or a parapet wall, (3) attachment to the ends of cantilevered outriggers. When using outriggers, the inboard ends must be positively secured against uplift by means of positive anchorage to the building or by counterweighting. Counterweighting methods used to be extremely hazardous since field personnel were often uninformed of the requirements of using a safety factor of 4 against overturning so that in many cases far too little counterweight was applied.

The preferred counterweighting method is to use steel or cast-iron counterweight blocking that can be positively attached to special brackets bolted to the outrigger beams. For handling ease, these blocks are generally made to weigh 50 lb (22.6 kg) each. The original conception of multiple, attachable counterweights used specially made blocks of poured concrete. An attendant problem with these is the hazard that with rough usage over a period of time they can be chipped and thereby become progressively lighter over the years. Strict inspection and control to maintain the original poured concrete weight is necessary. OSHA 1926.451(d)(3)(ii) states that counterweights shall be made of non-flowable material. Sand, gravel, and similar materials that can be easily dislocated shall not be used as counterweights. In addition, OSHA 1926.451(d)(3)(iii) states that only those items specifically designed as counterweights shall be used to counterweight scaffold systems. Construction materials, such as, but not limited to, masonry units and rolls of roofing felt, shall not be used as counterweights.

A development of the I-beam outrigger method is to mount pairs of them upon interconnected rolling towers, often known as "benchwork," so that the total rigging and counterweighting can be quickly relocated at successive lateral positions as required by the work. Such an application is shown in Fig. 14.21. It is important to follow the proper method of calculating the necessary counterweights that can use the self-weight of the benchwork to reduce the amount of separate counterweights to be placed at the rear of the towers. Seating of the outriggers can be done in two ways: U-bolted to the top head bars of the scaffold frames located in U heads or bolted concentrically over a frame leg. The latter must be carefully calculated to avoid overload of the caster supporting that leg. The head-bar method is generally preferred because the reaction loads at the fulcrum are shared by *two* legs and *two* casters if the outrigger is centrally placed on the frame.

Single-Point Suspension Bosun Chairs. For occasional spot work, a single-point suspension scaffold for use by one worker is often desirable. These generally contain a single powered winch mounted in a "work cage." These are often extended by adding additional components, often called "extension baskets" to both sides of the work cage for use by two workers—one in each extension. The weights of each worker plus tools should be carefully arranged to be approximately equal; otherwise imbalance will be experienced, causing the assembly to hang at an angle other than horizontal because the point of lifting of the winch is at the center. When raising or lowering the winch, one worker has to operate the controls at the winch center. To avoid an imbalance hazard, this means that both workers should relocate to the center section while raising or lowering the winch. There is economy of equipment involved here, to the detriment of safety. Realistically, if two workers are required, it could be done more safely by using a two-point suspension scaffold with a short platform. Some one-person cages are available using manually operated winches.

The original derivation of a one-person scaffold is a bosun chair, consisting of a wood seat, with rope slings attached to a multipart rope block and rope falls. This was used originally for ship work. Current technology generally uses a metal or plastic formed seat attached to a powered winch. Rigging for both types of single point should follow the same criteria as for two-point suspended scaffolds (swing stages).

Safety Considerations for Suspended Scaffolds.   Multiple-point suspended scaffolds are usually constructed using mesh between the guardrail and the toe board; OSHA and ANSI call for a screen mesh or solid panel capable of withstanding a force of 100 lb acting in any downward or outward direction. Plastics mesh screens are popular for this application. Full-body harness and lifelines are not required for workers on the scaffold. During erection of outriggers, safety harness and lanyards should be used in accordance with prevailing standard practice for structural steel work. During jumping and tying-off operations, it is recommended that harnesses and lanyards be tied off to the building or some part of the scaffold.

Full-body harness lanyards, and lifelines (safety lines) should be used at all times when working on two-point suspension scaffolds and bosun chairs. Multitiered two-point platform scaffolds present a problem; use of a safety harness and separate lifeline could result in an upper platform tier striking someone, should one end of the scaffold give way when a worker is being held at a relatively fixed height on the lifeline. For this use, the safety line is often allowed to be attached along the length of the scaffold, when doing so a separate independent suspension line equipped with an arresting device connected to the platform is required [see OSHA 1926.451(q)(3)(iii)].

Other Applications.   Many scaffolding applications involve the use of two or more different types of scaffold. Tubes and clamps can be used to greatly enhance the flexibility of sectional scaffolding when used in conjunction with them to provide lacing and ties between disconnected sections, to brace a high tower to a spread base, and to provide additional work levels not achievable with standard sectional scaffolding accessories.

Scaffolding winches suspended from roof steel or similar anchorage can be used to provide a suspended working platform with no support being required from the ground. The winches support bearers and runners of tubes and clamps or steel I beams which in turn support a planked platform. OSHA refers to this type of scaffold as an interior hung scaffold.

The use of rolling bench works to support swinging scaffolds has already been discussed. Human ingenuity is the only limiting factor in providing solutions to scaffolding problems, and it is important to keep in mind the facility of combining two or more disparate scaffolding products into a safe access or support system.

SAFE PRACTICES

We cannot cover all safety aspects of scaffolding, but some highlights are given in the following comments. The majority of scaffolding manufacturers through their direct operations and through their agents and distributors normally provide free safety rules and instructions with all shipments of equipment. These safety rules contain detailed instructions on safe use and proper construction. The Scaffolding, Shoring, and Forming Institute, Inc. (SSFI), besides providing safety rules through its industry members, also has available booklets on safe erection of certain products. The user should ensure that copies of such publications are obtained from a member supplier or directly from the institute and should most importantly ensure that they reach the field personnel who will be building and *using* the scaffolding.

An important reference for the industry is the *Safety Requirements for Scaffolding,* ANSI A10.8-2001, by the American National Standards Institute Inc., 1899 L Street, NW, 11th Floor Washington, D.C. 20036.

Guardrail Systems

As mentioned, for all scaffolds OSHA and ANSI call for use of 2×4-in (5×10 cm) guardrails and 1×6-in (2.55×15-cm) midrails. Attachment of wood guardrails to steel components should not be considered normal. In most cases installation can be achieved only by using (often inadequate) wire binding. Use the metal or other guardrails made for the particular brand of equipment being used.

Guardrailing can be relocated continuously to different working levels, provided that further work will not be necessary at a vacated level; if the planking is left in place, there is no collateral need for guardrails unless the platform is to be *reused* at a later time.

In the absence of guardrailing designed for use in conjunction with any particular type of scaffolding, expedient measures can be taken to ensure safe conditions and compliance. Tube and clamp scaffolds use tubes and clamps for guardrailing working levels; tubing can also be used for guardrailing of sectional steel scaffolding, being attached to frame legs with 1-$5/8$- $\times$ 2-in (41- $\times$ 50.8-mm) standard clamps. Similarly, manila, synthetic, and steel wire ropes can also be used for guardrails and midrails; ropes of any kind are easy to attach by simply making a loop or a half-hitch around an upright frame leg or post. To comply with the requirement for "minimum deflection," they must be strung tautly along the scaffold. Rope guardrails are particularly useful in guardrailing intermediate platform levels that may be occupied infrequently for uses such as inspection or light repairs. Most sectional scaffolding frames do not have built-in locking devices at levels suitable for attaching guardrailing for a simple reason: most scaffold frames are connected with cross bracing positioned in the vertical plane immediately *inside* the frame legs, thereby occupying the physical space in which a leg-attached guardrail would have to be located. However, separate, attachable devices are available for such use.

There exists an uncertainty in current regulations about guardrails being furnished and installed to comply with the *intent* of the regulations; minor differences of dimensions in specific instances should not be the subject of jobsite controversy.

Toeboards

Toeboards have been discussed. They are required wherever guardrails are installed. Many prefabricated planks and platforms have hinged toeboards or "drop-in" receptacles to receive them. For use with normal planking, various devices are available to facilitate toeboard positioning.

Planking

Planking has been covered; however, it is a regulation that, when overlapped on the scaffold support member, the overlap is required to be a minimum of 6 in (15 cm) and a maximum of 12 in (30 cm). Variously, some sections of OSHA require that planking be laid "tight," implying that there should be no openings through which tools and rubble could fall. Using five full or nominal 10-in (25-cm) planks with 5-ft (150-cm)-wide frames results in a *tightly* planked width between 51 and 46½ in (129 and 118 cm) so that to plank the full width of the scaffold frame [58-3/8 in (148 cm) between frame legs] there is between 7-3/8 and 11-7/8 in (18.7 and 30.1 cm) unplanked space. Where there is no work in progress beneath the platform, distributing this excess space by leaving gaps in the planking is reasonable; where there is need to protect persons or property under the platform, plywood sheets should be laid atop the planking. Prefabricated planks are designed to fill the whole 58-3/8-in, (148-cm) interior frame width, being from 19 to 19-3/8 in (48 to 49 cm) in width.

To provide continuous planking around corners, planks must be placed so as to avoid tipping; to avoid overloading the side planks of a platform, overlapped planks at right angles to each other should extend across the whole platform so that the load at the side is distributed over other planks when the most heavily loaded planks deflect. Where possible to lay planks diagonally across a corner, the diagonal planks should be placed first, running from bearer to bearer; thence the planking from the two adjoining elevations of scaffold are placed atop the diagonal planks. Under *these* conditions, the previous 6 to 12 in (15 to 30 cm) overlap does not apply and the diagonal planks must overlap their bearers by *at least* 12 in (30 cm).

If planks are to be butted to provide a smooth walking surface, ends of butted planks should be seated on separate bearers or equivalent solid seating. This is achieved simply by using additional bearers with tube and clamp scaffolds attached to runners. With sectional frame scaffolding various means of lumber packing can be used, or tube and clamp runners can be placed under the frame head bars. These, in turn, can support the additional required bearers. Where dislodgement of planks is possible, they should be cleated, nailed down, and/or prevented from uplift by wiring or some combination of all three. Where necessary to place planks at an angle other than horizontal, as for ramps or elevation changes of the platform for work on variable-height ceilings, any slope greater than 1:10 should have cleats nailed to the tops of the planks 12 in (30 cm) apart. Debris and rubble should not be allowed to accumulate on planking and should be removed as quickly as possible. Platforms should be cleared of ice or snow before being used for work.

Ladders

Ladders have been discussed previously; some additional comments are pertinent. Unless manufactured portable wood or metal ladders or job-built cleated ladders are used, there is no specific requirement that rungs be spaced 12 in (30 cm) apart. Variations from 12-in (30-cm) rung spacing are necessary with scaffolds for very simple reasons: frames may not be of even 1-ft (30.48-cm) heights, and where they are, the additional 1-in (2,54-cm) collar of the typical coupling pin adds 1 in (2.54 cm) to each frame tier. Therefore, three 5-ft (150-cm)-high frames give a height of 15 ft 2 in (4.62 m) to the top bearer; one 6-ft 6-in (1.98-m)-high frame plus two 5-ft (1.52-m)-high frames give a top bearer height of 16 ft 8 in (5.08 m). Common sense requires that rungs be relatively *evenly* spaced, but not necessarily exactly spaced, and in most cases they cannot be at an even 12 in (30 cm) unless some nonstandard increments are used from the ground to the lowest rung and from the uppermost rung to the platform. Pitched ladders are generally required to be placed with an overrun of 36 in (91 cm) above the top bearing support of the ladder; used at the side or end of a scaffold, this requires a person to have to jump down from the guardrail level from the platform or, as is unfortunately prevalent, the guardrails are *removed* at that point. The most important approach in applying finite restricting dimensions to scaffold access and guardrailing is to remember that by nature, all scaffolds are variable structures, with built-in flexibility for nonstandard and vastly varying circumstances and conditions. Common sense, practical measures, and broad interpretations of dimensional regulations should prevail. An excellent example is the 12-in (30-cm) ladder spacing mentioned before; there are no standards or studies that can show that 12-in (30-cm)-rung spacing is inherently more or less safe than, say, 10, 11½, or 13½ in (25, 29, or 34 cm). The 12-in (30-cm) spacing is convenient, traditional, and standard for building portable or fixed industrial ladders, which does not mean that scaffolds must be designed to comply with ladder regulations; rather they are products that must be evaluated on what they can achieve as a building tool having completely individual and separate form from any other product because there are no other comparable products that fulfill a similar function.

Why Scaffolds Collapse

The most important thing to consider when designing or providing a scaffold is *how it will be used*. The end use of the equipment can get lost between the customer, the salesperson, and the engineer. What the customer wants, what the salesperson sold, and what the engineer designs may have nothing to do with each other.

- A contractor rented a scaffold to fix a brick wall that was about to collapse. The wall collapsed, so did the scaffold.
- A mason contractor received a full truckload of brick. There was no place on the jobsite to store it so the foreman loaded it on the scaffold. "It's going to go there anyway"! The scaffold collapsed.
- A restoration contractor was to remove and replace the tiles on the face of a landmark building. The user intended to store the tile on a special section of the scaffold, and lower it to the ground at the end of each shift. The tile was never lowered. "They were only going to have to hoist it up and put it back on the scaffold eventually"; after three shifts, the scaffold collapsed.

A scaffold tower was provided to act as an access platform between a building and the hoists that brought material and personnel up to the floors. The diagonal braces, on the tower, were in the way. The scaffold user was told they could remove diagonals provided they put them back. The user didn't put them back. The scaffold collapsed.

BIBLIOGRAPHY

Occupational Safety and Health Standards for the Construction Industry (29CFR Part 1926) with Amendments as of November 29, 1996, U.S. Department of Labor.

Occupational Safety and Health Standards for the Construction Industry (29CFR Part 1910), July 1, 1992, U.S. Department of Labor.

CHAPTER 15
FALSEWORK/SHORING

Michael S. D'Alessio, P.E.

This chapter is intended to provide basic examples of the various types of shoring equipment available to the construction industry. The chapter also gives examples of efficient use, advantages and disadvantages, and special considerations related to shoring methods. All types of shoring can be efficient and cost-effective if matched to the proper job. Matching the most efficient shoring system to its most suitable application is the responsibility of the shoring layout design engineer.

Figure 15.1 shows a typical concrete floor supported by four-foot wide steel shoring towers spaced at seven-foot intervals. A sixteen-foot long steel I-beam spans over two seven foot shoring towers and supports aluminum joists that are fitted with wood nailer strips. The plywood sheet that comprises the form bottom is nailed to the joists.

All shoring erection must follow the shoring layout specifically and must be erected properly and in accordance with the layout. The reader should keep in mind that safety requirements are of primary importance regardless of the type of shoring being used.

Discussions of proprietary shoring systems are purposely avoided, so the following sections are intended as a general guide to the principles and procedures of safe shoring design and utilization. The manufacturer or agent should provide specific information on the characteristics and use of equipment. Under actual field conditions the manufacturer's design tables and instructions should be consulted and followed for the type of equipment in use.

FRAME SHORING

In the late 1930s and the 1940s welded-steel-frame scaffolding made its appearance as a more efficient and simpler replacement for the older, tube-and-coupler metal scaffolds. The concept of a welded frame taking the place of three or more pieces of tubing connected to

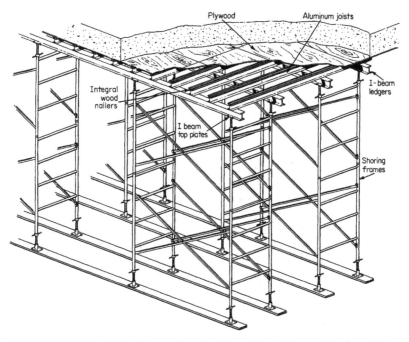

FIGURE 15.1 Typical shoring/framework section with steel-beam ledgers and aluminum joists.

each other by loose scaffold clamps changed the scaffolding practices in many areas of the world. Line drawings of tube-and-clamp scaffolding and frame-type shoring are shown in Figs. 15.2 and 15.3, respectively.

It was not until the 1950s that these laborsaving scaffolding "frames," as they generally were called, began to be used to provide vertical falsework/shoring support to horizontal formwork for slabs, beams, and other similar concrete construction. A frame width of 5 ft (1.524 m) became standard with 2 and 3 ft (0.61 and 0.91 m) widths available on a more limited basis.

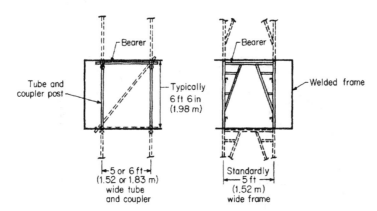

FIGURE 15.2 Cross section through tube-and-clamp and frame-type shoring.

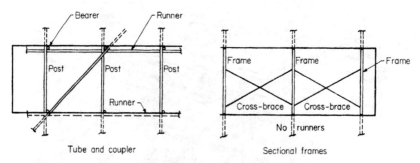

FIGURE 15.3   Side view of tube-and-clamp- and frame-type shoring.

The earliest applications used wooden ledgers (stringers) seated directly on the top, or "header," bars of the frames. This method was not satisfactory or efficient because releasing the shoring loads after concrete setting required the lowering of frames to break the bond to the concrete, as shown in Fig. 15.4. For heights of one or two frames this was a passable but awkward process; for higher work requiring multiple tiers of frames (lifts) it was very difficult to safely release the threaded screw legs at the base of the scaffold.

Attendant problems were soon discovered concerning the wooden ledgers. Unless 4-in (10-cm) wood was used, the ledgers were laterally unsupported and could not be loaded to their full strength. Also, the header bars were insufficiently strong to efficiently develop the load capacity of the frame legs. Efforts to reinforce the header bars of the frames did not eliminate the collateral problems of the local crushing of the lumber at bearing points on the tubular members. Reinforced or not, practical limitations resulted in the inability of the header bars to carry much more than the strength of *one* leg of the frames, that is half-capacity leg loading. This resulted in requirements of up to twice as many frames as theoretically necessary to support any given concrete load. Fifty percent inefficiency with unnecessarily high costs was the result. Many concrete contractors were therefore reluctant to try this and were unwilling to make additional investments to purchase 4×8 or 4×10 in (10×20 or 10×25 cm) ledgers that loaded the frames more efficiently. They were accustomed to using 3×4, 4×4, and 4×6 in (7.5×10, 10×10, and 10×15) joists and ledgers in conjunction with wood or metal single-pole shores. The scaffolding industry was, and still is, a rental-oriented one, but logically, rental of wooden joists and stringers is impractical.

However, some scaffolding manufacturers did make small I beams (14×7.7) available for rental as an efficient substitute for 4×8 and 4×10 in 10×20 and 10×25 cm) wood

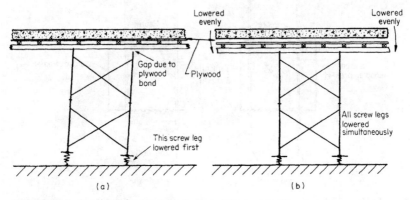

FIGURE 15.4   Stripping a shoring tower: (*a*) poor method and (*b*) better method.

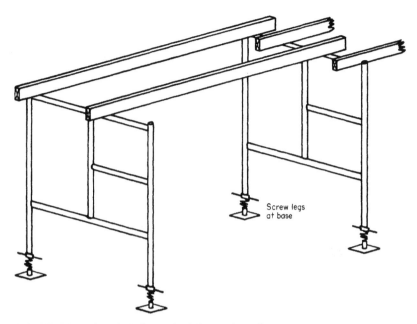

FIGURE 15.5 Early method of supporting ledgers on frames bars.

ledgers. While these became popular with many contractors and did expand the use of frame shoring, there still remained physical limitations and drawbacks such as necessary lateral overlapping of the beams that made positions on the frame header bars inefficient by having to position the ledgers away from the frame legs. (See Fig. 15.5.) Also, careless placement of such beams subjected the header bars to bending stresses and thus again failed to achieve optimum leg-load capacities of up to 5000 lb (22,240 N) per leg.

The Solution. To realize the frames' full potential, the shoring loads had to be applied *concentrically* to the frame legs. Accordingly, the next development was a historic one for both the construction industry and the scaffold manufacturers: the design and manufacture of specialized shoring accessories for use in conjunction with the standard manufactured scaffolding frames. The components were U heads or channels (Figs. 15.6 to 15.8) to

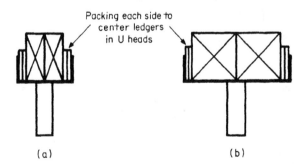

FIGURE 15.6 Ledgers in U heads: (*a*) two 2-in (5-cm)-wide ledgers in 4-in (10-cm) U head; (*b*) Two 4-in (10 cm)-wide ledgers in 8-in (20-cm) U head.

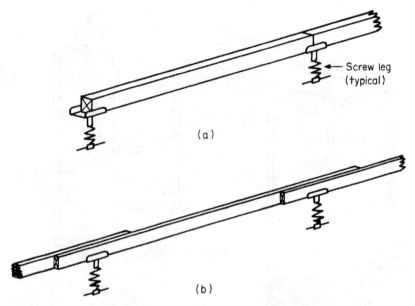

FIGURE 15.7   Ledger continuations: (*a*) ledgers butted in U heads (*b*) ledgers lapped in U heads.

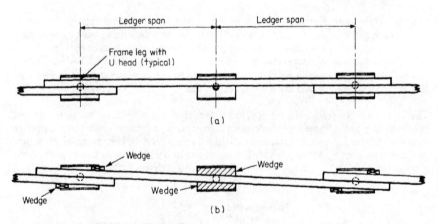

FIGURE 15.8   Ledger lapping: (*a*) ledgers lapped this way impose eccentric loading at center leg; (*b*) correct way to lap ledgers.

support and locate stringers and ledgers: adjustable threaded screw legs of up to 24-in (60-cm) open adjustment [although 12 to 16 in (30 to 40 cm) was normal], base plates with nail holes, and shorter diagonal bracing as small as 2 ft (60 cm) between frames. As these components became more plentiful, huge numbers of standard access scaffolding frames for shoring became available having safe load capacities between 3000 and 5000 lb (13,344 and 22,240 N) per leg. Thus, shoring concrete with scaffold frames became an everyday event instead of an occasional one. (See Fig. 15.9.)

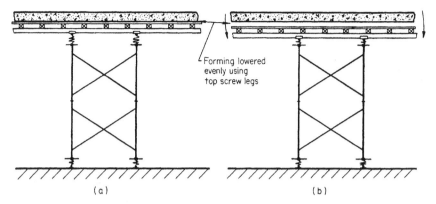

FIGURE 15.9 Four-leg shoring tower with screw legs at top and bottom.

Heavy-Duty Shoring Frames

The next step was another logical one: design frames *especially for shoring*. While doing this, why not also make them stronger?

Some premature efforts were made to manufacture frames with legs of nominal 1.9-or 2-in (4.8- or 5-cm)-OD tubing. Unfortunately, the resulting load capacity of 5000 to 7000 lb (22,240 to 31,136 N) per leg was not much greater than those obtainable with standard scaffolding frames. Frames generally had become standardized to use 1-5/8 -in (4.13-cm)-OD tubing with wall thicknesses of up to 0.125 in (3.2 mm).

Subsequently, the heavy-duty shoring frame made its logical appearance with tubular legs of 2-3/8 or 2½ in (6.0 or 6.35 cm) OD and with design shoring load capacities of 10,000 lb (44,480 N) or more per leg. Figure 15.10 shows typical heavy-duty shoring towers of various configurations.

The advantages and disadvantages of heavy-duty shoring frames were as follows:

Advantages	Disadvantages
Only one-half the number of frames for a given shoring condition	Weight increased to 70 lb (31.75 kg) for a 4 × 6 ft (1.22 × 1.83 m) high frame
Use of standard cross braces	Short supply while inventories were being enlarged
Simpler design procedures	
Sturdier accessories for rough handling	Market saturation (mid-1970s) took many years

From then on, the staple of the industry were frames with capacities of 10,000 lb (44,480 N) per leg. The 5-ft (1.524-m) width of the standard "access" scaffolding frames was changed to the more convenient modular 4-ft (1.22-m) width by 5 or 6 ft (1.524 or 1.83 m) high sizes; these quickly became the "bread and butter" items of shoring construction. Also, the shoring accessories were made of thicker and stronger materials to accept the larger vertical loadings involved with these heavy-duty frames. Frames of 10 kips per leg are heavy to move manually. A 4- × 6-ft steel frame weighs 69 lb, which can be easily carried by two men. The next logical development, of course, was to make them in aluminum. A comparable aluminum frame weighs less than 30 lb. The most popular shoring frame size is 10 kips and the first aluminum frames were produced with this capacity. They are for

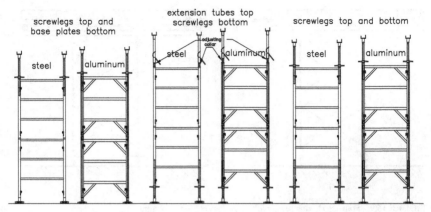

FIGURE 15.10   Various frame shoring tower configurations. (Bracing at right angles to frames not shown).

the most part interchangeable with their steel counterparts and some brands use the steel shoring components with their aluminum frames, but the frames are not intermixable with steel frames. Figure 15.10 shows a popular aluminum frame and some of the components that go with them. Larger U heads and screw legs and the availability of preexisting brace sizes made this type of shoring one of the most versatile and *adaptable* tools that the building contractor had ever known.

Another radical development was the telescoping tube used inside the frame leg. Its application was similar to that of a single-pole shore with a pin and pin-holes for rough adjustment: it had an adaptor with a short length of screw thread for fine adjustment. These telescoping tubes were named "extension" legs or tubes and had working lengths of up to 5 ft (1.529 m). With their rows of pinholes they were quickly christened "piccolos."

Today, there are many more types of equipment available for shoring, covered later in this chapter. However, the workhorse of the concrete construction industry is still the 4×5 or 4×6 ft (1.22×1.524 or 1.22×1.83 m) heavy-duty welded shoring frame.

Almost all makes of this frame are furnished with some form of quick-acting mechanical locking mechanism that enables fast attachment of the cross bracing to the frames.

Shoring Towers

A shoring tower is a modular assembly of single or multiple-tiered pairs of frames connected by pivoted diagonal cross bracing.

The historical importance of the scaffold shoring tower lies in the fact that it provided the facility of at least four single-pole adjustable steel shores with extremely rapid erection. Starting to shore a corner of an installation with single-pole shores puts great reliance on the moment strength of the ledger-shore top connection in which the only connection is nails. The use of wood bracing and lacing members to plumb vertical steel shores and tie them together is a time-consuming and difficult process for usually three or more workers if the shores are over hand-reaching height. A scaffold shoring tower, on the other hand, is quickly and safely erected by two workers, and, in extreme cases, one-frame-high towers can be erected by only one worker. The four legs of a tower give the equivalent facility of at least four shores, as mentioned, but they are usually of a higher load capacity, have built-in devices for easy attachment of bracing, are quick and easy to plumb and level, and can be erected to great heights.

The use of two or more tiers of single-pole shores is prohibited by many states. However, federal OSHA regulation [paragraph (b)(8) of Ref. 1] allows it but requires design and inspection of such installations by a structural engineer. OSHA 1926, 703(b)(8)(I) requires that "The design of the tiered post shoring shall be prepared by a qualified designer and the erected shoring shall be inspected by an engineer qualified in structural design." Sectional frame shoring does not have such a requirement unless called for by a state's OSHA regulations or if the job is unusually complex.

Sectional frame shoring must be installed and used in conformity with a shoring layout, generally designed and furnished by the concrete contractors, the shoring manufacturers, or their agents and distributors.

As the use of shoring frames expanded, the applications became more sophisticated and generally served to further reduce the use of single-pole shores.

The scaffolding industry makes cross braces available for spans as short as 2 ft (60 cm) and as long as 10 ft (3 m) between frames. When used with long-span horizontal shoring beams, bracing distances were expanded from 12 ft (3.6 m) up to 15 ft (4.5 m). Because of this, care must be taken to ensure that braces do not sag and pull the frames out of plumb. The combination of shoring frames and horizontal shoring beams is covered more fully later in this chapter.

Frame shoring towers are very flexible modules because they come in a wide range of bracing span lengths and frame widths. Although 4-ft (1.22-m) width is the standard—working nicely in conjunction with 4-ft (1.22-m)-wide plywood sheets—frames 2 and 3 ft (60 and 90 cm) wide are also popular because they are adaptable as shoring for beam soffits separately from slab shoring frames owing to soffit height differentials (Fig. 15.11).

Demand and supply developed enormously, peaking out in the early 1970s as other, proprietary shoring systems, such as column- and/or wall-supported flying deck systems, large "tables" with trusses and roof forms, also started to become popular. However, large shoring projects can require as many as 25,000 frames or more!

Accessories

By being able to combine various accessories and brace sizes, contractors now have more flexibility. Tower heights, taking advantage of standard 12- and 24-in (30- and 60-cm) extralength screw legs and piccolo extension tubes, can be easily assembled for any job height by using the relatively standard three frame heights of 4, 5, and 6 ft (1.22, 1.52, and

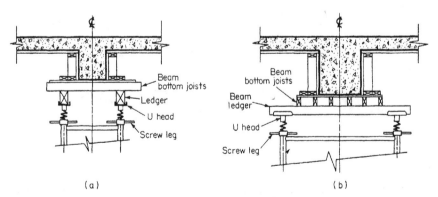

FIGURE 15.11 Beam shoring: (*a*) 2-ft (0.60-m)–wide frame beam shoring; (*b*) 5-ft (1.22-m)-wide frame beam shoring.

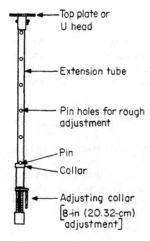

FIGURE 15.12 Extension tube (piccolo) with screw adjustment.

1.83 m) with which most shoring heights can be reached. Others, such as 3 ft 6 in (1.07 m) and 3 ft (0.91 m), are also available.

Typical accessories, shown in Fig. 15.12, provide threaded adjustment on tops, bottoms, or both ends of the tower. U heads are available for almost limitless ranges of size and types of ledgers, whether wood or steel (Fig. 15.13). The vertical sides of the U heads give support to wood ledgers during installation; however, they are not designed or intended to resist overturning of the ledgers under shoring loads.

A wide selection of steel I beams is popular and available for use as ledgers. These are generally used with "top plates" for attachment of the I beams to the frame legs. They are designed to permit lateral overlapping of ends of beams at frame leg junctions. Also available are rectangular top plates that permit attachment of other-sized I beams when the top plate is positioned at right angles to the normal. One of these is shown in Fig. 15.14.

STABILITY OF TOWERS

As a rule of thumb, any width-brace combination of tower size whose height is 4 or more times greater than its narrowest base dimension portends a stability hazard when not connected by ties to other towers; this hazard is substantially increased when people are working on the tower. If not laterally tied, 2-ft (0.61-m) wide frames and multiframe high towers having 2-, 3-, or 4-ft (0.61- 0.91- or 1.22-m) bracing are similarly unstable. The commonly accepted practice is to connect rows of towers to each other with tube-and-coupler horizontal lacing members, or where practical, with additional cross bracing so that rows of frames are continuously cross braced in one plane (Fig. 15.15).

With 4-ft (1.22-m)-wide frames having 4 ft (1.22 m) or longer brace lengths, towers should be tied (laced) to each other at a height of approximately 16 to 18 ft (4.9 to 5.5 m) from the base in line with the plane of the frames. If substantially high towers are involved, repeat the lacing-bracing at every third frame in height as work progresses.

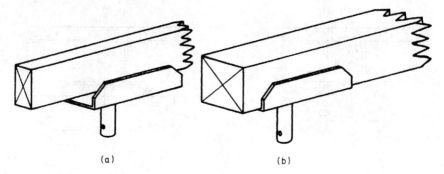

(a)   (b)

FIGURE 15.13 U-head details: (*a*) 4-in (10 cm) wide by 18 in (20 cm) long and (*b*) 8 in (20 cm) wide by 14 in (36 cm) long.

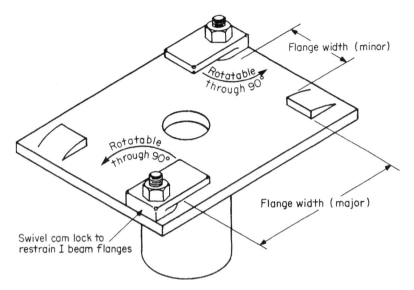

FIGURE 15.14 Top plate for securing I-beam ledgers.

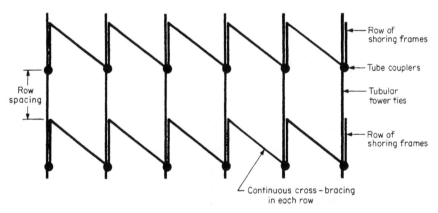

FIGURE 15.15 Lacing of shoring tower rows in one direction, continuous cross bracing in other direction.

Lumber is also used to effect lacing-bracing, utilizing specially shaped nailer plates (Fig. 15.16).

Lacing and bracing with 2-in (5-cm) nominal tubing 2-3/8- × 2-in (6- × 5-cm) clamps affords a significant degree of moment connection. This lacing should be installed in both horizontal planes at each three-frame level in the tower, unless continuous cross bracing can be used in one direction (plane at right angles to frames).

For relatively "clean" installations (i.e., having a flat, firm base), it is sufficient to attach these "lacing" members to the towers as continuous members with suitable scaffold

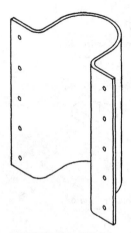

FIGURE 15.16   Nailer plate fits around frame leg, nailed to lumber brace of tie.

clamps between the tubing and *one* leg of each tower, even though the tube will pass by a second leg of the tower without being connected. (See Fig. 15.17.) Installations involving multivariable base and shoring height conditions should have each lacing member coupled to *two* legs of each tower that it passes. When shoring to or from sloping surfaces, each tower leg in contact with the forming of the sloping area should be tied to its adjacent, similar member with ties attached approximately parallel to the sloping surface(s). Figure 15.18 illustrates this, in addition to showing some recommended ways of proper fastenings to avoid slippage of the members in contact.

Continuity of these lacing-bracing members is quite important to the stability of the installation as a whole because each two or more connections develop a rudimentary moment connection that is a valuable stabilization feature. Also, screw legs can be tied together by these lacing-bracing members, thereby distributing the tendencies of one or more legs to deflect from the vertical owing to lateral forces.

In designing a shoring system it is very important that it be considered as a synergistic whole—possibly greater than the simple sum of its parts. The interface between ledgers-stringers and frame legs is an extremely important one and should take precedence over all considerations except overload of frame legs or ledgers. It is essential that ledgers be installed so as to (1) have maximum surface contact with the U head, (2) be suitably placed so that ledger reaction

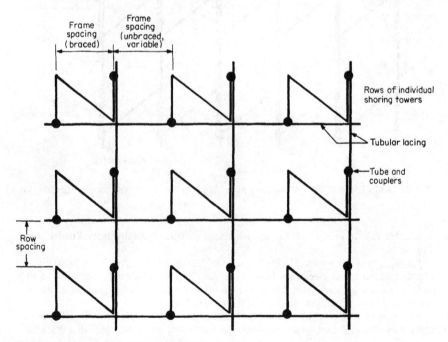

FIGURE 15.17   Shoring tower laced in both directions.

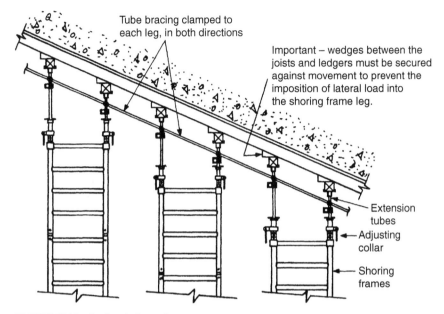

FIGURE 15.18   Shoring sloping surfaces.

loads are transferred *concentrically* to frame legs by the use of centering wedges and other suitable measures as shown in Fig. 15.8*b*, and (3) ensure that screw legs are not extended beyond the manufacturer's recommendations. When shoring to or from any sloping surfaces, it is important that wedges be used to alien the load with the shoring tower leg. It is also important that these wedges be secured in place (kept from sliding "uphill") to prevent the development of horizontal forces acting on the tower legs.

SCREW-LEG EXTENSIONS

Frame-leg capacity is generally lowered when using long screw-leg extensions. Variations in manufacturer's designs preclude generalizations as to the degree of load reduction. When a screw leg, completely closed, is inserted in a frame leg, it adds a certain, nonreducible extension to the leg. This is known as the "dead-leg" dimension and must be added to the frame and sprocket stackup height before adding the required screw-leg adjustment range for the situation. Several common screw-leg configurations are shown in Figs. 15.19 and 15.20.

As for type A and type B conditions shown in these figures, similar conditions exist at the tops of the towers between screw legs and U heads or top plates. Generally, total extensions up to 12 in (30 cm) do not significantly decrease the loading. Above 12 in (30 cm), the effect is variable and largely dependent on the unsupported column distance between the base of top plate and closest frame-leg structural connection, generally the cross-brace attachment. Based on this concept it is obvious that a type B screw leg will give a greater reduction of leg strength than a type A. Only the manufacturer will be able to give the precise allowable loading for various conditions of use for specific equipment.

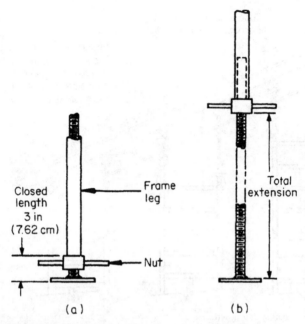

FIGURE 15.19   Type A screw legs: (*a*) closed and (*b*) open.

Extralong Screw Legs

Some manufacturers have available 24-in (60-cm) and longer extension screw legs for use with heavy-duty frames. Because proprietary equipment varies greatly, the specific manufacturer of any special legs should be consulted for allowable loads.

Other Extension Devices

The previously mentioned piccolos, also known as extension tubes, shore staffs, etc., give great additional flexibility to the shoring tower. Piccolos are tubes, generally 2-in (5-cm) nominal OD and 5 to 6 ft (1.52 to 1.83 m) long, which telescope inside the frame legs. They give a larger range of adjustment than the threaded screw leg, having pinholes at frequent intervals [3, 4, or 6 in (7.5, 10, or 15 mm)] to accept a hardened steel pin usually of ½ or 5/8 in (12.7 or 15.9 mm) diameter, which affords rough adjustment of extension. Fine adjustment is achieved with a relatively short length of a screw-threaded adaptor collar serving as a transitory connection between the extension tube and the frame leg. (See Fig. 15.12.) Depending on total length and the size of frame used in conjunction, extension-tube shoring lengths vary from about 1 to 5 ft (30 to 150 cm).

Certain proprietary manufacturers utilize a pair of extension tubes welded into a "head" frame (extension frame) configuration that can achieve high stability properties owing to the facility to brace the welded members together with pivoted cross bracing that is readily adaptable to the extension frame.

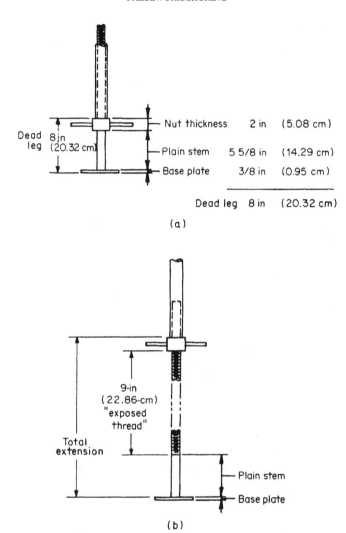

Nut thickness — 2 in — (5.08 cm)

Dead leg 8 in (20.32 cm)

Plain stem — 5 5/8 in — (14.29 cm)

Base plate — 3/8 in — (0.95 cm)

Dead leg — 8 in — (20.32 cm)

(a)

9-in (22.86-cm) "exposed thread"

Total extension

Plain stem

Base plate

(b)

FIGURE 15.20 Type B screw legs: (a) closed and (b) open. {*Important:* Most manufacturers call out the type of leg and *either* the fully extended leg (extension) from the frame leg, *or* give the amount of extension in terms of "exposed screw thread."}

Another means of bracing is by horizontal tube-and-clamp ties connecting each of four legs of a tower together in two planes. Where conditions allow, rows of towers can use long lengths of tubes in straight runs.

Nailer plates have also been developed to brace these legs with 2- × 4-in (5- ×10-cm) or 2- ×6-in (5- ×15-cm) wood lacing and bracing. The moment strength of nailer plates is not high, and generally stronger bracing is achieved using cross braces with special adaptors or tubes and clamps.

To illustrate the variables involved when using data published by manufacturers (Fig. 15.21), Table 15.1 shows a condensed comparison of two apparently similar types of extension tubes or staffs. To all intents and purposes the materials used are essentially the same. Manufacturer A does not publish recommendations on extension tube bracing, whereas manufacturer B does. Such data are usually supplied by the manufacturer or supplier of the equipment.

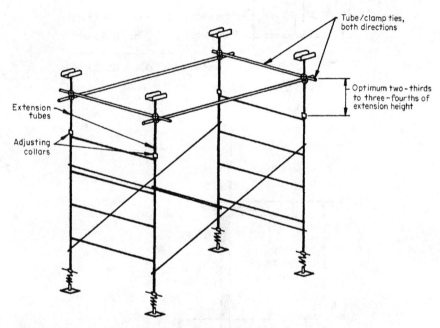

FIGURE 15.21   Shoring tower with braces extension tubes. (Refer to Table 15.1.)

TABLE 15.1   Leg Load Derating When Using Extension Tubes (Piccolos)
Allowable loads in pounds per leg

| | Manufacturer A | | | Manufacturer B | |
| | No. of frames high | | | Frame heights not specified | |
Extension, ft	1	2	3	Legs unbraced	Legs braced
Up to 2	9100	8600	8100	10,000	10,000
3	8500	7500	6900	8,000	10,000
4	7300	6500	5700	7,000	9,000
5	6500	5550	4600	6,000	8,000
6	6000	4750	3800	N/A	N/A

SELF-WEIGHT OF SHORING

The self-weight of heavy-duty shoring affects leg load design. Weights of frames, braces, and other appurtenances vary greatly among manufacturers. Taking an arbitrary heavy-duty shoring system, one section of a tower would weigh:

2 frames at 65 lb (29.5 kg) each

4 coupling pins at 2 lb (0.9 kg) each

2 cross braces at 20 lb (9 kg) each

Totaling 178 lb (80.6 kg) per section

A tower 10 frames high would weigh $10 \times 178 = 1780$ lb (806 kg)

Dividing by 4, the load per leg = 445 lb (1980 N)

It is a good rule of thumb that 5 percent or more should be deducted from the allowable leg load for self-weight. Because 445 lb (1980 N) is close enough to 5 percent of the 10,000 lb (44,480 N) allowable, it is recommended that design in this case be limited to 9500 lb (42,256 N) per leg. In general, for heavy-duty frames, deduct self-weight when towers are six or more frames high.

A similar but standard frame tower would be 10 to 20 percent lighter, say 140 lb (36.5 kg) per section or 35 lb (15.9 kg) per leg, per section. Because standard frames have a much wider range of allowable leg loads owing to size and configuration, the 5 percent rule is a good one to apply here too. For instance, a four-frame high tower with a nominal allowable leg load of 4000 lb (17,792 N) would have a self-weight per leg of $4 \times 35 = 140$ lb (63.6 kg) that is 3½ percent and that would be tolerable. Shoring is tested 3 frames high, so it is a good idea to deduct the self-weight of the frames for towers four or more frames high.

FRAME LAYOUTS

Frame towers on layouts may be shown with or without circles to indicate frame legs. Conventional ways of indicating shoring towers in plans are shown in Fig. 15.22.

It is difficult to design an efficient shoring layout unless the *total forming system* is considered as a whole. For any given concrete load, the items listed in Table 15.2 are also closely interrelated. Any one of these items, if not optimized, could possibly have an adverse effect on the design of other components of the system. Typical reasons for not using optimum size and spacing of components and their consequences could be

Design requirement	Reason for not using optimum size	Result
¾ - in (19-mm) plywood	Only 5/8 - in (16-mm) plywood is available	More joists at closer centers than need be*
4- × 4-in (10 × 10-cm) joists	Only 3- × 4-in (7.5 × 10-cm) joists available	More joists to supply and place at closer centers; inefficient use of plywood*
4- × 8-in (10- × 20-cm) ledgers	Only 4- × 6-in (10- × 15-cm) ledgers available	Frame legs at closer spacing than optimum and inefficiently loaded

*When these two results are combined, substantial excess handling occurs. This is discussed later in detail.

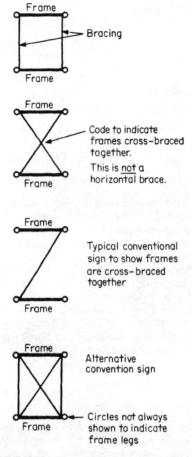

FIGURE 15.22 Convention for indicating shoring towers in plan view.

TABLE 15.2 Interrelated Items

Item	Optimum size and spacing	Depends on
Plywood	Optimum	Joist spacing
Joist spacing	Optimum	Joist span between ledgers
Ledgers	Optimum	Span between frame legs
Frame legs	Spacing	Finite maximum load/leg
Sills	Size/length	Leg loads

SHORING DESIGN EXAMPLE
(STANDARD FRAMES)

Various aspects of shoring design and comparisons of options will be explained through the use of an example.

Consider the forming and shoring of an 11-in concrete slab:

Plywood Available.   ¾ in (19 mm).

Ledgers Available. 4×8 in (10×20 cm).

Joists Available.   As required.

Design to determine whether standard or heavy-duty frames have best utilization.

Design Load.   11-in (28-cm) slab = 140 lb/ft^2 (6703 N/m^2)

$$\text{Live load} = 30 \text{ lb/ft}^2 \text{ (1436 N/m}^2\text{) (per SSI)}$$

$$\text{Total} = 170 \text{ lb/ft}^2 \text{ (8139 N/m}^2\text{)}$$

For ¾ - in (19-mm) plywood: With face grain at right angles to joists, maximum spacing is 21 in (53.34 cm) center to center [21 and 20 in (53.34 and 50.8 cm) are not evenly divisible into the 8-ft (2.44-m) length of a sheet of plywood].

Try: (a) 19.20-in (48.77-cm) spacing [one-fifth of an 8-ft (2.44-m) plywood sheet]: Load per foot of joist = 170 lb/ft^2 $\times$ 19.2 in/12 in = 272 lb/ft (3.97 kN/m) or (b) 16.0-in (40.64-cm) spacing: Load per foot of joist = 170 lb/ft^2 $\times$ 16 in/12 in = 227 lb/ft (3.31 kN/m)

Limitations: Limit the wood design to

$$\text{Maximum bending stress} = 1800 \text{ lb/in}^2 \text{ (1241 N/cm}^2\text{)}$$

$$\text{Horizontal shear} = 150 \text{ lb/in}^2 \text{ (10.34 N/cm}^2\text{)}$$

$$\text{Modulus of elasticity} = 1.76 \times 10^6 \text{ lb/in}^2 \text{ (1.21} \times 10^6 \text{ N/cm}^2\text{)}$$

$$\text{Limiting deflection} = \text{span}/360$$

These limiting stresses contain allowances of +25 percent for short-term loading for bending and shear and +10 percent for deflection.

Conclusions

1. Use 3- $\times$ 4-in (7.5 $\times$ 10-cm) joists at 19.2-in (48.77-cm) spacing for a 4-ft (1.22-m) span between ledgers if 4-ft (1.22-m)-wide heavy-duty frames are to be used, or
2. Use 4- $\times$ 4-in (10- $\times$ 10-cm) joists at 16-in (40-cm) spacing for a 5-ft (1.52-m) span between ledgers if 5-ft (1.52-m)-wide standard frames are to be used.

Ledgers—Standard Frames.   Using 4- $\times$ 4-in (10- $\times$ 10-cm) joists at 16-in (40-cm) spacing, the *ledger span* will govern the bracing length between frames.

Concentrated point load applied by joists to ledgers at 16 in (40 cm) is 227 lb/ft $\times$ 5-ft joist span = 1135 lb (5048 N) per point. Calculations show that a 4- $\times$ 8-in (10- $\times$ 20-cm) ledger has a maximum allowable span of 5.5 ft (1.68 m) with this loading.

Conclusions.   Use 5-ft (1.52-m) bracing to avoid design overload. The design for this example may now be finalized as

¾ - in (19-mm) plywood

4- × 4-in (10- × 10-cm) joists at 16 in (40 cm) center to center

4- × 8-in (10- × 20-cm) ledgers with a 5-ft (1.52-m) span leg to leg

5-ft (1.52-m)-wide standard frames braced, spaced 5 ft (1.52 m) apart

5-ft (1.52-m) spaces between rows of frames

Naturally, some dimensional concessions will require "closing up" some elements to allow for spandrels, columns, elevator shafts, floor openings, etc., where rows of elements cannot be practically reduced and require "expansion." There, single-pole steel shores can be used to good advantage. The layout will look as shown in Fig. 15.23. For a 100-ft (30-m) row of frames, there are twenty 5-ft (1.52-m) spaces, which equals nine 5- × 5-ft (1.52- × 1.52-m) single towers spaced 5 ft (1.52 m) apart, plus a double-braced 10-ft (3.04-m) tower at one end of each row. Each frame leg will support $170 \text{ lb/ft}^2 \times 5 \text{ ft} \times 5 \text{ ft} = 4250 \text{ lb}$ (18,900 N). All elements of this system would be efficiently loaded.

Although many standard frames can take 5000-lb (22,240-N) leg loads under optimum conditions, the type of frame available or long screw-leg thread extension may require a lower value and thus necessitate a load reduction by reducing one or more of the leading dimensions. The example uses a 4500-lb (20,016-N) limiting value.

Examination of Nonoptional Ledger Size.   Presuming the contractor did not have 4- × 8-in (10- × 20-cm) ledgers and wished to use or buy 4 × 6s (10 × 15 cm), let us examine what this would do to the efficiency of the system. With all other conditions the same 4- × 6-in (10- × 15-cm) ledgers could only span 3.5 ft (1.06 m) (a nonstandard brace length). There are few choices available to compensate for this change. One choice would be to use 3-ft (0.91-m) bracing for the towers with 3.5-ft (1.06-m) spaces, that is, 3-ft (0.91-m) braces alternating with 3.5-ft (1.06-m) spaces, between the towers. The 5-ft (1.52-m) spacing between rows would remain. The reduced leg load will now be

$$170 \text{ lb/ft}^2 \times \frac{5+5 \text{ ft}}{2} \times \frac{3+3.5 \text{ ft}}{2} = 2763 \text{ lb } (12,290 \text{ N})$$

$$\text{Efficiency} = \frac{2763}{4500} = 61 \text{ percent}$$

The frame quantity increase, with attendant waste of labor, is 100 percent/61 percent × 100 percent = 164 percent, that is, 64 percent more frames and necessary accessories, plus additional labor involved.

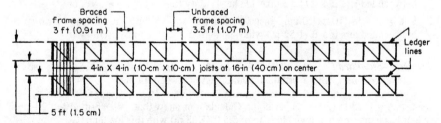

FIGURE 15.23   Initial layout of 5-ft (1.52-m)-wide shoring frames for numerical example.

Because many contractors are more disposed to use 4 × 6s (10 × 15 cm) for ledgers than 4 × 8s (10 × 20 cm), it is interesting to follow through the economics involved in comparing the two ledger choices because of their interrelation with amount of shoring equipment required. Assume a size for the building shored area to be three slabs, each 300 by 85 ft (91.2 × 95.8 m) and an average total *shored* period per slab of 21 days. Figure 15.24 shows the alternative frame layouts with the two ledger sizes. Figure 15.25 shows a view of an assumed shoring tower two frames high, with adjustable screw legs at top and bottom, using U heads. For cost comparison a rental period of 2 months is assumed. See Table 15.3a.

However, this example has not yet taken into account the *timing* of the job. Unless all shoring and formwork are relocated completely, the contractor will need equipment for *at least* 1½ floors to complete the work in the required time; many would use sufficient equipment to shore *two* slabs at one time. Obviously, this will increase rental costs; the labor cost would remain essentially the same except for additional material handling expenses. Therefore, increasing the rental costs by 50 percent for 1½ floors of equipment would increase the above shoring costs to those shown in Table 15.3b.

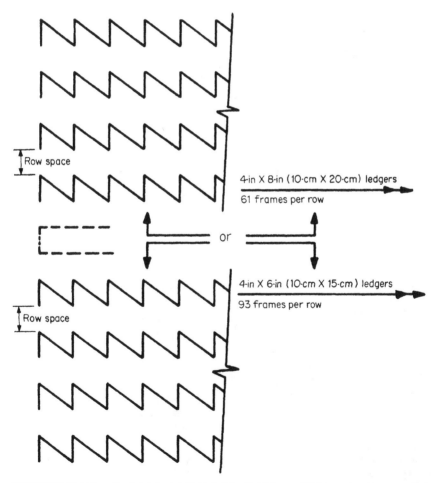

FIGURE 15.24   Alternative frame layouts for 11-in (28-cm) slab in numerical example.

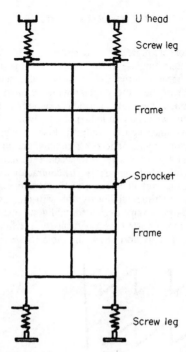

U head

Screw leg

Frame

Sprocket

Frame

Screw leg

FIGURE 15.25 Frame stack up for numerical example.

TABLE 15.3 Shoring Costs

	Layout with 4- × 8-in (10- × 20-cm) ledgers	Layout with 4- × 8-in (10- × 15- cm) ledgers
(a) Cost for one floor		
Frame positions per row	61	93
Number of rows	9	9
Total frames (two high)	1098	1674
Total screw legs	2196	3348
Total U heads	1098	1674
Total bases	1098	1674
Estimated 2-month rental at 5% of list price per month	$31,800	$48,400
Estimated shoring labor, place and remove	370 h	560 h
Labor costs for three slabs at $38.00/h	$42,180	$63,840
Shoring costs rental plus labor	$73,980	$112,240
Average cost per square foot		
76500 ft^2	$ 0.97	$1.43
(7107 m^2)	($10.41)	($15.80)
(b) Cost with rental overlap		
Rental plus labor	$89,880	$136,440
Cost per square foot	$1.17	$2.14
(Cost per square meter)	($12.64)	($19.20)

TABLE 15.3   Shoring Costs (*Continued*)

	Heavy-duty frames and accessories	
Frames and braces	576	
Screw legs	1152	
U heads	576	
Stacking pins (sprockets)	576	
Dollar values	High:	$150,400
	Low:	$110,000
Rental rate per month at 5%/month	High:	$7525
at 2%/month	Low:	$3000

Offset against the cost difference of $136,440 − $89,880 = $46,560 would be the initial cost of the 4- × 8-in (10- × 20-cm) ledgers for 1½ floors that assuming a cost of $500 per thousand board feet for Douglas fir select structural grade, would be

$$\frac{4 \times 8}{12} \times 300 \text{ ft} (9 \times 2 \times 1.5) \text{ lengths} \times 1.15 \text{ (waste)} \times \frac{\$1000}{1000} = \$24,840$$

It would not be realistic to write off this lumber expense against one small job such as this. If amortized over four jobs—a very conservative number—a realistic comparison would be based on one quarter of $24,840, which is $6210, that is

Extra cost of shoring and labor	$46,560
One-quarter cost of 4- × 8-in (10- × 20-cm) ledgers	−6,210
Saving	$40,350
Equivalent to cost difference of	$0.53/ft² ($2.83/m²)

This is a substantial saving over using 4- × 6-in (10- × 15-cm) lumber.

Similar economies can be applied to many shoring installations by optimizing the allowable strength relationships between the formwork and supporting shoring, especially when the shoring is to be totally disassembled before removal to other successive locations. The number of shoring uses involved will dictate the labor costs, and the potential labor and rental savings will be greater in a direct proportion to the number of uses.

SHORING DESIGN EXAMPLE USING ALUMINUM JOISTS AND LEDGERS (HEAVY-DUTY FRAMES)

The foregoing examples have been essentially based upon the historic use of sawn lumber joists and ledgers (stringers). Statements have been made earlier in this chapter regarding the impracticality of the "rental" of lumber joists and ledgers. Subsequently to these original writings, the design, manufacture, and rental availability of joists and ledgers of extruded (pultruded) aluminum members, having cross-sections of proprietary designs, have replaced the use of lumber to a large extent. Lateral spacing of aluminum joists is limited by plywood deflection; therefore, to load such joists efficiently one must utilize the maximum (deflection-limited) joist spans possible. Thereby, aluminum ledgers and vertical shoring components may be loaded to a higher level of optimal spacing, spans and consequently total shoring/forming system cost efficiency.

Various aspects of shoring design and comparisons of options will be explained through the use of an example. Consider the forming and shoring of an 11-in concrete slab:

Plywood Available.   ¾ in (19 mm).

Ledgers Available.   7½ in (19 cm) aluminum.

Joists Available.   6½ in (16.5 cm) aluminum.

Aluminum Beam Properties

	Moment of inertia, in^4	Section modulus (least), in^3	Shear area, in^2	Modulus of elasticity 6061-T6, psi
6½-in joist	16.64	4.52	0.4	10×10^6
7½-in ledger	35.0	9.07	0.765	10×10^6

Design Load.   11-in (28-cm) slab $= 140$ lb/ft^2 (6703 N/m^2)

$$\text{Live load} = 30 \text{ lb/ft}^2 (1436 \text{ N/m}^2) \text{ (per SSI)}$$

$$\text{Total} = 170 \text{ lb/ft}^2 (8139 \text{ N/m}^2)$$

For ¾-in (19-mm) plywood: With face grain at right angles to joists, maximum spacing is 21 in (53.34 cm) center to center [21 and 20 in (53.34 and 50.8 cm) are not evenly divisible into the 8-ft (2.44-m) length of a sheet of plywood].

Try: (a) 19.20-in (48.77-cm) spacing [one-fifth of an 8-ft (2.44-m) plywood sheet]:

$$\text{Load per foot of joist} = 170 \text{ lb/ft}^2 \times 19.2 \text{ in}/12 \text{ in} = 272 \text{ lb/ft} (3.97 \text{ kN/m})$$

$$\text{Limiting deflection} = \text{span}/360$$

Conclusions.   Use 6½ (16.5 cm) aluminum joists at 19.2 in (48.77 cm) spacing for a maximum span of 9 ft (2.47 m) between ledgers.

Ledgers—Heavy-Duty Frames.   Using 6½ in (16.5 cm) aluminum joists at 19.2-in (48.77-cm) spacing, the *ledger span* will govern the spacing between frames.

Concentrated point load applied by joists to ledgers at 19.2 in (48.77 cm) is 272 lb/ft × 9-ft joist span = 2448 lb (10,889 N) per point. Calculations show that a 7½ in (19 cm) aluminum ledger has a maximum allowable span of 7 ft (2.13 m) with this loading.

Conclusions.   Use 8-ft (2.44-m) bracing to avoid design overload. The design for this example may now be finalized as

¾-in (19-mm) plywood: 6½-in (16.5-cm) aluminum joists at 19.2 in (48.77 cm) spanning 8-ft (2.44-m) bracing and 9-ft (2.47-m) spacing between ledgers; 4-ft (1.22-m) heavy-duty frames in rows spaced 7 ft (2.13 m) apart

$$\text{Typical load per frame leg} = 170 \text{ lb/ft}^2 \times \frac{4 \text{ ft} + 7 \text{ ft}}{2} \times \frac{8 \text{ ft} + 9 \text{ ft}}{2}$$

$$= 7948 \text{ lbs} < 10,000 \text{ lbs}.$$

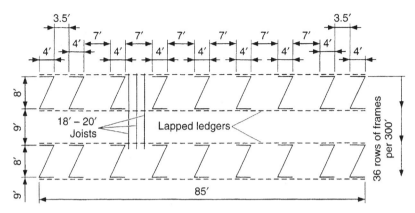

FIGURE 15.26 Heavy-duty frame layout for numerical example.

Laying out the heavy-duty frames to the slab size, some adjustment is necessary for fit because the 7- and 4-ft (2.13- and 1.22-m) dimensions are not evenly divisible into the width of 85 ft (26 m), and then the balance is subdivided into equal modules of frame-plus-rowspace modules. Because the 4-ft (1.22-m) frame plus 7-ft (2.13-m) space module is a most convenient one for 12-ft (3.66-m) aluminum ledger length, use it as much as possible and finish with one or two adjusting rows, that is, finalize the design to be

Nine 4-ft (1.22-m) rows of frames

Six 7-ft (2.13-m) spacing rows

Two 1.5-ft (0.46-m) spacing rows

as shown in Fig. 15.26.

Final adjustment of these dimensions would be to reduce one of the spaces if the shoring is within a confined area such as a basement or an area with spandrel beams. Generally, lay out the frame rows to get optimum economy and minimum quantities of joists. A careless choice of row spacing could result in a layout with many of the joists terminating in long, useless, cantilevered lengths that could result in excessive plywood deflection. Some additional cantilevered joist and ledger length is often necessary to provide bulkheads for slab edges.

ECONOMICS OF HEAVY-DUTY FRAMES

In the foregoing example, the 300-ft (91.4-m) building length is divisible into eighteen 8-ft (2.44-m) braces plus seventeen 7-ft (2.13-m) spaces with 3 ft (0.91 m) left over. Divide the 3 ft (0.91 m) into two 1.5-ft (0.46-m) cantilevers at the ends of each row, as shown in Fig. 15.26. Should such cantilevered lengths give undesirable, excessive deflection, their ends can be "picked up" or additionally supported with single-pole shores.

For this example, the bills of materials and their comparative costs for sufficient equipment for one floor are shown in Table 15.3c. The values shown are typical at the time of writing for the two types of equipment and necessary accessories.

The reader is reminded that such prices will escalate approximately in accordance with manufacturing material and labor costs. Therefore, an escalation allowance of 10 percent per year is recommended. Actual selling prices are normally determined by the manufacturer or distributor at the time and point of sale, and the above figures should be used only for comparison. Rental rates also vary, and therefore contractors are best advised to seek specific cost information from suppliers.

Both the selling and rental costs are for only *one* floor of equipment; they must be factored for job timing requirements as previously mentioned, that is factor by 1.5 for enough equipment for one and one-half floors. The rental values are per month, not per job duration. This factor is henceforth referred to as the "overlap factor."

For the purpose of further comparisons, consider these quantities from the previous example:

$$\text{Total area of three floors} = 76,500 \text{ ft}^2 \ (7100 \text{ m}^2)$$

$$\text{Three floors of concrete} = 2600 \text{ yd}^3 \ (1988 \text{ m}^3)$$

$$\text{Design load for three floors at } 170 \text{ lb/ft}^2 \ (8140 \text{ N/m}^2) = 13,005 \text{ kips} \ (57,846 \text{ kN})$$

With these quantities the purchase price range for *one* floor of shoring would be, using the given high and low values:

Heavy-Duty Frames, $		
	Purchase cost, $	Rental cost $/month
Cost per ft^2 of 11-in slab	3.18 → 4.34	0.18 → 0.44
(m^2 of 28-cm slab)	34.1 → 46.75	1.82 → 4.67
Cost per yd^2 of concrete	93.35 → 142.30	3.41 → 8.53
(m^3 of concrete)	122.01 → 166.85	5.12 → 12.76
Cost per kip of design load	18.34 → 25.49	0.71 → 1.71
(kN of design load)	4.19 → 5.73	0.024 → 0.057

The preceding costs must be multiplied by the overlap factor.

ESTIMATING RENTAL COSTS

It is most important to remember that rental rate per floor per month is usually a guide, and most equipment suppliers will quote lower monthly rates for long-term projects than for short-term ones. Also, longer-term jobs can often be completed in a shorter shoring time per floor by the use of more equipment. This serves the purpose of expanding the work to be done into more segments and enabling the various trade personnel to be more continuously employed with fewer layoffs or less downtime. Consequently, subject to *reshoring* requirements and procedures covered subsequently in this chapter, the use of two or even three floors of shoring equipment on multi-reuse work will increase the shoring cost per square foot by increasing the overlap factor to 2 and 3, respectively, but shortening the shored time per floor will give other economic advantages, such as earlier completion time and subsequently sooner building occupation.

These two variables, overlap factor and shored time per floor, must be carefully assessed when estimating shoring costs. A very common field error is to use the rental rate per month instead of the rental rate of equipment for the shoring duration per floor. The shoring cost per floor in the preceding example is, of course, derived from the total rental cost for a 2-month duration divided by the number of floors completed in that time, that is

$$\text{Rental} = 2 \text{ months}/3 \text{ floors}$$

Obviously, the miscalculation of using the rental rate per month for three floors will cause the false assumption that the job will take 3 months instead of the actual 2 months, resulting in 50 percent higher unit square foot and other costs.

The use of "flying shoring" techniques, utilizing steel-frame shoring, can further reduce labor costs for relocation of shoring and formwork from floor to floor with attendant time savings in job completion.

FRAME HEIGHT STACKUP

(The example is not metricated.) In the foregoing example we deliberately omitted any height calculations, preferring to do these separately.

Let us assume that the shoring is to rise from consolidated stone fill (rough grade) and that the height from grade to slab soffit is 14 ft 6 in.

The best way to determine frame combination is as follows:

1. Add up precise thickness of lumber. (¾ in for plywood + 5½ in for 4 × 6 joist + 7¼ in for 4 × 8 ledger + 2 in for sill + 15½ in.)

2. Add to this the amount of "dead leg" in the screw legs and stack up. (Assume 6 in at top + 6 in at bottom = 12 in.)

3. Subtract this total from shoring height. (14 ft 6 in – 2 ft 3½ in = 12 ft 2½ in.)

This is the height that must be filled by shoring frames plus variable adjustments on the screw legs. Let us assume that screw legs have a maximum of 10 in extension each.

At this height, two frames are certainly needed:

1. Try two 6-ft-high frames + 1-in coupling pins = 12 ft 1 in. Subtract 12 ft 1 in from 12 ft 4½ in = 3½-in adjustment called for. This will not be satisfactory unless one screw leg is dispensed with, not a practical solution because rough grade needs adjustment at the bottom for grade variations. At the top, at least 3- to 4-in adjustment must *always* be left to allow easy stripping of the form lumber before dismantling the scaffold.

2. Try two 5-ft frames + 1-in coupling pins = 10 ft 1 in. Subtract 10 ft 1 in from 12 ft 4½ in = 2-ft 3½-in adjustment needed. If screw legs are limited to 10-in thread extension (two of them 1 ft 8 in), we do not have sufficient height.

3. Try one 6-ft frame + 1 in = 11 ft 1 in. Subtract 11 ft 1 in from 12 ft 4½ in = 1-ft 3½-in adjustment needed. This combination leaves 15½ in for screw leg adjustment, say 8 in for bottom legs, and 7½ in for top legs. (Why should they be roughly equal? No reason except for symmetry.) A time-saving practice is to rough-set the screw legs before use (to approximately 8 in). Use a measuring stick or spacer and adjust legs until there is 8 in of screw leg extension showing between the underside of the handle nut and the base plate or head. It is faster to adjust the legs at ground level than while they are supporting the frames at the base and/or lumber at the top.

The average setting time is about 1 min each for clean, lubricated legs. Therefore, for 280 heavy-duty legs allow 4 to 5 labor-hours; for 432 standard legs allow 7 to 8 labor-hours.

There are two other items that should be considered: (1) grade settlement under load (should not exceed ½ in) and (2) compression of formwork lumber under load.

Grade settlement can be minimized using properly designed sills on compacted grade. Obtain from the soil consultant a ground settlement value for leg-base load intensity, or the per unit pressure that will be experienced. Single 2×10 plank sills are very local in their effect, so generally assume that their effective load distribution is only 1 ft^2 per leg. For a preliminary review of grade adequacy use the applied design load per leg as the pound per square foot intensity.

If a rough stone grade or compacted subgrade will withstand these pressures, all that remains is to place the sills so that they are in full contact with both base plates and grade. If the leg loads cause high settlement (i.e., more than a nominal ½ in), the sills must be designed for the actual conditions as described later in this chapter.

Consideration must also be given to local compression of the sills and form lumber. A good rule of thumb is to use 1/16 in for every lumber surface (see Fig. 15.27) in contact with another wood member or steel component loaded to more than 60 percent of capacity or allowable load; if loaded to less than 60 percent of capacity, use 1/32-in compression for each lumber surface in contact, except for plywood.

Shoring frame stackup charts are sometimes available from manufacturers to aid the contractor in selecting combinations of frame stacks. Table 15.4 is typical of such charts. It should be noted that although 6- and 5-ft-high frames are well standardized in size, the smallest height frames vary with the manufacturer and may be 3 ft, 3 ft 6 in, or 4 ft high. The table gives heights for the median 3-ft 6-in-high sizes; therefore, adjustment by ±6 in to the stackup heights is necessary in accordance with the size of frame to be used.

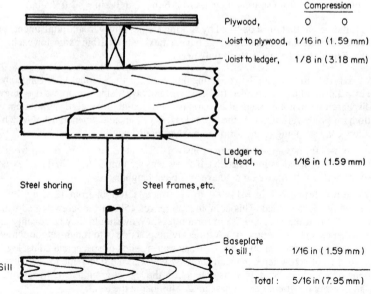

FIGURE 15.27   Compression at lumber surfaces in a shoring tower stack-up.

TABLE 15.4 Heavy-Duty Shoring Frame Stack up Chart

Combination	Dead screw leg[6] length at		Frame heights			Frame stack[3]	Shoring heights with 12 in adjustable screw legs[1]	
	Bottom, in	Top, in	3 ft 6 in	5 ft	6 ft		Minimum[2]	Maximum
1	4 N/A[4]	6		1		5 ft 0 in	6 ft 2 in	6 ft 10 in
2	6	6		1		5 ft 0 in	6 ft 4 in	8 ft 0 in
3	4 N/A[4]	6			1	6 ft 0 in	7 ft 2 in	7 ft 10 in
4	6	6			1	6 ft 0 in	7 ft 4 in	9 ft 0 in
5	6	4 N/A[4]	1	1		8 ft 7 in	9 ft 9 in	10 ft 5 in
6	6	6	1	1		8 ft 7 in	9 ft 11 in	11 ft 7 in
7	6	4 N/A[4]	1		1	9 ft 7 in	10 ft 9 in	11 ft 5 in
8	6	6	1		1	9 ft 7 in	10 ft 11 in	12 ft 7 in
9	6	4 N/A[4]		2		10 ft 1 in	11 ft 3 in	11 ft 11 in
10	6	6		2		10 ft 1 in	11 ft 5 in	13 ft 1 in
11	6	4 N/A[4]		1	1	11 ft 1 in	12 ft 3 in	12 ft 11 in
12	6	6		1	1	11 ft 1 in	12 ft 5 in	14 ft 1 in
13	6	4 N/A[4]			2	12 ft 1 in	13 ft 3 in	13 ft 11 in
14	6	6			2	12 ft 1 in	13 ft 5 in	15 ft 1 in
15	6	6	1	2		13 ft 8 in	15 ft 0 in	16 ft 8 in
16	6	6	1	1	1	14 ft 8 in	16 ft 0 in	17 ft 8 in
17	6	6	1		2	15 ft 8 in	17 ft 0 in	18 ft 8 in
18	6	6		3		15 ft 2 in	16 ft 6 in	18 ft 2 in
19	6	6		2	1	16 ft 2 in	17 ft 6 in	19 ft 2 in
20	6	6		1	2	17 ft 2 in	18 ft 6 in	20 ft 2 in
21	6	6			3	18 ft 2 in	19 ft 6 in	21 ft 2 in
22	6	6	1	3		18 ft 9 in	20 ft 1 in	21 ft 9 in
23	6	6	1	2	1	19 ft 9 in	21 ft 1 in	22 ft 9 in
24	6	6	1	1	2	20 ft 9 in	22 ft 1 in	23 ft 9 in
25	6	6		4		20 ft 3 in	21 ft 7 in	23 ft 3 in
26	6	6	1	3	1	21 ft 3 in	22 ft 7 in	24 ft 3 in
27	6	6		3	3	21 ft 9 in	23 ft 1 in	25 ft 9 in
28	6	6		2	2	22 ft 3 in	23 ft 7 in	25 ft 3 in

TABLE 15.4 Heavy-Duty Shoring Frame Stack up Chart (*Continued*)

Combination	Dead screw leg[6] length at Bottom, in	Top, in	Frame heights 3 ft 6 in	5 ft	6 ft	Frame stack[3]	Shoring heights with 12 in adjustable screw legs[1] Minimum[2]	Maximum
29	6	6		1	3	23 ft 3 in	24 ft 7 in	26 ft 3 in
30	6	6	1	4		23 ft 10 in	25 ft 2 in	26 ft 10 in
31	6	6			4	24 ft 3 in	25 ft 7 in	27 ft 3 in
32	6	6	1	3	1	24 ft 10 in	26 ft 2 in	27 ft 10 in
33	6	6	1	2	2	25 ft 10 in	27 ft 2 in	28 ft 10 in
34	6	6		5		25 ft 4 in	26 ft 8 in	28 ft 4 in
35[5]	6	6		4	1	26 ft 4 in	27 ft 8 in	29 ft 4 in
36	6	6		3	2	27 ft 4 in	28 ft 8 in	30 ft 0 in
37	6	6		2	3	28 ft 4 in	29 ft 8 in	31 ft 0in
38	6	6		1	4	29 ft 4 in	30 ft 8 in	32 ft 0 in
39	6	6			5	30 ft 4 in	31 ft 8 in	33 ft 0 in
40	6	6		6		30 ft 5 in	31 ft 9 in	33 ft 1 in
41	6	6		5	1	31 ft 5 in	32 ft 9 in	34 ft 1 in
42	6	6		4	2	32 ft 5 in	33 ft 9 in	35 ft 1 in
43	6	6		3	3	33 ft 5 in	34 ft 9 in	36 ft 1 in
44	6	6		2	4	34 ft 5 in	35 ft 9 in	37 ft 1 in
45	6	6		1	5	35 ft 5 in	36 ft 9 in	38 ft 1 in
46	6	6			6	36 ft 5 in	37 ft 9 in	39 ft 1 in
47	6	6		7		35 ft 6 in	36 ft 10 in	38 ft 2 in
48	6	6		6	1	36 ft 6 in	37 ft 10 in	39 ft 2 in
49	6	6		5	2	37 ft 6 in	38 ft 10 in	40 ft 2 in

[1]To use 24 in adj. legs add 12 or 24 in to maximum heights for one or two legs, respectively.
[2]Minimum height *includes* 4 in of screw adjustment at top, necessary for stripping.
[3]Coupling pins between frames assumed at 1 in each. Adjust for other dimensions.
[4]N/A means "no adjustment." Dead accessory length of 4 in: adjust for manufacturer.
[5]Short frames not included over 26 ft 4 in stack height; all heights reachable with 6- × 5-ft frames.
[6]Screw legs are assumed to be type B, Fig. 15.20.
NOTE: 1 ft = 0.305 m.

Adjustments may be made to the charts for use with extension tubes, extension frames, and similar accessories. Adjustments should be made to the dead leg lengths used in the charts for those corresponding to the design and/or manufacture of the shoring equipment being used.

For any required height, the combination having the least number of frames will be the most efficient. However, this must be balanced against sizes *available*. Generally, 3-ft- and 3-ft 6-in-high sizes are the least available; 4, 5, and 6 ft are readily available. It is noted that a 5-ft or 3-ft 6-in frame takes approximately the same time to erect as the 6-ft frame. It is also important to understand that when only one screw leg is required, the rule for using the screw legs at the top or at the bottom is of one's own choice.

Assume a building having soffit heights of 12 ft 6 in to the slab and 11 ft 8 in to the interior beams. The possibilities of frame combinations from Table 15.4 are the following:

	Slab	Beams
Soffit heights	12 ft 6 in	11 ft 8 in
Less lumber allowance	−1 ft 1½ in	−1 ft 1½ in
(assuming ¾ -in plywood, 4-in joist, 8-in ledger, and 2-in sill)		
Shore height	11 ft 4½ in	10 ft 6½ in

First Choice

Combination No. 6: 9 ft 11 in to 11 ft 7 in (slab)

Combination No. 6: 9 ft 11 in to 11 ft 7 in (beam)

Second Choice

Combination No. 8: 10 ft 11 in to 12 ft 7 in (slab)

Combination No. 6: 9 ft 11 in to 11 ft 7 in (beam)

Third Choice

Combination No. 9: 11 ft 3 in to 11 ft 11 in (slab)

Combination No. 6: 9 ft 11 in to 11 ft 7 in (beam)

Some combinations utilize one screw leg adjustment positioned at top or bottom of the tower. If support is from rough grade, the adjustment should be at the bottom. If support is from a prior concrete slab, use the single screw leg at the top because it makes for easier stripping. The third choice utilizes screw leg adjustment at both top and bottom. At this height economy would dictate use of the first or second choice, since the time needed to adjust screw legs is substantial and should be held to a minimum.

However, towers of *more* than two frames in height should always have adjustments both at top and bottom. Any irregularity in the support slab manifests itself as an "out-of-plumb" condition that becomes more serious with an increasing number of frames in the height. A ¼-in floor deviation over the 4-ft width of frame becomes 1¼ out of plumb at 20-ft height. This is an insidious condition that can be remedied only by using hardwood shims or blocking and only if discovered early enough. Once the shoring is three or more frames in height, it is best to consider the lower screw legs as "spacers" merely for height attainment and do all the adjusting with the top legs. The stripping will also go faster if the form lumber is not "racked" by premature lowering of the bottom legs. Generally, *never change lower leg adjustment from beginning to end of a job*, unless different height conditions such as a penthouse or machine-room floor are experienced.

MUD SILLS AND SHORING BASES

This is an area in which problems of excessive soil settlement often arise because of lack of specific instructions to field erection personnel. Without guidelines, there are often unforeseeable results despite workers doing their best based on prior expertise.

When, after having laid out a sill system and having erected the shoring, it is decided that the sills need "beefing up," it is very expensive or even impossible to do so.

This section offers some practical means to predetermine the sill design for a given soil or foundation condition. It must be emphasized that the settlement of soil under loads is a very complicated and technical subject involving many variables and factors. The investigation and calculations should not be attempted by unqualified personnel and should be requested from the engineer having responsibility to the contractor.

The information needed by the sill designer is the allowable load per unit area of bearing surface that will give a specified amount of settlement over a specified time. Assume that a recommended specification for a concrete bridge structure is ½-in (12.7-mm) settlement over a 7-day period. The sill designer and the contractor can agree to permit, say, any reasonable settlement over the time period such as 1 in (25.4 mm). This figure must be used to set the vertical shoring adjustment high by this specified amount. In this section on frame shoring/formwork, mention was made of calculating estimated shoring lumber compression, which must be added to the calculated soil settlement. The "time" factor in soil settlement is extremely important, and should be equated (by the engineer having responsibility) to the estimated time after which the structure will be self-supporting and will cease to deflect under its self-weight. Seven days is a typical time for this to happen, but it may vary up to 20 or 30 days depending on the design and job specifications. If soil settlement is negligible, the shoring will continue to carry the structure weight until it is "stripped," or removed. If soil settlement is of consequence, the soil settlement should virtually cease after it has deflected sufficiently to take up the structure's self-weight settlement. It should be obvious that self-weight settlement time and stripping time are related but different and are controlled by different factors.

For heavy structures such as (but not restricted to) bridges (see Fig. 15.28), the sill designer can begin after receiving the allowable pounds per square foot (newtons per square meter) psf load given by the engineer based on the specific *time* and settlement

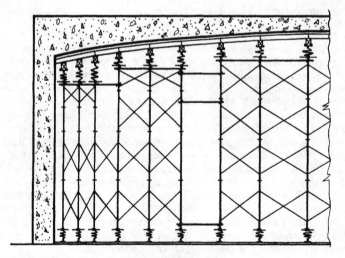

FIGURE 15.28   Bridge shoring on a continuous sill.

parameters required. The leg load divided by the allowable psf intensity will give the sill bearing area required per leg. The sill acts as an upside-down beam, with the point leg loads as beam support reactions and the soil as the uniformly distributed load. Unless this is truly uniformly distributed, the sill cannot be calculated as a beam. Variations in soil surface, if not leveled, or built up with such as level compacted sand or a light concrete "skim" slab, can cause the "sill beam" to deflect in a very nonuniform manner.

For relatively "soft" soils, heavy timbers such as 10×10 or 12×12 in (25×25 or 30×30 cm) can provide good bearing qualities because of their beam stiffness and the assumption that the loads will be uniformly spread over the undersides of them. Highly compacted sandy soil surfaces are also very suitable for long timbers, not necessarily as large as 12×12s (30×30 cm).

Many sill conditions must be treated more carefully, on a leg-to-leg basis. A good pragmatic assumption for these conditions is that most lumber will extend a stress influence line of approximately 45° outward and downward from the edges of a leg-applied load. If base plates are rectangular, always place them with the long side in the direction of the sill length. The influence area is primarily determined by the depth of the sill, bounded by any limiting dimension. In Fig. 15.29a, an example is shown using continuous or padded

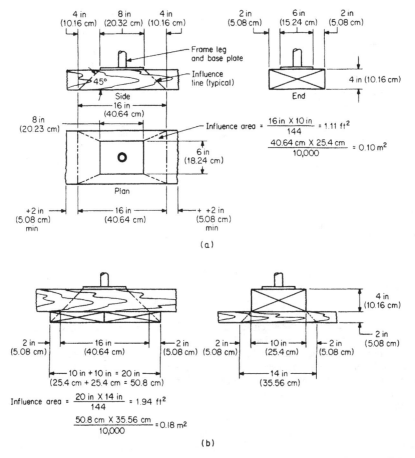

FIGURE 15.29 Various sill configurations: (a) single 4- × 10-in (10- × 25-cm) wood sill, (assumed full, rough), (b) 4- × 10-in (10- × 25-cm) sill plus two 2- × 10-in (5- × 25-cm) pads.

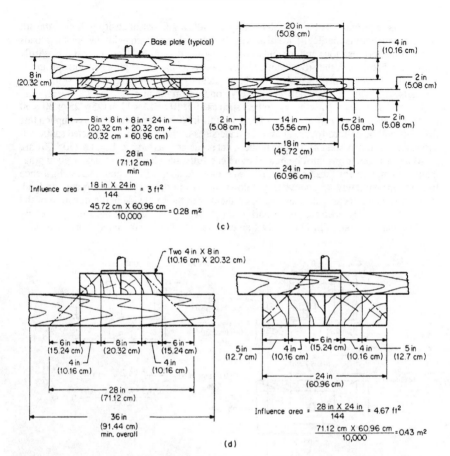

FIGURE 15.29 (*Continued*) Various sill configurations: (*c*) 4- × 10-in (10- × 25-cm) sill plus two levels of 2- × 12-in (5- × 50-cm) pads, (*d*) 4- × 8-in (10- × 20-cm) pads plus two 6- × 12-in (15- × 50-cm) sills (very soft condition).

4- × 10-in (10- × 25-cm) sills on the flat, which in conjunction with a 6- × 8-in (15- × 20-cm) base plate gives an influence area of $1.11\,\text{ft}^2$ $(0.1\,\text{m}^2)$. Figure 15.29b shows that by adding two 2- × 10- × 24-in (5- × 25- × 60-cm) pads under each 4×10 in $(10 \times 25\,\text{cm})$ the influence area is increased by 75 percent to $1.94\,\text{ft}^2$ $(0.18\,\text{m}^2)$. If the 2×10s $(5 \times 25\,\text{cm})$ were changed to 4×12s $(10 \times 30\,\text{cm})$ (or a double layer of 2×12s), the area would increase to $3\,\text{ft}^2$ $(0.18\,\text{m}^2)$ per leg per pad as in Fig. 15.29c.

The depth of the pads should be added to the influence line length at each end, that is, the 2- × 10-in (5- × 25-cm) pads should be $14 + 2 + 2 = 18$ in (46 cm) total length. For 4-in (10-cm) pads, the lengths would be $14 + 4 + 4$ in $= 22$ in (50 cm). Figure 15.29d shows a typical sill design for very soft soil and gives a bearing area of $4.67\,\text{ft}^2$ $(0.43\,\text{m}^2)$.

When erecting shoring from concrete slabs, it is important to use sills rather than seat base plates directly on concrete. During erection it is difficult to keep rows of frames properly in line because steel-concrete friction is low and the shoring "walks." Also, use of sills helps avoid "punch-through" on thin slab sections such as concrete dome, joist, and voided slabs. Normally 2-in (5-cm) sills are quite adequate for this purpose.

For average concrete construction from grade, especially gravel and rock base or hard undisturbed excavation, 2-in (5-cm) or thicker sills should always be used. If the ground is only roughly leveled, a single or doubled-up 2-in (5-cm) sill will act like the upside-down beams previously referred to with decidedly nonuniform load distribution. The sills will settle more under the legs and less between the legs, which will result in an incalculable amount of settlement. This, in turn, will affect leg-load reactions resulting in towers moving out of plumb. Therefore, unequal transmissions of loads throughout the complete shoring system could occur with obvious results of undesirable deflection and possibly overload and failure in some areas.

The maxim to be used here is to use 2-in (5-cm) sills very conservatively and, if in doubt, do not use them without approval of the engineer having responsibility.

Often in such cases, the contractor, rather than dealing with the vagaries of wood sills, elects to pour concrete pads under the shoring to better predict and to limit the effects of differential soil settlement.

RESHORING

When, in multistory buildings, a concrete floor is placed in forms supported by shoring, the total dead and live loads of concrete, falsework, workers, placing equipment, vibration, wind, and other forces must be supported and reshored from the lower floors of the structure until the poured concrete floor is self-supporting and sufficiently rigid to resist plastic deformation. The floors below the floor being poured are connected by various means of vertical shoring known as "reshores." New construction/pouring loads must be distributed among previously poured, reshored, lower slabs in some proportion to their relative flexural stiffnesses. After removal of reshores, the poured slabs become an integrated part of the total structure and thence perform their structural contribution as part of the total building design.

Various concrete behavioral phenomena, such as creep, excessive deflection, plastic deformation, temperature and concrete strength imbalances, varying compressions of reshores, hidden inadequacies and errors in the actual design or construction, are known to exist, but to a large extent these phenomena are not controllable in a predeterminable or precise manner. There is a wealth of available documented analytical research that is largely based upon *assumptions* that the reshored poured lower floors will behave under construction pouring loads with precision. This, unfortunately, is not true. The aforementioned behavioral variables do, indeed, adversely affect theory with occasionally disastrous results.

Often an "instinctive" approach based upon field experience can be successful by means of economic overkill, using far more reshores than necessary, even to the extent of carrying them all the way down to the ground. This can result in hazardous overloading of the lower reshore members, even on three- or four-story buildings. At the other extreme, when using *too few* reshoring members and reshored floors, it has been said that

> If the (reshoring) support is insufficient, the best to be expected is a series of dished slabs, deflected beams and radial cracks around the columns. The worst is a local collapse of a previously placed slab which triggers a chain reaction of collapse and may carry down for the full height. Since these are (usually) shear failures and there is no warning, the path of failure will search out the areas of deficiency and bypass those areas which, though weak, still have the minimum necessary factor of safety. In modern designs of apartment and office buildings there is little surplus strength in a floor. Even with no Live Load there is no strength available to act as a bumper capable of resisting the impact of a similar mass falling one story. And as the accumulation of falling mass comes down, there is no stopping the chain reaction.

From available articles and analytical studies of the problem of reshoring, it must be concluded that there is no one definitive source of guidance available. One of the main reasons for this lack of guidance is the often unpredictable behavior of poured concrete in structural form. We lack precise knowledge of creep, whether in the curing or aged state; we lack precise assurance of the quality of the concrete-steel bond, how much form-coating release agent got on the rebars, how many cups of coffee were poured into the forms (as was once discovered to be the cause of a "spongy" concrete column), how morning and night temperature shifts affect a monolithic pour, how seasonal variations affect checkerboard pours, etc.

Because of these factors, the builder is often unaware that crucial shear and flexural members are overloaded and their safety factors so diminished that a structure may be on the verge of a minor failure—or possibly a major one.

A contractor may take a job *including* responsibility for reshoring, without having a properly designed reshoring procedure to follow. A design could always be set out in the job specifications *by the only party having real knowledge of the parameters of the structural design: the designer.*

Perhaps, by examining what knowledge is available, we can propose some helpful suggestions.

To begin, there are three major approaches to the problem of reshoring: instinctive, analytical, and empirical. The empirical contains some of the best aspects of the instinctive and analytical approaches.

The Empirical Approach

A simplistic evaluation is available for the number of floor levels that must be connected with reshores to provide proper support. The following *assumptions* are made for this method:

1. Temporary load-carrying capacity of a slab is one-third above its design value.

2. Fresh wet concrete in the forms weighs 1.2 × dead load (allowing 20 percent for formwork and miscellaneous items).

3. Reshored floors all deflect equally; reshores do *not* shorten (or compress). The total load of system is divided into *shares* for respective floors in proportion to their stiffness.

4. Slabs are all equal in geometry and reinforcement; stiffness is in proportion to modulus of elasticity of the concrete. Because modulus increases more rapidly than does strength, stiffness is taken conservatively in proportion to concrete strength.

5. Concrete is up to its 28-day strength, and temperature conditions have been maintained at 50°F or above.

6. Concrete strengths (as percent of carrying strength) are listed in Table 15.5a with straight-line variation for intermediate ages.

7. Load-carrying capacity of floor slabs is shown in Table 15.5b.

TABLE 15.5a Concrete Strength Percent of 28-Day Strength

Age, weeks	Type I cement	Type III cement
4	100	100
3	85	100
2	70	90
1	50	75

TABLE 15.5b Carrying Capacity of Floor Slabs in Terms of Slab Weight

Ratio of LL/DL		0.50	0.75	1.00	1.25	1.50	1.75	2.00
carrying capacity in terms of dead load		2.00	2.33	2.67	3.00	3.33	3.67	4.00
Pour rate, floors/week	Type of cement	No. of levels below floor being poured that must be interconnected*						
	I	3	3	3	2	2	2	2
	III	2	2	2	1	1	1	1
	I	5	4	4	4	3	3	3
	III	4	3	3	2	3	2	2
	I	7	6	5	5	4	4	3
	III	5	4	4	3	3	3	3

*Includes number of floors of formwork still in place *below* slab being poured. Number of levels of *reshores is one less* than interconnected floors shown.

The requirements shown in Table 15.5*b* are minimal. They may not be sufficient to prevent plastic yield and permanent deflection owing to empirical assumptions. Wood reshores compress nonuniformly and allow additional, unpredictable slab deflections. Steel shores have negligible compression within the heights of typical building floors.

It is most inadvisable to remove reshores at any level within 2 days of the pouring of a new level. Reshore removal results in a *major readjustment of the loads in reshores* at all higher levels, and of course, possible changes in slab deflections, shear loads, and other interrelated forces. Part of the load readjustment will be in the reshores supporting the slab upon which the currently poured slab is supported. Mistiming will result in a readjustment of loading in the shores, and undesired deformation in plastic concrete will inevitably be experienced later. One is further advised against removal of shores under *any slab two floors below the one just concreted,* unless the schedule is so slow that 28-day strength data are available for the floor supporting the formwork.

The foregoing is a suggested empirical guide and thus, of necessity, must be treated strictly as such—not as a specification or a rigid procedure. Its application to any project must be made by the design or consulting engineer using job-based design criteria and procedures. The reader is urged to examine more exhaustive studies, such as Ref. 2.

The Analytical Approach

A most erudite and thorough analytical study of this subject was made by Paul Grundy and A. Kabaila in Ref. 3. Their conclusions are used by many as an authoritative guide on the subject, and the authors take into account a variety of factors and criteria that are too extensive for repetition here.

Some Practical Comments on Reshoring

1. Reshoring should be done on a one-for-one basis through as many floors as necessary to safely withstand the construction loads from the floors above. Slab-floor markings can be of great assistance in ensuring that vertical reshore continuity is maintained through *all* the shored and reshored levels. Reducing reshoring 2:1 or 3:1 at lower

reshored floors is a questionable practice. Such reduction is false economy that may result in punch-through, stress reversals, and redistribution of stresses that may be incompatible with the slab design.

2. Before the top construction floor is poured, the shored slab(s) by which it is supported should be allowed to take up its (their) own dead-load deflection(s). This is easily done by slightly releasing the adjustments of the shores. The shoring should then be retightened so as to snugly support the self-deflected slabs.

3. Shoring failure under vertical load is seldom experienced by frame shoring owing to the good predictability of the design loads and the built-in safety factors in the allowable strengths. Design error, bad sills, overloading, lateral instability, or lumber failure are generally to blame when shoring fails. Note, however, that frame shoring is normally not designed or intended for *reshoring* loads.

4. It is important to keep in mind that wood is not a homogeneous material. It cannot be batch-tested with assurance of consistency. Each piece has its own individuality, and test comparisons of apparently identical pieces of wood may give diverse results. When new, with standard safety factors in the allowable stresses, wood is relatively reliable. After multiple reuse, its behavior may change, therefore ample margin of safety should be allowed for wood that is used or aged.

5. Wood posts used for reshoring compress significantly. That may result in excessive deformations in the supported beams and slabs under their shore or reshore loads from upper floors. Solid lumber reshores of a type consisting of two 4 × 4's or 3 × 4's overlapped and spliced with mechanical hinges or splicing plates should be used with extreme caution and under strict control. Not only does the lumber compress but the splicing device may slip when the reshored slab is subjected to construction loads. Once this happens, the reshored slab will deflect, perhaps irreparably.

6. If the slab supporting the shoring is allowed to deflect soon after the next level pour is completed, the freshly poured slab will deflect in parallel. Once hardened in this position, the apparent deflection of the upper slab remains locked in.

7. Construction errors may create gross over-deflected slabs without collapse. Slabs of 30 ft span have been observed to "dish" or "sag" as much as 2 in or more without superimposed loads. The immediate emotional reaction is to blame the method and the installer of the shoring. However, if such deformations are experienced in the concrete pouring stage, it is seldom the fault of the shoring method or equipment but often the result of careless concrete pouring that may overload joists and ledgers.

8. Poor tightening or wedging of reshores should be guarded against.

9. Design specifications often state that after removal of shoring and forming materials beneath a freshly poured slab, reshores must be placed within a specified time limit. This may be stated in terms of hours, or quite vaguely, as "Reshores must be placed before termination of work on the same day as the shoring is struck." Good practice is to place reshores as quickly as possible. Do no wait for the time limit in the specifications. A semi-cured slab left without support for 14 h can be expected to deflect and creep severely.

10. The foregoing not withstanding, a slab area within its supports must have *all* of its shores removed or released before reshoring that area in order to allow the slab to assume its own dead load deflection. If some reshores are placed before all of the shoring is removed or struck, dangerous stress reversals may result, leading to cracking of the slab.

11. When using shoring that is adjustable by screw-threads, releasing or reducing all the screw supports simultaneously in a given bay will serve two purposes: break the

plywood-to-concrete bond and allow the released slab to acquire its own dead-load deflection. In this case, contrary to the advice in list 10, some reshores may be placed almost immediately, that is, before the physical removal of the released shores.

12. Avoid any type of construction load on shore-released slabs *prior* to placing of the reshores under it. Much of the deflection experienced by a slab at an early age may remain "locked in" permanently.

13. Good construction practices and timing of reshoring are the essence of "safe practice." Even though a collapse is not experienced, poor reshoring practices can result in badly deformed slabs and beams, and may generate time-consuming and expensive litigation to settle claims.

14. While reshoring load calculation methods have validity as guidelines, the myriad of possible construction activities reduce the reliability of theoretical calculations. Use sound engineering judgment and successful construction experience in developing and executing the reshoring.

Monitoring Reshoring Loads

Recent collapses of multistory apartment buildings point to the necessity for standards to guide the industry. One way to ensure adequate reshoring support for multilevel concrete loads is by monitoring the loads on shores and reshores with minihydraulic load cells. With experience and recorded data, concrete behavioral effects can thus be precisely measured even if not wholly and accurately predictable.

Obviously, the majority of multilevel concrete structures are completed without using monitoring gauges but with applying combinations of basic methods and common practices. However, because of unpredictable concrete behavior and the risk of construction deficiencies, such practices can give no real assurance of the efficacy of the shoring or reshoring system until the building is completed or suffers a failure. Monitoring is one way to be sure during construction even though it is not the only possible procedure.

STANDARDIZATION OF SHORING FRAME TESTS AND DETERMINATION OF SAFE ALLOWABLE LOADS

In the infancy of frame shoring, most manufacturers and users accepted load testing of individual frames to obtain allowable working loads. A commonly accepted value was 5000 lb (22,240 N) for each frame leg. The Scaffolding, Shoring, and Forming Institute, Inc., began, through the facilities of its members and those of certain professional laboratories, to research the behavior of scaffold frames in multitier frame configurations. Standard test procedures were developed as well as frames especially made for such tests. These frames were similar but not identical to those manufactured by member companies. Tests were made with four-leg towers, with the loading applied concentrically to each leg.

The results of this research had enormous impact on the use of scaffold frames for shoring. The frames were found to behave as four-legged columns when tested, with typical S-type buckling failure configurations similar to those of braced structural-steel columns. Unless the welded rigidity of a frame was very low, column buckling usually occurred in the plane of the cross bracing.

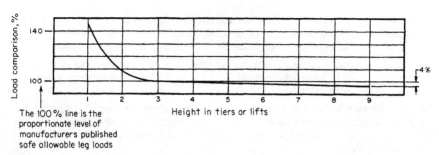

The 100% line is the proportionate level of manufacturers published safe allowable leg loads

FIGURE 15.30 Representative results of frame load tests by the Scaffolding, Shoring, and Forming Institute. Inc.

Of greatest importance was the finding that multitier shoring towers had significantly lower load capacities than one- or two-frame-high towers. Figure 15.30 illustrates these typical findings. The graph shows 40 percent greater capacity for one-frame-high towers than three-frame-high towers. There was a consensus that safest use would result from manufacturers' using their own individual allowable leg-load values based on three-frame-high tests.

Second, the amounts of extensions of screw legs and piccolos at tops and/or bottoms of the tested towers adversely affected frame strengths because they increased the unbraced column lengths of the frame legs from the highest or lowest cross-brace connection position. Third, the varying unbraced lengths between the positions of attachments of cross braces to the frame legs were found to result in important strength variations when comparing the relative performances of frames of different configurations and heights. Fourth, arch, or walk-through-type frames had lower capacities than frames having horizontal members at the tops and near the bottoms of frames. Fifth, narrow frames such as those 2 ft (60 cm) wide were not always as strong as those 4 or 5 ft (1.22 or 1.52 m) wide because of their lower cross-sectional stiffness when assembled into a four-legged braced column.

Traditionally the factor of safety for shoring is set at 2.5:1 against load refusal as determined at the time of testing (according to SSFI test procedures). The 2.5 is used to cover the effect of continuity on the loading of the frame leg, 2 for the structural safety factor, and 1.25 for the effect of a two span continuous beam. Thus no matter how the ledger beams are placed, the effect of continuity has been accounted for.

Users are cautioned that the test data were obtained from nontypical frames to standardize industry practice by using identical test criteria. Some examples in this chapter used 5000 and 10,000 lb (22,240 and 44,480 N per leg) for convenience only. For each type of frame combination, load data for specific conditions must be obtained from the manufacturer of the equipment or the authorized representative.

SINGLE-POST SHORES

There are many shoring applications where preengineered single-steel-post shores of adjustable height are used. They are normally constructed of a lower tube that contains a base plate, an adjusting nut to allow fine height adjustment, and a perforated telescoping staff member with a pin to allow large adjustments. (See Fig. 15.31.) The shore staff also includes a flat plate or a U head to which the formwork is attached.

FIGURE 15.31   Single-pole shore.

Applications where lateral loads are imposed on formwork supported by post shores must be designed with extreme care and caution. In addition to horizontal post bracing, as a minimum, each post-shore system must be braced as shown in Fig. 15.32. If bracing consists of lumber, timber bracing clamps, or nailing plates, it is imperative that the load-carrying capacity of these devices be known and that the quantity and frequency of bracing be adjusted to conform to these loads. If guy wires are used for system stability, the vertical compression effect on the load-carrying capacity must be taken into consideration.

Exercise extreme care and be sure all types of loads imposed are understood and resolved to ensure that the post shore is not overloaded.

In addition to the horizontal loads caused by wind, construction equipment movement, vibration of flowing concrete through pumping equipment, conveyors, or motorized concrete buggies, care must be given to lateral loads caused by deflection of horizontal shoring beams and truss members.

Double-tiered post shores should never be used. If the working height exceeds the maximum post-shore extension, an alternative shoring system, such as welded frame shoring, should be used.

Ledger Bracing.   Formwork loads are transmitted directly to the post shore through a system of joists and ledger beams (also known as stringer beams). Because the ledger beam is in direct contact with the post shore, it must be considered when bracing for stability. Ledger beams should not exceed a nominal height-to-width ratio of 2.5:1. If this ratio is exceeded, stabilization between ledgers is required to prevent rollover.

Layout.   Prior to developing a design layout, it is important to read, know, and understand all local, federal, and industry safety rules and regulations. In addition to the preceding design considerations, when making a shoring layout it is important that sufficient bearing under the post shore be available to safely support the design load. If ground conditions are not known at this time, they must be ascertained prior to post-shore installation. Avoid eccentric load placement on U heads or top plates by centering all ledgers and blocking and wedging them in place.

If post shores are used in conjunction with welded frame shoring, the post shores should be braced to an adjacent shoring frame.

Stripping Post-shore Formwork.   After the concrete has been cured or the engineer of record considers the concrete sufficiently cured to carry its own weight, post shores should be stripped by relieving the adjusting nuts in a planned sequence. Care must be exercised to avoid overloading individual posts. Once the entire load is relieved, the post shores and formwork can be removed safely. Post shores should not be struck at the base and allowed to topple.

Shoring low story heights with post and beam shoring has long been popular with American contractors. When the shoring consists of 4 × 4 wood posts on 4-ft centers, using

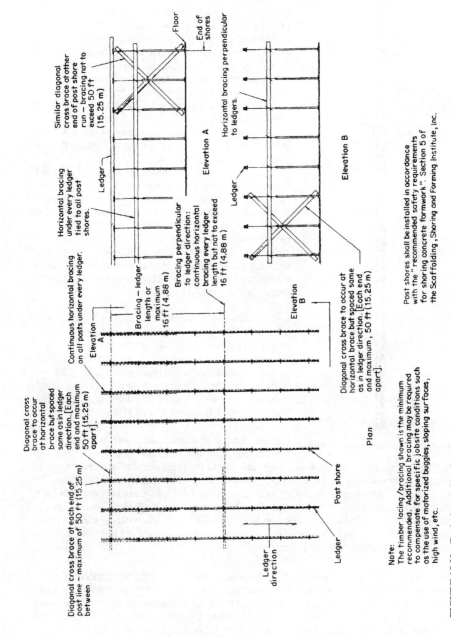

FIGURE 15.32 Typical post-shore layout.

Similar diagonal cross brace at other end of post shore run – bracing not to exceed 50 ft (15.25 m)

Horizontal bracing under every ledger tied to all post shores.

Bracing perpendicular to ledger direction: continuous horizontal bracing every ledger length but not to exceed 16 ft (4.88 m)

Floor

End of shores

Ledger

Elevation A

Horizontal bracing perpendicular to ledgers.

Ledger

Elevation B

Diagonal cross brace to occur at horizontal brace but spaced same as in ledger direction. [Each end and maximum, 50 ft (15.25 m) apart].

Continuous horizontal bracing on all posts under every ledger.

Bracing – ledger length or maximum 16 ft (4.88 m)

Elevation A

Elevation B

Diagonal cross brace to occur at horizontal brace but spaced same as in ledger direction. [Each end and maximum, 50 ft (15.25 m) apart].

Diagonal cross brace at each end of post line – maximum of 50 ft (15.25 m) between

Plan

Post shore

Ledger

Ledger direction

Note:
The timber lacing/bracing shown is the minimum recommended. Additional bracing may be required to compensate for specific jobsite conditions such as the use of motorized buggies, sloping surfaces, high wind, etc.

Post shores shall be installed in accordance with the "recommended safety requirements for shoring concrete formwork". Section 5 of the Scaffolding, Shoring and Forming Institute, Inc.

15.42

4×4 wood joists and ledgers (known colloquially as shoring with sticks and boards), the shoring setup is mindless. It uses the same pattern everywhere. It is easy to set up and tare down. To strip the forms, the posts are pulled out from the bottom; the formwork is allowed to fall to the deck; and the forms are broken-down, bundled, and sent up to the next floor—cheep and dirty.

Post and Beam Shoring System. The next logical step was the development of a post and beam shoring system that had advantages over the wooden shores. The crux of this approach is the drophead. As shown in Fig. 15.33*a*, the drophead allows the forms to be

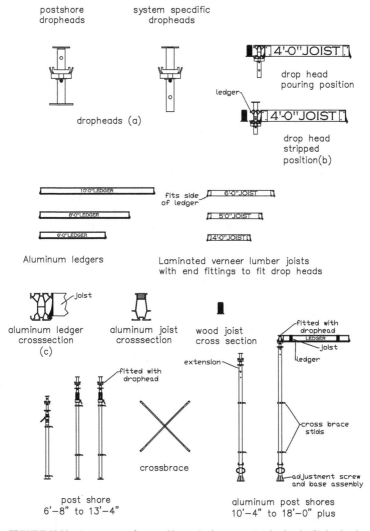

FIGURE 15.33 Components of post and beam shoring system (*a*) dropheads, (*b*) dropheads in stripped position, and (*c*) aluminum ledger cross section.

stripped, while leaving the posts in place to reshore the newly poured slab. This allows for better utilization of the formwork.

The formwork consists of an aluminum ledger (stringer) whose end hooks on to the drophead. The ledger is equipped with a groove into which the ends of the joists fit (see Fig.13.33b and 13.33c). The joists are aluminum extrusions, or they are made of wood; both have end pieces that fit into the side of the ledger. A typical partial layout, together with the explanation of some of the symbols is shown in Fig. 15.34. The cross braces are used to stabilize the posts. Usually one cross brace braces four post shores.

The slab is poured with the drophead in the up position. When the slab is sufficiently cured, the wedge holding the drophead in the up position is driven out. The ledgers and joists pull away from the underside of the plywood; they are removed, and sent up to the next deck. The posts are pulled down, one at a time, and the plywood is stripped. The posts are replaced and snugged up to the slab to act as reshoring.

The system is fast and efficient. It performs best when used to shore flat slabs that have regularly spaced beams and columns. The system uses wood shoring and frame shoring to deal with areas around beams and columns and to accommodate irregular shapes in the slab.

Post Ledger and Panel Shoring System. The next embodiment eliminates the joists and plywood and replaces them with forming panels. The example shown in Fig. 15.35a and 15.35b uses 2- × 6-ft and 3- × 6-ft panels, respectively. The panels can be either aluminum faced or plywood faced. With this system the posts are erected and the ledgers installed with the dropheads in the up position. The panels fit in the same ledger grooves as the joists in Fig. 15.33c. The posts are erected, the ledgers installed, and the panels are hooked into the ledgers, and the formwork is ready for the pour. Of course, this system requires "filling in" with wood and frame shoring.

When the slab is sufficiently cured, the dropheads are released, and the panes and ledgers removed. The posts are cracked loose and snugged to the slab. They are kept standing in place to reshore the green slab. Figure 15.35 identifies the components and shows a typical elevation and a partial plan.

Post and Panel System. The post and panel system eliminate the beams entirely. Forming panels fit directly on the dropheads. The panels are lightweight and handleable by one man. The erector stands up a pair of posts; the panel is hooked on to the drophead and rotated to the horizontal position using a special tool. The posts are placed in position under the panels and rotated into place. The process is repeated until the all the forms are up. When the slab is sufficiently cured, the drophead is released and the panels are removed and placed into a cart and moved to the next location. The post shores are cracked loose and snugged to the slab. This system underutilizes the post shores. The panels need to be light enough for a man to handle easily. Their size is about 5 ft × 2 ft 6 in—about 12½ ft². They weigh about 33 lb. The panels are delicate by construction standards. They must be handled carefully, for the system to be economically viable.

A ledger is available to allow better utilization of the post shores. When the ledger is used, the system reverts to a post, ledger, and panel system as described previously.

This system is fast, simple to use, and works well only because the aluminum panels and post shores are lightweight.

Modular Shoring System. Modular aluminum shoring is the most versatile system yet devised. It is built around a leg extrusion of 3½ in diameter (approximately) with longitudinal grooves around its periphery extending the length of the leg (see Fig. 15.36c).

Twist-bolts fit into this groove and are used to attach all manner of features, the most important of which are panels (Fig. 15.36a). The panels attach to the legs to form towers (Figs. 15.36c and 15.36d) with capacities up to 23,000 lb per leg, depending on the number and placement of the panels and the distance the inner extension leg is extended.

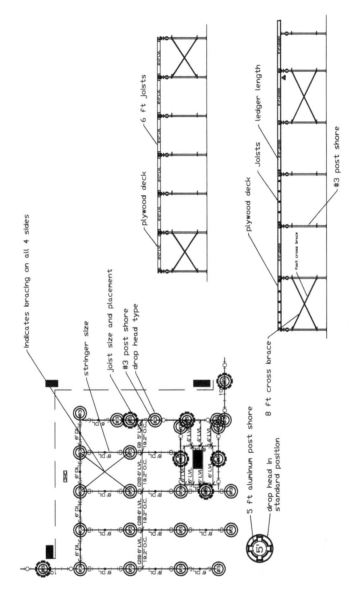

FIGURE 15.34 Typical post and beam shoring system layout.

15.45

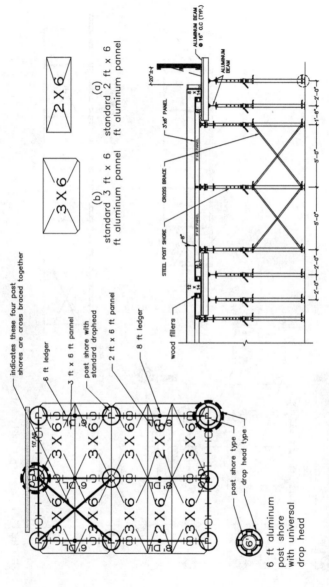

FIGURE 15.35 Post beam and deck system.

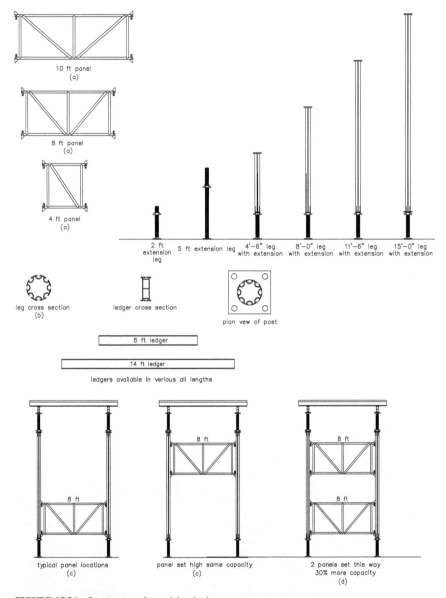

FIGURE 15.36 Components of a modular shoring tower.

A high-capacity extruded aluminum ledger is available to help utilize the high leg load capacity. Figure 15.36 shows some of the components of the system.

The fact that the panel braces can be set anywhere along the length of the leg allows easy accommodation to sloped surfaces, allows towers to clear obstructions, and provides ways to easily increase the shoring capacity. The legs may be used on their own as post shores or with drophead systems as described previously. This system can even be used to build

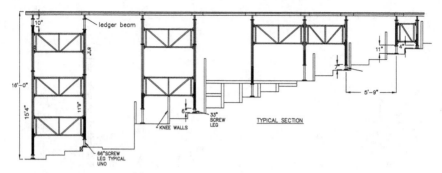

FIGURE 15.37   Typical section of a modular shoring application.

sidewalk canopies for the overhead protection of pedestrians. Figure 15.37 demonstrates the virility of the system. In this application it clears obstacles, climbs a slope, carries a heavy load with minimum pieces, and shores a flat deck.

FLYING SHORING SYSTEMS

The term "flying shoring" is generally used to describe the system of shoring and forming suspended reinforced concrete floors with large modular panels that are craned up from floor to floor as the building progresses (see Fig. 15.38).

FIGURE 15.38   Flying shoring table. Shown "flying out" from its slab shoring position. The table is constructed of two cross-braced aluminum trusses 55 ft (17 m) long supporting aluminum joists and plywood. The form deck area shored, and moved in one operation, is 100 ft² (93 m³). Tables of up to 2000 ft² (186 m²) are possible.

The first flying shoring system came into being in about 1960 and consisted of steel scaffold-type welded frames, braced together and stabilized with wood joists, wood stringers, and plywood on top making the forming surface. The complete unit, after the concrete was placed and set, was lowered by screw jacks in the scaffold frame legs. Roller attachments were added to the unit and rolled to the edge of the building, and the wire rope slings were attached, moved out, and flown by crane exactly like the flying truss system. This type of frame system is still being used today very successfully. It is an improved version with steel stringers and aluminum or steel joists replacing the wood stringers and joists.

Sometime later, the rolling shore bracket that supports the flying deck form came into being. There are two types of flying deck forms: the flying flat table, supported by columns or walls, and the flying truss. The rolling shore bracket consists of a roller assembly in the shore head and an adjustment screw below the head for final adjustment and for lowering the flying deck form a short distance after the concrete is placed and set.

The rolling shore brackets are bolted to columns or walls. The flying deck form consists of formwork stringers, usually steel joists that span from stringer to stringer, plywood, and the entire flying flat table tied together. The completed unit is then placed on the rolling shore bracket. After the concrete is placed and set, the table is lowered a few inches, rolled out toward the edge of the building, and again flown the same way as the flying truss or flying frame system. An advantage of this system is that it leaves the entire floor area below the formwork entirely free and open and reduces the need for reshoring. This system also is used to a great extent today with horizontal telescoping shoring beams as the joists in many installations. It should be pointed out that, in using the rolling shore bracket system, the concrete in columns must have sufficient strength to support the loads imposed by the bracket, and high early strength cement is commonly used.

Flying Truss Systems

Flying truss systems consist of a table supported by high-strength steel or aluminum trusses. Trusses are of various depths and extend the full length of the form. Depending on table width, there are two or three trusses per table. Trusses are held in a rigid position with lateral cross bracing of adjustable or preset steel tubing that allows the setting of variable truss spacing. Trusses are spanned laterally with joists of wood, steel, or aluminum and secured to the top chord with clips, screws, or steel banding. Joist spacing typically is 2 ft (60 cm) on center; however, this can vary according to slab thickness. Wood nailing strips are secured to the top flanges of some joists.

Joists constitute the basic support on which the plywood forming surface is mounted. Soffit boards may be incorporated into the support system lateral to the joists when additional nailing surface is needed for dome or pan forms.

Most trusses are supported on telescoping legs with vertical holes with pins to lock at the approximate height desired and with screw jacks that allow fine adjustment to final grade. Some systems may use only screw jacks for adjustment. See Fig. 15.38 for a typical installation.

The Forming Cycle

Stripping. To describe the forming cycle, it is best to start with stripping as the introduction to flying. Supplementary jacks are placed under the lower chord of the trusses.

Adjustable legs are telescoped back up into the trusses and pinned, and screw jacks are released and removed. The form is now quickly broken free.

Floor Movement of Forms.   If the table is to be moved straight ahead for flying, it is lowered onto sets of rollers positioned on the floor directly under the trusses. Rollers are sized so that their flanges are far enough apart to easily accommodate the bottom chords of the trusses.

If the form table is to be moved laterally before going out of the building, corner and middle leg positions may be fitted with casters and the form is moved manually to wherever it is to be exited from the building.

Depending on their size and weight, flying truss forms can be moved over floors on casters by crews of two to five workers. Many contractors, however, prefer to use power equipment for movement between stripping and flying.

Stripping a table, regardless of size, and readying it to fly can reasonably consume from 5 to 15 min. The form, with tag lines in place, is moved over the floor edge beyond the building line. Crane slings then are connected to the top chords of the trusses through small hatches opened in the form surface. The form is now ready to be flown.

Flying.   The form is swung clear of the building and up to the level it had recently supported. Usually on high-rise projects each table will occupy the same position on the new floor that it occupied below.

Flying forms generally are handled with medium capacity tower cranes. The operator often cannot see the form while booming it from the building side and will need direction to clear the floor edge, columns, or tables already set as the form is flown and brought in for a landing. Thus, usually the flying operation is accompanied by radio or hand communication between the rigger supervisor and crane operator.

Landing.   At the approach of the form for landing, four workers take it in hand and, together with the crane operator, guide the form to its next premarked position.

The four corner legs are extended first and locked into position at a predetermined setting, and the screw jacks are inserted. With the table in correct alignment, it is lowered into position on the slab.

The remaining legs are adjusted to the correct height at a later time as the crew now moves to position the next table.

SHORING DESIGN LOADS

Each structure must be thoroughly analyzed in terms of its physical configuration and content, relating these factors to the characteristics of the particular shoring and false-work to be used.

Vertical Loads.   It is customary to reduce the vertical load analysis to load per unit of projected area, such as pounds per square foot (newtons per square meter), so that the distribution of loads can be accurately determined when relating them to the supporting shores and falsework.

When determining this load, consideration must be given to all components, including such items as rebar, aggregate, and additives. These additives can vary the weight from as little as 100 lb/ft^3 (1600 kg/m^3) with lightweight aggregate to more than 300 lb/ft^3 (4800 kg/m^3) with heavy aggregate, high concentrations of rebar and lead additives as used in nuclear plants.

Consideration must also be given to the work practices anticipated during construction with due provision made in design calculations for these practices. Some examples are:

method of placement with attendant impact, motorized equipment such as buggies and vibrators, number of workers concentrated in one area, temporary storage of bundles of rebar, floor hoppers, and numerous other practices common during construction. Loads resulting from these conditions are customarily referred to as "live loads" in shoring calculations.

Scaffolding, Shoring, and Forming Institute, Inc., studies of actual field conditions provided data that were used to develop criteria for determining live and dead loads. These are as follows: A figure of 150 lb/ft^3 (2400 kg/m^3) should be used for calculating dead load where the concrete is commercial ready mix with normal rebar. An inverse relationship exists between dead and live loads when the dead load falls below 70 lb/ft^2 (3350 N/m^2). While a live load of 30 lb/ft^2 (1436 N/m^2) (958 N/m^2) for live load and 10 lb/ft^2 (479 B/m^2) for formwork dead load can be used for slabs over 5½ in (14 cm) thick, the live load must be progressively increased as the dead load becomes smaller so that the combined live and dead loads are not less than 100 lb/ft^2 (4788 N/m^2). When motorized carts are used for the placement of concrete, the live loads cited should be increased by not less than 25 lb/ft^2 (1197 N/m^2). Table 15.6 shows design loads for various slab thicknesses and weights of concrete.

In calculating the total design load for a beam, consideration must be given to increased formwork weight. It has to be added to the dead load and treated as a line load acting along the beam. As an example:

Depth of beam	Additional load, lb/ft (N/m)
Up to 3 ft (90 cm)	0 (0)
Up to 5 ft (150 cm)	5 (73)
Up to 8 ft (240 cm)	6 (87)

The figures cited in the table are guides to the most common methods used in shoring calculations and are minimum for that purpose. They are not intended to replace calculations based on actual conditions.

Once the basic design loads have been determined for those areas to be shored, it becomes necessary to support those loads with falsework, timber beams, horizontal shoring beams, steel frames or posts, single or in combinations in such a manner as to safely support those loads without failure or excessive deflection.

Horizontal Loads. Shoring and falsework structures must resist all foreseeable lateral loads such as wind, dumping of concrete on inclined surfaces, cable tensions, and stopping and starting of motorized placing or finishing equipment on deck forms. These temporary structures must also withstand the sideway effects caused by rapidly placed liquid concrete, vibration of concrete by power vibrator, and the effects that occur when concrete is unsymmetrically placed on slab or beam form.

In the absence of precise information on lateral loadings, it is suggested that the recommendations by the ACI in Ref. 3 for "Minimum Lateral Loads Acting in Any Direction" be used. This recommendation is as follows. Slab forms: 100 lb/lin ft (1459 N/m) of slab edge or 2 percent of total dead load on the form (distributed as a uniform load per lineal foot of slab edge) whichever is greater. "Consider only the area of slab formed in a single placement." These recommendations are only minimum requirements for slab form bracing, and a complete structural analysis of bracing requirements should be made when unusual construction conditions exist or are anticipated.

TABLE 15.6 Design Loads of Concrete Slabs Per Square Foot*
Allowance made for 10-lbf/ft2 forms and 20-lbf/ft2 live load

Wt. of concrete per ft³ slab thickness, in	150 lb/ft³ standard weight		120 lb/ft³ lightweight		100 lb/ft³ lightweight	
	Nonmotorized	Motorized	Nonmotorized	Motorized	Nonmotorized	Motorized
2 to 5.5	100	125	100	125	100	125
6.0	105	130	100	125	100	125
6.5	111	136	100	125	100	125
7.0	117	142	100	125	100	125
7.5	123	148	105	130	100	125
8.0	130	155	110	135	100	125
8.5	136	161	115	140	100	125
9.0	142	167	120	145	105	130
9.5	148	173	125	150	109	134
10.0	155	180	130	155	113	138
10.5	161	186	135	160	117	142
11.0	167	192	140	165	121	146
12.0	180	205	150	175	130	155
14.0	205	230	170	195	146	171
16.0	230	255	190	215	163	188
18.0	255	280	210	235	180	205
20.0	280	305	230	255	196	221
22.0	305	330	250	275	213	238
24.0	330	355	270	295	230	255

*For thicknesses of concrete greater than shown on the above chart, use dead weight of concrete plus 30 lb. If motorized, use dead weight of concrete plus 55 lb.
NOTE: 1 in = 2.54 cm; 1 lb/ft³ = 16.01 kg/m³; 1 lb/ft² = 47.9 N/m².

SHORING POSTTENSIONED CONSTRUCTION

Posttensioning is a complex and sophisticated procedure involving great accuracy of operation with high stresses and forces. There are occasions when these forces transmit additional loads to the shoring equipment over and above the dead loads and live loads for which they are normally designed. In almost all cases, the shoring remains in place until the posttensioning operation is completed.

One must therefore consider the effect of these loads and forces in all posttensioned jobs. The need for this consideration is clearly stated by the ACI in Ref. 3, Chap. 5, Sec. 5.5, "Forms for Pre-stressed Concrete Construction," Article 5.5.2.1, Design: "Where the side forms cannot be conveniently removed from the bottom or soffit form after concrete has set, such forms should be designed for additional axial and/or bending loads which may be superimposed on them during the pre-stressing operation." Because the shoring equipment supports these forms, the shoring layout must be designed giving consideration to any additional loads that may affect the equipment.

Procedures

First, determine if the construction uses the posttensioning method. The fastest way is to ask the contractor; otherwise a thorough study of the structural drawings is necessary.

1. Look for tendons in the shape of a parabola, as opposed to horizontal reinforcing bars, shown in longitudinal sections of beams and slabs. Posttensioned slabs are usually of the following types: (*a*) Deep, flat slab, generally with voids, such as used in bridge designs; (*b*) Pan-joist slabs, with tendons in the joists (generally long span); (*c*) Grid-dome slabs with tendons in the ribs (generally in long-span direction).

2. Look for anchorage points; find out whether these are at piers, abutments, columns, and/ or shear walls or at beams running at right angles to the tendons. This latter condition is of utmost importance *if you are shoring beams that carry anchorages from slabs or other beams.* The posttensioning forces applied at anchorage points are generally designated by "force vectors" at the anchorage point, or in tabular manner. They are generally shown in kips (thousands of pounds). It is important to know the values of these force vectors.

3. Obtain and check the job specifications, in addition to the drawings. Specifications generally devote several sections to describing the posttensioning system, end anchorages, and grouting and shoring requirements, in sections covering "posttensioning," "formwork," and "shoring" or "falsework."

4. Determine whether the shoring equipment will be left "in place" during the posttensioning operation (usually yes).

5. Determine the sequence of posttensioning; that is, what sections or areas will be posttensioned first.

6. Determine if the posttensioning method and sequence will superimpose additional loads on the equipment. Figures 15.39 and 15.40 illustrate two common conditions in which the tensioning will tend to "lift" the slab, ribs, or beams off the shoring support and transfer all or part of the "dead weight" of section A to section B. Section B is prevented from deflecting by the shoring.

   The persons to give a logical answer as to whether there will be superimposed loads, and if so in what amount, will generally be the structural design engineer, the contractor's engineer, or possibly the posttensioning subcontractor.

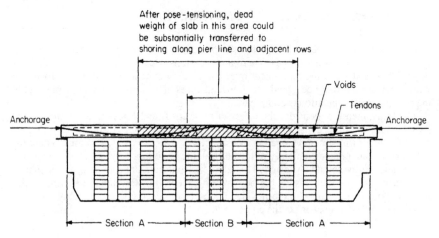

FIGURE 15.39   Shoring of posttensioned two-span slab.

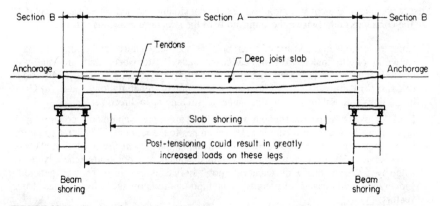

FIGURE 15.40   Shoring of posttensioned joist slab.

Failing to obtain satisfactory answers, the quotation *and* estimate layout should include the following notes: "This shoring is not designed to carry additional loads that may be superimposed upon it as a result of posttensioning operations. All reshoring and the reshoring of the superimposed posttensioning loads are the responsibility of others."

If information is obtained (preferably in writing or on a drawing) on the positions and amounts of the additional superimposed loads, the following notice should be used instead: "This shoring is designed on the consideration, as directed by the structural designer (note: or other appropriate party—substitute as necessary), that the posttensioning operation will induce additional loads on the shoring equipment in the amounts and locations indicated on this layout."

7. Obtain the contractor's consent to contact the parties named in step 6.

8. Have the contractor determine, or obtain the contractors permission to determine, whether the engineer has a special requirement concerning the procedure to follow for the removal of shoring, if this is not clearly spelled out in the drawings or specifications.

9. Similarly, determine if the engineer wants to review or approve the shoring layout. Although the design and safety of formwork is generally the responsibility of the contractor, the engineer or architect may wish to approve the formwork and shoring designs. If possible, it would be desirable in all cases to have a written or stamped approval as a confirmation of acceptance of any design assumptions made in the layout.

SHORING IN ALTERATION WORK

This subject is complex; it is therefore imperative to obtain complete information on loads and other pertinent factors. Based on this information, the shoring layout should be designed and shoring carried out under the supervision of a licensed structural engineer.

Shoring in alteration work differs from shoring a new building because each situation has to be studied individually to determine the expected loads. Factors to be considered are:

1. Occupancy, because whether the building to be altered will be occupied or not changes the live load to be considered on slabs and floors.

2. Structural characteristics: (*a*) slab thicknesses and strength, which will affect the size of bearing area for reshoring (similar to plywood thickness that affects joist spacing in placing fresh concrete); (*b*) direction of reinforcing steel rods; (*c*) whether the altered sections are load-bearing members or not (columns, main walls, etc.).

3. Single or multiple floors.

4. When alteration work is to be done inside the existing structure, some regular shoring considerations such as wind load and mud sills will not apply.

ERECTION TOLERANCES

All vertical shoring equipment must be erected and kept plumb in both directions. The maximum allowable deviation from the vertical is 1/8 in (3 mm) in 3 ft (90 cm) and it should never exceed 1 in (25 mm) in 40 ft (12 m). If this tolerance is exceeded, the shoring equipment should not be used until readjusted within this limit.

Eccentric loads on shore heads and similar members should be avoided. The capacity of a shoring leg or adjustable base is decreased by a large percentage when an eccentric load acts on it. [As an example, a jack that would support 33,000 lb (146 kN) when loaded concentrically will bend at 11,000 lb (49 kN) if the load is placed 2 in (5 cm) off center.]

FALSEWORK DRAWINGS

All jobs requiring vertical shoring should have a drawing prepared or approved by a qualified person. This drawing should have a complete plan of the area to be shorted along with elevations showing the makeup of the shoring equipment. Unusual conditions such as heavy beams, ramps, and cantilever slabs should be covered in detail to ensure a safe and proper installation. A copy of the drawing should be available and used on the jobsite at all times.

Shoring layout drawings should be available at the jobsite and should be strictly adhered to. All shoring layouts include general comments that provide additional information. Examples of these notes follow.

General Notes

This drawing is provided as a service to illustrate the assembly of the manufacturer's products only. It is not intended to be fully directive or cover engineering details of such products or equipment or materials not furnished by the manufacturer or the interconnection therewith. Inasmuch as the manufacturer does not control jobsite assembly or procedures, grade, or quality of material or equipment supplied by others, it is the responsibility of the contractor to integrate this drawing into a composite drawing suitably complete for construction purposes consistent with safe practice and overall project objectives. The following points should be covered:

1. All dimensions and details shown on this layout must be checked and verified by the contractor before proceeding with the work.

2. The concrete supported by the shoring shown on the layout is assumed to weigh _____ lb/ft^3 (kg/m^3).

3. The design layout includes a live load (including forms) of _____ lb/ft^2 (N/m^2) that does not include provisions for motorized concrete equipment. If motorized equipment is used, add 25 lb/ft^2 (1200 N/m^2) to the above figure.

4. Approximate amounts of screw jack extensions have been noted. These extensions may require adjustment due to field conditions. However, the maximum screw jack extensions for this layout are limited to _____ in (cm) top and _____ in (cm) bottom.

5. The contractor will design and provide suitable sills to properly distribute imposed shoring loads.

6. All stringers, ledgers, or other members resting on the manufacturer's equipment must be centered directly over the shoring legs.

7. Splicing of ledgers must occur at the centerline of jack screws.

8. Ledgers where supported by U heads will be equally packed or wedged.

9. In setting elevations, allow for lumber and soil compression.

10. *Timber Notes*

   a. Timber calculation and design criteria are based on dressed sizes conforming to the American Softwood Lumber Standard, Voluntary Product Standard 20-70, U.S. Department of Commerce, National Bureau of Standards, and to the *National Design Specifications for Stress-Grade Lumber and Its Fastenings,* 1968 edition, revised November 1970.

   b. The timber falsework details shown are manufacturer's suggestions and are based on the following minimum design values. Properties are for: Douglas Fir-larch No. 1 stress graded 1500 (per National Forest Product Association) size classified 2 to 4 in (5 to 10 cm) thick by 6 in (15 cm) and wider with a 25 percent increase for short-duration loading 10 percent for *E*.

Fiber stress in bending 1500 lb/in^2	1800 lb/in^2
(1034 N/cm^2) (engineered use)	(1290 N/cm^2)
Horizontal shear 95 psi (65.5 N/cm^2)	
[if checks are not in excess of width of piece,	150 lb/in^2
allow 33 percent increase to 126 lb/in^2 (86.9 N/cm^2)]	(103 N/cm^2)
Compression perpendicular to grain	480 lb/in^2
	(331 N/cm^2)
Modulus of elasticity *E*	1.76×10^6 lb/in^2
	1.21×10^6 N/cm^2

    c. Plywood design is based on American Plywood Association technical data (Concrete Formwork Brochure No. S71-90, January 1971). Grains of face plies are perpendicular to the supports. Plywood continuous over two or more spans. Values are for B-B plywood class I (exterior type), deflection limited to L/270. If B-B plyform class II is used, increase thickness of plywood shown on layout by 1/8 in.

11. The final responsibility for the formwork design and placement remains with the concrete contractor.

12. The shoring system, as shown, is designed on the assumption that formwork will be restrained from lateral movement by the contractor. Sufficient lateral support must be provided where necessary to prevent the imposition of lateral loads on the shoring system.

13. Shoring equipment should be erected and used per published safety rules and regulations of the Scaffolding, Shoring, and Forming Institute, Inc.

14. The print is the property of the manufacturer and is furnished for the exclusive use of the customer on the condition that it is not to be copied or used by others without written consent.

15. All beams must be secured to U heads with beam clamps or by another approved manner, as they are installed.

16. All lower frames of a tower must be plumb and level before erecting the remainder of the tower. Check again for plumbness prior to placement of concrete.

17. Reshoring design and procedures are the responsibility of others and should be thoroughly checked by the architect and/or engineer to determine proper placement and that sufficient capacity exists to support areas being reshored.

18. Tower leg loading should be as uniformly distributed as possible. Never load only one leg of a frame or one ledger of a tower.

19. *Horizontal Shore Notes*
    a. Job architect or engineer is to determine that the structural steel beams supporting horizontal shoring beams are capable of carrying construction loads during placement of concrete.
    b. An intermediate shore must never be placed under the lattice section of the horizontal shoring beam.
    c. An allowance of up to 5/8 in (16 mm) must be made for the depth of the horizontal shoring beam bearing prongs, in order to determine the correct elevation of the ledger.
    d. Do not nail horizontal shoring beam prongs to ledger.
    e. Care must be taken to follow the designed allowable clear span of the horizontal shoring beam.
    f. Beam hangers must be designed and loaded in accordance with manufacturer's specifications and designed to fit the shape of the horizontal shoring beam bearing prongs.
    g. All beam sides supporting horizontal shoring beams must have a minimum of 2- × 4-in (5.08 × 10.16-cm) vertical stud directly under each horizontal shoring beam or a design that takes horizontal shoring beam concentrated loads into consideration.
    h. The manufacturer assumes no responsibility for placement of beam ties or kickers for beamside forming.

20. Post shores, if shown, are to be the manufacturer's post shores. Any substitutions of other post shores are to be approved by the supplier.

21. The manufacturer does not provide all items illustrated on these drawings. Those items not supplied should be erected according to drawings furnished by their manufacturer or supplier. The final responsibility for design and placement of these items remains with the contractor.

22. Deviation from these layouts may be made only under the direction and supervision of a qualified person who, by possession of a recognized degree, certificate, or professional standing, or who by extensive knowledge, training, and experience has successfully demonstrated the ability to solve or resolve problems relating to the subject matter, the work, or the project; and/or in consultation of with the designer.

23. The customer or lessee bears the sole responsibility for ensuring that any erection or use of the goods shown on these drawings conforms to all laws, ordinances, and local codes and for checking the accuracy of field details and dimensions.

FIRE PROTECTION

The best fire protection is a good, clean, orderly jobsite with everything in its place and a place for everything. Flammable liquid storage and proper fireproof storage containers and a proper distance from buildings are a must. The proper fire control equipment in proper locations as prescribed in Federal OSHA Sections 1910 and 1926 and in National Fire Protection Association code rules must be followed. A thorough and proper survey of the jobsite, buildings, and adjacent areas by the job superintendent with the aid of a local fire protection agent can point out additional fire and safety precautionary measures and controls that must be taken. A fire drill must be run through to alert all to their responsibilities and assignments, as well as to the location of exits and fire control equipment. When space heaters are used, precautions must be taken in view of combustible materials on the jobsite. Any regulations covering the use of this type of equipment must be complied with.

Material shipped and stored on the jobsite must be analyzed for its flammable or inflammable qualities and stored and protected accordingly. Certification of any lumber or material as to its flame-retardant protection or combustible nature should be reviewed, noted, and tagged.

NEW PRODUCT TRENDS

Aluminum Shoring Trusses.   These are generally used on high-rise and other repetitive shoring work. They tend to eliminate the use of steel shoring frames for repetitive-use flying shoring.

Aluminum Shoring Joists.   Developed first for use with flying trusses, these joists are in popular use for all shoring applications. They are cost-effective; their light weight [average 4 lb/ft] (58.4 N/m) and high strength make them desirable to replace lumber joists on many jobs. Their ability to support concrete slabs over long spans enables shoring frames to be spaced farther apart and thus obtain higher leg loading closer to the shoring frames' safe allowable leg loads.

Aluminum Shoring Ledgers (Stringers).   These are stronger, heavier versions of aluminum joists and can replace steel I-beam ledger with weight and handling advantages. See Fig. 15.41.

Aluminum Shoring Frames.   These are relatively new, and their design is such that shoring loads of up to 15,000 lb (66,720 N) per leg can be accommodated safely, having a weight saving of about one-third over heavy-duty steel shoring frames that have average capacities of only 10,000 lb (44,480 N) per leg (see Fig.15.10).

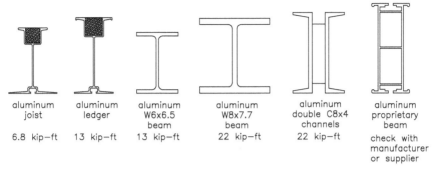

aluminum joist	aluminum ledger	aluminum W6x6.5 beam	aluminum W8x7.7 beam	aluminum double C8x4 channels	aluminum proprietary beam
6.8 kip–ft	13 kip–ft	13 kip–ft	22 kip–ft	22 kip–ft	check with manufacturer or supplier

FIGURE 15.41 Aluminum beams available for shoring, and their approximate moment capacities when fully laterally braced.

Single-Post Aluminum Shores.   The design advantages of aluminum are numerous; it enables almost unlimited flexibility of design shapes to obtain the most efficient strength-to-weight ratio. Aluminum extrusion dies are cheaper and simpler to build, whereas special shapes formed from steel are very expensive, resulting in the almost exclusive use of standard available tube, pipe, I-beam, and other rolled-steel shapes.

A typical section of shoring using steel frames, I-beam ledgers, aluminum joists, and plywood is shown in Fig. 15.1.

THE SCAFFOLDING, SHORING, AND FORMING INSTITUTE, INC.

In 1959, a group of scaffolding manufacturers got together and created an association with a mandate to promote safety and the safe use of their products by construction and other industries.

The Scaffolding, Shoring, and Forming Institute, Inc., publishes safety rules for the majority of uses of almost all types of scaffolding and shoring products. It also has available a series of booklets of safety requirements for major types of scaffolding. Another series of hip-pocket-sized booklets deals with how to erect steel frames (scaffolding and shoring) safely. Also available are a number of wall-poster-sized illustrations of various hazards and their prevention. Three groups of color slides are available at nominal cost; they come with descriptive booklets exhibiting dos and don'ts for scaffolding, shoring, and suspended powered scaffolds. The address of the institute is The Scaffolding, Shoring, and Forming Institute, 1300 Summer Avenue, Cleveland, Ohio 44115.

REFERENCES

1. Federal OSHA Regulation 29 CFR, part 1926. 701 Subpart Q.
2. *Lessons from Failures of Concrete Structures*, ACI monograph No. 1, American Concrete Institute.
3. *Recommended Practice for Concrete Formwork*, ACI Standard 347-88, American Concrete Institute, Detroit, 1988.

CHAPTER 16
CONCRETE FORMWORK*

John A. Brain, P.E.

*This chapter is updated and revised by John A. Brain, P.E., from the one in the second edition contributed by Kirk Gregory, P.E.

INTRODUCTION

Formwork is a classic temporary structure in the sense that it is erected quickly, highly loaded for a few hours during the concrete pour, and within a few days disassembled for future reuse. Also classic in their temporary nature are the connections, braces, tie anchorages, and alignment and adjustment devices, which forms need.

The term "temporary structures" may not fully connote, however, that some forms, tie hardware, and accessories are used hundreds of times, which necessitate high durability and maintainability characteristics and a design that maximizes productivity. Unlike conventional structures, the formwork disassembly characteristics are severely restricted by the concrete bond, rigidity, and shrinkage, which not only restricts access to the formwork structure but causes residual loads that have to be released to allow stripping from the concrete that initiates disassembly.

The general practice of using plywood and lumber with snap ties and brackets (Fig. 16.1) is well covered in the American Concrete Institute (ACI) special publication SP-4[1] and in ACI Standard 347-04.[2]

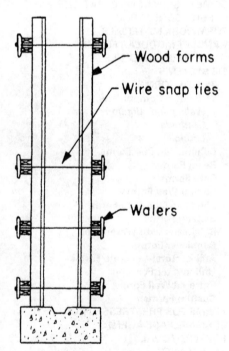

FIGURE 16.1   Wood forms and wire snap ties. (*From American Concrete Institute.*)

These set factor-of-safety criteria on form ties, anchors, and lifting inserts (Table 16.1) that were included by reference in the American National Standards Institute's ANSI A10.9[3] and thence by the U.S. Department of Labor, Occupational Safety and Health Administration's OSHA 1518.701.[4] The American Plywood Association (APA)[5] also publishes lumber-formwork guides. Furthermore, manufacturers of snap ties and tie hardware publish excellent design charts and load rating.

There is not enough space here to adequately cover the design of lumber formwork, so this chapter will only summarize the design criteria, types of forms, and jobsite conditions

TABLE 16.1 Design Capacities of Formwork Accessories*

MINIMUM SAFETY FACTORS OF FORMWORK ACCESSORIES

The safety factors shown below are based on ultimate strengths of accessories when new and comply with ACI's Guide to formwork for Construction (ACI 347-03) which equal or exceed OSHA 1926 SUBPART Q. User must determine which safety factor applies based on "Type of Construction" and other appropriate considerations. If there are unusual job conditions of shock, impact, or vibration, safely factors should be increased.

Accessory	Safety factor	Type of construction
Form tie	2.0	All applications
Form anchor	2.0	formwork supporting form weight and concrete pressure only
Form anchor	3.0	formwork supporting weight of forms, concrete, construction live loads and impact
Form hanger	2.0	All applications
Anchor inserts used as form ties	2.0	Precast concrete panels when used as formwork

*Design capacities guaranteed by manufacturers may be used in lieu of tests for ultimate strength.

that influence form selection and will describe special applications that present structural challenges.

DESIGN LOADS

Fresh concrete exerts pressure on formwork, similar to fluids or soils, that depends on the density of the concrete and the height of the pour. Unlike fluids, concrete consolidates because of vibration and the cement starts setting so that if poured slowly at a steady rate of rise, the maximum pressure can be controlled. ACI 347-04 contains" charts and formulas that give pressure as a function of rate of pour and temperature (Fig. 16.2). This maximum pressure is distributed as shown on the typical pressure diagram in Fig. 16.3. However, often overlooked are the conditions under which these charts and formulas are valid. Full liquid head pressure, 150 lb/ft³ (2400 kg/m³) × height for normal concrete, must be used when (1) slump exceeds 7 in (10 cm); (2) external vibrators are used; (3) retarder or workability admixtures are used; and (4) layers in excess of 4 ft (1.2 m) are poured.

When concrete is pumped *into the bottom* of the forms, the form pressure is the full liquid head plus additional pump pressure to overcome frictional flow resistance. This extra pressure depends on the concrete mix, the pump, and flow restrictions. Excessive pump speed can overpressure the forms, causing failure. Even with optimum conditions, 50 percent extra pressure at the pump inlet should be included in the design loads.

If there are significant restrictions to the flow of the concrete being pumped into the bottom, such as prepacked aggregate, precast elements inside the forms, large anchorage and embedments, or box-outs, the extra pressure should be 100 percent more than the full liquid head. Pressure acts perpendicular to the form surfaces. This causes uplift on battered wall forms or uphill thrust on sloping slabs, which therefore require positive anchors, braces, or dead loads to withstand the upward forces. Some of the pressure effects not readily apparent are:

1. A force imbalance when pouring against embankments, previous pours, and sheet piling requires anchors or braces.

2. The upward component of the reaction force from inclined wall braces when one-sided forms are poured (without ties) requires a tie-down anchor.

Table 1A — Unit weight coefficient C_w

INCH-POUND VERSION	
Weight of concrete	C_w
Less than 140 lb/ft³	$C_w = 0.5 [1+(w/145 \text{ lb/ft}^3)]$ but not less than 0.80
140 to 150 lb/ft³	1.0
More than 150 lb/ft³	$C_w = w/145$ lb/ft³

Table 1B — Chemistry coefficient C_c

CEMENT TYPE OR BLEND	C_c
Types I and III without retarders *	1.0
Types I and III with retarder	1.2
Other types or blends containing less than 70% slag or 40% fly ash without retarders *	1.2
Other types or blends containing less than 70% slag or 40% fly ash with retarders *	1.4
Blends containing more than 70% slag or 40% fly ash	1.4

TABLE 2: BASE VALUES OF LATERAL PRESSURE (LB/FT²) ON COLUMN FORMS: APPLY WEIGHT AND CHEMISTRY COEFFICIENTS TO OBTAIN PRESSURE FOR DESIGN

Rate of placement R, ft/h	p = [150 + 9000 R/T] for temperature indicated					
	90°F	80°F	70°F	60°F	50°F	40°F
1	600	600	600	600	600	600
2	600	600	600	600	600	600
3	600	600	600	600	690	825
4	600	600	664	750	870	1050
5	650	713	793	900	1050	1275
6	750	825	921	1050	1230	1500
7	850	938	1050	1200	1410	1725
8	950	1050	1179	1350	1590	1950
9	1050	1163	1307	1500	1770	2175
10	1150	1275	1436	1650	1950	2400
11	1250	1388	1564	1800	2130	2625
12	1350	1500	1693	1950	2310	2850
13	1450	1613	1821	2100	2490	3075
14	1550	1725	1950	2250	2670	3300
15	1650	1837	2078	2400	2850	3525
16	1750	1950	2207	2550	3030	3750

maximum pressure = *wh*

TABLE 3: BASE VALUES OF LATERAL PRESSURE (LB/FT²) ON WALL FORMS: APPLY WEIGHT AND CHEMISTRY COEFFICIENTS TO OBTAIN PRESSURE FOR DESIGN

Rate of placement R, (ft/h)	p = [150 + 43,400/T + 2800 R/T] for temperature indicated					
	90°F	80°F	70°F	60°F	50°F	40°F
1	663	728	810	920	1074	1305
2	694	763	850	967	1130	1375
3	726	798	890	1013	1186	1445
4	757	833	930	1060	1242	1515
5	788	868	970	1107	1298	1585
6	819	903	1010	1153	1354	1655
7	850	938	1050	1200	1410	1725
8	881	973	1090	1247	1466	1795
9	912	1008	1130	1293	1522	1865
10	943	1043	1170	1340	1578	1935
11	974	1078	1210	1387	1634	2005
12	1006	1113	1250	1433	1690	2075
13	1037	1148	1290	1480	1746	2145
14	1068	1183	1330	1527	1802	2215
15	1099	1218	1370	1573	1858	2285

Note: Values above are for walls where placement height does not exceed 14 feet. For walls where placement height does not exceed 14 feet and rate of placement is less than 7 ft/hr, use appropriate values from Table 2.

Do not use design pressures in excess of *wh* .

* Retarders include any admixture, such as a retarder, retarding water reducer, or retarding high-range water-reducing admixture, that delays setting of concrete.

FIGURE 16.2 Concrete pressure against forms. (*From American Concrete Institute.*)

16.4

3. In-plane forces due to pressure on bulkheads, which become severe on thick pours, such as bull-nose piers, require strong longitudinal connections.

4. In sloping tunnel roofs, the intensity of the pressure depends on the height of liquid head, which can be much greater than the slab thickness.

5. On thick circular walls, hoop-tension forces must be recognized and dealt with at panel joints and require positive anchorage.

Secondary load criteria relate to wind load on braces, motorized buggy turning and braking forces, walkway loads, lifting forces during tilt-up of gang forms as well as during moving them from pour to pour, and means of supporting the dead weight of gangs during erection and stripping.

While the concrete pressure is considered to be acting on the full height of the pour for design purposes (Fig. 16.3), it must be recognized that concrete has to be poured onto the bottom first, which tends to pivot the forms on the lower ties and kick the forms together at the top. Such partial loadings may govern the form design, so they must be considered.

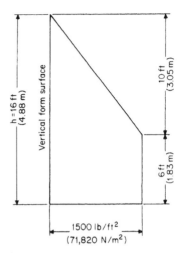

FIGURE 16.3 Typical pressure diagram. Maximum pressure under controlled rate of pour, such as 6 ft (1.83 m)/h at 40°F (4.4°C). (*From American Concrete Institute.*)

If at any one time during the pour the fresh concrete is higher along one side than along the opposite side within the formed area, a net horizontal thrust is created that can cause dislocation and sliding of the entire unit. This could also occur on a box culvert with inside bracing in lieu of ties between the wall forms.

Formwork cantilevers (and deckform cantilevers) where initial concrete load causes deflection more than *that* cantilever would finally deflect after all continuous spans are loaded should be avoided because as the cantilever tries to deflect into the concrete the high viscosity of the wet concrete, aggregate interlock, and its internal friction will not allow the cantilever to move inward. This in effect causes extra pressure on just the cantilever, a triangular load distribution, peaking at the tip, of about $150 \times L_c \times w$, plf. Here 150 is the density of concrete, pcf; L_c is the cantilever length, ft; and w is the width of the tributary area of the form member.

SOME DESIGN CONSIDERATIONS

The formwork structure must be integrated with the tie system that restrains the concrete pressure in a manner that allows quick erection, form-tie unloading, easy disassembly, and tie "removal." Because the ties positively anchor opposing forms, the proper locations of the ties at corners and pilaster, in relation to the formwork structure, are just as important as the strength and the modes of anchorage and removal of the ties.

Modular dimensions that easily combine into any size form assembly with minimum waste, good stackability, and adequate alignments make forming practical.

Formwork weight is always a factor whether hand-setting or crane-handling the gangs.

Multiple lift forms require anchors to carry dead weight, walkways for access, braces for wind loads, and plumbing.

Core forms, pilaster forms, etc., get severely "locked in" by the concrete pressures and its shrinkage. Some positive means to relieve this pressure greatly facilitates stripping, as discussed later in this chapter.

JOBSITE CONDITIONS

The nature of the job (including local conditions) is one of the primary factors in form-work selection. In addition to job specifications, variations in the shape of the concrete, reuse potential, form cycle time, gang-forming feasibility, proximity of form inventory and/or supplies, crew experience, crane capability and cost, inserts, penetrations and concrete pour rate are only some of the factors that influence the type and design of formwork systems.

If the concrete is to be exposed to the public, most of the handset modular systems will need liners owing to their grout leakage and the many "triple-joint" patterns that are ugly. The larger modular systems have higher pressure capacity and fewer, tighter joints to grind and finish, but the crane capacity, reach, and cost must be considered. However, if walls have many jogs, pilasters, or counterforts of varying sizes, modular systems, both handset and ganged, will fit better with fewer job-built fillers. This trade-off can be solved with intergraded systems comprising large crane set or handset rentable panels [4 × 10 ft, 8 × 9 ft, 4.5 × 10 ft, 4 × 9 ft, 4.5 × 9 ft (1.2 × 3 m, 2.4 × 2.7 m, 1.35 × 3 m, 1.2 × 2.7 m, 1.35 × 2.7 m)] that can easily form and fit pilasters, jogs, chamfers, and any other irregular shapes.

Construction sequence may be another factor. Buildings typically have high floor heights in the first few floors, so that a rentable formwork system will handle the first taller walls and columns, then reduce the rental equipment for the typical floor heights.

The architect may specify cone ties on 2-ft centers that fit older-style handset modular systems and job-built forms equally, but the newer modular clamp systems, both ganged and handset, have difficulty meeting such tie patterns because often the tie locations are spaced much farther (e.g., 4 × 5 ft) apart. Dummy cones can create the 2-ft (60-cm) cone patterns, but at extra cost. On the other hand, odd cone spacings, such as 16, 19.2, or 32 in (40.6, 48.8, or 81.3 cm), can be worked out with job-built forms.

Buildings and power plants usually have extensive electrical and mechanical requirements for conduit or plumbing form penetrations and/or box-outs for duct, window, and door openings. This requires nailing, drilling, and patching that is fast and easy with plywood form faces, but not with steel faces.

Most of the jobsite conditions have to be evaluated on their own merits, such as access, staging areas, contractor-owned forms, crew experience, weather, budget, and time schedules.

FORMWORK MATERIALS

Although lumber and plywood have been the usual materials because of their economy, availability, and jobsite workability, more durable materials are required for modern, large projects and for contractors having continuing repetitive formwork needs of a similar nature, such as large tracts of home basements, Table.16.2 is a list of form materials and their principal uses (it was reproduced in part from ACI 347-04).

Lumber and plywood formwork design has been well covered by ACI SP-4[1] (5th ed., now being updated) and APA,[5] including the load-span characteristics and the highly durable "overlaid" plywoods, phenolic-impregnated paper baked on as the plywood is laminated. Plastic laminates, such as fiberglass and HD polyethylene, have been developed for better durability and smoothness.

Engineered wood is higher-strength (215 percent) and stiffness (150 percent) as well as more durable owing to factory veneering and laminating with water-resistant glues Laminated veneer lumber (LVL), laminated strand lumber (LSL), parallel strand lumber (PSL), structural composite lumber (SCL), and glue laminated lumber (GLULAM) are

TABLE 16.2 Form Materials and Uses

Material	Principal use
Lumber	Form framing, sheathing, and shoring
Plywood	Form sheathing and panels
Steel	Panel framing and bracing
	Heavy forms and falsework
	Column and joist forms
	Stay-in-place forms
Aluminum	Lightweight panels and framing; bracing and horizontal shoring
Hardboard, particle board	Form liner and sheathing; pan forms for joist construction
Insulating board, wood, or glass fiber	Stay-in-place liners or sheathing
Fiber or laminated paper pressed tubes or forms	Column and beam forms; void forms for slabs, beams, girders, and precast piles
Corrugated cardboard	Internal and under-slab voids; voids in beams and girders (normally used with internal "egg-crate" stiffeners)
Concrete	Footings, stay-in-place forms, molds for precast units
Fiberglass-reinforced plastic	Ready-made column, dome, and pan forms; custom-made forms for special architectural effects
Cellular plastics	Form lining and insulation
Other plastics: polystyrene, polyethylene, polyvinyl chloride	Form liners for decorative concrete
Rubber	Form lining and void forms
Form ties, anchors	For securing formwork against placing loads and pressures
Plaster	Waste molds for architectural concrete
Coatings	Facilitate form removal
Steel joist	Formwork support
Steel frame shoring	Formwork support
Form insulation	Cold-weather protection of concrete

all processed wood materials with better structural properties up to 2800 psi (19,306 kPa) bending stress, yet are still nailable, cuttable, lightweight, etc.

Steel has high strength, stiffness, and durability (except light-gauge sheet metals), but as a form surface it loses nailability, light weight, and jobsite workability (other than by torch). However, for columns, dam faces, pier caps, and other surfaces where conduit or plumbing penetrations and box-outs are limited and where weight is manageable, all-steel formwork is practical if about 15 reuses are needed.

The most highly successful materials are steel frames with replaceable plywood faces. This combination affords the jobsite workability to minimize labor and yet gain large tie spacing as well as long-life structural frames, walers, and accessories. Overlaid plywood of high grade further extends the form-face wear and grout degradation of the plywood, yet retains nailability and workability.

The most successful of these systems embodies proprietary designs of high-carbon steels (to cut weight) that protect the edges of the plywood and absorb tie loads and stripping, racking, and lifting stresses. Because ties fit between panel joists (instead of through the plywood), the steel frame absorbs the tie loads and the wear. Only a few bolts or rivets attach the plywood to the frame; thus there are only a few overlay penetrations for grout to degrade the plywood. By patching the gouges and nail and conduit holes, and by faithful use of release agents, these panels are so durable that they are rented by form suppliers. Some suppliers maintain large inventories of various-size panels, fillers, corners, pilaster forms, chamfers, etc. They also offer assistance in planning and engineering layouts and provide special equipment for lifting, aligning, and bracing as well as cleaning and refacing.

Fiberglass-reinforced-plastic form materials produce such excellent as-cast concrete surfaces that they have found success as column forms, void forms for waffle slabs, and one-way concrete joists as well as for seamless gang forms, beam forms, and column capitals. Because fiberglass forms are made by a molding process, custom shapes are ideally formable with fiberglass materials.

Fiberglass underwater pier-repair sleeves offer excellent resistance to corrosion, erosion, chemical degradation, and marine borers.

Aluminum and magnesium also offer lightweight modular forms, but suffer in durability due to chemical attack of the alkaline concrete and due to large deflections and damage susceptibility indigenous to these materials. Brick-textured aluminum form-face systems are successful in the basement and housing formwork applications. Aluminum extrusions enable design flexibility to provide bolt slots, nailer pockets, and other special features. Beams, channels, tees, and tubes assemble into very strong but lightweight flying trusses. These nailable beams and double-channel walers provide large gang-wall forms that are exceptionally lightweight and straight owing to the extrusions.

Recently the structural characteristic of plastics have improved and today plastic facing materials as well as entire plastic sheathing materials have begun to replace plywood in modular-framed systems. Plastics have been used as form liners, water stops, rustication strips, and tie cones with varying success. "Structural-foam" plastic forms and domes have emerged overseas but have not found a domestic market.

Irrespective of the form material, ties and hardware are generally steel to resist the high loads and localized bearing stresses at the anchorages. Fiberglass ties have been tried, but low strength with high cost tend to offset the nonrust advantage. Sizes, designs, and special features of ties are described elsewhere in this chapter.

HANDSET MODULAR FORMS

While a 4- × 8-ft (1.22- × 2.44-m) sheet of plywood was the early module for forms, the term "modular form" has grown to refer to prefabricated all-metal or metal-supported-plywood systems of panels, fillers, corners, and ties whose integrated design of tie and connecting hardware is engineered to ensure dimensional control, speed of erection, and ease of stripping, as well as structural integrity and reliability.

The term "handset forms" is derived from the fact that these modular panel systems are capable of being erected or assembled by one or two men with common hand tools. These systems are often user friendly and have accessories that allow the system to handle most wall conditions. Currently there are two distinct styles of handset modular forms available within the domestic market.

The first generation of these systems consists of modular panels constructed of ½- or 5/8-in-thick plywood in a 2½-in-deep steel frame. The most common panel is 2 × 8 ft (6.1 × 2.44 m) in size, but a myriad of other sizes as well as various accessories are available

FIGURE 16.4   First-generation handset panels. (*From Symons Corporation.*)

depending on the manufacturer (Fig. 16.4). These systems utilize a tie system that consists of a small-diameter rod or flat bar that is attached using a simple wedge system. These systems have been in use for over 40 years and are used throughout the United States. Some of the benefits are that they are lightweight [a 2- × 8-ft (0.61- × 2.44-m) panel weighs approximately 80 lb (36 kg)], relatively simple to use, and can meet wall configuration without or minimal use of job-built fillers.

The second-generation systems were first developed by several European formwork manufactures in the late 1990s. These systems are handset-compatible versions of the large modular panel systems that utilize simple clamps and large-diameter ties. The most common panel is the 3 × 9 ft (0.91 × 2.7 m) size and is normally available in several other widths and heights depending on manufacturer. Some of the benefits of these systems are the reduction in labor to assemble, only two ties required for a 9-ft (2.7-m)-high panel as compared to seven ties on the first-generation panels along with reducing the need for the small wedges used to attach the panels together. These panels are slightly heavier [approximately 170 lb (77 kg) for a 3- × 9-ft (0.91- × 2.7-m) panel], but they still can be assembled by two men. In addition, these systems have various accessories to handle varying wall conditions as well as can be easily attached to the larger crane-handled modular systems with a transition panel (Fig. 16.5).

The availability of cranes or forklifts allows the handset modular systems to be built and used as large as are gang systems. This allows the users the flexibility to either gang and use as a larger panel or assemble by hand.

GANG FORMS AND TIES

This refers to any large formwork system that forms large areas at one time and is typically moved utilizing a crane. The four main types of gang form system are

1. Joist and waler systems
2. Large modular clamp systems (Euro-style forms)
3. Steel "girder" forms
4. Modular handset forms

These systems are assembled into large units and are designed to provide maximum reuse on a project. All these systems utilize tie systems that can be removed, released, or

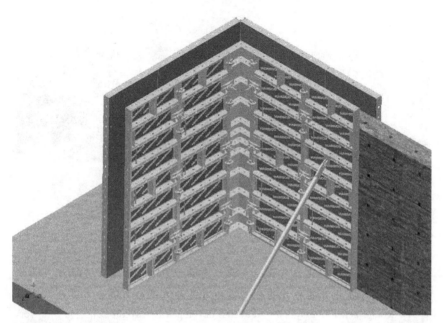

FIGURE 16.5 Second-generation handset panel system. (*From Harsco Infrastructure.*)

broken back to allow the gang to be stripped and reused. Some gangs have been known to have over 100 reuses without replacing the face sheet material.

Building shear walls and bridge-pier caps and stems are well suited for 8- × 12-ft (2.44- × 3.66-m) up to 8- × 20-ft (2.44- × 3.66-m) steel-faced "waler-less" forms that are structured so that the ties are spaced 4 ft (1.22 m) along the long edges. Thus the tie holes on shear walls are hidden behind the baseboards and there are very few joints to seal, align, and finish. Just the large 4- × 8-ft (1.22- × 2.44-m) tie spacing and high-pressure capacity of 1200 to 1500 lb/ft^2 (57,450 to 71,820 N/m^2) alone found tie-cost advantages and usefulness in spite of the high form cost and weight. One of the main features of these panels is the "girder action" that made pier-cap forming practical without all the high shoring and falsework. These big steel-faced forms have been called "girder forms" because they can be assembled into a boxlike structure for pier-cap construction wherein the panels function as girders to carry the dead weight to the piers without the normal shoring. To carry huge tensile forces across panel joints and to prevent buckling of the compression flanges requires an integrated system of accessories (girder bolts, braces, yokes, soffit angles, cover plates, and support brackets).

Initially gang forms were constructed by using dimensional lumber "joists" supported by double steel channels, commonly known as "wales" and heavy-duty reusable ties. These systems are known as "joist/wale systems" and are very common, especially when the concrete requires a smooth or architectural finish. As these systems developed, the dimensional lumber was replaced with extruded aluminum beams or "joists" that allowed the system to be designed more efficiently. These gangs are often designed and assembled for each project, so a large carpentry crew, as well as an area to assemble, is required.

Another common formwork system available today is the large modular clamp systems. These systems were originally developed in Europe during the early 1990s and often are known as "European clamp systems" (Fig. 16.6). These systems comprise modular panels constructed of heavy steel or aluminum frames that are connected together with a large clamp that both connects the panels and aligns them to keep the plumb and true.

FIGURE 16.6 Large-framed panel gang systems. (*From Harsco Infrastructure.*)

What makes these systems cost-effective is their modular design; all components are rentable with minimal site labor to assemble. This allows the forms to be assembled quickly and ready for use within a short time after delivery.

One of the shortcomings of all these bigger panel systems is fitting the odd dimensions—especially in between corners, pilasters, jogs, and counterforts. Therefore, it is wise to integrate the designs so that the handset system structurally connects to the big panels, thus enabling the big panels to fit all dimensions. Also, commingling the wood-faced panels where a cluster of penetrations or box-outs occurs is a benefit.

Culvert and room forms are gang forms that either roll or hoist to the next pour. Sometimes the slab and walls are poured monolithic, which tends to lock in the forms like elevator core forms. This requires a means to unload the forms and get clearance for movement, These applications are similar to the collapsible core forms (discussed elsewhere in this chapter) using the triple-hinge corner that eliminates slip joints, starter walls, and the cost of invert forms.

TIE SYSTEMS

Changes to the ACI Standard 347-04 require a tie factor of safety of 2.0 on pours that are "over 8 ft (2.44 m) above grade or unusually hazardous." This is intended to reduce the progressive failures that have occurred when only one tie is unlatched or fails.

Very high-strength steel wire and flat straps minimize the cost per tie; however, the tie anchorage and releaser must be quick yet reliable to minimize labor cost. Also, the tie design must provide for easy removal of protruding strubs that would rust, stain, be ugly, or would be safety hazards. Figures 16.7 through 16.10 depict some common ties.

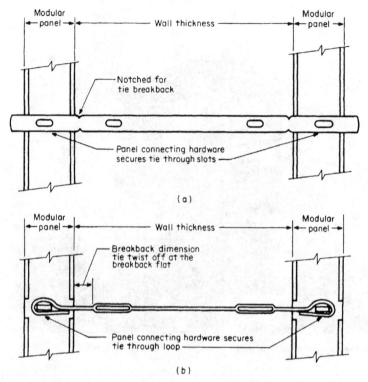

FIGURE 16.7   Two types of handset modular form wall ties: (*a*) flat tie and (*b*) wire panel tie. (*From Symons Corporation.*)

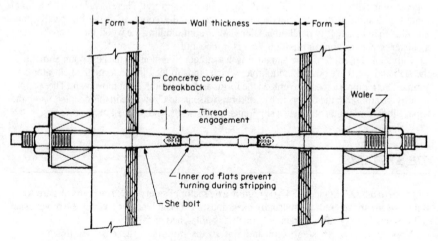

FIGURE 16.8   She-bolt tie. (*From Symons Corporation.*)

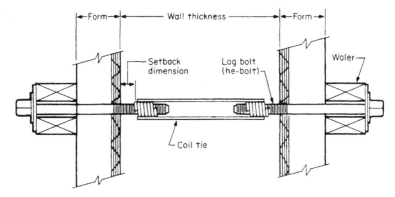

FIGURE 16.9 Coil tie. (*From Symons Corporation.*)

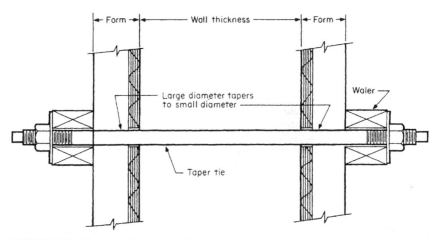

FIGURE 16.10 Taper tie. (*From Symons Corporation.*)

Note that many form and tie systems do not require separate load-gathering wales. The cone ties function as form spreaders and ensure a neat, adequate break-back.

A hingeable plastic cone evolved for field installation on wire ties; however, it does not spread the forms. Its cost was justifiable only in an emergency to avoid crew delays. While it may appear reusable, the labor economics and reuse life was not demonstrated.

Handset and gang-form "snap ties" are readily available for up to 4000-lb (17,800-N) tie loads (see ACI SP-4)[1], which limit tie spacings to about 2 ft (60 cm), or about 4 ft^2(0.37 m^2) of form area on each side. Larger ties have screw-type anchorages to carry the loads yet be readily releasable to strip the forms.

Snap ties, she-bolts, and coil ties leave a steel inner piece that must be recessed within the wall to prevent rust staining of the wall. She-bolts (Fig. 16.8) and coil ties (Fig. 16.9) have threadable inner units that are left in the concrete but can be reengaged for additional usage as anchors for subsequent pours.

She-bolts and taper ties (Fig. 16.10) will pass through both gang forms after setting. Figure 16.8 shows the tapered nose and female thread to engage the recessed inner unit yet allow easy stripping. The reusable she-bolts, taper ties, bolts, and bearing plates are often rentable from suppliers who generally offer assistance in form designs, tie selection, layouts, and logistics. Various sizes and strengths of she-bolts are available; however, the threads from various suppliers need to be checked for interchangeability.

Removable ties provide good long-term material and labor economics but leave holes all the way through the walls or columns. While these holes can be filled, the watertightness is dependent on the filler or membrane materials.

Threaded taper ties are the most common removable type and they come in various diameters, tapers, lengths, and strengths up to 96,000 lb (427,000 N) ultimate strength. A pullable flat tie is successful on thin (8-in) (20-cm) walls if pulled soon enough (about 2 days).

"Slim Jim" is a patented rubber-encased tie-rod that removes when pulled because the outside diameter of the rubber shrinks with tension.

Top-ties are external ties that attach to the forms or vertical walers extending above the forms.

TYPES OF FORMS

Wall-Form Components

Inside corners on wood forms are usually wrecked during stripping owing to the concrete pressure and shrinkage that creates residual loads on the forms even after the concrete is set. With most steel forms, these residual loads and stripping stresses are carried by the steel frames, not the plywood. There are also stripping corners that are specifically designed to relieve these residual loads (Fig.16.11). Many walls have pilasters, but because their dimensions vary widely, the use of adjustable pilaster forms is advantageous. They require minimal building and stripping labor and are of low cost because they are reused repeatedly without damage.

Outside corners are simply steel angles with holes that fit the connection bolt pattern.

Culvert forms, inside and outside bay corners, hinged corners, filler angles, bulkhead forms, stoop forms, brick ledges, step fillers, waler clamps, and haunch brackets are the many rental components that make a system complete enough to be practical. Versatility such as this is essential for today's complex forming to minimize high labor costs.

Waling and Aligning.   Double 2- × 4-in (5- × 10-cm) wales with snap ties passing between the 2-by-4s is the conventional formwork structure that is just as straight as the 2-by-4s available. However, alignment perpendicular to the waler requires additional "strong-backs," braces, and kickers to plumb and straighten the forms. Modular forms also require aligning in both directions; however, the clamping brackets for wood waler attachment to steel forms need to be positive and reliable, yet be quick to minimize labor costs. Steel alignment walers are also available in longer, straighter lengths and are much more durable than lumber.

Scaffolds.   Forms over 8 ft (2.4 m) high require a work scaffold and ladder access. The scaffolds are usually attached directly to the forms; however, if walers or strongbacks are well attached to the forms, they may support the scaffolds. Modular systems have scaffold brackets and attachments engineered for quick yet safe installation.

Lifting.   Lifting of gang forms requires proper lift brackets whose attachment points on the gang forms must withstand inclined loads from the slings during lift-up maneuvers.

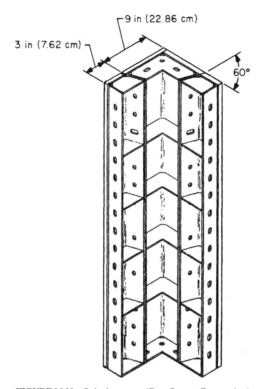

FIGURE 16.11   Stripping corner. (*From Symons Corporation.*)

When the gangs are set down on multilift walls, a wall bracket anchored to the previous pour is required to carry the weight of the outside gang. Since the previous pour is not fully cured concrete and the loads are eccentric, the inserts, anchors, and brackets must be carefully designed and tested.

Column and Pier Forms

Column forms are exposed to severe pressure because, while the amount of concrete required to fill them up is relatively small, the great height of fresh concrete creates large hydrostatic pressures at the base. This gave birth to adjustable steel-column clamps (Fig. 16.12) for wood forms about 75 years ago. Their high strength, reusability, speed of installation and removal, and better accuracy made these clamps the primary forming method for square and rectangular columns, until wood and labor costs became so high that crane-handled steel framed column forms became more economical. At first they were handled as two separate L-shaped assemblies, but hingeable forms with quick latches were developed subsequently, which require less labor and less highly skilled labor power to maintain square corners and dimensional accuracy. There were some 2000-lb/ft² (100-kN/m²) column forms for faster pours of tall columns.

Cylindrical column forms have been popular on bridge piers and parking garages where no wall framing or weather sealing was required. Waxed paper tubes spiral wound are available in many diameters and lengths; however, they have to be kept dry until used and

FIGURE 16.12 Adjustable steel-column form clamp.
(*From Symons Corporation.*)

cause a pronounced spiral-wound joint pattern and discoloration on the concrete. All-steel two-piece column forms are predominant in highway bridge-pier construction because many states have adopted standard diameters and because their long life provides good economics. They also function well with steel wall forms on bull-nose piers.

Fiberglass round forms evolved owing to the costs and stripping and handling problems of heavy steel-column forms on high-rise buildings. These are flexible enough to wrap around the column reinforcement to set and spring open to strip, leaving only one vertical joint to finish. These are now rentable in many diameters up to 36 in (90 cm) and lengths from 6 ft (1.8 m) up to any length, such as 40 ft (12 m), that can be shipped and handled on the job.

Custom-shaped columns, capitals, and column-spandrel beams are often part of the architectural design of buildings. These are available in steel or fiberglass, depending on the design. Generally radius corners or other nonflat shapes are economical in fiberglass if reused more than 10 times.

Underwater pile repair sleeves of fiberglass act as the form during pouring of concrete or epoxy filler, then stay in place to provide a smooth, chemically resistant, erosion-resistant shield.

Custom-automated 22-in (56-cm)-diameter cylindrical column forms designed to be pumped from the bottom to 42 ft (12.8 m) high with heavy piano hinges, ram-actuated latching wedges, and ram opening and closing achieve very high productivity and very low labor costs.

Beam Forms

Beam forms with form ties utilize conventional form design with studs and wales; however, when architectural finish or other factors preclude ties, outside means of support are required to withstand the lateral pressure.

The practice of transferring these lateral loads into the adjacent decking and/or deck joists takes special precaution in design to provide stripping relief of the decking. With flying deck forms, this practice is not recommended unless the beam form is part of the flying table.

The edge spandrel beam forms will still require kicker braces unless the beam form itself has been designed to withstand the lateral pressure, which is feasible with steel but more difficult with wood and fiberglass.

The outside of a spandrel beam receives the most severe brace load because the pressure is applied to the beam side that is also the slab edge form. This lateral load will cause overturning moments on the shoring that must be withstood. Whenever the spandrel beam is supported with a tall shoring tower, analysis of tower stability is critical to provide the following:

1. Wider towers so that the dead-load "righting" moment exceeds the overturning moment.

2. If the uncured columns are strong enough, the lateral load can be transferred with beams, trusses, and/or guy wires to the exterior columns.

3. Diagonal guys down to the previous deck, provided the shoring tower will take the additional load from the vertical component of the guy load.

4. Horizontal guys back to the top of interior concrete columns.

On sloping-deck garages the beams tend to be deep and narrow with large column spacings. Beam side pressure and lateral pressure on the sloping deck can accumulate from bay to bay, and the forms will move "uphill" on the slope unless the lateral loads are carried by kicker braces.

Core Forms

Elevator shaft and stairwell inside forms became known as core forms because concrete is usually poured on all four sides, one floor height at a time. Architects cluster utilities, elevators, and stairwells into a building "core" to enhance the architecture and broaden the views.

The concrete pressure and shrinkage locks in a core form on all four sides and creates severe stripping problems even with gang ties. The box-outs for door openings, inserts for elevator hardware, and conduit penetrations all add to the stripping problem. To avoid the high cost of wrecking and rebuilding wood core forms, stripping corners and inside-hinged corners evolved to achieve reusability. Crane handling an L-shaped gang or a collapsible core form was an improvement but still awkward.

Currently a specialized corner (Fig. 16.13) that is known as a stripping or striking corner allows the forms to be "stripped back" or released to allow the entire core form to be stripped and cycled as one entire unit. These corners function with an internal thread mechanism that allows the unit to slide back on itself, and thus allow the rigid core box to "collapse" inward, thus pulling the formwork away from the wall. A core form consisting of these corners is easy to rig, lift, and set into place for the next lift placement. Once the core formwork is lifted and is set for the next lift, the forms are ready for cleaning, oiling, and for insert and box-out installation without taking the crane time and storage space if the forms were required to be lifted to the ground after each use.

Wall-mounted brackets below the core form support its weight and, owing to the adjustable bracket design, provide the means of fine-grading the core form. Because of its rigid box configuration, this also plumbs the form-face surfaces. A safe falsework platform attached to these same wall brackets allows stripping and collapsing operations without the crane. The crane is needed only for lifting the core form and setting atop the core walls (for cleaning and oiling). Valuable crane time is minimized during both stripping and resetting. Haunch-corner core forms are now available and function in a manner very similar to the more common square-cornered cores.

FIGURE 16.13   Striking or stripping corner. (*From Harsco Infrastructure.*)

Core forms are also required inside of hollow bridge piers, and often have large chamfers, double cells, and/or tapered sides, which requires form panel adjustable widths and movable walers on outside as well as inside forms.

Curved-Wall Forms

Treatment plants often include large circular tanks with high walls. To form curved walls with 2-ft-(60-cm)-wide modular forms, there is only a slight departure from a truly circular surface (Fig. 16.14) and this is much less expensive because most of the equipment can be rented and suppliers have developed refined accessories and techniques.

There are several waling methods to hold the forms in the desired curve. Using plywood templates precut to the radius is the jobsite method, but this requires a lot of labor and special layout skills. Custom-rolled steel channels and angles are available from fabricating

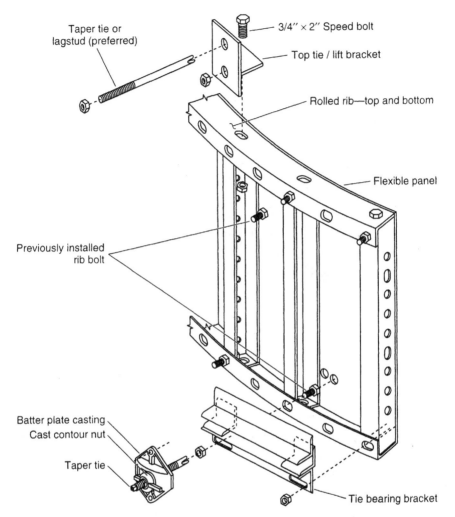

FIGURE 16.14 Flexible curved form. (*From Symons Corporation.*)

shops but get expensive if a lot of different radii are needed. Straight walers can be used with adjustable stand-off clamps. Curved pipe walers have limited bending strength but can be rerolled to a different radius if there is a sufficient quantity, say about 1000 ft (300 m); however, special attachment clamps are required.

Circular tank forming with 4- × 8-ft (1.22- × 2.44-m) modular flat forms and curved horizontal walers has found limited use owing to the high departure from true circular walls.

In addition to the systems in preceding paragraph, there are several other systems that have appeared within the domestic market over the last several years. Some formwork manufacturers have developed modular systems that have integral adjustment "knuckles" that will allow the panel to conform within a set range of diameters without the cost of special rolled angles or shim plates (Fig. 16.15). As with any of the modular handset or ganged systems, these are all rentable systems that allow cost-effective solutions to curved or radial structures.

FIGURE 16.15   Curved wall formed with flat modular forms. (*From Symons Corporation.*)

The third type of systems on the market are short steel walers that connect or have adjustment knuckles to the channels. Timber or aluminum joist and plywood must be attached to their front. These systems are quite commonly used when difficult forming challenges do not allow modular systems to meet the specified wall finish or appearance. Examples would be specific tie patterns or special architectural features.

Blind and One-Side Forms

In excavations when the concrete placement can be made against the sheet pile, slurry wall, or soldier-beam embankment, only the inside forms are required. This is one-sided forming whether the forms are tied with anchors in rock, stud-welded to the steel, or just braced.

Blind forming is a special situation where adjacent buildings or building-line restrictions prevent stripping access to the outside forming whether the forms are tied with anchors in rock, stud-welded to the steel, or just braced. The outside tie anchorage has to be removable from the inside so that the outside gangs can be stripped and lifted straight up. Threaded ties through the wall into captive nuts; such as taper ties, are 45° required.

In situations where ties cannot be used, such as when waterproofing is present between the existing surface at the concrete wall, external bracing methods are required. Many formwork manufactures can supply specialized support or braced frame systems. These systems are constructed from structural steel members into a vertical standing truss and anchored at the base through large cast in situ anchors (Fig. 16.16). These frames are

FIGURE 16.16   One-Sided Bracing Frames (*From Harsco Infrastructure*)

designed to brace formwork for walls up to 30 ft (9 m) in height, but higher walls can be attained by using these frames along with special components.

Sloping Forms

The large 45° haunches common in power-plant turbine pedestals and the girders on stadium bents experience severe lateral forces due to high concrete pressures and the steep angles. It must be recognized that concrete pressure acts perpendicular to the form surface. (This causes the well-known uplift on battered walls that requires tie-downs.) On haunches the vertical component is usually carried by the form ties on haunch forms; however, long-span plate-girderform stadium bents may need anchorage or special support brackets for this lateral (uphill) force.

Bulkheads with Waterstops

While bulkheads are not considered part of the formwork, on treatment-plant long walls and large tanks that use the "skip" pouring sequence to avoid later shrinkage cracking, the labor to form the keys, place and hold water stops and cut around the rebar penetrations, then to wreck it is very expensive.

New adjustable, reusable, laborsaving, steel split keyforms that clamp and hold the waterstop have proved to cut labor by a factor of 4.

Cantilever Forms

On structures that have multiple placements of tieless or one-sided walls, such as mass concrete pours on dams or those that do not allow double-sided formwork, there are specialized
systems that utilize heavy structural members constructed into frames that allow external
bracing of the formwork system. These systems are similar to the support frames used in
one-sided wall applications and are attached to the previous placement or lift utilizing large
anchors and reusable landing cones (Fig. 16.17). Because all the loads are transferred to

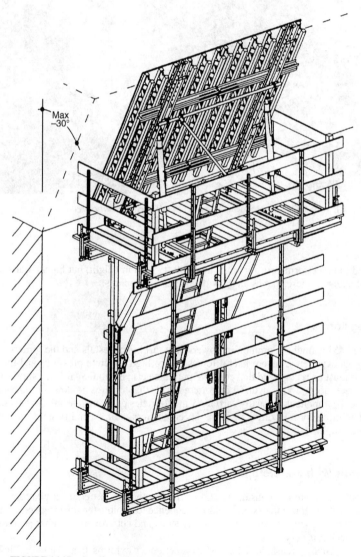

FIGURE 16.17   Cantilever Formwork System (*From Harsco Infrastructure*)

a single anchor point, these systems are often limited to a maximum pour height of 12 ft (3.6 m) and have limitations about the rate of placement of fresh concrete.

These systems are highly specialized and require highly detailed analysis and calculations to determine the frame spacing that will safely support not only the concrete load against the frames but also any live load from workers that use the working platforms during the cleaning, reinforcement setting, and placing of the concrete.

Architectural-Concrete Forms

Exposed architectural concrete has become popular with architects because it is a more creative material and less costly than brick and marble facings; however, it requires a wide range of special formwork details. Overlay plywood, tight form joists, well-sealed ties, gasketed joints on steel forms, foam-rubber-sealed laps to previous pours, and rustication strips are some of the steps necessary to minimize or mask unsightly form joints, tie holes, and construction joints. Overly deflected forms become readily apparent on flat architectural concrete, so special attention to form pressure and stiffness is required.

Form liners of various materials impart fins, grooves, brick, rock, and other textures to the concrete; hence the indigenous flaws in concrete (bugholes, discolorations, form bulges, and joints) are not so blatantly apparent on textured as on flat surfaces.

Fiberglass column and beam forms give very attractive architectural concrete; however, adequate aligning and bracing is required. Fiberglass column connectors and capitals, domes, and pans are also attractive owing to radius corners, seamless surfaces, and uniform concrete color.

Attractive concrete also requires carefully pretested mix design, proper selection of release agent, and thorough vibration and revibration to minimize honeycomb, bugholes, pour lines, and discolorations. A test wall and/or tests in unexposed areas (basement or garage walls) is the most reliable method of developing all the countertopposing factors into a harmonious construction plans. The test pour is also a crew training ground and visual aid to the architect.

Underwater Forming

Without caisson sinking and dewatering, the repair of underwater piles has become practical with flexible fiberglass sleeves. They snap around the old pile and interlock so they can be filled with epoxy or concrete and stay in place as barriers against erosion, borers, and chemical attack.

The sleeves require circumferential bands spaced along the sleeve to withstand the concrete pressure. Steel straps, "load-binder" belts, or fiberglass bands are only temporarily required during concrete placement and curing. The extent of the banding depends on the diameter, pressure, and type of interlock. New interlocks have been developed with a hook-type engagement that will carry much of the hoop tension created by the concrete pressure. This interlock reduces the number of bands required threefold, which thus reduces the diver labor to install and remove the bands. Pumping into the bottom of any form creates pressures much higher than the liquid head, which must be considered in the strength and spacing of the bands. Furthermore, these hook interlocks prevent future joint loosening due to wave action, temperature cycles, and mechanical impact.

Reusable fiberglass heavy-duty column forms have also been used underwater to form concrete encasement for deteriorated piles. Techniques have been developed to embed the permanent sleeves in the splash zone where they are most needed and, in the same pour, to form the concrete encasement below the sleeve to the mud.

Battered Wall Forms

If one or both wall forms have a slope to construct tapered walls, the inside concrete pressure has an uplift component. Tie-down anchors or dead-weight ballast are required to prevent the tendency of "floating" of the forms. Special spherical castings and nuts take tie angles, but the various tie lengths need preplanning.

Custom Forms

Custom steel tunnel forms and travelers, self-rising automated dam-face cantilever forms, integrated column-spandrel forms, hammerhead pier-cap and stem forms, and special column or wall forms, as well as custom fiberglass shapes too numerous to list are only a few customs that have been manufactured whenever the shape or the reuse potential justifies the cost. The dimensions, shapes, design pressure, stripping motions, traveling maneuvers, etc., are customized through normal proposal, drawing, and shop-drawing procedures.

FORMS FOR PRESTRESSED CONCRETE

Owing to the high-strength steel and concrete employed, the degree of precision in this formwork is higher in order to achieve correct dimensions of the structural member and accurate positioning of the tendons. This is critical because unbalanced cross sections or mislocated tendons induce eccentricities that are magnified by the high tendon stresses that may result in weakened or even unsafe structures. This accuracy is also required to meet the higher fit-up tolerances during assembly, especially where precambered and prestressed members fit side by side to build a level floor or roof or a load-bearing wall.

Precasting plants normally use permanent forms or casting beds with an anchorage abutment for a pretensioning type of manufacturing, which is beyond the scope of this handbook. However, there are limits to the size or weight that can be handled, hauled, or lifted into place, which has led to on-site and cast-in-place methods of posttensioning.

Circular poststressed tanks, silos, and chimneys employ very specialized patented methods and are outside the scope of this handbook.

Cast-in-place posttensioned members require shoring during pouring and curing until the tendons are pulled; however, this avoids the severe tolerance stackup, lifting requirements, and welding fit-up of on-site precast. On the other hand, when cast-in-place tendons are pulled, the dead load shifts away from the midspan shoring to the girder shoring. If the girder tendons can be partly stressed, then additional shoring may be avoided.

The formwork considerations for posttensioned members are very similar to those for conventional concrete, except for the following factors:

1. On-site precast posttensioned members require tighter tolerances so that the parts properly fit together when assembled and without excessive accumulations of multipart tolerances.

2. Form materials must be tougher and more durable to achieve the necessary reuses without creeping tolerances that cause fit-up problems later, which are almost impossible to correct.

3. Architectural surfaces require stiffer forms, tighter joints, smoother finishes, and more care to implement chamfers, rustications, and deep textures in such a manner as to prevent spalling during stripping and handling.

4. For some members such as long tees or piling, concrete shrinkage, foreshortening, or deflections during tendon stressing may induce additional loads on the formwork to be considered during design. This is in addition to the camber requirements.

5. Steam or other elevated-temperature cure methods may deteriorate some materials or induce differential expansion distortions to be considered. For example, unsealed wood rustication strips may swell and spall the edges unless kerfed and sealed. Void forms may swell or undergo thermal expansion to cause difficult stripping or damage to the concrete finish.

6. Interaction between chemicals, such as retarders, with plastic chamfers, formliners, spaces, and styrofoam void forms may be aggravated by higher-temperature cure methods.

7. Tendon drape, tie-downs, and/or flotation forces from embedded void forms may need to be considered in the form design.

8. Rebar or tendon congestion may prevent internal vibration, plus external vibration may be more productive, which induce severe localized forces and formwork resonances that require tough rigid-form materials and designs.

FORM-RELEASE AGENTS

A proper release agent not only makes stripping easier, but it prolongs the life of the plywood and reduces cleanup labor by 50 percent. Some agents chemically react with the concrete, leaving a soapy residue that may be objectionable because of discoloration, dusting, or preventing the bonding of curing compounds, paint, or paneling. All release agents are not alike; so pH balance, uniformity, and active ingredients should be checked before use.

A combination of a reactive release agent followed by a paraffin curing compound on round bridge piers has caused slippage of the friction collars which otherwise supported the pier cap pour.

Since most release agents have an oil base, they do pollute the air and ground, endangering aquifers and rivers, so EPA has clamped down some. While water-base release agents are available, freezing is a storage problem and stripping effort is noticeably more, so they are not too popular.

A form release made from seaweed has been used in China to save oil. Its processing gives a water-soluble white powder that is freezeproof owing to jobsite mixing but is not yet available in the West.

FORM REMOVAL

Formwork for columns, walls, and piers is generally removed in 12 to 48 h after placing of the concrete, depending on construction and wind loads, maturity for anchor bolts, concrete mix design, etc. Where textured architectural concrete is involved, and in elevated slabs and pier caps, where gravity loads are significant, a longer (3 to 7 days) cure of the concrete is required before form removal.

The following recommendations for form-stripping time are reproduced from ACI 347-04:

> When field operations are not controlled by the specifications, under ordinary conditions formwork and supports should remain in place for not less than the following periods of time....If high-early-strength concrete is used, these periods may be reduced as approved by the engineer/architect.

Conversely, if ambient temperatures remain below 50°F (10°C), or if retarding agents are used, then these periods should be increased at the discretion of the engineer/architect.

Walls	12 hours
Columns	12 hours
Sides of beams and girders	12 hours
Pan joist forms	
30 in (75 cm) wide or less	3 days
Over 30 in (75 cm) wide	4 days

	Where design live load is:	
	less than dead load	greater than dead load
Arch centers	14 days	7 days
Joist, beam, or girder soffits		
Under 10 ft (3 m) clear span between structural supports	7 days	4 days
10 to 20 ft (3 to 6 m) clear span between structural supports	14 days	7 days
Over 20 ft (6 m) clear span between structural supports	21 days	14 days
One-way floor slabs		
Under 10 ft (3 m) clear span between structural supports	4 days	3 days
10 to 20 ft (3 to 6 m) between structural supports	7 days	4 days
Over 20 Ft (6 m) clear span between structural supports	10 days	7 days.

TOLERANCES

ACI 117-06[6] covers concrete tolerances of all kinds and ACI 347-04 covers formwork aspects that affect deviations from print dimensions, cross sections, and grades, such as column alignment, plumbness or slab flatness, or wall straightness. Furthermore, ACI 347-04 establishes four classes of permitted irregularities in formed surfaces as checked with a 5-ft (1.5-m) template that is a straight edge, radius edge, radius arc, or other shape as specified:

Irregularity	Class A	Class B	Class C	Class D
Gradual	1/8 in (3 mm)	¼ in (6 mm)	½ in (12 mm)	1 in (25 mm)
Abrupt	1/8 in (3 mm)	¼ in (6 mm)	¼ in (6 mm)	1 in (25 mm)

Class A is a surface exposed to public view where appearance is of special importance.

Class B is intended for coarse-textured concrete intended to receive plaster.

Class C is a general standard for permanently exposed surface where other finishes are not specified.

Class D is a minimum quality for a surface where roughness is not objectionable.

The ACI classes provide a basis of quantifying tolerances for surface variations due to forming quality. Form offsets and fins are "abrupt," whereas warping, unplaneness, and deflection are "gradual" flaws.

DRAWINGS AND SPECIFICATIONS

Architect-engineer (A-E) structural drawings normally do not cover formwork or accessories, but they detail the end result required, such as the shapes and dimensions plus tie-cone size and spacing, rustication strips at construction joints, and the other specifics of the project. Their specifications also cover the concrete strength needed prior to stripping forms and decentering shores, the special tolerances and finishes.

Where architectural textures, treatments, or finishes are specified, a jobsite preconstruction mock-up should be constructed large enough to illustrate procedures and materials such as sealing form joints and ties, inside corners and core-form-stripping clearances, vibration technique, allowed tie-hole and honeycomb patching, sandblast or bush-hammer treatments, and release agents and curing compounds. This mock-up must be specified and detailed by the A-E so that it is included in all the bids and preconstruction preparations so that a full-scale acceptance standard is established.

Formwork drawings are prepared by contractors along with their suppliers to instruct builders and erectors so as to optimize equipment, reuse, labor power, materials, pour sequence, and other jobsite conditions. Such drawings show formwork elevation, plan, section views, tie size and location, formwork pressure limits, tie loads, bracing, work platforms, falsework, supports, pour sequence if required, as well as materials, dimensions, connections, and standard details. These are generally submitted to the A-E for review and approval prior to construction.

Formwork-drawing details for fabrication include member sizes and spacings as well as overall form dimensions, tie type and locations, wale size and spacings as well as overall form dimensions, tie type and locations, wale size and location; on elevated forms, the means of formwork support, work platforms, and wind-load bracing or guying; and other information to instruct field personnel how to erect, pour, strip, and cycle forms safely and efficiently.

For some projects, other information that may be valuable is:

1. Concrete pour sequence and timing of subsequent lifts, if required by the formwork design
2. Preattached inserts, anchors, block-outs, and rustications
3. Vibrator access and inspection ports
4. Bulkhead details for keyways and waterstops
5. Pour pockets and clean-out access
6. Formwork camber, if required
7. Special bracing for one-sided (pit) forms, including anchors and mudsills
8. Stripping relief between pilasters and inside core forms, such as crush plates or wrecking bars
9. Form coatings and when applied
10. Sequence of tie or form removal for worker safety as well as concrete strength

Standard panels or details are shown schematically or included by reference. Where field job building, such as around pipe penetrations and where large door openings are necessitated by nonrepetitive or field fit conditions, these areas should be so indicated.

Formwork drawings should be rechecked against structural drawings to verify the following:

1. Critical dimensions
2. All construction joint locations
3. Special tolerance on camber and deflection

4. Architectural concrete details and textures

5. Inserts for other trades

6. Precast and posttension coordination

SAFETY

The Scaffolding, Shoring, and Forming Institute publishes both field erection safety brochures and an 80-slide program of safety "do's and don'ts" that should be used in the planning stages and the crew training. These publications embody the latest ANSI and OSHA provisions plus practical safety pointers.

Design safety factors on form ties, anchors, hangers, and inserts are recommended by ACI 347-04, however, they depend on the type of construction (see Table 16.1). The responsibility for safety is the contractors,' because they control the crews, the material quality, the schedule, and all of the factors that determine tie and anchorage safety.

Design safety on formwork panels and walers is not specifically covered by OSHA as yet other than by reference to ANSI.

Allowable stresses of the materials are those used in standard structural design unless actual test data support higher values for a product design. Lumber and plywood stresses for short-term loading are permitted to be introduced owing to the brief duration of concrete pressure loading.

INSPECTION

It is essential that formwork be erected as designed so that formwork members are not overloaded inadvertently. If qualified supervisors are not available, the form designer should inspect the erection.

No field changes in the design should be permitted without approval of the form designer.

Inspection must ensure that all the form ties have been properly installed and latched, that threads are fully engaged and uniformly tightened, uplift anchors are provided, bracing is sound, connections are tight, and work platforms, guard rails, access ladders, and safety belts are provided. ACI suggests a three-stage inspection:

1. Preliminary inspection: after forms are built, but prior to oiling or rebar placement

2. Semifinal: just prior to final cleanup

3. Final inspection: immediately before concreting: check forms, spreaders, inserts, and fixtures for dislocation, and whether surfaces are clean, oiled, and if specified, wetted

FORMING ECONOMICS

Since formwork costs are the major portion of concrete costs that are controllable by the contractor (Fig. 16.18), success depends on thorough, objective analysis of the prevailing project and jobsite conditions and the resulting form design and selection.

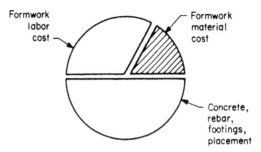

Formwork labor cost

Formwork material cost

Concrete, rebar, footings, placement

FIGURE 16.18 Concrete formwork labor and material costs. (*From Symons Corporation.*)

The Key to Material Cost: Reuse Life.   Job-built wood forms have a low first cost, but they can be reused at best only about 10 times. The labor cost to repair and reface job-built wood forms is so high that prefabricated steel modular forms have proved to be more economical because they last 10 to 20 times longer, repaying the higher initial cost many times.[7]

Steel-faced forms have a proven high-reuse life, if the face thickness is at least 10 gauge [1/8 in (3 mm)], preferably 3/16 in (5 mm), except for small items such as inside corners. However, nailable form faces save labor required for attaching inserts, box-outs, doors, windows, and ducts, and for penetrating the face with conduits, pipes, and other items. Therefore, specially developed coated plywood form faces, positively attached to steel frames, yet replaceable when needed and having the practicality of wood. Modular formwork systems, where the steel frame protects the edge of plywood from the inevitable wear and tear during stripping, jobsite handling, and shipping, have proven records of 100 to 200 reuses before plywood replacement. Since these steel frames last over 10 years, they have investment tax credit and depreciation tax incentives. These systems are so durable that even the plywood and all the connection hardware (except snap ties) are rentable.

Since these systems can be rented, they may be charged to the job under way; and after follow-up jobs are secured, the rental-purchase option can be exercised to convert the rental agreement to purchases. This reduces the up-front capital requirements.

Productivity: The Key to Labor Costs.   The rising cost of labor necessitates formwork designed to minimize crew size, delays, and wasted motion. The labor cost comes from actual experience with particular form systems and will vary by job type.

To determine true labor costs, time-and-motion studies have been conducted of actual jobs with time-lapse cameras that recorded every event on the job. From this film the productive labor power to accomplish each phase of the full formwork cycle was measured along with the square footage of formwork.

Formwork operations that were measured included five phases (including penetrations, insert or box-out attachment, bulkheads, and pilaster): (1) setting first side of formwork; (2) placing ties; (3) setting second side of formwork; (4) waling, aligning, and bracing; and (5) stripping, cleaning, and moving.

The cameras also identified nonproductive delays caused by material shortage, crane scheduling, improper sequence, and other inefficiencies that would have otherwise been misinterpreted as: "productive effort." Some of the delays were unavoidable results of weather or of equipment breakdown. But 60 percent of the delays were avoidable with better planning, training, standardization, and supervision. For example, it was found that 30 percent of the total effort was in plumbing, aligning, waling, and bracing, because scrap-lumber braces had been planned. Looking for such scrap and cutting and trying to make it work was twice as slow as using steel adjustable turnbuckles anchored to steel stakes. Since such items are rentable, why waste labor?

Two crews with the same contractor had 100 percent different productivity because one supervisor planned ahead better and had developed more effective techniques in eliminating delays.

Studying different types of jobs established practical (optimum could be 35 percent higher) productivity standards in square feet of contact area per labor-hour as follows:

On multilift industrial jobs, 37 ft²/labor-hour (3.5 m²/labor-hour)

On residential basement jobs, 72 ft²/labor-hour (6.7 m²/labor-hour)

On handset columns, 53 ft²/labor-hour (4.9 m²/labor-hour)

On gang-forming with steel walers, 37 ft²/labor-hour (3.5 m²/labor-hour)

On gang-forming with wood waters, 55 ft²/labor-hour (5.1 m²/labor-hour)

On handset curved walls, 34 ft²/labor-hour (3.2 m²/labor-hour)

Comparing this productivity with three well-known construction-estimating guides on job-built forming showed that modular-forms productivity was 67 to 190 percent higher than job-built. This translates into 30 to 50 cents/ft² ($3 to $5.50/m²) savings on each pour depending on carpenter rates and the type of forming.[7]

Rental + Productivity = Small-Job Success.   Combining formwork material and labor costs indicates that above 15 reuses, it is more economical to purchase modular forms than to job-build forms. For small jobs, rental of modular forms at about 50 cents/ft² ($5.50/m²) per month is more economical than $1.10/ft² ($11.80/m²) to job-build, especially if there are four reuses in that month [$12½ cents/ft² ($1.35/m²) per use]. Rental payments on modular forms can be applied to the purchase price as follows: After the first month's rent, apply 100 percent of rental to purchase; after the second month's rent, apply 90 percent; after the third, 80 percent; after the fourth, 70 percent; after the fifth, 60 percent, depending on the supplier.

Owing to re-plying, cleaning, and refurbishing costs, there are advantages to rental other than capital conservation; however, in general, for a job with 40 or more reuses, purchase of modular forms is more economical than rental (neglecting resale value).

Plywood and lumber prices have historically been more volatile than modular forms. The assurance of ready availability of modular forms combines with such price stability and with the above productivity standards to minimize the risk in bidding and schedule commitments.

Reuse the Ties to Control Tie Costs.   Bigger and stronger ties cost more per unit but less per square foot, are more reusable (thus rentable) and carry proportionately more area of forms, and they allow faster pours, which saves placement labor and crane time and often avoids overtime finish labor. The use of big ties is particularly advantageous on large gang forms because cranes are plentiful and are needed anyway to handle rebar bundles, concrete buckets, and other heavy items.

Gang-form snap ties spaced 2 ft (60 cm) on centers support 4 ft² (0.37 m²) on each side; thus they cost 4 cents/ft² (43 cents/m²) per use because the tie is not recovered; steel walers at 4 ft (1.22 m) on centers with taper ties at 4 ft (1.22 m) on centers cost more initially, but they last indefinitely (zero cost per use) and are rentable. A typical rental rate for such a waler and taper tie is about 30 cents/ft^2 per month. If the job allows the cycling of the forms four times per month, the cost per use is 7½ cents/ft^2 (81 cents/m^2). However, this taper tie leaves a 1¼ in (32-mm)-diameter hole all the way through the wall, which may be objectionable on some jobs such as treatment plants. In this case, she-bolts can be used that leave a ¾ in (190-mm)-diameter inside rod but no through-hole. Their rental cost is 6½ cents/ft^2 (70 cents/m^2) per use for the recoverable parts, plus 2 cents/ft^2 (22 cents/m^2) per use for the inside rod for an 8-in (20-cm) wall.

Furthermore, fewer ties to place, attach, and strip, and fewer tie holes to patch save labor cost. Assuming tie labor cost at $10 per tie, this amounts to $1.25/ft^2 ($13.50/m^2) for snap ties, only 32 cents/ft^2 ($3.50/m^2) for taper ties, and 50 cents/ft^2 ($5.50/m^2) for she-bolts. This comparison is oversimplified, since faster pour rates and gang-form movement offer additional labor savings with the big pass-through type of ties. The ultimate goal is faster cycle time (and maximum reuses), which cuts cost of rental equipment, overhead, and interim financing.

BRIDGE FORMWORK

Plate girder forms are large all-steel modular panels equipped with high-strength girder bolts and bolt bearing pads to interconnect the panels so that the panel joints are bridged to form long-span pier caps without shoring to the ground. The form face carries shear stress just like the web of a plate girder. These forms are called plate girder forms and are very useful on pier caps, especially hammerhead caps.

Box girder segmental bridges are typically precast segments lifted into place, tendons stressed sequentially to assemble a bridge span from the piers out to the midspan, then closure pours are cast-in-place to complete the span. Closures, diaphragms, guardrails, etc., require formwork. Some box girders are cast in place, either segmentally or totally, and then posttensioned in the air—all with custom forms.

Steel and concrete girders have an "I-shape" which requires cast-in-place diaphragms for stability and the slab overhangs and soffits between girders must be formed. Normally, these forms are job-built with lumber and plywood, but long bridges use rentable systems that form the bridge overhangs and the slab between girders, as shown in Fig. 16.19.

Either way the pier stems and caps are the domain of steel plate-girder form panels and/or round column forms, which are rentable. However, the elevated expressways in crowded cities typically have multilevels for ramps and cloverleafs requiring hammer-head pier caps. Tall tapered hollow pier stems are used on river bridges (especially navigable rivers) that must have experienced formwork engineers, rentable girder forms, and/or custom forms.

Bridge falsework induces such huge loads into the green concrete in the pier stems that the development of anchors, embedments, and bearings require extensive testing and analysis to avoid overstressing the concrete below. This becomes even more critical if the pier stem is hollow.

FORM LIFTING SYSTEMS

Form lifters are mechanical, hydraulic, electromechanical (or combinations thereof) power systems dedicated to lifting formwork, deckforms, and/or column and spandrel forms to the next pour—even though the structural support may be from the walls or columns below.

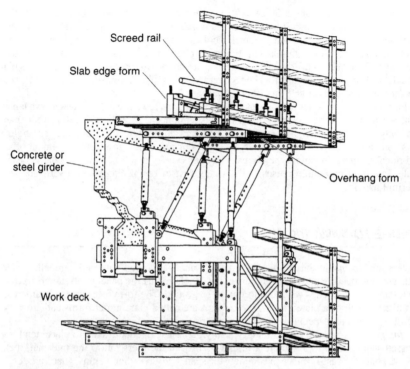

Screed rail

Slab edge form

Concrete or
steel girder

Overhang form

Work deck

FIGURE 16.19   Bridge deck overhang form. (*From Symons Corporation.*)

Historically the next pour is the next floor or lift above, as shown in Fig. 16.20; however, with the new top-down forming method the next pour may be the floor below.

SLIPFORMING

Slipforming or "sliding forms" differ radically from a "fixed" formwork system in that a fixed form is basically a mold into which the freshly placed concrete is cast and allowed to harden before the form is removed or "stripped," whereas the slipform is in reality a die through which the concrete is extruded. This means that the concrete emerging from the die must be at the right state of the setting process as it passes through the form. Since the "die" is often fairly extensive in area, it is obvious that a reasonably predictable uniform rate of concrete set is necessity to the operation. Herein lies the challenge. Fortunately, this requirement has sufficient tolerance to allow slipforming to be executed under conditions that are not normal in the construction industry.

Another basic difference between fixed and sliding forms lies in the integrity of the sliding forms. Fixed forms are made up of modular panels or other structural elements that work together to make a complete system. The slipform, on the other hand, is a single "structure." Whereas the fixed form will remain stationary during concrete placement, the slipform must be constantly moving during concrete placement. This means that the slipform must be designed to travel and act as one single unit.

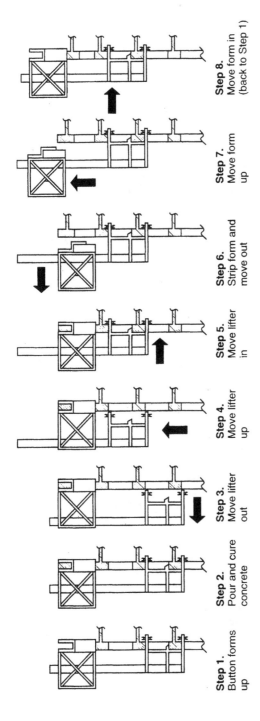

| **Step 1.** | **Step 2.** | **Step 3.** | **Step 4.** | **Step 5.** | **Step 6.** | **Step 7.** | **Step 8.** |
| Button forms up | Pour and cure concrete | Move lifter out | Move lifter up | Move lifter in | Strip form and move out | Move form up | Move form in (back to Step 1) |

Typical cycle for post-lifter form system involves two basic operations. First, forms which encase cured concrete are used to raise the lifter. Then the lifter is reattached to the building, and after forms are released it raises them to the next floor. Concrete is indicated by crosshatched lines.

FIGURE 16.20   Self-raising forms for high-rise buildings.   (*From Symons Corporation.*)

16.33

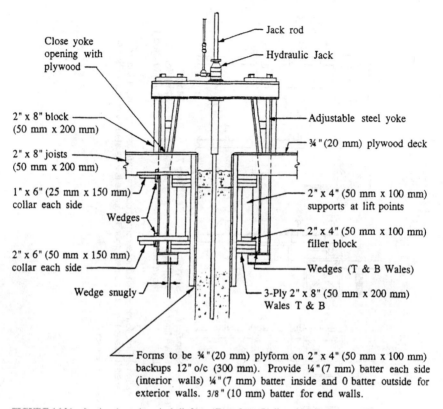

Close yoke
opening with
plywood

Jack rod

Hydraulic Jack

2" x 8" block
(50 mm x 200 mm)

Adjustable steel yoke

2" x 8" joists
(50 mm x 200 mm)

¾ " (20 mm) plywood deck

1" x 6" (25 mm x 150 mm)
collar each side

2" x 4" (50 mm x 100 mm)
supports at lift points

Wedges

2" x 4" (50 mm x 100 mm)
filler block

2" x 6" (50 mm x 150 mm)
collar each side

Wedges (T & B Wales)

Wedge snugly

3-Ply 2" x 8" (50 mm x 200 mm)
Wales T & B

Forms to be ¾ " (20 mm) plyform on 2" x 4" (50 mm x 100 mm)
backups 12" o/c (300 mm). Provide ¼ " (7 mm) batter each side
(interior walls) ¼ " (7 mm) batter inside and 0 batter outside for
exterior walls. 3/8 " (10 mm) batter for end walls.

FIGURE 16.21   Section through typical slipform (*From Leon Bialkowski P.E., Tuscon, AZ*)

Another very important function (and necessity) of the slipform is that it provides support for the necessary working platforms, storage facilities, and vertical transport equipment. Actually, when one designs slipforms, the design procedure has little in common with conventional wall form design. A sophisticated structure has to be designed that is capable of mobility and flexibility and yet must be able to carry without undue distortion vertical, lateral, and overturning loading. A section through a typical slipform is shown in Fig. 16.21.

Some common structures that utilize slipforming techniques are silos, elevator shafts or cores, offshore platforms, caissons, and nuclear containment vessels.

NEW CONSTRUCTION METHODS AND FORMWORK DEVELOPMENT

Garage Building System.   Parking garages with posttensioned beams and slabs have become so standard that a forming and shoring system evolved that formed the beams (supported by steel shoring towers having built-in horizontal and vertical adjusting screws) along with supporting steel beams that shored the wood I-beam deck soffit form. The deck panels and folded-leg shoring towers can on most garages roll up the ramps on dollies tugged by the forklifts that raise and lower the units. This is so fast and economical in labor and crane costs that the supplier rents the system. In some cases the savings pay for redesign of the garage structure to fit the garage forming system.

Room Tunnel System. In housing structures with hundreds of rooms the same size recurring systematically, such as condominiums in Florida, large motels and hotels and prisons, where the walls and slabs are poured simultaneously and then the forms can be recycled every day by heat curing the concrete all night, not only speeds up the construction time but also cuts costs.

Mole Method. This method employs piers placed underground down to bedrock, and the ground surface slab is poured on grade, after which the soil is excavated to the next lower level while above-ground construction proceeds upward. In the mole method the hanging formwork system is supported and lowered from above—either from the deck or from the columns.

While the recession stopped new product development in many areas, underground construction in major metropolitan areas where the population needs kept the urgency strong the mole method was refined so it would work below the water table or along waterfronts. This required the development of a new formwork hanging system (called "top-down") wherever unstable soils or water intrusion would not allow pouring each floor slab on grade. Whether the deckforms are hung from the preplaced piers or the slab above is a matter of design of the beam forms, capital forms, and deck forms, as well as the anchors, hanging devices, and recycling equipment. Very close coordination with the structural designer and contractor is essential because the upper slabs also become horizontal lateral bracing for the perimeter slurry walls, and because of the limited access, illumination, clearances, etc., as required for each site.

Automation

In the future there will be more automation in formwork, especially high-rise buildings and parking structures. Form lifters for shear-wall, core forms, and spandrel beams reduce labor and expensive crane time and enhance contract completion. High-strength steels and engineered wood laminates give higher-performance forms to meet the demand for faster pour rates and admixtures that result in higher concrete pressures. Bigger and stronger ties will support higher-pressure forms and/or wider tie spacings.

Composite Columns

Composite columns consist of steel pipe filled with reinforcing steel, studs, and concrete, which is a new ductile framing method that is fast and economical yet very seismic-resistant and wind-load resistant for high-rise buildings and parking structures. In fact they have already been used on the west coast. Such columns then engender connections such that deck forms can span from column to column for the so-called top-down hanging form method or for lifting the slabs for the roof first then the top floor, etc., so that the roofing, walls, windows, elevator equipment, mechanical/electrical/plumbing, and finish work is out of the weather and so completely unencumbered that work progress speeds upward and downward at the same time for early completion of ground floors, top floors, and garage sublevels.

Using such composite columns on high-rise buildings, the roof is poured first with deck forms supported on these composite columns. Then the deck forms are lowered from the slab pour cast previously right above or by hanging from the columns.

Composite columns and rentable deck forms provide the means, once contractors learn how to stage materials, preassemble components, and train their crews, to expedite high-rise completion by adding additional set(s) of deck form(s) and extra crews to form and pour multiple decks at the same time. Since the composite columns were pre-erected and cured, the deck progress could sequence from the bottom and middle levels upward using shores while at the same time progressing downward from both the top and middle levels using hanging deck forms hanging from the columns. All of this aboveground construction would not prevent underground "mole" construction of excavation beneath the ground floor slab to progress downward for the parking levels below ground. Furthermore, if the water table or unstable soils prevents shoring below ground, then hanging the deck forms from the predrilled piles will facilitate the progress of the parking levels. Such multiple levels of deck forms and multiple crews may only be justified when the economics of shorter interim financing periods and earlier usage of the building will pay off the added costs.

REFERENCE

1. Hurd, M. K.: *Formwork for Concrete*, 7th ed., ACI Special Publication No. 4, Farmington Hills, Michigan, 2005.

2. *Recommended Practice for Concrete*, ACI Standard 347-04, American Concrete institute, Detroit, Michigan, 2004.

3. *Concrete Construction and Masonry Work*, ANSI AI0.9, American National Standards Institute, New York, 1988.

4. *Forms and Shoring*, OSHA 1926, 701, U.S. Department of Labor, Occupational Safety and Health Administration, Washington, D.C., 2003.

5. *Plywood for Concrete Forming*, Brochure V345, American Plywood Association, Tacoma, Washington, 1998.

6. *Standard Tolerances for Concrete Construction and Materials*, ACT 117-06, American Concrete Institute, Detroit, Michigan, 2006.

7. *Steel-Ply System vs. Job-Built Forming Productivity and COSI*, Publication 77-I7, Symons Corporation, Des Plaines, Illinois, 1977.

CHAPTER 17
BRACING AND GUYING FOR STABILITY

James Scheld, P.E.
Lawrence K. Shapiro, P.E.

INTRODUCTION—THE ROLE OF BRACING AND GUYING IN TEMPORARY STRUCTURES

As a young field engineer, one of the authors of this chapter provided support for a crew erecting precast concrete framing for a cooling tower at a nuclear power plant. Upon erecting some slender columns and leaving them freestanding on the anchor bolts, he wondered to himself whether they should be left over the weekend without temporary guys. After some quick mental calculations, he decided that the columns should stand up in any wind that might arise over the next couple of days. That very night there was an earthquake severe enough that people in town ran from buildings out into the street. He immediately thought about those columns. In spite of his fears, they stood.

Broadly considered, temporary structures require bracing or guying for the same reasons as permanent structures, which is to impart stability and stiffness and to square up the structure. For a temporary structure that is homologous to a permanent structure except in duration of use, the demand is similar though the means perhaps differ in view of its ephemeral nature.

However, there are other considerations for a structure that is in a state of construction awaiting completion or curing, and the same goes for a weakened structure that has suffered damage or is undergoing demolition. An unbraced structure may be inherently unstable while it is in a temporary state. A temporary structure that is not plumb, that is degraded from damage, or that has imperfect ground support may need bracing or guying to compensate for these conditions. In some instances, guys or braces may be used to plumb up a structure that is leaning or to firm up a structure so that it remains square and plumb. These demands may be additional to potential lateral loads from wind, earthquakes, or equipment.

In the instance of those earthquake-shaken columns, the designers had seen fit to ensure that they could stand free with some lateral load.* With both a low probability of seismically induced toppling and a small risk to persons and property if it would occur, it is doubtful that an earthquake during erection was one of their concerns. More likely, they just figured it was good practice to ensure that the columns could stand on their own.

A program of temporary bracing should be built upon knowledge of anticipated loads and of the condition of the structure. In instances where potential loads are contingent or occasional, the probability of a potential loading should be weighed against the risk of ignoring it. In new construction, loads are more readily determined and the condition of the structure can be planned. When knowledge is incomplete, as it may be when the structure is old and perhaps deteriorated, the engineer may be compelled to factor commensurate caution into the bracing or guying design.

Differences between Bracing, Guying, and Shoring

The terms *shoring* and *bracing* are sometimes used interchangeably, but there is a functional difference. Shores are temporary compression elements that carry primary forces. In practice, these forces are usually gravity-induced, but they could arise from other sources such as earth pressure. Shores might be used to support primary loads during construction or while the structure is in a damaged or weakened state. Sometimes shores might also be used to achieve a temporary increase in the load-carrying capacity of a permanent structure, for instance, to support an overweight vehicle.

---

*This event occurred before OSHA mandated that columns must be placed on a minimum of four anchor bolts with an allowance for lateral load.

Whether they are in temporary or permanent construction, braces and guys support secondary loads. These loads are secondary in the sense that they are incidental to the primary load path, reacting against any tendency of the primary member to displace out-of-axis or out-of-plane. Relatively small in magnitude, the brace loads may be steady, occasional, or contingent upon outside events. Whereas guys are always in pure tension, a brace could act in tension or compression, or even in bending or torsion. In most applications, shores are vertical while braces and guys are lateral or diagonal. Bracing structures can also be quite elaborate as shown in Fig. 17.1.

FIGURE 17.1 Temporary steel braces preserving the exterior walls of a landmark building. The braces, preserving both the historic façade and party walls, were installed prior to the demolition of the floors. (*Photo and installation design by Benjamin Pattou.*)

Distinction between Temporary and Permanent Installations

Temporary works are distinct from permanent construction in several ways. A permanent installation may be intended to stay in place for 50 years or more, whereas the duration for temporary construction may be mere hours up to a few years. Permanent installations are exposed to the uncontrolled vagaries of nature and to random human activity, while braces and guys for construction are often in a more defined and managed environment.

The laws of statistics dictate that the opportunity for an exposed structure to be buffeted by extreme winds increases with the duration of its exposure. It stands to reason that a structure that will stand for a short duration can be designed for a lesser wind while maintaining the same probability of exceedance as a permanent structure. The ASCE 37 standard for design loads during construction incorporates this concept in its method for calculating wind loads.

A temporary installation is usually monitored and managed actively. The chance of exposure to excessive wind may be diminished if there is a possibility of securing or even dismantling it in advance. Braces or guys might be inspected or even enhanced in anticipation of a storm.

Permanent installations are generally designed and built for the project owner, based on governing codes and specifications. Temporary construction is usually considered means and methods to be thought out, implemented, and controlled by the contractor. Codes and specification may well apply to the temporary work, but, if so, they are likely to be broad brushed, leaving responsibility for execution with the contractor. In general, permanent work is designed on behalf of the project owner and temporary work on behalf of the contractor.

Materials used for temporary work are subject to different demands than those used for permanent construction. There are unlikely to be long-term concerns for corrosion and shrinkage. Where economy of weight and leanness of form may be important characteristics of permanent structural elements, temporary members might be oversized due to the availability of material or to compensate for crude installation. On the other hand, sometimes temporary elements must be carried through a building, installed and removed by hand. In such instances the weight and ease of handling are key considerations in the selection of materials.

During construction, constraints on the location of temporary bracing may be different from the demands when the structure is completed. The project-in-progress space is not filled with occupants, ductwork, pipes, and other building systems. On the other hand, guys and braces must not interfere with construction activities or with systems that will be installed before these temporary elements are removed.

Connections for temporary installations are often made more crudely than permanent connections because they need not be durable and they may benefit from being monitored and maintained frequently. A guy connection could be as basic as a cable wrap and a brace in compression as rudimentary as wedges or shims acting in simple bearing. As with the members themselves, long-term effects of corrosion, creep, and temperature movements are a lesser consideration if considered at all. Though a brace buried in a hidden part of a building might be left in place permanently, temporary braces and guys are usually removed. The remnant left behind should not diminish the strength, functionality, or aesthetics of the permanent structure.

APPLICATIONS OF TEMPORARY GUYING AND BRACING

Temporary bracing may be required for a specific step of a project—an element such as a wall, column, or deck may be plumbed and held until it is fully connected—or an entire structure might need to be stabilized in the absence of a permanent bracing system. The manner in which temporary braces or guys are applied to a structure depends on whether they are needed principally to impose geometric or force restraint, and also whether they are active or passive.

Geometric restraint is what occurs when a brace locks a structure into a shape or position. Braces or guys that square up or plumb a structure serve this function. Restraint can be applied actively as when guys are tensioned to plumb a column, or passively when bracing elements are installed to hold a structure in its preset shape.

Force restraint is what a brace provides to resist specific loads. These can be active loads needed to keep the structure in static equilibrium, occasional loads such as wind or notional loads derived by the designer to ensure structural stability.

In some applications forces must be actively transferred to temporary braces or guys. Guys can be preloaded (tensioned) with pulling devices and turnbuckles. Braces in compression might also be preloaded by driving wedges or with screw jacks. Conversely, when the work of these temporary elements is done, loads might need to be relieved from them before they are removed.

Comparison of Guys versus Braces

Guys and braces usually perform the same function, but the similarity ends there. Each has characteristics that can be an advantage or disadvantage in some applications:

- *Material.* Guys are made up of wire rope. Braces in temporary construction are usually wood or structural steel, though use of other materials is not precluded.

- *Behavior in tension and compression.* Unlike guys, braces might be designed to work in compression, tension, or both, but guys are limited to tension. Guys usually must be paired in opposition for balance and to brace a structure in opposing directions.

- *Rigidity.* Wire ropes sag and stretch, though this behavior can be overcome by preloading. Braces have a more manageable and determinate stiffness that makes it easier to predict deformation under load.

- *Length.* Cables have virtually no length limitation. Braces in compression lose strength with increased slenderness.

- *Planning and improvisation.* Guying is better suited to field improvisation. Cables can be spooled out and cut with rough measurement, while braces usually require cutting and fitting that is more suited to shop conditions.

- *Connections.* Wire rope ends are either shackled to a hole or fitting, or tied around the connected structure. Brace attachments are as varied as the materials used. Common connection means are nails, friction, clamps, anchors, bolts, and welding.

- *Adjustment.* Guys are often fitted with turnbuckles to make them adjustable. Braces usually have no such means, though it is possible to have adjustability built into some braces by means such as slotted holes, telescoping sections, or screw jacks. When a brace is not self-adjustable, an external device such as a chain fall, grip-hoist, or a hydraulic jack might be used to force the structure into alignment.

Guying Materials

Wire rope is available in a range of diameters from ¼ in (6 mm) to 3 in (75 mm) or more. Its construction is varied, too, with characteristics developed by rope manufacturers for a multitude of applications. Construction is designated in part by a pair of numbers: The first is the number of strands wound together to make up the rope and the second is the number of wires in each strand. For example, a common rope type used for guying is designated 6×19, with 6 strands and 19 wires in each strand. However, the latter number is nominal, as the actual number of wires per strand can vary. (For more information about wire-rope construction, see Chap. 20.)

Guying Behavior.   Some rope characteristics that are important for other applications are unimportant for guying; the ones that matter most are strength, flexibility, and elasticity. Strength is determined by the metallic cross-sectional area and the steel grade. Most rope types are available in several strength grades.

Good flexibility is helpful when a wire rope must be bent by hand around a fitting or around a structural element to make a connection. Ropes with fiber cores have less metallic area than ropes of the same diameter with wire cores, but fiber core ropes are more flexible. Ropes made up of smaller wires are more flexible than ones with larger wires. Thus a 6×36 rope would be more flexible than a 6×19 rope of the same diameter.

Elasticity is the propensity of a rope to stretch. A structure braced by a stiff (less elastic) rope will displace less under load. If there are other bracing elements, the stiffer braces will take a higher share of load. The modulus of elasticity (E) of a wire rope is nonlinear, increasing with load. As a rule of thumb, the value of E is taken as 90 percent of the nominal value when the rope is loaded below 20 percent of the breaking strength.

Cables sag from self-weight, a phenomenon known as *catenary action*. The catenary effect adds to elastic stretch in a nonlinear relationship. Preloading takes out the sag and effectively stiffens the rope. Tensioning a guy to at least 5 percent of its breaking strength is sufficient to reduce the catenary effect to a degree that it need not be considered in a typical guy analysis.[*] Rudimentary guying—most installations are in this category—preloading is done by feel, simply by taking the slack out. However, in critical applications cable tension can be set to a prescribed value using a tensiometer. Preload can also be correlated to cable sag measured at mid-span by applying the relationships for a uniformly loaded parabolic cable, as shown in Fig. 17.2.

Safety factors applied to wire rope vary with application. For temporary guying, a safety factor 3 (nominal rope breaking strength/applied load) is commonly used, though a different value may be used if deemed appropriate by code, specification, or the judgment of the design engineer.

Though there is no hard-and-fast rule to determine the proper amount of preload, for applications with opposing balanced cables the authors generally prefer a value of 10 to 17 percent of rope breaking strength. This level is sufficient to ensure that slack is taken

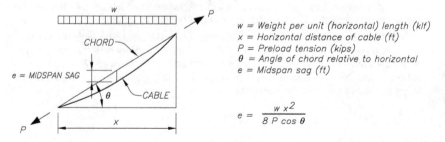

w = Weight per unit (horizontal) length (klf)
x = Horizontal distance of cable (ft)
P = Preload tension (kips)
θ = Angle of chord relative to horizontal
e = Midspan sag (ft)

$$e = \frac{w\,x^2}{8\,P \cos \theta}$$

FIGURE 17.2   Statics of parabolic cable subjected to uniform load.

---

[*]*Design of Guyed Electrical Transmission Structure*, ASCE, New York, 199, pp. 30–31.

out of the guys while avoiding an unnecessary increase in the overall compressive load on the guyed structure. See Example 1 for a guy design applying this concept.

Wire Rope Hardware. Wire-rope end terminations can be shop-made or field-made. Shop terminations, stronger and more reliable than field connections, are made with eyes or clevises that are swaged (pressed) onto the rope ends. Though sometimes it is possible to plan temporary guys with sufficient foresight that they can be shop-made, this degree of advance preparation goes against general intent: guys are usually improvised to suit field conditions.

The most common field-made rope termination is made by turning the rope back to form a loop (eye) and closing the loop with wire-rope clips. The correct number of clips must be installed in accordance with the manufacturer's instructions, but even so the strength of this connection is only about 80 percent of the rope strength. Swaged connections maintain the full strength of the rope.

The eyes at each end of the guy can be joined to the structure in a number of ways. A crude connection node can be made by choking a wire rope sling around or through the structure. Alternatively, a piece of wire rope can be wrapped around the structure in multiple loops with the ends held together by wire rope clips. The eye of the rope is attached to the choker or the wrap with a fitting, usually a shackle or a turnbuckle. In some applications, as illustrated in Fig. 17.3, the fitting is connected to a hole or pad-eye.

A slack guy is not an effective load-carrying member. Though sometimes it is possible to pull a guy sufficiently tight on initial installation, usually a turnbuckle is used.

Material for Bracing

The most common materials used for bracing are timber and steel. Both are readily obtained in a variety of shapes, grades, and sizes suitable for this use.

When timber is used for substantial braces, it is almost always in compression and field cut. The elements are often paired so that the compression is effective in opposing directions. Robust tension connections in timber are too difficult to make for most temporary applications. However, for minor braces that carry only incidental loads, side lap connections with nails or lag bolts are practical.

As timber is relatively inexpensive, braces may be oversized, often using readily available materials or scrap. Timber can carry surprisingly large loads in compression, though simple tension and shear connections are relatively weak. Long members might need to be built up from smaller ones to attain necessary lengths and to reduce column slenderness. Secondary braces might also be used to reduce slenderness of longer braces (Fig. 17.4).

Steel braces may be prepared in either the shop or the field with either bolted or welded connections. The connections are made in accordance with the same standards that apply to permanent construction, and the braces may be almost indistinguishable from permanent members. Single or double angles are frequently used because they are simple to connect. Pipe braces and stocky wide flange shapes are more structurally efficient, but the connections are most suited for applications that can be shop-fabricated. In applications requiring long member lengths, secondary braces may be needed to reduce unbraced lengths and slenderness. See Example 2 for a comparison of member sizes when secondary bracing is included in a long brace design.

A special category of pipe braces are those that are designed specifically for temporary service. Often, they have built-in screw jacks and may be made up of telescoping sections. These braces are sometimes designed to work only in compression, but they may have clevis ends that can push or pull.

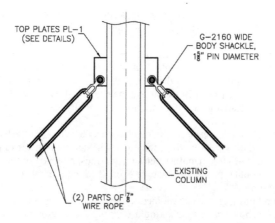

TOP PLATES PL−1
(SEE DETAILS)

G−2160 WIDE
BODY SHACKLE,
$1\frac{5}{8}$" PIN DIAMETER

EXISTING
COLUMN

(2) PARTS OF $\frac{7}{8}$"
WIRE ROPE

TOP OF COLUMN DETAIL
SCALE $\frac{1}{2}$"−1'

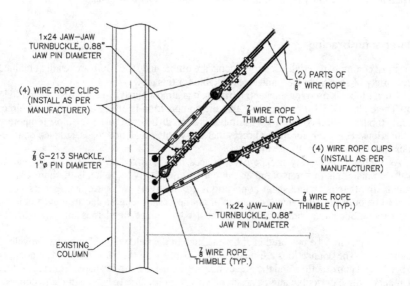

1x24 JAW−JAW
TURNBUCKLE, 0.88"
JAW PIN DIAMETER

(4) WIRE ROPE CLIPS
(INSTALL AS PER
MANUFACTURER)

$\frac{7}{8}$ G−213 SHACKLE,
1"∅ PIN DIAMETER

EXISTING
COLUMN

(2) PARTS OF
$\frac{7}{8}$" WIRE ROPE

$\frac{7}{8}$ WIRE ROPE
THIMBLE (TYP.)

(4) WIRE ROPE CLIPS
(INSTALL AS PER
MANUFACTURER)

$\frac{7}{8}$ WIRE ROPE
THIMBLE (TYP.)

1x24 JAW−JAW
TURNBUCKLE, 0.88"
JAW PIN DIAMETER

$\frac{7}{8}$ WIRE ROPE
THIMBLE (TYP.)

BOTTOM OF COLUMN DETAIL
SCALE $\frac{1}{2}$"−1'

FIGURE 17.3   Example of a cable connection to a gusset plates on a column.

FIGURE 17.4 Large-scale bracing of a distressed building leaning into a construction site. Slenderness of the long compression members is reduced by secondary braces.

Uses of Guys and Braces in Temporary Applications

Stabilization and Plumbing during Steel Erection. Whether it stands a few stories or many, a steel structure under construction is likely to need temporary braces or guys for stability (Fig. 17.5). This requirement is additional to guying to plumb up the structure, though the same cabling may be used for both plumbing and guying.

The permanent bracing system of the structure may comprise brace frames, moment frames, or shear walls. These may be completed floor by floor or in tiers of two floors. The metal deck diaphragms also participate in the bracing system. Until all these elements in the tier are fully made up with attendant bolting and welding completed, it will need temporary

(a)

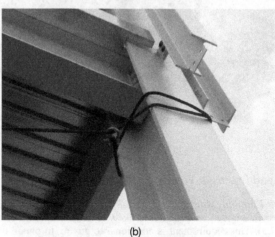

(b)

FIGURE 17.5 (*a*) Temporary guying of a steel structure under construction. The guys remain in place until the permanent bracing system of the building becomes effective. (*b*) A simple cable wrap may provide sufficient anchorage of a temporary guy to the steel framing. (*Falcon Steel, Inc.*)

lateral support. OSHA allows four floors or 48 ft (15 m) of uncompleted welding and bolting above the foundation or top secured floor.*

Temporary bracing or guying resists potential instability caused by load eccentricities, imperfections in plumbness and straightness of members, and second-order effects from deflection. A common design assumption is to apply a 2 percent lateral load to account for these effects, which is generally quite conservative. Wind forces and inertial loads from equipment are additional.

Long slender beam elements, trusses, and plate girders may also require bracing for stability. Guys are usually limited to 7/8 in (22 mm) diameter for ease of handling.

Cast-in-Place Concrete. Permanent lateral bracing of a reinforced concrete building is provided by shear walls, columns, moment frames, or some combination of these. A newly erected floor is not stable or self-supporting against lateral loads until its permanent bracing elements are poured and the concrete has achieved some measure of strength. Attainment of full design strength is not generally required. The decision of when temporary bracing can be removed is ultimately the design engineer's to make. Until the permanent bracing system is effective, temporary lateral support is needed.

Some formwork systems have bracing built-in, but in others it is job-built for each new floor. Shoring frames, assembled as towers, have inherent stability, requiring bracing only for uncommon conditions such as high floors or large lateral wind-catching surfaces. Formwork decks supported by individual posts have no inherent stability and must be braced, as do walls and columns. If walls and columns are poured on a previous day, the form deck may be braced in part by them.

Braces are usually light-duty pipes or dimensional lumber held by nails. The top end of a brace may be nailed to a stringer and the bottom end to a footblock that is secured to the slab. The strength of the brace is likely to be limited to a small value by the nails. For this reason, braces that are held by nails will be closely spaced. If higher-capacity braces and connections are used, the deck will act as a diaphragm spanning between the braced formwork bays.

The most common load on bracing is from wind. In some instance, inertial loads may be imposed by equipment. Wind loads can be calculated using ASCE 7 and ASCE 37, but minimum loads required by a code such as ACI 347 *Guide for Shoring/Reshoring of Concrete Multistory Buildings* may govern.

Stabilization during Demolition and Renovation. Bracing is used during partial demolition to maintain the stability and plumbness of the remaining structure (Fig. 17.6). A building being restored or adapted for a new use often has floors removed and walls maintained (Fig. 17.1). Sometimes new openings are made in existing walls. The interior space might be gutted while maintaining only the exterior walls or just the façade.

The means of support in older structures, especially masonry buildings, are often indistinct and concealed from easy scrutiny. Commonly, an urban building is braced by an adjoining building, or the two buildings share a party wall. A building may be a veritable patchwork of construction and alterations carried out over many generations. Elements may be degraded or structural systems compromised.

When floors or walls are removed, sometimes the unexpected can happen. A delicate equilibrium may be disturbed or an unanticipated load path interrupted. For this reason, bracing of existing structures sometimes should allow not only for known demands but sometimes also for the contingency of unknowns.

---

*OSHA *Health & Safety Policy and Procedures Manual,* June 2007 Section 27 (Steel Erection), p. 8.

FIGURE 17.6   Guying of structural steel columns to stabilize them during demolition. The guys are connected to the tops of adjoining columns.

A bracing program should take thorough consideration of existing conditions. In order to try to account both for those that are obvious and those that are subtle or hidden, the following measures may be useful:

- A survey to explore and document existing conditions before the start of work.
- Probing to uncover hidden conditions, reveal typical details of construction, and ascertain material properties.
- Gathering of local knowledge obtained from experience with similar structures.
- Monitoring of the structure. The monitors could include optical surveys to detect movement, crack monitors, and general inspections.

Emergency Stabilization.   Occasionally an incident occurs where a structure becomes distressed and requires emergency stabilization. In urban construction projects, excavation, underpinning, and demolition operations can create disturbances that upset the stability of a neighboring structure. Construction accidents, fires, and other disasters often demand immediate stabilization measures to prevent further collapse and safe-off the surrounding area.

Depending on the nature of the incident, a combination of shoring and bracing may be needed to stabilize the structure. These systems need to be conservatively designed and utilize materials that can be obtained and installed quickly. The level of urgency may curtail design time and only allow for quick-and-dirty "back of the envelope" calculations. Likewise, the materials called for must be readily available in sufficient quantity and suited to hastened field installation.

Timber is commonly used for shoring, cribbing, and bracing due to its wide availability and ease of installation. Wire rope cable systems, adjustable steel shoring towers, and pipe-and-clamp bracing are also used in emergency applications and are familiar to shoring contractors.

Structural steel has its place when the loads are large, the member lengths are long, or the abovementioned materials are not adequate for the job. Structural steel requires fabrication as well as field connection by welding or bolting, and may necessitate a crane for installation. However during an emergency, it may be the best bracing solution due to the exigency and accelerated pace of the work (Fig. 17.7).

FIGURE 17.7 Emergency bracing of a reinforced concrete parking garage that had collapsed during construction. The heavy steel braces are perhaps oversized, but the imprecise knowledge of conditions and danger of further collapse demanded a quick robust bracing solution.

Field improvisation may be necessary to create suitable brace anchorages. If conditions allow, braces might be anchored to an existing structure within the jurisdiction of the incident. If existing supports are not available, concrete deadman can be brought to the site or temporary concrete footings may be cast in-place. If time allows, anchors in confined spaces might entail the use of drilled micropiles, a solution that offers high-load capacity with minimal vibration.

Consideration should be given for the effect of the layout on future operations. For example, large rakers installed across an entire adjacent lot can partition the site and render it impassable to future equipment. Emergency shores or braces installed too close to a failed foundation or within the footprint of the replacement footing can complicate the ensuing excavation work. Obviously, the emergency takes precedent over all else; however, if there is an opportunity to plan ahead for the following phases of work, it may be beneficial down the road.

Several years ago, the authors partook in the emergency stabilization of a large-scale four-pole scaffold tower that buckled and nearly collapsed while being dismantled at a major bridge. Figures 17.8 and 17.9 illustrate the scale and dramatic distortion of the buckled tower.

FIGURE 17.8   A buckled four-pole scaffold tower. Flanking bays containing vertical x-bracing had been removed, leading to this outcome.

FIGURE 17.9   Pipe-and-clamp diagonal braces installed to stabilize a buckled scaffold tower.

Structurally, the four-pole scaffold is a three-dimensional grid with continuous vertical legs and pin-connected braces. As such, it relies on ties and diagonal braces to resist side sway in the two principal axes. The 320-ft-high scaffold comprised 86 bays that enclosed the two bridge towers. Approximately, every forth bay contain full-height vertical x-bracing. Outboard of this scaffold, but integral with the three remaining bays of scaffold as observed in Fig. 17.8, was a material and personnel hoist used to service the project.

An ambitious foreman thought it productive to dismantle the entire scaffold from top-to-bottom leaving in place the hoist and three bays of scaffold some 300 ft (90 m) high. Without an engineer's support and lacking an understanding of the how the system was braced, flanking bays containing the vertical x-bracing were dismantled along with the rest of the scaffold. On the morning of the incident, the crew rode the hoist to the top of the tower and stepped onto the landing. Within seconds, they heard a thunderous bang and felt the platform suddenly drop a foot. Fortunately there were no injuries or significant damage, just a jittery scaffold crew and a terribly distorted scaffold tower.

After the dust settled, an engineering assessment determined what had happened. With vertical x-bracing removed, the unbraced length of the scaffold legs was increased considerably; addition of a just small live load triggered global buckling.

Fortunately, the tower settled into a classic continuous S-shape without collapsing. Most of the leg sections even remained elastic and reusable. In the tower's new equilibrium, its sinuous legs were braced every 30 ft (9 m) by the hoist ties.

Emergency stabilization was achieved by suspending the weight of the scaffold from numerous cables anchored to the bridge tower. An array of pipe-and-clamp braces were then installed to the deformed scaffold legs to "lock" the structure in place. The structure was then dismantled conventionally from the top-down.

The incident underscores several points: Load path and bracing must be considered through the full life cycle of a temporary structure and an engineer's involvement may be needed through that life cycle as well.

LOADS AND LOAD COMBINATIONS

Design loads for temporary bracing and guying systems are principally lateral forces necessary to stabilize a structure during intermediate stages of construction or demolition. The usual design forces arise from wind, stability loads attributed to member out-of-plumbness, and lateral forces generated by construction cranes or erection activities. Occasionally, other loads such as foundation settlement or earthquake are also taken into account.

Design codes for permanent construction are useful for temporary works, but some adjustment is needed. Rational loads and load combinations may be different from permanent installations. The short duration of the installation may influence the designer to reduce consideration of loads due to natural events such as snow, wind, and earthquakes.

Codes, Standards, and References

The principal U.S. standard referenced by local building codes for determining gravity and environmental loads is ASCE/SEI 7-05 *Minimum Design Loads for Buildings and Other Structures*. The standard provides detailed and prescriptive methods for calculating dead load, live load, wind load, seismic loads, and other environmental loads.

A separate standard tailored to temporary structures and construction loads is SEI/ASCE 37-02 *Design Loads on Structures during Construction*. This standard considers loading specific to construction activities: live load during construction, wind load on temporary structures, equipment loading.

In addition to ASCE standards, several institutions have published useful design guides that cover the topic of bracing and guying. A list of references is given in the Bibliography.

In unusual cases, the designer must rationalize load and load combinations with little or no guidance from codes, standards, or references.

Wind Loads

Prescriptive methods for calculating service-level wind loads can be found in ASCE 7. The standard is dividing into two categories: (1) main wind-force-resisting systems for design of lateral support and stability of the overall structure and (2) components and cladding for design of wind effects imposed on the exterior envelope. Three design procedures are offered for main wind-force-resisting systems:

Method 1—Simplified procedure

For low-rise regularly shaped enclosed buildings 60 ft or less in height

Method 2—Analytical procedure

For regularly shaped structures not meeting the requirements of Method 1 or 3

Method 3—Wind tunnel procedure

For irregularly shaped structures or structures with unusual response characteristics in wind

Method 2 is applicable to most bracing and guying designs. The procedure begins with determining the basic wind speed for the location. The *basic wind speed* is defined as the wind velocity for 3-second gust at 33 ft above ground in exposure category C during a 50-year mean recurrence interval. Within the document, wind speed contour maps are provided for continental U.S., Alaska, and special wind regions in the Gulf of Mexico, Eastern Hurricane Coastline.

The 50-year wind can be excessive for temporary structures that may be in service for only weeks or a few years. There are procedures in both ASCE 7 and ASCE 37 that reduce the basic wind speed to normalize the risk for the duration of the structure's service life.

The 50-year wind can be modified for short-duration loading by scaling down the peak wind velocity using conversion factors for different recurrence intervals found in Table C6-7 of the ASCE 7 Commentary. These scale factors yield wind speeds that are calibrated to give approximately the same probability of being exceeded in a year as the 50-year event. For any of the mean recurrence intervals listed, there exists approximately a 2 percent probability that the wind speed will be equaled or exceeded during any given year.

ASCE 37 also provides scale factors compatible with ASCE 7 design wind speeds for various construction periods. Within that standard, factors listed for construction periods ranging between less than 6 weeks and 2 to 5 years.

Construction period	Factor
Less than 6 weeks	0.75
6 weeks to 1 year	0.80
1 to 2 years	0.85
2 to 5 years	0.90

Once the appropriate design wind speed V is established, the velocity pressure q_z can be calculated.

$$q_z = 0.00256 K_z K_{zt} K_d V^2 I \quad \text{[psf]}$$

Several factors are needed to define characteristics of the structure and surrounding features.

Exposure categories define the degree of ground-surface obstructions, vegetation, natural topography, which influence the intensity of wind.

Exposure B—Urban and suburban areas, wooded areas, or other terrain with numerous closely spaced obstructions having the size of single-family dwellings or larger.

Exposure C—Open terrain with scattered obstructions having heights generally less than 30 ft. Flat open country, grasslands, and all water surfaces in hurricane-prone regions.

Exposure D—Flat unobstructed areas and water surfaces outside hurricane prone regions. Smooth mud flats, salt flats, and unbroken ice.

Velocity pressure coefficient K_z or K_h takes into account above-ground height and exposure category.

Topographic factor K_{zt} considers influence of local topographic features such as escarpments, ridges, and hills.

Wind directionality factor K_d takes into account the reduced probabilities of maximum wind coming from any given direction and maximum pressure coefficient occurring for any given wind direction.

Importance factor I takes into account the level of structural reliability needed, depending on the occupancy category.

The service level design wind pressure, for main wind-force-resisting systems, is then

$$p = qGC_p - q_i(GC_{pi}) \quad \text{[psf]}$$

$q = q_z$ for windward walls evaluated at height z above the ground.

$q = q_h$ for leeward walls, sidewalls, and roofs evaluated at structure height h above the ground.

$q_i = q_h$ for windward walls, sidewalls, and leeward walls, and roofs of enclosed buildings and for negative internal pressure evaluation in partially enclosed buildings.

$q_i = q_z$ for positive internal pressure evaluation in partially enclosed buildings where height z is defined as the level of the highest opening in the building that could affect the positive internal pressure. For positive internal pressure evaluation q_i may conservatively be evaluated at height $h(q_i = q_h)$.

Gust factor G accounts for loading effects due to wind turbulence-structure interaction. Taken as 0.85 for rigid structures; it can also be calculated by a procedure.

External pressure coefficient C_p, empirically based coefficient, accounts for the variation in wind pressure based on the geometry of the structure, direction of wind, and the surface under consideration.

Internal pressure coefficient GC_{pi}, empirically based coefficient, accounts for internal pressure coefficient developed in enclosed and partially enclosed buildings. Internal pressure is not a consideration for most brace and guy design.

ASCE 7 also specifies a minimum design wind pressure of 10 lb/ft² for main wind-force-resisting systems.

The above procedure is applicable to enclosed, partially enclosed, and open buildings with a roof structure. During construction many structures are partially complete and are open and exposed. A variation to the above procedure is provided in ASCE 7 that addresses open structures. The design wind force is defined as

$$F = q_z G C_f A_f \quad \text{[lb]}$$

q_z = velocity pressure evaluated at height z of the centroid of area A_f using the exposure categories defined above.

G = gust-effect factor defined above.

C_f = force coefficient. Force coefficients or shape factors for numerous structure types and geometries are provided in ASCE 7.

A_f = projected area normal to the wind except where C_f is specified for the actual surface area, ft².

Open and exposed structures often comprise numerous elements or rows of elements that can shield or catch the wind. ASCE 37 addresses repetitive patterns unenclosed frames and structural elements in Section 6.2.2 "Frameworks without Cladding":

- The loads on the first three rows of elements shall not be reduced for shielding.

- The loads on the forth and subsequent rows shall be permitted to be reduced by 15 percent.

- Wind load on all exposed interior partition walls, temporary enclosures, signs, construction materials, and equipment shall be added to the total wind load on the structure.

Calculations shall be performed for each primary axis of the structure. According to ASCE 37, 50 percent of the wind load calculated for the perpendicular direction should be assumed to act simultaneously.

Structural Stability

Structural stability evaluation is fundamental to bracing or guying design. The stability of a member subjected to a compressive load is a function of the magnitude of the load, plumbness of the member, and the stiffness and strength of the brace.

A complex relationship exists between stiffness of the brace and its ability to restrain the braced member. The ideal brace stiffness is one that provides full restraint at the brace point. This results in an unbraced length equal to the distance between restraints and an effective length factor of $K = 1.0$. Flexible braces that do not adequately restrain the brace point produce longer effective lengths and diminished column strengths.

For new construction, out-of-plumbness is quantified through permissible construction tolerances. For concrete structures up to 1000 inches in height the allowable deviation from plumb is the lesser of 0.3 percent of the height or one inch. For steel structures the maximum out-of-plumbness permitted is $L/500$ or 0.2 percent. This translates to a lateral force equal to $0.002P$. Applied lateral loads further increase the lateral displacement.

Several methods are available that alleviate the complications of determining brace stiffness and out-of plumbness and thus simplify the problem for the designer.

A long-standing conservative rule of thumb for determining the lateral force for stability design is to use 2 percent of the total compressive load in the member. The 2 percent value is an empirical observation or engineering judgment that has been widely accepted. Though the assumption neglects the interaction of brace stiffness, it is assumed that practical brace sizes designed for the 2 percent value inherently possess the required stiffness. Care should be exercised when using the 2 percent value with flexible braces such as cables. Cables may provide sufficient strength, but at low tension values they may not provide sufficient restraint. Similarly, flexible connections may not impart adequate restraint and should be avoided.

Geometrically, the 2 percent value represents a lateral force corresponding to an out-of-plumbness equal to $L/50$. The degree of conservatism is underscored by comparing this value to the allowable construction tolerances for out-of-plumbness, say $L/500$. In an indirect sense, the 2 percent value covers the brace stiffness requirement and an amplification of lateral load due to member curvature, known as P-δ.

During demolition and emergency stabilization, the out-of-plumbness may sometimes exceed $L/50$. In these instances, 2 percent of the gravity load is unconservative and the designer should use the actual out-of-plumbness to derive the lateral force.

In the 13th edition of the AISC *Steel Construction Manual*, a simple rational method is provided that specifies the required brace strength and stiffness. In Appendix 6 "Stability Bracing for Columns and Beams," two types of bracing systems are considered for column and beam bracing.

Relative column bracing is another name for internal bracing. It controls displacement of a brace point by restraining it to another bracing point within the structural framework.

The required brace strength is

$$P_{br} = 0.004\, P_r \quad \text{[kips]}$$

The required brace stiffness is

$$\beta_{br} = 1/\phi(2P_r/L_b) \quad \text{[k/in]} \qquad \text{(LRFD)}$$

$$\beta_{br} = \Omega(2P_r/L_b) \quad \text{[k/in]} \qquad \text{(ASD)}$$

Nodal column bracing, also known as external bracing, controls displacement of a brace point by restraining it to an external restraint.

The required brace strength is

$$P_{br} = 0.01\, P_r \quad \text{[kips]}$$

The required brace stiffness is

$$\beta_{br} = 1/\phi(8P_r/L_b) \quad \text{[k/in]} \qquad \text{(LRFD)}$$

$$\beta_{br} = \Omega(8P_r/L_b) \quad \text{[k/in]} \qquad \text{(ASD)}$$

where $\phi = 0.75$
$\Omega = 2.00$
L_b = distance between braces, in
P_r = required axial compressive strength using appropriate load combinations

Relative beam bracing is lateral bracing that prevents relative twist of the section along the length of the member.

The required brace strength is

$$P_{br} = 0.008 M_r C_d / h_o \quad \text{[kips]}$$

The required brace stiffness is

$$\beta_{br} = 1/\phi(4M_rC_d/L_bh_o) \quad \text{[k/in]} \quad \text{(LRFD)}$$

$$\beta_{br} = \Omega(4M_rC_d/L_bh_o) \quad \text{[k/in]} \quad \text{(ASD)}$$

Nodal beam bracing is external lateral bracing that prevents the twist of the section along the length of the member.

The required brace strength is

$$P_{br} = 0.02M_rC_d/h_o \quad \text{[kips]}$$

The required brace stiffness is

$$\beta_{br} = 1/\phi \; (10M_rC_d/L_bh_o) \quad \text{[k/in]} \quad \text{(LRFD)}$$

$$\beta_{br} = \Omega(10M_rC_d/L_bh_o) \quad \text{[k/in]} \quad \text{(ASD)}$$

where $\phi = 0.75$
  $\Omega = 2.00$
  h_o = distance between flange centroids, in
  C_d = 1.0 for bending in single curvature; 2.0 for double curvature
  M_r = required flexural strength using appropriate load combinations

The specification includes additional provisions for torsional beam bracing.

Also in the 13th edition of the AISC *Steel Construction Manual*, an analytical method is provided that requires second-order analysis of the structural system to explicitly account for P-δ and P-Δ effects, initial out-of-plumbness, and reduced member stiffness due to yielding. In Appendix 7 "Direct Analysis Method," the details for the stability analysis are specified.

Equipment and Inertia Loads

Reactions from construction cranes, inertia forces from mechanical equipment, loads from temporary guys used for erection, and any other additional loading unique to an operation may need to be taken into account.

In Section 4.4 of ASCE 37, "Horizontal Construction Load," four criteria are provided to aid in accounting for horizontal loads:

1. For wheeled vehicles transporting materials, 20 percent for a single vehicle or 10 percent for two or more vehicles of the fully loaded vehicle weight. Said force shall be applied in any direction of possible travel, at the running surface.
2. For equipment reactions as described in Section 4.6, "Equipment Reactions," the calculated or rated horizontal loads, whichever are the greater.
3. 50 lb per person applied at the level of the platform in any direction.
4. 2 percent of the total vertical load. This load shall be applied in any direction and shall be spatially distributed in proportion to the mass. This load need not be applied concurrently with wind or seismic load.

Other Loads

There may be instances where environmental loads, other than wind, may need to be included in a brace or guy design. Snow loading, ice accretion, lateral earth pressure, hydrostatic pressure, and differential settlement are additional sources of load that may need to be considered.

In areas of high seismicity, temporary bracing structures of high value or long duration may warrant inclusion of earthquake loading. ASCE 7 has detailed provisions for determining these additional forms of environmental and earthquake loads. The seismic provisions are consistent with 2006 *International Building Code* and 2003 *NEHRP Recommended Provisions for the Development of Seismic Regulations for New Buildings and Other Structures*. AASHTO and Caltrans also have specific philosophies and methods for calculating earthquake loads on bridge structures.

Allowable Stress Design versus Load Factor Design

The development of loads in this discussion has been in terms of unfactored service loads. Depending on the project criteria or designer preference, either allowable stress design (ASD) or load and resistance factor design (LRFD) can be used. Appropriate combinations of loads shall be considered to determine the worst case effect. Both ASCE 7 and ASCE 37 provide load factors and load combinations suitable for ASD or LRFD.

ASCE 37 also includes load factors and load combinations specific to temporary structures. Additional factors are given for material loads, equipment loads, and other loads common to construction. As ASCE 37 is a general document that cannot possibly cover all scenarios, the designer may need to apply judgment to determine rational load factors and combinations for a particular problem.

EXAMPLES

Example 1—Guy Design for Shoring Tower

Design a cable guying system for an 80-ft-high shoring tower carrying a 400-k load. The tower is 8 ft × 8 ft in plan and constructed of W8 × 31 legs with square panels of L4 × 4 × 3/8 x-bracing on each side. The tower bears on a footing with no uplift resistance. Design the guys to resist the total lateral load. The site is a flat open area with a basic wind speed of 90 mph and the tower is expected to remain in service for a year. (See Fig. 17.10.)

1. Determine Design Loads

Wind load $= F = q_z\,GC_f A_f$ for on trussed tower

Flat open area = Exposure C

Wind velocity reduction factor due to short duration loading = 0.85

$V = 0.85\ (90\ \text{mph}) = 77\ \text{mph}$

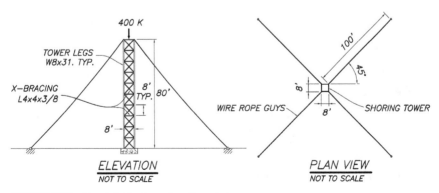

FIGURE 17.10   Guyed shoring tower from Example 1.

As a simplification, use K_z at 2/3 tower height.

$K_z = 1.10$ at 53 ft height

$K_{zt} = 1.0$ no influence by topographic features

$K_d = 0.85$ directionality factor for trussed tower

$I = 1.0$ ordinary structure

$q_z = 0.00256(1.10)(1.0)(0.85)(77 \text{ mph})^2(1.0) = 14.2$ psf

Gust factor was calculated separately using ASCE 7 procedure for flexible structures: $G = 1.3$.

Determine force coefficient.

Calculate percentage of solid for single face of 8 ft $\times$ 8 ft panel.

W8 $\times$ 31 legs	(2)(0.67 $\times$ 8.0 ft)	= 10.7 ft^2
L4 $\times$ 4 diagonal	(2)(0.33 $\times$ 10.6 ft)	= 7.0 ft^2
L4 $\times$ 4 horizontal	(1)(0.33 $\times$ 7.33 ft)	= 2.4 ft^2

$A_{solid} = 20.1$ ft^2/8-ft panel

$A_{gross} = 64.0$ ft^2/8-ft panel

$\varepsilon = 20.1/64.0 = 0.31$

$C_f = 4.0\varepsilon^2 - 5.9\varepsilon + 4.0 = 4.0(0.31)^2 - 5.9(0.31) + 4.0 = 2.6$

Check wind in the principal direction of the tower.

Neglect shielding of leeward face.

$A_f = (2 \text{ faces} \times 20.1 \text{ ft}^2)(10 \text{ panels}) = 402$ ft^2

$F_{w1} = (14.2 \text{ psf})(1.3)(2.6)(402 \text{ ft}^2)/1000 = 19.3$ k

Check wind on 45° skew.

Neglect shielding of leeward faces.

$A_f = (4 \text{ faces} \times 20.1 \text{ ft}^2 \times \cos 45)(10 \text{ panels}) = 569$ ft^2

$F_2 = (14.2 \text{ psf})(1.3)(2.6)(569 \text{ ft}^2)/1000 = 27.3$ k

ASCE 7-05 specifies that for trussed towers with square cross sections, wind forces shall be multiplied by $1 + 0.75\varepsilon$ but not greater than 1.2 when wind is directed along a tower diagonal.

$(1 + 0.75 \times 0.31) = 1.23$

$F_{w2} = 1.2(27.3 \text{ k}) = 32.8$ k controls

Stability loading: Design for equivalent lateral load equal to 2 percent of gravity load. The shoring load is 400 k and tower self-weight is approximately 24 k.

Equivalent lateral load induced by gravity $= F_G = 0.02 \ (424 \text{ k}) = 8.5$ k. See Fig. 17.11 for applied load diagram.

2. Design Cable Guys.   Apply wind load resultant at mid-height of tower. Apply equivalent lateral load for stability at the tower top. Preload guys such that the leeward guy just goes slack (zero tension) under maximum lateral load. Provide a safety factor 3 for the windward guy for the combined effects of preload and maximum lateral load.

Maximum tension for windward guy = 33 percent of breaking strength

Minimum tension for leeward guy = 0 percent of breaking strength

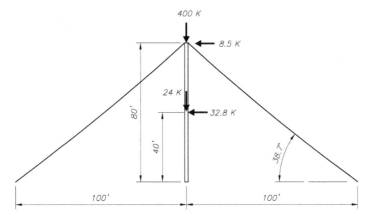

FIGURE 17.11   Applied loads on a guyed shoring tower.

Targeted preload for opposing guys = 1/2 (0 + 33 percent) = 16.5 percent of breaking strength

Usable cable strength for applied loads = 33 − 16.5 percent preload = 16.5 percent of breaking strength

Sum moments about base of tower.

T_{h1} = [8.5 k (80 ft) + 32.8 k (40 ft)]/2 opposing guys/80 ft = 12.5 k/guy horizontal component

T_{v1} = (80′/100′) T_{h1} = 10.0 k/guy vertical component

T_1 = sqrt $(T_{h1}{}^2 + T_{v1}{}^2)$ = 16.0 k externally applied cable tension

$T_u = T_1/0.165$ = 97 k targeted breaking strength required

Try (2) 3/4 in diameter 6 × 37 IWRC EIPS wire rope for each guy.

Rated breaking strength = (2 legs) 58.8 k = 118 k

Size preload based on leeward guy going slack under maximum lateral load.

Preload force, T_2 = 16 k/guy = 8.0 k/leg

Percentage of breaking strength = (16 k/118 k)(100) = 13.6 percent

$\theta = \tan^{-1}$ (80 ft/100 ft) = 38.7°

$T_{h2} = T_2 \cos \theta$ = 8.0 k cos (38.7) = 6.2 k/leg = 12.5 k/guy horizontal component

$T_{v2} = T_2 \sin \theta$ = 8.0 k sin (38.7) = 5.0 k/leg = 10.0 k/guy vertical component

Cable sag corresponding to preload, $e = w \, L^2/(8 \, T_{h2})$

S = sqrt (100 ft² + 80 ft²) = 128 ft approximate length of cable

L = 100-ft horizontal component of cable length

w_s = 1.04 plf weight of 3/4-in diameter cable

$w = w_s \, (S/L)$ = 1.04 plf (128 ft/100 ft) = 1.33 plf unit weight per horizontal length

e = (0.00133 klf)(100 ft)²/[8 (6.2 k)] = 0.27 ft = 3.2 in of mid-span drape

3. Check Additional Guy Forces Due to P-Δ.   It is assumed that under the specified preload initial elastic stretch and constructional stretch are removed from the system. Under maximum lateral load, leeward guy is unloaded and it provides zero lateral

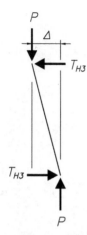

FIGURE 17.12 Freebody diagram of a tower with P-Δ effect.

restraint. Lateral resistance is provided only by the windward guy. See Fig. 17.12 for free-body diagram of tower with P-Δ effects.

$K_{lat} = AE \cos^2 \theta/L$

$A = 0.277$ in^2

$E = 14,000$ ksi

$L = 128$ ft $= 1536$ in

$K_{lat} = (2$ legs)$(0.277$ in$^2)(14,000$ ksi)$\cos^2(38.7)/(1536$ in$) = 3.1$ k/in

Lateral load acting at top of tower

$F = F_G + 1/2\ F_{w2} = 8.5$ k $+ 16.4$ k $= 24.9$ k

$\Delta = F/K_{lat} = 24.9$ k/3.1 k/in $= 8.0$ in $= 0.67$ ft

Additional windward guy forces due to P-Δ

$P = 424$ k load $+ (40$ k vertical components from 4 guys$) = 464$ k

$T_{h3} = (464$ k$)(0.67$ ft$)/80$ ft $= 3.9$ k

$T_{v3} = T_{h3} \tan \theta = 3.9$ k tan $(38.7) = 3.1$ k/guy

Windward guy

Total horizontal guy force

$T_h = T_{h1} + T_{h2} + T_{h3} = 12.5$ k $+ 12.5$ k $+ 3.9$ k $= 28.9$ k

Total vertical guy force

$T_v = T_{v1} + T_{v2} + T_{v3} = 10.0$ k $+ 10.0$ k $+ 3.1$ k $= 23.1$ k

Total guy force

$T = T_h/\cos \theta = 28.9$ k/cos $(38.7) = 37.0$ k

FS $= 118$ k/37.0 k $= 3.2 > 3$ ok

Leeward guy

Total horizontal guy force

$T_h = T_{h1} + T_{h2} = -12.5$ k $+ 12.5$ k $= 0$ k

Total vertical guy force

$T_v = T_{v1} + T_{v2} = -10.0$ k $- 10.0$ k $= 0$ k

Total guy force

$T = 0$ k

Provide (2) legs of 3/4-in diameter 6 × 37 IWRC EIPS wire rope for each guy. Preload each guy to 13.6 percent of rated breaking strength $= 2 \times 8.0$ kips per leg $= 16.0$ kips per guy.

Example 2—Stabilization of Distressed Building Using Long Braces

Design temporary rakers and corresponding lateral bracing to stabilize an existing 80-ft-long × 25-ft-wide × 60 ft-high five-story solitary masonry building leaning 6 in into a construction site during underpinning and excavation. The walls are constructed of 12-in-thick brick. The site is in urban area with a basic wind speed of 110 mph and the bracing is anticipated to remain in place for 1½ years. (See Fig. 17.13.)

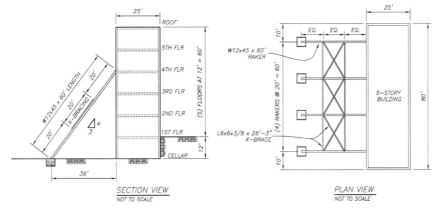

FIGURE 17.13 Stabilization of a distressed building using long braces used in Example 2.

1. Determine Design Loads

Wind load

Urban area = Exposure B

Wind velocity reduction factor due to short duration loading = 0.85

$V = 0.85 \, (110 \text{ mph}) = 94 \text{ mph}$

$K_z = 0.85$ at 60 ft height. As a simplification, use maximum K_z over entire height

$K_{zt} = 1.0$ no influence by topographic features

$K_d = 0.85$ directionality factor

$I = 1.0$ ordinary occupied building

$q_z = 0.00256(0.85)(1.0)(0.85)(94 \text{ mph})^2(1.0) = 16.3 \text{ psf}$

$G = 0.85$ for rigid structure

$C_p = 0.80$ for windward wall

$C_p = -0.50$ for leeward wall (L/B ratio = 0.31)

$p = q_z G C_p$ Treat as enclosed building

$p_{\text{windward}} = (16.3 \text{ psf})(0.85)(0.80) = 11.1 \text{ psf}$

$p_{\text{leeward}} = (16.3 \text{ psf})(0.85)(-0.50) = -6.9 \text{ psf}$

$p_w = p_{\text{windward}} - p_{\text{leeward}} = 11.1 \text{ psf} - (-6.9 \text{ psf}) = 18 \text{ psf}$

$F_{\text{wind}} = p_w B H = (18 \text{ psf})(80 \text{ ft})(60 \text{ ft})/1000 = 86 \text{ k}$

Stability loading

Equivalent lateral load induced by gravity = $F_{\text{gravity}} = (6 \text{ in}/720 \text{ in})W = 0.83$ percent W

Conservatively design for equivalent lateral load equal to 2 percent of gravity load

Weight takeoff

Floor loads: Dead load: 25 psf

Live load: 40 psf

Roof loads: Dead load: 25 psf

Live load: 30 psf

$W_{\text{floor}} = (65 \text{ psf})(78 \text{ ft} \times 23 \text{ ft})(5 \text{ floors})/1000 = 583 \text{ k}$
$\qquad\quad (55 \text{ psf})(78 \text{ ft} \times 23 \text{ ft})(1 \text{ roof})/1000 = 99 \text{ k}$
$W_{\text{walls}} = (2 \text{ walls})(80 \times 60 \text{ ft})(120 \text{ psf})/1000 = 1150 \text{ k}$
$\qquad\quad (2 \text{ walls})(23 \text{ ft} \times 60 \text{ ft})(120 \text{ psf})/1000 = 331 \text{ k}$

Neglect openings.

$W = W_{\text{floors}} + W_{\text{walls}} = 682 \text{ k} + 1480 \text{ k} = 2160 \text{ k}$
$F_{\text{gravity}} = 0.02 \, W = 0.02(2160 \text{ k}) = 43 \text{ k}$

Install top of raker equal with the fourth floor elevation (36 ft above grade). Sum moments about base of building.

$F_H = [(F_{\text{wind}} + F_{\text{gravity}}) \, H/2]/36 \text{ ft} = [(86 \text{ k} + 43 \text{ k})(30 \text{ ft})]/36 \text{ ft} = 107 \text{ k lateral load.}$

2. Size Raker Shore Using Wide Flange Shape.   Design four rakers spaced at 20 ft on center. Install the base of the raker 12 ft below grade to avoid interference during excavation of lot. Due to space constraints in the adjacent lot, the geometry of the raker can only assume that of a 3-4-5 triangle.

$P = 107 \text{ k}/4 \text{ rakers } (5/3) = 45 \text{ k/raker}$
$L = 60 \text{ ft} = 720 \text{ in}$

For initial member sizing, try independent W12 $\times$ 79 rakers without the use of secondary bracing. $F_y = 50$ ksi

Strong-axis compression:	Weak-axis compression:
$k = 1.0$	$k = 1.0$
$L = 720$ in	$L = 720$ in
$r_x = 5.34$ in	$r_y = 3.05$ in
$k \, L/r_x = 135$	$k \, L/r_y = 236$
$F_a = 8.2$ ksi	$F_a = 2.7$ ksi
$A = 23.2 \text{ in}^2$	$A = 23.2 \text{ in}^2$
$f_a = 1.9$ ksi	$f_a = 1.9$ ksi
$P_{\text{all}} = 190$ k	$P_{\text{all}} = 63$ k

Strong-axis bending: Include gravity effects on flexure of long member.

Span $L = 36$ ft
$w_{\text{dl}} = 79 \text{ plf beam} \times 60 \text{ ft length}/L = 132 \text{ plf}$
$M_x = w_{\text{dl}} \, L^2/8/1000 = 21.3 \text{ k·ft} = 256 \text{ k·in}$
$S_x = 107 \text{ in}^3$
$f_{\text{bx}} = M_x/S_x = 2.4 \text{ ksi}$
$F_{\text{bx}} = 12.0 \text{ ksi}$

Check combined stresses.

$\qquad C_{\text{mx}} = 0.85$
$F'_{\text{ex}} = 12\pi^2 E/23(kL_x/r_x)^2 = 12\pi^2(29000 \text{ ksi})/23(135)^2 = 8.2 \text{ ksi}$
$f_a/F_a + C_{\text{mx}} f_{\text{bx}}/[(1 - (f_a/F'_{\text{ex}}))F_{\text{bx}}]$
$1.9/2.7 + 0.85(2.4)/[(1 - (1.9/8.2))12.0]$

$0.70 + 0.22 = 0.92$ OK

Use (4) W12 $\times$ 79 $\times$ 60-ft raker shores spaced at 20 ft. on center.

Resize raker using secondary bracing to reduce slenderness ratio for weak-axis buckling and unbraced length for strong-axis bending. Place secondary bracing at third points. Try W12 $\times$ 45 F_y = 50 ksi steel.

Strong-axis compression:	Weak-axis compression:
$k = 1.0$	$k = 1.0$
$L = 720$ in	$L = 240$ in
$r_x = 5.15$ in	$r_y = 1.94$ in
$k\,L/r_x = 140$	$k\,L/r_y = 124$
$F_a = 7.6$ ksi	$F_a = 9.7$ ksi
$A = 13.2$ in^2	$A = 13.2$ in^2
$f_a = 3.4$ ksi	$f_a = 3.4$ ksi
$P_{all} = 100$ k	$P_{all} = 128$ k

Strong-axis bending: Include gravity effects on flexure of long member.

Span $L = 36$ ft
$W_{dl} = (45 \text{ plf beam} + 20 \text{ plf for bracing}) \, 60 \text{ ft} = 3900$ lb
$w_{dl} = 3900$ lb/L $= 108$ plf
$M_x = w_{dl}\,L^2/8/1000 = 17.6$ k·ft $= 211$ k·in
$S_x = 58.1$ in^3
$f_{bx} = M_x/S_x = 3.6$ ksi
$F_{bx} = 19.2$ ksi

Weak-axis bending: Carry in-plane reaction from internal bracing to end supports.

$P_{br} = 0.004 \text{ P} = 0.004(45 \text{ k}) = 0.2$ k
$L = 20$ ft
$M_y = P_{br}\,L = (0.2 \text{ k})(20 \text{ ft}) = 4$ k·ft $= 48$ k·in
$S_y = 12.4$ in^3
$f_{by} = M_y/S_y = 3.9$ ksi
$F_{by} = 37.5$ ksi
Check combined stresses.

$C_{mx} = C_{my} = 0.85$
$F'_{ex} = 12\pi^2E/23(kL_x/r_x)^2 = 12\pi^2(29000 \text{ ksi})/23(140)^2 = 7.6$ ksi
$F'_{ey} = 12\pi^2E/23(kL_y/r_y)^2 = 12\pi^2(29000 \text{ ksi})/23(124)^2 = 9.7$ ksi
$f_a/F_a + C_{mx}f_{bx}/[(1 - (f_a/F'_{ex}))F_{bx}] + C_{my}f_{by}/[(1 - (f_a/F'_{ey}))F_{by}]$
$3.4/7.6 + 0.85(3.6)/[(1 - (3.4/7.6))19.2] + 0.85(3.9)/[(1 - (3.4/9.7))37.5]$
$0.45 + 0.29 + 0.14 = 0.88$

3. Design Lateral Bracing.   Design diagonal lateral bracing to subdivide overall length into thirds for weak-axis bracing. Use AISC criteria for relative bracing to verify brace strength and stiffness.

Design brace as tension/compression brace for outer raker.

$P_{br1} = 0.004(45 \text{ k})/\cos(45) = 0.25$ k/brace

$\beta_{br1} = 2.0 \ (2 \times 45 \text{ k}/240 \text{ in})/\cos^2(45) = 1.5$ k/in/brace

Design brace to also laterally brace raker for gravity bending stresses.

$P_{br2} = 0.008 \ M_r \ C_d/h_o = 0.008 \ (211 \text{ k·in})(1.0)/(11.5 \text{ in})/\cos(45) = 0.21$ k/brace

$\beta_{br2} = 2.0(4 \ M_r C_d)/(L_b h_o)/\cos^2(45) = 1.2$ k/in/brace

Total demands on brace

$P_{br} = P_{br1} + P_{br2} = 0.25 \text{ k} + 0.21 \text{ k} = 0.46$ k

$\beta_{br} = \beta_{br1} + \beta_{br2} = 1.5 \text{ k/in} + 1.2 \text{ k/in} = 2.7$ k/in

Try L6 × 6 × 3/8 angle iron.

Compressive strength:	Axial stiffness
$k = 1.0$	$E = 29000$ ksi
$L = 340$ in	$L = 340$ in
$r_z = 1.19$ in	$A = 4.36 \text{ in}^2$
$k \ L/r_z = 285$	$E \ A/L = 372$ k/in
$F_a = 1.8$ ksi	
$A = 4.36 \text{ in}^2$	
$P_{all} = 7.8$ k	OK

Use (4) W12 × 45 × 60 ft raker shores spaced at 20 ft on center. Provide (3) panels of L6 × 6 × 3/8 secondary x-bracing at third points.

BIBLIOGRAPHY

Loads and Standards

1. ASCE/SEI 7-05: *Minimum Design Loads for Buildings and Other Structures*, Structural Engineering Institute of the American Society of Civil Engineers, 2005.
2. SEI/ASCE 37-02: *Design Loads on Structures during Construction*, Structural Engineering Institute of the American Society of Civil Engineers, 2002.
3. OSHA *Health and Safety Policy and Procedures Manual*, Section 27, Steel Erection, 2004.
4. ACI 347.2R-05 *Guide for Shoring/Reshoring of Concrete Multistory Buildings*, 2005.
5. *2006 International Building Code*, International Code Council, Inc. 2006.
6. *NEHRP Recommended Provisions for Seismic Regulations for New Buildings and Other Structures (FEMA 450)*, Building Seismic Safety Council for the Federal Emergency Management Agency, 2003 ed.

Structural Steel Buildings

7. *Steel Construction Manual*, 13th ed, American Institute of Steel Construction, 2005.
8. Fisher, James M. and West, Michael A.: *Erection Bracing of Low-Rise Structural Steel Buildings*, American Institute of Steel Construction, Steel Design Guide Series No. 10, 1997.
9. Fisher, James M. and West, Michael A.: *Erection Bracing of Structural Steel Frames*, Proceedings National Steel Construction Conference, American Institute of Steel Construction, 1995.

Guyed Towers

10. *Design of Guyed Electrical Transmission Structures*, ASCE Manuals and Reports on Engineering Practice No. 91, American Society of Civil Engineers, 1997.
11. ANSI/TIA-222-G-2: *Structural Standard for Antenna Supporting Structures and Antennas*, American National Standards Institute/Telecommunications Industry Association, 2009.
12. *Wire Rope Users Manual*, 3d ed., Wire Rope Technical Board, Woodstock, Md., 1993.

Bridge Falsework

13. *Falsework Manual*, State of California, Department of Transportation, Sacramento, Calif., 2010.
14. *Guide Design Specification for Bridge Temporary Works*, American Association of State Highway and Transportation Official, 1995.

Jake van Baarsel, P.E.

OVERVIEW

Falsework, which is the major temporary structure associated with bridge construction, may be defined in general terms as a temporary framework on which a permanent work is supported during its construction. Although steel structures occasionally require temporary support during construction, the term "falsework" is universally associated with the construction of cast-in-place concrete structures, particularly bridge structures. In this type of construction the falsework provides a stable platform upon which the forms may be built, and furnishes support for the bridge superstructure until the members being constructed have attained sufficient strength to support themselves.

Because falsework is a temporary structure subject to maximum load combinations for relatively short durations, many established standards and criteria applicable to design and construction of permanent work are neither technically appropriate nor economically feasible for bridge falsework. Thus the design and construction of the falsework can present an engineering challenge.

The design of falsework requires a basic knowledge of design principles for timber, steel, and concrete. In addition to these principles, the designers must also take into consideration the constructability of the falsework systems, including available material by the contractor, erection and removal methods, bracing and stability methods, and grade adjustment methods. The behavior of falsework should also be considered in the design. The falsework must be rigid enough to support the load, but also be flexible enough to allow for settlement, deflection, and grade adjustments.

This chapter presents an overview of the design and construction consideration and practices that are typical of falsework systems in general use for bridge construction.

FALSEWORK SYSTEMS

Falsework systems for cast-in-place bridges are not uniform and may come in many variations. There are no commercially available systems in the market that fit every type of bridge construction. Because of this, the falsework system must be able to adapt to the distinctions of each bridge. Bridges vary in length, width, depth, and height; they can be located over roadways, rivers, canyons, with ground conditions varying from rock to soft clay. Therefore falsework systems must be flexible, able to adjust to heights ranging from 0 to over 100 ft, and falsework span lengths may range from several feet to longer than 100 ft. Foundations may vary in size based on local soil-bearing values. A bridge over a future two-lane road with no obstacles to the layout, most likely calls for a different falsework system than a high connector ramp over a five-lane freeway with long falsework span (see Fig. 18.1).

Besides the variations of falsework systems, some common features are that the falsework system must be able to support the design loads, protect the public traveling through the falsework, and provide a safe working environment for the construction workers. Although typically there are no specifications for using a particular falsework system, it is expected that falsework systems conform to industry standard methods and procedures. The California Department of Transportation (Caltrans) developed the *Falsework Manual* that provides

FIGURE 18.1 Steel pipe falsework system at traffic openings.

guidance to the design and construction of falsework. This chapter reviews standard methods used for falsework construction. Individual components of the falsework, such as beams, posts, and foundation pads, must conform to standard design codes for steel and timber.

If standard falsework methods cannot be applied, alternative bridge construction methods may be considered as described later in this chapter.

The type of falsework system used for a particular type of bridge is usually determined by the contractor and the falsework design engineer. Their type selection may depend on the availability of falsework material in inventory, construction methods, and equipment use.

The most common cast-in-place concrete bridge in the western part of the United States is the prestressed box-girder bridge. The falsework is usually based on a typical system as shown in Fig. 18.2.

A typical conventional falsework system for a cast-in-place concrete bridge structure consists of soffit plywood and joists on top of stringer beams spanning between bents that are placed on a foundation. Falsework bents usually consist of a top cap beam, posts, braces, and a bottom sill beam. The foundation is made up of wedges on top of closely spaced corbels distributing the load to pads placed on stable ground.

Figure 18.3 shows some typical sections that can be used for bridge falsework for various heights. Falsework with timber post and braces (section 1) is usually more economical for falsework heights of less than approximately 30 ft due to the flexibility and ease of constructing with timber. When falsework exceeds a height of around 30 ft, steel post and bracing are typically used (section 2).

When the slenderness of the post exceeds the specified maximum ratio, a double-tiered bracing system (section 3) can be used to reduce the unbraced length of the post instead of selecting larger posts. Pony bent systems (section 4) can be used when the height of

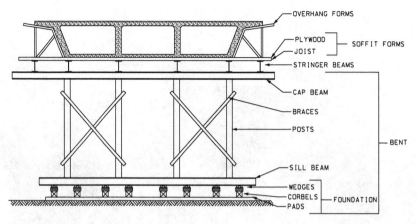

FIGURE 18.2   Typical falsework section.

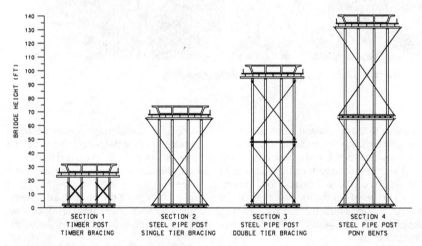

FIGURE 18.3   Various falsework sections.

the falsework makes a single level post system no longer economical or practical for the contractor.

The term "pony bent" describes a secondary support system on top of the main system or primary support system. At complicated falsework layouts requiring large skewed openings over traffic lanes, pony bents may be used to simplify the stringer layout, grade adjustment, and removal of the falsework.

Figure 18.4 shows a pony bent on top of beams spanning traffic in order to construct a new connecter ramp over an existing freeway and an existing connector ramp.

For falsework systems at traffic locations, the layout requirements are often such that typical falsework bents cannot be used due to large skewed openings or other obstructions.

Figure 18.5 shows several options that can be used for bridges that need to be constructed over live traffic at large skews where the span lengths exceed the option to use long stringer beams. Section 1 in Fig. 18.5 shows single post bents adjacent to traffic with

FIGURE 18.4   Falsework with "pony bent" over traffic.

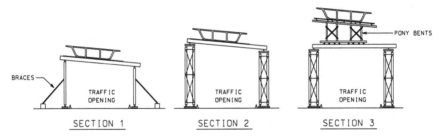

FIGURE 18.5   Falsework options over traffic openings with large skews.

external bracing to concrete deadmen. Section 2 uses braced towers with the stringer beams spanning perpendicular to the traffic opening with the soffit forms placed directly on the stringer beams. This system requires that the soffit forms support the relatively large bridge girder loads between the stringer beams. Section 3 can be used when a typical stringer and bent layout is desired for the ease of fine grading and removal of the falsework.

DESIGN LOADS

This section discusses the loads normally used for falsework design. The discussion is general as to the magnitude of the loads, since the actual design load values will be governed by applicable codes or will be included in the project specifications.

Vertical Loads

The design load for falsework includes vertical dead load, live load, and lateral loads. The dead load consists of the weight of the falsework, formwork, reinforcing steel, and the weight of the concrete. The weight of the concrete is usually assumed at 150 lb/ft³. For lightweight concrete, the actual unit weight of the concrete should be used. An additional 10 lb/ft³ is used for the weight of reinforcing steel and forms. Live loads include actual load of equipment supported by falsework and a minimum uniform load applied over the total area supported by the falsework. The uniform live load ranges from 20 to 50 lb/ft² depending on the specifications. The weight of bridge deck-finishing machine is taken as concentrated loads at the edge of bridge deck. Work bridge equipment behind the finishing machine should also be taken into consideration. The design load for equipment loads at the edge of deck is typically 75 lb/ft.

Live loads on walkways supported by the falsework are typically 100 lb/ft². Regardless of the actual vertical load imposed, the minimum design load for falsework members shall not be less than 100 lb/ft² for the combined live and dead load. Figure 18.6 shows a typical live load configuration for falsework construction.

Prestressing Forces.   Most prestressed concrete box-girder bridges are single frame bridges where the vertical prestress reactions are transferred to the columns and the abutments. At multi-frame bridges the prestress forces may also be transferred to hinges between frames. These hinges are often placed at 1/5 of the span length between columns where the bending moment in the superstructure is relatively small. The hinges are often constructed in stages and require to be supported on falsework during the prestressing operation. After the prestressing of the bridge is complete, the span on the long side of the hinge will be supported by the column on one side and the hinge on the other. Therefore the falsework at the hinge must be designed to support approximately half of the bridge span between the column and the hinge (see Fig. 18.7).

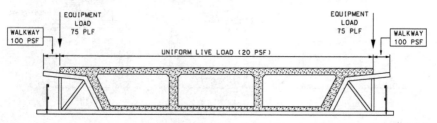

FIGURE 18.6   Typical live load on falsework.

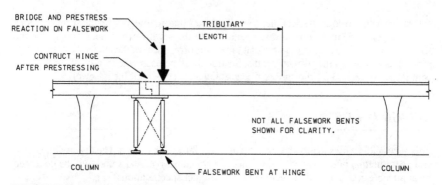

FIGURE 18.7   Falsework bents at hinge.

Horizontal Loads

All falsework design specifications include a requirement that the falsework be capable of resisting a horizontal design load. This requirement is included to provide a criterion for bracing design, and thus ensure the stability of the falsework system.

Typically, the horizontal design load will be either an assumed minimum load or the calculated wind load, whichever is greater. The minimum horizontal load is typically specified as not less than 2 percent of the total supported dead load at the location under consideration. Horizontal loads due to seismic activity are usually not considered. Falsework systems of all types are considerably more flexible and thus have inherently greater ability to accommodate ground motion than permanent structures. Because the falsework supports the full dead load of the superstructure for a relatively short time, the statistical risk is low, even in regions of known seismic activity. However, when a seismic event occurs, the falsework should be inspected to make sure the integrity of the falsework is not compromised.

Wind loads

In general, wind loads will not be a governing design consideration for falsework less than about 30 ft in height. Because of the number of variable factors involved, determining the actual force by wind on bridge falsework presents the falsework designer with an indeterminate problem. Factors affecting the wind forces include the true wind velocity, the "solidity ratio" or percentage of solid surface in a given frontal area, the downwind width of the falsework, the shading effect of and distance between successive downwind members, the drag or shape factor for the various members, and the height of the falsework above ground.

The wind load on the falsework can be determined using the *International Building Code* (IBC), but many public agencies have their own specifications for wind loading. The wind pressure values are often specified by height zone and location (at traffic openings or other locations). The wind pressure values should be based on wind speeds that are appropriate for the region where the falsework will be constructed. In general, the wind load for falsework is the product of the wind pressure value, the impact area, and the shape factor. The shape factor depends on the type of falsework system used. For steel pipe post the shape factor is 1, but, when shoring frame towers are used, the shape factor can be 2.2 or higher depending on the member size of the tower.

In Table 18.1 the drag coefficient, Q, shall be determined as follows: $Q = 1 + 0.2W$, but not more than 10, where W is the width of the falsework system measured in the direction of the wind being considered, ft.

The wind impact area is the gross projected area of the falsework, excluding the area between falsework bents or towers where diagonal bracing is not used. Areas where

TABLE 18.1 Wind Pressures on Falsework

Height zone (feet above ground)	Wind pressures, psf, on components other than post and heavy duty shoring							
	70 mph		80 mph		90 mph		100 mph	
	Typical	Traffic	Typical	Traffic	Typical	Traffic	Typical	Traffic
0 to 30	1.5 Q	1.5 Q + 5	2.0 Q	2.0 Q + 5	2.5 Q	2.5 Q + 5	3.0 Q	3.0 Q + 5
30 to 50	2.0 Q	2.0 Q + 5	2.5 Q	2.5 Q + 5	3.0 Q	3.0 Q + 5	3.5 Q	3.5 Q + 5
50 to 100	2.5 Q	2.5 Q + 5	3.0 Q	3.0 Q + 5	3.5 Q	3.5 Q + 5	4.0 Q	4.0 Q + 5
over 100	3.0 Q	3.0 Q + 5	3.5 Q	3.5 Q + 5	4.0 Q	4.0 Q + 5	4.5 Q	4.5 Q + 5

TABLE 18.2 Wind Pressure on Falsework Post and Heavy-duty Shoring

| Height zone (feet above ground) | Wind pressures, psf, on post and heavy duty shoring | | | | | | | |
| | 70 mph | | 80 mph | | 90 mph | | 100 mph | |
	Typical	Traffic	Typical	Traffic	Typical	Traffic	Typical	Traffic
0 to 30	15	20	20	25	25	30	30	35
30 to 50	20	25	25	30	30	35	35	40
50 to 100	25	30	30	35	35	40	40	45
over 100	30	35	35	40	40	45	45	50

flexible bracing systems are used, such as cable, steel rods or reinforcing steel bars, can also be excluded.

Table 18.2 shows wind pressures on falsework post and heavy duty shoring for various windspeeds and height zones.

The following example problem illustrates the wind pressure distribution on a falsework (see Fig. 18.8). The example assumes a falsework bent adjacent to traffic with the following dimensions:

- Falsework width is 40 ft.
- 4ea. 20-in-diameter steel pipe post per bent.
- Falsework bent is 105 ft tall (*H*).
- The stringer beams are 2 ft deep.
- Tributary span per bent is 50 ft.
- Wind velocity is 70 mph.

See Table 18.3 for the results of the example problem

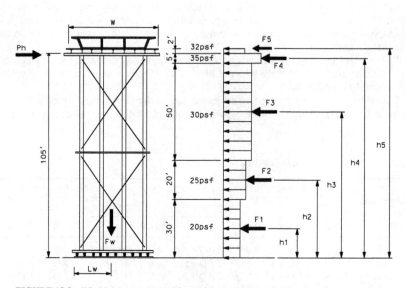

FIGURE 18.8 Wind load example on falsework bent.

TABLE 18.3 Results for Wind Load Example

Wind load on stringers per example:

a	b	W	Q = 1 + 0.2W	P = 3.0Q + 5	e	d	F = (P)(e)(d)	h = (b − a)/2 + a	M2 = (F)(h)
Height zone (feet above ground)		Falsework width ft	Drag coefficient	Wind pressure psf	Tributary span length lf	Stringer depth ft	Wind force lb	Arm lf	Moment ft-lb
105 to 107		40	9.00	32.00	50	2	3200	106.0	339200

Wind load on pipe post per example:

a	b		c	d	e	n	F = (b − a)(d)(e)(n)	h = (b − a)/2 + a	M1 = (F)(h)
Height zone (feet above ground)		Post type	Wind pressure psf	Pipe width ft	Pipe shape factor	No of pipe per bent EA	Wind force lb	Arm ft	Moment ft-lb
0 to 30		Pipe	20	1.67	1	4	4008	15.0	60120
30 to 50		Pipe	25	1.67	1	4	3340	40.0	133600
50 to 100		Pipe	30	1.67	1	4	10020	75.0	751500
100 to 105		Pipe	35	1.67	1	4	1169	102.5	119823
							Total overturning moment (ft-lb)		1065043

The total over turning moment: $M_{total} = M1 + M2 = 1,404,243$ ft-lb.
The horizontal design load is: $P_h = M_{total}/H = 13.374$ lb.

As shown in the example problem, when calculating the wind impact area of supported falsework, it is customary to neglect the formwork area above the bridge soffit. This is the case because, when subjected to a design wind velocity, it is generally assumed that the forms, because they are not designed to resist wind forces, would be blown of the falsework.

Impact Loads

Falsework design specifications at locations over or adjacent to public roadways or railroads usually include the following minimum requirements to resist impact loading:

- Vertical load on the falsework posts or towers shall be greater than 150 percent of the design load.
- Posts shall be mechanically connected to its supporting base to withstand a force of not less than 2000 lb, applied at the bottom of the post in any direction except toward the roadway or railroad track.
- The top of the post shall be mechanically connected to the cap beam capable of resisting a load in any horizontal direction of not less than 1000 lb.
- Mechanically connect to the falsework cap or framing, all exterior falsework stringers and stringers adjacent to the ends of discontinuous caps, the stringer or stringers over points of minimum vertical clearance and every fifth remaining stringer. The mechanical connections shall be capable of resisting a load in any direction, including uplift on the stringer, of not less than 500 lb.
- Connect all stringers to the caps for falsework spans over railroads.
- Connect timber bracing to falsework bents with 5/8-in-diameter or larger bolts.
- Timber posts shall have a minimum section modulus about each axis of 250 in^3.
- Steel post shall have a minimum section modulus about each axis of 9.5 in^3.
- Bracing at bents adjacent to railroads shall be designed for the required assumed horizontal load or 5000 lb, whichever is greater.

FALSEWORK STABILITY

In falsework design terminology, the term "stability" means resistance to overturning or collapse of the falsework system as a whole or an independent element of the system. Resistance to both overturning and collapse is provided by the falsework bracing system, which must be designed to resist all forces produced by the horizontal load.

It is important to recognize the technical distinction between "overturning" and "collapse" as these terms are used to describe the failure modes when the falsework is subjected to horizontal forces. Overturning describes the failure mode that occurs when the bracing system is of sufficient design strength to force the braced falsework to act as a single, rigid unit. In such cases the braced falsework will fail by overturning, or rotation about its base. If, however, the bracing system cannot prevent distortion of the falsework when it is subjected to horizontal forces, the system will collapse internally rather than overturn.

Falsework bracing whose purpose is to prevent collapse is customarily referred to as "internal" bracing, whereas bracing used to prevent overturning is described as "external" bracing.

The stability of pony bent systems should be given special consideration. Pony bents should be individually braced, and the bracing should be designed to resist the over-turning moment produced by the lateral design load acting at the top of the posts of the pony bent.

The overturning stability can also be compromised due to excessive stacking of material at the falsework bent foundation. This may occur when available posts are not quite long enough, adjustments for grading errors, or to provide a level sill beam at sloping ground. When the height-to-width ratio exceeds 2:1, the foundation assembly shall be externally braced to resist the greater of the horizontal wind, construction load, or a minimum 2 percent of the falsework dead load force applied to the top of the sill beam.

Transverse Stability

In a framed bent in a conventional falsework system, the internal bracing must be designed to resist the full horizontal design load. For analysis, the design load acts parallel to the direction of the bracing. In the previous wind load example, the lateral load must be resisted internally by the bracing system.

Even though the falsework is adequately braced to prevent collapse, a tall and narrow system may nevertheless fail by overturning, or rotation about its base, when the horizontal design is applied. Overturning may occur unless the falsework is inherently stable against overturning by reason of its shape and/or its dimensions, or is externally braced to prevent overturning.

For stability analysis, the horizontal design load produces a driving moment that acts to overturn the falsework system or element of the system under consideration. When calculating the overturning moment, the horizontal design load should be applied at the top of the vertical load-carrying members of the system, that is, the top of the posts or towers. The moment arm is the vertical distance between this location and the base of the falsework about which overturning rotation would occur.

The external stability must be provided by a resisting moment equal to the center of gravity of the weight of the falsework bent (including stringers, soffit, and rebar) multiplied by the distance between the falsework bent's rotation point and the center of gravity.

According to Fig.18.8 the resisting moment is as follows: $M_{resist} = (F_w)(L_w)$.

If the resisting moment is less than the overturning moment, the difference must be resisted by external bracing. Generally, cable guys are used for this purpose.

Longitudinal Stability

To ensure longitudinal stability, it is necessary to provide a system of restraint that will prevent the falsework bents from overturning when the horizontal design load is applied in the longitudinal direction. This can be accomplished by bracing between adjacent bents, or by transferring the horizontal load from unbraced falsework bents to a braced falsework bent. Longitudinal stability of falsework runs parallel with the bridge structure. Cable bracing can be used between bents to stabilize the bents. At a minimum a brace is required at each end of the falsework bent. Typically the cables are placed on either side of the top cap to the bottom sill beams of the adjacent bents. If adjacent bents are closely spaced, the cables may be too steep and should be placed to bents farther away. Falsework bents adjacent to columns or pier walls of the bridge structure can be braced against them for stability.

At traffic openings or other obstructions, it may not be possible to use cables to brace all falsework bents (Fig. 18.9, elevation 1). In this case other means can be used to transfer the

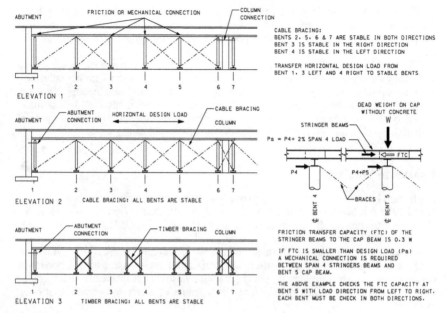

FIGURE 18.9 Longitudinal stability examples.

lateral design loads to a braced bent. In many cases, it is possible to take advantage of the frictional capacity between the stringer beams and the cap beam to transfer at least a part of the total longitudinal force acting at the bent. If frictional resistance alone is not sufficient to carry the load across the bent, a mechanical connection between the stringer beam and the cap beam must be provided to carry that portion of the total load in excess of the frictional transfer capacity at the location under consideration. Typically a friction factor of 0.3 is assumed for all contact surfaces, but this factor should be adjusted when wet conditions may occur. The transfer capacity based on friction is the total dead load (without concrete) on the falsework bent multiplied by the friction factor. The dead load used to generate the friction transfer capacity may include the weight of the soffit (plywood and joists), stringer beams, form panels, and rebar.

When mechanical connecting devices are used to transfer horizontal forces across an unbraced bent, at least two such devices should be used at each bent and they should be spaced far enough apart transversely to prevent eccentric loading on the connection.

Mechanical connections between stringer beams and cap beam can be made with a welded connection or bolted connection. "C" clamps can also be used to transfer lateral loads to stable bents (see Fig.18.10).

Longitudinal stability of the falsework system should be maintained during all phases of the falsework construction, from installation to grade adjustments until removal of the falsework is complete. Temporary braces must be used if permanent falsework braces are not yet installed or are no longer effective. Cable guys connected to concrete deadmen are often used for temporary bracing. Heavy-duty pipe braces are used for "push-pull"-type braces, resisting both tension and compression forces. Timber "A" frames can be used for low falsework with timber posts.

Timber braces can also be used between bents with timber post to provide longitudinal stability (see Fig. 18.9, elevation 3). Timber boards used for bracing are typically not longer

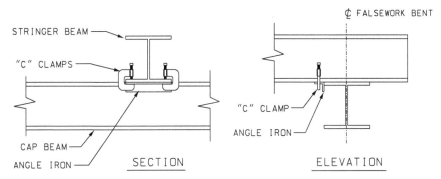

FIGURE 18.10 "C" clamp detail for transferring longitudinal loads.

than 20 ft, so falsework bents should be spaced relatively close together. Timber braces are applied at each post; this requires that posts of the braced bents are perpendicular to each other to ensure that the braces are placed flat against the post.

FALSEWORK DESIGN

As is the case with permanent work, a bridge falsework system and the individual components thereof should be designed using the formulas of civil engineering design applicable to statically determinate framed structures and, where appropriate, statically indeterminate structures as well. However, falsework is a temporary construction facility that differs from permanent work in both the way in which it is constructed and the manner and sequence in which it is loaded. These differences lead to a number of design considerations that are unique to falsework systems, as discussed in the following sections.

In most cases falsework systems are designed according to the agency's specifications. However, the specifications may vary between agencies or can be very general. The Caltrans *Falsework Manual* is often recognized by many agencies as industry standard for falsework design. The design approach in the *Falsework Manual* is often simplified and based on proven methods for temporary construction. Other design manuals, such as the AISC for steel design or the NDS for wood design, can also be used for design of falsework components. Some design methods in these manuals may vary significantly from the *Falsework Manual*, but must be used when specified or where the *Falsework Manual* does not apply.

The falsework system is subject to different load conditions during the various stages of construction. For a prestressed, cast-in-place concrete box-girder bridge, the following stages of falsework must be considered in the design to make sure that the falsework is stable during its "life":

- Falsework erection stage: The falsework bracing shall be designed to assist in the stability of the falsework during erection.
- Soffit and stem concrete placement: Falsework will see its first major vertical dead load.
- Deck concrete placement: Full vertical dead load is applied to falsework.
- Prestressing forces on falsework: Falsework at hinges must be able to support bridge reactions due to prestressing forces.

- Closure pours and construction joints.
- Falsework removal: During the removal the falsework shall remain stable. Removal equipment on top of the bridge must also be checked for its loading on the bridge deck.

Allowable Stresses

Unlike permanent work where construction is accomplished with new materials in virtually every instance, the typical falsework system will be constructed with a combination of new and used materials, or used materials exclusively. The type or grade for used material may not be known; therefore a careful evaluation of the material used in the falsework design should be done before establishing the allowable stresses for the design.

Agencies may specify allowable working stress values for unidentified material used in falsework. If needed, testing laboratories could also be used for establishing allowable stress values.

Structural Steel.   For standard rolled beam and other structural shapes of the type used in falsework, the grade of steel may vary from ASTM A36, ASTM 572 to ASTM A992. It is generally accepted to assume A36 grade steel when the grade of steel cannot be identified for used beams. ASTM A992 with yield strength of 50,000 to 65,000 psi was introduced around the year 2000 and has generally replaced the older ASTM grades A572 and A36 for wide-flange rolled beams.

Agencies such as Caltrans include maximum allowable working values for unidentified grades of structural steel based on A36 in their standard specifications.

Table 18.4 shows the maximum allowable working values for unidentified grades of structural steel:

In the Table 18.4 formulas, L is the unsupported length; d is the least dimension of rectangular columns, or the width of a square of equivalent cross-sectional area for round columns, or the depth of beams; b is the compression flange width and t is the thickness of the compression flange; r is the radius of gyration of the member. All dimensions are expressed in inches.

Most commonly used grades of steel for falsework pipe are the following:

- ASTM A252 Grade 2: minimum yield strength = 35,000 psi
- ASTM A252 Grade 3: minimum yield strength = 45,000 psi

TABLE 18.4   Working Values for Unidentified Steel per the Department of Transportation Standard Specifications of California, Nevada, and Washington State

Description	Unit	California	Nevada	Washington
Yield	psi	36,000	36,000	30,000
Tension, Axial, Flexural	psi	22,000	22,000	16,000
Compression, Axial	psi	$16000-0.38(L/r)^2$	$16000-0.38(L/r)^2$	$14150-0.37(KL/r)^2$
Shear	psi	14,500	14,500	9,500
Web Crippling	psi	27,000	16,000	22,500
Compression, Flexural	psi	$12 \times 10^6/(Ld/bt)$	$12 \times 10^6/(Ld/bt)$	$16000-5.2(L/b)^2$
Compression, Flexural (max)	psi	22,000	22,000	16,000
Modulus of Elasticity	psi	30,000,000	29,000,000	29,000,000

Timber. The strength characteristics of wood depend on two factors: the quality of the piece (i.e., the commercial grade) and the wood species. Design values for a given grade and species, used commercially in the United States, may be found in the *National Design Specification for Wood Construction (NDS)* published by the National Forest Products Association. Many timber design handbooks and other timber books incorporate material from the NDS specification, so that information about the appropriate working stress values to be used for timber is widely available for new material.

The lumber for falsework construction is often used lumber from previous falsework projects. Heavy falsework timbers such as posts, pads, or stringer beams can be reused many times without suffering any significant damage or loss of quality by reason of such reuse. Dimension lumber, because it often must be cut to fit a particular use, has a shorter life cycle, but it retains its quality until it is no longer of a size having a practical use as a load-carrying member.

Although used timber in falsework construction should not be a problem, the timbers should be inspected to ensure that the integrity of the timber falsework members is not compromised. Used timber may be warped, which does not reduce the inherent strength of wood, but may not provide sufficient bearing areas for joists and beams. Splits are also common in used wood, which is a separation of the fibers extending from one surface completely through a piece to another surface. Splits will reduce the shearing strength of members subject to bending, but have little effect on the strength of members subject to compression parallel to or perpendicular to grain.

The timber design in the *Falsework Manual* is based on using Douglas fir-larch. The appropriate values must be used when other species of wood are selected for falsework components.

Table 18.5 lists the values for maximum allowable stresses for timber when high-quality lumber is used. In the Table 18.5 formulas, L is the unsupported length; d is the least dimension of rectangular columns, or the width of a square of equivalent cross-sectional area for round columns.

Stringer Beams

Stringer beams are placed below the soffit to distribute the bridge load to the falsework bents (see Fig 18.11). At a box-girder bridge, the beams are typically placed at the girder line and at the edge of the bridge deck. The beams can also be placed between the girder

TABLE 18.5 Maximum Allowable Stresses for Timber per the Department of Transportation Standard Specifications of California, Nevada, and Washington State

Description	Unit	California	Nevada	Washington
Compression perp to grain	psi	450	450	450
Compression perp to grain when moisture content $\geq 19\%$	psi			300
Modulus of elasticity (E)	psi	1,600,000	1,600,000	1,600,000
Comp. par. to grain	psi	$0.3E/(L/d)^2$	1600, subject to column action correction	$0.3E/(L/d)^2$
Comp. par. to grain (max)	psi	1,600	1,600	1,500
Flexural stress when nominal depth > 8 in	psi	1,800	1,800	1,800
Flexural stress when nominal depth ≤ 8 in	psi	1,500	1,800	1,500
Horizontal shear	psi	140	140	140
Axial tension	psi	1,200	1,200	1,200

FIGURE 18.11   Stringer beam layout at connector ramp.

lines to reduce the span length of the joists. At solid concrete bridge bent caps or at concrete slab bridges, the beam spacing is often controlled by the maximum span length for the joist.

Timber beams can be used for relatively short spans, but rolled steel beams per the AISC manual are typically used for longer spans. For typical concrete box-girder bridges, falsework span lengths vary between 30 and 50 ft. Longer spans are used for spanning traffic lanes or other obstructions. Maximum span length for steel beams is around 100 ft. Spans longer than 100 ft may require steel trusses or deep plate girders that may make cast-in-place concrete bridges less economical than other types of bridges.

A large percentage of the falsework cost is the purchase of the steel or timber beams; therefore beam selection should be carefully considered. The weight of beam and therefore its cost will increase with the span length. But shorter beams require more falsework bents. In selecting the falsework beams, the following items should be taken into consideration:

- The falsework contractor's beam inventory and availability.
- Most economical falsework bent layout
- Spans over traffic may require shallow beams to provide the specified minimum vertical clearance.
- Deeper beams are more economical for longer spans to limit deflection.
- Deeper beams may require more braces to reduce the unbraced length.
- Available cranes sizes. Heavier beams require large cranes, trucks, etc.

The falsework beams are usually several feet longer than the span length between its supports. This means that the beams overlap at the cap beams and therefore they are not exactly placed directly below the bridge girder line. The soffit joists usually have sufficient capacity to handle the girder loads for the small offset between stringer beam and the bridge girder line, but in order to minimize the girder load on the joist, the beams should be placed

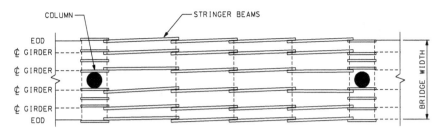

FIGURE 18.12   Typical stringer beam layout.

as close as possible to the girder line. Typically beams are placed on a skew to allow for overlap at the falsework bent locations (see Fig. 18.12).

Uniform beam sizes and lengths will provide the most economical falsework layout. However, often at large, traffic openings or other obstructions require the use of beams with different depths and lengths. For longer spans, beams may be deeper than the adjacent shorter spans or multiple beams can be used per girder line (see Fig. 18.13).

At traffic openings with large skews, and therefore large spans, the stringer beam may not be long enough to be placed along the girder line, so the beam layout may have to be skewed to the bridge to reduce the beam span (see Fig. 18.14). This type of layout with a

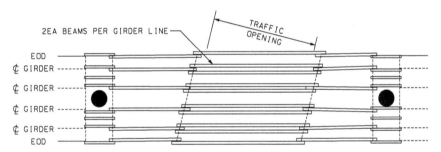

FIGURE 18.13   Stringer beam layout example at traffic opening.

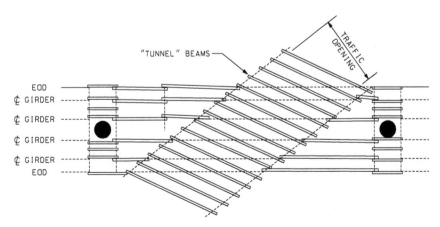

FIGURE 18.14   Stringer beam layout example at large skewed openings.

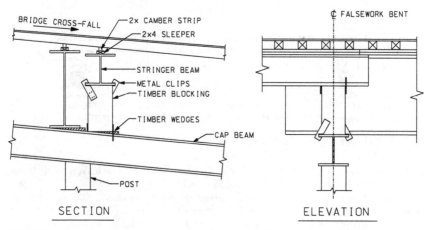

FIGURE 18.15   Stringer placement details with unequal depth beams.

"tunnel section" complicates the design of the stringer beam. Each beam at the tunnel section will have different loading, deflection, and camber requirements. Also, the joists must be able to support the girder weight between the stringer beams. To simplify the joist design and minimize its section, another layer of stringer beams can be installed along the girder line on top of the skewed stringers at the tunnel section.

When beams with different depths are used in adjacent spans, the deeper beams are placed directly on top of the cap beam and the shallower beams will have to be placed on blocking to make up the difference in beam height. Typically timber blocking is used for the support of the shallow beam (see Fig. 18.15).

The cap beam is typically placed at the same cross-fall as the bridge deck, which can be as much as 12 percent. When the cross-fall is small, 2 percent or less, the stringer beam may be placed directly on the cap beam without losing much capacity due to biaxial loading on the out of plumb beam. For larger than 2 percent cross-falls, the stringer beams must be analyzed for biaxial bending or placed on wedges to ensure that the beams are plumb. A wooden sleeper strip may be required at large cross-falls to provide clearance between the edge of the top flange and the soffit material.

Steel Beam Design.   The design load on stringer beams includes live load, weight of the supported load, and its own weight. The bridge load on the stringers is often not equal across the bridge section. Usually the stringer size is the same for a falsework span, but the loading for each stringer may vary. An analysis of the typical section must be done to determine the maximum stringer load. Figure 18.16 shows an example for a loading distribution to the stringers for a typical box-girder bridge.

The example in Fig. 18.16 uses a live load of 20 psf, an edge of deck load of 75 plf, and a 160 pcf for concrete, forms, and rebar. The sloping exterior girder requires a strut to support the formwork and girder concrete before the deck concrete is complete. The reaction in this strut should be applied to the edge of deck stringer and may be deducted from the load on the stringer at the exterior girder. In addition to the loading from the typical bridge section, other bridge components may add to the loading on the stringers, such as soffit and girder flares, bent cap loads, intermediate diaphragm loads, hinge loads, etc.

It is typical for deflection calculations to only include the weight of the concrete. For a box-girder bridge that is constructed in stages, the first stage of the bridge (i.e., soffit and girders without the deck) may have sufficient strength to support its own weight over

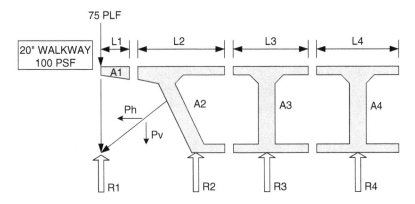

REACTIONS on STRINGER BEAMS:

- R1 = L1 x 20 psf + A1 x 160 pcf + 100 psf x 20" + 75 plf + Pv
- R2 = L2 x 20 psf + A2 x 160 pcf -Pv
- R3 = L3 x 20 psf + A3 x 160 pcf
- R4 = L4 x 20 psf + A4 x 160 pcf

FIGURE 18.16   Typical concrete box-girder load distribution.

a small span length before the deck concrete is placed. In general, when the stringer span length is less than 4 times the depth of the box-girder bridge, the weight of the deck concrete does not have to be included for the stringer design load.

The stringer beams used in falsework construction are checked for bending stress, shear stress, deflection, buckling, web crippling and biaxial bending.

The maximum deflection limit is specified by the agency, but it typically varies between L/240 and L/360. The beam design is typically done with beam analysis computer programs, but the biaxial bending may not always be included in the analysis. The *AISC steel manual* includes a formula for biaxial bending (see Fig 18.17).

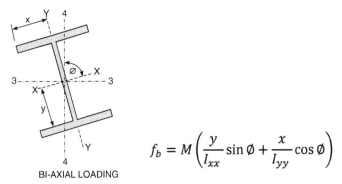

$$f_b = M \left(\frac{y}{I_{xx}} \sin \emptyset + \frac{x}{I_{yy}} \cos \emptyset \right)$$

Where: M is the bending moment due to
vertical loading on the beam.

FIGURE 18.17   Biaxial bending steel stringer beam.

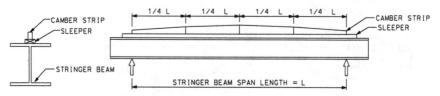

FIGURE 18.18   Typical camber strip configuration.

Camber Strips.   Camber strips are required to offset the deflection of the beam in addition to the required camber for the permanent structure. When superstructures are not placed in one pour, such as box-girder bridges, a modified beam deflection should be used to calculate the actual deflection. For a multiple section pour, the concrete structure itself may reduce the deflection of the stringer beam. The camber strip is often made of a 2× timber board and cut to the beam deflection at ¼ spans (see Fig. 18.18). The camber strip should be centered on the stringer beam to avoid undesirable torsional stresses. Although camber strips are typically not checked for crushing at the joists, it may need to be considered in sensitive areas of large load situations.

Beams are usually longer than the span length. When beams have long "tails" beyond their supports, uplift of the ends of the beam should be analyzed to make sure the ends will not lift up the soffit. A deeper camber strip or sleeper may be needed to avoid beam uplift from hitting the soffit.

Beam Continuity.   Continuous stringer beams over more than two supports is not recommended due to the sequence of the concrete placement. During a concrete pour, the loads will not be applied evenly to all spans at the same time but gradually from one end of the beam to the other end. Deflection of continuous beams will be uncertain and may create uplift in the unloaded span during the concrete pour. When continuous beams are used, the design should be such that no reduction in loads or stresses should be taken due to continuity, but increases in loads or stresses should be included. For example, the simple span condition is assumed when calculating the positive bending moments, whereas full continuity is assumed when calculating negative bending moments in these same stringers. In like manner, continuity is assumed when calculating beam reactions on interior supports, but the simple span condition is assumed when calculating the reactions at the end support.

Beam Bracing.   Beam bracing is used to prevent the compression flange from buckling. Beam bracing can consist of x-bracing between the stringer beams (see Fig. 18.19). The design load for the bracing can be taken as 2 percent of the maximum calculated compression force in the beam flange of the stringer beam.

The spacing and requirement for beam bracing is a function of the unbraced length of the beam. Typically the beam is designed for the required falsework span assuming the compression flange of the beam is braced. The brace spacing is determined to make sure that buckling does not govern the design of the beam.

Timber Beams.   For relatively short falsework spans a timber beam may be a cost-effective solution. Some falsework removal methods may be such that the stringer beams in some spans will fall to the ground while lowering the falsework beams and soffit. With this method the timber beams reduce the risk of damaging nearby columns or other structures. Common timber beam sizes used in falsework are 6×12, 6×16, and 12×12.

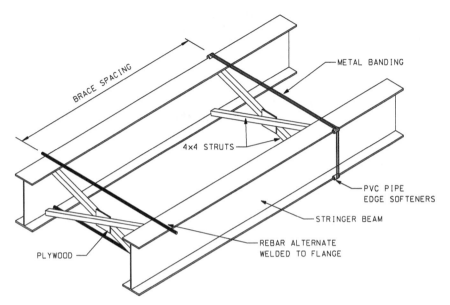

FIGURE 18.19 Typical stringer beam bracing.

The design of the stringer beams includes checking for flexural stress, shear stress, deflection, and beam stability.

Falsework Post

Posts used in falsework systems are the vertical support members between the foundation and the top cap of the falsework bent. Posts are the more critical components of the falsework system and therefore require careful attention during design and construction. Different types of post can be used, such as timber post, steel post, or proprietary systems.

The design load for the posts is the full dead load of the concrete bridge, the weight of the forms, stringer beams, and cap beams and live load. The design load is distributed through the stringer beams and cap beam to the post. Falsework posts can be placed anywhere along the falsework bent but are often placed at the bridge girder locations to minimize the bending stresses in the cap beam. A moment distribution analysis for the cap beam can be done to determine the post reactions.

For posts adjacent to traffic openings, agencies often specify that the post should be designed for 150 percent of the design load. The size of the post used in falsework is typically based on the contractor's available inventory. When the design load on the post is greater than the maximum allowable post load on the available posts, additional posts must be added to the falsework bent or larger post shall be selected.

Timber falsework posts are usually braced with timber braces with nailed or bolted connections directly to the post. When the braces are connected close to the top and bottom of the post, the horizontal force of the brace applied to the post is typically not considered in the post analysis. When transverse cable bracing is used for a timber bent,

FIGURE 18.20   Timber falsework post on sill beam.

the post may be subject to horizontal or eccentric loading due to elongation of the cable. When horizontal loads are present, the post must be checked for combined axial and bending stress.

Timber Post.   Timber posts are frequently used in falsework because of its durability and constructability. The most common timber post used in falsework in the western part of the United States is a 12 × 12 Douglas fir. A timber post can last for years when it is used and stored in dry condition and handled with care. Checks and splits are common in timber posts, but they do not greatly affect strength, unless they occur in zones of severe grain slope.

The flexibility and the great strength of timber make a wooden post a good choice for falsework design. The post can be cut to length with a chainsaw, and the desired cross-fall can be easily applied to the top of the post. The connection between the post and the cap beams is done with the use of metal beam clips (see Fig. 18.20).

A timber post has a high load capacity, is lighter than steel, and therefore easier to handle. The timber post is also easy to inspect for defects or other capacity-limiting factors. A disadvantage of a timber post is the limited height. Although splicing a timber post is possible, it is usually not economical to do. Therefore every time a post gets cut to length, the post will lose some of its value. Eventually the post becomes too short and will not be useful any longer.

The capacity of the timber post will decrease rapidly when the unbraced length is more than 20 ft. Therefore the use of timber post for bridges with falsework posts higher than 20 ft may no longer be cost-effective. The unbraced length of a post can be reduced by adding more transverse and longitudinal braces.

The maximum allowable load on the post is based on the cross-sectional area of the post, the effective length of the post, and the axial stress parallel to grain. Falsework posts are considered pinned-pinned-type columns where the effective length factor, $K_e = 1$, consequently the effective length is equal to the unbraced length of the post. Many agencies specify the following Euler formula to determine the maximum allowable compression stress for falsework posts:

$$F_c = \frac{0.3E}{\left(\dfrac{L_e}{d}\right)^2}$$

where: F_c = Allowable compression stress parallel to grain, psi
E = modulus of elasticity, psi
L_e = effective length of the compression member, in
d = least dimension of rectangular compression member, in

The slenderness ratio for solid columns L_e/d, shall not exceed 50.
    The maximum post load:

$$P = (F_c)(A)$$

where A = the cross-sectional area of the post, in^2.
    When a design formula is not specified, the allowable stresses of the post can be taken from the *National Design Specification for Wood Construction (NDS)*. According to the design method per the NDS 2005 edition (ASD), the compression design value parallel to grain F_c is multiplied with adjustment factors. The factors with the most impact for temporary construction are the duration factor (C_d = 1.25 for construction), wet service factor (C_m), and the column stability factor (C_p). Other factors such as size factor (C_f), temperature factor (C_t), and incising factor (C_i) typically have a value of 1. The column stability factor is calculated as follows:

$$C_p = \frac{1 + (F_{cE}/F_c^*)}{2c} - \sqrt{\left[\frac{1 + \left(\dfrac{F_{cE}}{F_c^*}\right)}{2c}\right]^2 - \frac{F_{cE}/F_c^*}{c}}$$

where: F_c^* = reference compression design value parallel to grain (F_c) multiplied by all applicable adjustment factors except C_p, psi
    c = 0.8 for sawn lumber
    c = 0.85 for round timber poles and piles
    c = 0.9 for structural glued laminated timber or composite lumber
    F_{cE} = critical buckling design value for compression members, psi

$$F_{cE} = \frac{0.822\ E'_{min}}{(L_e/d)^2}$$

where: E'_{min} = adjusted modulus of elasticity for column stability, psi
    L_e = effective length of compression member, in
    d = least dimension of rectangular compression member, in

$$E'_{min} = E_{min}(C_m)(C_t)(C_i)$$

Adjusted axial stress: $F'_c = F_c(C_f)(C_d)(C_p)(C_m)$
Axial force capacity: $P_a = (F'_c)(A)$

When a post is subject to axial and side loads, the following interaction equation must be satisfied:

$$\left[\frac{f_c}{F'_c} + \frac{f_b}{F_b}\right] \le 1$$

where: f_c = actual compression parallel to grain, psi
    F'_c = design value for compression parallel to grain, psi
    f_b = actual bending stress, psi
    F_b = design value for bending stress, psi

The modulus of elasticity for column stability (E_{min}) in the NDS is based on a safety factor of 1.66. The falsework design engineer may determine that a larger or smaller safety factor may be more appropriate for the type of falsework under consideration.

Steel Posts.   Steel pipe posts are often used for tall falsework where timber posts are no longer practical to use. Pipe posts can be used for falsework in excess of 100 ft tall. The slenderness of the post is usually limited by specifications to $KL/r = 120$. If the unbraced length exceeds the KL/r limit, additional braces at the posts need to be installed to reduce the unbraced post length or posts with large radius of gyration must be selected. Typically, there are no specification limits to the pipe size that can be used for falsework, but the pipe size usually ranges from 16 to 24 in with various wall thicknesses. It is recommended to use a minimum wall thickness of 0.25 in. Thinner walls may satisfy the design requirements but are easily damaged on construction sites due to handling.

Steel pipe can be spliced together with a welded connection using a splice ring as a backing plate for a single bevel groove weld butt joint. Pipes can be cut to length with a cutting and beveling machine (manual or motorized). Pipes can be welded directly to the cap and sill beam, but often an endplate welded on both ends of the pipe is used and clamped to the cap beam and sill beam. The top of the pipe must be cut on an angle to create the required cross-fall and profile of the bridge. Square cuts with swivel caps can also be used to create the required cross-fall on top of the pipe.

Pipe falsework bents are typically braced with rebar or cable braces. While rebar bracing is typically connected directly to the post, the cable bracing is usually connected to the ends of the cap beam and sill beam. The rebar bracing connections should be close to the end of the post to avoid large bending stresses due to lateral bracing loads unto the pipe.

When the grade of steel cannot be identified, Caltrans and other agencies specify the following formula to determine the maximum allowable axial stress on the steel post.

$$F_{ca} = 16000 - 0.38 \times \left(\frac{L}{r} \right)^2$$

where L = the lateral unbraced length of the post, in
      r = radius of gyration, in
      F_{ca} = allowable axial stress, psi

When the grade of the steel pipe is known, the post can be designed according to the agency's specification or the AISC manual. In chapter E of the *AISC Steel Manual* (13th ed.), the following formulas are used to determine the design value for axial stress:

$$\text{Column Slenderness} = \frac{KL}{r}$$

where K = the effective length factor, usually 1 for falsework application.
The elastic critical buckling stress, F_e (ksi):

$$F_e = \frac{(\pi^2)E}{\left(\dfrac{KL}{r} \right)^2}$$

where E = modulus of elasticity, ksi.

The flexural buckling stress, F_{cr} (ksi), is determined as follows:

When
$$\frac{KL}{r} \leq 4.71 \sqrt{\frac{E}{F_y}}$$

where F_y = specified minimum yield stress, ksi.

$$F_{cr} = [0.658^{Fy/Fe}]F_y$$

When
$$\frac{KL}{r} > 4.71 \sqrt{\frac{E}{F_y}}$$
$$F_{cr} = 0.877F_e$$

Allowable axial stress, F_{ca} (kips):

$$F_{ca} = \frac{F_{cr}}{\Omega_c}$$

where $\Omega_c = 1.67$ (ASD).
  The maximum allowable post load, P_a (kips):

$$P_a = F_{ca} \times A_g$$

where A_g = gross area of pipe post, in^2.
  When falsework pipe is subject to flexure and axial force, the following formula should be used:

$$\left| \frac{f_a}{F_{ca}} + \frac{f_b}{F_B} \right| \leq 1.0$$

where f_a = required axial stress, ksi
  f_b = required bending stress, ksi
  F_B = available bending stress, ksi

Proprietary System.   Proprietary systems consist of lightweight modular frames that can be assembled to form a four-leg shoring tower. Frames can be stacked to achieve the required height. Screw jacks on top or at the bottom of the assembled shoring tower allows for final grade adjustments.

  Most proprietary systems have capacities that range from 10,000 lb to 100,000 kips per leg. Allowable post reactions are usually provided by the manufacturer, which include taking in consideration the lateral internal stability of the shoring tower. The falsework design engineer must determine the bracing requirements for the external stability of the shoring tower. The required foundation for shoring towers must also be designed by the falsework engineer. Unequal loading on the legs of a modular frame tower must be limited to avoid unequal settlement of the tower assembly.

FIGURE 18.21   Stringer beams placed on cap beam.

Cap and Sill Beams

The cap beam, placed on top of the post, supports the stringer beams between the bents (see Fig. 18.21). The sill beam, supported on multiple corbels, distributes the post loads to the falsework foundation. The stringer beam reaction is the design load for the cap beam and the post reaction is the design load for the sill beam. The cap beam must be checked for bending stress, shear stress, web crippling, and deflection.

Bending stresses in the sill beam are usually small due to the closely spaced corbels. Sill beams are often stiff enough to where a moment distribution over corbels is not used to calculate the reactions to the corbels, but post reactions are evenly distributed to the corbels below the sill beam. High post reaction may require web stiffeners in the cap or sill beam to resist web crippling. Local flange stresses at the post due to uneven bearing are unpredictable but should be considered when selecting a beams size. It is recommended that the beam flanges of a cap and sill beam be at least 5/8 in thick to allow for uneven bearing. The overturning stability of the beam is also critical for selecting sizes for sill beams. It is recommended that the depth-to-width ratio of cap and sill beams is around 1, for example, HP14 × 117.

Falsework Bracing

The falsework bents must be braced to resist the lateral loads acting on the falsework in all directions. In general, the falsework system includes transverse bracing and longitudinal bracing. The transverse bracing can be done between individual posts or between cap beam and sill beam. Longitudinal bracing typically consists of cable braces between adjacent bents. Timber bracing can be used for longitudinal braces when timber posts are used.

The type of transverse bracing varies and depends on the falsework system. Proprietary tower systems have framed units that inherently provide the bracing system. Timber falsework systems typically use timber boards for their transverse bracing. The transverse bracing for steel pipe post may include

- Rebar bracing between exterior pipe posts
- Wire-rope cables between top cap beam and sill beam
- Prestressing strand between top cap beam and sill beam

Timber Braces. Timber braces usually consist of 2× boards connected to the timber post with nails or bolts at bents adjacent to traffic. Timber braces can be applied on one side of the bent or on both sides, depending on the required number of braces.

The design of timber braces with bolted connection, per the NDS manual, varies from the design method per Caltrans' *Falsework Manual.* The bolted bracing method in the *Falsework Manual* is simplified and can be used when allowed by the specifications. Also, the *Falsework Manual* is based on Douglas fir-larch values, so this method should be adjusted when other species of wood are used. The method in the NDS manual is based on yield modes for the side member, main member, and the bolts.

The following is a bolted brace example using the Caltrans *Falsework Manual* for a timber brace with 2ea 2 × 6 side members and a 12 × 12 main member (post) using 1-in-diameter bolts (see Fig. 18.22):

Given:

- Douglas fir-larch
- Brace width: $W = 10 \text{ lf}$
- Brace height: $H = 8 \text{ lf}$
- Modulus of elasticity: $E = 1{,}600{,}000 \text{ psi}$
- Depth of brace: $d = 1.5 \text{ in}$
- Area of brace: $A_s = 8.25 \text{ in}^2$

Side member capacity:

- Calculate brace angle.

$$\alpha = \operatorname{atan}\left(\frac{H}{W}\right) \qquad \alpha = 38.66°$$

- Calculate the effective brace length assuming that the side members are nailed together where they cross each other.

$$L = \frac{1}{2}\sqrt{H^2 + W^2} \qquad L = 6.4 \text{ ft}$$

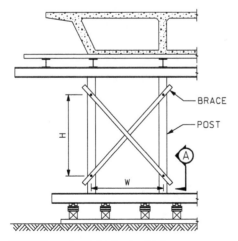

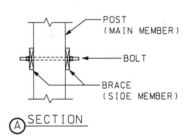

FIGURE 18.22 Timber bracing details.

- Calculate the allowable compression parallel to grain.

$$F_c = \frac{0.3E}{\left(\dfrac{12L}{d}\right)^2} \qquad F_c = 183 \text{ psi}$$

- Calculate lateral capacity for compression of side member.

$$C_s = (F_c)(A_s)\cos(\alpha) \qquad C_s = 1178 \text{ lbs}$$

- Determine the allowable bolt load parallel to grain.

For a side member loaded single shear parallel to grain, the allowable bolt load parallel to grain is 75 percent of the double-shear value for a member twice the thickness.

According to Table E-1 in *Falsework Manual*, the bolt design value for a 1-in bolt with a member thickness of 3 in is 3750 lb.

The allowable bolt load in the side member is

$$F_{ns} = 75\%(3750) \qquad F_{ns} = 2812 \text{ lbs}$$

Main member:

- From Table E-1:

    Allowable load parallel to grain (double shear): $F_p = 5080 \text{ lb}$

    Allowable load perpendicular to grain (double shear): $F_q = 2770 \text{ lb}$

- Angle between direction of grain and direction of load:

$$\theta = \operatorname{atan}\left(\frac{W}{H}\right) \qquad \theta = 51.34°$$

- Bearing strength of main member with Hankinson's formula:

$$F_{nm} = \frac{(F_p)(F_q)}{(F_p)\sin(\theta)^2 + (F_q)\cos(\theta)^2} \qquad F_{nm} = 3368 \text{ lbs}$$

- Assume single shear on both sides of posts, use 75 percent of double shear values:

$$F'_{nm} = 75\% F_{nm} \qquad F'_{nm} = 2526 \text{ lbs}$$

Total brace capacity:

- Determine duration factor for bolt capacity: $d = 1.25$
- Use smaller bolt capacity of side member and main member.

$$T = (\min(F_{ns}, F'_{nm}))(d)\cos(\alpha) \qquad T = 2465 \text{ lbs}$$

The compression capacity in the side member cannot exceed the bolt capacity.

$$C = \min(C_s, T) \qquad C = 1178 \text{ lbs}$$

- For two side members, the maximum allowable lateral load on the brace

$$P_h = 2(T + 50\%C) \qquad P_h = 6108 \text{ lbs}$$

Steel Bracing. Structural steel elements, such as angles, channels, or reinforcing bars, can be used for transverse bracing when steel beams or pipes are used for falsework post. The bracing can be connected directly to the steel posts. When steel pipe posts are used, the bracing is typically connected to "C" channel welded to the pipe post. The braces are usually connected only to the exterior post of the falsework bent. Steel braces are usually only considered in tension. Bolted connections are typically not used for brace connection. The strength of welded connections may be approximated by assuming a value of 1000 lb/in for each 1/8-in of fillet weld. This reduction in capacity is used because welding is not done in a controlled environment. If a higher weld value is required by the design, welding procedures must conform to the quality standards listed in the welding code, AWS d1.1.

Cable Bracing. Transverse bracing with cables can be done with wire-rope cables or pre-stressing strand. The wire-rope cables are connected to the cap and sill beams using traditional cable connection components such as shackles, thimbles, cable clips, and optional turnbuckles. All these elements are part of the design. The load in the cable is calculated as follows:

$$P_b = \frac{P_h}{(n)\cos(\alpha)}$$

where: P_b = load in cable, lb
P_h = lateral load applied to falsework bent, lb
n = number of cables per direction of brace
α = angle between cable and horizontal load

When cable bracing is installed on the falsework bent, a preload is applied to the cables in order for the cables to be sufficiently tight to ensure stability of the bent during the erection. The cables should not sag more than approximately 2 in after the falsework is erected. The *Falsework Manual* uses the following formula to calculate the sag:

$$A = \frac{(q)(x^2)}{8(P)\cos(\psi)}$$

where: A = sag, ft
q = cable weight, lb/ft
P = cable preload value, lb
x = horizontal distance between cable connection points, ft
ψ = angle between cable and sill beam, deg

A characteristic of wire-rope cable is that it will stretch when tension loads are applied to the cable. This stretch has two components: one is constructional stretch and the other is elastic stretch.

Constructional stretch happens when a load is applied to wire rope; the helically laid wires and strands act in a restricting manner, thereby compressing the core and bringing all the rope elements into closer contact. The result is a slight reduction in diameter and lengthening of the rope. The manufacturer will give recommended guidelines for constructional elongation in steel wire ropes as a percentage of rope under load. The percentage may vary from 0.25 to 0.50 percent based on the load in the cable. The Caltrans *Falsework Manual* uses the following formula to calculate the constructional stretch:

$$\Delta_{cs} = \frac{\text{Applied Load}}{65\% \text{ Cable Breaking Strength}} \times CS\% \times L$$

where: $CS\%$ = constructional stretch, % (provided by manufacturer)
  L = length of cable, ft
  Δ_{cs} = constructional stretch, ft

Elastic stretch is the actual physical elongation of the individual wires under load. It can be calculated by using the following formula:

$$e = \frac{W \times L}{A_c \times E}$$

where: e = elastic stretch, in
  W = load in cable, lb
  E = strand modules, psi
  A_c = cross section of the wire area, in^2

One of the consequences of the cable stretch is a small lateral displacement of the falsework cap beam (see Fig. 18.23). The cap movement introduces an eccentric load in the post and therefore a bending stress will be present in the post.

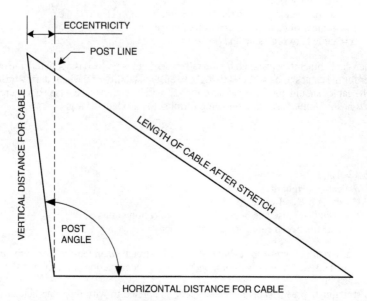

FIGURE 18.23 Falsework post eccentricity schematic due to cable brace elongation.

FIGURE 18.24 Transverse strand bracing with longitudinal wire-rope bracing connected to sill beam.

When prestressing strand is used for transverse bracing, the elongation behaves some-what different than wire rope but still has the same effect on the falsework bent. For strand the relaxation of steel causes a time-dependent loss in the initial prestressing force. The loss is generally assumed to be 2.5 percent when the initial prestress is 0.7 fpu (specified tensile strength of prestressing tendons). The actual load in the strand bracing is less than .7 fpu and therefore the relaxation will be less than 2.5 percent. Test information for relaxation of strand with small loads is not available; therefore 2.5 percent is conservatively used for relaxation loss in this design.

Caltrans limit the lateral movement of the cap to 1/8 in/ft of post height or ¼ of the post width or diameter, whichever is less. When the lateral movement is larger than specified, the post must be analyzed for combined bending and axial stress.

Wire-rope cable braces are also used for longitudinal bracing. The cables are typi-cally connected on either side at the ends of the cap beam and sill beam (see Fig. 18.24). A shackle through a drilled hole in the flange of the beam is a common connection detail. If more cable braces are required along the cap beam, the cable can be looped around the beam using PVC pipe at the flange edges of the beam to prevent damage to the cable.

An example of cable bracing calculation with prestressing strand (see Fig. 18.25) is discussed as follows.

Design loads:

- Vertical reaction in post: $P_v = 180,000$ lb.
- Horizontal load in brace: $P_h = 14,000$ lb.
- Apply preload in cable: $P_r = 2500$ lb.

Cable information (see Table 18.6):

- 0.5 in diameter low relaxation strand grade 270
- Cable section area: $A = 0.153$ in^2
- Specified yield: $f_{py} = 243,000$ psi

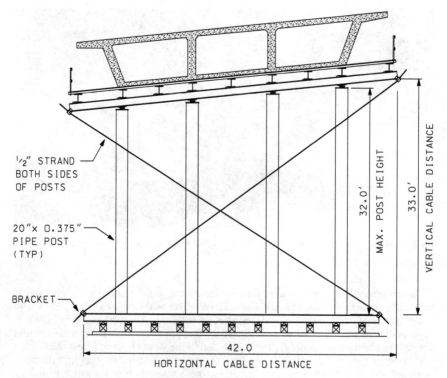

FIGURE 18.25 Falsework bent with prestressing strand bracing.

TABLE 18.6 Prestressing Strand Properties

Strand size in	Ultimate stress ksi	Yield stress ksi	Cross section area in²	Ultimate strength kips	Modulus of elasticity ksi	Nominal weight lbs/ft
0.5	270	243	0.153	41.3	28600	0.52
0.6	270	243	0.217	58.6	28600	0.74

- Modulus of elasticity: $E = 28.6 \times 10^6 \, \text{psi}$
- Relaxation loss: $RE = 2.5$ percent
- No. of cables per direction: $n = 2$

Brace analysis:

- Angle between cable and horizontal load of the longest bracing cable:

$$\alpha = \operatorname{atan}\left[\frac{L_V}{L_H}\right] \qquad \alpha = 38.16°$$

- Diagonal length of cable:

$$L_c = \frac{L_v}{\cos(\alpha)} \qquad L_c = 53.4 \text{ ft}$$

- Load in brace:

$$P_b = \frac{P_h}{\cos(\alpha)} \qquad P_b = 17805 \text{ lb}$$

- Load per cable:

$$P_c = \frac{P_b}{n} \qquad P_c = 8903 \text{ lb}$$

- Allowable load in cable: $\qquad P_a = 0.6 f_{py} A \qquad P_a = 22307 \text{ lb}$
- Add relaxation loss to cable load: $\qquad P_{cr} = P_c + 2.5\% P_c \qquad P_{cr} = 9125 \text{ lb}$
- Remaining load in cable: $\qquad P_{cn} = P_{cr} - P_r \qquad P_{cn} = 6625 \text{ lb}$
- Cable elongation:

$$e = \frac{(P_{cn})(12 L_c)}{(A)(E)} \qquad e = 0.97 \text{ in}$$

- Total cable length: $\qquad L_{cn} = L_c + e \qquad L_{cn} = 53.5 \text{ ft}$
- Top cap movement:

$$B = \text{acos} \left[\frac{(L_v^2 + L_h^2 - L_{cn}^2)}{2(L_v)(L_h)} \right] \qquad B = 90.18°$$
$$\theta = B - 90° \qquad \theta = 0.18°$$

- Cap displacement: $\qquad d = (L_v)(\sin(\theta)) \qquad d = 1.24 \text{ in}$

The post analysis must include the following adjustments due to cable bracing:

- Vertical cable reaction on post: $\qquad P_{cv} = (P_h)(\tan(\alpha)) \qquad P_{cv} = 11,000 \text{ lb}$
- Total post reaction: $\qquad P_s = P_v + P_{cv} \qquad P_s = 191,000 \text{ lb}$
- Bending moment in post: $\qquad M_b = d(P_s) \qquad M_b = 236,840 \text{ in.lb}$

Falsework Foundation

Depending on load and site characteristics, bridge falsework may be supported by a variety of foundation types. Concrete spread footings and/or pile foundations are often used to support relatively heavy concentrated loads. Pile foundations are used for structures over water or marshy ground, or where site or ground conditions preclude the use of spread footings. In most cases, however, the falsework will be supported on footings consisting of precast

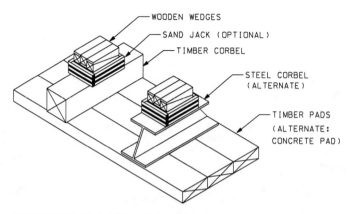

WOODEN WEDGES
SAND JACK (OPTIONAL)
TIMBER CORBEL
STEEL CORBEL (ALTERNATE)
TIMBER PADS (ALTERNATE: CONCRETE PAD)

FIGURE 18.26　Typical falsework foundations.

concrete or timber pads set directly on the ground surface (see Fig. 18.26). Precast concrete or timber systems are simple to construct, economical, and provide a safe and dependable method foundation support for the vast majority of falsework installations.

Although individual pads are occasionally used to support a single post, the most common arrangement consists of several posts supported by a continuous sill beam on top of multiple wedges, sand jacks, corbels, and pads.

Timber pads typically consist of 6×12 or 6×16 timbers that are placed flat on the ground parallel with the falsework bent line. The number of pads must be calculated based on the required pad area to resist the vertical load of the falsework bent. The length of the individual pad timbers varies but is usually 8 to 20 ft. It is good practice to stagger the pad joints and to make sure that the cantilever pad length between corbels does not exceed the maximum allowable. When using a multiple pad system, each individual pad should be supported by a minimum of two corbels. It is recommended to use a minimum pad thickness of 5.5 in and a minimum width of 2ea pads (6×12).

The corbels are short beams that distribute the post load on the sill beam to the individual pads in a multiple-pad system. Timber beams can be used for corbels at typical falsework systems; however, steel wide-flanged beams are used for corbels when the post loads are relatively high or when pads are relatively wide due to site restrictions requiring long corbels.

The wedges on top of the corbels are used to facilitate grade adjustment to the falsework. The wedges are typically cut from 4× timbers. The falsework grade adjustments are often done by using hydraulic jacks placed between the sill beam and the pads.

Soil-Bearing Value.　The soil-bearing capacity is usually assumed by the falsework design engineer. A typical range for soil-bearing values is between 2000 and 6000 lb/ft^2. Typical soil-bearing values are shown in Table 18.7. A field inspection of the soil is often

TABLE 18.7　Typical Soil-Bearing Capacity

Type of soil	Load bearing, psf
Rock w/ gravel	6000
Gravel	5000
Sand	3500
Clay	2000

sufficient to estimate a soil-bearing value. Bearing values based on field inspection should be conservative estimates. Soil-bearing tests must be done when higher or more accurate values are desired. The bearing capacity of the soil may also be determined from the soil data (borings) shown on the contract plans or in geotechnical reports.

Timber Pads. The design of the timber pad depends on several factors, such as soil-bearing value, size of timbers used for pads, and the corbel spacing. The size of the pads is typically determined by the available material in the contractor's inventory. The corbel spacing can vary based on the layout of the falsework system or the spacing can be determined based on the maximum span length of the pads between the corbels.

All timber pads are flexible to some degree, and thus in reality the soil-bearing pressure will not be uniform over the length of the pad. It is a common practice to limit the length of a cantilever pad at the exterior corbels to 80 percent of the calculated effective pad length.

The Caltrans *Falsework Manual* has an empirical procedure for analyzing timber falsework pads. The Caltrans method is suitable for evaluating the design of the pads, but trial-and-error must be used to determine the size and width of the pad system. Caltrans' effective pad length formulas are based on allowable bending stress in the pad and it also adds the width of the corbel to the pad length.

A simplified "beam analysis" method can also be used to calculate the required pad width by assuming the pad as a beam with a soil pressure acting as a uniform load (see Fig. 18.27). This method will directly determine the pad requirements without using the trial-and-error method. It is recommended to add 50 percent of the corbel width to the calculated span lengths.

The following design procedure can be used when the pad size is known:

- Calculate the maximum span length for the pad based on allowable bending stress for simple beam configuration.

$$L_{max} = \sqrt{\frac{(F_b)(S)(8)}{(W)(b)}} + \frac{t}{2}$$

where: F_b = allowable bending stress, psi
S = section modulus of an individual pad, in^3
b = pad width, in
t = width of corbel, ft
W = soil-bearing value, psf

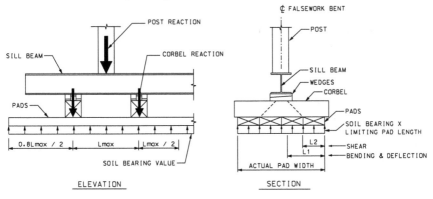

FIGURE 18.27 Falsework load distribution to the ground.

- Calculate the maximum allowable span length based on maximum allowable shear.

$$L_{max} = 2\left(\frac{2(F_v)(b)(d)}{3(W)(b)}\right) + 2d + \frac{t}{2}$$

where: d = depth of pad, in
F_v = allowable shear stress, psi

- Calculate the maximum pad length based on maximum allowable deflection (adjust formula when allowable deflection is different than $L/240$).

$$L_{max} = \sqrt[3]{\frac{384(E)(I)}{(240)(5)(W)(b)}} + \frac{t}{2}$$

where: E = modulus of elasticity, psi
I = moment of inertia, in^4

- Select a corbel spacing (s) that is less than L_{max}.
- Calculate post reaction on corbel based on selected corbel spacing. Although the reaction on posts within a bent varies and therefore the corbel reaction varies, typically the critical corbel load is used for determining the maximum pad requirement and applied to the entire falsework bent. Keep in mind that the cantilever pad length at the exterior corbel is reduced to 80 percent.
- Calculate minimum pad area at critical corbel.

$$A_{min} = \frac{P_c}{W}$$

where: P_c = reaction on the corbel, lb.
- Determine required minimum pad width.

$$B = \frac{A_{min}}{s}$$

where: B = minimum pad width, ft
s = corbel spacing, ft

- Determine number of pads.

$$n = int\left(\frac{B}{b}\right) + 1$$

where: b = actual individual pad width, in.
- Actual soil-bearing pressure, psf:

$$Q_a = \frac{P_c}{(s)(n)(b)}$$

Concrete Pads. An alternative to timber pads is precast concrete pads. The precast concrete pads can be made in small units to make it easy to handle. The layout of timber pads is more flexible than precast units, but timber is more expensive and more prone to deteriorate due to weather conditions. Precast concrete pads are typically not designed specifically for an individual falsework project, but they are made per the contractor's specifications. A frequently used precast pad size is 4 ft × 6 ft × 6 in thick. A maximum allowable soil-bearing pressure is assigned to the precast pad so that the falsework designer can determine the number of units required per falsework bent.

Corbels. The corbels can consist of timber or steel beams depending on the preference of the contractor. Where loads are relatively high and pads are wide, steel beams may be the better solution for a corbel beam. It is recommended that the corbel depth-to-width ratio should be close to 1 for lateral stability.

Design load on the corbel, W_c, lb/ft:

$$W = (Q_a)(s)$$

where Q_a = actual soil-bearing value, psf

When the actual pad width is known, the actual soil-bearing pressure can be determined. The actual soil bearing is used for the corbel design. The following is assumed before corbel design can begin:

- The post load is applied symmetrically and is uniformly distributed across the full width of the pad.

- When resisting the load applied by the pad, the corbel acts like a cantilever beam.

- For timber corbels, the point of fixity of the cantilever beam is located midway between the centerline and the edge of the sill beam.

- For steel beam corbels, the point of fixity is located at the edge of the sill beam.

Sand Jacks. The sand jacks consist of a timber or steel box filled with compacted sand (see Fig. 18.28). After the bridge superstructure is complete, the sides of the sand jack box are removed and the sand washed or blown out, which in turn lowers the falsework several inches and facilitates its removal. To prevent inadvertent settlement of the falsework due to

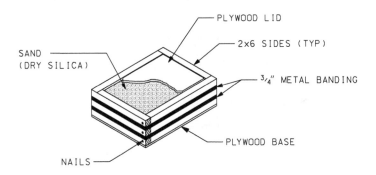

NOTE: INSIDE SAND JACK SHOULD BE LINED
WITH VISQUEEN ALL AROUND

FIGURE 18.28 Sand jack example.

premature erosion of the sand, care must be taken to ensure that the sand jack is protected from wind and water. As a general rule, sand jacks should be so constructed that the annular space between the top bearing plate and the side plates or frame does not exceed about ¼ in. Calculating the capacity of the sand jack is not straight forward. Capacities are often determined by load test. A typical sand jack has a capacity of about 60,000 lb.

Foundation Piles.   When falsework is constructed over waterways, soft ground, sloping ground, or other locations that make the use of pads not practical, piles can be used to support the falsework bent. In most cases timber piles will provide the most economical pile foundation. However, the design load on timber piles is usually limited to 45 tons (40,800 kg), or less by some specifications; consequently, steel piles may be more economical when large loads are to be carried. Regardless of other considerations, steel piles may be the better choice at any location where difficult driving conditions are anticipated.

Driven piles are usually cut off and capped near the ground line, in which case the vertical support system will consist of braced bents erected on top of the pile cap. For this arrangement, the piles will be supported throughout their length; therefore, they will be subjected to axial load only. Occasionally, site conditions will dictate the use of driven piles that extend a considerable distance above the ground surface. Such piles will be incorporated into free-standing pile bents that may be fully braced or only partly braced depending on site conditions.

When pile bents are only partly braced (see Fig. 18.29), as would be the case for a structure over a waterway where bracing is installed only above the water surface, the bracing must be adequate to resist the horizontal design load. Bracing so designed will provide rigidity to the system between the bent cap and the bottom of the bracing. Below the bracing, the individual piles will be subject to bending when the horizontal design load

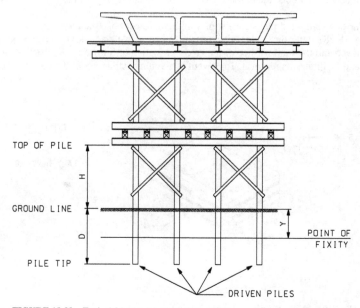

FIGURE 18.29   Typical falsework bent on timber piles.

is applied; consequently, both horizontal and vertical loads must be applied simultaneously and the resulting stresses combined to obtain the actual stress in the piles.

When free-standing pile bents are used, it is essential that the pile penetrate the subsurface soil to the depth necessary to develop a point of fixity. (The point of fixity is the location below the ground surface where the pile shaft may be considered as "fixed" against rotation when it is subject to a bending moment.)

Other factors being equal, the depth of embedment needed to develop pile fixity is a function of soil type. Obviously, soft soil requires a deeper penetration than firm soils, but determining the actual penetration needed is a matter of engineering judgment. One commonly used rule of thumb compares pile embedment to the height of the pile above ground and establishes a *D/H* (depth/height) ratio. Piles are considered fixed when the *D/H* ratio is 0.75 or more; however, this is only a guide. The important point is that the necessary penetration should be considered during the design stage, and a realistic value be assigned.

Assuming adequate penetration, the depth to the point of fixity is a function of soil stiffness and diameter of the pile at the ground line. The relationship is

$$y = (k)(d)$$

where: y = distance (depth) from the ground line to the point of fixity, ft
k = soil stiffness factor
d = diameter of the pile at the ground line, ft

A widely accepted rule of thumb assumes the point of fixity at about four pile diameters below ground surface (which corresponds to a k factor of 4) for soil conditions ranging from medium hard to medium soft, and this assumption has been verified by load tests.

The assumption that the depth of pile fixity is at four pile diameters below ground surface should be satisfactory for most soil types. For very soft yielding soils, this figure may be increased. Likewise, consideration may be given to raising the assumed point of fixity when piles are driven into very firm soil. However, caution is advisable because the driving of piles into any type of soil will tend to disturb the top few feet of the surrounding material.

For analyzing the adequacy of braced pile bents, the Caltrans *Falsework Manual* lists three categories, or bent types, depending on the L_u/d ratio of the pile.

All types of pile bents must be analyzed for bending stress due to vertical load eccentricity, eccentricity due to pile relaxation after pile pull, and axial compression stress (see Fig. 18.30).

FALSEWORK CONSTRUCTION

Falsework Erection

Falsework construction normally begins after the substructures such as abutments, columns, and pier walls are complete. Ground compaction at the falsework bents may be required, especially when bents are placed over recently backfilled footings or utility trenches. Falsework bent locations may be graded higher than the surrounding ground in order to provide sufficient drainage away from the falsework pads.

When all site work is complete, the installation of the falsework foundation (pads, corbels, sand jacks, wedges) can begin. When multiple timber pads are used at a bent, it is good practice to stagger the joints of the pads. Wedges may be required between the corbels and the timber pads to ensure sufficient bearing.

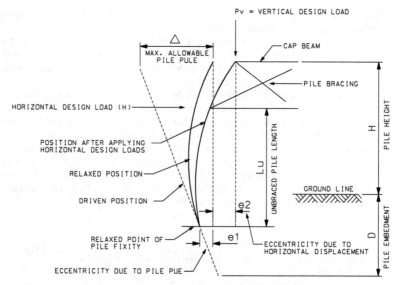

FIGURE 18.30   Timber pile schematic.

Bents are usually constructed on the ground where the posts are cut or spliced to the required length and connected to the sill beam and the cap beam. The transverse braces are installed before the bent is erected. Once the bent is complete, it will be hoisted on the falsework foundation and braced to a previous installed bent or to a temporary brace before it is released from the crane. Stringer beams are placed on top of the bents including the camber strip, and beam bracing is installed when required. Installation of the soffit can start as soon as all the stringer beams in a span are installed. It is recommended to start the falsework bent at a substructure so it can be braced to a stable bridge component.

After the falsework is complete and before the first concrete placement, the soffit elevations must be checked. Adjustments to the falsework shall be made by raising or lowering the bents. Hydraulic jacks are placed between the pads and the sill beam to assist in repositioning the wedges to the adjusted falsework grades.

The post length calculations must take into consideration the anticipated settlement off the falsework. The settlement occurs due to the joint take-up between all the falsework elements and the settlement of the ground. Settlement usually ranges from 3/8 to 1 inch, depending on the workmanship of the falsework crew, the number of joints between falsework bent material, and the ground conditions.

Telltales are used to measure settlement as concrete placement occurs. A "telltale" can be a thin strip of wood or metal banding attached to a joist as close as possible to the supporting bent and extending downward to a point near the ground. Before concrete placement begins, the bottom of the telltale is referenced by a mark on a stake driven into the ground. The number of telltales used must be sufficient to accurately determine the amount of joint take-up and settlement over the entire area of the concrete placement.

Falsework Removal

The removal of the falsework system is a critical operation and must be evaluated with great care. The removal methods and staging of the falsework removal should be determined before starting the falsework design in order to avoid complications during the removal stage. The falsework removal becomes even more critical when falsework is over traffic or railroads. The traffic closure windows are often limited and may only be done at night, so the removal may have to be performed in small segments in order to finish the removal operation before reopening traffic lanes. Railroads may have requirements that prohibit track closures; this would require falsework removal between rail traffic.

The removal of the falsework system can start after the bridge superstructure is complete. Releasing the load from the falsework is a high-risk operation. In order to remove the post of the falsework bent, sand jacks can be used to lower the falsework several inches to clear the falsework from the bridge superstructure. Hydraulic jacks placed between the pads and the sill beam can also be used to remove the wedges by lifting the sill beam slightly. After the wedges are removed, the falsework no longer supports the completed bridge structure.

After the falsework bent is lowered, the stringers and soffit can be removed with cranes, forklifts, or other equipment. During the stringer removal, the soffit can be temporarily suspended from the bridge or the soffit members, joists, and plywood can fall to the ground.

Another method to remove the stringers and soffit is by using winches (see Fig. 18.31). Winches are placed on top of the bridge deck and its cables are connected to the ends of the cap beam of the falsework bent. The winches suspend the cap beams with the stringer beams and soffit until the falsework posts below the cap beam are removed. Once the posts are removed, the winches lower the cap beams with the stringer beams and soffit to the ground.

The number of winches per cap beam is determined by the weight of the stringers, soffit, cap beam, and the capacity of the winch. A typical winch setup requires one winch at each end of the cap beam, but intermediate winches may be required for heavy loads.

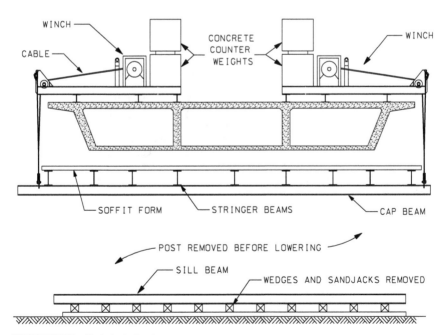

FIGURE 18.31 Typical winch setup.

Ideally the winch cable is placed outside the deck area, but openings through the bridge deck and soffit may be required when multiple cap beams per bent are used or when additional intermediate winches are needed. The location of the winches that require deck and soffit openings should be determined before the placement of the superstructure concrete in order to form block-outs for the cables to pass through.

Design of the falsework removal with winches shall include the design of the cap beam of the falsework bent, the load in the winches, and evaluation of the bridge deck at the winch locations. The cap beam must be able to support the stringer beams and the soffit weight between the winch connection points. The winches typically have rated load capacities and should be tested periodically according to their specifications. When the winch cable is placed outside the bridge deck area, counter weights are required to resist the overturning moment of the winch. The counter weights must be sufficient to resist the overturning moment due to the load in the winch cables.

FALSEWORK ALTERNATIVES

For concrete bridge structures over deep canyons, waterways, or other major obstructions, falsework may not be a viable construction option. Although conventional falsework systems are relatively economical to build, when bridge heights exceed 100 ft falsework systems may no longer be cost-effective or practical to use. In such cases, other construction alternatives, which eliminate the need for falsework, are available. This section briefly describes some of these alternatives.

Cast-in-Place Segmental Construction

For construction of cast-in-place concrete bridges, a frequently used alternative to falsework is the segmental cantilever method of construction. In this method a short section of the superstructure, which is commonly called the pier table, is constructed at the top of a completed pier using conventional falsework or brackets for support. A form traveler on both sides of the pier table is erected on the completed bridge deck, and forms are set for the first superstructure segment. By successive casting of segments and advancing of form travelers on opposite sides of the pier, a pair of cantilever segments is constructed. The superstructure is completed by connecting the cantilevers with a closure pour at the approximate midpoint of each span.

Typically, the length of a segment may be 16 ft or more, depending on the capacity of the form traveler. The work involved in constructing each segment, or cycle, consists of advancing the form traveler from the previously poured segment, placing reinforcing steel and tendons, placing and curing concrete, and prestressing tendons. The cycle of cast-in-place segmental is typically 1 week

Precast Segmental Construction

The construction sequence for precast segmental construction is somewhat similar to the cast-in-place construction method, but the super structure segments are precast instead of cast-in-place. The segments are made in a casting yard and transported to the bridge site where specialized equipment erects the segments in place. The precast segments are cast on one side against the previously cast segment to ensure a tightfit between the segments after being installed in its final position. The casting of the segments can start while the

FIGURE 18.32 Launching truss at precast segmental Otay River Bridge. (*Courtesy of International Bridge Technologies, Inc.*)

substructure is being built. This will make the construction time of the superstructure faster than the cast-in-place method.

The segment installation method may vary depending on project conditions and bridge design. One method is the span-by-span method where all segments of one span are installed and temporarily supported by a truss until all segments are stressed together to the pier caps. On average, two to three spans can be installed per week using the span by span method.

Another segment installation method is the balanced-cantilever method where segments are installed on both sides of the pier cap to maintain a balanced loading. On projects with good access at ground level, the segments can be installed with a crane. Other hoisting methods frequently used for balanced cantilever are the winch-and-beam method or an overhead launching gantry (see Fig. 18.32). The rate of segment installation may vary from two to six segments per day per cantilever, about one cantilever per 2 weeks.

BIBLIOGRAPHY

1. *California Falsework Manual*, issued by the California Department of Transportation.
2. *Standard Specifications*, California Department of Transportation, 2006.
3. *Standard Specifications*, Washington State Department of Transportation, 2010.
4. *Standard Specifications*, Nevada Department of Transportation, 2001.
5. *AISC Steel Manual*, 13th ed., 2005.
6. *National Design Specification for Wood Construction (NDS)*, 2005.
7. *Guide Design Specifications for Bridge Temporary Works*, AASHTO 1995.
8. *Construction Handbook for Bridge Temporary Works*, AASHTO 1995.

CHAPTER 19
TEMPORARY STRUCTURES IN REPAIR AND RESTORATION

Wilson C. Barnes, Ph.D., AIA, FCIOB
Kevin M. O'Callaghan

GENERAL

There are few generic endeavors in construction with such a potential for all the variations of temporary structures to occur as repair and restoration. Of these, repair is the broadest, for it must be construed to encompass renovation and the late twentieth century buzzword activity "recycling."

As advanced societies in the modern world continue to mature, we turn with greater frequency to the fixing up of what we have, to the conservation of our built environment rather than reduction of existing structures to rubble and starting over. There is no question that many factors enter into decisions of retain versus replace or ignore. These include factors of economics, technology, culture, and location, to name only a few.

Decisions to improve existing structures range beyond the simple and usually programmed decisions of routine scheduled maintenance, through the highly complex unprogrammed decisions of repair or restoration. No matter where a minor or major project lies in this spectrum of work, the need for temporary structures is virtually inescapable. Even the simple task of upgrading light fixtures usually requires temporary structure to bring a worker to the location of interest, or to extend the worker's reach.

The very nature of our built environment is such that it wears out or deteriorates in many different ways. Some parts are simply much more durable than others. Some parts are amenable to touch-up from time to time, and some have limited lifetime compared to other parts of the whole. When we reflect on the multitude of tasks that might be involved in the improvement of buildings or other constructed facilities, it readily becomes apparent that temporary structures are an important element in the conduct of improvement operations, and as such must be planned for properly in advance.

It is not uncommon for buildings to be designed with routine maintenance in mind. Almost always now, even such an infrequent activity as major HVAC equipment replacement is anticipated and appropriate accommodations are designed into the building. This can be seen as glass walls in equipment rooms or a bearing beam over a wall panel that could be a probable future opening. Also, some high-rise buildings have permanent cable-suspended platforms or staging that are used not just to transport window washers up, down, and around the structure but also to provide safe and functional working provisions for routine maintenance and repair of building envelope vertical surfaces. Most, however, use temporary suspension systems to facilitate the necessary work. Placement and securing of these systems must be properly planned and controlled; safety is of paramount concern.

In America especially, we turn increasingly to the past as we realize that many of our older buildings occupied prime sites with a scale and character that merits retention. Often these buildings need extensive repair and rebuilding. Masonry structures are among the most common in this regard. This is a different kind of work than new construction, with many hidden surprises in existing structures for the unsuspecting and unready. The improvement contractor soon becomes aware that extreme steps are sometimes necessary to reestablish structural viability while retaining the qualities that attracted notice in the beginning. Figure 19.1 shows the rebuilding of an entire exterior wall with the original brick and new lintels. Although the wall had not failed, it had bowed to a degree that called for drastic measures. The wall was rebuilt in sections with the entire elevation staged at once, and temporary shores installed throughout the floor systems to maintain structural integrity. This building was an important visual anchor on an urban space as well as being of historical interest.

FIGURE 19.1   Gerrish Hall, Chelsea, Mass. Wall reconstruction circa 1976. (*Photo by W. Barnes.*)

PLANNING

Review Documentation

In the beginning, when faced with an improvement project of any kind, it is best to have documentation. Whether it is drawings and specifications for a new project or simply old record drawings from file, it is essential to know what one has to work with before one can determine what to do. This means not just what the contract or documentation say, but also what they imply. Very few improvement instructions can be explicit enough to cover all eventualities. Sometimes essential information is in previous documentation, and sometimes it is not. Quite often it is necessary to look beneath the surface (both interior and exterior) of a building or other work to understand the nature and extent of what has to be done. The obvious implication is that we need this knowledge to plan our temporary structures. If no drawings exist, as is often the case, then some must be created to help analyze what is there and plan for repair and/or change.

Foundation and/or soil-bearing information can be of particular importance if extensive temporary structures are contemplated. The loads imposed by scaffolding, shoring, and even temporary rigging for the movement of materials and heavy equipment should be taken into consideration. Conservative analysis should be made of the existing structure, especially where basement slabs and assumed underlying soil are concerned. The original structural intent of an existing structure should be determined from the as-built documentation, and supplemented by conditions that can be observed on the site.

Where no documentation exists, it may be necessary for a structural engineer to recalculate the original intentions as well as the existing conditions. This caution applies beyond the foundations to an entire structure, especially where major repairs or modifications are contemplated. Safety is of paramount concern in work of this nature where removal or adjustment of a structural member can result in shifting of loads to other than originally intended paths.

Walk-through

An indispensable supplement to documentation review is a site visit and thorough walk-through of the project from the ground up. Not only does this provide a good visual familiarization with the existing conditions, but it provides an opportunity to note what might have been missed in drawings, specifications, and other formal instructions. It is well to remember that errors and omissions in documentation may be a legal concern after damage occurs, but more importantly this can result in the loss of a valuable building or life. It is not uncommon in improvement-type contracts to discover the unexpected as work progresses. Improvement-type contracts should contain a contingency provision for such eventualities. Discovery of the unexpected should not be discouraged by fear of reprisal for not finding it earlier. This necessitates a candid discussion and understanding with a building owner or project client before drafting a contract.

A straight-claw hammer and crowbar are frequently useful on design and preconstruction walk-throughs to expose structural members or other elements of the work that suggest an unknown condition. Depending on the nature of a project, it is well to take along subcontractors and engineers either on the initial or on a later visit. Although we all look to economy of time, it may be useful for a manager to get the big picture first and then bring others along for more detailed analysis. Getting the team together on the jobsite early can help coordinate temporary structure planning, especially the material requirements and logistical sequencing. The coordination between trades on this type of work is frequently more complex than in new construction and must be given appropriate consideration. This caution applies equally as well to safety concerns in both planning and conduct of operations.

Organize Job

After adequate evaluation of the documentation and physical features of a project, an organizational plan of feasible activity can be prepared. It is here that the nature and sequencing of required structures begins to emerge. What will be needed, at what point in the project, and for how long are all important aspects of the overall plan. The potential impact of weather, especially wind loads and stormwater control, must be provided for. Logistical planning strongly impacts the choice and deployment of temporary structures. Safety is a constant concern as are the disposal of waste and debris and the delivery of new materials. The logistical needs of a complete job are prime planning considerations and the requisite accommodations for pedestrian and traffic maintenance are long-term demands. In addition, not least of all, most building codes require that temporary structures meet the same code standards as more permanent installations for structural strength, fire safety, means of egress, light, ventilation, and sanitary requirements in terms of public health, safety, and general welfare [e.g., Building Official and Code Administrators (BOCA) *National Building Code*, Section 51 1 and IBC replacements]. Temporary structures can account for a significant portion of a project cost, and their use must be controlled as well as thoroughly thought out.

SUBSURFACE PREPARATION

Access and Stability

As with any project, smooth traffic circulation to, from, and around the site is essential to timely completion. This is of particular concern for projects in dense urban areas where congested street traffic can make movement difficult and excavation of any kind poses special stability problems. Access ramps should be properly engineered for anticipated loads. Lateral pressures when interior structure has been removed, excavation for the underpinning of footings, or simple deepening of a basement need to be compensated for with adequate bracing and/or sheeting. A prudent contractor will have such bracing and/or sheeting designed by a qualified engineer. This is especially critical where excavation occurs close to existing structures or trafficways. Modification or protection of utility feeder lines servicing a jobsite may require braced trenching covered with roadway plates for prolonged periods. The simplest trenches as well as deep excavation cuts must be shored, braced, and properly covered to accommodate traffic loads. The removal of an adjoining structure, nearby excavation, or site dewatering can all contribute to the destabilization of existing foundation soil, and consequent shifting or collapse of a column footing or bearing wall.

Utilities

All utilities need to be considered for complete shutdown or limited use, depending on the overall scope of activity. If utility usage is retained, protective covers for lines or stubs may be required. Feeder lines for power are especially vulnerable to activity movement around a project unless they are underground. Extended stubs inside an existing foundation need to be analyzed in terms of their possible differential movement should supporting or adjacent structures or soil move unexpectedly. The potential for damage to site utility lines and stubs is high. Such damage frequently generates costly and hazardous results. Obviously, coordination with local utility companies and departments is imperative.

Drainage

Effective perimeter drainage is especially important for urban projects with deep excavations such as sometimes found with facade retentions, foundation repair, or other work open to weather. If curb cuts are employed to facilitate vehicular movement to an excavation or to other depressed working area, ramp them appropriately inboard of the curb line, or set up diversion trenches or dikes to reduce the risk of flash flooding and washout of critical soil that may lead to the undermining of foundations or utility lines. This same eventuality should be considered where deep basements surrounded by adjoining bearing walls are left open for prolonged periods, Sump-type pumps already in place can avert catastrophic consequences when abnormal rainfall and runoff far above average occur.

PROTECTION

Adjacent Facilities

One quality that sets many contractors apart is the care that they exercise around existing buildings or other works whether they be adjacent to or part of a project. Not only care is necessary but also a careful wrapping or separation of what will remain as is, from what will be changed. Wrapping can range from polyethylene film taped in place to heavy blankets or quilts similar to movers' pads fixed firmly around objects subject to denting, scratching, or breakage. Sometimes, protective framing with dimension lumber and maybe plywood sheathing is necessary for a project duration.

Simple coverage with poly film is effective against liquids and washes of many kinds, especially water-based fluids and paints. However, before relying on poly to protect against chemical washes or acids, testing should be conducted on small areas to assure the poly's resistance to the fluid. Poly is also effective as a dust-control barrier around windows, doors, and similar openings. Quite large areas can be closed off or protected in this manner but need constant monitoring to assure the integrity of panel seams. Poly and duct tape should be in every improvement contractor's kit bag. They can provide quick, cheap, and versatile temporary protection. In addition to a good range of mil thicknesses for standard poly, fiber-reinforced film is also available for conditions of extreme wear and tear.

Bridging over Excavations

In urban areas, the provision of bridges or covers over excavated sidewalks and trenches, or protected walkways of an acceptable width in the street is generally a condition of permit and a matter of safety concern. Such bridges or walkways will be governed by local codes but as a general rule should be at least 4 ft (120 cm) wide. Both sidewalk and roadway bridges should be of sufficient strength to resist possible dynamic loading, and secured to preclude the shifting of components under load. Frequently, trench or other excavations may extend out into a street where traffic must be maintained. The bridges or covers there will need to be engineered for the possible vehicular loads and the accompanying dynamic conditions. Sometimes planking may be necessary over a finished surface in the road or on a sidewalk to protect it from heavy-wheeled or tracked equipment during the project conduct.

Pedestrians

Work on structures that are built up to or close to a street front or sidewalk line will need sidewalk sheds, exterior catch platforms, or thrust-out platforms to provide for safety of pedestrians below. Sidewalk sheds along the entire sidewalk facade of a structure are easily erected on sidewalk scaffolding, as discussed in Chap. 14. The shed roof should be made of heavy tongue-and-groove planking, crib supported, with planks tightly nested together to prevent the filtering through of debris from above. Plank thickness should be determined through analysis to resist anticipated impact loading. Dimensions of 1½ to 3 in (38 to 76 mm) are frequently used. Similar protection should be provided for rooftops of adjacent structures when work is being conducted at a higher level. This is of concern for steep roofs as well as flat roofs. Objects ranging from tools to entire sections of masonry wall have been known to fall onto lower roofs and damage not only the weather membrane but also supporting structure. Construction components for a typical heavy-duty sidewalk bridge, such as required in New York City, are discussed and shown in Chap. 21. Exterior catch platforms or outriggers are also diagrammed and discussed in that chapter. A simple, but at the time incomplete, installation of an outrigger catch platform is shown in Fig. 19.2.

Sidewalk sheds should have lighted walkways that are also of a minimum width to accommodate two people passing. Local codes may have specific requirements for this, as well as for all other temporary structures. but common sense and attention to safety are fundamental in developing the jobsite. Neither sheds nor higher platforms projecting from structures should have any materials stored on them. A supplement to sidewalk sheds for protection of pedestrians and comfort of workers is the combination safety net weather barrier panels used on exterior scaffolding (Fig. 19.3).

FIGURE 19.2 Urban project, Birmingham, England. Thrust platform at upper level. (*Photo by W. Barnes.*)

FIGURE 19.3 Urban project, London, England. Full staging with protective curtain panels. (*Photo by W. Barnes.*)

Fences are simple structures needed for almost all projects that are not completely self-contained. They are used for both security of the jobsite and a protective separation of pedestrians from work activity. In urban areas where building lines are close to street and sidewalk lines, fences are usually of boarding or plywood to inhibit movement of debris or pollutants through the fence. Equipment exhaust, pressure-line leaks or ruptures, and flying or falling debris are all potential irritants to pedestrians and need to be controlled. At the same time, appropriate opportunities for pedestrians to observe the jobsite activities should be provided. Openings or viewports should be of break-resistant transparent material of polycarbonate such as LEXAN or newer impact-resistant products now in the market.

Workers as well as passing pedestrians need protection. All temporary structures must be designed with sound rational analysis and erected in accordance with safe procedure. Not only must these structures be resistant to reasonable loadings, they must embody features to protect passersby and users from carelessness and discourage misuse. Such features include safety railings, toeboards, pulley covers and gear guards, ladder ties, safety nets, and so forth. A full definition of these requirements can be found in the OSHA Safety and Health Standards, 29 CFR Part 1926, for the construction industry. This reference contains much useful information to supplement the material in this chapter concerning demolition, excavation, ladders and scaffolding, floor and wall openings, stairways, and many other issues of interest.

Existing to Save

When working close to existing elements of a structure that are to be preserved or simply not worked on, sensible protective measures should be taken. These range from the sealing or cordoning off of areas with film, blankets, stand-off framing, or whatever is needed to preclude damage from work activity. Sometimes only a small area is affected; other times it may be major parts of the same structure or the entirety of an adjacent work.

DEMOLITION

In the repair and restoration arena, demolition is a far more selective procedure than wholesale destruction and removal of a structure. Sometimes demolition can be planned in advance fully; sometimes requirements are discovered or exposed as a project advances. In either case, the provisions to accommodate demolition are the same.

Types

A project may require one or more of four major types of demolition. (1) Cleaning can involve some very messy operations including stripping of finishes, washing of surfaces. and collection of accumulated trash and dirt. Cleaning may involve washing with water or chemical bases. It may involve blasting with abrasive material or brushing to remove dirt scale and grime. It may involve scraping, grinding, polishing, or etching. However it is done, cleaning generates a mess to be removed. (2) Exposing of structure or unsatisfactory service systems. This means getting into plenums; stripping, fireproofing, or other covering; and laying the primary structure bare. Unsatisfactory wiring or plumbing can be anywhere—in walls, in chases, and even sometimes threaded inside built-up or hollow structure. Almost the same can be said of ductwork except that horizontal runs are more prone to the opening of joints by sagging or to the growth of microorganisms due to collection of moisture in low spots or linings. (3) Removal of unwanted partitions, floors and flooring, ceilings, service systems, and structural elements that don't fit into the new scheme. Structural removal is the trickiest of all and should never be done without properly engineered transfer of the loads. Needling of masonry walls to permit an opening or rebuilding below the needle is a simple example of temporary removal. It is discussed more extensively later in the chapter. (4) Excavation. Lowering of basement floors or digging under foundations generates debris that must be lifted for removal from the jobsite. Most work of the four types shares common needs for temporary structures. Such excavation beneath or adjacent to foundations should be approached with caution and competent advice. Chapter 11 is a useful reference.

Dust Control

Measures for dust control during demolition are imperative, both to preclude undesirable contamination of other spaces and surfaces, and to contain dust as part of the debris. Such measures include the previously mentioned polyfilm closures or separations, and the misting or sprinklering of the dust clouds to reduce the potential for airborne movement of dust particles. It is also possible to simply close off or cordon off portions of an existing structure. This is a common technique where it is necessary to maintain normal occupant activity in portions of the structure not being worked on. As for misting or sprinklering, it is always important to consider protection of both vertical and horizontal surfaces before deciding on application of water as an abatement procedure. Water damage is water damage no matter how it is caused. Later, when demolition is complete, it is not uncommon to vacuum an entire remaining structure to complete the job of removing what is not wanted and to leave the surfaces clean and ready for further work. We see this as an OSHA requirement in asbestos removal along with very tightly prescribed filter types and monitoring of washing fluids and air in and around the project.

Debris

Removal of debris from the work spaces is the next major consideration. Both horizontal and vertical movement are involved. Horizontal movement of debris is best handled through appropriate combinations of carts, wheelbarrows, and simple hand-carry to a vertical transport device or opening. Temporary ramps and runways may have to be built to facilitate debris removal. In rare instances, a horizontal conveyor may be warranted; but usually the mixed nature of the debris material is not conducive to conveyor-belt transport.

Vertical movement is quite another matter. Large-scale removal generated by major renovations or recycling is frequently handled by special chutes tied to scaffolding or other convenient support, and terminating at the bottom above a collection point or truck loading dock where vehicles can be loaded to take the debris off site. Smooth rather than corrugated chutes should be used and care taken to keep them free of obstruction or jamming, especially at the bottom where many oversized objects will fail to clear a chute and clog up the opening. Some projects may permit the use of existing internal trash chutes or elevators. The OSHA Standards provide guidance here as well.

All debris should be removed from a jobsite as soon as possible. Accumulations of loose and unstructured material are a safety hazard as well as a breeding place for vermin and other unsanitary developments.

SCAFFOLDING, BRACING, AND SHORING

In almost all maintenance, restoration, and repair activity, access to the work surfaces, joints, and other areas of interest is a fundamental requirement. A very minor percentage of these needs are met through the use of ladders or simple elevated platforms. By far, however, the most universal temporary structure used for this purpose is scaffolding. It provides access inside and out of all kinds of structures. On occasion, a project will require temporary bracing to maintain stability of the existing structure, or shoring to keep the foundation soils and surrounding terrain from shifting until the work is completed. In such cases, combinations of scaffolding, bracing, and shoring are not uncommon, especially in congested urban areas.

Prefabricated scaffolding assemblies are the heart and soul of temporary structures for repair, restoration, and maintenance. These are used not only to stage entire projects by themselves, as in the world-renowned Statue of Liberty project, but also in concert with other specially designed temporary supports. This can be seen in Fig. 19.4, which illustrates renovation work of Madison Square Garden in New York City in the 1990s. The sketches

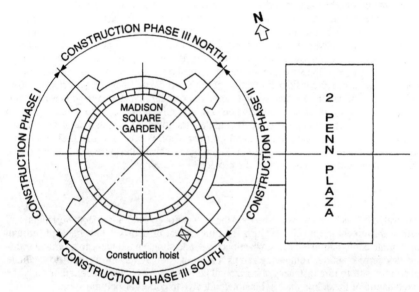

(a) PHASES FOR MODIFICATION OF EXISTING SKYBOXES AND CONSTRUCTION OF NEW EXECUTIVE SUITES

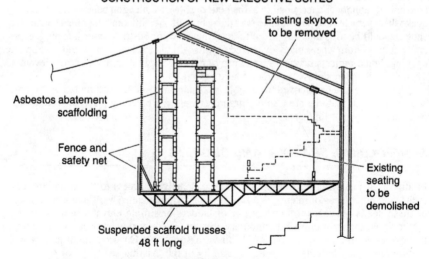

(b) SECTION THROUGH EAST SIDE OF MADISON SQUARE GARDEN

FIGURE 19.4   Madison Square Garden, New York City. Conceptual engineering sketches of temporary structures. (*Graphics by Universal Builders Supply Inc.*)

OUTER CIRCUMFERENCE OF SCAFFOLD   1276'-0"

TOTAL NUMBER OF SUSPENDED DECKS   48

PHASE I	WEST	452'-0" LG. SCAFFOLD × 37'-0" WIDE	17 TABLES
PHASE II	EAST	292'-0" LG. SCAFFOLD × 48'-0" WIDE	11 TABLES
PHASE IIIN	NORTH	266'-0" LG. SCAFFOLD × 37'-0" WIDE	10 TABLES
PHASE IIIS	SOUTH	266'-0" LG. SCAFFOLD × 37'-0" WIDE	10 TABLES

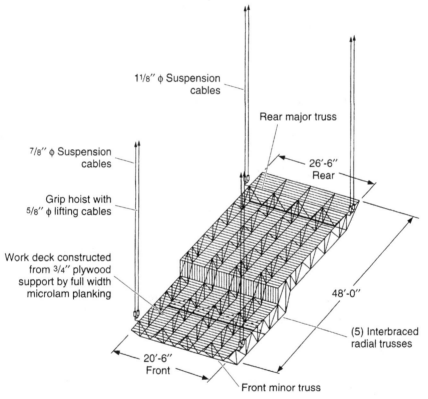

(c) TYPICAL SUSPENDED SCAFFOLD ASSEMBLY

FIGURE 19.4   *(Continued)*

illustrate conceptual engineering for the temporary structures to permit a variety of work inside the building without adversely impacting scheduled events. Notice that conventional prefabricated scaffolding frames are used in combination with specially designed and fabricated support trusses. Adjustable suspension of the total temporary assemblies was made possible by the overdesigned strength of the existing roof structure, which readily accommodated the temporary cable attachments.

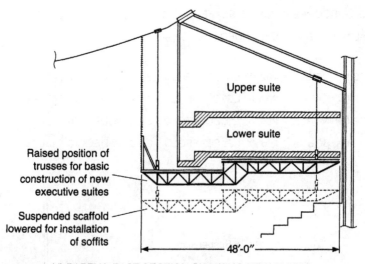

(d) PARTIAL EAST SECTION SHOWING NEW SUITES

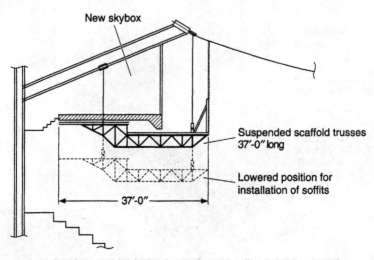

(e) PARTIAL WEST SECTION SHOWING NEW SKYBOX SUITE

FIGURE 19.4   *(Continued)*

Exterior Scaffolding

Engineered scaffolding using conventional components enables refurbishment and repair of all kinds. Many structures requiring external work are enclosed completely within creatively designed assemblies that rival great works of engineering in simplicity and sophistication. Notable among these efforts was the scaffolding to accommodate cleaning and maintenance repair of the Trinity Church steeple in New York City. Existing structure in the church tower was used to carry scaffold loads to the ground. At the same time, a unique tensile approach was used to convert objectionable lateral wind loads on the steeple scaffold wrap to vertical loads farther down the tower which could be tolerated. This is shown pictorially and diagrammatically in Fig. 19.5a and b. The 1000-ton weight of the tower and steeple helped to make this design scheme work.

Interior Scaffolding

Internal repair and restoration can be equally challenging. A case in point was the Pennsylvania State Capitol building, which had experienced long-term water damage to the structure and interior finishes of the central rotunda and complementary spaces. It was necessary for the building to continue normal operations and for any scaffolding to be completely free-standing without ties to the existing structure. The main scaffold loads were carried on triangular truss columns (which also supported a variety of other assemblies at the upper levels) to special footings in the basement. The conceptual engineering sketch in Fig. 19.6 shows how this was accomplished. Observe again, as in the Madison Square

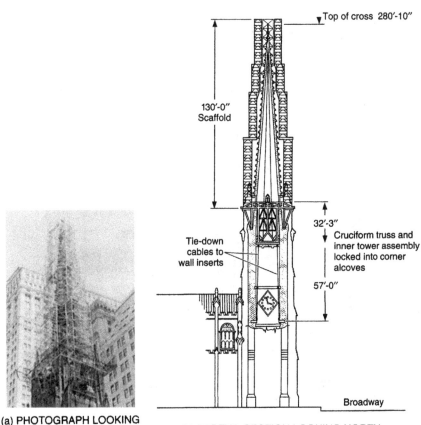

(a) PHOTOGRAPH LOOKING NORTHWEST

(b) PARTIAL SECTION LOOKING NORTH

FIGURE 19.5 Trinity Church, New York City. Conceptual engineering sketch of temporary structures and photo of implemented concept. (*Graphics by Universal Builders Supply Inc. Photo by Fred Tannery.*)

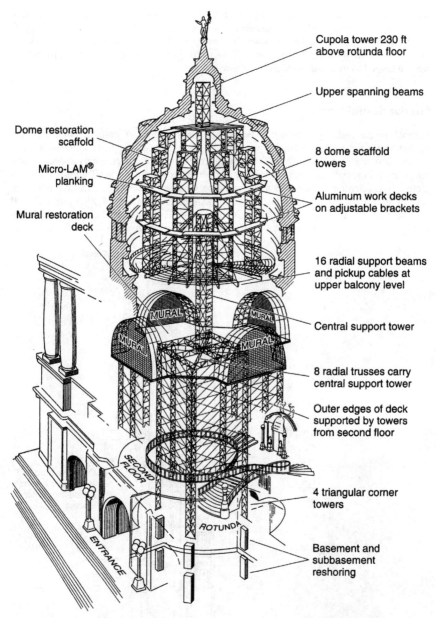

Cupola tower 230 ft above rotunda floor

Upper spanning beams

8 dome scaffold towers

Aluminum work decks on adjustable brackets

16 radial support beams and pickup cables at upper balcony level

Central support tower

8 radial trusses carry central support tower

Outer edges of deck supported by towers from second floor

4 triangular corner towers

Basement and subbasement reshoring

Dome restoration scaffold

Micro-LAM® planking

Mural restoration deck

FIGURE 19.6   Pennsylvania State Capitol, Harrisburg, Pa. Conceptual engineering sketch of temporary structures. (*Graphics by Universal Builders Supply Inc.*)

Garden example, how the mix of conventional frames and special fabrications enabled the total concept.

Rolling scaffolding is a work platform variation frequently seen in interior projects. Usually, the application is limited to assemblies of one or two frame sections high for obvious safety reasons. Wheels can be locked to keep an assembly in chosen position, but the wheels have a load capacity considerably less than the regular scaffold screw legs. On occasion, larger and higher scaffold assemblies will be mounted on wheels, allowing controlled movement where an extensive stationary scaffold assembly is impractical. Such assemblies should be conservatively designed and well braced with horizontal diagonals before applying any force due to moving.

Older Facades

Scaffolding is used extensively in facade restoration and preservation. Where the facade is an integral part of an existing structure being worked on, the scaffolding is relatively easy to design, erect, and use. Here, however, previously mentioned needs for sidewalk stability, sidewalk sheds, safety nets, weather wraps, lights, material hoists, and maybe stairs are all part of a total scaffold system to do the job. As a practical rule of thumb above the ground level, facade scaffolding should be tied into existing structure at every other level. For optimum security, such ties should be made to primary support structure or to the floor systems. Older buildings, especially of frame construction, frequently lack good structural integrity between the facade and the rest of the structure, even if the facade is a load-bearing wall.

Sometimes it is appropriate to stage a facade with cantilevered scaffolding to facilitate the movement of materials and provide a better working platform for craftspeople. A typical configuration might look like the sketch in Fig. 19.7. Many other conceptual approaches to facade scaffolds are workable. Frames of different widths, frames of different heights, and frame bay sizes of different lengths within the same job are feasible. Even scaffolding supported on one leg instead of two can work, if properly tied in and if of sufficient vertical rigidity. When working on existing structures, provision for easy passage of pedestrians nearby is usually a mandatory consideration and therefore may govern the scaffolding footprint. Proper bearing for scaffolding footings is extremely important. Scaffolds may be bracing structures as well as working platforms. In both cases they are subjected to a wide range of planned and unexpected live loads. Bearing points must have sufficient integrity and continuity of path to maintain position under imposed loads.

Facades and old party walls are especially susceptible to bowing, buckling, and collapse. Frequently an architecturally rich or historic facade will be retained with a completely new structure to be built behind it. In such situations, extensive pinning and bracing of the facade may be necessary to keep it from failing during construction until it can be secured to the new structure. A good example of this is in Fig. 19.8, a development behind historic facades in the Chelsea District of London. Note the liberal use of horizontal and vertical walers to bind the existing structure in place and preclude unnecessary separation of joints. The strong and protected diagonal bracing with buttresslike footings is intended to ensure the maintenance of the facade's vertical alignment and position on its foundation, while incidentally providing protection against errant vehicles. The materials in contact with a facade should be of good quality: stainless steel and protected metals, or inert blocking and padding of wood or synthetics intended to prevent staining as well as to maintain structural integrity. This type of protection, usually done prior to and during demolition behind the facade, preserves a heritage difficult to restore and impossible to recreate later.

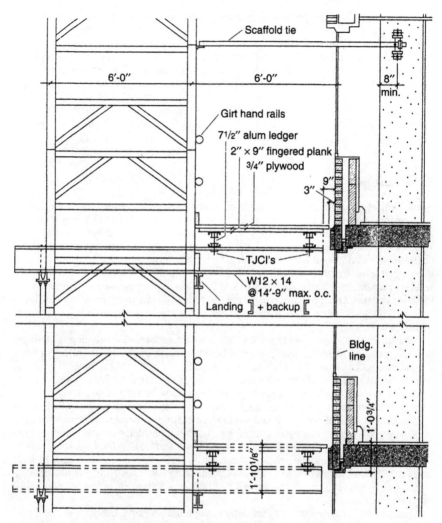

FIGURE 19.7 Cantilevered scaffold system showing tie to existing building. (*Graphics by Universal Builders Supply Inc.*)

Bracing and Shoring

More elaborate bracing behind facades may be necessary. This can take many forms of horizontal and vertical reinforcement to restrict movement of the facade: truss girders or extended bracing systems, or entire steel frames threaded through the existing structure before demolition. Some structures, including floors, may be left in position, or the entire space behind the facade may be empty. It may be necessary to brace the facade inside as well as outside. Bracing schemes, in general, must help the facade resist wind loads from both sides. They must contain the facade in its original position, and permit both demolition and then new construction operations to be conducted. Ties to the facade for bracing

(a)

(b)

FIGURE 19.8   Urban project, London, England. Preservation of historic facade for incorporation in new construction. (*Photos by W. Barnes.*)

(c)

(d)

FIGURE 19.8   (*Continued*)

should be designed to accommodate differential settlement as the new structure develops in position. Sometimes collars through window openings are used in lieu of ties. Sometimes bracing may be effected through a series of flying shores extending back to other temporary or permanent assemblies. Retained facades may be one or more floors in height, generally up to six or eight stories of older masonry. The configuration in plan, and the properties in section of existing structures usually will guide the temporary bracing design.

Shoring that is a combination of sheeting and bracing is used frequently in urban situations where it is important to maintain the integrity of surrounding soils and structures while work is being conducted. It may be necessary inside or outside of existing structures to permit other activity to take place. It may be left partially in place as permanent structure is positioned to take up the loads of concern. Diagonal shores alone also are used in a temporary manner to brace or stabilize existing older foundation walls exposed by excavation or by the removal of adjoining structure. Figure 19.8*b* and *c* illustrates a situation in the half block project in London where the below-grade walls have sufficiently strong engaged piers and buttresses to preclude shoring against streets and underground utilities on three sides. The fourth side, an existing structure not shown, was shored diagonally to maintain the wall foundation integrity until the project's new foundation and structure took over the task of soil containment and lateral bracing. Such shoring is common in construction of new structures in urban environments where the structures are built out to property lines. Extensive repair and renovation, especially that involving party walls, may need both underpinning and shoring. Timber shoring systems in trenches can include uprights, cross braces, and wales. Steel and aluminum are used in both fixed and hydraulic shoring and shield prefabrications for trenches that can readily be dropped into excavations or removed.

Shoring or falsework is also commonly thought of in connection with formwork and the temporary support of concrete while curing. That is a primary usage for columns, beams, and floor slabs above grade. Shoring has an equally important role in repair and restoration work. Shoring can support flat surfaces, arches, beams, girders, and load-bearing walls or assemblies of all kinds. It can support horizontal, angled, and vertical faces of formwork. Shoring can be simple adjustable posts or stronger structural members designed to carry significant loads. It can be single members or complex arrays of columns, grids, frames, trusses, and related support members. Load bearing or spreading is particularly important, especially at the bases or supporting locations for the shoring. In repair work, shoring is also sometimes used in concert with hydraulic jacks to realign assemblies and help maintain them in position until permanent fixing can be accomplished. Shoring is virtually indispensable in the repair of masonry bridges as well as in masonry and concrete buildings.

Masts, Mast Climbers, and Hoists

The use of masts and mast climbers has developed significantly since the 1980s, and they now are widely used in all types of new construction and repair/restoration projects, including routine maintenance options such as window washing. Masts and mast climbers are the new generation of scaffolding that provides quality vertical transportation for personnel, materials, and work platforms. Masts are generally rectangular or triangular in cross section depending on the planned loads and height on a project, and they come in a wide range of modular and special purpose vertical lengths. An excellent overview of the history and technology of masts and mast climbers appeared in *Masonry Magazine* in 2002. It can be found readily on the Internet at www.masonrymagazine.com/6-02/mastclimbers. html. Personnel and material hoists are the heavier-duty version of masts and mast climbers but generally with a different lifting and lowering technology. They have been on the construction scene for about 50 years and are constantly being improved in terms of capacity, versatility, and safety. A valuable reference for the do's and don'ts of this equipment can be found in OSHA Regulations (Standards—29 CFR), Standard Number 1926.552, "Material Hoists, Personnel Hoists and Elevators." There is a great variety of Internet entries for "masts and mast climbers" and "personnel and material hoists" where much useful information on these types of equipment can be obtained.

LADDERS AND TEMPORARY STAIRWAYS

Residential contractors and home owners who enjoy doing their own work are quite creative at finding ways to gain access for simple and short-term activity. However, even here the temporary structures should be chosen carefully. Ladders of a proper size and strength are entirely appropriate for certain work. However, ladders should be placed on even and stable surfaces and be braced or secured at the top. They should extend above an edge or parapet at least 42 in, if used to provide access to another level whether it be intermediary or where other work is located. Ladders are frequently subject to misuse through overload in terms of the design strength for possible spans. Most ladders that can be purchased today are manufactured to OSHA specifications (ANSI A14.2) and are so labeled with the rated capacities for their use to provide vertical access. They also are sometimes labeled with instructions for their use in a scaffold or horizontal bridge mode. They should not be used in the latter modes, however, unless they are so labeled, and never without scaffolding plank of the proper width to fill the rung space between the ladder rails.

Pairs of ladders are sometimes used to provide simple variable-height scaffolding. This can be seen in exterior situations where outrigger arms are attached to the ladders to support a span of planking; and sometimes both outside and inside where short-span planking or planking on a ladder bridge is supported by the rungs of two facing ladders. In the former case, ladder angle from the vertical is of consequence and should never be less than what is needed to maintain the centroid of a vertical load, imposed on the planking, inboard of the ladder base. Additional angle should be allowed to compensate for the natural pull on the ladders, away from the vertical resting surface, of persons climbing up. A horizontal setback that is one-fourth of the vertical reach is a good rule of thumb.

Ladders are frequently used in multilevel work up to several stories high, or work of greater height, until either temporary or permanent stairways can be erected. Local codes and OSHA standards prescribe allowable vertical distances for ladders to span. Where heights between permanent floors or other working levels are in excess of these limits, intermediary horizontal platforms should be built to accommodate transfer of movement from one ladder to the next. These platforms should be of dimension lumber or light-gauge metal per local codes, but capable of supporting no less than 50 lb/ft^2 (244 kg/m^2) for medium-duty foot traffic. Such platforms should not be used for storage of materials. They should be well secured to the permanent structure and accommodate secure attachment or bracing of the ladders. Proper securing of ladder feet to the supporting medium is frequently overlooked or disregarded. The use of wooden cleats or metal brackets securely fixing ladder feet to a platform is considered a good practice. Where it is difficult to mechanically anchor a brace to the horizontal surface, a stop weighted in excess of the horizontal component of the imposed ladder load may be considered. Friction should be taken into account and the center of gravity of the weight should be very low, for example, that provided by sandbags.

Temporary stairways are covered in most local codes; however, little is said about their design and construction. Where stairs are a part of commercial scaffolding assemblies, this concern is normally addressed by the scaffolding manufacturer. Where stairs are fabricated on site, materials should be of structural grade equivalent to that required for permanent use. Many contractors use stair runs that they shift from location to location or job to job. Where this is done, riser boards or plates should be included every 2 ft of vertical rise to provide lateral stability to the stringer-tread assemblies. Prefabricated runs must be secured at the top and bottom when installed. All temporary stairs must be equipped with handrails per OSHA requirements, and with lights for use under conditions of limited visibility.

Temporary ladders and stairs should both be used with caution, for personnel movement only, and not for transport of materials other than what a person can carry.

TEMPORARY OPENINGS

Temporary openings may occur in any plane of orientation from the vertical to horizontal. The two extremes are the most common and are treated quite differently.

Horizontal openings will be either in floor and ceiling or roof assemblies and rarely involve primary structural elements. When structure does need to be removed to accommodate passage of equipment or other major assemblies, temporary diagonal and header-type bracing are frequently used to recreate load paths that have been interrupted. Horizontal structure systems often provide lateral stability as well as carry both live and dead vertical loads, and this requirement should not be overlooked when making openings. Beyond the considerations of load transfer, openings need to be protected with both railings and toe boards or with a structurally sound temporary cover.

Vertical openings are quite different, but the same principles of load transfer apply. No matter what size of opening is made in a vertical surface, there is something above the opening of either a load-bearing or non-load-bearing nature that must be considered. The contractor must be very careful with vertical openings and be ready to react instantaneously to any sign of undesirable movement when conducting a breaching operation. The size of an opening is a major factor, as is the nature of the wall construction. Non-load-bearing walls often have sufficient internal integrity to permit temporary openings large enough to allow movement of personnel, and not exhibit distortion or damage beyond the opening. This is generally subject to a field judgment, and as long as the wall is not of masonry construction, temporary bracing may not be required. On the other hand, load-bearing walls of any kind and all masonry walls should be reinforced and/or braced securely before or during the opening operation. Headers and transfer beams should be considered for both distributed and concentrated loads. Temporary column support requires an adequate base to carry the shifted load. Adequately sized columns, beams, load spreaders, and base plates, for example, must be fixed securely in position before the existing load-bearing section is removed. Masonry walls can be needled and the load then shifted to ground or other support before openings are cut. A simple needling operation is shown in Fig. 19.9 where the vertical loads were carried and spread to firm bearing both outside and inside. (The needle beams in the photograph are the two horizontal timbers through the masonry wall, transferring the vertical loads to temporary posts.) Wider openings, including those needed for insertion of permanent headers, can be accomplished with multiple needles picked up on both sides by temporary transfer beams or girders. Temporary openings in load-bearing and in all masonry walls must be reconstituted to properly support loads after completion of all other work.

FIGURE 19.9   Urban project, Chelsea, Mass. Simple needling operation in restoration of existing masonry facade. (*Photo by W. Barnes.*)

NOTABLE RECENT PROJECTS

Grand Central Terminal

The great train terminals of the 19th and 20th centuries have long proven their worth and provided a handsome return on investment to their cities of location. Those still in use have experienced their share of repair and restoration. Two features have characterized such projects: First is the sheer monumental size of the structures and second is the necessity to maintain a high volume of pedestrian traffic through and often under the principal areas of work. Both of these applied to Grand Central Terminal where, in addition to cleaning the historic barrel-vaulted ceiling of the main concourse, it was also necessary to install ductwork, lights, and a sprinkler system in the crawl space between the ceiling and the main supporting vault.

To stage the space of approximately 100 ft × 300 ft in the most economical manner, an easily moveable system of temporary structure was devised to provide worker access to a reasonable strip of the vaulted ceiling across the space at one time. The engineering concept is illustrated in Fig. 19.10a where we see the great trusses (four of them) spanning the space and carrying the 30-ft-wide worker platforms (shown below in plan at different

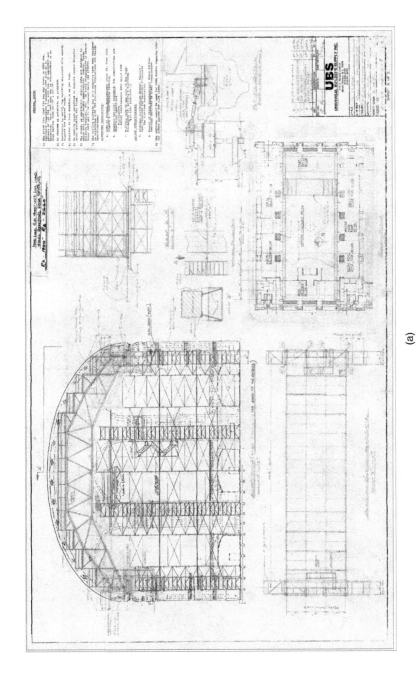

FIGURE 19.10 (a) Grand Central Terminal, New York City. Great hall engineering concept drawing—section, plan, and details of rolling truss assembly to access ceiling. (*Graphics by Universal Builders Supply Inc.*)

19.23

levels) across the ceiling. This whole double debris netted assembly rolled along a double I-beam track supported by scaffolding on either side of the great hall as seen in the drawing. Scaffolding shown under the trusses at the east end of the hall was used to facilitate truss assembly and dismantling. Lower in the drawing, we see the great express concourse in plan where greater than 500,000 people would traverse daily. The high-level tracks were supported on standard scaffolding running parallel to the long sides of the hall and braced to the giant masonry columns supporting the vaulted ceiling structure. Details of scaffolding ties to the giant columns, as well as an interesting shock absorber mechanism to buffer dynamic loading on the double I-beam tracks, can be seen. Figure 19.10*b* shows the hall with temporary structures actively in place ready for both ceiling work and traffic below.

(b)

FIGURE 19.10   (*b*) Grand Central Terminal. Photo of great hall with temporary structures and safety devices in place and ready for business. (*Photo by Universal Builders Supply Inc.*)

Baltimore Basilica

In 2004 a complete restoration was undertaken of the largest Catholic church in the United States. Known as the Basilica of the Assumption, it was begun in 1806 and, apart from its religious history and significance, it is also a widely known and respected extant work of Benjamin Latrobe, who, through his projects of that time, earned a reputation as the father of today's American architectural profession. While earlier repairs and restorations had served to maintain the structure in its essential form and décor, the advanced stage of deterioration and undesired deviations from the original design called for a massive restoration of the interior and exterior surfaces of the building. The planning and staging for this operation involved both conventional and innovative approaches.

The most innovative part of the restoration project in terms of temporary structures was accommodation for roof replacement east and west of the main dome. These roofs covered the main nave and ancillary areas on the west and the entire altar area on the east. The original design called for these roofs to be hidden behind or with eaves just set back from a stone masonry parapet. Repairs over the years resulted in replacement roofs that overlapped the stone parapet and altered the original intent to maintain a consistent visual parapet line and further diverted rainwater control to the exterior of the shell.

To maintain a secure weather envelope while rebuilding the roofs with integrated flashing and water control, the existing roofs were cut loose from their structural ties and raised 4 ft by hydraulic jacks and hand-adjusted tension devices, commonly known as hand-tuggers, and then were closed with a perimeter weather sheathing and temporary flashing to provide worker access and a protected workspace to frame and complete new roofs according to the original design.

Figure 19.11a shows one of the roofs being lifted from the stone masonry parapet; visible are the hydraulic jacks and tuggers. Figure 19.11b is a view from the inside showing

(a)

FIGURE 19.11 (a) Baltimore Basilica, Baltimore, Md. Roof in lift from parapet. (*Photo by Universal Builders Supply Inc.*)

(b)

FIGURE 19.11   (*b*) Baltimore Basilica, Baltimore, Md. Interior with typical brace and aluminum beams. (*Photo by Universal Builders Supply Inc.*)

(c)

FIGURE 19.11   (*c*) Baltimore Basilica, Baltimore, Md. Roof view with lifting structure. (*Photo by Universal Builders Supply Inc.*)

(d)

FIGURE 19.11   (*d*) Baltimore Basilica, Baltimore, Md. Staging on Byzantine tower. (*Photo by Universal Builders Supply Inc.*)

one of the many stay-braces around the perimeter to stabilize the lifted roof. Also visible here are aluminum beams carrying joist loads, cross beams to facilitate new roof construction, and a severed section of the old roof framing to free the roof for lifting. Figure 19.11*c* shows the temporary structure set up for the tuggers to work against in complement to perimeter hydraulic jacks. Figure 19.11*d* shows more typical staging around one of the Byzantine towers with work platforms at various levels. Both towers were repaired and rehabilitated. Swing staging was used around the exterior and full scaffolding gave access to all interior surfaces. The project was completed in 2006.

Augusta Sportswear

Faced with the need for expanded warehouse and operating space, R. W. Allen of Augusta, Georgia, undertook a raising of the roof on the existing 50,000 ft^2 sportswear facility. The roof was raised by 18 ft of clear height to more than double the existing working capacity of the company. With the help of the subcontractor Roof Lifters of Toronto, Allen maintained roof integrity and weather protection while raising the whole roof uniformly using synchronized hydraulic jacks in extendable masts. With the jacks in place, the roof perimeter was cut all around and existing columns were cut free at the base plates. Electric lines, fire protection lines, and roof drainage systems were rerouted to maintain operation during the lift. At the final height, stud columns and beams were placed and pre-cast wall panel extensions doweled on top of the existing. All mechanical systems, roofing, lighting, and fire protection were left in place, lifted and reused. The job was completed in 7 months with maximized cost savings for the client. Figure 19.12*a* through 19.12*d* shows various stages of project completion.

1717 Rhode Island Ave, Washington, D.C.

This 10-story class A office building in Washington, D.C. was built in 2003. During occupancy, problems became apparent and extensive replacement of elements in the envelope was called for requiring removal of significant portions of the brick facing. This reinforced concrete-framed structure overlaid with brick veneer on metal studs was sited adjacent to a historic church on one side and behind historic buildings on the street frontage with associated roof setbacks that required atypical approaches to support work platforms because much of the scaffolding could not be supported directly from grade.

A frontal view of 1717, as shown in Fig. 19.13*a*, exhibits both grade-supported and support arm/flying beam-supported scaffolding to carry the work platforms across the face of the building. The debris/dust curtain is standard in projects of this type. The support arm/flying beams (e.g., Fig. 19.13*b*) were of lightweight aluminum secured on husky temporary angle brackets anchored to the original reinforced concrete frame (e.g., Fig. 19.13*c*). This

(a)

FIGURE 19.12 Augusta Sportswear, Augusta, Ga. Views of hydraulic jack roof raising. (*Photos by R.W. Allen.*)

(b)

(c)

FIGURE 19.12 (*Continued*)

(d)

FIGURE 19.12   (*Continued*)

(a)

FIGURE 19.13   (*a*) 1717 Rhode Island Ave., Washington, D.C. Frontal view. (*Photo by Universal Builders Supply Inc.*)

(b)

FIGURE 19.13 (*b*) 1717 Rhode Island Ave., Washington, D.C. Flying beam support arm. (*Photos by Bruce Wilkinson.*)

(c)

FIGURE 19.13 (*c*) 1717 Rhode Island Ave., Washington, D.C. Husky angle bracket. (*Photos by Bruce Wilkinson.*)

technique, creating terraces to support conventional scaffolding, facilitated work above surfaces impractical to support scaffolding on and in most cases requiring protective cover of framed plywood covered by an impervious membrane. An arrangement protecting part of the church property is shown in Fig. 19.13*d* and above flat roofs where sometimes dimension planking may be called for. This latter protective technique is common above sidewalks or other close-in traffic ways where the supporting structure may be conventional scaffold frames or special towers as in Fig. 19.13*e* that, in this case, support the scaffold platform as well as facilitate traffic movement underneath. The support arm/flying beams also enabled suspension of swing staging in a situation where hanging from a roof/ parapet location was not practical or, as in the subject situation, the roof configuration was not conducive. An upper hook for the swing staging is seen in Fig. 19.13*f*. Where masonry removal was necessary to expose shelf angles and associated flashing, miniature strut braces were used to maintain integrity of the masonry envelope left in place as shown in Fig. 19.13*g*.

Westin Hotel, Atlanta, Georgia

In 2008 a rogue tornado swept through downtown Atlanta and sucked out or cracked some hundreds of windows in this 74-story cylindrical tower. Located in the midst of heavy-vehicle and pedestrian traffic patterns and relatively narrow streets, repair of the damaged glazing posed a considerable logistic and safety problem, typical in today's urban conditions. Built in 1976, the hotel still is tallest in the Western Hemisphere and has enjoyed a strong occupancy profile as well as a significant revenue and publicity source in a rotating restaurant on top of the structure where one has sweeping views of the surrounding urbanscape and beyond. Due to the age of the building, replacement of the damaged glazing alone was rejected in favor of a complete reglazing responding to present-day energy and wind loading standards. The entire reglazing of the cylinder and its scenic elevator shaft

(d)

FIGURE 19.13   (*d*) 1717 Rhode Island Ave., Washington, D.C. Protective coverings. (*Photos by Bruce Wilkinson.*)

(e)

FIGURE 19.13 (*e*) 1717 Rhode Island Ave., Washington, D.C. Protected traffic-way bearing scaffolding and work platforms above. (*Photo by Bruce Wilkinson.*)

(f)

FIGURE 19.13 (*f*) 1717 Rhode Island Ave., Washington, D.C. Swing staging hook. (*Photo by Bruce Wilkinson.*)

(g)

FIGURE 19.13 (*g*) 1717 Rhode Island Ave., Washington, D.C. Miniature strut braces. (*Photo by Bruce Wilkinson.*)

involved 6350 window panels. This total external refitting of the building envelope and subsequent replacement of fan-coil units in the guest rooms presented a client-requested material movement requirement that well exceeded capacity of the existing loading dock, two freight elevators, and connecting corridors, which also had to maintain servicing of the guest rooms and restaurant while the repair work proceeded.

A complete operational analysis comparing usage of the internal material movement paths described previously versus an external system of temporary structures convinced the client to choose the latter proposal. Conceptually, the external scheme involved four different types of temporary structures: (1) A scaffold-based loading dock area that facilitated material movement both in and out, and stabilized the buck hoist and its landing on the street side of the 16th-floor level; (2) a hoist that climbed from the 16th-floor level to the 70th-floor level for the movement of materials in and debris out at every floor; (3) three mast climbers around the semicircular external panorama viewing express elevator shaft to the 70th-floor restaurant and the roof beyond at the 75th-floor level; and (4) swing staging encircling the entire cylinder to facilitate external assistance to crews removing old and installing new glazing from the inside. See Fig. 19.14*a* through 19.14*e* illustrating these structures.

The setback of the tower cylinder from its base at street edge necessitated a two-stage hoist to facilitate total vertical movement. This is shown in the conceptual sketch (Fig. 19.14*a*) with a larger-scale diagram of the transfer point at level 16 with its cantilevered support at level 15 (Fig. 19.14*b*) that also picked up the inboard end of a short bridge from the buck hoist landing at the upper level of the loading dock area base scaffolding. Vertical bracing to secure the interior end of the level 15 cantilever extending into the tower spanned two-floor systems. The buck hoist upper landing with the main materials hoist mast behind is shown in Fig. 19.14*c* where mast stabilizing ties to the base scaffolding and the main tower can be seen.

Rigging of the mast climbers for the external elevator shaft is shown in Fig. 19.14*d* with debris netting visible below. This netting is also visible in Fig 19.14*c*. The crowning

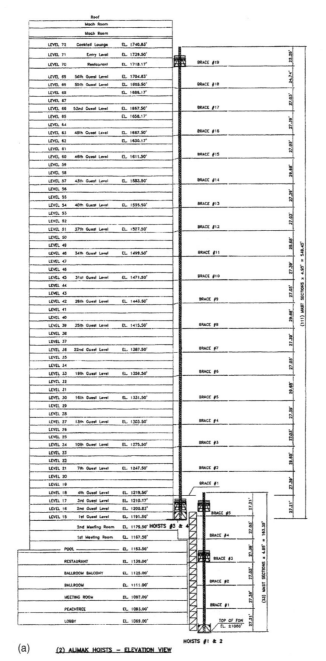

(a)   **(2) ALIMAK HOISTS — ELEVATION VIEW**

FIGURE 19.14 (*a*) Westin Hotel, Atlanta, Ga. Engineering drawings/ elevation views of hoists. (*Graphics by Robert Posch, P.E.*)

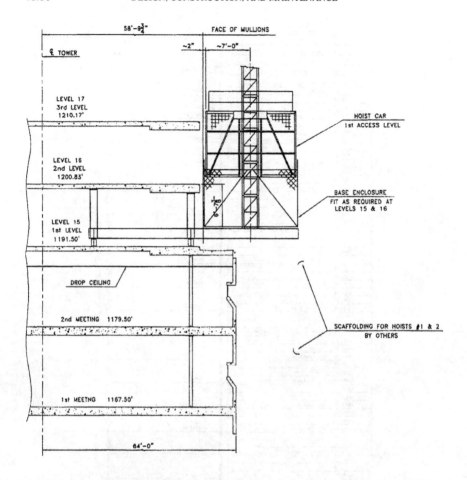

(b) **HOISTS #3 & 4 — ELEVATION VIEW (BASE LEVELS)**

FIGURE 19.14(b) (*Continued*)

element of temporary structure in this project is the swing staging or high-stage climbers suspended from the roof that enabled safe conduct of external operations over the entire cylinder envelope from top to bottom. See Fig. 19.14*e* and note the full deployment of suspension cables and safety railings.

Fine work and a satisfied client led to another contract to refit the main central utility room in the pedestal structure. Work included stripping out 3800 ft^2 of electric and HVAC primary equipment and replacing the functions with new and energy-efficient installations. Figure 19.14*f* shows the scaffolding complex with related hoists, ramps, and platforms to facilitate removal and installation. Most materials could be broken down for movement

(c)

FIGURE 19.14 (*c*) Westin Hotel, Atlanta, Ga. Level 16 buck hoist landing and connection to main personnel/material hoist. (*Photo by David Luffel.*)

(d)

FIGURE 19.14 (*d*) Westin Hotel, Atlanta, Ga. Mast climber rigging. (*Photo by David Luffel.*)

(e)

FIGURE 19.14 (*e*) Westin Hotel, Atlanta, Ga. High stages and climbers. (*Photo by David Luffel.*)

(f)

FIGURE 19.14 (*f*) Westin Hotel, Atlanta, Ga. CPR scaffold. (*Photo by W. Barnes.*)

(g)

FIGURE 19.14 (*g*) Westin Hotel, Atlanta, Ga. Boiler insertion into CPR. (*Photo by David Luffel.*)

with the exception of the main boilers weighing 10,000 lb each that had to be extracted en-masse through a lateral opening. The new boilers were inserted in the same manner by use of a crane (see Fig. 19.14*g*). Maintenance of boiler operation for hot water and heat was accomplished through a temporary boiler situated on an adjacent rooftop supported by a temporary platform. This upgrading of a major HVAC central plant is representative of many such projects in the future that will be called for to refit existing commercial and institutional buildings and bring them up to current code and energy standards.

Perry Memorial

This 352-ft-tall monument was built on an island in Lake Eire just off the coast of Ohio in 1900. It commemorated the victory on Lake Eire of Commodore Oliver Hazard Perry in the War of 1812. Over time, this fluted Doric column developed considerable deterioration on the upper-level observation deck and its soffit casings that capped the top of this massive stone masonry structure. To avoid ground-supported scaffolding in excess of 300 vertical ft, a set of trusses were designed to provide suspended work platforms under and around the observation deck casings and transfer all loads back to pedestal points directly above the main column structure. Hoist trusses were configured to similarly transfer their dead and live loads back to pedestal points that would bear directly on the column. The hoists were in constant use to assist in the removal of casing stones around and under the deck, position them for cleaning and then replacement and proper joint treatment. As each stone weighed 1 to 3 tons, considerable dynamic load was imposed on the entire temporary structure during the course of the project.

Figure 19.15a shows the monument in full height with all trusses, work platforms, and hoist in place. Figure 19.15b focuses in to the trusses and hoist mast, and Fig. 19.15c is a view of the top showing all assemblies in position, including a debris curtain at a point of facing stone hoist.

(a)

FIGURE 19.15   (a) Perry Memorial, Put-in-Bay, Ohio. Full-height view with temporary structures. (*Photo by Universal Builders Supply Inc.*)

(b)

FIGURE 19.15 (*b*) Perry Memorial, Put-in-Bay, Ohio. Trusses, hoist mast, and work platforms. (*Photo by Universal Builders Supply Inc.*)

(c)

FIGURE 19.15 (*c*) Perry Memorial, Put-in-Bay, Ohio. All assemblies in place. (*Photo by Universal Builders Supply Inc.*)

SUMMARY

Most activities in and around maintenance, repair, and restoration projects that require temporary structures to be built can be classified under one or more of the headings discussed in this chapter. This is a broad subject area with many variations, permitting many approaches to problem solving. What has been presented is illustrative of what can be done.

The principles of sound engineering practice should be applied in developing solutions to temporary structure requirements. Some allowance should always be made for the unknown when dealing with existing structures, especially older ones. One has to be very conservative in judgments and estimates; and one has to be conservative in analysis and design. It is imperative to understand the original design before devising changes and to keep in mind the load paths and stability during alterations. Existing conditions in a structure may result in quite different structural behavior than was intended or foreseen in the original design.

Repair and restoration is challenging work. It is made more so if one considers that existing buildings emptied for work may undergo considerable environmental change. The normal pitfalls of construction combined with hidden surprises of existing structures characterize repair and restoration work. This chapter has focused on qualitative aspects of repair and restoration operations. The bibliography is rich in references augmenting chapter commentary and containing quantitative guidance. It has been said that a good rehabilitation contractor can probably be good at anything in the construction field. However, more important than the relative ability of successful practitioners is the caution that must be exercised by those uninitiated in the art. Caution and sound engineering practice will generally succeed where novices fear to tread.

ATTRIBUTIONS

Some information on projects in the "Notable Recent Projects" section was provided by the following individuals:

1717 Rhode Island Ave., *Bruce Wilkinson* of Brisk Waterproofing and *Mark Tsirigos* of Universal Builders Supply Inc., 2010–2011.

Augusta Sportswear, *John Martin, Laura Berry* of R.W. Allen, 2011.

Baltimore Basilica, *Mark Tsirigos* of Universal Builders Supply Inc., 2010–2011.

Grand Central Terminal, *Chris Evans* of Universal Builders Supply Inc., 2010–2011.

Perry Memorial, *Mark Tsirigos* of Universal Builders Supply Inc., 2011.

Westin Hotel, *Graeme Kelly, David Luffel, John Reyhan* of Skanska USA, 2010–2011.

BIBLIOGRAPHY

"All the Arena Is a Stage," *ENR (Building Team Edition)*. McGraw-Hill, Aug. 12, 1991.

Bernstein, David: "Tackling the Older Types of Building," *The Architects' Journal*, Feb. 10, 1993.

Burns, John A.: "Measuring and Documenting Existing Buildings," *Progressive Architecture*. June 1992, pp. 39–46.

Chandler, Ian: *Repair and Refurbishment of Modern Buildings*. B. T. Batsford Ltd., London, 1991.

Code of Federal Regulations. Title 29. Chapter XVII, Occupational Safety and Health Administration, Department of Labor (Continued). Part 1926.

"Construction and Demolition Operations—Concrete and Masonry Work—Safety Requirements," ANSI A10.9-2004, American National Standards Institute, New York.

"Construction and Demolition Operations—Temporary Floor and Wall Openings, Flat Roofs, Stairs, Railings, and Toeboards—Safety Requirements," ANSI A10.18-2007, American National Standards Institute, New York.

Dyton, Francis: "A Contractor's Thoughts on Heavy Structural Repairs," *Construction Repairs and Maintenance,* London, November 1985, pp. 19–20.

"Engineering Skills Meet Challenge of Capitol Restoration," *Newsletter,* Scaffold Industry Association, Van Nuys, March 1986.

"Glaze of Glory." *International Construction Review,* The Chartered Institute of Building, Ascot, 2d Quarter 2010.

Goodchild, S. L.: "Facade Retention Projects—2," *Structural Survey,* H. Stewart, London, vol. 3, no. 3, 1984, pp. 231–242.

Haylor. J. Denise: "Interior Conservation: Looking inside Historic Buildings." *Construction 77.* CIOB, Englemere, Ascot, September 1990, pp. 6–9.

Highfield, David: *The Construction of New Buildings behind Historic Facades,* E & FN Spon, London, 1991.

Hill, W. F.: "Facade Retention Projects—I," *Structural Survey,* H. Stewart, London, 3, no. 1, 1984, pp. 12–23.

International Building Code, International Code Council, Inc., Country Club Hills, Ill., 2009.

Keller's Official OSHA Safety Handbook, J. J. Keller & Associates, Inc., Neenah, Wis. 1993.

Lee, David: "The Role of the Structural Engineer in Repairs and Maintenance," *Construction Repairs and Maintenance,* Palladian Publications, London, March 1986, p. 3.

Lindsay, R. H. W.: "Access Scaffolding," *Structural Survey,* H. Stewart, London, vol. 1, no. 2, 1982, pp. 140–145.

"Mast Climbers." *Masonry Magazine,* June 2002.

"R. W. Allen Goes Vertical at Sportswear Factory." *Georgia CONSTRUCTION TODAY,* 4th Quarter 2010.

"Safety Requirements for Demolition Operations," ANSI A10.6-2006, American National Standards Institute, New York.

"Scaffolding—Safety Requirements," ANSI A10.8-2011, American National Standards Institute, New York.

"Scaffolding—Trinity Church Braces Itself against the Winds of Change," *ENR,* McGraw-Hill, Sept. 28, 1989.

Seelye, Elwin E.: *Foundations—Design and Practice,* 2d printing, Wiley, New York, 1966, pp.10–13.

Spradlin, W. H. (ed.): *The Building Estimator's Reference Book,* Frank R. Walker Company, 28th ed. 2008.

Universal Builders Supply Inc. *Website,* 2010, 2011.

CHAPTER 20
CRANES

Lawrence K. Shapiro, P.E.

CRANE AND RIGGING BASICS

From the time humankind coalesced into civilizations there has been a need to lift heavy objects. Ancient constructors relied on inclined planes. Greek builders may perhaps have been the first to advance to the use of devices that would be recognized today as cranes and rigging. Crane-like equipment was used in the Roman Empire and in medieval Europe. From the Industrial Age to the present, crane technology has tracked the general progression of technology, utilizing materials, power transmission and control systems of the day.

Definitions

A *crane* is a machine designed to lift heavy loads and reposition them by traveling, trolleying, booming, or swinging. An archetypical crane uses ropes and pulleys for hoisting, but some modern crane-like machines manage without ropes by use of articulating extensions. Whether or not such devices should be classified as cranes is a debatable question. This author favors using the term "crane" only to machines that lift with ropes and pulleys.

Cranes in common use for construction come in several form, the most prevalent being *mobile cranes* and *tower cranes*. Each of these types, in turn, is available in various forms and configurations described in this chapter. A crane is a self-contained piece of equipment, whereas the power plant of a *derrick* is a separate unit. The derrick and its power source are brought together for each particular job.

Out of many hundreds of words in the crane and rigging lexicon, the following basic terms are broadly used and useful for comprehension of this chapter:

All-terrain crane. A mobile crane with a truck chassis that is suitable both for highway travel and driving on rough terrain (Fig. 20.1).

Boom. A strut or spar used to project the upper end of the hoisting tackle; also called jib (European).

Crawler crane. A crane with a base mounting that incorporates a rigid central *car body* supporting a pair of side frames and tracks. Each side frame is an armature carrying a continuous belt of track pads to travel the crane forward or rearward (Fig. 20.2).

Critical lift. A lifting operation judged to carry a potentially high level of risk due to factors such as load weight, complex procedures, high value, or presence of hazards. Though no uniform standard exists for identifying a critical lift, there is a consensus that risk can be mitigated by engineering, planning, and field controls (Fig. 20.3).

Derrick. A lifting device that uses ropes for hoisting, with or without a boom, utilizing a mast or equivalent member held at the head by guys or braces. Unlike a crane, a derrick does not have an integral winch (Fig. 20.4).

Friction crane. A mobile crane that relies on mechanical power transmission; the operator engages clutches and brakes to engage power and stop hoist motion (Fig. 20.2).

Hammerhead jib. A tower-crane jib (boom) fixed at a horizontal or near-horizontal angle, equipped with a *trolley* to support the hook block and traverse loads radially from the tower (Fig. 20.5).

Hydraulic crane. The vernacular name for a crane for a telescoping cantilevered boom crane; the telescoping of the boom and the various other motions of the crane are typically powered by hydrostatic (hydraulic) drives (Fig. 20.6).

Jib. (1) In U.S. practice, an extension to the boom mounted at the boom tip, in line with the boom longitudinal axis or offset to it (Fig. 20.6). (2) On tower cranes, the structural

FIGURE 20.1 The telescopic boom on this *all-terrain crane* carries both a main load fall and an auxiliary tip for a whip line. The carrier, with a separate driving cab, is suited both for highway driving and off-road travel. Note that all wheels are turned to maneuver the crane into operating position, demonstrating the operation of crab steering. This crane carrier has five axles but machines with heavier road weight have as many as nine. (*Paul Yuskevich.*)

member extending forward from the mast to support the lifting trolley, sheaves, hook block, and load. (3) In Europe, the term is used for a crane boom. Correspondingly, the extension to the main boom is known as a fly jib.

Latticed boom. A boom constructed of four longitudinal corner members, called *chords*, assembled with transverse and/or diagonal members, called *lacings*, to form a truss work in two directions. The chords carry the axial boom forces and bending moments, while the lacings resist the shear forces (Fig. 20.2).

Lifting capacity. The maximum gross load weight that a crane manufacturer has determined a crane can safely handle under specified conditions as stipulated in the load chart.

Lift crane. A crane configured for lifting, booming, swinging, or traveling with loads attached to the crane hook block as contrasted to a crane configured for lifting material with a bucket or grapple.

Load. The suspended weight applied to the crane, including the weight of lifting hardware such as hook block, shackles, and slings.

Load rating chart. A tabulation of load ratings provided in a formal document.

Luffing. Changing the boom angle by activating the boom hoist winch; also called luffing, booming, and topping.

FIGURE 20.2 A *crawler crane* may be costly to bring to the site, but, once assembled, it can be moved about on its tracks, sometimes even on bare ground with a load on the hook. The offset tip seen here keeps the load away from the delicate latticed boom. This crane is being used for dynamic compaction of the soil, a duty-cycle operation requiring a *friction crane* allowing the load to freefall. (*Paul Yuskevich.*)

Luffing tower crane. A crane with a boom pinned to the superstructure at its inner end and containing load hoisting tackle at its outer end and with a hoist mechanism to raise or lower the boom in a vertical plane to change load radius (Fig. 20.7).

Mast. (1) An upright load-bearing component of a crane or derrick. (2) The tower of a tower crane.

Mobile crane. A crane capable of either traveling or transiting under its own power. In Europe, a crane mounted on a truck carrier.

Outriggers. Extensible arms attached to a crane base mounting that include means for relieving the wheels (crawlers) of crane weight; used to increase stability (Fig. 20.8).

Overhaul weight. Weight added to a load fall to overcome resistance and permit unspooling at the rope drum when no live load is being supported; also called headache ball. (Fig. 20.9).

FIGURE 20.3 *Critical lift* involving multiple cranes and complex procedures. Hazards of this hangar dismantling were mitigated by engineering, planning, and field controls. (*Valentin Ciocoi*).

FIGURE 20.4 In a historic photograph, a *guy derrick* lifts a truck crane onto a high-rise steel structure.

FIGURE 20.5 A multitude of hammerhead tower cranes sweep this large casino development in Macau, China. The cranes above the high-rise building are the saddle jib style. The ones nearer the camera are the flattop style, selected for these locations because they do not project vertically into airspace swept by other cranes. (*Lawrence Shapiro.*)

Parts of line. A number of *running lines* (ropes) supporting a load, generally running through two opposing blocks (Fig. 20.9).

Radius. Horizontal distance from the axis of rotation to the center of gravity of a freely suspended load.

Rated load. The maximum allowable working load on a crane or rigging component designated by the manufacturer under specified working conditions; expressed in pounds, kilograms, short tons, or metric tons.

Reeve/reeving. (1) The pattern followed by a rope over pulleys or sheaves and drums. (2) The act of pulling a rope over pulleys or sheaves and drums.

Rough-terrain crane. A mobile crane with a chassis that has oversized tires for driving over uneven terrain. The carrier is incapable of driving at full highway speed (Fig. 20.6).

Self-erecting crane. A tower crane that assembles and raises itself on the site, usually with no assist crane (Fig. 20.10).

Shackle. A component of rigging hardware consisting of a U-shaped steel clevis with a pin passing through the open ends of the U. A shackle is used to temporarily attach other rigging components such as wire rope or chain slings to fixed lugs or fittings.

Sheave. A wheel or pulley with a circumferential groove designed for a particular size of wire rope; used to change direction of a running rope.

Sling. A rope or strap with integral loops at both ends used for supporting a load.

Superstructure. The upper portion of the crane that rotates.

FIGURE 20.6 A *hydraulic crane*, more properly called a *telescoping cantilevered boom crane*, set up to lift from its swingaway jib. An unused extension piece of the jib is mounted to the side of the boom. (*Paul Yuskevich.*)

FIGURE 20.7 A *luffing tower crane* is a fixture of dense urban high-rise districts and industrial plants. (*Dominique Singh.*)

FIGURE 20.8 *Outriggers* extend the tipping fulcrum of a truck-mounted crane. The load is distributed over the supporting roadway by *cribbing* and the jacks will be extended so that the wheels are fully clear of the ground as in Fig. 20.6. The immense size of this crane is evidenced by its eight axles and the robust cribbing required under the outriggers. (*Mathieu Chaudanson.*)

FIGURE 20.9 An *overhaul weight* overcomes rope and sheave resistance to allow an empty hook to be lowered to the ground. The one on the left overhauls a single-part whip line. The larger weight on the right must compensate for the resistance from the three *parts of line* of the main fall.

20.8

FIGURE 20.10 A *self-erecting crane* operates like a tower crane but travels and sets up like a mobile crane. This unusual installation is supported on a platform above the street. (*Morrow Equipment Company.*)

Swing. A crane or derrick function where the boom or load-supporting member rotates about a vertical axis (axis of rotation); also called slewing.

Tower crane. A lift crane with the working boom mounted on a vertical mast; the working boom slews about a rotation center that is usually fixed at the center of the mast (Figs. 20.5 and 20.7).

Trolley. A carriage carrying the hook block for radial movement along the lower *chords* of a horizontally mounted tower-crane jib.

Truck crane. A crane superstructure and boom mounted on a rubber-tired carrier that can be driven over the road from a driver's station remote from the crane cab (Fig. 20.1).

Weathervaning. Allowing an *out-of-service* crane to rotate freely in response to wind forces so as to expose a minimal area to the wind.

Winch. A power-driven drum capable of winding rope for pulling or for lifting and lowering loads.

Wire rope. A flexible rope composed of multiple-wire steel strands helically wound around a core.

Temporary versus Permanent

By nature construction is a temporary condition, a fact that makes construction cranes temporary installations by definition. This is a key consideration that distinguishes construction machines from those used in general industry and marine service, shaping how construction cranes are designed, tendered for service, manned, maintained, regulated, and inspected.

The nature of "temporary" is quite variable. A telescopic truck crane may be on a jobsite for an hour to make one pick and then leave, while a *tower crane* on an infrastructure project might stay put for 3 or more years.

In the first example, the crane might be rented out by the hour and furnished with an operator who also drives the crane to the site. The crane is designed to be deployed for "taxi service." Maintenance and inspections would take place in a shop remote from the jobsite.

In the latter example, a crane erected to work in one place for 3 years is subjected to a different set of circumstances. Economics may favor that it be purchased. The operator is likely to be an employee of the user. Maintenance and inspection are performed on site, and parts need to be available—on site or on call—to keep the crane working if it breaks down. The costs of shipping this crane to the site and installing it (as well as removing it) are great, but they are amortized over its long duration of use.

Temporary service also implies that a crane is less likely to be exposed to severe environmental conditions than a permanent installation. A mobile crane, for example, might be removed from the site or the boom placed out of harm's way in advance of a severe storm wind. Even a construction tower crane is not ordinarily designed to withstand the same magnitude of wind as a permanent structure, as its service time of a few months to a couple of years is still relatively curtailed.

The Basic Hoisting System

At the heart of every crane is a winding drum, hoisting rope and one or more pulleys, or *sheaves*.

Figure 20.10 shows a crane with a load attached to the lower block, or *overhaul weight* (Fig. 20.9), and the block in turn supported by two ropes, or *parts of line* suspended from the upper block. Each rope will carry half the weight of the load, giving the system a mechanical advantage of two. Mechanical advantage is governed by the number of ropes actually supporting the load. As parts of line are added, the force needed to raise or lower the load decreases and load movement speed decreases proportionally.

The block is an assemblage of sheaves. The rope is strung in one continuous piece from the winding drum through the blocks to a dead end on either the lower or the upper block. When the winding drum is not in motion, the load is uniform on all the parts of line. The load acting on any one part is found by dividing the weight of the lifted load by the mechanical advantage. In Fig. 20.3 the lifted load of each crane would include the lower block, sometimes called the hook block. When the distance between the upper and lower blocks is great, the rope weight can be a significant part of the load.

Friction comes into play when the hoist drum is in motion. The major part of friction loss occurs in the sheave shaft bearings. As each sheave retards the motion of the rope, there is a small difference in load from one rope segment to the next, that is, sheave to sheave. The loss coefficient can vary from a high of about 4½ percent of rope load for a sheave mounted on bronze bushings to as little as 0.9 percent for a sheave on precision ball or roller bearings. An arbitrary value of 2 percent is a reasonable approximation for sheaves on common ball or roller bearings when the rope makes a turn of 180°. Due to friction at the sheaves, the pattern of rope tension changes when going from raising to lowering the load.

The lowering of an unloaded hook is resisted by friction, by the weight of the rope opposing the block-and-tackle and by the inertia of the winding-drum mass. Weight at the hook must exceed the rope weight multiplied by the mechanical advantage plus an allowance to overcome friction and inertia. If the weight at the hook is insufficient, the hook will not lower and might even rise on its own until it strikes the upper block. To prevent this action, it is necessary to have a lower block with adequate weight or to add an overhauling weight so that the hook will descend on command. Because the overhaul weight becomes part of the dead weight of the mechanism and remains in place throughout operations, it must be added to the lifted load when planning a lift.

Figure 20.3 shows the hoisting mechanisms at work on four cranes lifting simultaneously. The lower block on each crane is provided with heavy side plates for overhauling multiple parts of line. On the two outer cranes a single-part auxiliary fall, the *whip line*, is overhauled by a lighter cast weight irreverently called a *headache ball.*

Winches, Ropes, and Sheaves. An assemblage of one or more winding drums mounted on a frame is called a hoist or *winch.* A free-standing winch unit with integral controls and power plant is called a *base-mounted drum hoist.* Power sources, power transmission, control systems, and drum characteristics vary.

A winding drum spools and stores the wire rope, and it also transmits power from the drum motor to the rope. Each turn of the rope around the full circumference of the drum is called a wrap. Rope is helically wrapped around the drum, starting at one end flange and progressing toward the opposite flange. A series of wraps extending from flange to flange is referred to as a layer. Spooling continues after completion of a layer, proceeds back toward the starting flange in a second layer, and so forth. Storage capacity is limited by the width of the drum and its flange height.

Drums are often grooved to guide the first layer. A good first layer is a necessity if succeeding layers are to wrap properly; the rope itself creates the groove for subsequent layers. Grooves are cut to suit a particular rope diameter, and grooved drums can be properly used only for that diameter of rope.

For proper spooling, the angle at which the rope leads onto the drum, called the fleet angle (Fig. 20.11), must be kept within controlled limits. Fleet angles should be no less than 0.5° and no more than 1.5° for smooth drums or 2° for grooved drums when the lead

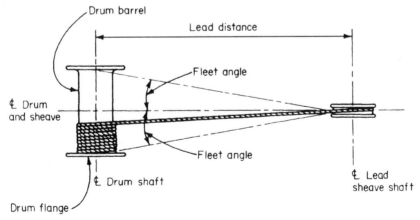

FIGURE 20.11 Fleet angle. (*From Shapiro, Lawrence K. and Jay P. Shapiro: Cranes & Derricks, 4th ed., McGraw-Hill, New York, 2010, Fig. 1.4.*)

sheave is centered on the drum. Where there is insufficient space to allow an optimal fleet angle, one of several alternative arrangements such as a pivoted lead sheave may be used.

A traditional friction-drive winch is powered by a gasoline or diesel engine with a mechanical gear train and two or three drums. A brake band is built around the periphery of one flange on each drum. For additional safety, a ratchet-and-pawl system can be included at each drum, providing a means for positively locking the drum against inadvertent spooling out of rope.

With electric controls and hydraulic or electric power transmission, modern hoist machines only superficially resemble their rudimentary predecessors. Freefall lowering is replaced by full control under power or retarding by a torque converter.

Winch units for use in a crane are tailored to the specific requirements of the crane and are integrated with the control and mechanical systems. Hydraulically driven winches and some electric winches are furnished with automatic brakes that are normally engaged. When either power-up or power-down control signals are initiated, the application of power to the drum triggers release of the brake. Tower cranes and many contemporary mobile cranes have electric or hydraulic winches that do not permit free-fall lowering. Powered lowering provides good load control for precision work such as placing machinery on anchor bolts.

Sheaves are used to change the direction of travel of wire ropes. Assembled in multiples, in the form of blocks, they are able to provide almost any required mechanical advantage. Sheaves rotate about their mounting shafts on bushings or bearings. Grooves are shaped to provide some tolerance for misalignment.

Wire Rope. The contemporary crane cannot exist without wire rope. Ordinary fiber rope is so limited in strength that it is generally used only for unpowered applications, and chains are both awkward and heavy. High-performance fiber ropes are used for slings and pendants but are not yet considered generally useful for hoist ropes.

Wire ropes are constructed of individual wires laid together into strands with the strands in turn laid over a core to form the rope. The number of wires in a strand, the number of strands in a rope, and the nature of the core vary. Ropes are categorized by classes, such as 6×19, which give the number of strands to the rope and the nominal number of wires in the strand. The class designations have a basis in tradition rather than in precision; thus a 6×19 class rope may have anywhere from 15 to 26 wires in its strands. Figure 20.12 shows a few of the many styles of rope available.

Rope cores are of several types, namely, fiber core (FC), wire strand core (WSC), and wire rope (independent wire-rope core, IWRC). The core acts to support the strands. A wire core adds strength but reduces flexibility; however, it generally increases resistance to crushing and bending fatigue.

Wire ropes are manufactured in several grades. Minimum required safety factors, or strength factors vary with rope application and sometimes the rope type. For most crane uses, four rope properties are of key importance:

- *Strength* is controlled by the size, grade, construction, and core type.
- *Flexibility* and *fatigue resistance* are improved by strands with a large number of small wires and by preforming.
- *Abrasion resistance* is enhanced by large outer wires or by Lang lay construction.
- *Crushing resistance* is improved with IWRC or WSC, large outer wires, and regular-lay rope.

The lay of the rope refers to the direction of rotation of the wires and the strands. Regular-lay (right or left) ropes are made with the wires in the strands laid in one direction and the strands laid in the opposite direction, so that individual wires have the appearance of running

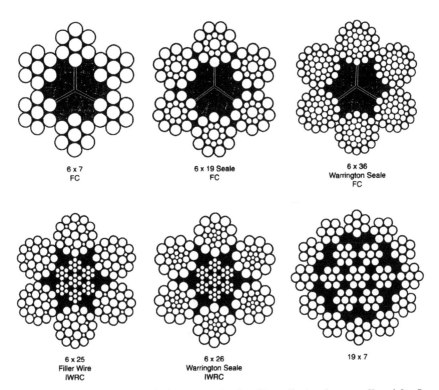

FIGURE 20.12 A few styles of wire-rope construction. (*From Shapiro, Lawrence K. and Jay P. Shapiro: Cranes & Derricks, 4th ed., McGraw-Hill, New York, 2010, Fig. 1.9.*)

parallel to the long axis of the rope. Regular-lay ropes are easy to handle, resist kinking and twisting, and are stable. Lang lay ropes have wires and strands laid in the same direction, offering greater surface exposure and hence greater abrasion resistance. They are more susceptible to untwisting and kinking than regular-lay ropes and have less crushing resistance, but they are unsuited to applications where one end of the rope is free to rotate as would occur with a swivel installed. However, ropes with Lang lay are more flexible and fatigue-resistant than regular lay.

Preformed wire ropes have strands shaped to in helical form before being laid into the rope, a process that produces stable rope that will return to its original form and not unwind when cut. These ropes also have improved fatigue resistance. Most wire rope is now preformed.

The arrangement of the wires in a strand contributes to the properties of the rope. Some common arrangements are

- *Simple.* All wires of the same size
- *Seale.* Large wires on the outside for abrasion resistance and small wires inside to increase flexibility
- *Warrington.* Alternative wires large and small to combine flexibility with abrasion resistance
- *Filler wire.* Very small wires placed in the spaces between the wires in the inner and outer layers of wire for increased fatigue resistance

A complete specification for a particular wire rope construction could read 6×25 filler wire, preformed, IPS IWRC, right-regular-lay rope; if not designated, right regular lay is assumed.

A loaded rope tends to unwind and stretch from its original helical shape, a phenomenon called constructional stretch. If the rope ends are unrestrained, some unwinding, or spin, will occur, marked by end rotation and additional elongation. A load at the end of the rope will spin if not restrained. Taglines, usually manila ropes connected to the load and controlled by workers on the ground, are used to prevent load spin while they also orient and steady the load.

Rotation-resistant ropes are often used for single-part applications. In these ropes, the tendency of one layer of strands to rotate is counteracted to some degree by the opposite propensity of another layer. The advantage in using rotation-resistant ropes is that under ideal conditions the load will not spin and a tagline may not be necessary.

In U.S. practice the working load of the rope is the rope-breaking strength divided by a safety factor, usually 3.5 to 5.0 for running lines and 3.0 for standing lines or guys. A 5.0 minimum factor is used for rotation-resistant ropes, which must only be used as running lines. In European practice, the basis of rope selection includes factors for the intensity of use and the number of sheaves the rope will have to pass over.

Design factors for slings are generally set higher than those set for running ropes and standing ropes on cranes. Because slings are often used repeatedly under particularly severe and loosely controlled conditions, provision is made for their inevitable deterioration and for potential misuse. Regulations such as those from OSHA and CEN (European Committee for Standardization) specify sling capacities in various configurations and conditions of use. Actual load capacity values can be found in rigging handbooks and vendor literature.

Wire Rope Fittings.   Various types of fittings are used to attach wire ropes to structural connections and to each other. The strength of a fitting is expressed as *efficiency*, which is actually the percentage of the strength of the attached rope.

Rope-end fittings provide the rope end with a loop, eye, pinhole, or hook for attachment. Permanent fittings, which require cutting the rope for removal, are most efficient, often matching the strength of the rope. Removable fitting types, wedge sockets, and clip and thimble loops develop about 80 percent of rope strength. Figure 20.13 illustrates some common rope fittings. More detailed information can be found in a wire-rope handbook.

With superior strength and reliability, permanent rope fittings should be used where feasible. There are, however, many purposes for which permanent fittings are impractical such as the dead-end attachment of a hoist rope. In design practice the efficiency of the rope-end fittings is commonly ignored, though this practice should be used judiciously.

Shackles are used to connect a rope-end loop or eye at a structural pinhole, lug, or bail. To accommodate varied conditions of use, the mouth opening is made quite wide. The connecting lug plate is commonly built up or the shackle packed out with washers to create a reliable concentric connection.

LIFTING MACHINES AND THEIR ACCESSORIES

Cranes and other lifting appliances are deployed in construction, maritime commerce, and general industry. Though there are large overlaps among all of these uses, construction has its own set of demands and culture of practice that set it apart. The machines described in this section, in keeping with the subject matter of the handbook, are the ones most commonly used in construction.

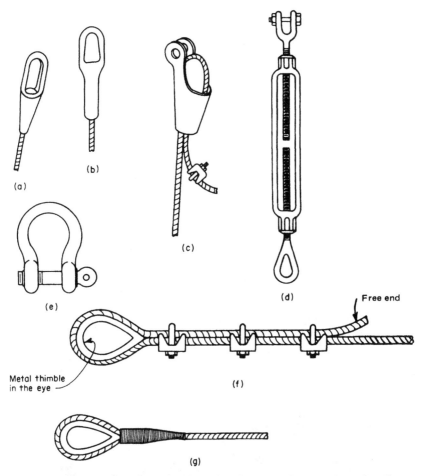

FIGURE 20.13   Wire-rope fittings: (*a*) spelter socket, (*b*) swage socket, (*c*) wedge socket, (*d*) turn-buckle, (*e*) shackle, (*f*) wire rope clips and thimble, (*g*) hand-spliced eye and thimble. (*From Shapiro, Lawrence K. and Jay P. Shapiro: Cranes & Derricks, 4th ed., McGraw-Hill, New York, 2010, Fig. 1.10.*)

Derricks

A derrick is a device employing ropes and tackle for hoisting loads and moving them later-ally but connected to a hoisting engine that is not an integral part of the machine. Though a derrick usually makes use of structural parts such as a boom or mast, a basic derrick could be made of nothing more than ropes and fittings, relying on the structure on which it is mounted for rigidity and stability. In the not too distant past, booms and masts of derricks were wooden poles. Most of these now are angle iron or tubular latticed structures.

There are various derrick forms that serve a multitude of uses. Just a few are presented here:

A *Chicago boom* (Fig. 20.14) can be mounted on a building frame during or after con-struction. Simple in concept, though infrequently used nowadays, it is useful to illustrate

FIGURE 20.14   A *Chicago boom* mounted on the side of a building.

basic lifting mechanisms and underlying principles. Raising and lowering of the boom is actuated by the topping lift. Swing guys often are fitted to the boom tip and are run laterally on each side to a point of anchorage. Not a high-production machine, its uses are "on again, off again" operations where the load placement is above the reach or capacity of mobile cranes.

A *guy derrick* (Fig. 20.4) was at one time the only practical means to erect steelwork on a high-rise structure, work mostly done now with tower cranes. Guy derricks are now unfamiliar to most ironworkers, and the skill set needed to operate them has been largely lost.

A guy derrick can be described as a Chicago boom with its own integral column, called a mast, held vertically by six or more guy ropes, some of which are visible in Fig. 20.4. Guys, anchored to the building frame, radiate in a horizontal circle about the derrick. When a guy

FIGURE 20.15   A stiffleg derrick on the rooftop of a building in New York City. (*Budco Rigging Co.*)

derrick swings, the boom and mast move together. The top pivot itself is called a gudgeon pin connected to a fitting to which the guys are attached, called a spider. Swinging, or slewing, is accomplished by using a large horizontal wheel, called a bull wheel, which is fitted at the bottom of the mast. Wire rope is run around the bull wheel, and winch power is used for swinging.

Guy derricks are practical for steel erection work because it is possible for the derrick to lift itself, or jump, as the height of the work increases. A steel grillage is placed under the mast to transfer the loads to the host structure. The winch is left at base level as the derrick jumps; the winch operator works blind, receiving all instructions by signal or voice communication.

A *stiffleg derrick*, like a guy derrick, has a boom supported from a vertical mast. However, instead of guys, it has two rigid inclined legs supporting the mast (Fig. 20.15). Operation is restricted by the stifflegs to an arc that is between 2/3 and ¾ of a full circle. Reassembled in the field from small components, it is often a good choice as a heavy-duty lifting apparatus to be mounted on a rooftop. For special applications, it can be mounted on a tower or a portal frame either fixed or traveling.

Mobile Cranes

The utility of cranes has long been enhanced by making them mobile, whether by mounting on rails, watercraft, wheels, or tracks. Today what is known as a mobile crane in the United States is on tires or crawler tracks, though the Europeans restrict the definition to those cranes that are mounted on truck carriers. Most twentieth-century mobile cranes were built with latticed booms of tubular steel or angle iron, but the dominant form now is the telescoping cantilevered boom crane, commonly called a hydraulic crane (Fig. 20.6). Latticed booms continue to be used, mostly on crawler cranes (Fig. 20.2). Whether telescopic or latticed, the boom is pinned to a revolving *superstructure* mounted on the truck or crawler base.

With few exceptions, a telescoping boom is carried on a wheeled chassis with outriggers. It utilizes hydraulic (hydrostatic) power transmissions for extending (telescoping) and elevating (luffing) the boom, and vice versa. The boom is raised and lowered with one or a tandem pair of hydraulic cylinders acting on its base section. Telescoping of the boom sections is powered by one or more hydraulic rams hidden within the closed sections, sometimes assisted by ropes and sheaves. In keeping with the relatively delicate nature of the equipment, these machines do not allow the load on the hook to free fall (i.e., lower against the brake). Likewise, users must take care to avoid duty-cycle work and other harsh service applications.

Power is mediated from the engine through a gearbox to hydraulic pumps and motors that drive the various crane motions. A small crane model is equipped with one engine that powers both the crane and the carriage that drives it, with fluid passing through a hydraulic swivel between the crane superstructure and the carrier. A larger crane has separate engines for driving the vehicle and operating the crane.

Mobile-crane load ratings are mostly determined by the ability of the machine to resist overturning. The machine weight itself provides overturning resistance, but counterweights may be added at the rear of the superstructure to improve lifting capacity. Backward tipping limits the amount of counterweight that can be added. However, lifting capacity is sometimes determined by other factors such as strength and thus not always enhanced by adding counterweights. When required by road limitations, counterweights are detachable for transport. Some cranes are rated for different counterweight options so that the user is spared the cost of shipping more than is needed.

Travel refers to movements about the jobsite while transit is movement of a crane over the highway. A crawler crane under its own power only travels on site. A wheeled crane is capable of both.

Wind is an important operational consideration, as it can compromise the crane structurally, reduce stability, or rob the operator of the ability to maintain control of the load. Lifting must be stopped when the wind reaches an excessive level, though judging what is excessive can be difficult. A maximum permissible operating wind speed might be stipulated in a code or in the manufacturer's literature, but lifting a load with a large surface area might be problematic in a lesser wind. Guidance is sometimes provided by the manufacturer, but sometimes left to the user's judgment. Jibs may be particularly susceptible to overstress from side loading wind.

Preparing a mobile crane to be taken out of service for a severe wind is easy when the boom can be readily telescoped down or lowered to the ground. Long booms and complicated configurations pose a larger challenge. A mobile crane should not be placed on a site without having a means to take it out of harm's way in advance of a storm. Manufacturer's literature can be consulted to determine wind limitations for leaving a boom in the air and the means for securing the crane.

Crawler Cranes

As a crawler crane requires assembly on site, the additional work piecing together the boom may be justified by the capabilities gained. Some situations where crawler cranes have a distinct advantage are

- Duty-cycle and repetitive cycle applications such as pile driving and bucketing concrete
- Pick-and-carry operations where the crane must travel with the load on the hook
- Long-term assignments where the lower monthly cost of the crane offsets the assembly, dismantling, and shipping costs
- Extreme lifting assignments that are beyond the capability of truck-mounted cranes
- Projects that require frequent repositioning from one working location to another

The robust construction, cable-suspended boom, and generously sized power plant of a crawler crane make it suited to duty-cycle work. Such tasks as lifting and releasing a concrete bucket or clamshell in a steady cycle can be harsh. Delicate equipment on a telescopic crane cannot endure the repetitive impact, reversing loads, and heat buildup. Wear components such as clutches, brakes, bushings, and slide bearings deteriorate quickly. Many crawler cranes are designed with this harsh exposure in mind. Some manufacturers divide their crawler crane product lines to separate duty-cycle machines from lift cranes, while others offer the similar machines in different versions.

While pick-and-carry operations are extremely limited for truck-mounted cranes, the same cannot be said for crawler cranes. Nonetheless there are limitations: the tracks must always be on firm-level support and load pendulation must be held in check.

Highway moves of truck cranes are constrained by dimensional and weight restrictions, but crawler cranes are broken down into as many pieces as necessary for transport. Designers have leveraged this distinction by developing crawler cranes into some of the largest lifting machines in existence.

Truck Cranes

A truck carrier for a crane must be stronger and stiffer than a conventional truck unit. The largest carriers have as many as nine axles (Fig. 20.8). Road speeds of 35 to 50 mi/h (56 to 80 km/h) are common. Lifting loads while supported on tires is done occasionally, but substantial lifting requires that outriggers be extended and tires lifted clear of the ground

Identified easily by oversized tires and a single cab for driving and operating, a *rough-terrain* carrier has two axles and a telescoping boom (Fig. 20.6). These carriers perform transit moves at about 30 mi/h (48 km/h), with some discomfort to the operator. For long moves they are transported on low-bed trailers.

All-terrain carriers (Fig. 20.1) combine the high road speeds of truck carriers with some of the off-road capabilities of the rough-terrain crane. To achieve maneuverability, these cranes typically have all-axle drive and steering as well as crab steering. Many machines are furnished with suspension systems that maintain equalized axle loading on uneven surfaces while the crane is in motion or is static. The largest of these cranes require some dismantlement to travel legally over highways.

Attachments and Accessories

A mobile crane has no utility without an attachment, the most basic of which is a boom. Over the century or since when mobile cranes have been around, manufacturers and inventive users have put into use a wide variety of devices that vary or augment the basic configuration. Attachments at the front end, modifying or extending the boom, improve lifting performance or reach, or adapt the crane to carry out specialized tasks. The most common front-end attachments are jibs. Attachments at the rear end provide for increasing counterweights and raising the topping lift so that the crane can lift greater loads.

Jibs.   A *fixed jib* increases the reach of a crane. (The U.S. jib is called a fly jib in the European Union.) A relatively lightweight structure mounted at the boom tip, it can be in line with the boom or offset as much as 45°. A typical crane is arranged so that it can carry the main hook suspended from the boom tip while a second light-duty hook is suspended by a *whip line* from the jib.

FIGURE 20.16   A swingaway jib in the process of being unfolded on a street. (*Paul Yuskevich.*)

There are two fixed jib styles: suspended and stowaway, associated with lattice booms and telescopic booms, respectively. A *suspended jib* is supported from a jib strut by fixed-length forestay ropes. The jib strut is then held in place by backstay ropes anchored to the boom. A *swingaway jib* is mounted with brackets on the side of a telescopic boom when it is not in use. Sufficient space must be made available at the work site to allow unfolding (Fig. 20.16).

A *luffing jib* is an articulating extension to the main boom. In place of the stationary backstays of a fixed jib, it has a suspension of running cables that allow the jib to luff. The lattice-boom crane will need four drums to make full use of the boom and jib: one to operate the main boom, another for luffing the jib, a winch for a boom fall, and another for the jib fall.

Whether mounted on a conventional or telescopic boom, a luffing jib configuration can dramatically expand the versatility of a mobile crane, creating a large working envelope with robust lifting capacities. On an open site, it can sweep at long radii. In congested urban or industrial sites the main boom can be set at a high angle to clear obstructions such as buildings whereas the jib reaches to set loads over those obstructions, as in Fig. 20.17.

The luffing jib can be longer than the boom. It is assembled on the ground in line with the boom and raised up like a jackknife. The space needed for assembly is slightly longer than the combined length of the crawler or carrier plus the boom and jib. Space must be allocated to allow assembly and dismantling; the boom-jib combination might also need to be set down for contingencies such as inspection, repair, or high winds.

FIGURE 20.17 A luffing jib dramatically increases the reach and capability of a mobile crane. (*Mathieu Chaudanson.*)

Heavy-Lift Attachments. Mobile-crane manufacturers have devised various schemes that boost the lift capacities and reach beyond the envelope of conventional machines, to as much as 2000 tons (1800 t) and in excess of 700 ft (213 m). These arrangements require special attachments that are not part of the everyday equipment set. Substantial space and site preparation may be needed. Extraordinarily heavy lifts demand large numbers of parts of line in the boom suspension and the load fall. Operations are correspondingly slow, but slowness can be an advantage when lifting massive loads of high value.

Two general principles are behind these arrangements, either singly or jointly that transform mobile cranes, principally crawlers, into lifting leviathans: overturning resistance is improved by adding counterweight set far back from the tipping fulcrum, and stress on the boom is reduced to allow an increase to its working capacity.

Boom stress may be reduced by raising the boom hoist suspension on a high live *mast*. Stability against overturning is enhanced by supplementary counterweights placed behind the superstructure. The tray carrying the supplementary counterweights might roll on the ground or on a circular track (Fig. 20.18). As the crane swings, the counterweight tray and superstructure rotate together about the center pin. Though there are some cranes designed from the ground up with these characteristics, more commonplace are proprietary kits called *heavy-lift attachments* that adapt conventional crawler cranes to the same end.

FIGURE 20.18   A heavy-lift attachment on this crawler crane converts it into a lifting behemoth. Auxiliary counterweights on the trailing wagon follow the swing motion. (*Mathieu Chaudanson.*)

Tower Cranes

Much like a mobile crane, a tower-mounted crane moves loads by executing three motions: the hook is raised and lowered by means of a winch and fall, carried in a circular path by the swing gear, and carried in a radial motion by either luffing the jib or rolling a trolley carriage on its underside. The simplest of tower cranes have only these motions, but more complicated arrangements include mechanisms that allow the base to roll on a track or the crane to change elevation by climbing.

Tower cranes are the lifting machines of choice worldwide for most mid- and high-rise building construction. They are used also on expansive sites where the broad hook sweep and the relative ease of coordinating multiple-tower cranes is an advantage. In most of the world outside the United States, small-tower cranes are used for modest-size residential and commercial projects. Many of these rigs are self-erecting machines that are pulled to the site by truck.

Freestanding hammerhead tower cranes range up to about 300 ft (91 m) in height; for luffing tower cranes the limitation is less. Though most tower cranes free-stand and remain at a fixed height, various self-climbing arrangements permit a tower crane to attach to a building under construction and rise with it. With such supplemental means of support, a tower crane can ascend to any building height.

Tower cranes must ride out storms. Design practices with respect to wind vary with the manufacturer or country of origin. However, the storm demands have all to do with the place where a crane is installed and not the place where it was designed, manufactured, or where its owner resides. Some areas are prone to more extreme winds. In the United States, these regions are on the Atlantic, Gulf, and northwest coasts as well as some river valleys and mountain passes. Installation practices should be adjusted for those regions. A heavier mast and foundation might be used or the freestanding height reduced.

Lift capability of tower cranes is gauged by a moment rating expressed as tonne-meters, obtained by multiplying rated capacity in metric tons by the working radius in meters. The smallest machines used for light construction have ratings of about 20 tonne-meters and the very largest in production exceed this by a factor of about a hundred. Most used for heavy construction are in the range of 150 to 650 tonne-meters.

The cost associated with installing and removing a tower crane is small for the self-erecting type but can be considerable for most others. At minimum, those costs would include trucking, hiring a rigging crew and an assist crane, construction of a foundation, electrical hookup, and the services of a trained technician. More complicated installations such as those that climb have considerable additional expenses. The high costs of installation, as well as the considerable investment of time and planning, make all but the smallest tower cranes a tool for longer-term projects where these expenditures can be amortized.

Excluding self-erectors, tower cranes are broken down into relatively small components for shipping and erection. These parts are connected by pins or bolts. Bolts must be preloaded to carefully controlled levels. Failure to establish and maintain the preload can result in bolt fatigue failure.

With the exception of a few diesel-powered machines with hydrostatic drives, tower cranes are fully electric. The hoist motors on older machines have high- and low-speed ranges with stepped increments in each range. More recent models have variable-frequency drive or other forms of continuously adjustable speed motors with friction or eddy-current brakes and creep speed. Automatic acceleration for all motions is typical on many cranes. Remote controls are sometimes offered.

A tower crane is said to be top-slewing if the swing circle is mounted near the tower top as in Fig. 20.7, and it is said to be bottom-slewing if the swing circle is near the base as in Fig. 20.10. Among contemporary cranes bottom-slewing is associated with self-erecting machines.

Hammerheads and Luffers.   In the U.S. tower crane lexicon, the terms *boom* and *jib* are used more-or-less interchangeably, though *jib* is more likely to be applied to a hammerhead style and *boom* to a luffing crane. Most everywhere else, *jib* is the common term for both.

A tower crane with a *hammerhead jib* carries a trolley that traverses the bottom of the jib to change the hook radius. Available in a broad range of sizes and relatively low in cost, it is the dominant tower crane type worldwide. The traditional hammerhead form is called a *saddle jib*, illustrated in Fig. 20.19. Its operator's cabin is typically mounted just below the jib but above the *slewing* circle, allowing the operator a full view of the load and the trolley. An opposing strut, called a *counterjib*, projects opposite the jib to carry the counterweights while also supporting the load winch, power plant, and control panels. The *flattop* style shown in Fig. 20.5 is preferred where maintaining low headroom is a priority such as at airports or where another crane passes overhead.

No obstruction can be permitted to prevent a hammerhead crane from *slewing* through a full 360°. The crane must swing freely with the wind, or allow *weathervaning*, when out of service. At the completion of a work shift, the operator must leave the crane with the *slewing* brakes disengaged. Where multiple hammerheads have overlapping swing circles, they are offset at different elevations to avoid interference.

A variety of *operational aids* and *limiting devices*, mandated by codes and standards, are provided with hammerhead tower cranes. Devices prevent excessive line pull, excessive load moment, and over-travel of the trolley. Another prevents the load block from striking

FIGURE 20.19   A saddle jib tower crane in the process of top climbing. The tower section on the hook is used to balance the superstructure of the crane, a requirement for climbing. The suspended tower section next to the mast is about to be inserted into the climbing frame to increase the mast height. (*Morrow Equipment Company.*)

the upper block, an occurrence known as *two blocking*. None of these devices should substitute for attentiveness and good judgment by an operator.

The shortcoming of a hammerhead crane is that it requires a wide swath of space for the boom to operate, and when the crane is not operating, the boom must have clearance over a full circle so that it can act like a weathervane. Built-up areas of cities and industrial sites sometimes do not have adequate clear space. Increasingly, moreover, authorities enforce "air rights" that prevent a crane user from occupying the space over adjoining land. These difficulties have provided a market opportunity for tower cranes with luffing jibs.

A luffing jib (Fig. 20.7) has several distinct advantages over a hammerhead. It resembles a latticed boom mobile crane superstructure mounted on a tower. With its high vertical reach, it can accomplish the same task as a hammerhead with a lesser tower height. When arranged for climbing, its climbs do not need to be scheduled to synchronize with other cranes that overlap. The mechanical works of the luffer are more complex, and its jib presents a more formidable task for erection and dismantling.

Just as with a hammerhead, a luffer must act like a weathervane when out of service, though there are a very few that are permitted to have the swing locked. An installation planner must be cognizant of the need to allow adequate space for the boom to clear a full swing circle at a suitable luffing angle, which is to say an angle that is not too high to allow the boom to blow over backward or to catch the leeward wind. The governing clearance might be between the boom and a nearby tall building, or it might be crane to crane.

Base Mountings. Almost every tower crane is freestanding, if not for the duration of a project then at least at the starting phase. A freestanding tower requires a base mounting that is either weighted with *ballast* or anchored to a massive structure that can resist overturning moment. There are three common types of mountings: static, traveling, and climbing.

A *static base* has the bottom of the tower firmly affixed to an immovable concrete mass or to structural framing. This static base might be an undercarriage carrying ballast blocks, a concrete spread footing, a pile cap, or a frame anchored to another structure (Fig. 20.20).

The crane imposes vertical load, lateral load, and moment to the foundation. Vertical force is the weight of the crane plus the hook load. Lateral load on a freestanding crane arises strictly from wind, though slew torsion also acts on the anchorage and foundation in a horizontal plane. Overturning moment on the tower derives from the weight of the jib and the counterjib, ropes, trolley, the suspended load, and the wind. It may apply in any azimuth angle as the crane slews. While the loads acting on tower legs and anchors are maximal when the overturning moment acts on a diagonal with respect to the tower, the least stable direction on a ballasted base is directly over the side of the base.

A *travel base* is a mobile chassis on which ballast weights are stacked (Fig. 20.21), supported on bogies, usually four sets of electrically driven wheels guided on steel rails. The rails must be installed at a level to keep the tower plumb and well supported so that they will not bend or settle excessively under loads. A section of track, called a *parking track*, is set up with anchorage arrangements and added strength to resist the loads produced by storm conditions. Travel bases enable one or more cranes to be arranged to provide hook coverage throughout a large construction site, industrial facility, shipyard, or port.

A *climbing base* is a steel frame, part of a bottom climbing system that allows the tower crane to raise itself on the building as construction progresses (Fig. 20.22). The initial installation may be freestanding on a static base, transferring to the climbing base after the structure has been erected high enough to support a jump. The crane can sometimes be located in an elevator shaft, but tower size or scheduling of the elevator installation usually makes this choice impractical. In most instances, temporary floor openings are made.

FIGURE 20.20   A fabricated steel base supports a freestanding tower crane being erected on the roof terrace of a reinforced concrete building. The corners of the base frame are anchored to columns. (*Lawrence Shapiro.*)

FIGURE 20.21   A traveling tower-crane base with ballast blocks mounted on rails. Tie-downs to anchor the crane in a storm are visible. (*Manitowoc Crane Corp.*)

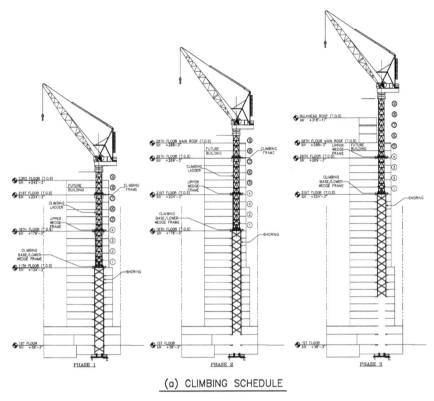

(a) CLIMBING SCHEDULE

FIGURE 20.22 A typical internal climbing arrangement with the tower crane supported on a climbing base and wedged at the floors of the building. (*a*) The climbing schedule provides direction for shoring and wedging.

Self-Erecting Cranes. *A self-erecting crane* fits into tight spaces such as a narrow street, alleyway, or courtyard. Functionally it is a tower crane but deploys like a mobile crane (Fig. 20.10). A typical machine of this type is driven to the construction site and erected rapidly on its outrigger base with a small crew. The telescoping mast, composed of tubular or latticed sections, gives it the ability to operate close to buildings. When the work is done, it folds up and is hauled away. The self-erecting crane can weathervane in the wind but for an extreme storm might be folded up.

INSTALLATION AND OPERATION

Contractors with moderate lifting needs are fortunate that the world is full of small mobile cranes and boom trucks available on demand that are easily dispatched to the site and that can be set up for work with minimal effort. These contractors need not usually suffer for lack of planning. For those with greater needs, lack of foresight will most assuredly lead to trouble.

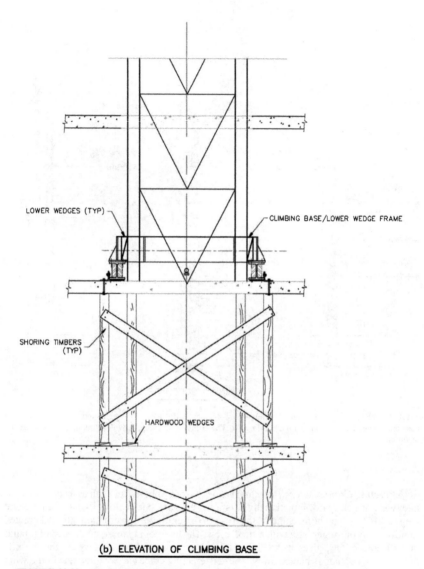

LOWER WEDGES (TYP)

CLIMBING BASE/LOWER WEDGE FRAME

SHORING TIMBERS
(TYP)

HARDWOOD WEDGES

(b) ELEVATION OF CLIMBING BASE

FIGURE 20.22 (*Continued*) A typical internal climbing arrangement with the tower crane supported on a climbing base and wedged at the floors of the building. (*b*) The climbing base transmits the weight of the crane to shores and lateral loads to the floor.

Mobile Crane Deployment and Use

Planning for large mobile cranes starts with a look at the means of getting the machine to the work location. Dimensions, turning radius, gross weight, and axle loads come up against transit route restrictions. Crane components carried on flat bed or lowboy trailers are often wide or high loads. Large cranes may require 10 or more truckloads of components. A proposed route should be examined for width and height limitations as well as bridge or culvert capacities. Power lines running over the route require close scrutiny and might need to be turned off, relocated, or shielded.

Construction site gates and internal access roads are often too constricting for a large crane. Abrupt changes in grade can cause the chassis of a truck crane or the bed of a lowboy haul trailer to hang up. If the site area is limited, space may need to be found to marshal trucks in the neighborhood until their loads are needed for assembly of the crane.

A crawler crane is capable of self-powered travel about the jobsite, but for transit it must be trucked or carried by rail. As weight and width increase, it becomes necessary to remove more and more components in order to remain within road or rail limitations. For rail transit, the carrier needs to be consulted, as limitations vary with the routing.

Travel within the Site. Haul roads to remote sites and internal construction site roads may be in poor shape and badly graded. A travel path should be defined and surveyed before attempting it; corrective earthwork might be needed. *Rough terrain cranes* or *all terrain cranes* are inherently capable of negotiating uneven and soft ground. A crawler crane with a short boom may also manage with the boom at a low angle to avoid bouncing backward into the boom stops. Existing ground might be improved by compaction or by firming up with crushed stone. Alternatively, the existing ground could be overlaid with mats.

Crane travel with a load on the hook is a special operation that requires a high standard of ground evenness and levelness. Avoidance of side slope and side wind is of particular importance.

Whenever possible a long-boom truck crane should be assembled in the spot where it is to be operated, but that is not always possible. While traveling, a crane with a long boom or an extensive front-end attachment is sensitive to side slope and side wind. With all components intact, axle loads are optimized by varying the boom angle and orientation. Many truck cranes are built with more axles supporting the rear than the front and the optimum loading will be proportioned to axle capacities. In some instances, counterweights must be removed for travel to protect axles or tires from overload. When in doubt, the manufacturer should be consulted.

Space is often at a premium, particularly at urban sites, making assembly a challenge. Latticed booms and luffing jibs require a level stretch of ground somewhat longer than the member to be assembled. Consideration should be given to the space required for the trucks delivering the sections and perhaps also for an assist crane to assemble boom and counterweights. If power lines are nearby, proper distances must be allowed for both the assist and main cranes.

An assembled crawler crane is designed for traveling onsite, but it cannot do so unrestricted. With its wide stance instability is not a threat, but side load from out-of-levelness can overstress the boom and jib. A greater concern is instability fore-and-aft caused by deficient ground support. The tracks must always have firm support, particularly at the heavy end, which is usually the rear end under the counterweights. Heavy ground pressure at the rear end is advantageous for travel; when the leading edges of the tracks are lightly loaded they advance without digging into the soil.

Clearances and Reach. It is embarrassing, to say the least, to send a crane to a job only to find that it is incapable of placing the loads where needed. Embarrassment may be only the first problem in a series, as some accidents come about because field crews try to improvise and work around an unexpected limitation.

There are numerous ways that a crane might come up short. Shortness of reach or lifting capacity is an obvious example. This is best averted by diligent review of loads, radii, and crane capabilities. Limitations apply not only to how far a crane can reach but how close, too, as each crane has a minimum working radius. A crane with a large front-end attachment such as a boom with a luffing jib has a correspondingly large minimum working radius.

FIGURE 20.23 A long telescopic boom with a fixed jib nestled between buildings. A clearance study must consider the deflection of the jib, the swing of the jib, the load path swinging past adjoining buildings and clearance of the bottom of the jib to the parapet of the roof where the load is being set. (*Mathieu Chaudanson.*)

Less obvious shortcomings, as debilitating to an operation as the straightforward ones, might be ferreted out only by careful study with the benefit of an experienced eye. Some subtle shortcomings involve interferences between the moving crane or the load and other objects (Fig. 20.23). These clearance problems can come up in various ways:

• The lifted load is at risk of fouling the boom or jib. A wide load lifted close to the head machinery at a high boom angle could be at risk. This is sometimes a dilemma faced by riggers placing unwieldy vessels or machinery. Several styles of boom heads such as a hammerhead tip or offset tip are configured to lessen the possibility of fouling. The head sheaves of telescopic booms are often similarly offset, too.

- The suspended load or the boom fouls a nearby building or another obstruction. In tight quarters, the boom might not be capable of being raised high enough to swing past the obstruction. Urban and industrial sites can be places of heightened exposure to this peril. A crane that starts a project with a free swing might not have clear space by the time the erection work is completed.

- Reaching past the leading edge or parapet of a building, the underside of the boom or jib cannot clear it.

- The hook cannot reach sufficient height to place the load. This is known as a drift deficiency. It is sometimes caused by an underestimation of the length of slings or the height of the rigging supporting the load.

- The aft end of the crane superstructure cannot clear an obstruction. The counterweights, for instance, interferes with a tree or a live mast fouls a building. The tail-swing radius may encroach over an active traffic lane. A large crane can have a rear-projecting gantry, strut, backstays, or a live mast capable of fouling trees, buildings, or electric power lines.

- Any part of the crane or the suspended load encroaches within a restricted zone surrounding a power line.

Clearance problems can be complex and difficult to visualize even for experienced planners. Basic checking can be done with two-dimensional scale diagrams and layouts. When conditions appear to be too close to permit reliance on the diagram, calculations or a more comprehensive graphical approach might be used. In many instances, a conventional computer-aided design (CAD) drawing, accurately dimensioned and detailed, is a suitable tool for clearance checking. Building information modeling (BIM) or other forms of three-dimensional modeling may be justified sometimes by high value or complexity.

Drift. A load aloft can be only raised as high as the maximum attainable hook elevation minus the height of the load and rigging. In addition, some "wiggle space" is needed for maneuvering. What is enough wiggle space allowance is a matter of judgment. For a carefully measured rigging job 10 ft (3 m) could be acceptable, but that figure might be doubled or more depending on the reliability of the planning information and the need for production speed.

A load block cannot be allowed to strike the head sheaves. When this does happen, it is called *two-blocking*. A contemporary mobile crane is required to have an *anti-two-block device*. However, even when such a device is installed, sometimes the inertia of a rapidly raised load block can overrun it. An operator should behave as if the device is not there, slowing down the lift speed as the block approaches the upper limit.

Drift is defined as the vertical clearance between the top of a lifted load and the crane hook when the hook is in its highest possible position; it is a measure of the gap available for rigging and for maneuvering the load. The drift is a useful piece of information for the planner to make decisions about the rigging and handling of the load.

Ground Support. Reliable information for determining mobile crane support reactions is not always available. Rough conservative approximations are sometimes used. Most crane manufacturers, however, supply software and tables that can give the user accurate and comprehensive answers when the resources are properly used.

Presumptive soil-bearing capacities used for building foundation design are often conservative when applied to crane supports. These capacities sometimes reflect a measure of long-term settlement control not necessary for mobile crane use, as mobile cranes can be leveled from time to time. Practical bearing limitations can be based on avoidance of abrupt settlement or soil shear failure. Evaluation by a qualified person, preferably a licensed professional engineer, is certainly advisable if any doubt exists. Cribbing, usually timber or steel, is needed when the ground pressure directly below the crane exceeds the ground-bearing capacity.

Operating track pressures are relatively low, usually not in excess of 6 tons/ft^2 (575 kN/m^2). Often, a 1- to 1½-ft (300- to 450-mm)-thick layer of road base or crushed stone will make a satisfactory crane support.

On most ground, a crawler crane is not likely to develop enough pressure beneath the tracks to induce soil failure. Settlement is a different question, however, as it can cause the crane to go out of level or lose the firm support that is necessary for its stability. Often, the solution requires nothing more than placing small-sized scrap lumber beneath the tracks and using crane weight to press the lumber into the soil. It may take a few repetitions before the spreading effect of the scrap will furnish a firm-level base. Good support under the ends of the track is essential to the stability of the crane and should be checked frequently.

When soils are very good, it is possible that reliable support and a level crane can be attained with the tracks resting directly on the ground (Fig. 20.2).

Nonetheless, mobile cranes are furnished with support elements that are too small to bear directly on most soil. Cribbing is placed under the supports to spread the load. Ground surfaces can conceal places of potential peril. Pavements may cover vaults, poorly consolidated backfill, or voids where fine-grained soils have washed out. The variety of soil conditions encountered in crane work is as wide as the field of soil mechanics. Where there is complexity or doubt, a geotechnical engineer might be consulted. Here are some general lessons from the authors' experience:

- Recently backfilled areas should be approached with caution, unless the crane user knows that the fill is a select material that has been placed and tamped in layers then tested for compaction. Utility trenches are often hastily backfilled in spite of a surface that may look deceptively sound. Backfilling around footings is also frequently loose and uncontrolled.

- Differential settlement, especially dangerous for crawler cranes, can be caused by support that is partly on soft backfill and partly on hard ground.

- Pavements conceal manhole chambers and utility vaults that are not fully revealed at the surface. Cellars of urban buildings sometimes extend out under the sidewalk or even under the roadway.

- Heavily trafficked areas around a construction site are often well compacted and thus suitable for cranes. If the ground is churned up and muddy, the surface can be improved with crushed stone or brickbats.

- Old urban streets sometimes have sewers and gas mains below that are not in good condition. The problem for cranes may be exacerbated by the disturbance caused by new cellar excavation. Sheeting and bracing may bring about lateral displacement of utility lines or the washout of fine grained soils around them. Utility lines have been known to rupture from this combination of soil disturbance and crane loading.

Cribbing.   A theme of the bulleted points is the need to survey a site to evaluate surface and below-grade conditions before a crane is brought in. In some instances it is not possible to determine observationally whether backfill has been prepared adequately to support a crane. Some load testing might be advisable, say by running trucks or heavy equipment across a backfill to see its effect. The crane itself could be used for a self-check by swinging the counterweight over the backfill with the boom at minimum radius and with no load on the hook.

Supporting Crawler Cranes.   Side frames that carry the crawler crane tracks, being very rigid, can arch over minor void spaces or low points if they are situated near the center of a track. However, all four track ends need to be well supported to avert tipping. On any corner, loosely draped track shoes may deceptively lay in contact with the ground while the side frame is unsupported. The only way to detect that the side frame is not supported is to observe that one or more track rollers are not in contact with the shoes. When the crane swings or picks a load under this condition, it will rock like a table with one short leg. The tipping fulcrum will have dramatically foreshortened and the crane can overturn.

During operation, the greatest pressure exerted on the ground occurs when the superstructure swings over a corner of the crawler base. The maximal loading may occur either with a load on the hook or with the boom at minimum radius and nothing on the hook. Yet greater pressure may be exerted when raising a long boom off the ground. Performed with the boom facing directly over the track end, the entire weight of the crane can concentrate on a small surface. For this reason, manufacturers sometimes specify blocking the ends of the tracks when lifting a long boom. Some models, however, lift the boom over the side of the track using supplemental outriggers.

Mobile Crane Picks. A mobile crane can have a small margin between a hook load that is acceptable and one that will tip the crane. For that reason, the concept of a *critical lift* has particular relevance for mobile crane operations. A load that approaches the full-rated capacity should give pause to the planner and the operator, compelling either of them to give a second look at the load and the radius, and perhaps to other facets of the lift as well. Rated loads include the weight of the hook block, slings, and other lifting accessories; a reliable accounting of those weights will be needed for a critical pick.

A mobile crane should be sized to avoid near-capacity picks whenever possible, most particularly for repetitive work. Though there is no rule that disallows capacity picks, they do by nature engender elevated risk. A rule of thumb is to limit the crane to about 75 percent of capacity for work that is repetitive or not tightly controlled.

Multiple Crane Lifts. When two or more cranes are used to lift a single object, the risk increases. The cranes interact with one another, and a failure of one is likely to cause all to fail. By nature, a pick involving more than one crane carrying the load simultaneously is a *critical lift* (Fig. 20.3). Multiple crane lifts require formal planning. Risk control measures keep the lifting operation within limits established by the crane manufacturer and the operation planners.

Though a fuller discussion of multiple crane lifts is beyond the scope of this text, here are some common-sense rules of thumb:

1. A single-crane lift at full capacity is usually better than a multiple crane lift at less than full ratings.

2. Planners, field supervisors, and crane operators should be experienced.

3. Use as few cranes as possible.

4. A formal, written lift plan is advisable. The plan should set out the position of each crane, loads and radii, the movements to be made by each crane including sequence and radius changes, and the operational and risk control measures to be used. The crew should be thoroughly briefed.

5. The weight of the lifted load, and its center of gravity (CG), should be determined. The distribution of load to each crane should be determined for each phase of the operation. Weights of hook blocks and lifting accessories must be considered.

6. Crane load lines must be kept plumb within set tolerances at all times throughout the course of every multiple crane lift. Simultaneous hoisting and swinging should be avoided as should hoisting and luffing. No motion should be combined with travel. The

cranes can hoist together as needed, and any time when one crane swings, travels, or luffs, it will be necessary for the other crane(s) to move synchronously in order to keep the load lines plumb.

7. In most instances, only one crane should have a locked swing with the others freewheeling. This action will limit side loading.

8. In typical circumstances, loads should be limited to 75 percent of rated load or less. However, with rigorous controls, cranes can be allowed to operate closer to full capacity.

9. A jib, especially one with a substantial offset angle, is not generally suitable to be used in a multiple crane lift because of its heightened vulnerability to side loading failure.

Tower Cranes That Climb

The majority of tower cranes in the world are freestanding arrangements to construct buildings of moderate height. However, on a high-rise project, the freestanding crane does not have sufficient height to complete the structure. This limitation is overcome by climbing. There are various climbing systems, but they broadly fall into two methods: *top climbing* makes the tower grow in height by adding sections to the top and *bottom climbing* uses the rising host structure to support the rising crane.

Climbing requires skill and attentiveness because improper procedures can lead to disaster. An experienced and knowledgeable crew is highly recommended; there should always be at least one key person on the crane with expert knowledge and familiarity with the manufacturer's manual. The actual jacking should proceed slowly and deliberately with eyes watching all potential snag and distress points.

Before a climb, the climbing apparatus should be visually inspected by someone familiar with the system. The hydraulic mechanism needs to be in good working order; the climb should not proceed otherwise. Only a trained hydraulics mechanic can be allowed to make valve adjustments.

The crane must be *balanced* before the climb; this usually requires that a specified hook load be held at a specified radius, although some cranes are balanced by setting counterweight or boom positions. If a weight is suspended from the hook for balancing, the space below it must be cleared of pedestrians, workers, and traffic, as with any load on the hook. Field adjustment of the radius is almost always needed to achieve a fine balance. It is important that the crew recognizes the appearance and behavior of a balanced crane.

Climbing should not be done in high winds, as a tower crane is in a balanced stance that may be particularly sensitive to its disturbance. In order to ensure that the climbing operation is well controlled, the wind limitation is often set particularly low, usually 20 to 25 mi/h (9 to 11 m/s).

Top Climbing. By means of a climbing frame employing hydraulic rams, top climbing is accomplished by raising the upper works of the crane to allow a new section to be inserted into an opening on one face. The procedure is repeated until the desired tower height is achieved (Fig. 20.19). As the tower rises in height, the mast is periodically guyed or tied to a host structure. There is virtually no limit to the crane height that can be erected, though cost or availability of tower sections can be a prohibitive factor for a very tall building (Fig. 20.24).

When the building is completed, the top climbing procedure is reversed and the tower is reduced in height for dismantling by a mobile crane. Whereas the crane was free to swing during construction of the building, during the climb-down and dismantling there is a building in the way. Planners must be sure to allow space for the tower sections to be loaded out during the climb-down and for the mobile crane to reach both the jib and counterjib of the tower crane.

FIGURE 20.24   A top-climbing crane mounted to the exterior of a high-rise building, with periodic ties to the floors. (*Paul Yuskevich.*)

Bracing and Guying.   A laterally braced mast utilizes the strength of an adjoining structure. Where there is no suitable structure alongside but there is sufficient space on four
sides, guying to anchorage points on the ground might be considered instead.

Guying and bracing systems give rise to large forces that must be resisted by both the
crane mast and a supporting structure. Guys must be anchored either to the earth or to suitably massive elements. While brace loads can readily be sustained by the wind-resisting
systems of most high-rise buildings, the local area of brace attachment may have to be
reinforced to accommodate these loads.

Guying or bracing operations are commonly coordinated with the addition of tower
sections to increase crane height. The combined work of raising and securing the crane is
outlined in a climbing schedule that also links the construction schedule to the operation.
If the building is permitted to move ahead of the crane, the crane will be obstructed by the

rising structure. If the crane moves too far ahead of the building, there will not be structure high enough to receive the bracing. Devising a *climbing schedule* requires knowledge of crane limitations and familiarity with the pace and practices of the construction operations.

Economics and engineering considerations dictate that ties should be spaced apart as generously as conditions allow. First, ties are relatively expensive to install. Second, close spacing of ties tends to magnify counterflexure forces causing one brace to "fight" against the next.

The principal elements of a mast tie are a *collar* and *struts*. Most tie strut arrangements have three struts because this is statically determinate (Fig. 20.24). Struts are not connected directly to the crane mast, as the mast framing is not robust enough to sustain the strut loads. Instead, they are pinned into the collar that has been fitted tightly around the mast. A collar is a stiff "picture frame" that distributes brace forces concentrically to the legs and inhibits racking distortion of the mast cross section. A collar may weigh several thousand pounds, and the weight of the tie struts bears on it as well. The collar must be adequately supported on the mast; often it is hung by cables with turnbuckles or by chain blocks.

On a concrete building, the floor is usually used to secure tie struts because of its diaphragm strength and ease of access. Loading is out of the plane of the floor diaphragm and induces slab bending. Shoes are secured with high-strength through-bolts. Friction-grip connections may be preferred to minimize strut movement, but, if not, lining the holes with pipe sleeves reduces the otherwise high bearing stress on the concrete. Shear walls and columns are also used sometimes for tie attachment points.

On a steel building, tie struts may be connected into the framing system with details and fabrication having been modified for that purpose. The cost to accommodate the crane ties is lowest when introduced early in the detailing and procurement schedule.

The top mast tie experiences the highest level of loading with the load on the second tie somewhat less and acting in the opposite direction. Permissible tower height above the top tie is controlled by out-of-service storm wind loading.

Bottom Climbing. A bottom climbing crane, mounted inside the building, supports the weight of the crane on the host structure. In most systems, the climbing base is a detachable component that also serves to support climbing apparatus and wedges. It disassembles easily so that the pieces can be passed up for reuse at a higher floor.

In the common climbing arrangement, support framing is needed at three different elevations (Fig. 20.22a), two for supporting the operating crane and the third for supporting the next climb. Frames at all levels are identical; the climbing frame for one phase becomes the climbing base in the next. At the completion of a climb, the tower is wedged horizontally to the climbing frames at two levels at a prescribed minimum distance apart. Wedges transfer horizontal forces and resist overturning moment.

Climbing requires a means to raise the balanced crane vertically from one support level to the next. Sometimes hydraulic rams are provided with sufficient stroke to accomplish this in a single stroke. More often, a *climbing ladder* is used in conjunction with rams or some contrivance that serves the same purpose. Climbing the ladder is carried out by hydraulic rams and dogs or pawls that enable the crane to advance up the rungs. Crane weights are too substantial for the climbing-frame and climbing-base reactions to be supported on a typical floor; temporary shores are installed to distribute the loads to additional floor levels.

The topmost floor, after a climb, is often a wedging floor. With a fast construction sequence, it is critically important that the structure be developed to support the upper wedging loads. On a reinforced concrete building, this is a question of having developed sufficient concrete strength, while on a steel structure the floor framing must be completed for these lateral loads.

When the structure is completed, there is a tower crane on the roof and no way for the crane to disassemble or lower itself. If a mobile crane can reach, it is usually the most economical tool for removal. An alternative is to install a derrick, usually a stiffleg derrick, on the roof for dismantling and lowering the crane components piece by piece. Helicopters are occasionally used to remove tower cranes. However, components of large cranes are generally too heavy for helicopter lifts. Aside from the rigging problems, a safe flight path and a close-by staging area are needed.

SAFETY, RESPONSIBILITIES, AND REGULATIONS

Accidents that had at one time been considered a cost of progress are no longer treated so matter-of-factly by government, industry, or the public. Cranes are far from the biggest source of workplace accidents and fatalities, but crane failures are often dramatic, newsworthy, and destructive.

Private industry has responded to demands for greater safety by improving training and standards. Governments have tightened regulations and increased the cost of accidents by imposing harsher penalties (fines and shutdowns) and stricter accountability.

The United States has historically relied on voluntary standards such as the ASME B30 *Safety Standards for Cableways, Cranes, Derricks, Hoists, Hooks, Jacks, and Slings*. However, the implementation of new OSHA construction crane regulations in 2010 is emblematic of an increase in government involvement. These rules, emphasizing account-ability and responsibilities, include new provisions about operator certification, site hazard recognition, and inspections. Canada also has voluntary standards, with regulation occur-ring at the provincial level. E.U. crane regulations are issued by CEN and enforced by national governments.

In both government and private industry, there is a recognition that an important ele-ment of crane safety is to have a clear delineation of responsibilities. Key parties and their responsibilities are

Crane owner must provide a crane with documentation, which meets manufacturer's specifications and that has been properly inspected, and maintained.

Crane user must provide qualified operator, personnel, and supervision; ensure daily inspections, maintenance, and repairs; conform to applicable regulations; and verify that the crane has adequate lifting capacity.

Site supervisor coordinates crane activity at the site, ensuring that the site is prepared, hazards mitigated, and designated and qualified persons performing the work.

Lift director oversees the crane and the rigging crew, ensuring that the crane and work area are prepared and that personnel understand their responsibilities.

Crane operator is responsible to understand the crane functions and limitations, stop the operation when there are doubts about its safety, perform frequent inspections, and report items requiring adjustment or repair.

These descriptions of responsibilities are summaries. For more comprehensive descrip-tions, other sources such as the ASME B30 standards should be consulted.

Reducing Risks. Though an accident may occur during the most mundane and seemingly minor crane operation, some activities obviously entail more risk than others. An elevated risk could come about from the complexity of the operation or the presence of hazards. A lift close to the rated capacity of the crane or a load of great value also may be judged to have a high level of risk. A clear assignment of responsibilities reduces the chance that some important element of planning or execution will be overlooked.

The concept of a *critical lift* is generally understood but not universally defined. The division of all lifts into this binary classification is an oversimplification. Every crane move requires some level of diligence. In the best run operations there is recognition, though it may be implicit, of a gradation of risk. Skilled personnel are a finite and expensive resource. A realistic assessment of risk allows for these resources to be allocated to the best advantage.

BIBLIOGRAPHY

Dickie, D. E.: *Crane Handbook*, Construction Safety Association of Ontario, Canada, 1975.

Dickie, D. E. and Roland Hudson: *Mobile Crane Manual*, Construction Safety Association of Ontario, Canada, 1985.

Shapiro, Lawrence K. and Jay P. Shapiro: *Cranes & Derricks*, 4th ed., McGraw-Hill, New York, 2010.

Rossnagel, W. A., Lindley R. Higgins, and Joseph A. MacDonald: *Handbook of Rigging for Construction and Industrial Operations*, McGraw-Hill, New York, 2008.

Wire Rope Users Manual, 3d ed., Wire Rope Technical Board, Woodstock, Maryland, 1993.

CHAPTER 21
PROTECTION OF SITE, ADJACENT AREAS, AND UTILITIES

Richard C. Mugler III
Richard C. Mugler, Jr. (dec.)

GENERAL PRACTICES, LIABILITIES

The problems of protection at construction sites are immense. The one statistic, provided by the Bureau of Labor Statistics, that most clearly gives the problem its dimension is the fatality rate at construction sites. An average of nearly two people are killed every single day in the construction industry. This number has decreased substantially in the last two decades, primarily as a result of a commitment to improving construction site safety and safety education.

The word "protection" clearly implies a liability, and the bulk of litigation derived from construction sites alleges improper protection of lives and property. Over a lifetime in construction one is likely to spend many days at examination before trial and then in the courtroom defending the adequacy of the protection provided at the construction site. Keep in mind that most construction damage claims involve more than one contractor; therefore, if an award is made, it is often apportioned among the defendants on the basis of degree of liability. Because this is often a very subjective area, you will find that a claim will turn against you on seemingly unimportant evidence such as having used the wrong color danger sign. A clever lawyer working with a minor technicality will have a jury thinking you regularly worked with total disregard for safety. The main point to remember is that form may be as important as substance in a legal sense. In other words, it is not enough to put up a required danger sign; it must be the prescribed color and size. It is on such seemingly trivial details that major damage suits are won and lost.

With this legalistic point of view in mind, we will review the various considerations in the protection of a construction site.

OSHA AND OTHER SAFETY REGULATIONS

Every contractor must retain a copy of the OSHA manual for construction titled Code of Federal Regulations, Part 1926, *Safety and Health Regulations for Construction,* U.S. Department of Labor, OSHA.

Many of the OSHA requirements, as will be seen in the following pages, are nothing more than good housekeeping and require little or no expense for compliance. Experienced contractors treat all OSHA citations seriously due in part to the fact that each citation becomes part of the contractors' permanent records regardless of where they work in the United States. These records become especially important in the event of serious accidents. Many prior OSHA citations will imply a pattern of disregard for safety that can prejudice the contractor's position in defenses against liability and property damage claims, plus the ever-present threat of criminal negligence indictments. In the 1978 collapse of the cooling-tower scaffold in West Virginia that resulted in the death of 51 men, it was immediately reported in the newspapers that the *site* had received numerous OSHA citations many months before the accident. This clearly implied to the public that there was some connection between the collapse and bad safety procedures. In reality the citations were against other contractors not connected with the scaffold collapse, but it serves to show how a bad record will compound the problems when a serious accident occurs.

Aside from the OSHA regulations that apply uniformly across the country, each state and locality may have differing building codes and labor laws that will affect construction in their jurisdictions. Most seasoned contractors when working in new localities will make personal contact with the building departments and agencies that will oversee construction at their sites. This is the kind of courtesy and diplomacy that can turn a potential adversary into a valuable ally. The personnel of the various building departments and agencies tend to be quite protective of their powers and responsibilities. A contractor who ignores, snubs, or bruises these sensibilities may be in for a rude awakening.

SIGNS

Proper signs are an important part of a contractor's legal obligation to protect the public and workers from the hazards that may exist around the construction site. Signs scribbled on plywood and boards are an invitation to allegations of negligent protection of the site.

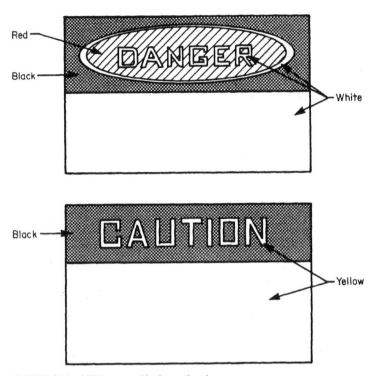

Red

Black

White

Black

Yellow

FIGURE 21.1   OSHA-approved basic warning signs.

OSHA outlines very simple sign requirements in part 1926.200. Approved signs are available at most contractor supply houses. The two basic signs are "Danger" and "Caution," as shown in Fig. 21.1

Traffic control signs become more complicated, and, because moving vehicles are involved, it is very important to comply with regulations for directing traffic. Traffic control signs have been standardized nationally and are detailed in U.S. Department of Transportation Federal Highway Administration (USDOT FHWA) *Manual on Uniform Traffic Control Devices for Streets and Highways*, 2009 edition. Local highway departments usually have copies of this manual.

When conditions are such that a flagman is needed to direct traffic, the degree of liability escalates. The contractor's employee is actually directing traffic, and if an accident occurs, it will usually be alleged to have been caused by improper traffic direction. For this reason it is especially important that the flagman be properly equipped, using a red flag of at least 24 in (60 cm) square, using red lights at night, and wearing a fluorescent red-orange or yellow-green vest with retro reflective striping.

A few general rules should be observed in the use of traffic control signs. Traffic signs should be mounted at right angles to traffic and about 2 ft (60 cm) from the roadway. Advance warning signs should begin at least 1500 ft (450 m) before the construction site and then be followed by repeat signs at 1000 and 500 ft (300 and 150 m) from the site. Typical signs would be diamond-shaped with an orange background and black lettering, as shown in Fig. 21.2.

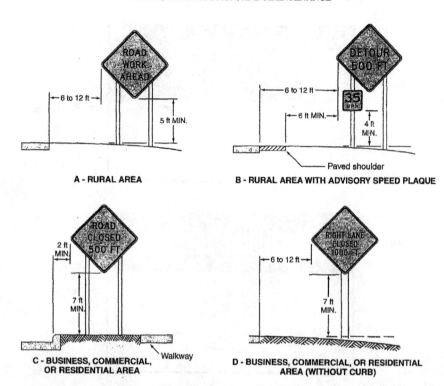

FIGURE 21.2   Basic traffic control signs. *Source:* USDOT FHWA, *Manual on Uniform Traffic Control Devices for Streets and Highways*, Section 6F, 2009.

Suggested speed limits, when posted by the contractor in the prescribed orange background and black lettering, are not intended as enforceable speed limits. The suggested speed limits should be reasonable for existing conditions. If they are too low, they will be ignored. The contractor should use only the orange and black signs that are now used universally for construction warning signs. The use of standardized regulatory signs of the red-and-yellow and black-and-white types normally placed and enforced by police and highway departments should be avoided. If they are to be used, it should be with the approval and supervision of the public agency having jurisdiction. Such signs are not just warning signs, as are the typical construction site signs; they impose legal obligations on the public. Unauthorized use of regulatory signs will cause unnecessary legal complication in the event of an accident.

BARRICADES AND FENCES

Barricades take many forms and are fully covered in the USDOT FHWA *Manual on Uniform Traffic Control Devices for Streets and Highways*. Figure 21.3 illustrates a typical heavy-duty street barricade that is both strong and movable. Barricades have been generally standardized into three types as shown in Fig. 21.4

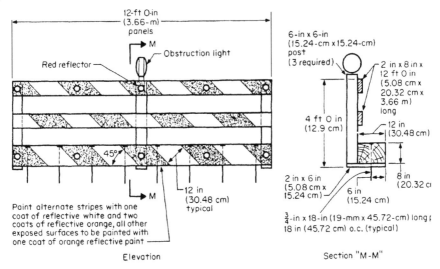

FIGURE 21.3   Heavy-duty timber barricade type III.

Drums (see Fig. 21.4) of the 30-gal (0.11-m³) and 55-gal (0.21-m³) size are very good traffic guides when they are closely spaced. They should be striped with 6-in (15-cm) orange stripes and 4-in (10-cm) white stripes using reflectorized paint. Plastic drum shapes are now sold for traffic control by building supply houses.

Cones, as shown in Fig. 21.4, are another good traffic control device, especially for short-duration jobs of less than a day. They should be at least 18 in (45 cm) high and preferably orange in color.

Drums and cones should never be used in the road without advance-warning signs as described in the previous section.

Figure 21.5 is a sample of a good jobsite-condition report used by many contractors for closing down the job each day. When faithfully used, it will make a real contribution to safety. In the event of accident, it becomes a powerful piece of evidence to refute claims of negligence.

Fences are important not only to protect equipment and materials but also to reduce the contractor's exposure to public-liability claims. Construction sites have repeatedly been held by the courts to be attractive nuisances for children. The primary burden seems to be on the contractor to keep them out by virtually impenetrable fences.

Chain link is in common use on large, open sites, but it is not suitable for confined urban sites. A plywood fence is the most practical for city construction. It reduces the invitation to thefts and offers better protection to pedestrians. This is especially important if blasting will occur at the site. A typical fence would be constructed of ½-in (12-mm) plywood mounted on an 8-ft × 8-ft (2.5-m × 2.5-m) frame. The plywood should be mounted on frames to facilitate alterations to the fence line, which will certainly occur on confined urban sites. The fence posts should be a minimum of 4-in × 4-in (10-cm × 10-cm) posts set at every 8 ft (2.5 m) or less in 2-ft (60-cm)-deep holes. It is a good practice to brace the fence with 2-in × 4-in (5-cm × 10-cm) braces back to stakes driven into the ground at every 8 ft (2.5 m) to stabilize the fence against wind. A poorly built fence will blow down without warning and can cause severe pedestrian injuries.

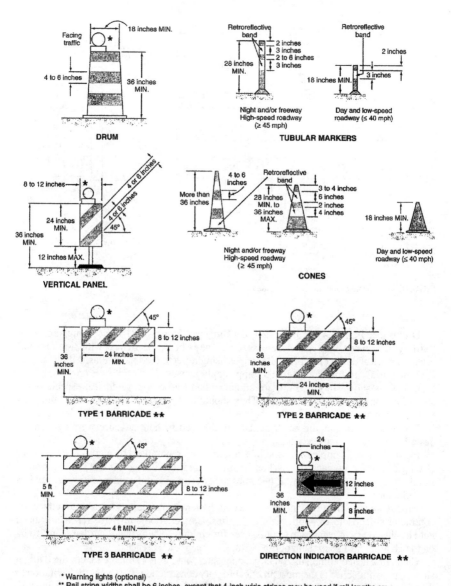

FIGURE 21.4 Barricade Devices. *Source:* USDOT FHWA, *Manual on Uniform Traffic Control Devices for Streets and Highways*, Section 6F, 2009.

Gates for truck entry are often a neglected item at construction sites. A typical truck entrance is 16 ft (5 m) wide. Swinging gates are not suitable for an opening of this size, due to the wind hazard. They also pose a maintenance problem and are difficult to secure well enough to prevent entry by children. The safest and most secure gate is a chain-link sliding gate, which can be built inexpensively using low-cost tracks and rollers available at most building supply houses.

JOB SITE CONDITION REPORT _____

_____ for signs, barricades, and lighting

Date _____ Time _____ M Weather _____

Job Location _____

	Yes	No
1. Are signs posted in advance of job limits and hazards		
2. Are flares and flashers lit		
3. Are signs clean and in good repair		
4. Is job site kept free of holes and ruts		
5. Are tools, equipment, etc. off roadway		
6. Any signs no longer needed		
7. Are detours well identified and lighted		
8. Is road kept free of rocks and spillage		
9. Are shoulders well identified		
10. Dust control precautions taken		

*Note: All unsafe conditions are to be corrected before leaving job site.

Unsafe conditions corrected:

1. _____
2. _____
3. _____
4. _____
5. _____

Signed _____

Reviewed by _____ Title _____

Date _____

Instructions:

This report is to be kept with our permanent records

Standards for signs and barricades, along with proper methods for constructing detours are available at the field office.

FIGURE 21.5 Typical jobsite condition report.

The use of spectator viewing openings in the fence should be considered carefully. On major projects the owners often specify that the fence will have a certain number of viewing locations on the theory that it is good public relations. This may be good for the owner, but it creates an added liability for the builder and subcontractors. The fewer people attracted to the site, the better, especially if blasting is in progress. The average pedestrian does not respond to the blasting whistle. If viewing windows are unavoidable, they should be covered with heavy shatterproof plastic and located away from compressor hoses. The compressor hoses in any case should be kept as far from the fence as possible, since they do occasionally rupture and thrash violently.

One final admonition in the maintenance of good fences and barricades: a persistent cause of pedestrian and vehicular accidents at construction sites is the failure to properly close fences and place barricades at the end of the day. Everything is there to make the job safe for the night, but someone neglects to place them properly. This is more likely to happen on evenings when a crew that does not normally close down the job is working late. In any case, the safe closing of a job is the direct responsibility of the construction superintendent.

HEAT PROTECTION

The proper use of temporary heat at the construction site is regulated by a great deal of state and local law as well as by detailed OSHA regulations. Strict compliance is important for safety reasons as well as legal reasons. Serious fires due in part to violation of fire regulations fall in the criminal negligence area. Inspections for compliance with fire regulations by various agencies are quite frequent, and noncompliance tends to be quite obvious.

Almost every construction site will store a certain amount of flammable or combustible liquids and liquid petroleum gas (LP gas). Flammable liquids are those that will ignite below 140°F (60°C), and combustible liquids are those that will ignite above 140°F (60°C). Flammable and combustible liquids must never be stored in hallways, on stairs, or at exits. OSHA regulations permit storage of up to 25 gal (0.1 in^3) of flammable or combustible liquids within a building without special storage cabinets. Quantities above 25 gal (0.1 m^3) must be stored in an approved metal cabinet. A given storage cabinet cannot have over 60 gal (0.22 m^3) of flammable liquid or 120 gal (0.45 m^3) of combustible liquids. Liquefied petroleum gases (LP gas) such as butane and propane are widely used on construction sites for temporary heat. The fuel containers or cylinders must never be stored inside the building. This is a serious and dangerous violation in every jurisdiction due to the explosive nature of these gases. The OSHA regulation for storage of containers outside of buildings is shown in Table 21.1.

TABLE 21.1 Storage Regulations for LP Gas in the Vicinity of Buildings

Quantity of LP gas stored		Distance from nearest building	
lb	kg	ft	m
500 or less	228	0	0
501 to 6000	229 to 2736	10	3
6001 to 10,000	2737 to 4560	20	6
Over 10,000	Over 4560	25	7.6

Approved fire extinguishers must always be available within 25 ft (7.62 m) of stored flammables. Regulations vary widely as to the approved number, type, and size of extinguishers for a given condition. Local fire departments are the most readily available source for such information. Water-type extinguishers are never to be used on flammable or combustible liquids and electrical fires; only foam, carbon dioxide, and dry chemicals are suitable and effective.

Temporary heaters used to protect concrete pours during freezing weather are usually LP gas or solid fuel salamanders. LP gas heaters should not be used without first consulting with the local fire department. There are many regulations controlling the use of these heaters due to the ever-present danger of explosion. One of the basic OSHA rules requires that the heater be at least 6 ft (1.8 m), from the LP gas container. Individual heaters and containers should be kept at least 20 ft (6 m) away from other heater units.

All heaters must be kept at least 10 ft (3 m) from tarpaulins to avoid being turned over by a thrashing tarpaulin in high winds.

NOISE CONTROL

The problems of noise control have received a great deal of attention in recent years from the public and various governmental agencies. The recognition of the need for more control came very slowly, probably for two reasons. First, the equipment in use gradually became more powerful over a long span of time. Second, the damage to the human ear is itself a function of time. Hearing damage tends to be gradual, and the person affected may not be aware of a change until an appreciable loss has occurred. This would be typical of heavy-equipment operators over many years of operation without protection.

This new awareness has resulted in a steep rise in compensation claims for loss of hearing. Awards in most states are for "loss of function," meaning loss of hearing, even though there may be no loss of earning power. The contractor therefore has a strong financial as well as legal obligation to enforce noise-control regulations.

Noise-related ear damage is almost always irreversible. For this reason noise-control programs must be effective and rigorously enforced. Noise or sound is measured in decibels (dB). OSHA has established that for noise levels above 90 dB for an 8-h working day, ear protection must be provided. A worker may be exposed to higher decibels without protection, providing it is for a limited period of time, as shown in Table 21.2 reprinted from the OSHA *Occupational Safety and Health Standards*.

TABLE 21.2 Permissible Noise Exposures

Duration per day, h	Sound level slow response, dBA
8	90
6	92
4	95
3	97
2	100
1 ½	102
1	105
½	110
¼ or less	115

Source: CFR Part 1910, *Occupational Safety and Health Standards*, US Department of Labor, Occupational Safety and Health Administration, Washington, D.C. 1974, Rev 2009.

Table 21.3 shows the noise levels for various types of construction equipment and will help alert the reader to the danger areas.

The types of recognized ear protection fall into two categories: earplugs and ear covers or muffs. Cotton is not an acceptable earplug; it will give some comfort but very little

TABLE 21.3   Construction Equipment Noise Levels

Equipment Description	Avg. Reading (dBA)	Equipment Description	Avg. Reading (dBA)
Auger Drill Rig	84	Man Lift	75
Backhoe	78	Mounted Impact Hammer (hoe ram)	90
Boring Jack			
Power Unit	83	Pavement Scarifier	90
Chain Saw	84	Paver	77
Clam Shovel (dropping)	87	Pickup truck	75
Compactor (ground)	83	Pneumatic Tools	85
Compressor (air)	78	Pumps	81
Concrete Mixer Truck	79	Refrigerator Unit	73
Concrete Pump Truck	81	Rivit Buster/Chipping Gun	79
Concrete Saw	90	Rock Drill	81
Crane	81	Roller	80
Dozer	82	Sand Blaster (single nozzle)	96
Drill Rig Truck	79	Scraper	84
Drum Mixer	80	Sheers (on backhoe)	96
Dump Truck	76	Slurry Plant	78
Excavator	81	Slurry Trenching Machine	80
Flat Bed Truck	74	Vacuum Excavator (Vac-truck)	85
Front End Loader	79	Vacuum Street Sweeper	82
Generator	81	Ventilation Fan	79
Gradall	83	Vibrating Hopper	87
Grapple (on backhoe)	87	Vibratory Concrete Mixer	80
Horizontal Boring		Vibratory Pile Driver	101
Hydraulic Jack	82	Warning Horn	83
Impact Pile Driver	101	Welder/Torch	74
Jackhammer	89		

Source: Construction Noise Handbook, Sect. 9, *Highway Traffic Noise*, USDOT FHWA, 2010.

protection. Earplugs of the soft rubber type are effective when properly fitted. They do cause irritation to some people and are not convenient for continuous removal and reinsertion; they must also be kept clean and cannot be used by people with ear infections or drainage. The earmuffs which cover the ear have none of the earplug disadvantages and are readily removable for conversations. In extreme situations plugs and muffs can be used together for maximum protection.

There are several steps that can be taken at the construction site to reduce the level of noise emanating from the site. The construction fence made with plywood is the single most effective noise barrier that is readily available to the contractor. Wood is an excellent noise attenuator and is used extensively in the control of sound. Noise attenuation is a function of material and mass; therefore, the heavier the plywood, the more effective the noise barrier. The height of the fence is also important in reducing the amount of noise that leaves the site. As a practical matter, fences over 12 ft (3.6 m) pose stability and wind problems, although with proper engineering anything is possible in this area.

Special enclosures around excessively noisy machines are often appropriate. This option is particularly important during alterations inside existing buildings. Noise levels from machinery within a building can reach dangerous levels very quickly. The use of plywood panels sandwiched with insulation is an effective do-it-yourself enclosure; however, the fact that plywood is combustible may make its use dangerous or illegal within buildings where fireproof materials are required. There are available today a variety of prefabricated fireproof shelters and enclosures that will provide a noise reduction of 25 to 40 dB. The basic component is a galvanized metal panel with either double or triple walls packed with a high-density mineral fiber. These panels are interlocking tongue-and-groove design and self-supporting. A typical panel might be 2 ft × 8 ft (60 cm × 240 cm) and weighs 6 to 7 lb/ft^2 (30 to 35 kg/m^2). Window and door panels are also available.

Sound absorbers may be useful to temporarily reduce the construction noise leaving an enclosed area. There are two main types available. One is a cylinder of absorbent material the size of a 50-gal (0.2-m^3) drum designed to be hung from the ceiling, and the other is a sound-absorbent panel for wall and ceiling mounting.

SIDEWALK SHEDS

Sidewalk sheds as typically used in all major cities are primarily for protection of the public from falling construction debris and secondarily for the storage of construction materials. The standard heavy-duty bridges as required in cities such as New York are sturdy structures capable of withstanding great impact and sustaining loads of 300 lb/ft^2 (14 kN/m^2) (see Fig. 21.6). For construction more than six stories above the sidewalk, anything less than the heavy-duty bridge does not offer good pedestrian protection.

For low-rise construction sites, a light-duty sidewalk bridge is used which is essentially a heavy-duty scaffold. This is available from many equipment-rental companies and scaffolding contractors. The sidewalk bridge should be built with the following minimum dimensions: clear height for pedestrians, 7 ft (2.1 m); width of walkway, 5 ft (1.5 m); distance from curb, 2 ft (0.6 m) but not over 4 ft (1.2 m); protected lights every 8 ft (2.4 m).

Fences should not be attached to the bridge on the curb side. This creates a concealed tunnel which is attractive to muggers.

There are conditions under which a sidewalk bridge should not be built. If a building develops structural problems such as a buckling curtain wall or failing cornices or parapets where a major collapse of material is possible, the only safe procedure is to close the sidewalk and even the street if necessary. The use of a sidewalk bridge under such conditions would be false security. There is an element of judgment involved, but the decision is rarely the contractor's alone, since the local building department would certainly be on the scene.

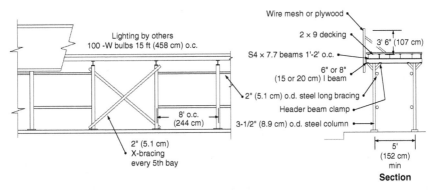

FIGURE 21.6 Heavy-duty sidewalk bridge (shed).

CATCH PLATFORMS AND SCAFFOLDS

Catch Platforms and Safety Nets

Catch platforms, outriggers, safety nets, canopy structures, or debris nets, as they are sometimes called, are discussed in section 1926, subsections 105, subpart M, 759 and 760 of the OSHA construction manual.

A sturdy, well-built catch platform can be a very effective form of protection in high-rise construction. A great deal of material inadvertently falls from a skyscraper during construction. Quite naturally, the higher the building, the less of this material falls conveniently on the sidewalk bridge. The traditional way to overcome this problem is to follow below the construction deck with a catch platform or netting. The first catch platform or netting should be built at about the tenth floor. To be effective, it should be raised as construction progresses so as not to be more than three or four floors below the construction deck.

Because a catch platform is a cantilevered structure, it is subject to great stress and must be built of sound design. In addition, it must be able to withstand impact and severe winds. OSHA permits the use of 2-in × 10-in (5-cm × 25-cm) lumber for light-duty catch platforms, but it is good practice not to use less than 3-in × 10-in (7.5-cm × 25-cm) lumber for the outrigger beams. This will reduce the risk of a failure during the construction of the platform when the carpenters must be on the platform. Safety harnesses are a requirement for those building the platform. Figure 21.7 details a heavy-duty catch platform.

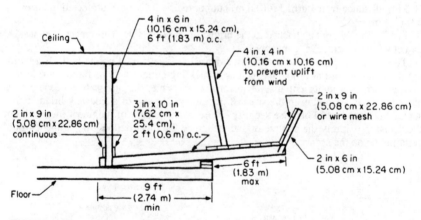

FIGURE 21.7   Heavy-duty catch platform (outrigger).

Safety nets have become more popular in recent years because of the ease with which they can be installed and dismantled. The outriggers consist of telescoping aluminum poles, fastened to the building, at the top and bottom with steel brackets. A heavy-duty catch-all net is supported by wire rope running from the top bracket to the tip of the outrigger. The retractable outrigger can then be adjusted accordingly. Figure 21.8 details a common safety net installation. Due to the extreme hazard of an improperly designed catch platform or safety net, all designs should be reviewed by a professional engineer.

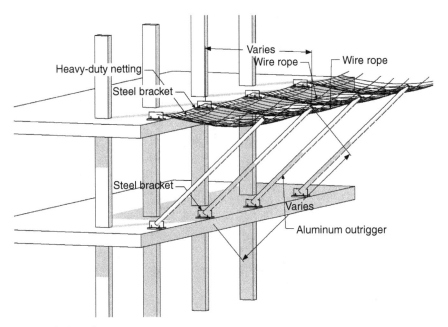

FIGURE 21.8 Safety net.

Scaffolds

Scaffolds are one of the primary causes of injury to workers and the public. Scaffolds are constantly being built and used by workers not necessarily versed in the basic safety principles of scaffold construction. OSHA covers scaffold requirements in section 1926, subpart L and this is a must reading for every contractor and superintendent.

There are several fundamental precautions that must be reviewed constantly to reduce the likelihood of scaffold accidents. OSHA requires training in scaffold erection and scaffold use. Most contractors require proof of training before any installation. All workers working above 6 ft must have a body harness. Guard rail systems, safety nets, and/or harnesses are required on all scaffolds. Toe boards and handrails must always be in place. Cover the scaffold with nylon netting or wire mesh when working over the sidewalk. Quantities of materials, tools, and hardware fall through the scaffold planking. For this reason it is also a good rule not to allow one trade to work above another on a scaffold even with overhead protection. Improper planking is another area of great abuse. Planking should be scaffold-grade full-thickness undressed lumber usually 2 in × 9 in (5 cm × 22 cm). Spans over 10 ft (3 m) are not permitted for any scaffold plank. A typical scaffold on the side of a building should be secured to the building no more than 30 ft (9 m) horizontally and 26 ft (8 m) vertically. These ties must be checked daily due to the problem of careless workers removing ties to facilitate their work and then not replacing them. Many scaffolds have blown off buildings such as shown in Fig. 21.9.

Observe the height rules for a given make of scaffold. Most steel-frame scaffolds are designed for heights up to 125 ft (38 m), but not all. Ask your contractor for a copy of the manufacturer's specifications or a state approval. If the planned height of the scaffold exceeds either state or manufacturer's specifications, do not allow the scaffold to be built without modifications approved by a professional engineer.

FIGURE 21.9   High winds tore scaffold away from building. (*Courtesy of Brian Christopher*)

Scaffold accidents that can be traced to a violation of any one or more of the basic safety rules are a particular threat to builders and/or their superintendents. When serious injury or death is involved, there is a strong possibility that criminal indictments will follow for those responsible for having permitted or overlooked the unsafe use of a scaffold.

DAMAGE DUE TO CONSTRUCTION OPERATIONS

Blasting

The largest single risk associated with blasting is the accidental discharge of electric blasting caps. Current can be induced in blasting caps by radar, microwaves, radio transmitters, lightning, adjacent power lines, dust storms, or any other source of electricity.

Due to the ever-present risk or accidental detonation, it is essential that OSHA-required warning signs be prominently displayed as illustrated in Fig. 21.10.

The most likely source of extraneous electricity is radio transmission and cellular telephones, which have greatly increased in recent years. The risk of

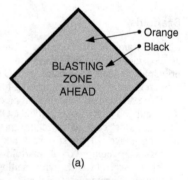

FIGURE 21.10   OSHA – required warning signs for blasting: (*a*) 48 × 48 inches +/– (122 × 122 cm +/–) and (*b*) 42 × 36 inches +/– (107 × 91 cm +/–).

picking up energy from these sources can be considerably reduced by making certain that all detonating wires are kept on the ground and are not allowed to go over fences or materials. An elevated wire has a much greater ability to receive radio signals.

During the preconstruction survey of the jobsite where blasting is planned, it is essential to determine if any adjacent buildings house radio transmitters or cellular telephone towers. Arrangements can usually be made with amateur operators, but commercial users may present very difficult and costly legal problems. This kind of problem would have to be resolved even before an excavation contract is let. To proceed without a resolution will ultimately lead to either the involuntary shutting down of the commercial transmitter or the cessation of the blasting operation. Either one can be very costly.

Careful planning of the blasting operation can reduce the possibility of damage to adjacent structures to a minimum. This can be accomplished by first obtaining a designed blasting pattern that will assure that the shock waves transmitted to adjacent buildings will be of low enough magnitude not to cause damage. Specialists in blasting techniques can make on-site field tests to determine the transmission factor of the rock at the site. This test will determine the energy ratio being transmitted through the rock. From these data a blasting pattern can be designed that will assure a safe blasting operation.

It is worth mentioning that the key to designing safe but effective blasting is the use of proper delays in detonation. The U.S. Bureau of Mines estimates that an instantaneous detonation of a given amount of dynamite will produce an energy ratio 11 times greater than a timed delay detonation. The work produced in breaking the rock is the same in either case, but the power of the shock wave jumps dramatically in an instantaneous blast.

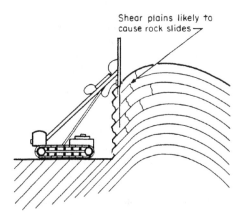

Shear plains likely to cause rock slides →

FIGURE 21.11   Rock drilling under conditions likely to cause rock slides.

In supervising a blasting operation there are several precautions to monitor on a day-to-day basis. Check the blasting mats to be certain the blaster is using the proper number of mats and is not using torn mats. Observe the strike of the rock to determine the possibility of a rock slide. Folded rock, common in the eastern United States, can be extremely dangerous under the condition illustrated in Fig. 21.11 and may require pinning or bracing.

Drilling should not be permitted within 50 ft (15 m) of a loaded hole in the event of accidental detonation.

As the drilling and blasting approaches adjacent buildings, recheck elevations to be certain the cellar of the adjacent building is not behind the rock that will be blasted. As obvious as this may sound, careless excavators are occasionally blasting in foundation walls.

The storage of explosives and related hardware is strictly regulated by federal and state law. Violations in this area will usually involve criminal law; therefore, strict adherence to the law is essential.

Every construction site using explosives is required to have two very secure magazines, one for explosives and one for blasting caps, detonating primers, and primed cartridges. The owner of the explosives must keep an accurate inventory and use record of all explosives. The quantity of explosives stored at the construction site should be kept to a minimum for safety and theft reasons. Terrorists and criminal elements have always been tempted by construction site magazines. Any thefts, discrepancies in inventory, or attempted break-ins must be reported immediately to the local police and the FBI.

In the event of a fire close enough to the magazines to possibly cause detonation, the generally recommended procedure is to evacuate personnel rather than attempt to fight the fire.

Pile Driving

Damages due to pile driving are, on average, not difficult to control. The operation itself is noisy and conspicuous, thus attracting the attention of the public and adjacent owners. Human sensitivity to vibration is very acute and even more so when accompanied by noise. People are aware of vibrations of an energy level only 1/100 of that required to do damage to a structure. Even vibrations perceived to be severe by people are only one-fifth of the vibrations necessary to damage a sound structure.

The only professional way to control pile-driving vibrations is to retain a consulting engineer to monitor the site with vibration monitors. This instrument is a velocity seismograph which measures vibrations in three directions (one vertical and two horizontal) and records the maximum disturbance that was measured in any one of three directions in inches per second of particle velocity—more commonly referred to as the energy ratio. For energy ratios up to 3, damage is unlikely to occur. For energy ratios between 3 and 6, slight damage is possible. In general, for a distance of 20 ft (6 m) from a pile driver, energy ratios of 1.0 or less will develop in most soils. This level of vibration would be comparable to a train passing at the same distance. Soil type has a direct bearing on the transmission of vibratory energy. In general, the energy ratio in sand will drop to almost zero at a distance from the pile equal to the length of the piles. The energy ratio increases somewhat as the sand grades off to gravel. In clay soils measurable energy will be transmitted a distance of 2 or 3 times the pile length. The above generalizations assume dry conditions. A high water table may cause a much higher-energy ratio for a given distance from the pile. Backfilled areas, rubbish, and swamps are special conditions where high-energy ratios may be encountered. Vibration recordings are an invaluable tool under these conditions.

The above generalizations are only a guide and an aid in avoiding problems. Pile driving is fraught with dangers due to the wide range of variables. Vigilance is the best protection against damages from the unpredictable.

A word of caution in using energy ratios: The safe ranges assume sound structures. In most of our urban centers, this is often not the case. In the eastern United States it is very likely that some adjacent building will be 100 years old and constructed of lime mortar that has been thoroughly leeched out. Walls, chimneys, and parapets will often be found out of plumb, bulging, and in various stages of incipient collapse. Even the slightest vibration can be damaging to such buildings. A pre-job survey should reveal such conditions and must be done well in advance of construction to allow time for remedies. Many owners of such rundown buildings are often unaware of their true condition. The local building department will usually issue a violation to the owner when defects are pointed out. This removes some of the responsibility from the shoulders of the builder and makes it a matter of public record that an adjacent building is unsound. The problem of making it safe enough for the builder to begin work can become very complicated legally. Many owners are absentee or destitute or both. In any case, they often cannot be induced to remedy their own problems. Builders are then forced to brace, tie, or in some manner shore up the buildings at their own expense. Most state and city codes allow for this problem to the extent that the project cannot be stalemated by an uncooperative adjacent owner. Legal counsel is essential in these matters to avoid problems of trespass and alleged damages.

The most hazardous part of pile driving is to the pile-driving crew. The equipment is heavy and there is a great deal of hoisting and moving about. The builder can contribute little to improving the safety of the crew, because the equipment, crew, and operations are under the control of a pile-driving contractor. The builder can, however, monitor and control two important aspects of the operation. The site itself must be made safe for a pile rig to move about. The rig must not be required to work too close to the top of excavated banks or under banks that do not have a safe angle of repose. The second area of responsibility for the builder is to be certain that the rig does not have to work within 10 ft (3 m) of a power line carrying up to 50 kV. For power lines over 50 kV, greater distances are required. Consultation with the power company is essential. It is possible to eliminate all hazard by temporarily shutting down power.

Dewatering

The subject of dewatering is covered in depth in Chap. 9. This section will be confined to a few remarks on the potential hazards of dewatering.

The average dewatering problems encountered will usually be overcome without any detectable damage to surrounding ground and structures. When possible, it is best to pump continuously with automatic sumps rather than allowing the water to rise and fall. Not allowing the water table to fluctuate up and down reduces the degree of consolidation of the soil and also reduces the flow of fine-grained material to the pumps. There are, however, conditions under which an improperly conceived dewatering plan may cause massive damage to surrounding buildings, roads, and utilities. Compounding the problem is the likelihood that a site requiring heavy dewatering will also require piles. The net effect is that the water is withdrawn from the soil, permitting consolidation, and the vibration of pile driving further encourages the consolidation. The use of vibratory pile-driving hammers can aggravate the consolidation process to an even greater degree.

The best course to follow when heavy dewatering is encountered is to put the problem in the hands of a soils engineer and dewatering expert.

Excavation

Safe excavation is a matter of constant vigilance and adherence to several elementary rules. Most problems that develop during excavation are not really accidents, but would more properly be called gambles taken and lost. The temptation to dig a little further without sheet piling or closer to a building than is safe seems irresistible, but the consequences can be costly.

The most elementary rule to observe in excavation is to maintain 30° slopes when the depth of the excavation exceeds 5 ft (1.5 m). If these slopes cannot be maintained, sheet piling should be used for earth banks and underpinning in the case of adjacent buildings. As far as safe slopes are concerned, OSHA does allow variations in the 30° rule if the soil type is clearly a more compacted material. Table 21.4 is a summary from the OSHA *Construction Safety and Health Regulations,* Part 1926.

Wood sheet piling should never be used in lieu of underpinning for supporting adjacent buildings. The 30° rule should always be observed regardless of soil type when excavating toward adjacent buildings. The exact method of underpinning is a matter to be determined by a professional engineer.

If groundwater is encountered, the OSHA table does not apply and excavation near adjacent buildings becomes a special condition requiring the judgment of a professional engineer.

A problem frequently encountered when excavating is a change in soil type. This can be dangerous when the excavation begins in a compacted material and then abruptly changes

TABLE 21.4 Approximate Angle of Repose for Sloping of Sides of Excavations

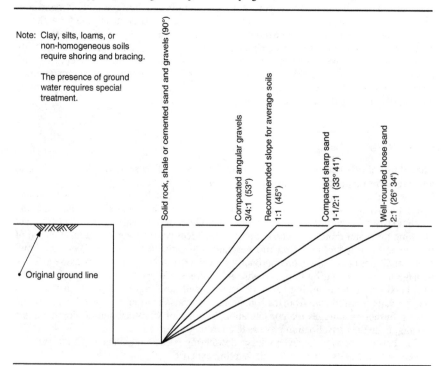

Note: Clay, silts, loams, or non-homogeneous soils require shoring and bracing.

The presence of ground water requires special treatment.

Solid rock, shale or cemented sand and gravels (90°)

Compacted angular gravels 3/4:1 (53°)

Recommended slope for average soils 1:1 (45°)

Compacted sharp sand 1-1/2:1 (33° 41')

Well-rounded loose sand 2:1 (26° 34')

Original ground line

to a loose sandy material overlain by a compacted material. The loose sand will have a much lower angle of repose and result in undermining and collapsing of the compacted material above, as shown in Fig. 21.12. If this condition is encountered, the bank above must be cut back or sheet piling must be installed.

Special attention must be given to the operation of heavy equipment above the banks next to the excavation. Table 21.4 does not take into account superimposed loads of equipment or materials. This becomes especially critical if sheet piling is supporting the banks. The load of equipment can easily exceed the soil pressures for which the sheeting was originally designed. Unless the sheeting is designed to support the additional load of equipment, it is good practice to fence or barricade the sheeting for a distance equal to approximately 45° back from the base of the sheeting, as shown in Fig. 21.13.

Excavation of a site between two existing buildings poses serious potential risk to the adjacent buildings. A typical condition develops from the demolition of a building in the middle of a block and the subsequent backfilling of the cellar hole up to grade. The property often becomes a parking lot while plans are completed for a new building. The backfilling of the cellar hole creates a heavy surcharge against the foundation walls of the adjacent buildings, which were never designed for earth loads. Over time, the foundation walls begin to creep into the building. In dark and cluttered cellars this danger is often undetected. Prior to excavation these walls must be examined and braced if necessary. The use of front-end loaders near such walls is particularly dangerous. Figure 21.14 is a photograph taken in a cellar illustrating the foundation wall collapsing into the cellar from the weight

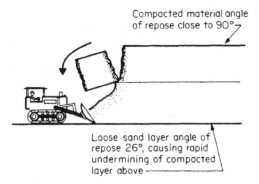

FIGURE 21.12   Dangers inherent in excavation of materials of sharply varying compactness.

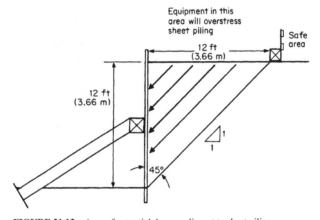

FIGURE 21.13   Area of potential danger adjacent to sheet piling.

of backfill and equipment against the outside of the wall. This wall was more than 12 in out of plumb. Less than 72 hours after this photograph had been taken, the building partially collapsed (Fig. 21.15).

The excavation of pits is especially hazardous to those excavating the pits. Many needless fatalities occur every year from careless pit operations. The basic rule is to sheet-pile pits over 5 ft (1.5 m) deep, regardless of how compact the soil seems. Pits that might otherwise seem safe are always subject to the movement of heavy equipment close to them, causing a sudden surcharge that will shear the soil without warning and bury the workers below. For this reason pits should always be barricaded as well. For small pits up to 4 ft × 4 ft (1.2 m × 1.2 m) square, 2-in × 9-in (5-cm × 22-cm) undressed lumber is usually adequate for sheet-piling a pit. For larger pits the variations in conditions are so numerous that generalizations are dangerous; therefore, the sheeting should be designed by a professional engineer. For pits over 10 ft (3 m) deep, special care must be taken to ensure constant clean air in the pit. Carbon monoxide from nearby gasoline engines is the greatest danger. Smoking, although prohibited on most construction sites, should never be allowed in pits, due to the possibility of trapped gases being encountered.

FIGURE 21.14 Collapsing foundation wall due to backfilling of adjacent cellar following demolition of the building. (*Courtesy of Richard C. Mugler III.*)

PROTECTION OF NEARBY STRUCTURES

The best tool for protecting nearby structures is the pre-job survey. When this job is done by an engineering firm specializing in surveys, very little will be left to chance. The physical integrity of nearby structures can directly affect such operations as pile driving and blasting. When adjacent buildings are sound, these operations can be carried on with little danger of damage, providing energy levels are controlled and monitored with accelographs as discussed earlier. If, on the other hand, walls, parapets, and chimneys are out of plumb, bulging, and in various stages of collapse, extensive shoring and reinforcement will be required.

Bracing

The most direct technique available for supporting weak or failing walls is the common brace. To be effective, the brace should not be steeper than 2 ½:1 or about 22° above horizontal. Figure 21.16 shows a typical bracing system using a heavy timber on the wall to distribute load at floor level and a solid heel to develop sufficient bearing value in the soil. This bracing system is not suitable for all conditions. The type, size, and extent of bracing for each condition should be determined by a professional engineer. Figure 21.17 is an example of timber bracing. When conditions do not allow for timber bracing, steel bracing is a suitable substitute and in many cases the preferred choice. The versatility of steel allows the bracing to take on many forms. From stabilizing a building (Fig. 21.18) to preserving

FIGURE 21.15 Partial collapse due to backfilling of adjacent property. (*Courtesy of Richard C. Mugler III.*)

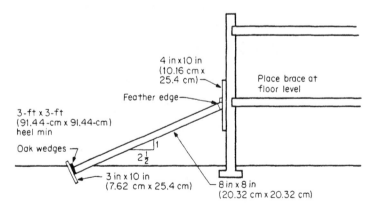

FIGURE 21.16 Bracing of unstable buildings adjacent to construction sites.

FIGURE 21.17   Timber bracing using (2) 3 × 10 strapped and nailed together. (*Courtesy of Richard C. Mugler III.*)

FIGURE 21.18   Steel bracing of building to prevent collapse. (*Courtesy of Richard C. Mugler III.*)

a facade (Figs. 21.19*a* and 21.19*b*), the possibilities are limited only by one's imagination and the cost of implementation.

Wall Ties

Some of the problems of protecting nearby structures develop during the demolition of adjacent buildings. Party walls left standing on adjacent buildings, while quite safe when supported from both sides, become unstable when required to stand alone with support from only one side. A typical procedure to reinforce such walls is to attach channels to the floor beams as shown in Fig. 21.20. Conditions under which this type of reinforcing is done vary so widely that it is best to have the system designed by a professional engineer.

Old foundation walls left standing on property lines with earth loads behind them usually become very unstable when the floor is removed that retained the top of the wall. Here again, conditions vary widely and a professional engineer should be consulted to design a safe bracing system.

FIGURE 21.19   Steel bracing of façade: (*a*) front view. (*Courtesy of Richard C. Mugler III.*)

FIGURE 21.19 (*Continued*) Steel bracing of façade: (*b*) side view. (*Courtesy of Richard C. Mugler III.*)

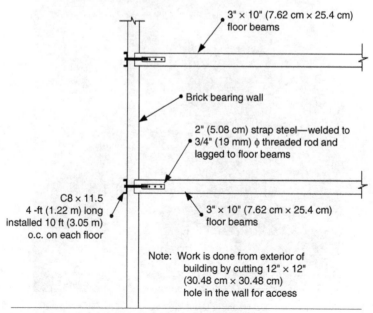

3" × 10" (7.62 cm × 25.4 cm) floor beams

Brick bearing wall

2" (5.08 cm) strap steel—welded to 3/4" (19 mm) φ threaded rod and lagged to floor beams

C8 × 11.5 4 -ft (1.22 m) long installed 10 ft (3.05 m) o.c. on each floor

3" × 10" (7.62 cm × 25.4 cm) floor beams

Note: Work is done from exterior of building by cutting 12" × 12" (30.48 cm × 30.48 cm) hole in the wall for access

FIGURE 21.20 Reinforcing old brick walls using channels and rod ties.

An old free-standing building adjacent to an excavation often requires reinforcement of all four walls. A system of channels and rods as shown in Fig. 21.21 is a typical solution, but again the system requires careful engineering. Figures 21.22 and 21.23 offer some examples of the sketches described in Figs. 21.20 and 21.21. Note the "Z" pattern in Fig. 21.23. The channels are installed horizontally and diagonally across two existing chimneys. Historically, these chimneys, whether active or not, have little or no reinforcing. They are only one or two wythes thick. The mortar joints have had to withstand countless heating and cooling cycles, not to mention water penetration.

The chimneys, and just as often, stairwells, are easily overlooked and can be the first part of a wall to collapse, especially with the vibration from demolition and/or excavation.

The picture in Fig. 21.24a is graphic evidence of the failure to properly tie in an old wall. The building on the left lost its wall when settlements began during excavation

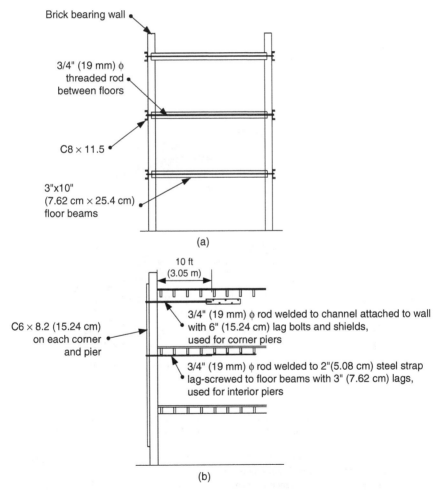

FIGURE 21.21 A strong tie-in system for all four walls of a weakened building: (*a*) Side bearing walls and (*b*) Front and rear walls.

FIGURE 21.22   Wall tie system similar to Figure 21.21b. (*Courtesy of Richard C. Mugler III.*)

FIGURE 21.23   Wall tie system across chimneys or stairs. (*Courtesy of Richard C. Mugler III.*)

for a new office building on the right. Note that complete collapse of the building was avoided by timely shoring of all floors with 8-in × 8-in (20-cm × 20-cm) timber posts and headers.

Figure 21.24*b* is a photograph of an old brick building that has lost its front and side wall simultaneously during excavation for a new building on the left, behind the plywood fence. The three-story building is balanced precariously on its own partitions. The building was too dangerous to save and was demolished. A properly designed tie-in system as shown in Figs. 21.21 and 21.23 would have prevented collapse of the walls.

(a)

(b)

FIGURE 21.24 (*a*) Failure to reinforce the bearing wall resulted in total collapse of the wall, (*b*) a three-story brick building after collapse of the front and side walls.

Nogging and Brickface

Nogging and brickface, when used in old buildings, conceal a hidden danger. Nogging is loose brick placed between the studs of a wood frame building to provide fire protection and insulation (Fig. 21.25). This was a popular construction technique in many cities during the nineteenth century. Brickface is a stucco coating troweled on the walls of old wood and masonry buildings to give the appearance of brick and also to waterproof the old wall. The danger results from the weight of these materials. The worst combination is a wood frame building with nogging in the walls and brickface on the exterior (Fig. 21.26). The wood

FIGURE 21.25   Loose bricks known as nogging between wood studs. (*Courtesy of Richard C. Mugler III.*)

FIGURE 21.26   Wood frame building overloaded with brickface and nogging begins to collapse. (*Courtesy of Richard C. Mugler Jr.*)

frame is clearly overloaded but at first glance appears as a solid brick building. The building in Fig. 21.26 began to buckle from the vibration of adjacent excavation. Emergency shoring prevented complete collapse.

Roof Protection

After the new building comes out of the ground, there are still precautions that must be taken with adjacent buildings. Roof damage is a primary concern when the new building rises any distance above the adjacent property. The most severe claims actually come from water damage due to clogged drains caused by debris from the new construction. Sawdust, paper, and wood chips can clog a roof drain in minutes. A good precaution is to circle the roof drains with ½-in (12-mm) wire mesh at least 2 ft (60 cm) from the drain. The protection of the roof itself will vary with the height of the work above the roof. Certainly ½-in (12-mm) plywood placed on the adjacent roof would be a minimum protection. For five floors or more above adjacent roofs, serious consideration has to be given to impact from falling debris. The kind of roof construction will have a bearing on the amount of protection. Obviously, a concrete roof is better protection than a typical wood beam and board construction. The occupancy of the structure to be protected is also a factor. A building used for dead storage is clearly less of a concern than a crowded apartment building.

When heavy protection is required, a typical design would be 3-in (7.5-cm) lumber on the flat laid at right angles to the existing roof beams, followed by a 3 in × 10 in (7.5 cm × 25 cm) on edge covered with at least 2-in (5-cm) lumber. This pattern could be repeated again for added protection (see Fig. 21.27).

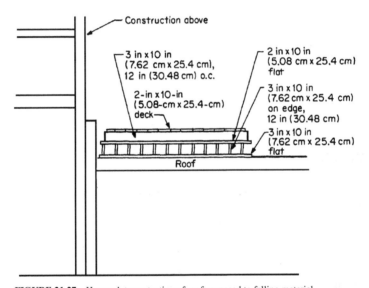

FIGURE 21.27   Heavy-duty protection of roofs exposed to falling material.

Shoring of the roof from below may be required due to the weight of the protection itself and to ensure its ability to withstand impact.

The photograph in Fig. 21.28 shows a heavy cribbing section that has just withstood the impact of a 10-ft × 15-ft (3-m × 4.5-m) section of brick wall that fell 12 stories onto the cribbed roof. There was no roof damage. The cribbing had been placed there when a buckle developed in the wall of a completed office building. The wall was too dangerous to work on and was allowed to collapse.

The distance to carry protection out from the new building is a matter of judgment based on the relative height of the new construction over adjacent buildings. Certainly 10 ft (3 m) out would be a minimum for any condition. Where the new building is hundreds of feet higher, protection is often carried out 50 ft (15 m) and more.

FIGURE 21.28 Heavy-duty roof protection after collapse of a brick wall above. (*Courtesy of Richard C. Mugler Jr.*)

After completion of the new building, it is a good practice to have another survey of adjacent buildings. Damage, if any, should be carefully documented and certified by a professional engineer. The after-job survey is particularly important in protecting the contractor from questionable or outright fraudulent claims. Many an owner sees adjacent construction as an opportunity to get a new roof, make plaster repairs, and repaint it at the builder's expense. The survey reports, if professionally done, will carry great weight in any future litigation for damages.

PROTECTION OF UTILITIES

Temporary support of utilities must be approached with extreme care. The utility company or agency controlling the utility in question should always be consulted. All temporary supporting systems should be designed by a professional engineer and approved by the controlling agency. The location of various shutoffs for the utilities in the vicinity of the jobsite should be marked in the field and posted in the construction office to assist emergency crews in the event of an accident.

The first step in protecting utilities is to mark their approximate location, keeping in mind that utility drawings are notoriously inaccurate.

The utilities of concern to the builder are those just outside the foundation walls and utilities that may cross the construction site. The utilities just outside the excavation but within the angle of repose will have to be supported by sheet piling. The strength and type of sheeting will depend on the proximity of the utilities, the depth of excavation, and soil material. There is always reconsolidation of the material behind the sheeting; therefore, some movement of the ground is to be expected. Solid-steel sheeting, where conditions permit its use, offers the most positive retention of the soil. If H beams are to be driven for the sheeting, numerous test holes should be dug to positively locate the utilities. Blind driving, depending on utility drawings, is an invitation to disaster.

If plans call, for the utilities to cross the construction site and remain in place, a careful plan of support and/or protection must be developed. Utilities that cross the site at the bottom of the excavation are relatively easy to protect. Barricades are necessary to keep machinery away. Hoisting over the utilities should be avoided. Bridges over the utilities for the movement of equipment are usually necessary. Earth bridges are very risky because the

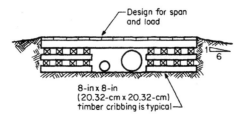

FIGURE 21.29 Construction ramp over utilities.

FIGURE 21.30 Temporary hanging of concrete—encased conduit at an excavation. (*Courtesy of Richard C. Mugler Jr.*)

utility must withstand a good percentage of the equipment load that passes over it. The best bridge is one that permits no load on the utility, as illustrated in Fig. 21.29.

Utilities that must be undermined present a more difficult problem of support. The system used will depend on the weight of the utility, its own structural strength, and its joint interval. For lighter utilities a single H beam driven next to it at appropriate intervals with a carrying beam is usually adequate. The photograph (Fig. 21.30) shows a concrete-encased conduit hung from W12 × 53s, 20 ft (6 m) above an excavation.

For heavier utilities a frame must be built, again using H beams driven on both sides, as shown in Fig. 21.31.

On some sites utilities are relocated and temporarily hung from brackets on the sheet piling. This may be convenient, but it is a dangerous practice. Sheet piling is under great stress and subject to sudden movements due either to a failure or to equipment knocking down a brace.

Utilities that come through foundation walls into the excavation are of particular concern. This is most common where a new wing is being added to an existing building. If the utility is temporarily supported and later backfilled, again great care must be taken to prevent a shear at the foundation wall. The ground below the utility must be fully compacted, or the backfill will move down the wall with glacierlike force and take the utility with it. The photograph in Fig. 21.32 shows sheet piling for an addition to a hospital. The gas, water, and sewer pipes are all hung from the sheet piling as they leave the existing building. The new wall for the addition went up about 4 ft (1.2 m) from the sheet piling. When the space between the sheeting and the new wall was backfilled, all three utilities were ripped from the sheeting and sheared off at the wall of the existing building.

Utility poles are often found along the curb just outside foundation walls. Excavation for a new building will likely disturb the anchors, requiring temporary guying of the pole. A suitable anchor for the guy is often a problem. If H beams are being driven for sheet piling, a cable can be attached to a beam with ease because the sheet piling is usually adjacent to the poles, it is a good practice to drive one or two long H beams for cabling to the pole, as shown in Fig. 21.33.

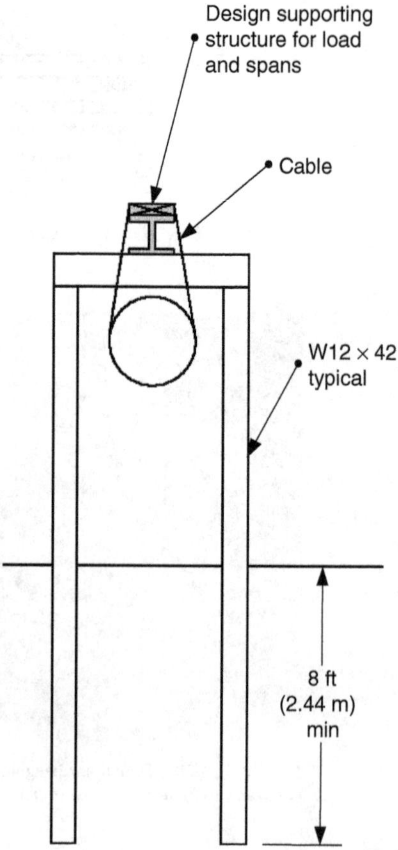

MONITORING SURROUNDING GROUND AND NEARBY STRUCTURES' CONDITIONS

This section deals mainly with spotting the problems that develop during the excavation phase of a building project.

Probably the best early warning of impending trouble is simply constant visual inspection. Many of the points made here will seem simplistic; yet, over and over again, perfectly obvious signs are overlooked in their early stages.

A well-designed sheet-piling system will not show any visible signs of movement. If, however, a failure is developing, the first

FIGURE 21.31 Hanging heavy utilities during excavation.

signs of movement will usually show behind the sheet piling—in the earth which is actually held in place by the sheet piling. Cracks will start to form a few feet back from the sheeting and parallel to the line of sheeting. These cracks will grow measurably each day if movement is taking place, and new lines will form farther back from the sheeting.

If these cracks are along a curb line of a roadway, they must be filled daily until movement stops. Water from heavy rains follows the curbs and will run down these cracks, causing a dramatic increase in sheeting load and probable loss of material from behind the sheeting. Soil and road cracks can in some cases result from reconsolidation of the soil, due possibly to loss of material while installing the sheeting. If this is so, the movement will stop, but in any case the appearance of such cracks requires thorough examination of the sheeting itself. Check for bending boards, braces, or walers. The heels for the braces may be failing or may have been disturbed by machinery. All these signs call for attention and review by a professional engineer. Sheet-piling failures usually have a domino effect in that they trigger water, gas, and sewer main breaks and the loss of job momentum.

Buildings are relatively easy to monitor visually if done daily by the same person. If a building starts to settle or a wall moves laterally, it will show almost immediately on the

FIGURE 21.32 A potentially dangerous method of supporting utilities. (*Courtesy of Richard C. Mugler Jr.*)

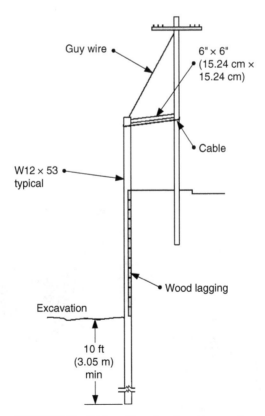

FIGURE 21.33 Cabling utility poles adjacent to excavation.

inside by cracks in the corners and separation of the floors from the walls. Interior inspection is vastly superior to exterior inspection. A crack developing in plaster on the inside may not show in the brick on the outside until considerably more movement has taken place.

The use of the level and transit to monitor roads, buildings, and utilities is the best legally acceptable record of job conditions when kept in log form. Most of the larger projects shoot levels and lines on a daily basis and record them in a log in order to detect vertical or lateral movements in ground and structures. Many projects seem to put too much reliance on shooting levels, probably because it is easier to do. Detecting lateral movement can in some cases be much more significant than detecting vertical movement. Sheet piling and supported utilities almost always fail laterally.

The older, structurally unsound masonry buildings will often begin failure by the walls bulging out in the middle. The bulge could grow several inches before the level readings would pick it up, if at all. Lateral readings will pick up the movement immediately and allow valuable time for bracing or tying in of the walls.

Concrete underpinning requires special attention to prevent and detect lateral movement. It is the nature of the underpinning operation that the underpinning must for a period act as a retaining wall until the adjacent walls of the new building are installed against the underpinning. Most underpinning is designed to only take the vertical load of the structure above. It has basically no lateral stability; therefore, if the earth loads behind it are significant, it must be braced until the new foundation wails are built. During this period while waiting for the new walls to be built, the entire stability of the building is dependent on these braces. Lateral readings are especially important during this period to assure that the bracing system is adequate and holding. There have been many dramatic collapses in the past due to failure to properly brace the underpinning. Underpinning when it collapses appears to be instantaneous, but it is almost always preceded by several days of lateral creep which would easily be detected by lateral readings.

Figure 21.34 shows the total collapse of a portion of a building when the unbraced underpinning rolled into the excavation. The collapse occurred on a Sunday, or there would

FIGURE 21.34 Collapse of underpinning causing collapse of building. (*Courtesy of Richard C. Mugler Jr.*)

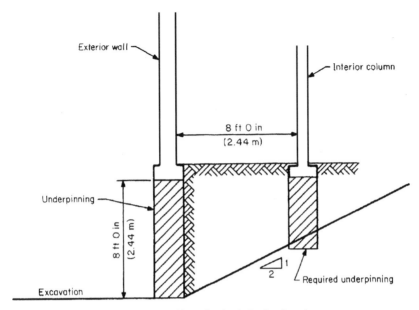

FIGURE 21.35   The often-neglected problem of underpinning interior columns.

have been a large loss of life. After the collapse, the remaining underpinning was properly braced, as shown on the left side of the photograph.

Under certain conditions the monitoring of interior columns of adjacent buildings can take on great significance. The operations that can affect interior columns and footings are pile driving, blasting, dewatering, and underpinning. The detection of movement by level readings is the same regardless of the cause. The control of settlements due to excessive vibration was discussed earlier in this chapter in reference to the use of vibration monitors and proper design of blasting patterns.

The chapter on construction dewatering in this book details the implementation of a safe dewatering system that will greatly reduce the risk of settlements from improper dewatering.

Underpinning is a special condition under which one can inadvertently undermine interior footings. If foundation plans are not available, which is the usual case, the pre-job survey will indicate whether the interior footings will be undermined. If there is doubt as to the depth of interior footings, arrangements must be made for a test pit to determine exact footing elevations. Problems with adjacent owners can complicate work inside the building, but there is no way for the builder to avoid the responsibility of underpinning interior footings if required. The general rule governing the slope between underpinning and interior columns is 2:1, which is the same as in most footing relationships. Figure 21.35 illustrates the application of the 2:1 rule.

BIBLIOGRAPHY

"Blasting Vibrations and Their Effects on Structures," *Bulletin 656*, U.S. Bureau of Mines, Washington, D.C., 1971.

"Census of Fatal Occupational Injuries Summary," Bureau of Labor Statistics, U.S. Department of Labor (USDL-10-1142), Washington, D.C., 2009, 8/19/10.

"Commerce in Explosives," IRS Publication 26 CFR 181, Internal Revenue Service, Washington, D.C.

Construction Noise Handbook, Section 9, *Highway Traffic Noise*, U.C. Department of Transportation, Federal Highway Administration, Washington, D.C., 2010.

Excavation of Trenching Operations, OSHA 2226, Rev. 1985.

Flammable and Combustible Liquids Code, NFPA 30-93, National Fire Protection Association, Boston, Mass., 1993.

LePatner, Barry B., and Sidney M. Johnson: *Structural and Foundation Failures*, McGraw-Hill, New York, 1982.

Maintenance and Use of Portable Fire Extinguishers, NFPA no. 10A-94 National Fire Protection Association, Boston, Mass., 1994.

Occupational Safety and Health Administration Safety and Health Standards, CFR Part 1910, *Occupational Safety and Health Standards*, U.S. Department of Labor, Washington, D.C.

Occupational Safety and Health Administration Safety and Health Standards, CFR Part 1926, *Safety and Health Regulations for Construction*, U.S. Department of Labor, Washington, D.C.

Uniform Traffic Control Devices for Streets and Highways, U.S. Department of Transportation, Federal Highway Administration, Washington, D.C., 2009.

CHAPTER 22

LEADING EDGE VERTICAL CONTAINMENT SYSTEMS

Chris Evans, P.E.

OVERVIEW

Two of the most challenging aspects to any construction project, particularly high-rise construction in congested urban areas, are protection of the workers and protection of the public.

Safety regulations typically require the protection of the public in close proximity to structures under construction. This can be in the form of sidewalk bridges (Fig. 22.1) or other overhead protection systems (Fig. 22.2) to protect pedestrians on the sidewalk adjacent to a construction site, or roof protection on adjoining properties. The rationale behind these systems is to provide protection at the location of the element to be protected.

FIGURE 22.1   Custom sidewalk bridge

FIGURE 22.2 Overhead netting.

Taken a step further, the traditional construction of high-rise buildings typically also requires the use of a horizontal net system, which is required to stay a prescribed distance below the top of the building under construction (Fig. 22.3). The rationale behind this system is to catch falling personnel and/or material before reaching the ground or adjoining structures.

A new way of thinking in the construction industry is to prevent material from falling off of the building altogether by providing protection at the top most floors under construction (Fig. 22.4) to not only "contain" material and debris, but also to provide fall protection

FIGURE 22.3 (*a*) Illustration of horizontal netting.

FIGURE 22.3 (*b*) Illustration of horizontal netting.

(a)

(b)

(c)

FIGURE 22.4   Illustrations of the Cocoon® systems at the top of buildings: (*a*) 42nd Street Hotel, NY, NY (*b*) 150 Amsterdam Ave., NY, NY (*c*) 10 East 102nd Street, NY, NY.

for those working on these floors. To those ends, innovative new vertical containment systems have been developed in order to encapsulate, or "cocoon," the top floors of buildings under construction.

In the New York area, where much of the development of these enclosure systems is taking place, concrete construction typically follows a very aggressive 2-day cycle in which a new concrete floor is poured every second day. This presents two key challenges to any system proposed to protect the top floors.

One is that the system must be able to be raised, or "jumped" at the same rate the building is constructed without interfering with and slowing down the construction activities themselves. Another is that because the concrete placement schedule is so rapid, the system must be able to cantilever, or project, above the uppermost floor to sufficient heights to protect not only the floor above that is currently being constructed, but also the floor above that which is going to be constructed next.

Figures 22.5(a) and (b) illustrate the location of the Cocoon system during construction.

PHASE 1: PROTECTION AT THE FORMING FLOOR AND FLOOR ABOVE

The system attaches to two floors, and cantilevers above the top tie floor by two floors plus a minimum height for the handrail.

After the system is initially installed, work platforms are lowered into place at the location of the two tie floors, providing protection against falling debris as well as acting as work platforms for the construction workers. Handrails must be present at the outside face of the system to provide the required fall protection. Below the bottom tie floor is a material-handling net with fine mesh liner to further contain any small debris.

This configuration allows for the concrete in the next floor above to be placed, as well as for the forming and shoring that will support the floor above that will be started within the confines of the already installed vertical protection system. This is critical where the fast pace of construction can see the forming for the next floor starting the same day as the floor that has just been poured.

PHASE 2: PROTECTION AT THE STRIPPING AND FORMING FLOORS

The platforms at the tie floors are typically at the "stripping floor," where the forming and shoring systems are being removed after the concrete slab has gained sufficient strength to allow for the shoring removal.

After a new floor is placed, an additional outrigger is installed above the current top tie floor. The work platforms are then raised up to pass by the existing slab edge formwork, and the system is raised one floor with the secondary material-handling net left in place to provide protection during the raise. Once the system has been raised one floor and the work platforms secured in their new locations, the netting is raised to the new bottom tie floor. This process of "jumping" is repeated during the construction of the building, providing continuous protection at the currently uppermost and stripping floors.

Fall protection needs to be provided for the construction workers, not only at the upper most floors where the new building is being constructed, but also at the lower floors.

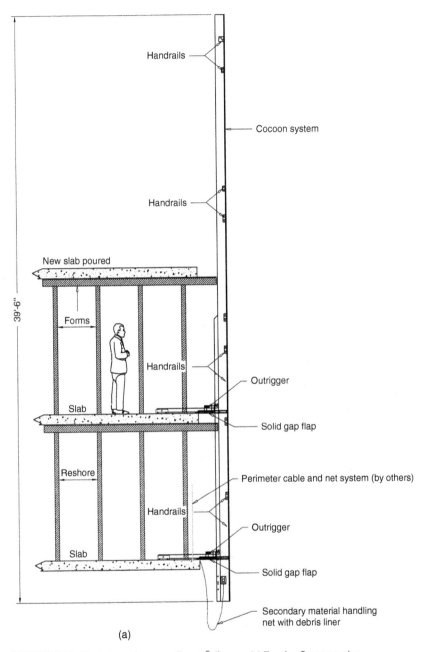

(a)

FIGURE 22.5   Vertical containment or Cocoon® diagram: (*a*) Forming floor protection

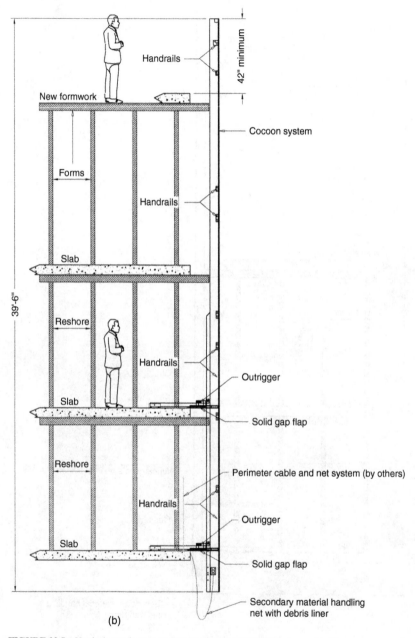

(b)

FIGURE 22.5 Vertical containment or Cocoon diagram: (*b*) Striping floor protection.

In concrete construction the forming system typically extends beyond the slab edge, and is constructed close to the containment system structure. With steel construction the pour stop is typically extended out beyond the steel structure and is also close to the containment structure. At these top floors, handrails need to be provided. Traditionally these handrails would be constructed out of wood or some other material as the forming and shoring is installed. With a vertical containment system this protection is in place prior to work commencing. At the two tie floors a solid platform must be installed in addition to the handrails to close the gap between the building and the containment structure (Fig. 22.6). Figure 22.7 illustrates the top floor under construction, both with, and without, the Cocoon system in place.

There are many benefits from using a vertical containment system:

Debris and building materials are contained within the perimeter of the building. There is a great risk of falling debris and material, particularly during high-rise construction that is often exposed to high winds. There is also the human element, and accidents may happen. With the pace of construction often demanded by developers, owners, and construction managers, the top floor of a high-rise building is a very busy and congested place. Preventing the accidental dropping of tools and material is a priority.

Fall protection for the workers. Traditional perimeter fall protection consists of handrails constructed in one of several methods, but typically connected to the forming system. This requires work to be done at an unprotected edge. If a cantilevered vertical containment system is used, the protection is in place prior to the work reaching it.

FIGURE 22.6   Work platforms at lower tie floors.

(a)

(b)

FIGURE 22.7   Illustration of top floor under construction: (*a*) with the Cocoon system and (*b*) without the Cocoon system.

Protection of the public below and adjoining properties. Accidents or injuries related to the public and damage caused to adjoining properties can be very costly, often resulting in higher insurance premiums. Project delays resulting from stop-work orders or investigations can also result in significant cost impacts. The natural progression in the thinking of how protection should be approached, at the source, greatly reduces the exposure of the owner and building to these costs.

Protection at the stripping floor. The area with the greatest exposure for falling debris and material during concrete construction is the stripping floor. With the increasing demand for speed, the work proceeds at a rapid pace, and with a vertical containment system, it is completed safely in a protected environment.

Reduced insurance costs. As this style of protection become widely adopted, and the benefits become more and more apparent, there is a great likelihood that insurance costs on projects will be reduced.

These systems are gaining widespread support and adoption as the benefits are becoming apparent. From the general contractors and construction managers to the various governmental agencies to those performing the work, the unanimous opinion is that the systems enable the project to be built faster, safer, and more economically. Many feel it is only a matter of time before these systems are not only recommended, but required.

P · A · R · T · 3

FAILURES

CHAPTER 23

FAILURES OF TEMPORARY STRUCTURES IN CONSTRUCTION

Robert T. Ratay, Ph.D., P.E.

CONSTRUCTION FAILURES

More structures fail during construction than in service after completion, and many, if not most, of the construction disasters occur as the result of the failure of temporary structures. A tacit attitude seems to prevail in the design-construction industry: "these things" are temporary only, hence generally less important; therefore greater risks are acceptable than in permanent structures. However, a dollar or a life lost at a construction site is no less valuable or less tragic than its loss elsewhere. More than 1200 construction workers lose their lives and many others are injured each year in the United States. According to some estimates, the direct and indirect cost of construction injuries in this country is more than $20 billion annually. No one appears to have estimated and published the total value of property losses in construction failures. These occurrences are not unique to the United States but are experienced in other countries as well, especially in those making great strides in construction fueled by the developing global economy.

In the United States, as well as throughout the world, disturbingly large numbers of structural failures occur during construction. Advances in construction technology, newly developed materials, increasingly fine-tuned designs, and the construction of more daring structures, as well as the pressure of time and cost cutting driven by competition, are all contributing factors. However, the most frequent direct causes of failures are human factors: oversight; carelessness; incompetence; breakdown of organization; poor management and communication; disregard of codes, standards, and specifications; and general non-adherence to good practice.

There appears to be a disconnect between the practices of one group, the designers-of-record—who by necessity distance themselves from the construction of the project—and the objectives and capabilities of the other group, the constructors—who by contract must perform under the constraints of agreed time and money.

Safety of construction, hence temporary structures, is the concern of the designer, the contractor, the building official, and the insurer, as well as the workers on the job and the general public.

Yet, this most important component of the construction process is still not a "field" of practice but a neglected stepchild, at times claimed, at other times disclaimed by both designers and contractors, and almost totally neglected by researchers and educators.

Failures of unbraced excavations, scaffolding, falsework, formwork, excavation supports, and temporary erection shoring, bracing, and guying—in approximately this order—are the most frequent occurrences of temporary structure failures. Often, it is the total absence of some of these, such as excavation supports, shoring, bracing, or guying that is the proximate cause of a disaster.

Reasonable and clearly written codes, standards, and regulations that emerged in recent years improve construction safety, but the best ways to mitigate failures obviously are competent designs, good construction practices, utmost care, strict inspection, and unwavering enforcement of high standards.

CASE HISTORIES OF TEMPORARY STRUCTURE FAILURES

Construction failures caused by defective performance or complete absence of temporary structures in construction are an almost daily occurrence. Just about every step along the design-construction process includes hidden risks and has been shown to be prone to errors or omissions that result in subsequent construction failure. Failures of excavation supports, scaffolding, falsework, formwork, and temporary shoring, bracing, and guying (in approximately this order) are the most frequent occurrences of temporary structure failures. Often, the total absence of some of these—such as excavation supports, shoring, bracing, and guying—is the proximate cause of a disaster.

With very few exceptions that involve large fatalities, temporary structure failures do not make as much news as the collapse of a building or bridge. They may happen away from the public eye, at an isolated construction site, or behind solid fences, and they are often kept hidden—for everyone's benefit. The author has observed that whenever a construction failure is reported in *ENR: Engineering News-Record*, in books, or in other technical publications, it is nearly always the permanent structure that is described, with little or no discussion of the details of the temporary structure even if *it* was the thing that actually failed.

It is the author's opinion that almost always one of three reasons is the underlying cause of all temporary structure failures: one is the willingness, indeed deliberate choice, to accept greater risks; another is error or omission out of oversight, carelessness, or ignorance; and the third is an unanticipated confluence of events or conditions. All are human failings and all can be averted. Therefore—at the risk of being simplistic—construction failures can be prevented.

An interesting fallout from failures and their investigations is that regardless of what is found to have been the proximate cause of the failure and who is found to have the immediate responsibility, often several other errors or defects related or unrelated to the proximate cause of the failure are discovered as well, which might never have come to light if not for the failure. If you scratch hard enough, you will find many problems and several culprits—often more than first meet the eye.

It is believed by some that the best way to learn how to prevent future failures is by studying past failures. In this author's opinion such studying is often superficial and largely useless unless one delves really deeply into the design and construction details of the project. Few construction failures are so simple that all the pertinent information can be discussed adequately in a paragraph, in a page, or even in two or three pages. Nevertheless, for the sake of illustration of the nature and extent of temporary structure failures, and at the risk of leaving questions hanging in the air, 13 cases of temporary structure failures are described here very briefly. It is pointed out that the conclusions offered are those that have been formulated and/or published by one or more investigators and that appear to have "carried the day." Different investigators may come to different conclusions for the same event from the same data, and even trials by jury can come to surprising decisions.

Some of these cases were publicized catastrophes with tragic fatalities; others were obscure events known only to those involved. They were selected for inclusion here to illustrate the kinds of errors, omissions, and goofs that can cause temporary structure failures. Four of the cases were summarized from reports of investigations by National Institute of Standards and Technology (NIST) [formerly National Bureau of Standards (NBS)] and OSHA; two from the records of the Worker's Compensation Board of British Columbia; two were contributed by individuals; and four came from the author's job files of his own forensic consulting practice.

A reader with particular interest in reading case histories will find a number of construction failures discussed in the recently published book that is a collection of forensic case studies (Delatte, 2008).

It is noted once again that reciting the highlights of failures and the summaries of conclusions of their investigations in rudimentary fashion may be meaningless, and may even be misleading, without the reader being given all of the details and without the reporter himself or herself having been involved in the case. It is also noted that the causes of failures stated in these case histories may not have been agreed upon by all the investigators. In some failures, there is no consensus even years or decades later. Nearly all failures lead to disputes, nearly all disputes lead to claims and forensic investigations, and many forensic investigations lead to differences in the conclusions of the opposing parties. Recitation of case histories are usually based on one investigator's opinion and are often written so as to lead up to or justify his or her conclusion—without rebuttal or challenge. Therefore, be careful with what you "learn" from a failure.

Case History 1: Riley Road Interchange Falsework

Three spans of an elevated highway, the Riley Road Interchange Ramp C (also known as the Cline Avenue Bridge) in East Chicago, Indiana, failed during its construction on April 15, 1982. The entire 180 ft of one span, and 160 ft and 135 ft, respectively, of two adjoining 180-ft spans, collapsed, killing 13 and injuring 18 workers. On the day of the collapse, workers had been casting concrete in one of the spans. The failure occurred before posttensioning of the cast-in-place concrete superstructure when all the construction loads were still carried by the falsework/shoring.

At the request of OSHA, NBS conducted an investigation to determine the most probable cause of the collapse. NBS researchers concluded that the collapse most likely was triggered by the cracking of a concrete pad under a leg of one of the shoring towers that were to support the ramp during construction. This initial failure occurred because the pads did not have an adequate margin of safety to support the loads.

The initial failure of the pads caused additional tower components to fail, leading to the collapse of the support system as well as major segments of the partially completed ramp. The support tower location where the collapse most likely began was pinpointed by NBS engineers, who also reported the most likely sequence of failure.

Three other deficiencies were identified that did not trigger the collapse but contributed directly to it:

- Specified wedges that were to have been placed between steel cross beams and stringer beams at the top of the support system (or falsework) to compensate for the slope of the roadway were omitted, thus increasing the load on critical pads.
- The tops of the shoring towers were not adequately stabilized against the longitudinal movement that occurred when the concrete pads cracked and the tower frames dropped slightly.
- The quality of welds in the U-shaped supports for the falsework's cross beams at the top of the towers was poor, thus making them unable to resist the forces resulting from the longitudinal movement.

"Had any one of these deficiencies not existed, it is unlikely that the collapse would have occurred," the NBS report concludes.

Additional deficiencies contributed to the collapse of an adjacent ramp unit, which failed about 5 min after the first unit:

- Specified 1-in bolts to connect certain stringer beams to cross beams were not used; frictional clips were used instead.
- Special overlap beams at the ramp's supporting piers were not constructed as specified.
- The construction sequence deviated from that specified in the construction drawings, so that concrete for one span was placed before it should have been (see Fig. 23.1).

(a)

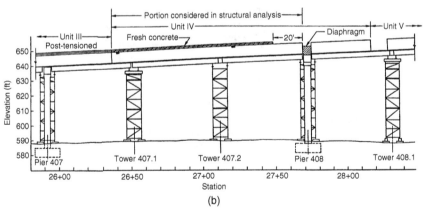

(b)

FIGURE 23.1   Riley road interchange. (*a*) Shoring towers before failure. (*From John Duntemann, Wiss, Janney, Elstner Associates, Inc., Northbrook, Ill.*) (*b*) State of construction at the time of collapse. (*From NBSIR 82-2593, October 1982.*)

(c)

(d)

FIGURE 23.1 (*Continued*) Riley road interchange. (*c*) View of base of shoring towers. (*From John Duntemann, Wiss, Janney, Elstner Associates, Inc., Northbrook, Ill.*) (*d*) Typical shoring tower screwleg. (*From NBSIR 82-2593, October 1982.*)

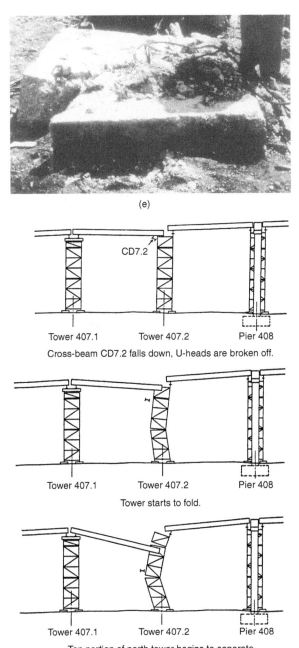

(e)

Tower 407.1   Tower 407.2   Pier 408

Cross-beam CD7.2 falls down, U-heads are broken off.

Tower 407.1   Tower 407.2   Pier 408

Tower starts to fold.

Tower 407.1   Tower 407.2   Pier 408

Top portion of north tower begins to separate.

(f)

FIGURE 23.1 (*Continued*) Riley road interchange. (*e*) Failed concrete foundation pad. (*From John Duntemann, Wiss, Janney, Elstner Associates, Inc., Northbrook, Ill.*) (*f*) Possible collapse sequence of falsework at Tower 407.2. (*From NBSIR 82-2593, October 1982.*)

(g)

(h)

FIGURE 23.1 (*Continued*) Riley road interchange. (*g*) View of collapsed unit IV from the southeast. (*From John Duntemann, Wiss, Janney, Eltsner Associates, Inc., Northbrook, Ill.*) (*h*) View of top of pier 407 and buckled shoring towers from the northwest. (*From John Duntemann, Wiss, Janney, Eltsner Associates, Inc., Northbrook, Ill.*)

Case History 2: Baltimore-Washington Expressway Shoring Towers

A highway bridge built to carry Maryland Route 198 over the Baltimore-Washington Expressway collapsed during construction on August 31, 1989, injuring five workers and nine motorists. The structure was designed as five contiguous posttensioned box girders spanning 100 ft between simple-support abutments. The bottom slab was cast in July, and the box girder webs were cast between July 21 and August 4. On the day of the collapse, workers were pouring the 8-in deck slab. The collapse occurred 5 hours into the pour, when 120 of the 160 ft^3 of concrete were in place. The shoring collapsed "in a flash" without warning, landing all the formwork and concrete on the roadway below. The formwork had been used earlier to cast the westbound structure.

The Federal Highway Administration (FHWA) and private investigators found that the most likely factor causing the failure of one of the shoring towers was the use of 10-kip rather than the 25-kip screw jacks, shown on the approved drawings, on the tops of the shoring towers supporting the bridge. A review board found no evidence that the FHWA (the owner) had not lived up to its contract responsibilities, and ruled that the assembly of the falsework system in accordance with the approved design was the responsibility of the contractor. The FHWA requires a contractor's engineer to certify that falsework has been assembled according to approved drawings before it is loaded. A fine of more than $900,000 was levied by the state against the contractor.

The FHWA report also said that "the top screw jacks were rusty" and that much of the cross bracing had "large amounts of rust and heavy pitting." In one section, the cross brace pieces were connected by nail instead of the required bolt (see Fig. 23.2).

(a)

FIGURE 23.2 Baltimore-Washington parkway. (*a*) Aerial view of collapse. (*From The New York Times, September 1, 1989.*)

(b)

(c)

FIGURE 23.2 (*Continued*) Baltimore-Washington parkway. (*b*) Collapsed falsework previously supported construction of bridge's twin. (*From Kate Patterson, Springfield, Va., and ENR,
September 7, 1989.*) (*c*) End view of collapsed falsework. (*From Kate Patterson, Springfield, Va.,
and ENR, September 7, 1989.*)

Case History 3: 14th & H Streets, Washington, Excavation Supports

Portions of a 150-ft by 208-ft by 47-ft-deep braced open excavation for an office building in Washington, D.C., collapsed on November 19, 1990. Because it occurred in the late evening, after construction work stopped, no one was injured, although there was a significant potential for casualties.

Due to the size of the excavation, external support with wales and tiebacks was estimated to be more economical than internal support with wales, struts, diagonals, and rakers. However, the constraints by the adjacent UPI Building basement, the existing utility lines, and the underground right-of-way permit prohibited the installation of tiebacks in certain locations. Thus, the west portion of the excavation was supported by an external tieback system, while the east portion was supported by an internal structural steel bracing system consisting of diagonal struts, cross-lot bracing, and cross-corner bracing. Based on the original design drawings, both the tiebacks and the bracing were at three levels or three tiers.

For the protection of adjacent building foundations and utilities, the support system was designed to maintain a preload in the system and to minimize the stress change in the adjacent ground mass in order to limit its deformation until the permanent structural concrete gained sufficient strength to maintain this preload force. Thus, any tier of the internal support system could not be removed before the adjacent concrete gained sufficient strength. In addition, to build the basement wall to coincide with the property line to obtain the maximum usable space, it was elected to place the temporary soldier beam and lagging wall just outside the property line, and use this wall as the outside half of the formwork for concrete. To accomplish these two goals, sufficient clearance (3 ft) had to be maintained (by 3-ft-long spacers or "outlookers") between the soldier beam and wale, to accommodate the thicknesses of the perimeter wall of the permanent basement between the lagging and the wales, and to provide for removing the wale and cutting off the protruding outlookers from inside the finished basement wall when the concrete wall gained adequate strength. The outlookers were welded to the inside face of the soldier beams and the outside face of the wales.

The initial collapse was followed 17 hours later by a second additional collapse of a portion adjacent to the first one, resulting in a great deal of damage in and around the excavation.

OSHA investigators concluded that the collapse of the excavation occurred due to the failure of certain structural members of the internal support system along the north and south walls, and that the external support tieback system did not fail. They also found that the failure load, the soil pressure immediately preceding the collapse, was lower than the load for which the temporary structure was designed by the shoring subcontractor.

It was apparently overlooked in the design and/or in the detailing that the force component of a 45° corner brace in the direction of the longitudinal axis of the wale in the north and south wall did not have a physical reaction in the system. The "outlookers" could transmit the bracing force component (earth pressure) *across* the plane of the wall, but had no ability to resist the east-west force component *in* the plane of the wall (see Fig. 23.3*f* showing "plan") and just "folded" over in a horizontal direction like a two-hinged linkage.

(a)

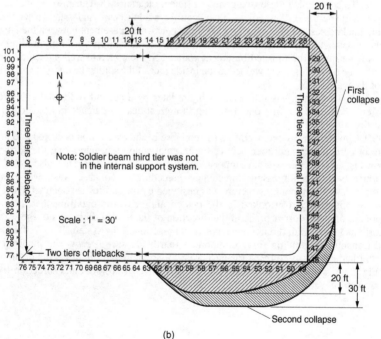

(b)

FIGURE 23.3 Washington excavation. (*a*) Overall view of the collapsed site—looking toward the east. (*From OSHA Investigation of November 19, 1990 Excavation Collapse at 14th & H Streets, N.W., Washington, D.C. report, May 1991.*) (*b*) Extent of initial and second collapse. (*From OSHA Investigation of November 19, 1990 Excavation Collapse at 14th & H Streets, N.W., Washington, D.C. report, May 1991.*)

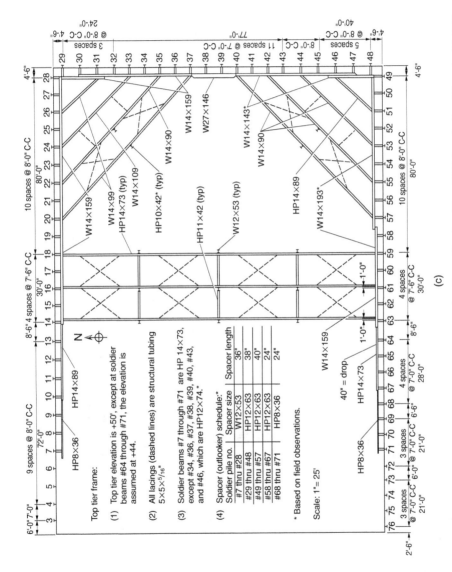

FIGURE 23.3 *(Continued)* Washington excavation. (c) Top tier of internal bracing. *(From OSHA Investigation of November 19, 1990 Excavation Collapse at 14th & H Streets, N.W., Washington, D.C. report, May 1991.)*

Top tier frame:

(1) Top tier elevation is +50', except at soldier beams #64 through #71, the elevation is assumed at +44.

(2) All lacings (dashed lines) are structural tubing 5×5×5/16".

(3) Soldier beams #7 through #71 are HP 14×73, except #34, #36, #37, #38, #39, #40, #43, and #46, which are HP12×74.*

(4) Spacer (outlooker) schedule:*

Soldier pile no.	Spacer size	Spacer length
#7 thru #28	W12×53	36"
#29 thru #48	HP12×63	38"
#49 thru #57	HP12×63	40"
#58 thru #67	HP12×63	24"
#68 thru #71	HP8×36	24"

* Based on field observations.

Scale: 1"= 25'

40" = drop

N

W14×159
W14×99
HP14×73 (typ)
W14×109
HP10×42" (typ)
HP11×42 (typ)
W12×53 (typ)
W14×159
W14×159
W14×90
W14×159
W27×146
W14×143*
W14×90
HP14×89
W14×193*
HP14×73
W14×159
HP8×36
HP14×89
HP8×36

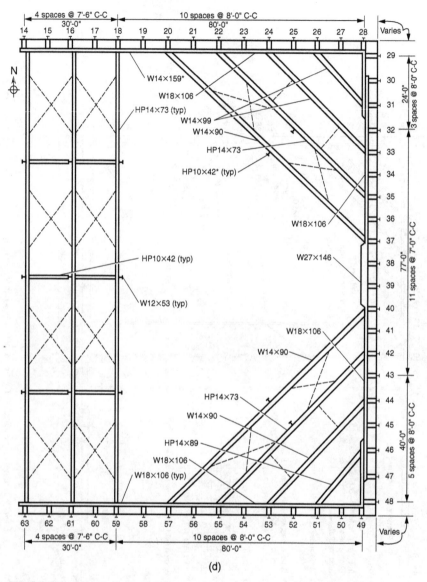

FIGURE 23.3 (*Continued*) Washington excavation. (*d*) Middle tier of internal bracing. (*From OSHA Investigation of November 19, 1990 Excavation Collapse at 14th & H Streets, N.W., Washington, D.C. report, May 1991.*)

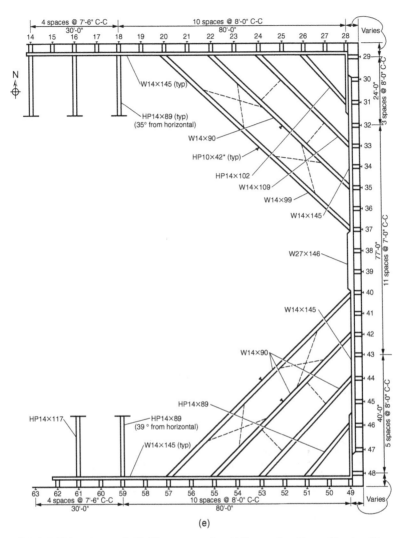

(e)

FIGURE 23.3 *(Continued)* Washington excavation. *(e)* Bottom tier of internal bracing. *(From OSHA Investigation of November 19, 1990 Excavation Collapse at 14th & H Streets, N.W., Washington, D.C. report, May 1991.)*

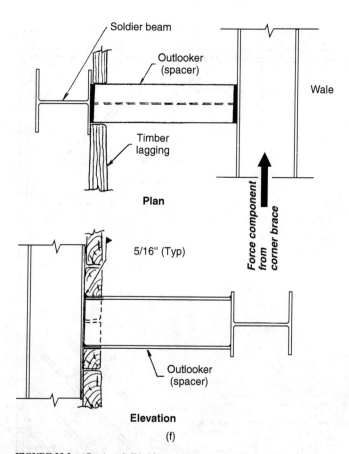

FIGURE 23.3 *(Continued)* Washington excavation. *(f)* An "outlooker/spacer" between a soldier column and a wale. *(Based on OSHA Investigation of November 19, 1990 Excavation Collapse at 14th & H Streets, N.W., Washington, D.C. report, May 1991.)*

(g)

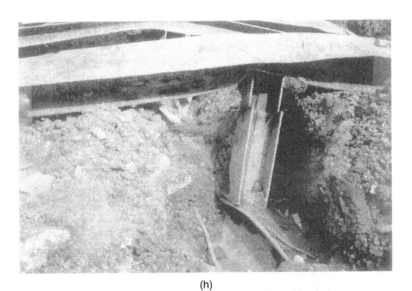

(h)

FIGURE 23.3   (*Continued*) Washington excavation. (*g*) End view of a bottom-tier wale, out-looker, and soldier beam. (*From OSHA Investigation of November 19, 1990 Excavation Collapse at 14th & H Streets, N.W., Washington, D.C. report, May 1991.*) (*h*) A toppled middle-tier out-looker welded to a wale and a soldier beam. (*From OSHA Investigation of November 19, 1990 Excavation Collapse at 14th & H Streets, N.W., Washington, D.C. report, May 1991.*)

Case History 4: Denver I-70 Overpass Scaffolding/Shoring Tower

A scaffolding/shoring tower collapsed, and two workers were badly injured in 1986 at the construction of the Wadsworth exit overpass bridge over Interstate I-70 in Denver, Colorado. The portion of the bridge under which the tower collapsed was to be of three horizontally curved steel tub girders running parallel and connected to each other with vertical diaphragm braces, working in composite action with the cast-in-place concrete deck on top of them. The girders were continuous over three spans of 150, 180, and 160 ft. The average radius of horizontal curvature of the bridge in the middle span, under which the shoring tower collapsed, was about 800 ft. As a tub girder was lifted into place, it was connected at its end to the previously set girder with the permanent bolts only finger-tight, and the diaphragms were installed. When all the girders over the three continuous spans were in place, the formwork for the concrete deck was erected on top of them. Before pouring the concrete deck, the elevations of all the girder end-to-end splice connections had to be adjusted and the bolts tightened, or "driven home." This was done with the aid of a 22-ft-tall temporary scaffold/shore tower erected on the ground directly below the connection.

The tower was assembled from standard 4-ft-wide by 6-ft-high scaffold frames: four parallel planes of frames stacked three-high, with screw jacks at the base and at the top, connected by diagonal braces, so that the assembled tower measured 4 ft by 7 ft in plan. A rectangular grid of four 8-in wide flange beams was placed on top of the screw jacks on the top of the tower to receive the hydraulic jacks. Four hydraulic jacks were placed between the top of the tower and the underside of the two butting tub girders.

Two men positioned themselves on a work platform near the top. Each man was pumping a pair of jacks, one pair under each of the two abutting girder ends, until the steel reached the proper elevation signaled by a surveyor. Several of the connections had already been completed when, during the jacking of a connection (Marked 3 in Fig. 23.4b), the tower collapsed and, with the two men and all equipment, fell to the ground. The connection being jacked when the collapse occurred was in the middle span of a three-span outside girder.

Investigators for the plaintiffs (the injured workers) and the defendant (the scaffold/shoring rental company) never came to an agreement on the cause of the collapse. An out-of-court settlement was made between the defendant and the more seriously injured worker; a significant award after trial by jury was made to the other worker.

It is important to know that the contractor did not specify the type, size, or shape of the tower, but showed the bridge structural drawings to the rental agency's engineer and asked him to furnish whatever was needed.

Plaintiffs' expert, this author, found that both the rated and the calculated capacities of the tower were less than the load to be lifted at the collapse location. (The actual mode of initial failure of the tower could not be determined from the wreckage, so all possible modes of failure had to be evaluated.) He further found, from the calculations of the rental agency's engineer, that the engineer calculated the load to be lifted as the weights of the girders from the connection to halfway to the next bridge pier, as if the girders were simply supported at both ends. That is, he did not consider resistance against lifting that would be created by the continuity of the girder into the neighboring span. Nor did the rental agency's engineer consider loads on the lifted girder from the weight of the adjacent parallel girder tied to it through the connecting diaphragm. The difference between the assumed simple-beam and the real continuous-beam reactions against jacking was very significant.

Defendant's expert contended that the reason for the collapse was not the miscalculation of the load and the consequent inadequacy of the tower but the workers' uneven pumping of the jacks, hence unbalanced loading of the tower legs. He built a full-size replica of the tower and load-tested it under various top support conditions.

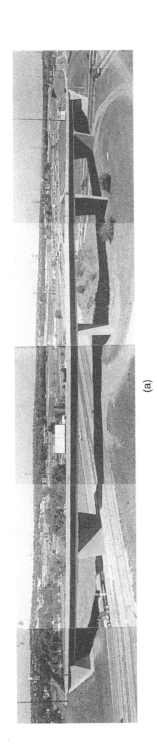

(a)

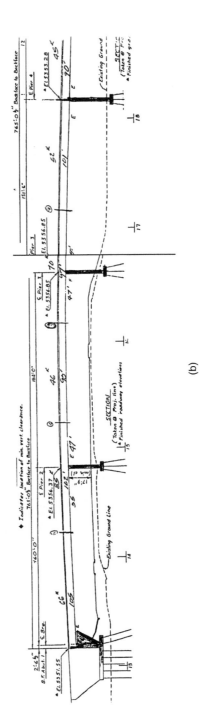

(b)

FIGURE 23.4 Denver I-70 overpass. (*a*) View of completed interchange overpass. (*From R. T. Ratay Engineering, P.C.*) (*b*) Longitudinal section with weights of girder segments. (*From R. T. Ratay Engineering, P.C.*)

23.21

(c)

(d)

FIGURE 23.4 (*Continued*) Denver I-70 overpass. (*c*) Part of collapsed scaffolding/shoring tower toppled over the embankment. (*From R. T. Ratay Engineering, P.C.*) (*d*) Close-up of collapsed scaffolding/shoring tower toppled over the embankment. (*From R. T. Ratay Engineering, P.C.*)

(e)

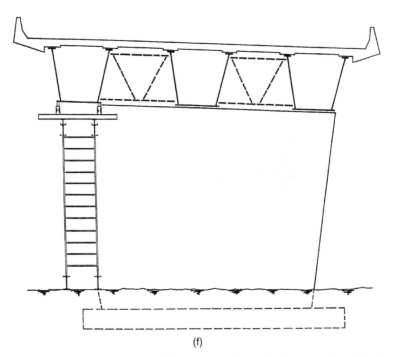

(f)

FIGURE 23.4 (*Continued*) Denver I-70 overpass. (*e*) Reassembled upper tier of scaffolding/
shoring tower. (*From R. T. Ratay Engineering, P.C.*) (*f*) Schematic cross section of bridge with
scaffolding/shoring tower in place. (*From R. T. Ratay Engineering, P.C.*)

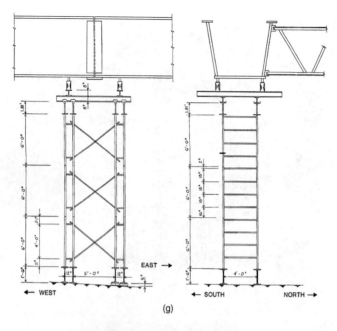

(g)

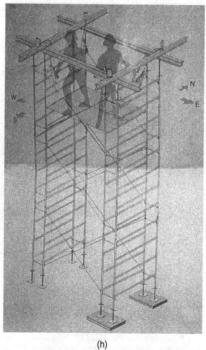

(h)

FIGURE 23.4 (*Continued*) Denver I-70 overpass. (*g*) Dimensions of scaffolding/shoring tower. (*From R. T. Ratay Engineering, P.C.*) (*h*) Presentation board showing tower and two workers prepared for the jury trial. (*From R. T. Ratay Engineering, P.C.*)

Defendant's expert further contended that the rental agency's young engineer, even though a licensed professional engineer, should not be expected to have the sophistication of all these experts with doctorates to understand all the complexities arising from the continuous, curved, interconnected, and composite beam action. Plaintiffs' expert contended that a licensed professional engineer must be expected to at least recognize when there is so much complexity that he is getting in over his head and needs to seek competent advice. Based on its verdict, the jury appears to have agreed with the plaintiff.

An interesting question in the forensic investigation and litigation was whether the tower was scaffolding, because it provided access for the workers, or shoring, because it supported superimposed loads, or both a scaffolding and a shoring at different times. Design safety factors and allowable height-to-width ratios are different for scaffolding and shoring, as per OSHA regulations (see Fig. 23.4).

Case History 5: West Coast Hotel Scaffolding*

During the 1992 construction of a west coast hotel, three construction workers were seriously injured when a scaffold collapsed, causing them to fall about 30 ft along with hundreds of pounds of scaffold parts and building materials. The scaffold platform was erected in the lobby of the hotel to provide access for the installation of the gypsum wallboard ceiling and lighting fixtures. The platform was 32 ft above the lobby floor and covered a 30-ft by 80-ft area. The temporary structure included two towers constructed from pipe frame scaffolding, with a platform using steel putlogs spanning between the towers. The putlogs spanned 18.5 ft and had knee braces at their third points; 2-by-10 wood planks spanned between putlogs to form the platform.

On the day of the collapse, 80 to 100 sheets of gypsum wallboard were stacked in four or five piles at various locations on the platform. On the morning of the collapse, four workers were on the platform. One of them approached a stack of gypsum wallboard that was near the middle of the platform. As he got close to the stack, a sound similar to a gunshot rang out and the center of the platform dropped 2 to 3 in. The initial sag was followed quickly by the total collapse of a central putlog. Within seconds, the four center sections of the platform were on the lobby floor along with several seriously injured workers among the pile of scaffold parts, planks, and gypsum wallboard.

The postcollapse investigation associated with the litigation initiated by the injured parties included observations of hundreds of photos, hours of videotape, and samples of the scaffold parts; metallurgical studies of the metal at the failure location; and analyses of forces in the structure, stresses at the failure location, and progression of the collapse. The review of the photos and videotape that were taken after the collapse immediately identified shortcomings, including missing ties between the scaffold towers and the building and missing or incorrectly installed braces between adjacent putlogs. The photos of the arrangement of the fallen structure and the order of the pieces in the pile on the floor also helped to identify the region where the collapse initiated.

The photos showed that one putlog near the center of the platform had an end sheared off. This was the only scaffold member that broke; almost all the other collapsed members were bent and twisted without breaking. In addition, the putlogs and towers collapsed in toward the center of the platform, suggesting that the collapse initiated in the general area of the putlog that sheared off during the collapse.

Visual inspection of the failed scaffold members revealed that the broken putlog had one location in the bottom chord that had a welded repair before use on this job. The weld

---

*This case history was contributed by Raymond W. LaTona, Simpson Gumpertz & Hager, Inc., San Francisco, California.

metal was deposited on the outside surface of the pipe, but it was not fully fused to the metal of the pipe wall. Metallurgical studies showed that the section at the point of the break had been broken previously, and the bottom chord pipe had cracked around 75 percent of its circumference. The weld that was intended to repair the putlog only covered the crack and did little to reestablish the strength of the cross section. The metallurgical examination also revealed that the repaired bottom chord location failed in a brittle manner (consistent with the sound like a gunshot and the platform dropping a few inches while the loads transferred from the putlog bottom chord to the top chord), and that the top chord of the broken putlog failed in a ductile manner (consistent with the pause between initial failure and total collapse while the top chord yielded).

Analysis of the platform using the scaffolding supplier's load tables showed that the dead weight alone resulted in the putlogs being loaded to the published capacity. Therefore, based on the supplier's tables, the allowable superimposed loading on the platform supported by the putlogs was essentially zero, and the platform was grossly underdesigned for its intended use. Even though the platform was improperly designed, that condition did not explain the failure because the published capacities appropriately include factors of safety. A structural analysis showed that although loaded beyond the allowable loadings, these putlogs should still take the additional load of the gypsum wallboard and a worker without failing. The putlogs under the stacks of gypsum wallboard were nearly at yield stress under the applied load. Using the reduced capacity for the damaged putlog based on the conclusions of the metallurgical analysis and the computational results for the applied loads, the repaired putlog was loaded to the ultimate strength of the repaired location under the weight of the gypsum wallboard. The weight of the worker approaching the stack of gypsum wallboard was enough additional load to break the repaired putlog.

In conclusion, the combination of the seriously underdesigned scaffold platform and the weight of the stack of gypsum wallboard plus the worker on an area supported by a poorly repaired putlog caused the failure. The collapse spread to a larger area beyond the broken putlog because the platform was not properly braced and tied into the building.

How did this happen? The platform design was inappropriate; the scaffolding supplier and the contractor never discussed the intended purpose of the platform, so the scaffolding supplier did not know what load the platform was to carry. Also, the supplier's designer said the design was based on "experience," and he never looked at load tables or performed calculations of any kind. This resulted in a scaffold platform with no capacity for any superimposed loads. Another contributing factor for the collapse was that some of the scaffold components were not properly handled and maintained. The supplier's policy stated that damaged components were to be discarded and not repaired; consequently, the putlog with the broken chord should have been discarded. Finally, the scaffold supplier did not follow the company bracing and tie requirements, which led to a general collapse of the temporary structure (see Fig. 23.5).

(a)

FIGURE 23.5 A west coast hotel. (*a*) Collapsed scaffolding. (*From R. A. LaTona Simpson, Gumpertz & Hager, Inc., San Francisco, and Civil Engineering, April 1998.*)

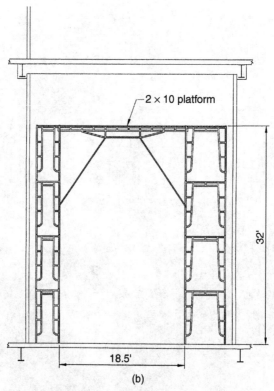

(b)

FIGURE 23.5 (*Continued*) A west coast hotel. (*b*) Typical elevation of the scaffolding. (*From Raymond A. LaTona, Simpson Gumpertz & Hager, Inc. San Francisco.*)

(c)

FIGURE 23.5   (*Continued*) A west coast hotel. (*c*) Break at a defective repair weld in the bottom chord. (*From Raymond A. LaTona, Simpson Gumpertz & Hager, Inc. San Francisco.*)

Case History 6: Masons' Scaffold

A seven-frame-high masons' scaffold collapsed during construction of a multistory brick veneer facing on a building in British Columbia, Canada. Three workers who were working on the fifth scaffold level rode down with the frames and other materials. At the time of the failure in addition to the three workers and their materials on the fifth level, bricks were stockpiled on the seventh level. The weight of the stockpiled materials was approximately double the safe working load of the scaffold. Investigators of the Worker's Compensation Board of British Columbia (similar to a state OSHA in the United States) found that the horizontal bracing members of the scaffold were incomplete and the scaffold's ties to the permanent structure were inadequate. In addition, they indicated that the contractor was not aware of the maximum allowable load that could be placed on the scaffold, nor did he have manufacturer's documentation available to perform reasonable assessment of the maximum allowable superimposed load (see Fig. 23.6). Figures 23.6*a* and *b* are fair illustrations of the "mess" that is often created by a structural collapse.

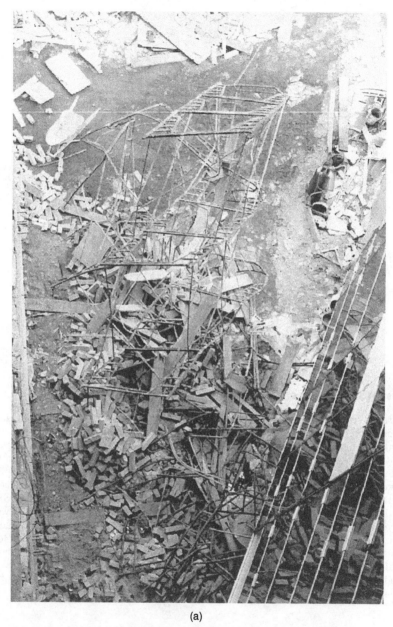

(a)

FIGURE 23.6   Mason's scaffold. (*a*) Aerial view of collapsed scaffold. (*From Jozef Jakubowski, Workers' Compensation Board of British Columbia, Vancouver, B.C., Canada.*)

(b)

FIGURE 23.6 (*Continued*) Mason's scaffold. (*b*) Collapsed scaffold. (*From Jozef Jakubowski, Workers' Compensation Board of British Columbia, Vancouver, B.C., Canada.*)

Case History 7: Indianapolis Structural Steel Frame Guying*

A major section of the structural steel frame collapsed during erection of a large warehouse-type building in Indianapolis, Indiana, in April 1992. There was no loss of life, but there were injured ironworkers, subsequent lawsuits, and out-of-court settlements. As a result of the settlements, there were no trials and neither side's case was fully developed or challenged. The case, however, is instructive because of the relationships among the various parties, the allegations made, the expert opinions offered, and some pertinent temporary structure information that was "lost in the shuffle."

The essentially one-story structure with extensive mezzanine areas was approximately 220,000 ft² T-shaped in plan, with two expansion joints separating the arms of the T from the stem. It was one of the approximately 60,000-ft² arms that collapsed during erection.

The structure consisted of steel roof deck on steel joists and joist girders supported by hot-rolled structural steel beams, girders, and columns on shallow spread footings. Lateral load resistance was provided by the steel deck diaphragms, rigid frames, and precast concrete shear walls.

The project was a design-build type with a construction manager. The architect retained a structural engineer to design the foundations and exterior walls. A steel fabricator was responsible for the structural design, fabrication, and erection of the steel frame, but subcontracted both its design and erection. Because of this division of design responsibilities, the designer of the steel frame had to communicate essential features of the design, such as foundation and shear wall reactions, to others. In addition to providing memoranda listing these reactions, a diagram of diaphragm reactions was provided on the structural steel construction documents. Following the *AISC Code of Standard Practice for Structural Steel Buildings*, the structure was identified as "non-self-supporting." This means that elements not characterized as being of structural steel (such as precast concrete shear walls) are needed for the lateral stability of the completed structure. The column reactions furnished to the foundation designer were identified as being only those associated with the behavior of the completed building.

At the time of the collapse, the steel columns, joist girders, and nearly all joists had been set. The joists between those on the column lines had been spread out in each bay but not yet welded to their supports. The columns were resting on leveling nuts beneath the base plates; the base-plate grout had not yet been installed. There was little temporary cable guying in place. It was reported that there was a significant increase in wind velocity before the collapse.

The expert retained by the injured worker plaintiffs cited insufficient guying as contributing to the collapse, but also named design and construction conditions as contributors. Some of the expert's contentions are listed as follows: (Positions of the defense are shown in parentheses.)

- The designer of the steel frame was responsible for the design of its temporary support. (Such design was excluded by contract. The *AISC Code of Standard Practice* in paragraph 7.9.1 states that "Temporary supports such as temporary guys, braces, falsework, cribbing, or other elements required by the erection operation will be determined and furnished and installed by the erector.")

- The notices of "non-self-supporting" frame in the construction documents were not explicit enough.

---

*This case history was contributed by Michael A. West, Computerized Structural Design, S.C., Milwaukee, Wisconsin.

- The method of temporary support of the columns on four leveling nuts was improper, and pregrouted leveling plates should have been used. (The use of leveling nuts was a common practice in the industry.) The anchor bolts and piers should have been designed for loads resulting from erection operations. (It was the steel erector's responsibility to make this analysis.)
- In at least one instance the concrete pier was constructed too low, resulting in a relatively long projection of the anchor bolts. This condition was directly contributing to the collapse.
- The owner and construction manager had responsibility for site safety. They failed to observe the long anchor bolt projection and the lack of temporary guying of the steel frame.

Although there were differing and unresolved expert opinions as to the cause of the collapse, it was the structural designer's belief that the collapse could have been prevented by the erector's preparation and implementation of a proper temporary erection bracing and guying system. This would have resulted in the necessary temporary bracing and guying materials being on site as they were needed, and in instructions for their timely installation and removal as the permanent lateral load-resisting systems were completed (see Fig. 23.7).

(a)

FIGURE 23.7   Indianapolis steel frame. (*a*) "Orderly line-up" of structural steel framing. (*From Michael A. West, Computerized Structural Design, Milwaukee, Wisc.*)

(b)

(c)

FIGURE 23.7 (*Continued*) Indianapolis steel frame. (*b*) Partially toppled steel columns. (*From Michael A. West, Computerized Structural Design, Milwaukee, Wisc.*) (*c*) Pulled-out and broken column base anchor bolts. (*From Michael A. West, Computerized Structural Design, Milwaukee, Wisc.*)

Case History 8: Brooklyn Wall Bracing

A 15-ft-high, 93-ft-long, 12-in-thick load-bearing unreinforced concrete masonry block wall collapsed during its construction in Brooklyn, New York, on March 23, 1990. The wall was being built as the south side of a second-floor addition to a one-story commercial building. The wall toppled away from the mason's scaffold, fell onto and broke through the roof of the adjacent one-story building, killing 2 and injuring 14 people below. The mason's scaffold was not connected to the wall, so rather grotesquely it remained standing in the middle of nowhere.

Several improprieties and violations were alleged by OSHA and other investigators soon after the collapse. These included no building permit, no seal on the architectural/structural drawings, no shop drawings for the wall support beams, inadequate and unstable support at the base of the wall, no inspection, and, perhaps most importantly for construction safety, no temporary lateral bracing of the wall. Because of the fatalities and the obvious negligence, criminal proceedings were started against the contractor and the engineering investigations were halted. OSHA did complete its report but has not released it for the public.

While this appears to be an extreme case, collapses of masonry walls during construction are chronic occurrences and the cause is nearly always the total absence of temporary bracing.

Note, however, that a contractor wanting to actually design temporary wall bracing would have a difficult time determining what the required lateral design load should be. Where would he or she find that information? (See Fig. 23.8.)

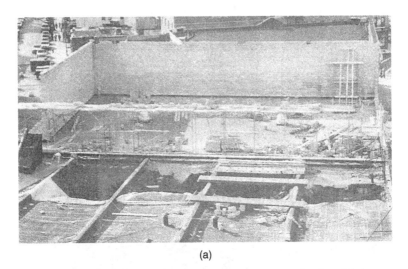

(a)

FIGURE 23.8   Brooklyn wall. (*a*) Unbraced second-story masonry wall collapsed during construction, breaking through adjacent roof. (*From ENR, April 5, 1990.*)

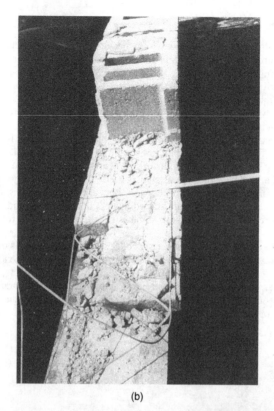

(b)

(c)

FIGURE 23.8 (*Continued*) Brooklyn wall. (*b*) Remnant of concrete block wall built on widened top flange of wide flange beam. (*From R. T. Ratay Engineering, P.C.*) (*c*) Web of wide flange beam bent and top flange rotated under the wall. (*From R. T. Ratay Engineering, P.C.*)

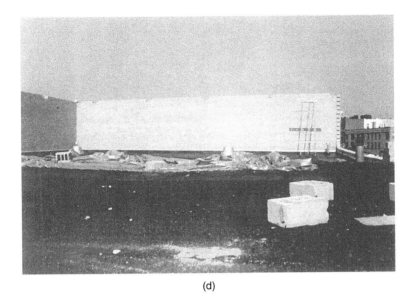

(d)

FIGURE 23.8 (*Continued*) Brooklyn wall. (*d*) Another wall received temporary lateral bracing after the catastrophic collapse. (*From R. T. Ratay Engineering, P.C.*)

Case History 9: Huntington Wall Bracing

An unfinished 18-ft-high, 180-ft-long hollow-core concrete masonry bearing wall collapsed on February 25, 1996, several months after construction had been halted in Huntington, New York. The wall had Durowall horizontal wire reinforcement at every third course, 16-in-wide by 8-in-deep unreinforced pilasters at 18-ft intervals, and the wall was standing on an 8-in-wide reinforced-concrete foundation wall. The 180-ft length was broken into three segments by two vertical control joints that separated the wall from the adjacent pilaster at those two locations. Before the collapse, the wall was allegedly braced, probably from one side only, by an unknown number of 16-ft-long (or so), 9- by 2-in wood planks. During the night of the collapse, the fastest-mile wind speed, according to the National Weather Service records, was 45 mph (creating a calculated stagnation pressure of 5.4 lb/ft^2). The 50-year design wind speed by the *New York State Building Code* for the subject location was 75 mph (translating to 15 lb/ft^2 stagnation pressure).

According to the engineering investigator retained by the insurance company, while the 9-by-2 in diagonal bracing planks when straight would provide some support, they were bent so much (about 2-in snag at mid-length), hence having so little axial stiffness, as to be ineffective to restrain lateral movement of the top of the wall (see Fig. 23.9).

(a)

(b)

FIGURE 23.9 Huntington wall. (*a*) Inadequately braced unfinished masonry block wall toppled in strong wind. (*From R. T. Ratay Engineering, P.C.*) (*b*) Remnant of collapsed wall. (*From R. T. Ratay Engineering, P.C.*)

(c)

FIGURE 23.9   (*Continued*) Huntington wall. (*c*) "Orderly line-up" of collapsed masonry on the ground. (*From R. T. Ratay Engineering, P.C.*)

Case History 10: Wood Truss Erection

Dozens of 42-ft-span wood trusses collapsed during construction of a major addition to a church in British Columbia, Canada. One worker fell with a truss and was slightly injured. The trusses were approximately 5 ft deep with sloping parallel chords peaking at midspan, spanning north-south 42 ft between two 22-ft-high load-bearing stud walls of 2-by-6 studs at 8 in on center. The causes of the collapse appeared to have been inadequate temporary bracing of the trusses and the incomplete construction of the load-bearing stud wall on the south side of the building. The wall was not fully sheathed, the sheathing that was in place was only partially nailed, and it is likely that this wall was not adequately braced. No truss drawings were available at the site. According to the workers, "the bracing didn't make any sense, the other guys on the other site were responsible for the bracing" (see Fig. 23.10).

(a)

(b)

FIGURE 23.10 Wood trusses. (*a*) Wood roof toppled over during construction. (*From Jozef Jakubowski, Workers' Compensation Board of British Columbia, Vancouver, B.C., Canada.*) (*b*) Close-up view of collapsed wood trusses. (*From Jozef Jakubowski, Workers' Compensation Board of British Columbia, Vancouver, B.C., Canada.*)

Case History 11: Column Reinforcing Steel Bracing

A tall reinforcing bar cage for a concrete column collapsed in Vancouver, British Columbia, in the morning of November 2, 1993, seriously injuring one worker. The column rebar cage consisted of 44 bars of 25 mm diameter and 38 ft length each from the ground up, to which an additional 40 bars of 25 mm diameter and 24 ft length were spliced starting at 9 ft above ground, or 16 in above the top of the previous pour. The collapsed cage was reported as weighing approximately 14,000 lb. The column was to be part of a ductile moment-resisting frame. The reinforcing steel requirements and lap locations were detailed on the structural engineering drawings. Five guy cables were stabilizing the cage. At the time of the accident, the injured worker was on top of a previously poured concrete wall, rigging a section of gang form to be moved by a tower crane. He was positioned such that two of the five guylines were on either side of him, to his east and west.

Two possible scenarios were developed by investigators of the Worker's Compensation Board of British Columbia: "Either the moving gang form panel tripped the guyline to the west of the worker, initiating the rebar cage's collapse, or someone had undone one of the guylines to the east, which sent the cage over in a northwesterly direction."

They further opined, "Either scenario could have caused the other guylines to go down, with the one behind the worker (to the east) catching and flinging him approximately 32 ft across the site. Had the rebar cage been adequately braced, it would not have collapsed." Following this incident, and others in which reinforcing steel or its temporary supports had collapsed, the Worker's Compensation Board issued a Technical Commentary, stating that

"The contractor is required to address the support and stability of reinforcing steel in concrete structures..." and suggested to "consider retaining an engineer to design the temporary support and stability of rebar column cages, wall panels and [elevated] reinforcing steel mats in [thick] foundation slabs." (See Fig. 23.11.)

(a)

FIGURE 23.11   Column reinforcing steel. (*a*) Tall reinforcing steel cage collapsed before erecting formwork. (*From Jozef Jakubowski, Workers' Compensation Board of British Columbia, Vancouver, B.C., Canada.*)

(b)

FIGURE 23.11 (*Continued*) Column reinforcing steel. (*b*) Column rebar cage bent over at top of previously poured concrete. (*From Jozef Jakubowski, Workers' Compensation Board of British Columbia, Vancouver, B.C., Canada.*)

Case History 12: Bridge Demolition

The State of Connecticut contracted to demolish the old Sikorsky Bridge on State Route 15 over the Housatonic River, and construct a new bridge. The demolition work was done with the use of cranes on barges anchored in the river below.

The bridge to be removed was a 61-ft-wide, 1824-ft-long, 12-span carbon steel highway bridge on reinforced concrete piers constructed in 1939. It consisted of an open-grid steel roadway deck on longitudinal stringers that were supported on transverse floor beams at 32-ft spacing. The floor beams framed into three longitudinal plate girders, along the center and along the two edges of the bridge spanning pier to pier.

A catastrophic failure occurred on February 17, 2004 during the demolition of span 8. During its removal, the north girder buckled, dragging and toppling one of the cranes off the barge into the water, killing the crane operator who was entrapped in the cab.

The following description of the incident is from the July 14, 2004, four-page memorandum by OSHA, Investigation Report of the February 17, 2004 Fatal Crane Incident at Sikorsky Bridge, Stratford, Connecticut.

The incident occurred on February 17, 2004, at approximately 12:00 PM. The contractor decided to employ two cranes, each mounted on a barge, to hoist the north girder [in Span 8] after removing the bridge deck, stringers, and floor beams. The contractor removed the metal deck and stringers. But before the floor beams could be cut from their connections to the north girder, the girder was rigged to the two cranes. The American crane had applied a lifting force of 70,000 pounds at the east [cantilever] end of the girder, while the Manitowoc crane applied a force of 68,000 pounds at the west end [over Pier 7]. At this moment, the girder was still partially supported and seated on the piers in Span 8. The contractor then cut the remaining four floor beams from the north girder and the cranes were ready to perform tandem lifting.

According to the interview statements, the last floor beam was cut near the west end of the girder. Both cranes were directed to lift the load. The Manitowoc crane was directed to lift the load slightly. However, before the Manitowoc crane operator could lift the load, the operator of the American crane had already lifted the east [cantilever] end of the girder between 6 in to 2 ft above the pier cap. The girder reportedly oscillated in the north-south direction and then buckled at a point near the east [cantilever] end. The boom of the American crane buckled and failed, dragging the crane off the barge and plunging the crane into the river. The crane operator was killed. The west end of the Manitowoc crane tilted up 4 to 5 feet from the barge. Fearing that the crane could overturn, the Manitowoc crane operator immediately released the load. The load dropped and the boom snapped backwards overtop of the crane.

The OSHA memorandum did not include illustrations.

A similar description and additional details of the incident are in the illegibly dated untitled seven-page accident investigation supplementary narrative report of the State of Connecticut Department of Public Safety. Of particular interest are the following two paragraphs:

Prior to the accident which occurred on February 17, 2004, [the contractor] had a demolition crew remove the necessary steel grid deck, stringers and some of the floor beams in preparation to removing the north girder of span 8. On the morning of February 17, 2004, two cranes were employed to lift the 192 foot long north girder. An American 931 OA crawler crane equipped with 180 feet of lattice boom was positioned on a barge by the east [cantilever] end of the north girder. This American 931 OA was rigged to the north girder by a shackle and a 20 foot endless sling approximately 10 feet, 6 inches from the east [cantilever] end. A Manitowoc 4100 W crawler crane equipped with 200 feet of lattice boom was positioned on a barge by the west end [over Pier 7] of the girder by a shackle and a 20 foot endless sling about 3 feet in from the west end of the girder. Both cranes were equipped with a load cell which enabled each operator to know the load or force that he was applying to an accuracy of plus or minus 1 percent as stated by the manufacturer.

With both cranes rigged to the 192 foot long girder, a lifting force of 70,000 pounds was applied by the American crane positioned on the east [cantilever] end of the girder, while the Manitowoc crane applied a lifting force of 68,000 pounds at the west end [over Pier 7]. At this point in time, both cranes were just securing the girder so that the demolition crew could complete the remaining cuts on the four floor beams. With this force applied by each of the cranes, the girder was still resting on its bearing seat. Iron workers of the demolition crew had severed the last floor beam which was attached by the west end of the north girder [over Pier 7]. The iron worker who made the last cut on the floor beam maintained radio communication with the crew (which included both crane operators). Upon finishing his last cut he made his way to a safe location and instructed the Manitowoc crane to "cable up" (to lift). However, this ironworker noted that the American crane [at the cantilever end], not the Manitowoc crane [over Pier 7], had lifted the girder by the east [cantilever] end off its seat or pier cap. It appears that the operator of the American crane [at the cantilever end] acted on the ironworker's command meant for the operator of the Manitowoc crane at [Pier 7].

The girder that failed during removal was the Outside North Girder in span 8, cantilevering into span 9. Before its removal, the west end of the 192-ft girder was supported on pier 7; it spanned 160 ft across span 8 to pier 8, and then continued east as a 32-ft-long cantilever beyond pier 8 into span 9. It was a built-up plate girder of carbon steel. The web plate was 91 in tall by 7/16 in thick with $6 \times 4 \times 3/8$ vertical stiffeners on both sides at approximately 6-ft 5-in spacing, and a pair of $8 \times 8 \times \frac{3}{4}$ flange angles running the length of the girder along its top and bottom edges. All top and bottom flange cover plates were 24 in wide by 5/8 in thick, varying in number from one to three layers along the length of the girder—appearing to be consistent with the design moment diagram. The girder had a field splice shown at 70 ft from the cantilever (east) end and 122 ft from pier 7 (west end) on erection drawings.

Before disassembly of the bridge, transverse floor beams and horizontal braces had framed into the interior (south) side of the north girder transmitting vertical loads and providing lateral bracing. After cutting away the floor beams and braces, 1- to 2-ft-long stubs of these members appear to have remained attached to the interior (south) side of the girder. Being off-center, these became eccentric loads that created torsion in the girder—twisting and laterally bending it toward the south.

Before the pick, the 192-ft girder was a propped cantilever with a 160-ft main span (between piers 7 and 8) and a 32-ft cantilever (from pier 8 toward pier 9). After the pick, when the cantilever (east) end of the girder was lifted up from pier 8, the girder became a propped cantilever with a longer 181.5-ft main span (192 ft – 10.5 ft = 181.5 ft) between pick points and a shorter 10.5-ft cantilever. The entire moment diagram and the location of the maximum moment changed, resulting in larger moments and higher bending stresses at weaker sections of the variable cross-section girder.

A simple structural analysis of the girder in its position when the cantilever end of it was lifted was performed. The purpose of the analysis was to examine the approximate available ultimate strength of the girder in bending as limited by lateral-torsional stability when lifted up off pier 8 and supported on pier 7 at its west end and on the crane line near its cantilever east end, and to compare that strength to what was required of the girder.

The available ultimate strengths, that is, ultimate moments controlled by lateral torsional buckling at selected cross sections of the girder, were calculated by hand in accordance with the AISC and AASHTO LRFD design specifications. The required strengths (bending moments due to the self-weight) of the variable cross-section girder were calculated with the aid of a computer using the ANSYS structural analysis software. For simplicity, all vertical loads were assumed to act in the plane of the web. The results of the simple analysis indicated that the girder had inadequate strength and stability when lifted. This was the result even without considering the twisting and destabilizing effect of the eccentric weights on the south side of the girder. More accurate stability analyses could have been performed using a sophisticated nonlinear finite-element computer software, but in light of the convincing results of the simple analysis, the time and expense of greater sophistication were not warranted.

It was the author's opinion with reasonable engineering certainty that the girder failed in a lateral-torsional buckling mode. The reasons for the failure were the girder's instability on account of its slenderness (due to the narrowness of its top flange relative to the long length between the pick locations). The eccentric weights of the floor beam and brace stubs that remained attached to the south side of the web plate created twisting moments and consequent lateral bending in the girder immediately upon lifting, which further compromised its stability. A secondary contributing factor to the failure was the nature of the suspended support at the end of the crane line that allowed lateral displacement (swaying) of the east end when lifted off the pier. Other possible contributing factors may have been lateral pressure from wind, or attachment of the lifting shackle in a manner that created tilting of the girder when lifted.

Engineering intuition and judgment, results of the structural analysis, and observations in the OSHA and State DOT reports all are consistent with the mode of failure of the girder, and point to the validity of the above listed causes. The failure could possibly have been prevented by (a) installing lateral bracing along the top flange of the girder, or by (b) lifting the girder at two points closer together (see Fig. 23.12).

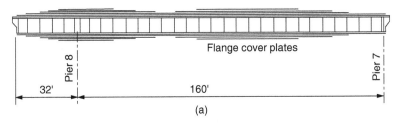

Flange cover plates

32'   160'

Pier 8   Pier 7

(a)

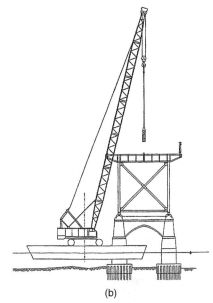

(b)

(c)

FIGURE 23.12 Bridge demolition. (*a*) Schematic elevation of north girder. (*b*) Disassembly by crane on barge. (*c*) Girder collapsed when lifted for removal.

23.45

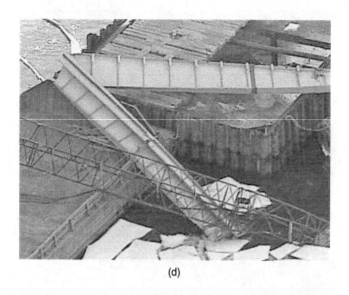

(d)

(e)

FIGURE 23.12 (*Continued*) Bridge demolition. (*d*) Close-up of collapsed girder. (*e*) Collapsed girder in storage.

Case History 13: A Tower-Crane Collapse*

On a calm night of November 16, 2006, a 210-ft-tall freestanding tower crane toppled at a construction site in Bellevue, Washington (Fig. 23.13a). The operator, who was just ending his shift in the cab at the top of the crane, remarkably suffered only minor injuries. However, the crane collapsed onto several adjoining buildings and in one a young professional, a recent arrival to the city, died when the boom crashed through the ceiling of his apartment.

The crane, constructing a high-rise tower, had been supported on a fabricated steel base mounted to columns and shear walls in the underground garage. After initial blame focused on the crane mast, the steel base quickly became the focus of investigators.

This custom steel base had been devised as the solution to a construction dilemma. The project was a restart on a site that had been abandoned a few years earlier. Foundations and several underground parking levels had been constructed previously but the old tower-crane footing was not in a workable location for construction of the redesigned building. The construction team determined that digging out a crane footing in a new location was unfeasible because existing posttensioned parking decks could not sustain cutting of openings for the mast. After brainstorming and weighing a number of alternatives, the construction team held a meeting during which they chose the steel base. It would be mounted above the existing parking decks with the framing designed by the building structural engineer using loads provided by the crane manufacturer.

(a)

FIGURE 23.13   Tower crane. (a) Toppled due to a failure of its fabricated steel base.

---

*This case history was contributed by Lawrence K. Shapiro, P.E., of Howard I. Shapiro Associates, Lynbrook, New York.

(b)

FIGURE 23.13 (*Continued*) Tower crane. (*b*) Failure occurred at the base frame connections due to high stresses and cyclical loading. Notches in the beams, made because of an unnecessary offset in the height of the connecting girder, created a stress concentration that hastened the failure.

Unfortunately the two key parties in this meeting, the construction manager and the structural engineer, came away with different ideas about how the tower crane was to be initially supported. The engineer assumed the crane would be erected with a tie connecting its mast to the building core, but the CM had determined that a tie was not possible at the time of initial installation because the core would not be built to the required elevation. The engineer proceeded to design the support frame using the wrong loads, assuming overturning moment would be resisted by this nonexistent tie, making no allowance for a tall freestanding crane to be carried fully on its base.

Hence the base was fabricated and installed using severely understated load values, and the crane was erected upon it. It stood and worked for several months until the November night when it collapsed. During its erection a snapping noise prompted the rigging crew to scurry down the mast. The engineer was called to the site to review the base, but, failing to note that the crane mast was not tied, he gave his approval for the crane to go to work.

Other telltale signs of deficiency were ignored. Excessive leaning of the mast should have prompted a survey of base deflection and tower plumbness, but that never was ordered. The base was also notorious for loud metallic groaning.

The failed base revealed deficiencies in detail design of connections and in welding (Fig. 23.13*b*). Though neither was a prime agent of the collapse, these additional lapses reinforced the notion of a systemic breakdown of responsibilities. Procedures followed in carrying out the design, procurement, and implementation were all flawed.

Investigation uncovered a lack of checks and balances, and corporate cultures that accepted perfunctory performance of duties. Information was compartmentalized. Key tasks were handled by inexperienced staff without adequate supervision. Nobody was in overall charge.

Before installation, the conceptual error at the root of the failure might have been detected by common management practices such as dissemination of meeting minutes or formal inspections. After installation, the behavior of the crane and the base gave ample reason for additional scrutiny. A collapse could have been prevented at either stage.

There was also a managerial failure to recognize and prioritize risk. Obviously a tower-crane base is critical to safety, but this support solution engendered an additional layer of risks due to its complex interaction with the crane and the dependency of its design factors on operational considerations. The crane was not adequately understood by the engineers or construction managers; they did not compensate for their lack of expertise by bringing in additional parties for consultation or review.

Assigning a third-party reviewer would have been a wise choice. As a result of this accident, the State of Washington amended its industrial code to require an independent engineering review for all nonstandard tower-crane bases.

HOW TO AVOID TEMPORARY STRUCTURE FAILURES?

Take them seriously: Consider them just as important as the permanent facility they help build!

ATTRIBUTION

Much of this chapter is reproduced from Chap. 10, "Temporary Structures in Construction," from the book *Forensic Structural Engineering*, 2d ed., Robert T. Ratay (ed.), published by McGraw-Hill, 2010.

Some of the case studies were extracted in part from publications, such as the *ENR: Engineering News-Record*; some others were extracted in part from books, such as *Construction Failure*, 2d ed., by Feld and Carper; *Design and Construction Failures*, by Dov Kaminetzky; and *Why Buildings Fall Down*, by Matthys Levy and Mario Salvadori. Several were summarized from reports of investigations by NBS (now NIST) and OSHA; two from the records of the Worker's Compensation Board of British Columbia; four were contributed by individuals; and four came from the author's job files of his own forensic engineering practice.

BIBLIOGRAPHY

ASCE/SEI 37-02 *Design Loads on Structures During Construction*. American Society of Civil Engineers, Reston, Virginia, 2002 (Next edition 2012)

Campbell, P. (ed.): *Learning from Construction Failures: Applied Forensic Engineering*, John Wiley & Sons, New York, 2001.

Delatte, N. J.: *Beyond Failure: Forensic Case Studies for Civil Engineers*, ASCE, Reston, Va., 2008.

Duntemann, J. R., et al: *Construction Handbook for Bridge Temporary Works*, FHWA Publication No. FHWA-RD-93-034, Washington, November 1993.

Feld, J. and K. L. Carper: *Construction Failure*, 2d ed., John Wiley & Sons, Inc., New York, 1997.

Lapping, J. E.: "OSHA Standards That Require Engineers," ASCE/SEI *Proceedings, Construction Safety Affected by Codes and Standards*, Robert T. Ratay (ed.), October 1997.

Peraza, D. P.: "Avoiding Structural Failures during Construction," *STRUCTURE Magazine*, November, 2007.

Ratay, R. T.: "Temporary Structures in Construction—US Practices," *Structural Engineering International 4/2004*, Journal of the International Association for Bridge and Structural Engineering (IABSE), 2004.

Ratay, R. T.: "Temporary Structures Failures—Designers Beware!," *STRUCTURE Magazine*, Editorial, December, 2006.

Ratay, R. T.: "Changes in Codes, Standards and Practices Following Structural Failures—Part 1, Bridges," *STRUCTURE Magazine*, December, 2010.

Ratay, R. T.: "Changes in Codes, Standards and Practices Following Structural Failures—Part 2, Buildings," *STRUCTURE Magazine*, April, 2011.

Ratay, Robert T. (ed.): *Forensic Structural Engineering*, 2d ed., McGraw-Hill, New York, 2010.

INDEX

Note: Page numbers followed by *f* denote figures and *t* denote tables.

Printed in the USA
CPSIA information can be obtained
at www.ICGtesting.com
JSHW010227280923
49197JS00003B/57